Student Companion for Stryer's BIOCHEMISTRY

Student Companion for Stryer's BIOCHEMISTRY

Fourth Edition

Richard I. Gumport
Ana Jonas
Richard Mintel
Carl Rhodes

Department of Biochemistry
College of Medicine at Urbana-Champaign
University of Illinois

Expanded Solutions to Text Problems contributed by

Roger E. Koeppe

Department of Biochemistry
Oklahoma State University

W. H. Freeman and Company
New York

ISBN: 0-7167-2560-6

Excerpt from "Stopping by Woods on a Snowy Evening" by Robert Frost copyright 1923; © 1969 by Holt Rinehart and Winston; copyright 1951 by Robert Frost. Reprinted from "The Poetry of Robert Frost," ed. Edward Connery Latham, by permission of Henry Holt and Company, Inc.

Printed in the United States of America

First printing 1995, VB

Contents

PART III METABOLIC ENERGY: Generation and Storage

PART IV BIOSYNTHESIS OF BUILDING BLOCKS

PART V GENES: Replication and Expression

Acknowledgments

The *Companion* has its origins in a curriculum guide that was initiated over twenty years ago for first-year medical students studying biochemistry at the College of Medicine of the University of Illinois at Urbana-Champaign. A number of colleagues have contributed to the *Companion* over the years. We wish to thank John Clark, Lowell Hager, Walter Mangel, William McClure, and Robert Switzer for their efforts in helping to develop the initial curriculum guide. Special thanks go to George Odal and James Kaput, our fellow teachers of biochemistry at the College of Medicine; and Dale Kaiser and Robert Lehman, in the Department of Biochemistry at Stanford. We thank Gordon Lindberg for his careful reading and insightful contributions to the book. We also thank William Sorlie for educating us to the value of learning objectives. Thanks are also due to the students who took the time to criticize the *Companion*. Finally, we appreciate the support provided our teaching efforts by the Department of Biochemistry and the College of Medicine. In particular, we thank Cheryl Steinkamp and the Word Processing Center for their careful and cheerful preparation of our written materials.

Richard I. Gumport
Ana Jonas
Richard Mintel
Carl Rhodes

Acknowledgments

To the Student

The woods are lovely, dark, and deep,
But I have promises to keep,
And miles to go before I sleep,
And miles to go before I sleep.

ROBERT FROST

Opening a comprehensive biochemistry text for the first time can be a daunting experience for a neophyte. There is so much detailed material that it is natural to wonder if you can possibly master it in one or two semesters of study. Of course, you can't learn everything, but experience indicates that you can, indeed, learn the fundamental concepts in an introductory biochemistry course. We have written this *Student Companion for Stryer's Biochemistry* to ease your entry into the exciting world of biochemistry.

Your goal is to "know" and "understand" biochemistry. Unfortunately, awareness of these grand goals offers no practical help in reaching them, because they represent such high-level and complex intellectual processes. In addition, it is difficult for you to know to what extent you have attained them. We have found that, by subdividing these goals into simpler ones and expressing them in terms of demonstrable behaviors, you can begin to approach them and, in addition, can readily assay your progress toward reaching them. Thus, a part of each chapter consists of Learning Objectives that ask you to do things that will help you to begin to understand biochemistry. When you can master the objectives, you are well on your way to learning the material in the chapter. It is important to add a cautionary note here. Being able to respond to all the objectives adequately does not mean that you know biochemistry, for they are a limited sampling of all the possible objectives, and more to the point, they do not explicitly require such higher level activities as creation, analysis, integration, synthesis, problem solving, evaluation, application, and appreciation. These more advanced skills will develop to varying levels as you continue your studies of biochemistry beyond the introductory stage.

Each chapter in the *Companion* consists of an introduction, Learning Objectives, a Self-Test, Answers to Self-Test, Problems, Answers to Problems, and Expanded Solutions to Text Problems. The introduction sets the scene, places the chapter material in the context of what you have already learned, and reminds you of material you may need to review in order to understand what follows. The Learning Objectives are presented in the order that the information they encompass appears in Stryer's *Biochemistry.* Key words—important concepts or vocabulary—are italicized in the objectives. Self-Test questions, requiring primarily information recall, are followed by the answers to the questions. A Problems section, in which more complex skills are tested, is followed by answers to the problems. Finally, expanded solutions to Stryer's end-of-chapter problems are presented.

There are many ways to use the *Companion,* and as you begin your studies you will develop the "system" that is best for you. Over twenty years of experience teaching introductory biochemistry to first-year medical students has suggested one pathway that you should consider. Start by reviewing the prerequisite chapters mentioned in the introduction and skim the Learning Objectives to obtain an overview of what you are to learn. Some students like also to skim the Self-Test questions at this time to form an impression of the levels of difficulty and the kinds of questions that will be asked. Next, read the chapter in *Biochemistry,* using the Learning Objectives to help direct you to the essential concepts. Note the key words and look up any you don't know. Then attempt to meet the objectives. When you cannot satisfy an objective, reread the relevant section of the text. You should now take the Self-Test to check your ability to recall and apply what you have learned. Finally, solve the Problems, which have been designed to further test your ability to apply the knowledge you have gained. It is not sufficient simply to read the problems and look at the Answers to see if you would have done them the same way. You must struggle through the solutions yourself to benefit from the Problems. As you are using the *Companion,* you will, of course, be integrating what you have learned from your studies and your lectures or laboratory exercises.

Besides helping you to learn biochemistry, you will find the *Companion* useful in studying for examinations. Go over each of the Learning Objectives in the chapters covered by an exam to ensure that you can respond to them knowledgeably. Similarly, review the key words. Decide which chapter topics you feel uncertain about and reread them. This protocol, coupled with a review of your lecture and reading notes, will prepare you well for examinations.

It is important to talk about biochemistry with others in order to learn the pronunciation of scientific terms and names and to help crystallize your thinking. Also, realize that although biochemistry has a sound foundation and we understand much about the chemistry of life, many of our concepts are hypotheses that will require modification or refinement as more experimental evidence accrues. Alternative and sometimes contradictory explanations exist for many biochemical observations. You should not regard the material in *Biochemistry* or the *Companion* as dogma, and you should, wherever possible, attempt to read about any given topic in at least two sources. Try to follow up topics that particularly interest you by reading about them in the scientific literature. References are given in *Biochemistry,* and your instructor can help you locate research and review articles. In this way, you can begin to appreciate the diversity of opinion and emphasis that exists in the field of biochemistry.

The authors welcome reader's comments, especially any drawing our attention to errors in the text. Comments should be sent to:

Professor Richard I. Gumport
Department of Biochemistry
University of Illinois
600 S. Mathews Avenue
Urbana, Illinois 61801-3792

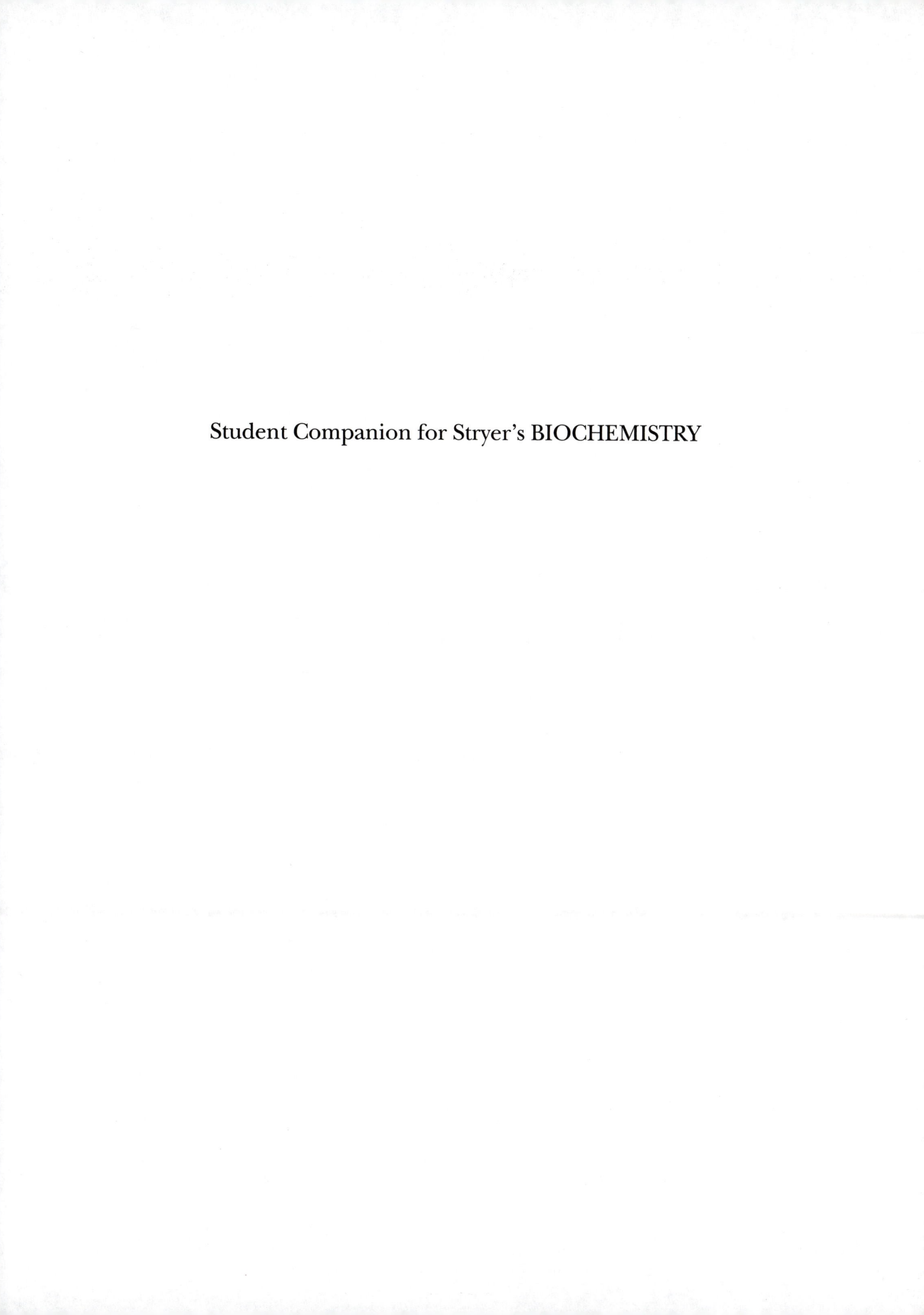

Student Companion for Stryer's BIOCHEMISTRY

CHAPTER

1

Prelude

The introductory chapter of *Biochemistry* begins by highlighting the impact of biochemistry on modern biology and medicine, which sets the stage for the study of this exciting branch of science. Stryer then presents the three types of atomic models that will be used to depict molecular structure. Since biochemistry is the study of the molecular basis of life, molecular dimensions, energies, and time scales are important to know; they are presented next. Then follows a discussion of the forces involved in the interactions between biomolecules. These interactions are responsible for the enormous variety of biological structures and biological processes found in nature. Because the great majority of biochemical processes occur in water, the properties of water and their effects on biomolecules are also described. Finally, Stryer outlines the five-part organization of the text.

LEARNING OBJECTIVES (Stryer pages 3–16)

When you have mastered this chapter, you should be able to complete the following objectives:

1. Discuss the most important achievements of *biochemistry* in the elucidation of the *molecular basis of life* and in the advancement of modern biology and medicine.

2. Explain the uses of different *molecular models.*

3. Correlate the metric scale with the *dimensions* of *biomolecules, assemblies* of biomolecules, and *cells.*

4. Give the ranges of the *time* and the *energy scales* relevant to biochemical processes.

5. List the three kinds of *noncovalent bonds* that mediate interactions of biomolecules and describe their characteristics.

6. Describe how the properties of *water* affect the interactions among biomolecules.

7. Explain the origin of *hydrophobic attractions* between *nonpolar molecules* and give examples of their importance in biochemical interactions.

SELF-TEST

1. Which of the following molecular patterns or processes are common to both bacteria and humans?

(a) Development of tissues
(b) Information flow from proteins to DNA
(c) The "energy currency"
(d) Genetic information flow
(e) Similar biomolecular composition

2. Match the types of molecular models in the left column with the appropriate application in the right column.

(a) Space-filling model ________
(b) Ball-and-stick model ________
(c) Skeletal model ________

(1) Shows the bond framework in macromolecules
(2) Indicates the volume occupied by a biomolecule
(3) Shows the bonding arrangement in small biomolecules

3. Hydrogen atoms are frequently omitted from ball-and-stick models and skeletal models of biomolecules. Explain why.

____

____

____

4. Assume that amino acids, the building blocks of proteins, can be represented by spheres having a radius of 3 Å. Approximately how many amino acids could make up a spherical protein having a radius of 2 nm?

(a) 3 amino acids
(b) 30 amino acids
(c) 300 amino acids
(d) 3000 amino acids

5. If the protein in question 4 were extended so that the amino acids were in linear sequence, how long would it be in nanometers?

(a) 90 nm
(b) 180 nm
(c) 900 nm
(d) 1800 nm

6. Could one see the extended protein in question 5 with a light microscope? Explain.

____

____

____

7. Could one see the protein in question 5 in an electron microscope? Explain.

____

____

____

8. The shape of a red blood cell can be approximated by a disk 7 μm in diameter and 3 μm thick. If the cell density is close to 1.30 g/ml, how much does a single red blood cell weigh?

(a) 1.5×10^{-10} g
(b) 1.5×10^{-8} g
(c) 1.5×10^{-6} g
(d) 1.5×10^{-4} g

9. Most enzyme-catalyzed reactions occur in

(a) picoseconds.
(b) nanoseconds.
(c) microseconds.
(d) milliseconds.
(e) seconds.

10. How long would it take 1 picomole of enzyme to convert 1 micromole of substrate to product if the turnover number of the enzyme (the time required for one enzyme molecule to convert one substrate molecule to one molecule of product) were 1 per millisecond?

(a) 10^{-1} sec
(b) 10 sec
(c) 10^{1} sec
(d) 10^{2} sec
(e) 10^{3} sec

11. What would you expect the generation time of a virus that infects a bacterium to be? Explain.

____

____

12. Ultraviolet light, which has an energy of around 100 kcal/mol, can be damaging to tissues. Explain why.

____

____

13. Arrange the following in order of increasing energy content:

(a) Protein molecule
(b) Glucose molecule
(c) Covalent C–C bond
(d) Hydrogen bond

14. Convert the usable energy present in 1 mole (mol) of ATP (~12 kcal/mol) into joules.

15. For the bonds or interactions in the left column, indicate all the characteristics in the right column that are appropriate.

(a) Electrostatic interaction ____________

(b) Hydrogen bond ____________

(c) van der Waals bond ____________

(d) Hydrophobic interaction ____________

(1) Requires nonpolar species
(2) Involves charged species only
(3) Requires polar or charged species
(4) Involves either O and H or N and H atoms
(5) Involves polarizable atoms
(6) Is also called a *salt bridge*
(7) Exists only in water
(8) Is optimal at the van der Waals contact distance
(9) Has an energy between 3 and 7 kcal/mol
(10) Has an energy of around 1 kcal/mol
(11) Is weakened in water

16. The properties of water include

(a) the ability to form hydrophobic bonds with itself.
(b) a disordered structure in the liquid state.
(c) a low dielectric constant.
(d) being a strong dipole, with the negative end at the O atom.
(e) a diameter of 5 Å.

17. Biological membranes are made up of phospholipids, detergentlike molecules with long nonpolar chains attached to a polar head group. When isolated phospholipids are placed in water, they associate spontaneously to form membranelike structures. Explain this phenomenon.

____

____

18. If two molecules had a tendency to associate with each other because groups on their surfaces could form hydrogen bonds, what would be the effect of putting these molecules in water? Explain.

____

____

ANSWERS TO SELF-TEST

1. c, d, e

2. (a) 2, (b) 3, (c) 1

3. The ball-and-stick model and skeletal model best show the bonding arrangements and the backbone configurations of biomolecules; the inclusion of the numerous hydrogen atoms would obscure the very features revealed by these models.

4. c. The volume of the protein is estimated with the equation $\frac{4}{3}\pi r^3$. This volume is then divided by the average volume of an amino acid to provide an approximate value. The volume between spheres is neglected in this approximation.

5. b. The average diameter of an amino acid is multiplied by the number of amino acids in the protein.

6. No. Although the resulting length of 1800 Å is close to the resolution limit of the light microscope, the width of only 6 Å would not permit resolution.

7. Yes. The resolution of the electron microscope is in the nanometer range and could probably reveal the protein, if it were collapsed into a globular (spherical) conformation with a diameter of 4 nm.

8. a. The volume of a cylinder is $\pi r^2 \times h$, so the volume of the red blood cell is 115 μm^3. Converting to cubic centimeters (which are the same as milliliters) gives 1.15×10^{-10} cm^3. To obtain the weight, multiply the volume by the cell density.

9. d

10. e. The number of molecules is directly proportional to the number of moles, so that 1 mole of substrate is converted per mole of enzyme in 1 millisecond. Therefore, 1 picomole (pmol, 10^{-12} mol) of enzyme forms 1 pmol of product in 1 ms (10^{-3} s). Since 1 μmol of substrate must be converted, 10^{-6} mol divided by 10^{-12} mol/ms = 10^6 ms, or 10^3 s.

11. The generation time of a virus has to be longer than the synthesis time of a typical protein, since many proteins (and other macromolecules) must be made to form a virus. The time must, however, be shorter than the generation time of a bacterium, which is a much more complex entity, because division of the bacterium would disrupt the formation of the virus. Therefore, one would expect the generation time of a virus to be between 10 seconds and 20 to 30 minutes—the generation time of a rapidly growing bacterium.

12. An energy of 100 kcal/mol is in the range of the energy of many covalent bonds; therefore, ultraviolet light can break covalent bonds and, consequently, damage the structures of essential biomolecules.

13. d, c, b, a

14. 50,200 joules (J)/mol. Since 1 J equals 4.184 cal, the product of 12,000 and 4.184 is the answer.

15. (a) 2, 6, 9, 11 (b) 3, 4, 9, 11 (c) 5, 8, 10 (d) 1, 7. Hydrophobic interactions are stengthened in water.

16. d

17. When the nonpolar chains of the individual phospholipid molecules are exposed to water, they form a cavity in the water network and order the water molecules around themselves. The ordering of the water molecules requires energy. By associating with one another through hydrophobic interactions, the nonpolar chains of phospholipids release the ordered water by decreasing the total surface area and hence reduce the energy required to order the water. Such coalescence stabilizes the entire system, and membranelike structures form.

18. Because of the high dielectric constant of water and its ability to form competing hydrogen bonds, the interaction between the molecules would be weakened.

PROBLEMS

1. As found in the answer to question 8 of the self-test, the volume of a red blood cell is 1.15×10^{-10} cm^3, or 1.15×10^{-16} m^3. A hemoglobin molecule has a diameter of 65 Å.

   (a) If a red blood cell were completely filled with hemoglobin molecules, how many hemoglobin molecules would it hold?
   (b) Referring again to question 8 of the self-test, the weight of the red cell is 1.5×10^{-10} g. The molecular weight of a hemoglobin molecule is 65,000 g. Using your answer for part (a), estimate the weight of a red blood cell completely filled with hemoglobin.
   (c) To calculate the number of hemoglobin molecules in a red blood cell in a more realistic way, use the molecular weight of hemoglobin and the weight of the red blood cell from question 8 to calculate the number of hemoglobin molecules in a cell, assuming that 80% of the weight of the cell is due to hemoglobin.

2. As will be seen in succeeding chapters, enzymes provide a specific binding site for substrates where one or more chemical steps can be carried out. Often these sites are designed to exclude water. Suppose that at a binding site, a negatively charged substrate interacts with a positively charged atom of an enzyme.

   (a) Using Coulomb's equation, show how the presence of water might affect the interaction. What sort of environment might be preferable for an ionic interaction?
   (b) How would an ionic interaction be affected by the distance between the oppositely charged atoms?

3. The melting points of straight-chain hydrocarbons increase in rough proportion to their length. Relate these observations to the three types of noncovalent interactions discussed in Chapter 1.

4. In some proteins the contact distance between an amide hydrogen and a carbonyl oxygen that are participating in hydrogen binding is somewhat less than expected from adding their respective van der Waals contact distances. What feature of hydrogen bonding allows the two atoms to be closer to each other?

5. Consider a spherical eukaryotic cell that is 20 μm in diameter and a virus that is a cube with sides of 200 Å.

(a) Estimate the maximum number of virus particles that could adhere to the surface of the cell.
(b) Most viruses interact with host cells by binding to specific receptor molecules on the cell surface. If there are 1000 receptors that are uniformly distributed on the cell, what is the maximum separation distance between each of the receptors?

6. Oxidation of glucose ($C_6H_{12}O_6$) in a cell or by combustion in a calorimeter yields about 686 kcal per mole. In a normal 70-kg male after an overnight fast, the concentration of glucose in the plasma is 5 mM, and a total of about 20 grams of free glucose is circulating in the blood.

(a) How many molecules of glucose are found in 1 liter of blood?
(b) How many kcal of energy would be produced by the oxidation of the circulating glucose?
(c) Calculate the approximate volume of blood in a normal 70-kg male.

7. The oxygen-carrying protein myoglobin is composed of 153 amino acids, linked by covalent bonds into an unbranched polymeric chain. If all amino acids in the chain assume a regular and periodic conformation in which each residue is separated from the next by a distance of 1.5 Å, then the molecule could be as long as 230 Å (153 residues × 1.5 Å per residue). Analysis of the myoglobin molecule in solution reveals that it is no more than 45 Å in length. What does this observation tell you about how a linear polymer of amino acids might behave in solution?

8. Water molecules have an unparalleled ability to form hydrogen bonds with one another. Water also has an unusually high heat capacity, as measured by the amount of energy required to increase the temperature of a gram of water by 1° C. How does hydrogen bonding contribute to water's high heat capacity?

9. Plants harness light energy to convert carbon dioxide to glucose. A key catalytic protein in this process is called *rubisco;* it is perhaps the most abundant protein in the living world. Rubisco catalyzes a chemical reaction that links carbon dioxide with another biological molecule, forming a new compound that can enter the pathway to glucose production. Compared with most enzymes, a rubisco molecule works rather slowly, catalyzing the linkage of carbon dioxide at the rate of only 24 times per second (most enzymes can work up to 1000 times faster). However, the concentration of rubisco molecules in plants is very high, as much as 4 mM in some tissues. The molecular weight of the enzyme is 560,000, and the volume of a plant cell is 1×10^{-13} L.

(a) How many moles of carbon dioxide can a plant cell fix per hour?
(b) What is the concentration of rubisco protein, in grams per liter, in a plant cell?

10. The single molecule of DNA that comprises the chromosome of the bacterium *E. coli* contains 4×10^6 nucleotides paired in a duplex helical molecule. The molecule is about 20 Å thick, with a potential length of 1.4 mm, based on a distance of 3.4 Å between each base pair in the linear, unbranched molecule.

(a) The limit of resolution of the most powerful light microscope is 2 μm. How many DNA molecules must be laid side by side in order to be visible?
(b) If the bacterial cell is a sphere 2 μm in diameter, how many times must the chromosome be folded in order to fit into the cell?

ANSWERS TO PROBLEMS

1. (a) The volume of a sphere is $\frac{4}{3}\pi r^3$, so the volume of a hemoglobin molecule is

$$1.33 \times 3.142 \times (3.25 \times 10^{-9}\text{ m})^3 = 143.5 \times 10^{-27}\text{ m}^3 = 1.43 \times 10^{-25}\text{ m}^3$$

To find the number of hemoglobin molecules in the red blood cell, we divide the volume of the red blood cell by the volume of a hemoglobin molecule:

$$\frac{1.15 \times 10^{-16}\text{ m}^3}{1.43 \times 10^{-25}\text{ m}^3} = 8.0 \times 10^8 \text{ molecules}$$

(b) To find the weight of a molecule of hemoglobin, we divide the molecular weight of hemoglobin by Avogadro's number:

$$\frac{65{,}000\text{ g}}{6.023 \times 10^{23}} = 1.07 \times 10^{-19}\text{ g}$$

Now, 8×10^8 molecules would weigh about 8×10^{-11} grams!

(c) Another estimate of the number of hemoglobin molecules in a red blood cell is found by first calculating the weight of the cell due to hemoglobin:

$$(1.5 \times 10^{-10}\text{ g}) \times 0.8 = 1.2 \times 10^{-10}\text{ g}$$

The number of molecules in the cell is

$$\frac{1.2 \times 10^{-10}\text{ g}}{1.07 \times 10^{-19}\text{ g/molecule}} = 1.12 \times 10^9 \text{ molecules}$$

2. (a) The magnitude of the electrostatic attraction would be diminished by the presence of water because *D*, the dielectric constant, is relatively high for water. Inspection of Coulomb's equation

shows that higher values of D will reduce the force of the attraction. Lower values, such as those for hydrophobic molecules like hexane, allow a higher value for F. We shall see that many enzyme active sites are lined with hydrophobic residues, creating an environment that enhances ionic interaction.

(b) Inspection of Coulomb's equation also reveals that the force between two oppositely charged atoms will vary inversely with the square of the distance between them.

3. Straight-chain hydrocarbons, like hexane and decane, do not form hydrogen bonds, nor are they normally ionized, but they do interact through van der Waals forces. Because a homogeneous group of straight-chain molecules provides many opportunities for steric complementarity, a large number of bonds are formed. The longer the chains are, the larger the number of bonds that are formed per molecule. Thus, for longer molecules, greater amounts of thermal energy are required to disrupt the bonds and cause the molecular assembly to melt.

4. Both atoms have partial charges that attract each other. The single electron of the hydrogen atom is partially shifted to the nitrogen atom to which the hydrogen is covalently bound. As a result, the distance between the electronic shells of the hydrogen and the carbonyl oxygen is reduced, allowing them to approach each other more closely.

5. (a) The surface area of a sphere is πd^2, so the surface area of the eukaryotic cell is

$$3.14 \times (20 \times 10^{-6}\ \text{m})^2 = 1.26 \times 10^{-9}\ \text{m}^2$$

The surface area of a face of the virus cube is

$$(200 \times 10^{-10}\ \text{m})^2 = 4 \times 10^{-16}\ \text{m}^2$$

To find the number of virus particles that could adhere to the surface of the cell, we divide the surface area of the cell by the surface area of one face of the virus cube:

$$\frac{1.26 \times 10^{-9}\ \text{m}^2}{4 \times 10^{-16}\ \text{m}^2} = 3.15 \times 10^{6}\ \text{virus particles per cell}$$

(b) The maximum area available for each receptor is the surface area of the cell divided by the number of receptors:

$$\frac{1.26 \times 10^{-9}\ \text{m}^2}{10^{3}\ \text{receptors}} = 1.26 \times 10^{-12}\ \text{m}^2\ \text{per receptor}$$

Assuming that each maximum area is a circle, then the diameter of the circle gives the maximum distance between neighboring receptors:

$$\tfrac{1}{4}\pi d^2 = 1.26 \times 10^{-12}\ \text{m}^2$$

$$d = \left(\frac{1.26 \times 10^{-12}}{0.785}\right)^{1/2} = 1.27 \times 10^{-6}\ \text{m}$$

6. (a) The number of glucose molecules in a liter of blood is

$$(5 \times 10^{-3}\ \text{mole})\ (6.023 \times 10^{23}\ \text{molecules/mole}) = 3.01 \times 10^{21}\ \text{molecules}$$

(b) 20 total g of circulating glucose is equivalent to

$$\frac{20\ \text{g}}{180\ \text{g mol}^{-1}} = 0.111\ \text{mole glucose}$$

The number of kilocalories produced by the oxidation of 0.111 mole is

$$(0.111\ \text{mole})\ (686\ \text{kcal/mol}) = 76.2\ \text{kcal}$$

It is interesting to note that from 2200 to 2800 kcal in fuel molecules is needed daily to provide energy for normal activities. Thus, circulating gluose reserves are not nearly sufficient for the daily requirement. Later we will see that the liver continuously supplies glucose from its fuel reserves to the blood for distribution to other tissues, including muscle and brain.

(c) The total blood volume can be estimated from the total amount of glucose present in the circulation and the normal glucose concentration in a liter of blood:

$$20\ \text{g circulating glucose} = 0.111\ \text{mole total} = 1.11 \times 10^{-1}\ \text{mole of glucose}$$

$$\frac{1.11 \times 10^{-1}\ \text{mole glucose}}{5 \times 10^{-3}\ \text{mole glucose/liter}} = 22.2\ \text{liters blood volume}$$

7. Because the length of the myoglobin molecule in solution is much less than its extended length, it is likely that the polymeric chain is folded into a compact structure. This conclusion was first reached in the 1930s when studies on the radius of gyration of certain proteins showed that they are shorter than their predicted length. The globular structure of a soluble protein was visualized in detail by John Kendrew in 1957 when he used x-ray analysis to show that myoglobin is an assembly of rodlike chains with overall dimensions of 45 × 35 × 25 Å. It is now well established that most soluble proteins fold into globular, compact structures in solution. Discussion of those folded structures as well as how they undergo folding will form much of Part II of Stryer's text.

8. When water is heated, considerable energy is required to break the hydrogen bonds. Only after a large percentage of bonds are broken are the molecules more mobile and the temperature raised. This buffering capacity of water is very important to cells, which can resist changes caused by increases in temperature because of water's high heat capacity.

9. (a) First calculate the number of rubisco molecules in a plant cell:

$$(4 \times 10^{-3}\ \text{mole/liter of protein})\ (6.023 \times 10^{23}\ \text{molecules/mol}) = 2.41 \times 10^{21}\ \text{molecules/liter}$$

$(2.41 \times 10^{21}$ molecules/liter$)$ $(1 \times 10^{-13}$ L$) =$
2.41×10^{8} molecules in a plant cell

After calculating how many carbon dioxide molecules are fixed per second in the cell, calculate the number fixed in an hour and then convert the number of carbon dioxide molecules to moles.

2.41×10^{8} enzyme molecules
$\times$ 24 CO_2 molecules fixed per second per enzyme molecule $=$
5.78×10^{9} molecules CO_2 fixed per second

or

5.78×10^{9} molecules/sec $\times$ 3.6×10^{3} sec/hr $=$
20.1×10^{12} molecules per hour

$$\frac{20.1 \times 10^{12} \text{ molecules per hour}}{6.023 \times 10^{23} \text{ molecules per mole}} =$$

3.34×10^{-11} mole carbon dioxide fixed by one plant cell per hour

(b) The molecular weight of rubisco is 5.6×10^{5} g/mol and its concentration in plant cells is 4×10^{-3} mole per liter.

$(5.6 \times 10^{5}$ g/mol$)$ $(4 \times 10^{-3}$ mole/liter$) =$
22.4×10^{2} g rubisco protein per liter

10. (a) 0.2 μm $= 2 \times 10^{-7}$ m, and 1 Å equals 1×10^{-10} m, so the limit of resolution of the light microscope is 2×10^{-7} m $\times$ 1 Å$/10^{-10}$ m $=$ 2000 Å. This means that about 100 DNA molecules would have to be laid side by side in order to be visible by light microscopy.

(b) To fit inside the bacterial cell, a DNA molecule that is 1.4×10^{-3} m in length would have to be folded to form 700 chains no longer than 2×10^{-6} m in length.

$$\frac{1.4 \times 10^{-3} \text{ m}}{2 \times 10^{-6} \text{ m required to fit}} = 700 \text{ folded chains}$$

CHAPTER

2

Protein Structure and Function

Proteins are macromolecules that play central roles in all the processes of life. Stryer begins Chapter 2 with a description of the chemical properties of amino acids, the building blocks of proteins. It is essential that you learn the names, symbols, and properties of the 20 common amino acids at this point, as they will recur throughout the text in connection with protein structures, enzymatic mechanisms, metabolism, protein synthesis, and the regulation of gene expression. It is also important to review the behavior of weak acids and bases, either in the appendix to Chapter 2 or in an introductory chemistry text. Following the discussion of amino acids, Stryer turns to peptides and to the linear sequences of amino acid residues in proteins. Next, he describes the folding of these linear polymers into the specific three-dimensional structures of proteins. You should note that the majority of functional proteins exist in water and that their structures are stabilized by the forces and interactions you learned about in Chapter 1. Stryer concludes this chapter with examples of how protein conformations determine their functions.

LEARNING OBJECTIVES

When you have mastered this chapter, you should be able to complete the following objectives:

Introduction (Stryer pages 17–18)

1. List examples of *protein* functions.

Amino Acids (Stryer pages 18–23)

2. Draw the structure of an *amino acid* and indicate the following features, which are common to all amino acids: *functional groups, side chains, ionic forms,* and *isomeric forms.*

3. Classify each of the 20 amino acids according to the side chain on the α *carbon* as *aliphatic, aromatic, sulfur-containing, aliphatic hydroxyl, basic, acidic,* or *amide derivative.*

4. Give the name and one-letter and three-letter *symbol* of each amino acid. Describe each amino acid in terms of *size, charge, hydrogen-bonding capacity, chemical reactivity,* and *hydrophilic* or *hydrophobic* nature.

5. Define *pH* and *p*K. Use these concepts to predict the *ionization state* of any given amino acid or its side chain in a protein.

Amino Acid Sequences of Proteins (Stryer pages 24–27)

6. Draw a *peptide bond* and describe its *conformation* and its role in *polypeptide* sequences. Indicate the *N-* and *C-terminal residues* in *peptides.*

7. Define *main chain, side chains,* and *disulfide bonds* in polypeptides. Give the range of *molecular weights* of proteins.

8. Explain the origin and significance of the unique *amino acid sequences* of proteins.

9. List examples of the *modification* and *cleavage* of proteins that expand their functional roles.

Three-Dimensional Structure of Proteins (Stryer pages 27–39)

10. Explain the importance of protein conformation to *biological activity.*

11. Differentiate between two major *periodic structures* of proteins: the α *helix* and the β *pleated sheet.* Describe the patterns of hydrogen bonding, the shapes, and the dimensions of these structures.

12. Describe α*-helical coiled coils* in specialized proteins and the role of β *turns* or *hairpin turns* in the structure of common proteins.

13. Describe the biological significance, unique amino acid sequence, and helical structure of *collagen.*

14. List the types of interactions among amino acid side chains that stabilize the *three-dimensional structures* of proteins. Give examples of *hydrogen bond donors* and *acceptors.*

15. Using *myoglobin, ribonuclease,* and *integral membrane proteins* as examples, describe the main characteristics of native folded protein structures.

16. Describe the *primary, secondary, supersecondary, tertiary,* and *quaternary structures* of proteins. Describe *domains.*

17. Using *ribonuclease* as an example, describe the evidence that the information needed to specify the three-dimensional structure of a protein is contained in its amino acid sequence.

Specific Binding by Proteins (Stryer pages 39–40)

18. Explain the ability of proteins to interact with diverse molecules.

19. Give examples of protein interactions in enzymatic reactions, the transmission of signals, and the conversion of one form of energy into another.

SELF-TEST

Introduction

1. Match the proteins in the left column with the appropriate function in the right column.

(a) Hemoglobin ________	(1) Enzymatic catalysis
(b) Ribonuclease ________	(2) Transport
(c) Acetylcholine receptor ________	(3) Transmission of nerve impulses
(d) Myosin ________	(4) Immune protection
(e) Collagen ________	(5) Coordinated motion
(f) γ-Globulins ________	(6) Control of growth
(g) Nerve growth factor ________	(7) Mechanical support

2. Explain how proteins can have the diverse functions they exhibit.

____

____

Amino Acids

3 (a). Examine the four amino acids given below:

A: COO^- — ^+HN—C—H (ring: H_2C, CH_2, CH_2)

B: COO^- — ^+H_3N—C—H — CH_2 — (benzene ring) — OH

C: COO^- — ^+H_3N—C—H — CH_2 — CH — H_3C CH_3

D: COO^- — ^+H_3N—C—H — CH_2 — CH_2 — CH_2 — CH_2 — NH_3^+

Indicate which of these amino acids are associated with the following properties:

(a) Aliphatic side chain ______

(b) Basic side chain ______

(c) Three ionizable groups ______

(d) Charge of +1 at pH 7.0 ______

(e) pK ~10 in proteins ______

(f) Secondary amino group ______

(g) Designated by the symbol K ______

(h) In the same class as phenylalanine ______

(i) Most hydrophobic of the four ______

(j) Side chain capable of forming hydrogen bonds ______

(k) May cross-link polypeptide chains of some specialized proteins ______

3 (b). Name the four amino acids:

A ______

B ______

C ______

D ______

3 (c). Name the other amino acids of the same class as D:

______

______

4. Draw the structure of cysteine at pH 1.

5. Match the amino acids in the left column with the appropriate side chain types in the right column.

(a) Lys ______	(1) Nonpolar aliphatic
(b) Glu ______	(2) Nonpolar aromatic
(c) Leu ______	(3) Basic
(d) Cys ______	(4) Acidic
(e) Trp ______	(5) Sulfur-containing
(f) Ser ______	(6) Hydroxyl-containing

6. Which of the following amino acids have side chains that are negatively charged under physiologic conditions (i.e., near pH 7)?

(a) Asp
(b) His
(c) Trp
(d) Glu
(e) Cys

7. Why does histidine act as a buffer at pH 6.0? What can you say about the buffering capacity of histidine at pH 7.6?

____

____

8. Where stereoisomers of biomolecules are possible, only one is usually found in most organisms; for example, only the L amino acids occur in proteins. What problems would occur if, for example, the amino acids in the body proteins of herbivores were in the L isomer form whereas those in a large number of the plants they fed upon were in the D isomer form?

____

____

Amino Acid Sequences of Proteins

9. How many different dipeptides can be made from the 20 L amino acids? What are the minimum and the maximum number of p*K* values for any dipeptide?

____

____

10. For the pentapeptide Glu-Met-Arg-Thr-Gly,

(a) name the carboxyl-terminal residue: ______________

(b) give the number of charged groups at pH 7: ______________

(c) give the net charge at pH 1: ______________

(d) write the sequence using one-letter symbols: ______________

(e) draw the peptide bond between the Thr and Gly residues, including both side chains:

11. If a polypeptide has 400 amino acid residues, what is its approximate mass?

(a) 11,000 daltons
(b) 22,000 daltons
(c) 44,000 daltons
(d) 88,000 daltons

12. Which amino acid can stabilize protein structures by forming covalent cross-links between polypeptide chains?

(a) Met
(b) Ser
(c) Gln
(d) Gly
(e) Cys

13. Several amino acids can be modified after the synthesis of a polypeptide chain to enhance the functional capabilities of the protein. Match the type of modifying group in the left column with the appropriate amino acid residues in the right column.

(a) Phosphate ______	(1) Glu
	(2) Thr
(b) Hydroxyl ______	(3) Pro
	(4) Ser
(c) γ-Carboxyl ______	(5) N-terminal
(d) Acetyl ______	(6) Tyr

14. How can a protein be modified to make it more hydrophobic?

______

______

Three-Dimensional Structure of Proteins

15. Which of the following statements about the peptide bond are true?

(a) The peptide bond is planar because of the partial double-bond character of the bond between the carboxyl carbon and the nitrogen.
(b) There is relative freedom of rotation of the bond between the carboxyl carbon and the nitrogen.
(c) The hydrogen that is bonded to the nitrogen atom is trans to the oxygen of the carboxyl.
(d) There is no freedom of rotation around the bond between the α carbon and the carboxyl carbon.

16. Which of the following statements about the α helix structure of proteins is correct?

(a) It is maintained by hydrogen bonding between amino acid side chains.
(b) It makes up about the same percentage of all proteins.
(c) It can serve a mechanical role by forming stiff bundles of fibers in some proteins.
(d) It is stabilized by hydrogen bonds between backbone amino groups and carboxyl groups of polypeptide chains.
(e) It includes all 20 amino acids at equal frequencies.

17. Which of the following properties are common to α-helical and β pleated sheet structures in proteins?

(a) Rod shape
(b) Hydrogen bonds between main-chain CO and NH groups
(c) Axial distance between adjacent amino acids of 3.5 Å
(d) Variable numbers of participating amino acid residues

18. Explain why α helix and β pleated sheet structures are often found in the interior of water-soluble proteins.

______

______

19. Which of the following amino acids may alter the direction of polypeptide chains and interrupt α helices?

(a) Phe
(b) Cys
(c) Trp
(d) His
(e) Pro

20. Explain why Gly occurs every third residue in the sequence of collagen.

____

____

21. Approximately 1000 amino acid residues constitute each chain of the three-stranded collagen helix. How long is each molecule in Å, μm, and cm?

____

____

22. Which of the following amino acid residues are likely to be found on the inside of a water-soluble protein?

(a) Val
(b) His
(c) Ile
(d) Arg
(e) Asp

23. Which of the following statements about the structures of water-soluble proteins, exemplified by myoglobin and ribonuclease A, are not true?

(a) They contain tightly packed amino acids in their interior.
(b) Most of their nonpolar residues face the aqueous solvent.
(c) The main-chain NH and CO groups are often involved in H-bonded secondary structures in the interior of these proteins.
(d) Polar residues such as His may be found in the interior of these proteins if the residues have specific functional roles.
(e) All these proteins contain β sheet structural motifs.

24. Match the levels of protein structures in the left column with the appropriate descriptions in the right column.

(a) Primary __________
(b) Secondary __________
(c) Super-secondary __________
(d) Tertiary __________
(e) Quaternary __________

(1) Association of protein subunits
(2) Aggregate of α-helical and β sheet structures
(3) Linear amino acid sequence
(4) Spatial arrangement of amino acids that are near each other in the linear sequence
(5) Necessary for the catalytic activity of an enzyme

25. The hydrophobic interactions of the aliphatic side chains of amino acids help stabilize the folded structure of proteins. What purpose is served by the variety of sizes and shapes of aliphatic side chains?

26. Which of the following statements are true?

(a) Ribonuclease (RNase) can be treated with urea and reducing agents to produce a random coil.
(b) If one oxidizes random-coil RNase in urea, it quickly regains its enzymatic activity.
(c) If one removes the urea and oxidizes RNase slowly, it will renature and regain its enzymatic activity.
(d) Although renatured RNase has enzymatic activity, it can be readily distinguished from native RNase.

27. When most proteins are exposed to acidic pH (e.g., pH 2), they lose biological activity. Explain why.

Specific Binding by Proteins

28. Which of the following amino acids has a lone electron pair at one of the ring nitrogens that makes it a potential ligand in binding the iron atoms in hemoglobin?

(a) Lys
(b) Trp
(c) His
(d) Pro
(e) Arg

29. Explain how a protein may act as a molecular switch.

ANSWERS TO SELF-TEST

Introduction

1. (a) 2 (b) 1 (c) 3 (d) 5 (e) 7 (f) 4 (g) 6. You will learn later that myosin, a major protein of muscle, also has an enzymatic activity.

2. Proteins are made up of 20 amino acids arranged in unique sequences that are folded to form specific three-dimensional structures. Therefore, there are a great variety of distinct protein conformations that are especially adapted for diverse biological functions.

Amino Acids

3. (a): (a) C (b) D (c) B, D (d) D (e) B, D (f) A (g) D (h) B (i) C (j) B, D (k) D. (b) A is proline, B is tyrosine, C is leucine, and D is lysine. (c) Histidine and arginine (basic amino acids).

4. See the structure of cysteine in the margin. At pH 1, all the ionizable groups are protonated.

5. (a) 3 (b) 4 (c) 1 (d) 5 (e) 2 (f) 6

6. a, d

7. Histidine acts as a buffer at pH 6.0 because this is the p*K* of the imidazole group. At pH 7.6, histidine is a poor buffer because no one ionizing group is partially protonated and therefore capable of donating or accepting protons without markedly changing the pH.

8. All metabolic reactions in an organism are catalyzed by enzymes that are generally specific for either the D or the L isomeric form of a substrate. If an animal (an herbivore in this case) is to be able to digest the protein from a plant and build its own protein from the resulting amino acids, both the animal and the plant must make their proteins from amino acids having the same configuration.

COOH
$^{+}H_3N$—C—H
CH_2
SH

Cysteine

Amino Acid Sequences of Proteins

9. The 20 L amino acids can form 20 × 20 = 400 dipeptides. The minimum number of p*K* values for any dipeptide is two; the maximum is four.

10. (a) Glycine
  (b) 4
  (c) +2, contributed by the N-terminal amino group and the arginine residue
  (d) E-M-R-T-G
  (e) See the structure of the peptide bond in the margin.

11. c

12. e

13. (a) 2, 4, 6 (b) 3 (c) 1 (d) 5

H O H O
~N—C—C—N—C—C—O^-
H—C—OH H H
CH_3

Peptide bond

Three-Dimensional Structure of Proteins

14. The attachment of a fatty acid chain to a protein can increase its hydrophobicity and promote binding to lipid membranes.

15. a, c

16. c, d

17. b, d

18. In both α helix and β sheet structures, the polar peptide bonds of the main chain are involved in internal hydrogen bonding, thereby eliminating potential hydrogen bond formation with water. Overall the secondary structures are less polar than the corresponding linear amino acid sequences.

19. e

20. Since there are three residues per turn of helix, every third residue points inward. The interior residues are Gly because Gly is the only residue small enough to fit inside the superhelical cable.

21. Since the rise of the collagen helix is 2.9 Å per residue, the length of the molecule is 2900 Å, 0.29 μm, or 2.9×10^{-5} cm.

22. a, c. Specific charged and polar amino acid residues may be found inside some proteins, in active sites, but most polar and charged residues are located on the surface of proteins.

23. b, e. Statement (b) is incorrect because globular, water-soluble proteins have most of their nonpolar residues buried in the interior of the protein. Statement (e) is incorrect because not all water-soluble proteins contain β sheet secondary structures. For example, myoglobin is mostly α-helical and lacks β sheet structures.

24. (a) 3 (b) 4 (c) 2 (d) 5 (e) 1, 5

25. The different sizes and shapes of aliphatic side chains fill the space in the interior of proteins and promote contacts that increase van der Waals interactions.

26. a, c

27. A low pH (pH 2) will cause the protonation of all ionizable side chains and will change the charge distribution on the protein; furthermore, it will impart a large net positive charge to the protein. The resulting repulsion of adjacent positive charges and the disruption of salt bridges often cause unfolding of the protein and loss of biological activity.

Specific Binding by Proteins

28. c

29. In proteins, particularly those that contain subunits, the specific binding of a molecule or ion at one site can elicit conformational changes that are propagated over significant distances to produce a change in another site, often in another subunit. The conformational changes in the second site then result in a change of its biological activity.

PROBLEMS

1. The hydration of carbon dioxide by carbonic anhydrase requires the participation of a zinc ion. Why would you not expect to observe enhancement of bicarbonate formation in an aqueous solution that contains only CO_2 and Zn^{2+}?

2. Suppose you are studying the conformation of a monomeric protein that has an unusually high proportion of aromatic amino acid residues throughout the length of the polypeptide chain. Compared with a protein containing many glycine residues, what would you observe for allowed ranges of ϕ and ψ? What can you say about the relative α-helical content in each of the two types of proteins?

3. Glycophorin A is a glycoprotein that extends across the red blood cell membrane. The portion of the polypeptide that extends across the membrane bilayer contains 19 amino acid residues and is folded into an α helix. What is the width of the bilayer that could be spanned by this helix? The interior of the bilayer includes long acyl chains that are nonpolar. Which of the 20 L amino acids would you expect to find among those in the portion of the polypeptide that traverses the bilayer?

4. Iodoacetate reacts with cysteine side chains in proteins to form *S*-carboxymethyl derivatives. Why is it a good idea to treat a protein first with β-mercaptoethanol before carrying out carboxymethylation of cysteine residues? Why might treatment with urea also be useful?

5. The net charge of a polypeptide at a particular pH can be determined by considering the p*K* value for each ionizable group in the protein. For a linear polypeptide composed of 10 amino acids, how many α-carboxyl and α-amino groups must be considered?

6. For the formation of a polypeptide composed of 20 amino acids, how many water molecules must be removed when the peptide bonds are formed? Although the hydrolysis of a peptide bond is energetically favored, the bond is very stable in solution. Why?

7. In a particular enzyme, an alanine residue is located in a cleft where the substrate binds. A mutation that changes this residue to a glycine has no effect on activity; however, another mutation, which changes the alanine to a glutamate residue, leads to a complete loss of activity. Provide a brief explanation for these observations.

8. Before Anfinsen carried out his work on refolding in ribonuclease, some scientists argued that directions for folding are given to the protein during its biosynthesis. How did Anfinsen's experiments contradict that argument?

9. Many types of proteins can be isolated only in quantities that are too small for the direct determination of a primary amino acid sequence. Recent advances in gene cloning and amplification allow for relatively easy analysis of the gene coding for a particular protein. Why would an analysis of the gene provide information about the protein's primary sequences? Suppose that two research groups, one in New York and the other in Los Angeles, are both analyzing the same protein from the same type of human cell. Why would you not be surprised if they publish exactly the same primary amino acid sequence for the protein?

10. Wool and hair are elastic; both are α-keratins, which contain long polypeptide chains composed of α helices twisted about each other to form cablelike assemblies with cross-links involving Cys residues. Silk, on the other hand, is rigid and resists stretching; it is composed primarily of antiparallel β pleated sheets, which are often stacked and interlocked. Briefly explain these observations in terms of the characteristics of the secondary structures of these proteins.

11. What is the molarity of pure water? Show that a change in the concentration of water by ionization does not appreciably affect the molarity of the solution.

12. When sufficient H^+ is added to lower the pH by one unit, what is the corresponding increase in hydrogen ion concentration?

13. You have a solution of HCl that has a pH of 2.1. What is the concentration of HCl needed to make this solution?

14. The charged form of the imidazole ring of histidine is believed to participate in a reaction catalyzed by an enzyme. At pH 7.0, what is the probability that the imidazole ring will be charged?

15. Calculate the pH at which a solution of cysteine would have no net charge.

16. In his discussion of protein modification and cleavage, Stryer refers to the synthesis and cleavage of a large polyprotein precursor of the proteins of poliovirus, as well as to the synthesis of multiple polypeptide hormones from a single polypeptide chain. Is there an advantage to synthesizing a large precursor chain and then cleaving it to create a number of products?

17. In normal people collagen provides a matrix for bone, which is then solidified by the deposition of crystals of calcium hydroxyapatite. A connective tissue disease called *osteogenesis imperfecta* is caused by defective collagen. Among the symptoms of this disorder are skeletal deformities caused by multiple fractures due to brittle bones. Analysis of the gene coding for one of the collagen chains in an affected person shows that the glycine residue normally located at position 988 in the primary sequence is changed to a cysteine residue. The alteration disrupts the collagen helix and causes it to unfold at room temperature. How could the substitution of cysteine for glycine cause helix disruption?

18. More complex integral membrane proteins are composed of a number of membrane-spanning segments, which form bundles of α helices which are packed closely together. In some cases these bundles can form a membrane channel or pore. Examination of the amino acid sequences of these segments shows that the percentage of hydrophobic side chains is not as large as that found in a protein with a single membrane-spanning segment. In some cases, every third or fourth amino acid residue in a membrane-spanning bundle may have a polar side chain. Suggest how such bundles could be accommodated in the hydrophobic core of a membrane.

19. A survey of the location of reverse turns in soluble proteins shows that most reverse turns are located at the surface of the protein, rather than within the hydrophobic core of the folded protein. Can you suggest a reason for this observation?

20. The enzyme collagenase was first characterized in tissues from the tail of tadpoles. It cleaves a collagen triple chain at a single site, between residues 775 and 776 in a 1056-residue sequence. The cleaved helical chain unfolds and is then hydrolyzed by other proteases. Collagenase has been found in human uterine tissue after pregnancy. What is the functional significance of collagenase? Why is it important to regulate its synthesis and activity?

ANSWERS TO PROBLEMS

1. In carbonic anhydrase, the zinc atom is located in a deep pocket in coordination with three histidine side chains. Water bound to the fourth coordination position is deprotonated to form hydroxide ion, which is stabilized at pH 7, where OH^- does not usually exist. Other nearby sites on the enzyme bind CO_2, which then reacts with OH^- to form bicarbonate. Zinc ion in solution would be less likely to be found in a tetrahedral configuration with other ligands, and even if hydroxyl ion were generated, its effective concentration in solution would be low. Carbon dioxide would seldom be in proximity for the reactions with OH^- to occur.

2. The larger the side chain, the more likely it is that steric hindrance will restrict the range of ϕ and ψ. Since aromatic side chains, like the phenyl group of tyrosine, are much bulkier than the hydrogen atom of glycine, the range of ϕ and ψ values will be lower. The relative levels of glycine and the aromatic amino acids in α-helical regions are low; hence, one would expect both types of proteins to have relatively low α-helical content. Presumably the large side chains of aromatic residues interface with establishment of the regular structure of the α helix. Although there seems to be ample room for glycine residues to fit within an α-helical structure, its frequency in such chains is almost as low as that of proline, a fact that is not yet understood.

3. Since each residue in the α helix is 1.5 Å from its neighbor, the length of the chain that spans the membrane bilayer is 19×1.5 Å = 28.5 Å, which is also the width of the membrane. One would expect to find nonpolar amino acid residues in the polypeptide portion associated with the membrane bilayer. These would include Ala, Val, Leu, Ile, Met, and Phe. The actual sequence of the buried chain is

I–T–L–I–I–F–G–V–M–A–G–V–I–G–T–I–L–L–I

4. Many proteins contain disulfides (cystine), which must be reduced by thiol reagents to sulfhydryls (cysteine) before they can react with iodoacetate. Urea promotes the unfolding of the protein, providing exposure of otherwise buried disulfides to β-mercaptoethanol.

5. Only the N-terminal α-amino group and the C-terminal α-carboxyl group will undergo ionization. The internal groups will be joined by peptide bonds and are not ionizable.

6. For a peptide of n residues, $n - 1$ water molecules must be removed. A significant activation energy barrier makes peptide bonds kinetically stable.

7. Both alanine and glycine are neutral nonpolar residues with small side chains, whereas the side chain of glutamate is acidic and bulkier than that of alanine. Either feature of the glutamate R group could lead to the loss of activity by altering the protein conformation or by interfering with the binding of the substrate.

8. The fact that ribonuclease folded in vitro to yield full activity indicated that the biosynthetic machinery is not required to direct the folding process for this protein.

9. Because the sequence of DNA specifies, through a complementary sequence of RNA, the amino acid sequence of a protein, knowledge about any one of the three types of sequences yields information about the other two. One would also expect the coding sequence for a particular protein to be the same among members of the same species, allowing for an occasional rare mutation. For that reason, the published primary amino acid sequences are likely to be the same.

10. When the α helices in wool are stretched, intrahelix hydrogen bonds are broken as are some of the interhelix disulfide bridges; maximum stretching yields an extended β sheet structure. The Cys cross-links provide some resistance to stretch and help pull the α helices back to their original positions. In silk, the β sheets are already maximally stretched to form hydrogen bonds. Each β pleated sheet resists stretching, but since the contacts between the sheets primarily involve van der Waals forces, the sheets are somewhat flexible.

11. The molarity of water equals the number of moles of water per liter. A liter of water weighs 1000 grams, and its molecular weight is 18, so the molarity of water is

$$\text{M} = \frac{1000}{18} = 55.6$$

At 25°C, K_w is 1.0×10^{-14}; at neutrality, the concentration of both hydrogen and hydroxyl ions is each equal to 10^{-7} M. Thus, the actual concentration of H_2O is (55.6 – 0.0000001) M; the difference is so small that it can be disregarded.

12. Because pH values are based on a logarithmic scale, every unit change in pH means a tenfold change in hydrogen ion concentration. When pH = 2.0, $[H^+] = 10^{-2}$ M; when pH = 3.0, $[H^+] = 10^{-3}$M.

13. Assume that HCl in solution is completely ionized to H^+ and Cl^-. Then find the concentration of H^+, which equals the concentration of Cl^-.

$$\begin{aligned} \text{pH} = -\log [H^+] &= 2.1 \\ [H^+] &= 10^{-2.1} \\ &= 10^{0.9} \times 10^{-3} \\ &= 7.94 \times 10^{-3}\text{M} \end{aligned}$$

Thus, $[H^+] = [Cl^-] = [HCl] = 7.94 \times 10^{-3}$ M

14. Use the Henderson-Hasselbalch equation to calculate the concentration of histidine, whose imidazole ring is ionized at neutral pH. The value of pK for the ring is 6.0.

$$\text{pH} = \text{p}K + \log \frac{[\text{His}]}{[\text{His}^+]}$$

$$7.0 = 6.0 + \log \frac{[\text{His}]}{[\text{His}^+]}$$

$$\log \frac{[\text{His}]}{[\text{His}^+]} = 1.0$$

$$\frac{[\text{His}]}{[\text{His}^+]} = 10$$

At pH 7.0, the ratio of uncharged histidine to charged histidine is 10:1, making the probability that the side chain is charged only 9%.

15. To see which form of cysteine has no net charge, examine all the possible forms, beginning with the one that is most protonated:

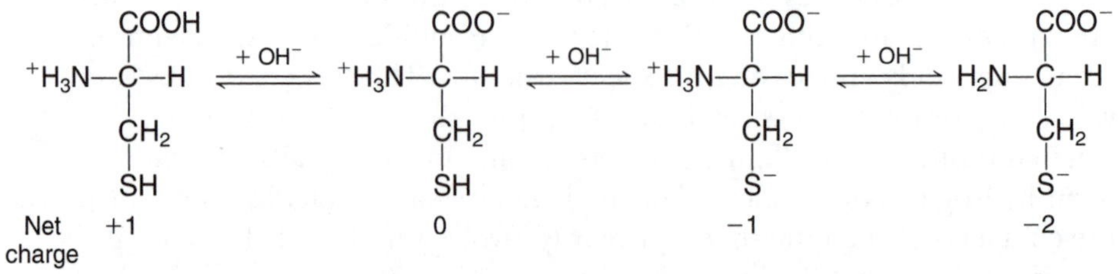

$$\begin{array}{c} COOH \\ | \\ {}^+H_3N{-}C{-}H \\ | \\ CH_2 \\ | \\ SH \end{array} \xrightleftharpoons{+\,OH^-} \begin{array}{c} COO^- \\ | \\ {}^+H_3N{-}C{-}H \\ | \\ CH_2 \\ | \\ SH \end{array} \xrightleftharpoons{+\,OH^-} \begin{array}{c} COO^- \\ | \\ {}^+H_3N{-}C{-}H \\ | \\ CH_2 \\ | \\ S^- \end{array} \xrightleftharpoons{+\,OH^-} \begin{array}{c} COO^- \\ | \\ H_2N{-}C{-}H \\ | \\ CH_2 \\ | \\ S^- \end{array}$$

Net charge: +1, 0, −1, −2

The pH of the cysteine solution at which the amino acid has no net charge will be that point at which there are equal amounts of the compound with a single positive charge and a single negative charge. This is, in effect, the average of the two corresponding p*K* values (see p. 43 of the text), one for the α-carboxyl and the other for the α-amino group. Thus, $(1.8 + 10.8)/2 = 12.6/2 = 6.3$. This value is also known as the *isoelectric point.*

16. The primary advantage of precursor chain synthesis is that the production of related proteins can be coordinated. This could be important in poliovirus infection, where 60 copies of each of four proteins associate to form a new virus particle. It may also be important for coordinated synthesis of hormones with related activities. It is worth noting that there are other reasons for the synthesis of polyprotein precursors. The genome of poliovirus consists of a single RNA molecule that acts as a messenger on entering the cytoplasm of the host. In eukaryotic cells a messenger RNA molecule can be translated into only one polypeptide chain. Therefore the poliovirus can reproduce only by synthesizing its proteins by sequential cleavages.

17. In the helical region of collagen, glycine occupies every third position in a sequence of just over 1000 amino acids that constitute the typical collagen chain. Each glycine residue is located on the inside of the three-chain helix, where there is no room for a larger side chain. The substitution of a cysteine residue at one location probably disrupts the helix because the larger side chain ($-CH_2-SH$) cannot fit inside.

It is remarkable that an alteration at only 1 of over 300 glycine residues can so dramatically disrupt the helix. There is good evidence that disruption of the helix at position 988 causes excessive hydroxylation of neighboring prolines and an increased level of glycosylation of lysine residues. These changes make it impossible for the chains to form highly ordered arrays and also interfere with cellular secretion of newly synthesized collagen protein.

18. The essential requirement for a membrane-spanning segment is that most side chains that face the lipid core be hydrophobic. However, closely packed bundles of α helices will have side chains that face the interior of the bundle assembly. These side chains could be polar, forming interchain hydrogen bonds. Examination of the basic structure of an α helix shows that the side chain of every third residue (more precisely, every 3.6 residues) is on the same side of the main chain. Thus in a membrane-spanning segment, each third polar side chain could face the inside of the bundle array, away from the hydrophobic core.

19. Figure 2-37 in Stryer's text shows that in a reverse turn the CO group of residue 1 is hydrogen-bonded to the NH group of residue 4. However, there are no adjacent amino acid residues available to form intrachain hydrogen bonds with the CO and NH groups of residues 2 and 3. These groups cannot form hydrogen bonds in the hydrophobic environment found in the interior portion of a folded protein. They are more likely to hydrogen-bond with water on the surface of the protein.

20. Collagenase probably functions to help with the rebuilding and reorganization of tissues. Recall that collagen is the most abundant protein in connective tissues. During metamorphosis, amphibians undergo dramatic morphological changes involving the appendages as well as other organs.

Collagenase assists in removing areas of tissue that are no longer needed (the tail of the tadpole during metamorphosis, for example). It is of course desirable to control collagenase synthesis and activity in order to avoid destroying vital tissue structures at an inappropriate time. Human collagenase is first produced as an inactive precursor that can be cleaved to yield an active enzyme. It is also subject to control by inhibitory proteins.

EXPANDED SOLUTIONS TO TEXT PROBLEMS

1. (a) Since tropomyosin is double-stranded, each strand will have a mass of 35 kd. If the average residue has a mass of 110 d, there are 318 residues per strand (35,000/110), and the length is 477Å (1.5 Å/residue × 318).
   (b) Since 4 of the 40 residues formed the hairpin turn, 36 residues formed the antiparallel β pleated sheet which is 18 residues long (36/2). In β pleated sheets, the axial distance between adjacent amino acids is 3.5 Å. Hence, the length of this segment is 63 Å (3.5 × 18).

2. Branching at the β-carbon of the side chain (isoleucine), in contrast to branching at the γ-carbon (leucine), sterically hinders the formation of a helix. This fact can be shown with molecular models.

3. Changing alanine to valine results in a bulkier side chain, which prevents the correct interior packing of the protein. Changing a nearby, bulky, isoleucine side chain to glycine apparently alleviates the space problem and allows the correct conformation to take place.

4. The amino acid sequence of insulin does not determine its three-dimensional structure. By catalyzing a disulfide-sulfhydryl exchange, this enzyme speeds up the activation of scrambled ribonuclease because the native form is the most thermodynamically stable. In contrast, the structure of active insulin is not the most thermodynamically stable form. The three dimensional structure of insulin is determined by the folding of preproinsulin which is later processed to mature insulin.

5. Appropriate hydrogen-bonding sites on the protease might induce formation of an intermolecular β pleated sheet with a portion of the target protein. This process would effectively fully extend α helices and other folded portions of the target molecule.

6. Being the smallest amino acid, glycine can fit into spaces too small to accommodate other amino acids. Thus, if sharp turns or limited spaces for amino acids occur in a functionally active conformation of a protein, glycine is required; no substitute will suffice. An easily visualized example is the triple helix of collagen (Figure 2-41 in Stryer). In view of the above, it is not surprising that glycine is highly conserved.

7. To answer this question one needs to know some of the characteristics of the guanidinium group of the side chain of arginine and of the other functional groups in proteins. Most of the needed information is presented in Figure 2-37 in Stryer; note that the guanidinium group has a

positive charge at pH 7 and contains several hydrogen bond donor groups. The positive charge can form salt bridges with the negatively charged groups of proteins (glutamate and asparatate side chains and the terminal carboxylate). As a hydrogen bond donor the guanidinium group can react with the various hydrogen bond acceptors shown in Figure 2-37 (glutamine, asparagine, glutamate, and the main chain carbonyl). It can also hydrogen bond with the hydroxyl group of serine and threonine (not shown in Figure 2-37). Hydroxyl groups accept hydrogen bonds much like water does.

8. The keratin of hair is essentially a bundle of long protein strands joined together by disulfide bonds. If these bonds are broken (reduced) by the addition of a thiol and the hair curled, the keratin chains slip past each other into a new configuration. When an oxidizing agent is added, new disulfide bonds are formed, thus stabilizing the new "curled" state.

CHAPTER

3

Exploring Proteins

Chapter 3 extends Chapter 2 by introducing the most important methods used to investigate proteins. Many of these were essential in discovering the principles of protein structure and function presented in the preceding chapter. These methods also constitute the essentials of the armamentarium of modern biochemical research and underlie current developments in biotechnology. First, Stryer describes methodological principles for the analysis and purification of proteins. He then describes methods of sequencing the amino acids in proteins, and explains why the knowledge revealed by these techniques is so important. He continues with a discussion of antibodies as highly specific analytical reagents. Optical methods that reveal information concerning protein conformation are presented, followed by a discussion of the use of x-ray crystallography and nuclear magnetic resonance spectroscopy to determine the three-dimensional structures of proteins. To close the chapter, he describes the uses of peptides of defined sequence and how they are chemically synthesized.

LEARNING OBJECTIVES

When you have mastered this chapter, you should be able to complete the following objectives:

Separation and Purification of Proteins (Stryer pages 45–53)

1. Describe the principle of *electrophoresis* and its application in the separation of proteins.

2. Explain the determination of *protein mass* by *sodium dodecyl sulfate–polyacrylamide gel electrophoresis.*

3. Define the *isoelectric point* (*pI*) of a protein and describe *isoelectric focusing* as a separation method.

4. List the properties of proteins that can be used to accomplish their *separation* and *purification,* and correlate them with the appropriate methods: *gel-filtration chromatography, dialysis, salting out, ion-exchange chromatography,* and *affinity chromatography.* Describe the basic principles of each of these methods.

5. Compare the factors that affect the *velocity of migration* of a particle during *ultracentrifugation* and during *electrophoresis.*

6. Define the *sedimentation coefficient S,* and give its common name. Note the range of *S values* for biomolecules and cells.

7. Relate the *sedimentation velocity* of a particle to the *hydrodynamic properties* of the particle and to the density of the solution. The hydrodynamic properties of a particle include its *size, shape,* and *density.*

8. Describe *zonal centrifugation* and *sedimentation equilibrium* and explain their applications to the study of proteins.

9. Outline the application of *electrospray mass spectrometry* to the analysis of proteins and compare the merits of the various methods of determining the molecular weights of proteins.

Determination of Amino Acid Sequences (Stryer pages 53–60)

10. Outline the steps in the determination of the *amino acid composition* and the *amino-terminal residue* of a peptide.

11. Describe the sequential *Edman degradation method* and the automated determination of the amino acid sequences of peptides.

12. List the most common reagents used for the *specific cleavage* of proteins. Explain the application of *overlap peptides* to protein sequencing.

13. Describe the additional steps that must be used for sequencing *disulfide-linked polypeptides* and *oligomeric proteins.*

14. Explain how *recombinant DNA technology* is used to determine the amino acid sequences of *nascent* proteins. Note the differences between a nascent protein and one that has undergone *posttranslational modifications.*

15. Give examples of the important information that amino acid sequences reveal.

Use of Specific Antibodies (Stryer pages 60–63)

16. Define the terms *antibody, antigen, antigenic determinant (epitope),* and *immunoglobulin G.*

17. Contrast *polyclonal antibodies* and *monoclonal antibodies* and describe their preparation.

18. Outline methods that use specific antibodies in the analysis or localization of proteins.

Determination of Protein Structure (Stryer pages 63–68)

19. Explain the principles of *circular dichroism* and describe its applications to determining the *α helix content* and *composition* of proteins.

20. Describe the fundamentals of the *x-ray crystallographic analysis* of a protein, and give the basic physical principles underlying this technique.

21. Provide a similar description of the method and basic physical principles underlying *nuclear magnetic resonance spectrometry* as applied to protein structure determination.

22. Compare the relative advantages and disadvantages of x-ray crystallography and NMR spectroscopy for protein structure determination.

23. Note how determining the three-dimensional structure of proteins provides fundamental insights into their functions, interactions, and evolution.

Synthesis of Peptides (Stryer pages 68–70)

24. List the most important uses of *synthetic peptides*.

25. Outline the steps of the *solid-phase method* for the synthesis of peptides.

SELF-TEST

Separation and Purification of Proteins

1. The following five proteins, which are listed with their molecular weights and isoelectric points, were separated by SDS–polyacrylamide gel electrophoresis. Give the order of their migration from the top (the point of sample application) to the bottom of the gel.

	Molecular weight (daltons)	pI
(a) α-Antitrypsin	45,000	5.4
(b) Cytochrome *c*	13,400	10.6
(c) Myoglobin	17,000	7.0
(d) Serum albumin	69,000	4.8
(e) Transferrin	90,000	5.9

Top ——————————— Bottom

2. Which of the following statements are not true?
   (a) The pI is the pH value at which a protein has no charges.
   (b) At a pH value equal to its pI, a protein will not move in the electric field of an electrophoresis experiment.
   (c) An acidic protein will have a pI greater than 7.
   (d) A basic protein will have a pI greater than 7.

3. If the five proteins in question 1 were separated in an isoelectric-focusing experiment, what would be their distribution between the positive (+) and negative (–) ends of the gel? Indicate the high and low pH ends.

Cathode (–)___________________________(+) Anode

4. SDS–polyacrylamide gel electrophoresis and the isoelectric-focusing method for the separation of proteins have which of the following characteristics in common? Both

(a) separate native proteins.
(b) make use of an electrical field.
(c) separate proteins according to their mass.
(d) require a pH gradient.
(e) are carried out on supporting gel matrices.

5. Before high-performance liquid chromatography (HPLC) methods were devised for the separation and analysis of small peptides, electrophoresis on a paper support was frequently used. Separation was effected on the basis of the charge on a peptide at different pH values. Predict the direction of migration for the following peptides at the given pH values. Use C for migration toward the cathode, the negative pole; A for migration toward the anode, the positive pole; and O if the peptide remains stationary.

	pH 2.0	4.0	6.0	11.0
(a) Lys-Gly-Ala-Gly				
(b) Lys-Gly-Ala-Glu				
(c) His-Gly-Ala-Glu				
(d) Glu-Gly-Ala-Glu				
(e) Gln-Gly-Ala-Lys				

6. How would the time required for the separation described in question 5 be changed if all the solutions that were used during the electrophoresis contained 100 g/L of table sugar (sucrose)?

____

____

7. Match the properties of proteins listed in the left column with the appropriate separation or purification methods listed in the right column.

(a) Size __________
(b) Charge __________
(c) Specific binding __________
(d) Solubility __________

(1) Electrophoresis
(2) Gel filtration
(3) Salting out
(4) Immunoprecipitation
(5) Isoelectric focusing
(6) Affinity chromatography
(7) Ion-exchange chromatography
(8) Zonal ultracentrifugation

8. Which of the methods listed in question 7 are sufficiently specific to isolate a particular protein from a complex mixture of proteins in one step? Explain why.

____

____

9. The sedimentation velocity of a protein in a centrifuge does not depend on the

(a) density of the solution.
(b) density of the protein.
(c) charge on the protein.
(d) shape of the protein.
(e) mass of the protein.

10. The molecular weight of a protein can be determined by SDS–polyacrylamide gel electrophoresis or by sedimentation equilibrium. Which method would you use to determine the molecular weight of a protein containing four subunits, each consisting of two polypeptide chains cross-linked by two disulfide bridges? Explain your answer.

11. After isolating and purifying to homogeneity a small enzyme (110 amino acids long) from a culture of bacteria, you are confused as to whether you grew wild-type bacteria or a mutant strain that produced the enzyme with a valine residue at position 66 instead of the glyine found in the wild-type strain. How could you quickly determine which protein you had?

Determination of Amino Acid Sequences

12. Which of the following are useful in identifying the amino-terminal residue of a protein?

(a) Cyanogen bromide
(b) Fluorodinitrobenzene
(c) Performic acid
(d) Dabsyl chloride
(e) Phenyl isothiocyanate

13. Which of the following statements concerning the Edman degradation method are true?

(a) Phenyl isothiocyanate is coupled to the amino-terminal residue.
(b) Under mildly acidic conditions, the modified peptide is cleaved into a cyclic derivative of the terminal amino acids and a shortened peptide (minus the first amino acid).
(c) Once the PTH amino acid is separated from the original peptide, a new cycle of sequential degradation can begin.
(d) If a protein has a blocked amino-terminal residue (as does *N*-formyl methionine, for example), it cannot react with phenyl isothiocyanate.

14. When sequencing proteins, one tries to generate overlapping peptides by using cleavages at specific sites. Which of the following statements about the cleavages caused by particular chemicals or enzymes are true?

(a) Cyanogen bromide cleaves at the carboxyl side of threonine.
(b) Trypsin cleaves at the carboxyl side of Lys and Arg.
(c) Chymotrypsin cleaves at the carboxyl side of aromatic and bulky amino acids.
(d) 2-Nitro-5-thiocyanobenzoate cleaves on the amino side of cysteine residues.
(e) Chymotrypsin cleaves at the carboxyl side of aspartate and glutamate.

15. What treatments could you apply to the following hemoglobin fragment to determine the amino-terminal residue and to obtain two sets of peptides with overlaps so that the complete amino acid sequence can be established? Give the sequences of the peptides obtained.

Val-Leu-Ser-Pro-Ala-Lys-Thr-Asn-Val-Lys-Ala-Ala-Trp-Gly-Lys-Val-Gly-Ala-His-Ala-Gly-Glu-Tyr-Gly-Ala-Glu-Ala-Thr-Glu

16. The following reagents are often used in protein chemistry:

(1) CNBr	(7) 6 N HCl
(2) Urea	(8) Fluorescamine
(3) β-Mercaptoethanol	(9) Phenyl isothiocyanate
(4) Trypsin	(10) Chymotrypsin
(5) Dicyclohexylcarbodiimide	(11) Dilute F_3CCOOH
(6) Dabsyl chloride	

Which of these reagents are best-suited for the following tasks?

(a) Determination of the amino acid sequence of a small peptide:

(b) Identification of the amino-terminal residue of a peptide (of which you have less than 10^{-7} grams):

(c) Reversible denaturation of a protein devoid of disulfide bonds:

(d) Hydrolysis of peptide bonds on the carboxyl side of aromatic residues:

(e) Cleavage of peptide bonds on the carboxyl side of methionine:

(f) Hydrolysis of peptide bonds on the carboxyl side of lysine and arginine residues:

(g) Reversible denaturation of a protein that contains disulfide bonds (two reagents are needed):

(h) Activation of carboxyl groups during peptide synthesis:

(i) Determination of the amino acid composition of a small peptide:

(j) Removal of *t*-Boc protecting group during peptide synthesis:

17. Which of the following techniques are used to locate disulfide bonds in a protein?

(a) The protein is first reduced and carboxymethylated.
(b) The protein is cleaved by acid hydrolysis.

(c) The protein is specifically cleaved under conditions that keep the disulfide bonds intact.
(d) The peptides are separated by SDS–polyacrylamide gel electrophoresis.
(e) The peptides are separated by two-dimensional electrophoresis with an intervening performic acid treatment.

18. In spite of the convenience of using recombinant DNA techniques for determining the amino acid sequences of proteins, chemical analyses of amino acid sequences are frequently required. Explain why.

____

____

19. Which of the following are important reasons for determining the amino acid sequences of proteins?

(a) Knowledge of amino acid sequences helps elucidate the molecular basis of biological activity.
(b) Alteration of an amino acid sequence may cause abnormal functioning and disease.
(c) Amino acid sequences provide insights into evolutionary pathways and protein structures.
(d) The three-dimensional structure of a protein can be predicted from its amino acid sequence.
(e) Amino acid sequences provide information about the destination and processing of some proteins.
(f) Amino acid sequences allow prediction of the DNA sequences encoding them and thereby facilitate the preparation of DNA probes specific for the regions of their genes.

Use of Specific Antibodies

20. Match the terms in the left column with the appropriate item or items from the right column.

(a) Antigens ____________
(b) Antigenic determinants ____________
(c) Polyclonal antibodies ____________
(d) Monoclonal antibodies ____________

(1) Immunoglobulins
(2) Foreign proteins, polysaccharides, or nucleic acids
(3) Antibodies produced by hybridoma cells
(4) Groups recognized by antibodies
(5) Heterogeneous antibodies
(6) Homogeneous antibodies
(7) Antibodies produced by injecting an animal with a foreign substance
(8) Epitopes

21. The methods used to localize a specific protein in an intact cell are
(a) Western blotting.
(b) solid-phase immunoassay.
(c) enzyme-linked immunosorbent assay.

(d) immunoelectron microscopy.
(e) fluorescense microscopy.

22. Explain why immunoassays are especially useful for detecting and quantifying small amounts of a substance in a complex mixture.

___

___

Determination of Protein Structure

23. Which of the following statements concerning x-ray crystallography is not true?

(a) Only crystallized proteins can be analyzed.
(b) The x-ray beam is scattered by the protein sample.
(c) All atoms scatter x-rays equally.
(d) The basic experimental data are relative intensities and positions of scattered electrons.
(e) The electron-density maps are obtained by applying the Fourier transform to the scattered electron intensities.
(f) The resolution limit for proteins is about 2 Å.

24. How can x-ray crystallography provide information about the interaction of an enzyme with its substrate?

___

___

25. Match each term, property, or requirement in the right column with the more appropriate method of structure determination in the left column.

(a) Circular dichroism ___
(b) Nuclear magnetic resonance spectroscopy ___

(1) Peptide bond
(2) Provides atomic detail
(3) Radiofrequency region of electromagnetic spectrum
(4) Optical activity
(5) Protein in solution
(6) α-Helical content of a 10-kd protein
(7) Nuclear Overhauser effect (NOE)
(8) Far-ultraviolet region of electromagnetic spectrum
(9) α-Helical content of a 100-kd protein
(10) Atomic nucleus

Synthesis of Peptides

26. The amino acid sequence of a protein is known and strong antigenic determinants have been predicted from the sequence; however, you do not have enough of the pure protein to prepare antibodies. How could you circumvent this problem?

___

___

27. Which of the following is commonly used as a protecting group during peptide synthesis?

(a) *tert*-Butyloxycarbonyl
(b) Dicyclohexylcarbodiimide
(c) Dicyclohexylurea
(d) Hydrogen fluoride
(e) Phenyl isothiocyanate

ANSWERS TO SELF-TEST

Separation and Purification of Proteins

1. Top $\xrightarrow{\text{e d a c b}}$ Bottom

2. a, c. Regardless of the pH, a protein is never devoid of charges; at the pI, the sum of all the charges is zero.

3. High pH (−) $\xrightarrow{\text{b c e a d}}$ Low pH (+)

4. b, e

5.

	pH 2.0	4.0	6.0	11.0
(a)	C	C	C	A
(b)	C	C	O	A
(c)	C	C	A	A
(d)	C	O	A	A
(e)	C	C	C	A

For example, peptide b carries a net charge of +1.5 at pH 2.0 (Lys side chain, +1; α-amino group, +1; Glu side chain, 0; and terminal carboxyl, −0.5, since the pH coincides with its pK value). At pH 4.0, the net charge is +0.5; the Glu side chain is half ionized (−0.5), but the terminal carboxyl is almost completely ionized (−1). At pH 6.0, the net charge is 0 due to a +2 charge contributed by the Lys residue and a −2 charge contributed by the Glu residue. At pH 11.0, the α-amino group is deprotonated (charge of 0) and the Lys side chain is half-protonated (charge of +0.5); thus, the net charge is −1.5. The same answer for peptide b can be given graphically:

		Net charge
pH 2.0:	$^{+}H_3N$—Lys—Gly—Ala—Glu—COOH (COO^-); Lys side chain NH_3^+; Glu side chain COOH	+1.5
pH 4.0:	$^{+}H_3N$—Lys—Gly—Ala—Glu—COO^-; Lys side chain NH_3^+; Glu side chain COOH (COO^-)	+0.5
pH 6.0:	$^{+}H_3N$—Lys—Gly—Ala—Glu—COO^-; Lys side chain NH_3^+; Glu side chain COO^-	0
pH 11.0:	H_2N—Lys—Gly—Ala—Glu—COO^-; Lys side chain NH_2 (NH_3^+); Glu side chain COO^-	−1.5

6. More time would be required for the separation, since the velocity of movement of the compounds would be slowed because of the increased viscosity of the solution. The velocity at which a molecule moves during electrophoresis is inversely dependent on the frictional coefficient, which is, itself, directly proportional to the viscosity of the solution.

7. (a) 1, 2, 8 (b) 1, 3, 5, 7 (c) 4, 6 (d) 3. Note that some electrophoretic methods, such as SDS–polyacrylamide gel electrophoresis, separate proteins according to their size, although their migration on the gel is due to the overall charge. Salting out may depend on the charge of the protein.

8. Immunoprecipitation and affinity chromatography depend on the specific interaction of a protein with an antibody and a binding substance, respectively; therefore, these methods should be sufficiently specific to isolate a particular protein. Isoelectric focusing can also be quite specific, but it is impractical for complex mixtures.

9. c

10. Determinations of mass by SDS–polyacrylamide gel electrophoresis are carried out on proteins that have been denatured by the detergent in a reducing medium; the reducing agent in the medium disrupts disulfide bonds. Therefore, to determine the molecular weight of a native protein containing subunits with disulfide bridges, you must use the sedimentation equilibrium method. Another nondenaturing method, gel-filtration chromatography, can be used to obtain approximate native molecular weights.

11. Electrospray mass spectrometry could easily distinguish a protein of this approximate mass (101 amino acids × approximately 110 d per amino acid = 12 kd) that contained an extra 41 atomic mass units as a result of the substitution of a valine for a glycine residue.

Determination of Amino Acid Sequences

12. b, d, e

13. a, b, c, d

14. b, c, d

15. The amino-terminal residue of the hemoglobin fragment can be determined by labeling it with fluorodinitrobenzene or dabsyl chloride or by analyzing the intact fragment by the Edman degradation method; this shows that the amino-terminal residue is Val. Trypsin digestion, separation of peptides, and Edman degradation give

Val-Leu-Ser-Pro-Ala-Lys

Thr-Asn-Val-Lys

Ala-Ala-Trp-Gly-Lys

Val-Gly-Ala-His-Ala-Gly-Glu-Tyr-Gly-Ala-Glu-Ala-Thr-Glu

Chymotrypsin digestion, separation of peptides, and Edman degradation give

Val-Leu-Ser-Pro-Ala-Lys-Thr-Asn-Val-Lys-Ala-Ala-Trp

Gly-Lys-Val-Gly-Ala-His-Ala-Gly-Glu-Tyr

Gly-Ala-Glu-Ala-Thr-Glu

16. (a) 9 (b) 6, 7 (c) 2 (d) 10 (e) 1 (f) 4 (g) 2, 3 (h) 5 (i) 7, 8 (j) 11

17. c, e. The performic acid oxidizes the disulfide bonds to SO^{3-} groups and releases new peptides.

18. The amino acid sequence derived from the DNA sequence is that of the nascent polypeptide chain before any posttranslational modifications. Since the function of a protein depends on its mature structure, it is often necessary to analyze the protein itself to determine if any changes have occurred after translation.

19. a, b, c, e, f. Answer d may be correct in some cases where homologous proteins are compared in terms of amino acid sequences and known three-dimensional structures.

Use of Specific Antibodies

20. (a) 2 (b) 4, 8 (c) 1, 5, 7 (d) 1, 3, 6

21. d, e. Immunoelectron microscopy provides more precise localization than does fluorescence microscopy.

22. Because the interaction of an antibody with its antigen is highly specific, recognition and binding can occur in the presence of many other substances. If the antibody is coupled to a radioactive or fluorescent group or an enzyme whose activity can be detected in situ, then a sensitive method is available for the detection and quantitation of the antigen-antibody complex.

Determination of Protein Structure

23. c

24. If an enzyme can be crystallized with and without its substrate and the three-dimensional structures of both are obtained using x-ray crystallography, the difference between the two structures should reveal how the substrate fits in its binding site and which atoms and what kind of bonds are involved in the interaction.

25. (a) (1), (4), (5), (6), (8), and (9)

(b) (2), (3), (5), (6), (7), and (10)

Synthesis of Peptides

26. You could synthesize peptides containing the putative antigenic determinants, couple these peptides to an antigenic macromolecule, and prepare antibodies against the synthetic peptides. If the same antigenic determinants are present in the protein of interest and are not occluded by the structure of the protein, then the antibodies prepared against the synthetic peptides should also react with the protein.

27. a

PROBLEMS

1. A glutamine residue that is the amino-terminal residue of a peptide often undergoes spontaneous cyclization to form a heterocyclic ring; the cyclization is accompanied by the release of ammonium ion. Diagram the structure of the ring, showing it linked to an adjacent amino acid residue. How would the formation of the ring affect attempts to use the Edman procedure for sequence analysis? A similar heterocyclic ring is formed during the biosynthesis of proline; in this case, *glutamate* is the precursor. Can you propose a pathway for the synthesis of proline from glutamate?

2. Mass spectrometry is often used for the sequence analysis of peptides from 2 to 20 amino acids in length. The procedure requires only microgram quantities of protein and is very sensitive; cationic fragments are identified by their charge-to-mass ratio. In one procedure, peptides are treated with triethylamine and then with acetic anhydride. What will such a procedure do to amino groups? Next, the modified peptide is incubated with a strong base and then with methyl iodide. What groups will be methylated? Which two amino acids cannot be distinguished by mass spectrometry?

3. Many of the methods described in Chapter 3 are used to purify enzymes in their native state. Why would the use of SDS–polyacrylamide gel electrophoresis be unlikely to lead to the successful purification of an active enzyme? What experiments would you conduct to determine whether salting out with ammonium sulfate would be useful in enzyme purification?

4. A peptide composed of 12 amino acids does not react with dabsyl chloride or with phenyl isothiocyanate. Cleavage with cyanogen bromide yields a peptide with a carboxyl-terminal homoserine lactone residue, which is readily hydrolyzed, in turn yielding a peptide whose sequence is determined by the Edman procedure to be

E-H-F-W-D-D-G-G-A-V-L

On the other hand, cleavage with staphylococcal protease yields an equivalent of aspartate and two peptides. Use of the Edman procedure gives the following sequences for these peptides:

G-G-A-V-L-M-E and H-F-W-D

Why does the untreated peptide fail to react with dabsyl chloride or phenyl isothiocyanate?

5. Most enzymes are relatively stable in the pH range from 5.5 to 8.5 or 9.0. Proteins with few histidine residues exhibit relatively stable affinities for ion exchangers over that pH range, whereas proteins with significant percentages of histidine residues vary greatly in their affinity for ion exchangers over the same pH range. Why? Suppose you have an active protein in a crude extract that, at various pH ranges, adsorbs neither to cation nor to anion exchangers. Why should you continue to investigate the use of ion exchangers as a means of purifying the protein?

6. A laboratory group wishes to prepare a monoclonal antibody that can be used to react with a specific viral coat protein in a Western blotting procedure. Why would it be a good idea to treat the viral coat protein with SDS before attempting to elicit monoclonal antibodies?

7. Of the techniques for analyzing proteins discussed in Chapter 3 of Stryer's text, which one would be the easiest to use for accurately determining the molecular weight of a small monomeric protein? Comment on the standards you would wish to use in this technique. What types of proteins might not be analyzed accurately by your suggested method?

8. In gel-filtration chromatography, porous gel beads are packed into a glass column. The total volume of the column V_t is equal to V_o, the volume outside the beads; plus V_i, the volume of the pores inside the beads; and V_b, the volume occupied by the beads themselves:

$$V_t = V_o + V_i + V_b$$

Usually V_b is less than 1% of V_t and can be ignored in many applications.

When a solution containing various sizes of molecules passes through the column, the molecules are distributed between V_o and V_i, depending on their ability to enter the beads, which in turn depends on their size. For a molecule of a particular size that can penetrate the beads, K_D, the distribution coefficient, is defined as

$$K_D = (V_e - V_o)/V_i$$

where V_e is the elution volume of the molecule, or the volume required to effect the passage of the molecule through the column.

(a) What is K_D for a molecule that is too large to enter the beads in the column?
(b) If a solution contains a number of molecules of different sizes, all of which can enter the beads, in what order would you expect them to emerge from the column? What is the relationship between the size of a molecule and its corresponding K_D?
(c) Usually a very small molecule (often a colored dye molecule) can be used to determine V_t. Why?
(d) As is the case for SDS–gel electrophoresis, gel-filtration chromatography can also be used to estimate molecular weights. In this case, K_D, the distribution coefficient, is plotted against the logarithm of the molecular weight, and a roughly linear plot can be obtained for many types of proteins. Given that proteins are not denatured when their molecular weights are estimated through gel chromatography, what key assumption is made in order for the estimates to be valid?

9. A hexapeptide that is part of a mouse polypeptide hormone is analyzed by a number of chemical and enzymatic methods. When the hexapeptide is hydrolyzed and analyzed by ion-exchange chromatography, the following amino acids are detected:

Tyr Cys Glu
Ile Lys Met

Two cycles of Edman degradation of the intact hexapeptide released the following PTH–amino acids:

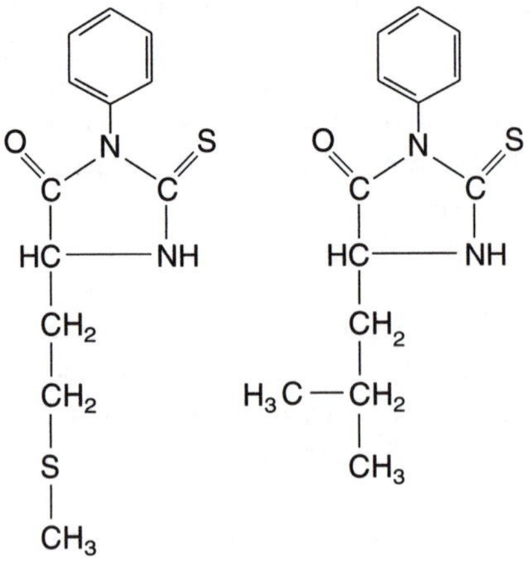

Cleavage of the intact protein with cyanogen bromide yields methionine and a pentapeptide. Treating the intact hexapeptide with trypsin yields a dipeptide, which contains tyrosine and glutamate, and a tetrapeptide. When the intact hexapeptide is treated with carboxypeptidase A, a tyrosine residue and a pentapeptide are produced. Bearing in mind that the hexapeptide is isolated from a mouse, write its amino acid sequence.

10. A map of the electron density is necessary for the determination of the three-dimensional structure of a protein, but other information is also needed. Hydrogen atoms have one electron and cannot be visualized by x-ray analyses of proteins. Bearing this in mind, compare the structures of amino acids like valine, threonine, isoleucine, and valine and then describe what additional information would be needed along with an electron-density map.

11. Some crystalline forms of enzymes are catalytically active, and thus are able to carry out the same chemical reactions as they can in solution. Why are these observations reassuring to those who are concerned about whether crystallographic determinations reveal the normal structure of a protein?

ANSWERS TO PROBLEMS

1. Glutamine cyclizes to form pyrrolidine carboxylic acid:

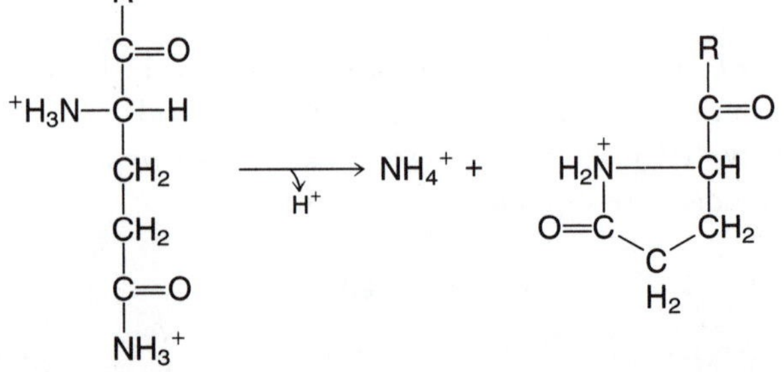

N-terminal glutamine residue

Pyrrolidone carboxylate residue

The Edman procedure begins with the reaction of phenyl isothiocyanate with the terminal α-amino group of the peptide. In the pyrrolidine ring, that group is not available. Therefore, the cyclized residue must be removed enzymatically before the Edman procedure can be used. During the biosynthesis of proline, glutamate undergoes reduction to form glutamate γ-semialdehyde; this compound cyclizes, with the loss of water, to form Δ^1-pyrroline-5-carboxylate, which is then reduced to form proline.

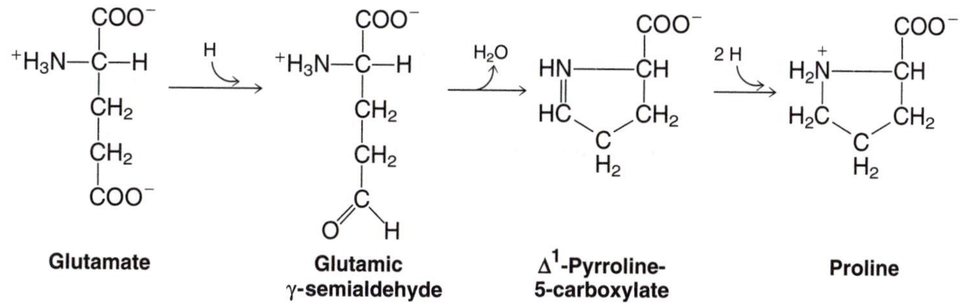

Glutamate → (H) → Glutamic γ-semialdehyde → (H_2O) → Δ^1-Pyrroline-5-carboxylate → (2 H) → Proline

2. Treatment with triethylamine and then with acetic anhydride will yield acetylated amino groups. The strong base removes protons from amino, carboxyl, and hydroxyl groups. These groups would then be methylated with methyl iodide. Leucine and isoleucine have identical molecular weights, so they cannot be distinguished by mass spectrometry.

3. SDS disrupts nearly all noncovalent interactions in a native protein, so the renaturation of a purified protein, which is necessary to restore enzyme activity, could be difficult or impossible. You should therefore conduct small-scale pilot tests to determine whether enzyme activity would be lost upon SDS denaturation. Similarly, when salting out with ammonium sulfate is considered, pilot experiments should be conducted. In many instances, concentrations of ammonium sulfate can be chosen such that the active enzyme remains in solution while other proteins are precipitated, thereby affording easy and rapid purification.

4. The sequences of the peptides produced by the two cleavage methods are circular permutations of each other. Thus, the peptide is circular, so it has no free α-amino group that can react with dabsyl chloride or with phenyl isothiocyanate.

Circular peptide: L-M-E-H-F-W-D-D-G-G-A-V (-L)

Cyanogen bromide → E-H-F-W-D-D-G-G-A-V-L + M (as homoserine lactone)

Staphylococcal nuclease → G-G-A-V-L-M-E + H-F-W-D + D (as free aspartate)

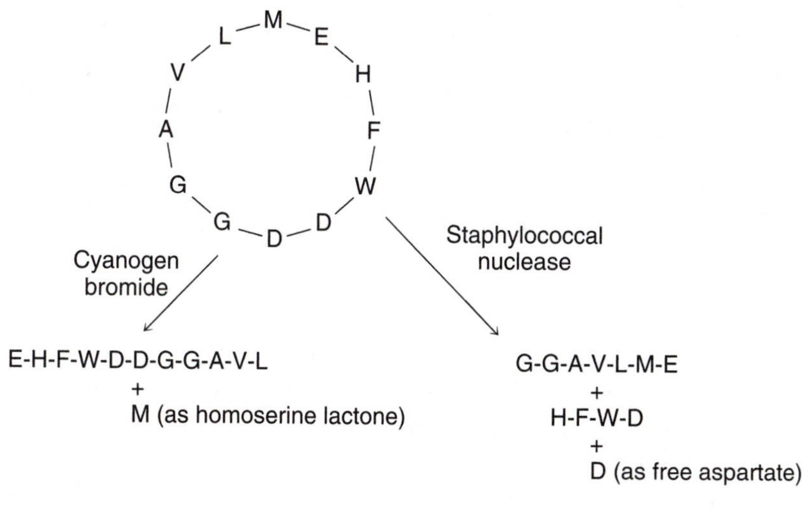

5. Because the pK for the imidazole group of histidine is approximately 6.5, the charge on the residue can vary considerably over the pH range from 5.5 to 9.0. The greater the number of histidine residues in the protein, the greater the protein charge varies when pH changes. Although the desired protein may not be adsorbed by either ion exchanger, other proteins in the extract may be adsorbed; this increases the specific activity (grams of enzyme per grams of total protein) and thereby helps to purify the protein of interest.

6. Samples for assay by Western blotting are separated by electrophoresis in SDS before blotting and antibody staining, so the reacting proteins are denatured. The use of an SDS-denatured antigen to generate the monoclonal antibody response to the viral coat protein could assure that similar specificities are achieved in the test.

7. SDS–polyacrylamide gel electrophoresis, a sensitive and rapid technique which takes only a few hours and which has a high degree of resolution, is probably the easiest and most rapid method for providing an estimate of molecular weight. Small samples (as low as 0.02 μg) can be detected on the gel. For standards or markers on the gel, you should use two or more proteins whose molecular weights are higher than that of the protein to be analyzed, as well as two or more whose molecular weights are lower. The relative mobilities of these markers on the gel can then be plotted against the logarithms of their respective molecular weights (see Figure 3-5 in Stryer's text), providing a straight line that can be used to establish the molecular weight of the protein to be analyzed. Proteins which have carbohydrate molecules covalently attached, or those which are embedded in membranes, often do not migrate according to the logarithm of their mass. The reasons for these anomalies are not clear; in the case of glycoproteins, or those with carbohydrate residues, the large heterocyclic rings of the carbohydrates may retard the movement of the proteins through the polyacrylamide gel. Membrane proteins often contain a high proportion of hydrophobic amino acid residues and may not be fully soluble in the gel system.

8. (a) When the large molecule is excluded from the beads, its elution volume will be the same as the volume of solute outside the beads. Thus, $V_e = V_o$ and $K_D = 0$ for an excluded molecule.

(b) You would expect the largest molecules to emerge first from the column, since they will be unable to penetrate the beads as much as smaller ones. The smallest molecules would emerge last. The value of K_D depends on the difference between elution volume and the V_o. For a small molecule, the elution volume will be greater, and K_D will be larger as well.

(c) A small molecule will easily penetrate the pores of the gel beads, so that its elution volume equals V_o plus V_i, which is approximately equal to V_t.

(d) The key assumption is that all molecules have the same shape, so that they penetrate the pores of the beads in the same way. Problem 6 on page 73 in Stryer's text provides an example of differences in shape that could lead to inaccurate estimates of molecular weights. Troponin, a 93-kd rod-shaped molecule, could be expected to be partially excluded from pores, while hemoglobin

(65 kd) is roughly spherical and can enter beads more readily. The value of K_D for rod-shaped molecules is often lower than that for spherical molecules of the same molecular weight because of the inability of the rods to penetrate the beads.

9. The sequence of the mouse hexapeptide is Met-Ile-Cys-Lys-Glu-Tyr, or MICKEY. Cyanogen bromide treatment cleaves methionine from one end, and the PTH–Met derivative places Met at the N-terminal end, with Ile next in the sequence. Trypsin treatment cleaves on the carboxyl side of Lys, so that Lys is on the C-terminal end of the tetrapeptide, next to Cys. Tyrosine is located on the C-terminal end, as shown by the observation that it is released as a single amino acid when the intact hexapeptide is treated with carboxypeptidase A. Glutamate must therefore be located between Lys and Tyr.

10. Even though individual atoms can be delineated at a resolution of 1.5 Å, the structure of individual side chains that are similar in shape and size cannot be clearly established. The primary structure of the polypeptide chain must be available. The path of the polypeptide backbone can be traced and the positions of the side chains established. Those that are similar in size and shape can be distinguished by using the primary structure as a guide as they are fitted by eye to the electron-density map.

11. As Stryer will point out in future chapters, the chemical reactions carried out by proteins like hemoglobin and lysozyme depend on the precise orientation of atoms involved in binding and acting on substrates. The fact that catalysis can occur in enzyme crystals argues that these same orientations are preserved and that the structure of the enzyme must be the same as that found in solution.

EXPANDED SOLUTIONS TO TEXT PROBLEMS

1. (a) The Edman method is best because it can be used repeatedly on the same peptide; hence, phenyl isothiocyanate.

   (b) Since you have a very small amount of sample, sensitivity is important. Hence, dabsyl chloride or dansyl cloride is the reagent of choice over FDNB (Sanger's reagent).

   (c) Reversible denaturation is usually achieved with 8 M urea. However, if disulfide bonds are present, they must first be reduced with β-mercaptoethanol to obtain a *random coil* by urea treatment.

The known cleavage specificities of chymotrypsin (d), CNBr (e), and trypsin (f) provide easy answers.

2. One can rearrange the Henderson-Hasselbach equation to the following:

$$\text{pH} - \text{p}K = \log \frac{[\text{base}]}{[\text{acid}]}$$

Substituting pH = 4 into this equation shows that the log of the [base]/[acid] is −2; therefore, the ratio is 1/100. By similar reasoning, at pH 5, 6,

7, and 8 the respective ratios are 1/10, 1/1, 10/1, and 100/1. Note the logic: the lower the pH, the lower the [base]/[acid]; the higher the pH, the higher the [base]/[acid].

3. Whereas the hydrolysis of peptides yields amino acids, hydrazinolysis yields hydrazides

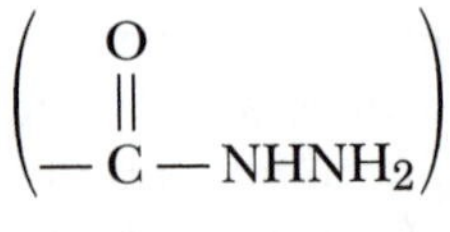

$$\left(-\overset{\overset{\displaystyle O}{\|}}{C} - NHNH_2 \right)$$

of all amino acids *except* the carboxyl-terminal residue. The latter can be separated from the hydrazides by the use of anion exchange resin. (The hydrazides of aspartic and glutamic acids would also be picked up by the anion exchange resin; thus, a further purification step might be necessary.)

4. Ethyleneimine reacts with cysteine to form *S*-aminoethylcysteine, which has the following structure:

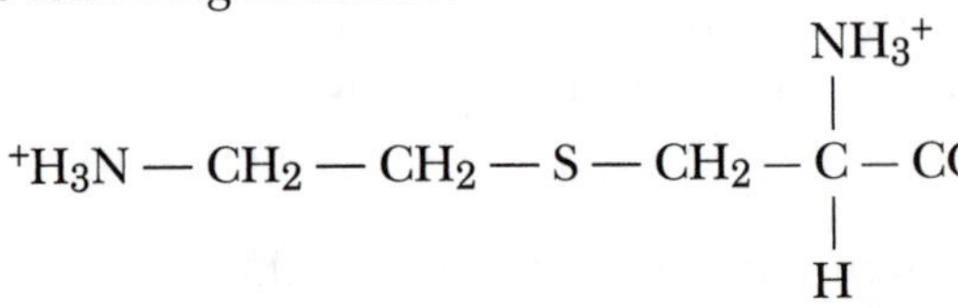

$$^{+}H_3N - CH_2 - CH_2 - S - CH_2 - \underset{\underset{\displaystyle H}{|}}{\overset{\overset{\displaystyle NH_3^{+}}{|}}{C}} - COO^{-}$$

Note that the cysteine side chain has increased in length and has added a plus charge. It closely resembles lysine in both size and charge, and therefore, its carboxyl peptide bonds are susceptible to hydrolysis by trypsin.

5. A 1-mg/ml solution of myoglobin (17.8 kd) is 5.62×10^{-5} M (1/17,800). The absorbance is 0.843 ($15{,}000 \times 1 \times 5.62 \times 10^{-5}$). Since this is the log of I_0/I, the ratio is 6.96 (the antilog of 0.843). Hence, 14.4% (1/6.96) of the incident light is transmitted. Note that when we say the ratio is 6.96, we are really saying 6.96/1. Inverting this ratio gives the 1/6.96.

6. Rod-shaped molecules have larger frictional coefficients than do spherical molecules. Because of this, the rod-shaped tropomyosin has a smaller (slower) sedimentation coefficient than does the spherical hemoglobin, even though it has a higher molecular weight. Imagine a metal pellet and a nail of equal weight and density sinking in a syrup. The pellet will sink in an almost straight line, whereas the nail will twist and turn and sink more slowly.

7. From equations 2 and 5 in Stryer, we can derive the expression $s \propto m^{2/3}$, where s is the sedimentation coefficient and m is the mass, if we assume that the buoyancy ($1-v\rho$) and viscosity (η) factors are constant. Then the mass (m) of a sphere is proportional to its volume (v). Since $v = \frac{4}{3}\pi r^3$, m is $\propto r^3$ and $r \propto m^{1/3}$. Also, if $f = 6\pi\eta r$ (equation 2, p. 46), $f \propto r \propto m^{1/3}$.

When the buoyancy factor is constant, s is proportional to m/f (equation 5, p. 51) $\propto m/m^{1/3} \propto m^{2/3}$. Note that this says that the sedimentation coefficient is proportional to the two-thirds root of the mass.

Therefore, $s(80\text{ kd})/s(40\text{ kd}) = 80^{2/3}/40^{2/3} = 2^{2/3} = 1.59$.

8. Electrophoretic mobilities are usually proportional to the log of the molecular weight (see Stryer Figure 3-5, p. 48). Note in Figure 3-5 that

$$\text{Slope} = \frac{\Delta \log \text{MW}}{\Delta \text{mobility}} = \frac{\log 92000 - \log 30000}{0.41 - 0.80} = \frac{\log x - \log 30000}{0.62 - 0.80}$$

where x is the molecular weight of the unknown. Solving,

$$\log x = \frac{(4.964 - 4.476)(-0.18)}{-0.39} + 4.476 = 4.701$$

and

$$x = \text{antilog of } 4.701 = 50 \text{ kd}$$

9. Compare the *diagonal electrophoresis* (p. 58) patterns obtained with the normal and the mutant proteins. If these patterns are essentially identical, the disulfide pairing is the same in both proteins; if they are not, the new cysteine residue is probably involved in a new disulfide pairing in this mutant.

10. (a) At pH 7 all lysine side chains carry a plus charge, whereas at pH's above 10 a substantial portion of the side chains are uncharged. Since like charges repel, helix formation is inhibited at pH 7 but not at the higher pH's.

(b) The p*K* of the glutamate side-chain carboxyl is approximately 4.3. Therefore, we predict that at pH 5 or above, where virtually all the side chains are ionized, helix formation will be inhibited owing to the repulsion by like charges (COO^-). Around pH 4, most side chains will be protonated (uncharged) and will not inhibit helix formation.

11. Assuming that cells with receptors that bind bacterial degradation products may also bind fluorescent derivatives of those products, you could synthesize a fluorescent-labeled derivative of a degradation product (perhaps some peptide of interest) and use this derivative to detect cells having receptors for this peptide.

12. To answer this question you must remember that enantiomers (mirror images) have identical physical properties (solubility, etc.) and differ only in the *direction* in which they rotate polarized light and in their reaction to *optically active* reagents. In contrast, diastereoisomers (optically active isomers that are *not* mirror images) do not have identical physical properties. Since the synthetic proteases are enantiomers, they will have identical sedimentation and electrophoretic properties. The circular dichroism spectra are expected to have the same magnitude but be of opposite sign or direction. You would also predict that the D protease would hydrolize only peptides of D amino acids, and vice versa for the L enzyme. This expectation has recently (1992) been proved. For reference see Stryer p. 1016.

13. Light was used to direct the synthesis of these peptides. Each amino acid added to the solid support contained a photolabile protecting group

instead of a *t*-Boc protecting group at its α-amino group. Illumination of selected regions of the solid support led to the release of the protecting group, thus exposing the amino groups in these sites and making them reactive. The pattern of masks used in these illuminations and the sequence of reactants define the ultimate products and their locations. [See S.P.A. Fodor, J.L. Read, M.C. Pirrung, L. Stryer, A.T. Lu, and D. Solas, *Science* 251(1991):767, for an account of light-activated, spatially addressable parallel chemical synthesis.]

CHAPTER

4

DNA and RNA: Molecules of Heredity

Chapters 2 and 3 introduced you to proteins. Stryer now turns to a second class of macromolecules, the nucleic acids. First, he describes the structures of the nucleotide building blocks of DNA and the phosphodiester bond that links them together. He then discusses the experimental bases for the conclusion that genes are DNA. Following this, the Watson-Crick DNA double helix is presented, an overview of how DNA is replicated is given, and DNA polymerase is introduced. Stryer concludes the chapter by describing viruses in which the genetic material is not double-helical DNA but rather is single-stranded DNA or RNA. He describes the way in which these viruses replicate through double-stranded nucleic acid intermediates formed by specific base pairing. The replication of one of the RNA viruses is described to illustrate the reversal of the usual direction of flow of information from DNA to RNA.

LEARNING OBJECTIVES

When you have mastered this chapter, you should be able to complete the following objectives:

DNA Composition (Stryer pages 75–77)

1. Recognize the structural components of *DNA,* that is, the *nitrogenous bases,* the *sugar,* and the *phosphate group.* Recognize the various conventions used to represent these components and the structure of DNA.

2. Distinguish between *purines, pyrimidines, ribonucleosides, deoxyribonucleosides, ribonucleotides,* and *deoxyribonucleotides.* Recognize the *deoxyadenosine, deoxycytidine, deoxyguanosine,* and *deoxythymidine* constituents of DNA, and describe the *phosphodiester bond* that joins them together to form DNA. Relate the direction of writing a DNA sequence to the *polarity of the DNA chain.*

Genes Are DNA (Stryer pages 77–80)

3. Describe the major experiments that indicated DNA is the genetic material.

DNA Structure (Stryer pages 80–83, 87–88)

4. List the important features of the *Watson-Crick DNA double helix.* Relate the pairing of *adenine* with *thymine* and of *cytosine* with *guanine* to the *double-helical (duplex) structure* of DNA and to the replication of the helix. Explain the determinants of the specific base pairs in DNA.

5. Compare the lengths of DNA molecules in viruses, bacteria, and eukaryotes.

6. Describe *supercoiling* and state its biological consequences.

DNA Replication (Stryer pages 83–87, 88–89)

7. Describe the *Meselson-Stahl experiment* and relate it to *semiconservative replication.* Define the *melting temperature* (T_m) for DNA and relate it to the separation of the strands of double-helical DNA. Describe *annealing.*

8. List the substrates and the important enzymatic properties of *DNA polymerase I* as they relate to replication. Distinguish between a *primer* and a *template* and describe their functions.

RNA and Single-Stranded DNA Viruses (Stryer pages 90–92)

9. Describe the data that led to the conclusion that bacteriophage ϕX174 contains single-stranded DNA. Explain how single-stranded DNA replicates via a *replicative form.*

10. Contrast the composition and structure of *RNA* and DNA. Distinguish between *uracil* and thymine and between *ribose* and *deoxyribose.*

11. Outline the experiments with tobacco mosaic virus (TMV) that led to the conclusion that RNA can function as a gene. Describe the replication of TMV, including the function of *RNA-directed RNA polymerase.*

12. Relate the catalytic activity of *reverse transcriptase* (an *RNA-directed DNA polymerase*) to the replication of *retroviruses.*

DNA Composition

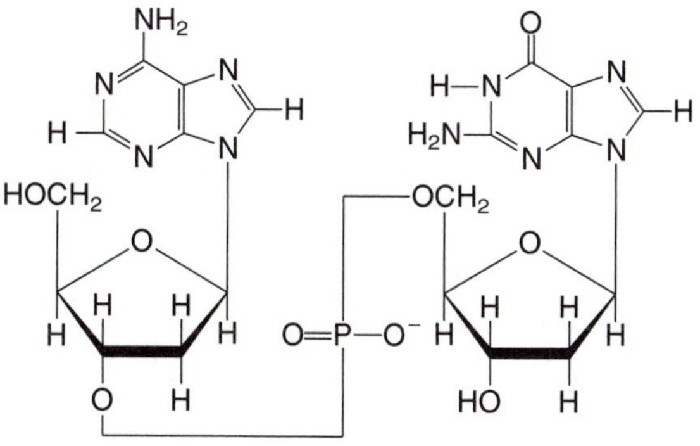

A

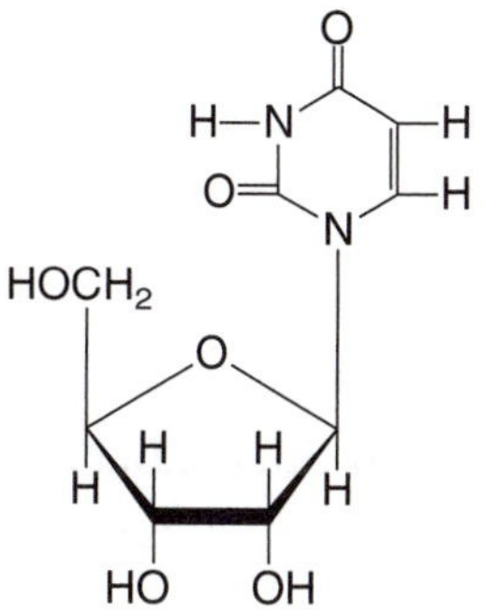

B

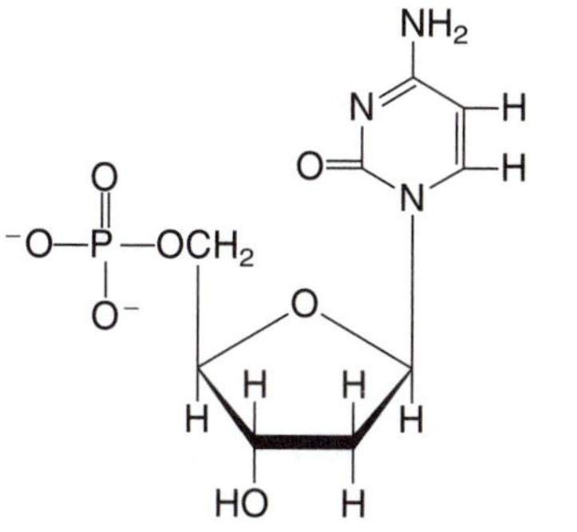

C

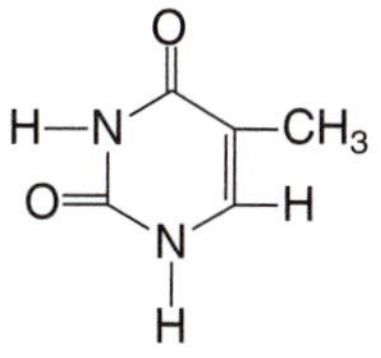

D

1. Which of the preceding structures

(a) contains ribose? ______

(b) contains deoxyribose? ______

(c) contains a purine? ______

(d) contains a pyrimidine? ______

(e) contains guanine? ______

(f) contains a phosphate monoester? ______

(g) contains a phosphodiester? ______

(h) is a nucleoside? ______

(i) is a nucleotide? ______

(j) would be found in RNA? ______

(k) would be found in DNA? ______

Genes Are DNA

2. The transforming principle of pneumococcus found by Avery, MacLeod, and McCarty was

(a) DNA.
(b) mRNA.
(c) polysaccharide capsular material.
(d) protein.

3. The transforming principle of pneumococcus had which of the following properties?

(a) Its activity was not affected by ribonuclease.
(b) Elemental chemical analysis of it was consistent with that of DNA.
(c) Elemental chemical analysis of it was consistent with that of RNA.
(d) Its activity was affected by deoxyribonuclease.
(e) Its activity was affected by protease.

4. Why did the radioisotopes ^{32}P and ^{35}S specifically label the DNA and the protein coat, respectively, of bacteriophage T2 in the Hershey-Chase experiment?

DNA Structure

5. Which of the following are characteristics of the Watson-Crick DNA double helix?

(a) The two polynucleotide chains are coiled about a common axis.
(b) Hydrogen bonds between A and C and between G and T hold the two chains together.
(c) The helix makes one complete turn every 34 Å because each base pair is rotated by 36° with respect to adjacent base pairs and is separated by 3.4 Å from them along the helix axis.
(d) The purines and pyrimidines are on the inside of the helix and the phosphodiester backbones are on the outside.
(e) Base composition analyses of DNA duplexes isolated from many organisms have shown that the amounts of A and T are equal, as are the amounts of G and C.
(f) The sequence in one strand of the helix varies independently of that in the other strand.

6. If a region of one strand of a Watson-Crick DNA double helix has the sequence ACGTAACC, what is the sequence of the complementary region of the other strand?

---

7. Explain why A·T and G·C are the only possible base pairs in double-stranded DNA.

---

8. Match the appropriate characteristics in the right column with the structures of double-stranded or single-stranded DNA.

(a) Double-stranded DNA ______________

(b) Single-stranded DNA ______________

(1) Is a rigid rod
(2) At a specific temperature shows a hyperchromic effect
(3) Reacts readily with formaldehyde
(4) Contains equal amounts of A and T bases
(5) Contains different amounts of C and G bases
(6) Contains U rather than T bases

DNA Replication

9. DNA polymerase I activity requires

(a) a template.
(b) a primer with a free 5′-hydroxyl group.
(c) dATP, dCTP, dGTP, and dTTP.
(d) ATP.
(e) Mg^{2+}.

10. Derive the polarity of the synthesis of a DNA strand by DNA polymerase I from the mechanism for the formation of the phosphodiester bond.

____

____

11. You are provided with a long, single-stranded DNA molecule having a base composition of C = 24.1%, G = 18.5%, T = 24.6%, and A = 32.8%; DNA polymerase I; [α-^{32}P]dATP (dATP with the innermost phosphate labeled), dCTP, dGTP, and dTTP; a short primer that is complementary to the single-stranded DNA; and a buffer solution with Mg^{2+}. What is the base composition of the radiolabeled DNA product after the completion of one round of synthesis?

____

____

12. You are given two solutions containing different purified DNAs. One is from the bacterium *P. aeruginosa* and has a G+C composition of 68%, whereas the other is from a mammal and has a G+C composition of 42.5%.

(a) You measure the absorbance of light of each solution as a function of increasing temperature. Which solution has the higher T_m value and why?

____

____

(b) After melting the two solutions, mixing them together, and allowing them to cool, what would you expect to happen?

(c) Would appreciable amounts of bacterial DNA be found associated in a helix with mammalian DNA? Explain.

13. Outline the basic process by which a Watson-Crick duplex replicates to give two identical daughter duplexes. Explain the reasons for the accuracy of the process.

14. The DNA in a bacterium is uniformly labeled with ^{15}N, and the organism is shifted to a growth medium containing ^{14}N-labeled DNA precursors. After two generations of growth, the DNA is isolated and is subjected to density-gradient equilibrium sedimentation. What proportion of light-density DNA to intermediate-density DNA would you expect to find?

15. Haploid human DNA has 2.9×10^6 kilobases (a kilobase, abbreviated *kb,* is 1000 bases or base pairs). What is the length of human DNA in centimeters?

16. Purified double-helical DNA molecules can be

(a) linear.
(b) circular and supercoiled.
(c) linear and supercoiled.
(d) circular and relaxed.

RNA and Single-Stranded DNA Viruses

17. For the viruses in the left column, indicate the appropriate characteristics from the right column.

(a) Tobacco mosaic virus ____________
(b) Bacteriophage T7 ____________
(c) Rous sarcoma virus ____________

(1) Linear genome
(2) Genome contains U rather than T
(3) Single-stranded nucleic acid genome
(4) DNA intermediates are involved in replication
(5) Uses RNA-directed RNA polymerase to replicate
(6) Uses RNA-directed DNA polymerase to replicate

18. Describe how a single-stranded DNA virus could replicate by incorporating semiconservative replication into the process.

____

____

____

____

19. From the following nucleic acids, select those that appear during the infection of a cell with a retrovirus and place them in the order in which genetic information flows during the process.

(a) Double-stranded DNA–RNA helix in the cell
(b) Single-stranded RNA in the virus
(c) Single-stranded RNA in the cell
(d) Double-stranded DNA in the cell
(e) Double-stranded RNA in the virus
(f) Double-stranded RNA in the cell

ANSWERS TO SELF-TEST

DNA Composition

1. (a) B (b) A, C (c) A (d) B, C, D (e) A (f) C (g) A (h) B (i) C, but strictly speaking, A is called a *dinucleotide,* not a *nucleotide;* (j) B (k) A, C, D

Genes Are DNA

2. a

3. a, b, d

4. The radioisotope ^{32}P labeled the DNA because there is one phosphorus atom per phosphodiester bond linking the deoxynucleosides of DNA to each other. The radioisotope ^{35}S labeled the protein coat because sulfur is a constituent of the amino acids cysteine and methionine in the protein coat. There is no sulfur in DNA and no phosphorus in the protein coat of bacteriophage T2.

DNA Structure

5. a, c, d, e. Answer b is not correct, because A pairs with T and G pairs with C. Answer f is not correct because the sequence of one strand determines the sequence of the other by base pairing.

6. GGTTACGT. The convention for indicating polarity is that the 5′ end of the sequence is written to the left. The two chains of the Watson-Crick double helix are antiparallel, so the correct complementary sequence is *not* TGCATTGG.

7. The space between the two ribose-phosphodiester strands is precisely defined. This space is not big enough for two purines to hydrogen-bond. On the other hand, two pyrimidine bases would not be close enough to form a hydrogen bond. Furthermore, in the double-stranded structure the hydrogen donor and acceptor groups are not properly aligned for the G·T or A·C pairs.

8. (a) 1,2,4 (b) 3,5

DNA Replication

9. a, c, e. Answer b is not correct because, although the enzyme requires a primer, the state of its 5′ end is irrelevant, since dNMP residues are added to the 3′ end. Answer d is not correct because the enzyme uses dNTP and not NTP molecules, where N means A, C, G, T, or U.

10. To form the phosphodiester bond the 3′-hydroxyl of the terminal nucleotide of the primer makes a nucleophilic attack on the innermost phosphorus atom of the incoming dNTP that is appropriate for Watson-Crick base pairing to the template strand. As a result, a dNMP residue is added onto the 3′ end of the primer, with the concomitant release of PP_i, and the chain grows in the $5' \rightarrow 3'$ direction.

11. After the completion of one round of synthesis, the template strand will have directed the polymerization of a complement in which C = 18.5%, G = 24.1%, T = 32.8%, and A = 24.6%. The complementary strand will be labeled with ^{32}P on the 5′ side of A. Since the primer is short with respect to the template, its contribution to the composition of the product strand can be neglected.

12. (a) The bacterial DNA solution has the higher T_m because it has the higher G+C content and is therefore more stable to the thermally induced separation of its strands.

(b) The complementary DNA strands from each species will anneal to form Watson-Crick double helices as the solution cools.

(c) No; each strand will find its partner because the perfect match between the linear arrays of the bases of complementary strands is far more stable than the mostly imperfect matches in duplexes composed of one strand of bacterial and one strand of mammalian DNA would be.

13. When replication occurs, the two strands of the Watson-Crick double helix must separate so that each can serve as a template for the synthesis of its complement. Since the two strands are complementary to each other, each bears a definite sequence relationship to the other. When one strand acts as a template, it directs the synthesis of its complement. Thus, the product of the synthesis is a duplex molecule that is identical with the starting duplex. The process is accurate because of the specificity of base pairing and because the protein apparatus that catalyzes the replication can remove mismatched bases.

14. After two generations, one would expect to find equal amounts of light-density DNA, in which both strands of each double-helical DNA were synthesized from ^{14}N precursors, and intermediate-density DNA, in which each double-helical DNA consisted of a heavy ^{15}N strand paired with a light ^{14}N strand.

15. 98.6 cm. The math is as follows:

$$(2.9 \times 10^6 \text{ kb}) \times 10^3 \text{ bases/kb} \times 3.4 \text{ Å/base} \times 10^{-8} \text{ cm/Å} = 98.6 \text{ cm}$$

16. a, b, d. Answer d is correct because if at least one discontinuity exists in the phosphodiester backbone of either chain of a circular duplex molecule, the chains are free to rotate about each other to assume the relaxed circular form. Answer c is incorrect because supercoiling requires closed

circular molecules. In a linear molecule, the ends of each strand are not constrained with respect to rotation about the helical axis; therefore, the molecule cannot be supercoiled.

RNA and Single-Stranded DNA Viruses

17. (a) 1, 2, 3, 5 (b) 1, 4 (c) 1, 2, 3, 4, 6

18. The single-stranded DNA penetrates the cell, where it is converted enzymatically to a double-stranded replicative form through Watson-Crick base pairing. The replicative form is then reproduced by a mechanism similar to that used for the semiconservative replication of the duplex chromosome of the host cell. Finally, after this stage of replication, the mechanism shifts to one in which the replicative form serves as a template to produce copies of the single-stranded DNA found in the mature virus.

19. Starting with the single-stranded RNA in the virus and ending with the single-stranded progeny viruses, the order in which genetic information flows during the infection of a cell with a retrovirus is b, c, a, d, c, b. Retroviruses use the enzyme reverse transcriptase to convert their single-stranded genomes into a DNA-RNA replicative form that is subsequently converted into a duplex DNA replicative form before insertion into the host chromosome and ultimate reconversion into the single-stranded viral RNA.

PROBLEMS

1. The genome of the mammalian virus SV40 is a circular DNA double helix containing 5243 base pairs. When a solution containing intact DNA molecules is heated, one observes an increase in the absorbance of light at 260 nm. When the solution is then cooled slowly, a decrease in absorbance is observed. If one or more breaks are made in the sugar-phosphate backbones of the SV40 double-stranded circles, heating causes a similar hyperchromic effect. However, when the solution of nicked molecules is cooled, the reduction in absorbance is much slower than that observed in the solution containing intact molecules. Why do the two types of molecules behave differently when they are cooled after heating?

2. A number of factors influence the behavior of a linear, double-stranded DNA molecule in a 0.25 M sodium chloride solution. Considering this, explain each of the following observations:

(a) The T_m increases in proportion to length of the molecule.
(b) As the concentration of sodium chloride decreases, the T_m decreases.
(c) Renaturation of single strands to form double strands occurs more rapidly when the DNA concentration is increased.
(d) The T_m is reduced when urea is added to the solution.

3. (a) Many proteins that interact with double-stranded DNA bind to specific sequences in the molecule. Why is it unlikely that these enzymes operate by sensing differences in the diameter of the helix?

(b) What other features of the double-stranded helix might be recognized by the protein?

4. You have a double-stranded linear DNA molecule, the appropriate primers, all the enzymes required for DNA replication, four ^{32}P-labeled nucleoside triphosphates, Mg^{2+} ion, and the equipment needed to detect newly synthesized radioactive DNA. Why is this system not sufficient to distinguish between conservative and semiconservative replication of the DNA molecule?

5. Certain deoxyribonuclease enzymes cleave any sequence of single-stranded DNA to yield nucleoside monophosphates; these enzymes do not hydrolyze base-paired DNA sequences. What products would you expect when you incubate a solution containing a single-strand-specific deoxyribonuclease and the following oligonucleotide?

5′-ApGpTpCpGpTpApTpCpCpTpCpTpApCpGpApCpTp-3′

6. Formaldehyde reacts with amino groups to form hydroxymethyl derivatives. Would you expect formaldehyde to react with bases in DNA? Suppose you have a solution that contains separated complementary strands of DNA. How would the addition of formaldehyde to the solution affect reassociation of the strands?

7. When double-stranded DNA is placed in a solution containing tritiated water, hydrogens associated with the bases readily exchange with protons in the solution. The greater the percentage of A·T base pairs in the DNA, the greater the rate of exchange. Why?

8. At the same time that many experiments were suggesting that DNA in chromosomes is very long and continuous, it was established that DNA polymerase I adds deoxyribonucleotides to the 3′-hydroxyl terminus of a prime chain and that a DNA template is essential. Why was a great deal of attention then given to determining whether chromosomal DNA contains breaks in the sugar-phosphate backbone?

9. The value of the T_m for DNA in degrees Celsius can be calculated using the formula

$$T_m = 69.3 + 0.41\ (G+C)$$

where G+C is the mole percentage of guanine plus cytosine.

(a) A sample of DNA from *E. coli* contains 50 mole percent G+C. At what temperature would you expect this DNA molecule to melt?
(b) The melting curves for most naturally occurring DNA molecules reveal that the T_m is normally greater than 65°C. Why is this important for most organisms?

10. During early studies of the denaturation of double-stranded DNA, it was not known whether the two strands unwind and completely separate from each other. Suppose that you have double-stranded DNA in which one strand is labeled with ^{14}N and the other is labeled with ^{15}N. If density-gradient equilibrium sedimentation can be used to distinguish between both double- and single-stranded molecules of different densities, how can you determine whether DNA strands separate completely under denaturation?

11. Under strongly acidic conditions, several atoms of DNA bases are protonated; these include the N-1 and N-7 of adenine, the N-7 of guanine, the N-3 of cytosine, and the O-4 of thymine. Relate these observations to the stability of double-stranded DNA at low pH.

12. Sol Spiegelman found that some types of single-stranded RNA can associate with single-stranded DNA to form double-stranded molecules. What is the most important condition that must be satisfied in order to allow the formation of these hybrid molecules?

13. Many cells can synthesize deoxyuridine 5′-triphosphate (dUTP). Can dUTP be utilized as a substrate for DNA polymerase I? If so, with which base will uracil pair in newly replicated DNA?

14. The DNA of bacteriophage λ is a linear double-stranded molecule that has complementary single-stranded ends. These molecules can form closed circular molecules when two "cohesive" ends on the same molecule join, and they can form linear dimers, trimers, and longer molecules when sites on different molecules are joined.

(a) What conditions should be chosen to ensure that λ phage DNA molecules form closed circular monomers?

(b) Under certain conditions, λ phage DNA molecules are infective. When a very low concentration of λ phage DNA is incubated with DNA polymerase I and the four deoxyribonucleoside triphosphates, the infectious activity of λ phage DNA is destroyed. Brief treatment of λ phage DNA with bacterial exonuclease III, an enzyme that removes 5′-mononucleotides from the 3′ ends of double-stranded DNA molecules, also destroys infectivity, but subsequent brief treatment of that DNA with DNA polymerase I and nucleotide substrates can restore infectivity. Exhaustive treatment of the exonuclease III–digested DNA with DNA polymerase I and nucleotide substrates will completely destroy infectivity. Describe more completely the structure of λ phage DNA, and provide an interpretation of the action of the two enzymes on the molecule.

15. The isolation of viral DNA from animal cells that have been infected with adenovirus yields linear double-stranded molecules that, when denatured and allowed to reassociate under conditions favoring intramolecular annealing, form single-stranded circles. Although the circles can be visualized using the electron microscope, resolution is not sufficient to visualize the ends of the molecule. Other analyses of the single-stranded molecule show that each end has a sequence that allows the structure shown in Figure 4-1 to be formed.

(a) Suppose that the base sequence at one end of the single-stranded molecule is 5′-ACTACGTA . . . . What is the corresponding sequence at the other end? Show how these sequences would allow full-length, double-stranded, linear molecules to be formed.

(b) An alternative suggestion for the formation of the single-stranded molecules was also proposed; it is shown in Figure 4-2. Why is this proposed pairing scheme unlikely?

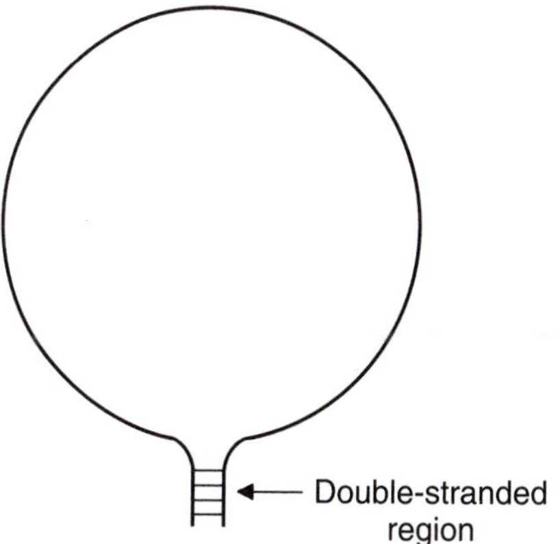

Figure 4-1
A single-strand circle formed by intra-molecular annealing of adenovirus DNA.

16. Thermoacidophilic bacteria can grow in volcanic sulfur springs at pH 2 and at temperatures as high as 85°C. DNA polymerase purified from the thermophile *Sulfolobus acidocaldarius* has an optimal activity at 70°C and is stable at 80°C. When incubated with a circular DNA template at 100°C, the isolated polymerase can extend a 20-nucleotide primer by more than 100 nucleotides. These experiments require that enzyme-to-primer

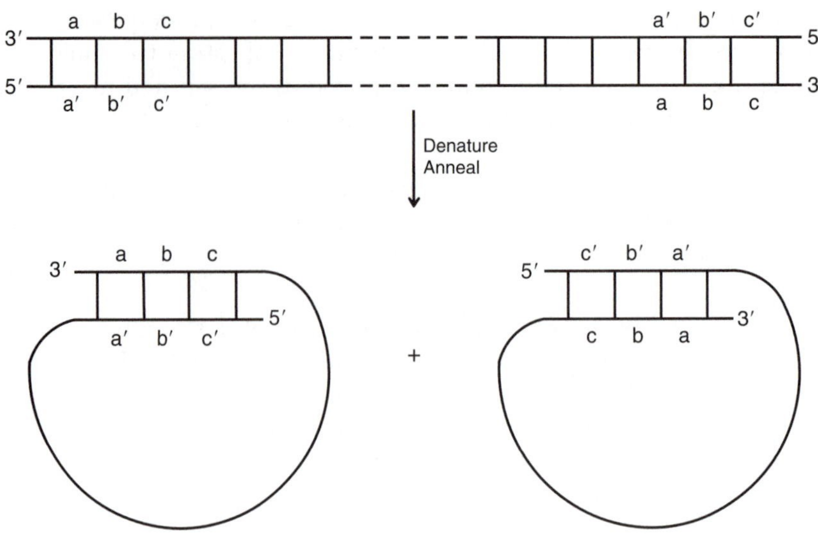

Figure 4-2
Another proposal for formation of single-stranded molecules of adenovirus DNA.

concentration be about 1:1. T_m for the double-stranded DNA used in the experiment is about 60°C. Unlike DNA polymerase I from *E. coli,* DNA polymerase from *S. acidocaldarius* has no demonstrable ability to correct mistakes in DNA by removing mismatched nucleotides (such an enzyme activity is often referred to as *proofreading*).

(a) Why should the ratio of enzyme to primer be 1 in order for primer extension to take place at 100°C?
(b) Would you expect to find an auxiliary proofreading enzyme in *S. acidocaldarius*? Why?
(c) Would you expect DNA from the genome of *S. acidocaldarius* to have a G + C content higher or lower than that from a bacterium that grows at a more normal temperature? Why?

17. Terminal deoxynucleotidyl transferase (TdT), an enzyme found in bone marrow and thymus tissue, can extend a DNA primer by 5′ → 3′ polymerization using deoxynucleotide triphosphates as substrates. The primer must be at least three nucleotides in length and must have a free 3′-OH end. The enzyme does not require a template nor does it copy one.

(a) Compare TdT with DNA polymerase I.
(b) Would TdT be useful for synthesizing DNA molecules that carry genetic information? Why?

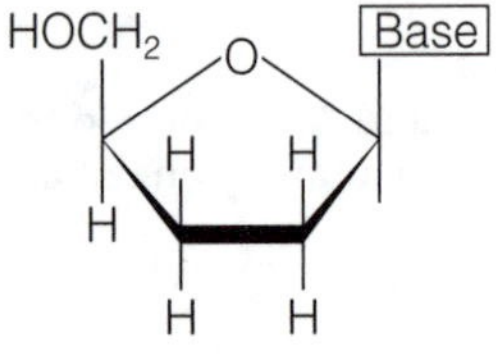

2′, 3′ - Dideoxynucleoside

18. The 2′,3′-dideoxynucleosides can be used as reagents to inhibit the action of DNA polymerase. These analogs must be converted to dideoxynucleoside triphosphates in order to have a measurable effect on DNA synthesis. When incorporated into a growing DNA chain, a single dideoxynucleotide residue can effectively block subsequent chain extension.

(a) Why must a 2′,3′-dideoxynucleoside be converted to a dideoxynucleoside triphosphate to be incorporated into DNA?
(b) What feature of a 2′,3′-dideoxynucleoside is most likely to account for inhibition of DNA chain extension?

19. In each chain-elongation reaction catalyzed by DNA polymerase, a phosphodiester bridge is formed and pyrophosphate is concomitantly released. Hydrolysis of pyrophosphate to two molecules of inorganic phosphate occurs rapidly because most cells have a potent pyrophosphorylase. By removing one of the products of the chain-elongation reaction, pyrophosphate cleavage in the cell is partially responsible for the forward progress of polymerization. However, isolated DNA polymerases can efficiently carry out chain extension in the absence of pyrophosphate cleavage, so long as the double-stranded helix is allowed to form during elongation. What forces in a DNA helix might contribute to driving polymerization forward?

20. In his studies of DNA in the late 1940s, Chargaff established that DNA from all organisms studied has equal numbers of adenine and thymine bases and equal numbers of guanine and cytosine bases. Considering that for this problem thymine and uracil are both equivalent in their abilities to form hydrogen bonds with adenine, state whether you would expect similar constraints on base composition to be found in the following:

(a) single-stranded RNA from tobacco mosaic virus.
(b) the DNA-RNA hybrid molecule synthesized by reverse transcriptase.
(c) RNA from a virus in the reovirus family, which have large genomes composed of double-stranded RNA molecules.

ANSWERS TO PROBLEMS

1. When the intact double-stranded circular DNA molecule is heated in solution, its base pairs are disrupted and an increase in the absorbance of light at 260 nm is observed. However, the two resulting single-stranded circles are so tangled about each other that they remain closely associated. When the molecules are cooled, the interlocked strands move relative to each other until their base sequences are properly aligned and a double-stranded molecule is reformed. This molecule absorbs less light at 260 nm than does the pair of denatured single strands. Breaks in one or both strands of a double-stranded DNA molecule allow them to separate completely from each other during denaturation. In order to form a double-stranded molecule, the separate strands collide randomly until at least a small number of correct base pairs are formed; then the remaining base pairs form to generate a double-stranded molecule. Reassociation of a pair of separate strands in solution is slower than that of a pair of interlocked circles, so a corresponding difference in the reduction of absorbance will be observed.

2. (a) The longer the DNA molecule, the larger the number of base pairs it contains. As a result, more thermal energy is required to disrupt the helical structure of the longer DNA molecule. Experiments show that such a relationship is true for molecules up to 4000 base pairs in length.

(b) Sodium ions neutralize the negative charges of the phosphate groups in both strands. As the concentration of NaCl decreases, repulsion between the negatively charged phosphate groups increases, making it easier to separate the two strands. The tendency for the strands to separate more easily means that dissociation occurs at a lower temperature, which means that the T_m of the molecule is lower.

(c) The reassociation of single strands begins when a short sequence of bases in one strand forms hydrogen bonds with a complementary sequence in another. Once a short stretch of base pairs is formed, reassociation to form the longer double-stranded molecule occurs rapidly. The higher the concentration of DNA, the greater the number of complementary sequences there are in the solution, and thus the quicker the complementary sequences will find and pair with each other.

(d) Urea in the solution disrupts hydrogen bonds. Because hydrogen bonds are partly responsible for the stability of the double helix, the disruption of these bonds makes the structure more sensitive to denaturation by thermal energy and thereby reduces the T_m. In addition to hydrogen bonding, the tendency of bases to stack also contributes significantly to the stability of the helix. Base stacking minimizes the contact of the relatively insoluble bases with water, and it also allows the sugar-phosphate chain to be located on the outside of the helix, where it can be highly solvated. Urea may also cause destabilization of the helix by allowing bases to associate more readily with water.

3. (a) Figure 4-11 in Stryer's text shows that each of the purine-pyrimidine base pairs found in the DNA double helix are almost identical in size and shape, so the diameter of the double helix is essentially uniform all along its length. It is therefore unlikely that a protein can identify a specific sequence by sensing differences in the diameter of the helix.

(b) Proteins that interact with specific sequences might do so by forming hydrogen bonds with the bases; in some cases, it would be necessary for the double strand to undergo local unwinding or melting in order for the bases to form hydrogen bonds with a protein. However, hydrogen bond donors and acceptors are also found in the grooves of the helix. A protein could also bind to a specific location on DNA by forming hydrogen bonds with a particular group of atoms in one of the grooves of the helix. (See pages 962–963 in Stryer's text.)

4. Although the system described could yield ^{32}P-labeled daughter DNA molecules, chemical methods cannot distinguish DNA in which both strands are radioactively labeled from DNA in which one strand is labeled and one strand is unlabeled. In their experiments, Meselson and Stahl used a physical technique, density-gradient equilibrium sedimentation, to separate the labeled molecules according to their content of ^{14}N and ^{15}N.

5. In solution, the oligonucleotide forms an interchain double-stranded molecule with flush ends and a small single-stranded loop containing the sequence 5′-pTpCpCpTpCp-3′. The deoxyribonuclease hydrolyzes the nucleotides in this single-stranded region to form nucleoside monophosphates, leaving a small double-stranded linear molecule containing seven base pairs.

6. Formaldehyde could react with the amino groups on the C-6 carbon of adenine, the C-2 of guanine, and the C-4 of cytosine to form hydroxymethyl derivatives. Because these derivatives cannot form hydrogen bonds with complementary bases, formaldehyde-treated single strands would reassociate to a lesser extent than would untreated single strands. The actual sites of the reaction of formaldehyde with DNA are not precisely known; these sites may also include the ring nitrogen atoms in pyrimidines.

7. This experiment suggests that the hydrogen bonds of base-paired regions of double-stranded DNA may undergo reversible dissociation to form single-stranded regions, often known as *bubbles*. The transient disruption of these hydrogen bonds allows the exchange of protons with the tritiated water. A·T pairs open more easily than G·C pairs. Thus, the greater the percentage of A·T pairs, the greater the rate of proton exchange.

8. A continuous double-stranded DNA molecule has only two 3′-OH groups available for the initiation of DNA synthesis by DNA polymerase I; because each is located at an opposing end of the molecule, no template sequence is available. In order to construct a relatively simple mechanism for chromosomal replication, one could postulate that the enzyme initiates DNA replication at a number of breaks along the chromosome, with each of the breaks offering the 3′-OH group required for the initiation of the new DNA strand. The template required for replication would then be located on the strand opposite the break, thus ensuring that DNA synthesis could continue. It is now well established that DNA in chromosomes is very long and continuous. The fact that there are no breaks in the molecule makes the mechanism of replication more complex. It involves a number of enzyme activities, as well as the use of RNA to prime the synthesis of DNA. For details, see page 805 and succeeding pages of Stryer's text.

9. (a) The expected melting temperature for *E. coli* containing 50% GC base pairs is

$$\begin{aligned} T_m &= 69.3 + 0.41(\text{G+C}) \\ &= 69.3 + 0.41(50) \\ &= 69.3 + 20.5 \\ &= 89.8°\text{C} \end{aligned}$$

(b) Most organisms live at temperatures that are considerably lower than 65°C. Because both the transmission and expression of genetic information depends on the integrity of the double-stranded DNA molecule, it is important that the molecule not be disrupted by thermal energy.

10. First, you must determine the temperature at which the hydrogen bonds are disrupted and single strands are formed. You can do this by heating the double-stranded DNA to various temperatures and measuring the extent of hyperchromicity. Once the DNA has been melted, centrifuge the sample using the density-gradient equilibrium sedimentation technique to attempt to separate the ^{14}N-labeled DNA strands from the ^{15}N-labeled DNA strands, which will be the denser of the two. If you are successful, this would suggest that the strands separate completely during thermal denaturation.

11. The protonation of the N-1 and N-7 of adenine, the N-7 of guanine, the N-3 of cytosine, and the O-4 of thymine makes hydrogen bonding at these locations impossible. Therefore, at low pH, where proton concentrations are high and protonation is likely to occur, double-stranded DNA is less stable than at higher pH values.

12. The association of a molecule of RNA with a molecule of DNA to form a hybrid molecule depends primarily on the two molecules' having complementary sequences of bases. The formation of hydrogen bonds between complementary bases will allow the formation of a double helix composed of RNA and DNA.

13. The deoxyribonucleotide dUTP can be used as a substrate for DNA polymerase I during DNA replication because the structure of uracil is very similar to that of thymine. When incorporated into a double-stranded DNA polymer, uracil pairs with adenine, as does thymine. For a discussion of the reasons uracil is not normally incorporated into DNA, refer to page 812 of Stryer's text.

14. (a) To ensure that λ phage DNA molecules form closed circular monomers, the concentration of λ phage DNA should be relatively low so that the intrachain formation of hydrogen bonds is favored. At higher concentrations, the probability of interchain joining is enhanced. At very low DNA concentrations, the molecules will remain as linear monomers.

3′ ———————— 5′
5′ ———————— 3′

(b) The most reasonable model for the structure of the λ phage DNA molecule is a double-stranded molecule having single-stranded protrusions at the 5′ ends, as illustrated in the margin. The 3′ ends have hydroxyl groups, which allow them to serve as primers for DNA replication catalyzed by DNA polymerase I. This enzyme fills in the single-stranded regions of the molecule, perhaps producing a molecule with flush ends. Such a molecule no longer has cohesive ends, which could be required for infectivity.

Molecules treated with exonuclease III have a longer single-stranded sequence at either end; they may not be infective because the newly exposed bases may not be complementary to each other, which would mean that the ends could no longer be joined. When the exonuclease-treated DNA is treated with DNA polymerase I, the single-stranded regions are filled in until the molecule has protruding single strands that are approximately the same length as those in the native molecule. Hence, the molecule once again becomes infective. Further treatment of the molecule with DNA polymerase I will once again produce a molecule that has flush ends and is no longer infective.

15. (a) The sequence at the other end of the single-stranded molecule must be composed of complementary bases. It must therefore be

. . . TACGTAGT-3′

The structure of the full-length, double-stranded, linear molecule would be

5′-ACTACGTA ———— TACGTAGT-3′

3′-TGATGCAT ———— ATGCATCA-5′

Each single strand has a pair of inverted repeats.

(b) The formation of double-stranded helical segments depends on hydrogen bond formation between bases in nucleotide chains that are *antiparallel,* as follows:

5′ . . . PuPyPuPyPu . . . 3′
3′ . . . PyPuPyPuPy . . . 5′

where Py = pyrimidine and Pu = purine. When the suggested structure is labeled using this scheme, as in Figure 4-3, it can be seen that it would require the formation of base pairs in *parallel* chains, and such pairing cannot take place.

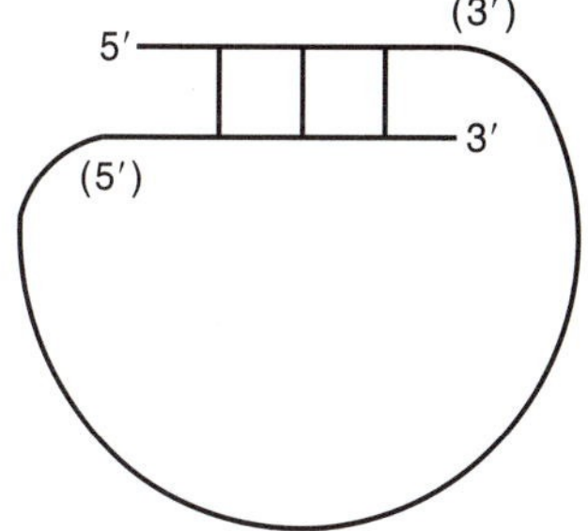

Figure 4-3
Base pairing between parallel nucleotide chains, required to form the circular structures shown in Figure 4-2, is not known to occur in DNA.

16. (a) Each enzyme molecule probably tightly clamps the primer as well as the newly synthesized chain, thereby protecting the helix from degradation at high temperature. The association between primer and enzyme may protect the enzyme from thermal denaturation as well. There are many examples in biochemistry where substrate binding can stabilize enzyme protein structure.

(b) During polymerization of a new DNA chain, the chance that base incorporation errors can occur will increase at high temperatures. Even though a proofreading or error-correcting activity is not found in the *S. acidocaldarius* polymerase polypeptide, you would expect an auxiliary enzyme to be present in the cells in order for the bacterial genome to be accurately replicated.

(c) The higher the G+C content of double-stranded DNA, the higher the melting point at which the helix is denatured. Therefore you might expect DNA from a thermophile to have a higher G+C composition; surprisingly, DNA base ratios are not very different from those in bacteria living at lower temperatures, so that DNA isolated from thermophilic bacteria melts at temperatures lower than those encountered in hot springs. There must be proteins or other molecules in the bacterial cells that protect the genomic DNA of thermophiles from thermal denaturation.

17. (a) Like DNA polymerase I, TdT can extend a DNA primer by using nucleoside triphosphates as a substrate. However, TdT does not use a template and cannot copy from one, so that the base composition of the newly synthesized single strand of DNA will depend solely on the relative concentrations of bases in the nucleoside triphosphate substrates. For chains synthesized by DNA polymerase I, the base composition will be complementary to that of the template strand. While DNA polymerase has an activity that removes mismatched bases from newly synthesized strands, TdT has no such activity and does not need one.

(b) DNA molecules that carry genetic information must be synthesized as faithful copies of template strands; TdT cannot copy a template and would not be useful in genomic DNA synthesis.

18. (a) DNA polymerase requires nucleoside triphosphates as a substrate for DNA chain extension. Nucleosides like the 2′,3′-dideoxy analogs must be converted to nucleoside triphosphates in order to serve as a substrate for DNA polymerase. Studies on inhibition of DNA synthesis in living cells include incubating those cells with nucleoside forms of the analogs instead of nucleoside triphosphates, because negatively charged phosphate anions cannot pass across the plasma membrane, while relatively neutral

nucleotides can. Once inside the cell, nucleoside analogs are phosphorylated by cellular enzymes that normally function to "salvage" nucleosides generated by turnover of nucleotides from DNA and RNA.

(b) Dideoxynucleosides lack a free 3′-hydroxyl group, which would normally serve as a primer for incorporation of the next nucleotide into the growing polynucleotide chain. The lack of a 3′-OH group also interferes with excision by the error-correcting activity of DNA polymerase, so that chain extension is blocked.

19. Noncovalent forces could also contribute to helix formation. Hydrogen bonds form between opposing A and T bases as well as between G and C molecules in the antiparallel chains. There are also significant hydrophobic interactions between adjacent bases on the same strand, and these stacking interactions may in fact contribute more significantly to helix formation and stability. In vivo, it is likely that these noncovalent forces, along with the cleavage of pyrophosphate, together account for the forward progress of chain elongation.

20. (a) The ratios observed by Chargaff can be attributed to the requirement that in a double-stranded polynucleotide, only certain bases can form hydrogen bonds with each other. while a single-stranded polynucleotide might form some hydrogen bonds between bases as it folds, the overall base ratios will not conform to Chargaff's established rules. This is also true for single-stranded DNA in a virus like ϕX174.

(b) Because hydrogen bonding between A and T (or U, in the case of RNA) and between G and C can occur in a duplex molecule formed by one strand of DNA and a complementary strand of RNA, you would expect to see base ratios like those observed by Chargaff.

(c) Double-stranded RNA molecules form hydrogen bonds between bases in a manner similar to those in DNA helices. Therefore, bearing in mind that U, not T, would normally be found in RNA, the number of uracil residues would equal the number of adenines, and the number of guanine bases would be expected to be the same as those of cytosine.

EXPANDED SOLUTIONS TO TEXT PROBLEMS

1. By convention, when polynucleotide sequences are written, left to right means $5' \rightarrow 3'$. Since complementary strands are antiparallel, if one wishes to write the complementary sequence without specifically labeling the ends, the order of the bases must be reversed.

(a) TTGATC
(b) GTTCGA
(c) ACGCGT
(d) ATGGTA

2. (a) Since [A] + [G] account for 0.54 mole-fraction units, [T] + [C] must account for the remaining 0.46 (1 – 0.54). However, the individual mole fraction of [C] or [T] cannot be predicted.

   (b) Due to base pairing (A:T, G:C) in the complementary strand, [T] = 0.30, [C] = 0.24, and [A] + [G] = 0.46.

3. To answer this question one must know that $2\mu m = 2 \times 10^{-6}$, that one Å $= 10^{-10}$ m, and that the distance between the base pairs is 3.4 Å. The length of a DNA segment (in this case 2×10^{-6} m) divided by the distance between the base pairs (3.4×10^{-10} m) gives the answer, 5.88×10^3 base pairs.

4. After 1.0 generation, one-half of the molecules would be ^{15}N-^{15}N, the other half ^{14}N-^{14}N. After 2.0 generations, one-quarter of the molecules would be ^{15}N-^{15}N, the other three-quarters ^{14}N-^{14}N. Hybrid ^{14}N-^{15}N molecules would not be observed.

5. Under intracellular conditions, melting would be more difficult because the bacterial chromosome is circular and stabilized by DNA-binding proteins. If it occurred, the mobility of the single strands would be restricted, thus enhancing reannealing to the original molecule.

6. Competent strands of bacteria are able to take up foreign DNA and incorporate portions of it into their genome. Non-competent strands are unable to do this. Why? Apparently, they are unable to take up DNA, destroy the foreign DNA after taking it up or are unable to integrate fragments of DNA into their genome. Any one of these deficiencies would render a cell noncompetent.

7. The Hershey and Chase experiments would have been indecisive if performed with M13 because both ^{35}S (in protein) and ^{32}P (in DNA) would have been found in the bacteria, giving little information as to which component carried genetic information.

8. (a) Thymine is the molecule of choice since it occurs in DNA, is not a component of RNA, and is not readily converted to cytosine or uracil. If they enter the cell, labeled deoxythymidine or dTTP would also be useful molecules. Its large negative charge prevents dTTP from entering most cells.

   (b) During DNA synthesis, the β- and γ-phosphorus atoms of the nucleoside triphosphates are lost as pyrophosphate. Since the α-phosphorous atom is incorporated into DNA, one should use dATP, dGTP, dTTP, and dCTP labeled with ^{32}P in the α position.

9. Only (c) would lead to DNA synthesis since (a) and (b) have no primer or open end to build on and (d) has no template extending beyond a free 3′-OH. Note: single-stranded linear DNA can be used as a template for DNA synthesis because it can prime synthesis through hairpin formation at its 3′ end.

10. A short polythymidylate chain would serve as a primer since T base pairs with A. Radioactive dTTP labeled in any position except the β- and γ-phosphates would be useful for following chain elongation.

11. After the synthesis of the complementary (–) DNA on the RNA template, the RNA must be disposed of by hydrolysis prior to the completion of the synthesis of the DNA duplex.

12. One should treat the infectious nucleic acid with either highly purified ribonuclease or deoxyribonuclease and then determine its infectivity. RNAse will destroy the infectivity if it is RNA; DNAse will destroy it if it is DNA.

13. Ultimately, this mutation results in half the daughter DNA duplexes being normal and half having a TA pair that had been CG. The first two rounds of replication at the mutant site will be as follows:

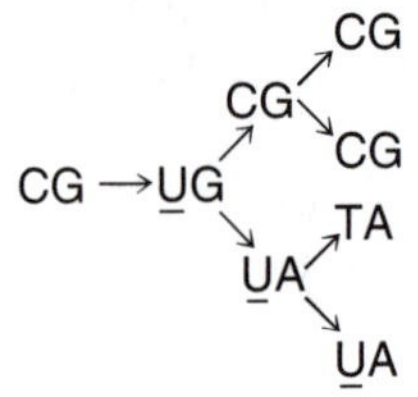

$$CG \rightarrow \underline{U}G \begin{cases} CG \begin{cases} CG \\ CG \end{cases} \\ \underline{U}A \begin{cases} TA \\ \underline{U}A \end{cases} \end{cases}$$

14. Since HCN is the only molecule containing carbon, hydrogen and nitrogen, it is the obvious choice. Also, note that adenine (see Stryer page 76) has the empirical formula $(HCN)_5$ and purine itself is $(HCN)_4 + C$. Although it is interesting to speculate, in fact we do not know how early biomolecules were formed from simpler molecules in the primordial atmosphere.

15. (a) From the 4 mononucleotides one can formulate 16 different dinucleotides. If you don't believe it, try it! From these you can make 64 different trinucleotides. Note that 64 is 4^3. There will be 4^4 (256) tetranucleotides. Proceeding in this manner we get to 4^8 (65,536) different octonucleotides (8-mers).

(b) A bit specifies two bases (say A and C) and a second bit specifies the other two (G and T). Hence, two bits are needed to specify a single nucleotide (or base pair) in DNA. An 8-mer stores 16 bits ($2^{16} = 65{,}535$), the *E. coli* genome (4×10^6 bp) stores 8×10^6 bits, and the human genome (2.9×10^9 bases) stores 5.8×10^9 bits of genetic information.

(c) A high-density diskette stores about 1.5 megabytes, which is equal to 1.2×107 bits. A large number of 8-mer sequences could be stored on such a diskette. The DNA sequence of *E. coli,* once known, could be written on a single diskette. Nearly 500 diskettes would be needed to record the human DNA sequence.

CHAPTER

5

Flow of Genetic Information

In Chapter 4, Stryer introduced you to DNA and RNA as the storage materials of genetic information. In this chapter, he begins a discussion of how this information is expressed in the proteins of a cell. Stryer starts with descriptions of the basic structures and kinds of RNA and an explanation of their central roles in the overall flow of genetic information. He then presents the functions of messenger RNA, transfer RNA, and ribosomal RNA in protein synthesis, along with a description of the enzyme RNA polymerase that synthesizes all cellular RNA. The experiments that elucidated the genetic code and its general features are described. The colinear relationship between the sequences of nucleotides in the DNA and the amino acids of the encoded protein is compared in prokaryotes and in eukaryotes, in which some genes are interrupted by noncoding sequences. Stryer next describes the mechanism by which the noncoding sequences are removed from the initial transcript to form messenger RNA and the biological consequences of such splicing. He concludes the chapter with a discussion of catalytically active RNA and the implications of its existence with respect to hypotheses concerning molecular evolution.

LEARNING OBJECTIVES

When you have mastered this chapter, you should be able to complete the following objectives:

Introduction (Stryer pages 95–97)

1. State the role of DNA in protein synthesis. Outline the flow of genetic information during gene expression.

2. Define the terms *transcription* and *translation* and relate these processes to the *flow of genetic information.*

3. Describe the structure of the *phosphodiester bond* that links the *ribonucleosides* of *RNA* to one another. Recognize the four major bases of RNA. Contrast the composition and structures of RNA and DNA.

4. Name the *three major classes of RNA* found in *E. coli* and outline their functions. Compare their sizes and their relative proportions by weight.

Messenger RNA and Transcription (Stryer pages 97–99)

5. Identify the primary cellular locations of RNA and protein synthesis in eukaryotes and relate these locations to the concept of *messenger RNA (mRNA)*. List the properties of mRNA and describe its role in the flow of genetic information.

6. Describe the experiments that indicated mRNA was an informational carrier in the conversion of the genetic information in DNA into proteins.

7. Describe the *ribosome* and explain its role in *gene expression.*

8. Describe the technique of *hybridization* and explain how it was used to show the relationship of mRNA, *transfer RNA (tRNA),* and *ribosomal RNA (rRNA)* to their *DNA templates.*

9. Contrast the roles of mRNA, tRNA, and rRNA in genetic information flow.

RNA Polymerases and Transcription (Stryer pages 100–102)

10. List the substrates, products, and important enzymatic properties of *RNA polymerases (DNA-dependent RNA polymerases).* Explain the roles of the *DNA template,* the *promoter,* and the *terminator* in transcription.

11. Describe the *transcription* of duplex DNA to form single-stranded RNA and relate this phenomenon to the *conservative* nature of RNA synthesis.

12. Compare DNA-dependent RNA polymerases and *RNA-dependent RNA polymerases.*

The Genetic Code and Protein Synthesis (Stryer pages 102–112)

13. Describe the role of tRNA as the *adaptor molecule* between mRNA and amino acids during protein synthesis. Outline how specific amino acids are covalently attached to specific tRNA molecules. Explain the relationship of the *codon* and *anticodon* to the specific interaction between mRNA and tRNA.

14. Explain what the *genetic code* is and list its major characteristics. Define the terms *degenerate, ambiguous, synonyms, triplet,* and *nonoverlapping* as they apply to the genetic code. Identify the *termination codons* and the *initiation codon.*

15. Recount the historical development of the concepts of information flow and name the major contributors.

16. Using the genetic code, predict the sequence of amino acids encoded by any template DNA or mRNA sequence.

17. Discuss the near universality of the genetic code.

Introns and Splicing (Stryer pages 113–115)

18. Describe the *colinear relationship* between the sequence of DNA in a bacterial gene and the sequence of the amino acids in the protein it encodes. Relate this concept to the *intervening sequences,* or *introns,* that frequently occur in eukaryotic genes.

19. Recount a hypothesis relating *exons* and *functional domains* to the generation and evolution of protein diversity. Differentiate between *genetic recombination* and *RNA splicing* in this process.

Catalytic RNA and Molecular Evolution (Stryer pages 115–116)

20. List the properties of the self-splicing precursor ribosomal RNA (*ribozyme*) in *Tetrahymena.* Explain how the existence of *catalytically active RNA* molecules fits into a hypothesis concerning the evolution of molecules prior to the existence of DNA and proteins.

SELF-TEST

Introduction

1. Transcription is directly involved in which of the following possible steps in the flow of genetic information?

(a) DNA to RNA
(b) RNA to DNA
(c) DNA to DNA
(d) RNA to protein
(e) Protein to RNA

2. Translation is involved in which of the following possible steps in the flow of genetic information?

(a) DNA to RNA
(b) RNA to DNA
(c) DNA to DNA
(d) RNA to protein
(e) Protein to RNA

3. Answer the following questions about RNA:

(a) What is the name of the bond joining the ribonucleoside components of RNA to one another? ____________________
(b) Is this bond between the 2′- or the 3′-hydroxyl group of one ribose and the 5′-hydroxyl of the next? ____________________

(c) Intramolecular base pairs form what kinds of structures in RNA molecules?

___

(d) What bases pair with one another in RNA?

___

(e) What are the three major classes of RNA in a cell, and which is most abundant by weight?

___

(f) Which major class of RNA is most abundant in terms of different molecular species?

___

Messenger RNA and Transcription

4. After the infection of *E. coli* with T2 bacteriophage, which of the following events occur?

(a) New proteins are synthesized.
(b) New RNA is synthesized.
(c) New ribosomes are synthesized.
(d) New RNA becomes associated with new ribosomes.
(e) New RNA becomes associated with preexisting ribosomes.
(f) New RNA is formed that has a long half-life.

5. The following are parts of the eukaryotic cell:

(1) Lysosomes
(2) Nucleus
(3) Plasma membranes
(4) Cytoplasm
(5) Nucleolus

(a) In which part does the synthesis of mRNA usually occur? ___

(b) In which part are most proteins synthesized? ___

6. If you have samples of pure RNA and duplex DNA, how can you tell whether they contain complementary nucleotide sequences?

___

___

7. If all the RNA referred to in question 6 turns out to have sequences that were complementary to the DNA, will its percentage of G and C be identical with that of the DNA? Explain.

___

___

8. If samples of each of the three major classes of RNA found in a cell were hybridized to denatured DNA from the same cell and the RNA–DNA hybrids were filtered through a nitrocellulose membrane, which of the classes would be retained on the filter?

(a) mRNA
(b) rRNA
(c) tRNA

RNA Polymerases and Transcription

9. Which of the following are required for the DNA-dependent RNA polymerase reaction to produce a unique RNA transcript?

(a) ATP
(b) CTP
(c) GTP
(d) dTTP
(e) UTP
(f) DNA
(g) RNA
(h) Mg^{2+}
(i) Promoter sequence
(j) Operator sequence
(k) Terminator sequence

10. What is the sequence of the mRNA that will be synthesized from a template strand of DNA having the following sequence:

. . . ACGTTACCTAGTTGC . . .

____

11. Describe the mechanism of chain growth in RNA synthesis. What is the polarity of synthesis, and how is it related to the polarity of the template strand of DNA?

____

____

The Genetic Code and Protein Synthesis

12. Which of the following are characteristics or functions of tRNA?

(a) It contains a codon.
(b) It contains an anticodon.✓
(c) It can become covalently attached to an amino acid.✓
(d) It interacts with mRNA during the process of transcription.
(e) It can have any of a number of different sequences. ✓
(f) It serves as an adapter between the information in mRNA and an individual amino acid. ✓

13. What is the minimum number of contiguous nucleotides in mRNA that can serve as a codon? Explain.

____

____

14. What is the sequence of the polypeptide that would be encoded by the DNA sequence given in question 10? Assume that the reading frame starts with the first nucleotide given. The genetic code is given on page 109 of Stryer.

____

____

15. The following is a partial list of mRNA codons and the amino acids they encode:

AGU = serine AGC = serine
AAU = asparagine AAC = asparagine
AUG = methionine AUA = isoleucine

Based on this list, which of the following statements are correct?

(a) The genetic code is degenerate.
(b) The alteration of a single nucleotide in the DNA directing the synthesis of these codons could lead to the substitution of a serine for an asparagine in a polypeptide.
(c) The alteration of a single nucleotide in the DNA directing the synthesis of these codons would necessarily lead to an amino acid substitution in the encoded polypeptide.
(d) A tRNA with the anticodon ACU would be bound by a ribosome in the presence of one of these codons.

16. Explain why mitochondria can use a genetic code that is different from the standard code used in the nucleus.

____

____

Introns and Splicing

17. Explain how genetic techniques and amino acid sequence analyses were used to show the colinear relationship of a prokaryotic gene and the protein it encodes.

____

____

____

18. Answer the following questions about what was discovered when the sequence of the gene for the β chain of hemoglobin was determined:

(a) What was the major finding when the sequence of the gene and the amino acid sequence of the β chain were compared?

____

____

(b) What did hybridization between the partially denatured DNA and the mRNA for β-globin show?

____

____

(c) What must happen to the primary transcript from the β-globin gene before it can serve as an mRNA for protein synthesis?

____

____

19. How might the fact that some exons encode discrete functional domains in proteins be related to the evolution of new proteins?

____

____

Catalytic RNA and Molecular Evolution

20. What would you predict would happen if the gene for the precursor of rRNA found in *Tetrahymena,* which contains an intron, were introduced

into *E. coli* and converted into a primary transcript by the DNA-dependent RNA polymerase of *E. coli*?

____

____

ANSWERS TO SELF-TEST

Introduction

1. a, b. Answer b is correct because the reverse transcription of RNA sequences into DNA sequences occurs during the replication of retroviruses (see Stryer, Chapter 4).

2. d

3. (a) The bond is called the *phosphodiester bond.*
   (b) The bond joins the 3′-hydroxyl to the 5′-hydroxyl to form a $3' \rightarrow 5'$ phosphodiester bond.
   (c) Hairpin loops are formed when the RNA chain folds back on itself and base pairs become hydrogen-bonded.
   (d) A pairs with U, and G pairs with C; G can also pair with U, but the association is weaker than that of the G · C base pair.
   (e) The three major classes of RNA found in a cell are mRNA, rRNA, and tRNA; the most abundant is rRNA.
   (f) mRNA constitutes the major class of cellulsar RNA containing the most molecular species.

Messenger RNA and Transcription

4. a, b, e

5. (a) 2 (b) 4

6. You could sequence the RNA and DNA and compare the specific sequences of each to see if the two are complementary; this method provides definitive evidence of identity. An easier but less precise way would be to use the hybridization technique. You would mix the samples, heat the mixture to melt the double-stranded DNA, slowly cool the solution, and examine it to see if it contains any double-stranded DNA–RNA hybrids. If it does, this would indicate that the RNA and DNA sequences are complementary.

7. Not necessarily; RNA synthesis is asymmetric, and only one strand of any region of the DNA usually serves as a template. Asymmetric transcription can lead to RNAs with G and C compositions different from those of the DNA from which they are copied.

8. a, b, c

RNA Polymerases and Transcription

9. a, b, c, e, f, h, i, k. Answer f is correct because DNA is needed to serve as the template. Answers k and i are correct because the promoter and terminator sequences are needed to specify the precise start and stop points, respectively, for the transcription.

10. The mRNA sequence will be . . . GCAACUAGGUAACGU . . . , written in the 5′ → 3′ direction. Sequences written without explicit designation of their ends as 3′ or 5′ follow the convention that the 5′ end is leftmost.

11. The 3′-hydroxyl terminus of the growing RNA chain makes a nucleophilic attack on the α phosphate (the innermost phosphate) of the ribonucleoside triphosphate that has been selected by the template strand of the duplex DNA. RNA polymerase catalyzes the reaction. A ribonucleoside monophosphate residue is added to the chain as a result, and the chain has grown in the 5′ → 3′ direction; that is, the chain has grown at its 3′ end. As with all Watson-Crick base pairing, the stands are antiparallel; that is, the RNA chain is assembled in the 3′ → 5′ direction with respect to the polarity of the template strand of the DNA.

The Genetic Code and Protein Synthesis

12. b, c, e, f. Answer d is incorrect because the interaction of tRNA with mRNA takes place during translation, not transcription.

13. Three contiguous nucleotides are the minimum number that can serve as a codon. There are four kinds of nucleotides in mRNA. A codon consisting of only two nucleotides (either of which could be any of the four possible nucleotides) allows only 16 possible combinations ($4 \times 4 = 16$). This number would be insufficient to specify all 20 of the amino acids. A codon consisting of three nucleotides, however, allows 64 possible combinations ($4 \times 4 \times 4 = 64$), more than enough to specify the 20 amino acids.

14. The sequence of the polypeptide would be Ala-Thr-Arg. The reading frame is set by the nucleotide at the 5′ end of the mRNA transcript; the fourth codon of the mRNA transcript is UAA, which is a translation termination codon.

15. a, b, d. Answer a is correct because both AGU and AGC specify serine; since more than one codon can specify some amino acids, the genetic code is said to be degenerate. Answer b is correct because the alteration of a single nucleotide in the DNA could change a codon on the mRNA transcript from AGU, which specifies serine, to AAU, which specifies asparagine. Answer d is correct because the anticodon ACU would base-pair with the codon AGU. Answer c is *not* correct, because the alteration of a single nucleotide in the DNA could result in another codon that specifies the same amino acid; for example, a codon changed from AGU to AGC would still specify serine.

16. Mitochondria can use a genetic code that differs from the standard code because mitochondrial DNA encodes a distinct set of tRNAs that are matched to the genetic code used in their mRNAs.

Introns and Splicing

17. The mutations in a given gene of *E. coli* were mapped by recombination analysis. The proteins encoded by the wild-type and the mutant genes were then sequenced, and the location and nature of the amino acid substitution for each mutation was identified. The order of the mutations on the genetic map was the same as the order of the corresponding changes in the amino acid sequence of the polypeptide produced by the gene, thus establishing that genes and their polypeptide products are colinear.

18. (a) The number of nucleotides in the gene was much greater than three times the number of amino acids in the protein. There were two stretches of extra nucleotides between the exon sequences that encode the amino acids in the β chain.

(b) The mRNA hydridized to the DNA under conditions where DNA–RNA hybrids are more stable than DNA–DNA hybrids, but there were sections of duplex DNA between the hybrid regions. This indicated that there are intron sequences in the DNA that have no corresponding sequences in the mRNA. (See Figure 5-19 on p. 112 of Stryer.)

(c) The intervening sequences in the nascent, or primary, transcript, which are complementary to the template strand of the DNA of the gene but do not encode amino acids in the protein, must be removed by splicing to generate the mRNA that functions in translation.

19. The shuffling of exons that encode discrete functional domains, such as catalytic sites, binding sites, or structural elements, preserves the functional units but allows them to interact in new ways, thereby generating new kinds of proteins.

Catalytic RNA and Molecular Evolution

20. Because the sequence of ribonucleotides is catalytically active and can splice itself to remove its own intron, it would do so in the *E. coli* that produced the transcript.

PROBLEMS

1. Certain DNA endonucleases degrade double-stranded DNA to yield mononucleotides and dinucleotides, but these enzymes do not degrade those sequences to which other proteins are bound.

(a) How can you use a DNA endonuclease and RNA polymerase to locate a promoter site?

(b) Why should this process be performed in the absence of ribonucleotide triphosphates?

2. The amino acid at position 102 in the primary sequence of a bacterial enzyme is valine, and the corresponding codon in the mRNA sequence for the enzyme is GUU. Suppose a mutation that alters the codon to GCU has no effect on the activity of the enzyme, but another mutation that changes the codon to GAU completely inactivates the enzyme. Briefly explain these observations.

3. It is essential for spliceosomes to remove introns precisely, that is, between the terminal nucleotide of an intron and the first nucleotide of an exon. To see why, suppose that the sequence at the junction between an intron and an exon is

...UUAG | GCUAACGG...

Intron   Exon

Suppose further that a spliceosome occasionally cleaves the RNA transcript between the C and U residues in the exon sequence to yield the following two molecules:

. . . UUAGGC UAACGG . . .

What would be the consequence of this cleavage?

4. Although nearly all the proteins synthesized by a bacterial cell after it has been infected with T2 bacteriophage are determined by the viral genome, some bacterial proteins are also required for successful infection. What bacterial enzyme is needed to initiate viral infection when T2 DNA first enters the cell?

5. (a) The genome of bacteriophage ϕX174 is a single strand of DNA containing 5386 nucleotides. If only one AUG in the genome were used as an initiation signal, how many amino acids could be encoded by the genome? If the average molecular weight of an amino acid is 112, what is the maximum molecular weight of protein encoded by the genome?

(b) Studies have shown that the ϕX174 genome can encode a larger number of proteins than expected. One reason for this increased encoding capacity is that some of the genes overlap one another. For example, the coding sequence for gene B is located entirely within the sequence that codes for gene A. However, the amino acid sequences of the two proteins specified by these genes are entirely different. How is this possible?

6. In contrast to DNA polymerase, RNA polymerase has no nuclease capability to excise mismatched nucleotides. Suggest why the two enzymes are different in this respect.

7. The genome of bacteriophage G4 is a small, single-stranded circle of DNA. Replication of the circle is initiated when an RNA polymerase, a product of the *dnaG* gene of *E. coli,* synthesizes a small segment of RNA that binds to a unique sequence on the G4 chromosome. Initiation of G4 DNA synthesis does not occur in bacterial *dnaG* gene mutants, which have an inactive RNA polymerase. Suggest a function for the small segment of RNA.

8. (a) In *E. coli,* a tRNA that carries tyrosine is composed of 85 nucleotides. However, transcription of the gene that codes for tyrosine tRNA yields an RNA molecule consisting of 350 nucleotides. At least three ribonuclease enzymes cooperate in removing a 41-base segment on the 5′ side of the tRNA sequence and a 224-base segment that extends from the 3′ terminus of the tRNA sequence. The tRNA sequence in the primary transcript is continuous, and no nucleotides are removed from that part of the transcript during processing. How does this type of RNA processing differ from the splicing operations described in Chapter 5 of the text?

(b) Another primary transcript that is synthesized in bacteria contains 6500 nucleotides, including sequences for the 23S, 16S, and 5S RNA molecules found in ribosomes. This primary transcript has sequences on either side of the set of rRNA sequences, as well as “spacer” sequences between each of them. Suggest a reason for the synthesis of a transcript containing all three rRNA sequences.

9. As discussed on page 108 of the text, the codons UAA, UAG, and UGA are signals for chain termination in protein synthesis because none of these codons are read by tRNA molecules. These codons are normally found at the ends of coding sequences for proteins. However, single-base mutations in certain codons can also cause premature termination of the protein chain.

(a) Which codons can be converted to the chain-termination codon UAA by a single base change?
(b) Suppose a mutation creates a UAA codon that is three codons away from the 3′ end of the normal mRNA coding sequence. Why might you assume that the prematurely terminated protein could still be functional?
(c) Revertants of chain-termination mutants include those in which a single base substitution changes a termination codon to one that can again be read by a tRNA molecule. For example, a UAG codon can mutate to UCG. What amino acid would then be found at the corresponding position in the protein?
(d) Other revertants retain the original termination codon at the premature termination site, but an amino acid is inserted at the corresponding site in the protein so that the protein has the same length as the nonmutant protein would have. These revertants are due to another mutation in which the anticodon of a tRNA molecule is altered so that the tRNA molecule can read a termination codon. These tRNA molecules are called *suppressor tRNAs* because they suppress the effect of a chain-termination mutation. Suppose you have a chain-termination mutation that is due to the presence of a UAG codon in the normal coding sequence. If the effect of the UAG codon is suppressed by a tRNA mutation, which amino acids could be found at the site corresponding to the premature termination signal? Assume that a single base change occurs in each case.

10. Polynucleotide phosphorylase was used in the laboratory to synthesize polyribonucleotides that were useful in determining the genetic code. Why is it unlikely that this enzyme synthesizes RNA in the cell? Suggest how the cell uses this enzyme.

11. (a) How did Spiegelman's experiments with T2 RNA suggest that bacterial transcription takes place on only one of the two DNA strands at any location on the chromosome?

(b) Later the principle of transcription on only one strand of DNA was established more firmly by studies with DNA from the virus SP8, which infects the bacterium *Bacillus subtilis.* Because the two complementary strands of SP8 DNA have very different base compositions, they can be easily separated by density centrifugation. How could you use these separated strands to learn more about the transcription of SP8 DNA?

12. You are studying the effects of amino acid replacements on the stability of a particular α helix that is buried in the protein myoglobin. You carry out a series of replacements of a particular leucine residue located in the helix, using site-specific mutagenesis, a technique described in detail in Chapter 6 of Stryer's text. The replacements are as follows:

(i) Leucine ⇒ arginine
(ii) Leucine ⇒ valine
(iii) Leucine ⇒ proline
(iv) Leucine ⇒ glycine
(v) Leucine ⇒ alanine

(a) For each replacement, state whether the change would stabilize, destabilize, or have no effect on the structure of the α helix. Briefly explain each of your answers.

(b) For each replacement, write the most likely mRNA codon required to code for the particular amino acid. Which of the replacements can be carried out by single-base changes? Which of the replacements can be produced only by altering two bases in the mRNA codon?

13. Cordycepin (3′-deoxyadenosine) is described in Stryer's text (p. 118, problem 4) as a compound that can block the synthesis of RNA, because a cordycepin residue in an RNA chain lacks the 3′-OH end needed for chain extension by RNA polymerase. The structure of cordycepin is shown below.

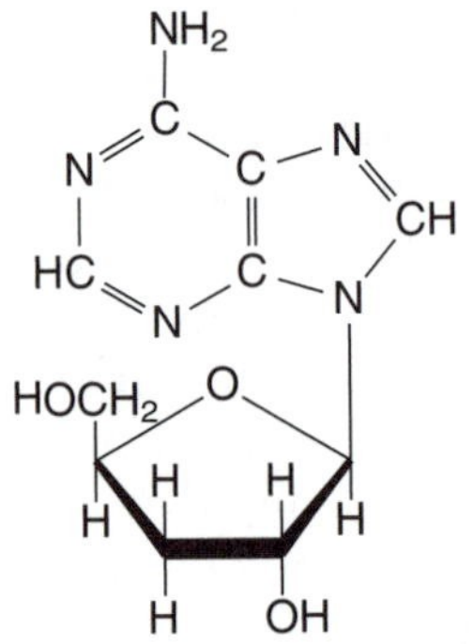

NH_2, N, C, C, N, CH, HC, C, N, N, $HOCH_2$, O, H, H, H, H, H, OH

(a) Cordycepin does not inhibit the growth of bacteria, but it does inhibit growth and division of mammalian cells. Consider the reactions that are required for cordycepin to be converted into a substrate for RNA polymerase and then propose a reason for its ineffectiveness in bacteria.

(b) Would you expect cordycepin to block DNA synthesis as well? Why?

14. Problem 10 on page 118 of Stryer's text describes experiments using Raney nickel to convert cysteinyl-tRNA^{cys} to alanyl-tRNA^{cys}. When this altered aminoacyl-tRNA is used in a protein-synthesizing system, alanyl residues are placed in the position normally occupied by cysteinyl residues in the protein. What does this experiment tell you about the ability of the protein-synthesizing machinery to recognize an inappropriate aminoacyl-tRNA like alanyl-tRNA^{cys}?

15. The products of the cleavage of RNA in dilute alkali include 2′- and 3′-monophosphates. What does this observation reveal about the mechanism of cleavage of RNA? How might sensitivity to base-catalyzed cleavage have affected the choice of DNA or RNA as the prime carrier of genetic information?

16. Many steps in the flow of genetic information are subject to regulation. Stringent control of the production of macromolecules limits expenditure of energy by the cell, permitting the synthesis of particular proteins only as they are required. Consider the steps in storage and transmission of genetic information, and describe which one, when regulated, makes it possible to achieve the greatest economy in energy expenditure by a mature cell.

ANSWERS TO PROBLEMS

1. (a) To locate a promoter site, you would first incubate the double-stranded DNA with RNA polymerase; the RNA polymerase will bind to the promoter site. Next, you would add the DNA endonuclease, which will degrade the DNA that is not protected by RNA polymerase. Electrophoresis can be used to determine the size of the protected fragments of DNA, and the base sequence can be determined using methods discussed in Chapter 6 of the text.

   (b) Ribonucleotide triphosphates are substrates for the transcribing activity of RNA polymerase. If present when this process is performed, they will allow the polymerase molecule to move from the promoter site to the site on the template where transcription begins, as well as onward as transcription progresses. As a result, the promoter site would no longer be protected from endonuclease degradation.

2. The GCU codon in the first mutation corresponds to a substitution of alanine for valine at position 102. Although the side chain of alanine is smaller than that of valine, both are aliphatic amino acids, so the alteration in the structure of the enzyme does not affect the enzyme activity. The GAU codon of the second mutation specifies the substitution of aspartate for valine at position 102. This substitution could have a detrimental effect because the side-chain carboxyl group of aspartate has a negative charge at neutral pH. The charged group could disrupt the native conformation of the enzyme, thereby inactivating it.

3. If the spliceosome cleaves the transcript within the normal exon sequence as shown, the exon coding sequence will be altered because of the loss of two bases. Instead of beginning with the codon GCU in the normal exon, the reading frame will begin with the codon UAA in the altered exon. This codon is, in fact, a termination signal for protein synthesis, which means that the translation of the polypeptide specified by the spliced messenger RNA will be terminated prematurely, indeed, before it even starts.

4. Because only T2 DNA enters the cell, the synthesis of viral proteins cannot begin until T2 messenger RNA has become available. Transcription of the T2 DNA must therefore be carried out by bacterial RNA polymerase using ribonucleotide substrates synthesized by the bacterial cell.

5. (a) Three consecutive bases are required to encode an amino acid, so up to 1794 amino acids could be specified by the ϕX174 genome. The molecular weight of this much protein would be approximately 201,000.

(b) Overlapping genes can yield proteins with very different primary amino acid sequences only if each of the protein coding sequences is read in a different reading frame. The critical feature for the establishment of the proper reading frame is the site of the AUG initiation signal. As an example, consider the following mRNA sequence, which contains two AUG codons that are in overlapping but different reading frames.

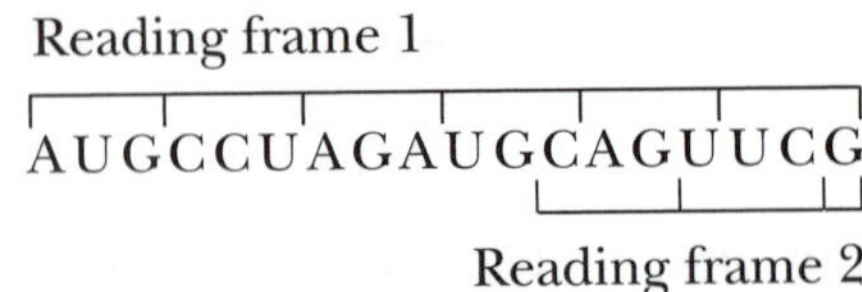

When an initiator tRNA binds to the first AUG codon, reading frame 1 is established; similarly, reading frame 2 is established when an initiator tRNA binds to the second AUG codon. The polypeptides specified by the two different mRNA sequences will necessarily have different amino acid sequences.

6. DNA polymerase is responsible for the duplication of chromosomes, which are the repositories of the genetic information that is passed on to the progeny cells. Any error that occurs in the copying of a DNA template will be transmitted not only to the duplicated chromosome but also to the messenger RNA molecules transcribed from the miscopied DNA template. Therefore, it is very important that DNA polymerase be able to correct errors that would result from mismatched nucleotides. RNA polymerase can make many copies of mRNA, but these molecules have very brief lives in the cell, and very few are passed on to progeny cells. Occasional errors in the transcription can result in the production of defective proteins, but it appears that the cell can tolerate such errors, provided not too many occur.

7. The small RNA molecule serves as the primer for the intiation of DNA synthesis by DNA polymerase. As described in Chapter 4 of the text, DNA polymerase requires a primer nucleotide with a 3′-hydroxyl terminus along with a template. Studies show that either a deoxyribonucleotide or a ribonucleotide can serve as a primer. DNA polymerase cannot synthesize a primer sequence on a closed DNA circle, but the RNA polymerase produced by the *dnaG* gene can and does synthesize a short primer sequence, which is then extended as DNA is synthesized by DNA polymerase. When replication has extended around the circle and the 5′ terminus of the RNA primer is reached, the primer is hydrolyzed and the small gap is filled in with a DNA sequence. Bacterial *dnaG* mutants cannot support G4 infection because they cannot synthesize the RNA primer.

8. (a) The processing of the primary tRNA transcript removes RNA on either side of the uninterrupted tRNA sequence of 85 nucleotides, whereas splicing operations remove RNA sequences that are located within regions that code for continuous polypeptide; these sequences must be removed to ensure that the protein specified by the RNA will have the correct amino acid sequence.

(b) The most obvious reason for transcribing all three rRNA sequences simultaneously is that it ensures that an equal number of the various ribosomal RNA molecules will be available for the assembly of ribosomes. In addition, only one promoter, rather than three, is required for rRNA synthesis.

9. (a) The protein-encoding codons that could be mutated to UAA by a single base change are CAA, GAA, AAA, UCA, UUA, UAU, and UAC.

(b) Chain termination near the 3′ end of the normal coding sequence could allow the prematurely terminated protein to be functional because most of the polypeptide sequence would be intact. Removing a few amino acids from the C-terminal end of many, but not all, proteins does not appreciably affect their normal function.

(c) When a UAG codon reverts to UCG, a serine residue will be found at the site on the protein that corresponds to the chain-termination site; it will be linked by a peptide bond to the next amino acid in the polypeptide.

(d) To determine which amino acids would be carried by suppressor tRNAs to a UAG codon site, you should identify all those tRNA molecules having an anticodon that, by a single base change, can read a UAG codon. Each such tRNA molecule would suppress the premature termination of the chain by inserting an amino acid at the corresponding site in the protein. The amino acids that could be found at the site (along with their codons are Glu (GAG), Gln (CAG), Leu (UUG), Lys (AAG), Ser (UCG), Trp (UGG), and Tyr (UAC and UAU). When a cell contains a suppressor tRNA, proteins whose mRNA sequence normally ends with a single stop codon may not be terminated. Although the extension of such proteins could be lethal, most cells are able to tolerate suppression. One explanation is that other proteins involved in chain termination may recognize a stop codon even though a tRNA that reads the codon is present. More work is needed to develop a full understanding of the reasons for toleration of suppression.

10. Equilibrium for the reaction catalyzed by polynucleotide phosphorylase lies toward the direction of RNA degradation rather than synthesis. High concentrations of ribonucleotides are required to achieve the net synthesis of RNA; and it is likely that the concentrations of ribonucleotides in the cell are not sufficient to drive net polynucleotide synthesis. Also, polynucleotide phosphorylase does not use a template, so the polyribonucleotides it synthesizes contain random sequences, which makes them of no value for protein synthesis. The cell may use polynucleotide phosphorylase as a degradative enzyme in conjunction with other nucleases that regulate the lifetimes of RNA molecules, including mRNA, which in bacteria are relatively short.

11. (a) If transcription occurred simultaneously on both DNA strands, the complementary RNA molecules synthesized from the two templates would be double-stranded. Spiegelman found no evidence of such molecules in his studies.

(b) To establish whether one or both strands of SP8 DNA are used for transcription, you can carry out hybridization experiments with the separate strands using radioactive RNA synthesized during the infection of Bacillus with SP8. The results show that such RNA hybridizes to only one of the two strands, which means that only one of the two strands of the DNA of the SP8 virus is transcribed. In most other organisms, parts of both strands are used for transcription; SP8 virus is exceptional in that only one strand is used for all mRNA synthesis.

12. (a) (i) Leucine ⇒ arginine destabilizes the helix because you are substituting a hydrophilic residue for a hydrophobic residue (most likely in a hydrophobic region of the helix).
(ii) Leucine ⇒ valine has no effect (a conservative replacement of one hydrophobic residue for another).
(iii) Leucine ⇒ proline destabilizes the helix because proline does not allow rotation about its peptide bond and no hydrogen atom in this imino acid is available for hydrogen bonding.
(iv) Leucine ⇒ glycine destabilizes because glycine is a very flexible residue and can act as a swivel, disrupting the helix (the frequency of glycines in an α helix is as rare as the frequency of prolines).
(v) Leucine ⇒ alanine has no effect (a conservative replacement of one hydrophobic residue for another).

(i) CUU ⇒ CGU Single-base change required
(ii) CUU ⇒ GUU Single-base change required
(iii) CUU ⇒ CCU Single-base change required
(iv) CUU ⇒ GGU Two-base alteration required
(v) CUU ⇒ GCU Two-base alteration required

13. (a) Cordycepin is a nucleoside, and it must be converted to the triphosphate form before it can be incorporated (as cordycepin monophosphate) by RNA polymerase into a growing polynucleotide chain. The conversion of cordycepin to the triphosphate form is carried out by a number of kinase enzymes (see Stryer's Chapter 29 on nucleotide metabolism) that utilize ATP as a phosphate donor. Bacteria probably cannot phosphorylate cordycepin efficiently, which makes them less susceptible to inhibition of RNA and DNA synthesis.

(b) You would expect cordycepin to inhibit DNA polymerase because a cordycepin residue in DNA would also lack a 3′-OH end required for primer extension.

14. The ribosomal complex that carries out protein synthesis is unable to recognize alanyl-tRNA^{cys} as an inappropriate or foreign form of tRNA. An amino acid that is attached to a transfer RNA molecule will be transferred into a growing polypeptide chain solely on the basis of recognition between the anticodon in tRNA and the codon in the messenger RNA molecule. Once an aminoacyl-tRNA has been formed, accurate translation does not depend on recognition of the attached amino acid. This important point was established by Dintzis and von Ehrenstein, who carried out the incisive experiments using Raney nickel to convert the cysteinyl residue on cys-tRNA^{cys} to an alanyl residue, then analyzing the results using an in vitro hemoglobin-synthesizing system.

15. The base-catalyzed generation of 2′- and 3′-monophosphates argues that a cyclic 2′,3′-phosphodiester is formed during cleavage of RNA. It is likely that a hydroxyl anion abstracts the hydrogen atom of the 2′-OH of RNA, leaving a 2′-O$^-$ that attacks the phosphorus atom, cleaving the 5′ phosphodiester bond and generating a cyclic 2′,3′-phosphodiester. This cyclic derivative is unstable and can decompose to form either a 2′- or a 3′-phosphate ester. DNA has no 2′-OH groups and is therefore not susceptible to alkali degradation. Because it is more stable than RNA, it may be selectively appropriate as the carrier of genetic information to future generations.

16. The step that would afford the maximum economy is probably transcription, through the control of the activity of RNA polymerase. Recall the discussion of Monod and Jacob's experiments with lactose uptake and utilization in *E. coli;* they found that lactose-utilizing proteins are produced only when lactose, the substrate for those enzymes, is present in the culture medium, and synthesis of those proteins stops shortly after the sugar is exhausted. It follows that controlling messenger RNA production, by stimulating or inhibiting the activity of RNA polymerase, allows the cell to make certain types of mRNA and to synthesize the corresponding proteins only when required by the cell. A cell that cannot regulate RNA polymerase activity will produce unneeded mRNA molecules, and the energy required to produce those polynucleotides will be wasted, even if translation is stringently regulated.

EXPANDED SOLUTIONS TO TEXT PROBLEMS

1. (a) Deoxyribonucleoside triphosphates versus ribonucleoside triphosphates.

(b) $5' \rightarrow 3'$ for both.

(c) DNA serves as the template for both polymerases. During DNA replication by polymerase I each parent strand acts as a template for the formation of a new complimentary strand. Since each daughter molecule receives one strand from the parent DNA molecule, the template is said to be semiconserved. However, after guiding the synthesis of RNA by RNA polymerase, the DNA double helix remains intact. Hence the template is said to be conserved.

(d) DNA polymerase I requires a primer, whereas RNA polymerase does not.

2. (a) Because mRNA is synthesized antiparallel to the DNA template and A pairs with U and T pairs with A, the correct sequence is 5′-UAACGGUACGAU-3′.

(b) Since the 5′ end of an mRNA molecule codes for the amino terminus, appropriate use of the genetic code (Stryer, p. 109) leads to Leu-Pro-Ser-Asp-Trp-Met.

(c) Since one has a repeating tetramer (UUAC) and a 3-base code, repetition will be observed at a 12-base interval (3 × UUAC). Comparison of this 12-base sequence with the genetic code leads to the conclusion that a polymer with a repeating tetrapeptide (Leu-Leu-Thr-Tyr) unit will be formed.

3. The instability of RNA in alkali is due to its 2′-OH group. In the presence of OH^- the 2′-OH group of RNA is converted to an alkoxide ion (RO^-) by removal of a proton. Intramolecular attack by the 2′-alkoxide on the phosphodiester in RNA gives a 2′,3′-cyclic nucleotide, cleaving the phosphodiester bond in the process. Further attack by OH^- on the 2′,3′-cyclic nucleotide produces a mixture of 2′ and 3′-nucleotides. Note that the mechanism for ribonuclease action is quite similar (see Figure 9-18, p. 216). Since DNA lacks a 2′-OH group, it is quite stable in alkali.

4. Apparently cordycepin is converted to its 5′-triphosphate and incorporated into the growing RNA chain. This chain containing cordycepin now lacks a 3′-OH group; hence, RNA synthesis is terminated.

5. Only single-stranded mRNAs can serve as templates for protein synthesis. Since poly(G) forms a triple-stranded helix, it cannot serve as a template for protein synthesis.

6. Note that each complimentary strand is missing one of the four bases; d(TAC) lacks G and d(GTA) lacks C. Thus, incubation with RNA polymerase and only UTP, ATP, and CTP led to the synthesis of only poly(UAC), the RNA complement of d(GTA). When GTP was used in place of CTP, the complement of d(TAC), poly(GUA), was formed.

7. (a) Changing the codon for lysine to one for aspartate requires two mutations, and the probability of getting two mutations in a three-base sequence is very small.

   (b) The more palatable substitutions are those that can result from a *single* base change in the lysine codon AAG. By consulting the genetic code (Stryer, p. 109), one can see that the lysine codon (AAG) can be changed to the codon for Arg, Asn, Gln, Glu, Ile, Met, or Thr by changing a single base.

8. Since three different polypeptides are synthesized, the synthesis must start from three different reading frames. One of these will be in phase with the AAA in the sequence shown in the problem and will therefore have a terminal lysine, since UGA is a stop signal. The reading frame in phase with AAU will result in a polypeptide having an Asn-Glu sequence in it, and the reading frame in phase with AUG will have a Met-Arg sequence in it.

9. Highly abundant amino acid residues have the most codons (e.g., Leu and Ser each have six), whereas the least abundant ones have the fewest (Met and Trp each have only one). Degeneracy (a) allows variation in base composition and (b) decreases the likelihood that a substitution of a base will change the encoded amino acid. If the degeneracy were equally distributed, each of the 20 amino acids would have three codons. Benefits (a) and (b) are maximized by assigning more codons to prevalent amino acids than to less frequently used ones.

10. During peptide synthesis, mRNA codons recognize the anticodons of tRNAs, *not* the amino acid the tRNA is carrying. Therefore, the cysteine tRNA (with an anticodon for UGU) carrying alanine will be recognized for UGU on the mRNA during peptide synthesis. The product will therefore be Phe-Cys-His-Val-*Ala*-Ala. All the ^{14}C should be in the first alanine (the fifth amino acid.)

11. Organisms that survive at elevated temperatures are expected to have more GC in their DNA, since this raises the melting temperature. The reverse might be expected in colder temperatures. Therefore, one would predict that the hot spring alga has a higher frequency of the GUC and GUG codons for valine, whereas that from the Antarctic has a higher frequency of GUU and GUA.

12. The degeneracy of the genetic code makes it possible for comparable DNA molecules to have less homology than the proteins they code for, since a base change in DNA need not cause a change in the amino acid sequence. It is possible for DNA molecules having very limited sequence homology to code for identical or nearly identical proteins.

CHAPTER

6

Exploring Genes

The nature of hereditary material and the flow of information from DNA to protein via RNA were outlined in Chapters 4 and 5. Review these chapters in preparation for studying Chapter 6.

In this chapter, Stryer presents the methods and techniques that are used to analyze and manipulate DNA. He begins with a description of restriction endonucleases and their products—specific DNA restriction fragments. He then presents three methods for the sequencing of DNA and an automated method for synthesizing DNA by chemical means. A more detailed description of restriction enzymes and DNA ligase follows; these enzymes make possible the precise production and joining of DNA fragments. Next, the self-replicating carriers of the target genes are discussed, and the most widely used of these vectors are described. The problems of locating specific genes in the genome are considered. The polymerase chain reaction for amplifying specific DNA sequences is described. The special role in recombinant DNA technology of cDNA, which is produced from mRNA, is discussed, followed by consideration of the technology by which foreign genes are inserted and then expressed in eukaryotic cells. Stryer then describes how site-specific mutagenesis and gene splicing can produce new proteins. Finally, he describes various applications of these powerful new biochemical technologies.

LEARNING OBJECTIVES

When you have mastered this chapter, you should be able to complete the following objectives:

Introduction (Stryer pages 119–120)

1. Define *recombinant DNA*.

Restriction Enzymes and Restriction Fragments (Stryer pages 120–122)

2. Describe the reaction of *restriction enzymes* and the sequence characteristics of the sites they recognize.

3. Explain why *gel electrophoresis of DNA* is essential to recombinant DNA technology. Describe how DNA molecules can be detected in gels.

4. Contrast the *Southern, Northern,* and *Western blotting* techniques.

DNA Sequencing and Chemical Synthesis (Stryer pages 122–126)

5. Outline *DNA sequencing* by *specific chemical cleavage* (the *Maxam-Gilbert method*). Describe the function of *polynucleotide kinase* in this process.

6. Outline DNA sequencing by the *controlled interruption of enzymatic replication* (the *Sanger dideoxy* method).

7. Compare and contrast the Maxam-Gilbert and Sanger dideoxy DNA sequencing methodologies.

8. Describe how *oligonucleotides* are chemically synthesized.

9. List the common uses of oligonucleotides. Note the use of *DNA chips* in sequencing.

Recombinant DNA Construction and Cloning (Stryer pages 126–135)

10. List the desired characteristics of a *vector.* Outline the major steps in *cloning* a DNA molecule.

11. Name the substrates of *DNA ligase,* and describe the reaction catalyzed by it. Describe the termini of the DNA fragments joined by DNA ligase.

12. Explain how *linkers* are used in recombinant DNA technology.

13. Name some common vectors used in prokaryotes and eukaryotes and compare their properties and relative merits.

14. Describe the preparation and *screening* of a *genomic library.*

15. Describe what a *probe* is and explain the biochemical basis of its specificity. Describe how probes are obtained and used in identifying DNA segments.

16. Explain how to design a probe by converting an amino acid sequence into a nucleotide sequence using the genetic code.

17. Outline *chromosome walking* and *jumping* and state the kinds of information that these methods reveal.

18. Outline the *polymerase chain reaction* (*PCR*) steps and give examples of PCR applications.

Expression of Recombinant DNA (Stryer pages 135–139)

19. Define *cDNA*. Distinguish between *genomic* and *cDNA libraries.*

20. Describe the reactions catalyzed by *reverse transcriptase* and *S1 nuclease.* Outline the process for converting the information in mRNA into duplex DNA.

21. Explain how the presence of *introns* in some eukaryotic genes prevents the expression of these genes in prokaryotes.

22. List the properties of *expression vectors.*

23. Contrast screening for recombinant cells using *nucleic acid probes* versus *immunochemical methods.*

24. List the methods by which DNA can be incorporated into plant and animal cells. Define *transgenic organism* and give an example.

Protein Engineering and Site-Specific Mutagenesis (Stryer pages 139–141)

25. Outline the processes of *oligonucleotide-directed mutagenesis* and *cassette mutagenesis.* Define *stringency* as it relates to *nucleic acid hybridization.*

26. Describe the process of protein engineering and explain its value. List the different kinds of mutational changes that can be engineered into genes. Define *chimeric protein.*

27. List applications of recombinant DNA technology.

28. Reflect on the implications of recombinant DNA technology with respect to moral and social values.

SELF-TEST

Restriction Enzymes and Restriction Fragments

1. Which of the following portions of a longer double-stranded (duplex) DNA segment is likely to be the recognition sequence of a restriction enzyme?

(a) 5′-AGTC-3′
3′-TCAG-5′
(b) 5′-ATCG-3′
3′-TAGC-5′
(c) 5′-ACCT-3′
3′-TGGA-5′
(d) 5′-ACGT-3′
3′-TGCA-5′

2. Which of the following reagents would be useful for visualizing DNA restriction fragments that have been separated by electrophoresis in an agarose gel and remain in the wet gel?

(a) $^{32}P_i$
(b) [α-^{32}P]ATP
(c) Diphenylamine
(d) Ethidium bromide
(e) DNA polymerase
(f) Polynucleotide kinase

3. Which blotting technique is used for the detection of DNA that has been separated from a mixture of DNA restriction fragments by electrophoresis through an agarose gel and then transferred onto a nitrocellulose sheet?

(a) Eastern blotting
(b) Northern blotting
(c) Southern blotting
(d) Western blotting

DNA Sequencing and Chemical Synthesis

4. Which of the following reagents would be useful for labeling the oligonucleotide d(GGAATTCC)?

(a) [γ-^{32}P]ATP
(b) $^{32}P_i$
(c) DNA-dependent RNA polymerase
(d) Polynucleotide kinase
(e) DNA ligase

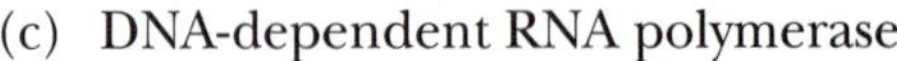

5. In producing fragments for sequencing, which of the following reagents is used to modify the purines in a DNA molecule so that they become labile to alkali?

(a) Dimethylsulfate
(b) Hydrazine
(c) NaCl
(d) Piperidine
(e) Phosphite triester

6. Complete the following statements about the Sanger dideoxy method of DNA sequencing:

(a) The incorporation of a ddNMP onto a growing DNA chain stops the reaction because

____

____

(b) The fragments are usually labeled with ^{32}P by

____

____

(c) A "universal" primer may be used when sequencing any insert cloned into M13 because

____

____

(d) It is preferable to label the oligonucleotide primer with a fluorescent rather than a radioactive group because

____

____

7. Which of the following statements are correct? Chemically synthesized oligonucleotides can be used

(a) to synthesize genes.
(b) to construct linkers.
(c) to introduce mutations into cloned DNA.
(d) as primers for sequencing DNA.
(e) as probes for hybridization.

8. Which of the following statements are correct? The successful chemical synthesis of oligonucleotides requires

(a) high yields at each condensation step.
(b) the protection of groups not intended for reaction.

(c) a single treatment for the removal of all blocking groups.
(d) methods for the removal of the blocking groups that do not rupture phosphodiester bonds.
(e) a computer-controlled, automated "gene machine."

Recombinant DNA Construction and Cloning

9. State whether each of the following is or is not a desired characteristic of vectors and explain why or why not:

(a) Autonomous replication ______________________________

____

(b) Unique restriction sites ______________________________

____

(c) Genes that confer antibiotic resistance ____________________

____

(d) Small size ______________________________________

____

(e) Circularity ______________________________________

____

10. You have been supplied with the oligonucleotide d(GGAATTCC) and an isolated and purified DNA restriction fragment that has been excised from a longer DNA molecule with a restriction endonuclease that produces blunt ends. Which of the following reagents would you need to tailor the ends of the fragment so it could be inserted into an expression vector at a unique EcoRI cloning site?

(a) DNA polymerase
(b) All four dNTPs
(c) EcoRI restriction endonuclease
(d) ATP
(e) DNA ligase
(f) Polynucleotide kinase

What would happen if the restriction fragment had an internal EcoRI site?

____

____

11. Which of the following statements are correct? Inserting a long DNA fragment into the middle of a vector gene that specifies an enzyme that hydrolyzes an antibiotic and incorporating the altered vector into a bacterium

(a) leads to transfer of drug resistance.
(b) is called *insertional inactivation.*
(c) renders the cell sensitive to that antibiotic.
(d) can be used to identify bacteria that contain the vector with the DNA fragment.
(e) is a method of destroying pathogenic bacteria.

12. Briefly describe genomic and cDNA libraries.

____

____

13. Which of the following partial amino acid sequences from a protein whose gene you wish to clone would be most useful in designing an oligonucleotide probe to screen a cDNA library?

(a) Met-Leu-Arg-Leu
(b) Met-Trp-Cys-Trp

Explain why.

____

____

14. Which of the following statements are correct? Very long DNA fragments (> 100 kb) from eukaryotic genomes

(a) can be efficiently packaged in a λ phage vector.
(b) can be propagated in yeast artificial chromosomes.
(c) must have cohesive ends for cloning.
(d) can be analyzed by chromosome walking.
(e) can be separated by polyacrylamide gel electrophoresis.

15. Match the conditions of the PCR reaction, in the left column, with the appropriate reaction.

(a) Cooling abruptly to 54°C __________	(1) DNA synthesis by Taq DNA polymerase
(b) Heating to 72°C, 30 s __________	(2) Hybridization of primers
(c) Heating to 95°C, 15 s __________	(3) Strand separation

16. Using the information in question 15, give the sequence of the PCR reaction steps, and explain the rationale for each step.

____

____

____

17. Which of the following are possible applications of the PCR technique?

(a) Detection of very small amounts of bacteria and viruses
(b) Introduction of a normal gene into animals containing the corresponding defective gene
(c) Amplification of DNA in archaeological samples
(d) Monitoring of certain types of cancer chemotherapy
(e) Identification of matching DNA in forensic specimens

Expression of Recombinant DNA

18. Explain how the presence of introns in eukaryotic genes prevents the production of the protein products they encode when expression is attempted in bacteria. How can this problem be circumvented?

____

____

19. Which of these reagents would be required to perform an immunochemical screen of a population of bacteria for the presence of a particular cloned gene?

(a) [γ-^{32}P]ATP
(b) Polynucleotide kinase
(c) DNA polymerase
(d) All four dNTPs
(e) A radioactive antibody to the protein encoded by the cloned gene

20. The gene for a eukaryotic polypeptide hormone was isolated, cloned, sequenced, and overexpressed in a bacterium. After the polypeptide was purified from the bacterium, it failed to function when it was subjected to a bioassay in the organism from which the gene was isolated. Speculate why the recombinant DNA product was inactive.

___

___

Protein Engineering and Site-Specific Mutagenesis

21. Which of the following statements are correct? Oligonucleotide-directed site-specific mutagenesis

(a) depends on having an oligonucleotide with a sequence completely different from that of the target gene.
(b) can be used to produce deletion, insertion, and point mutations.
(c) is a good method for identifying the functional domains of an enzyme.
(d) is useful for determining the involvement of a particular amino acid in the catalytic mechanism of an enzyme.
(e) can involve the use of the same oligonucleotide to produce the mutant and to detect it.

22. Outline the steps necessary to synthesize a gene. Be explicit about the information and reagents, including enzymes, you would need.

___

___

ANSWERS TO SELF-TEST

Restriction Enzymes and Restriction Fragments

1. d. It has twofold rotational symmetry; that is, the top strand, 5′-ACGT-3′, has the same sequence as the bottom strand. Many restriction enzymes recognize and cut such palindromic sequences.

2. d. Ethidium bromide intercalates into the DNA double strand, and its quantum yield of fluorescence consequently increases. Upon uv irradiation, it fluoresces with an intense orange color wherever DNA is present.

3. c

DNA Sequencing and Chemical Synthesis

4. a, d. The reaction of the oligonucleotide with these reagents would yield [5′-^{32}P]d(pGAATTCC).

5. a. Dimethylsulfate methylates guanine at N-7 and adenine at N-3. Hydrazine reacts with the pyrimidines, and piperidine is the base that catalyzes the cleavage of the phosphodiester backbone at apurinic (a site lacking a purine) or apyrimidinic sites.

6. (a) The newly synthesized ddNMP terminus lacks the requisite 3′-hydroxyl for continuation of DNA synthesis.

   (b) Using an [α-^{32}P]dNTP in the reaction mixture because DNA polymerase incorporates the [5′-^{32}P]dNMP portion of the nucleotide into the growing DNA chain.

   (c) A single primer sequence that is complementary to a region adjacent to the site of insertion can be used to sequence any DNA segment cloned into that site (see Stryer, p.130).

   (d) It avoids the use of radioisotopes and could allow the automated detection of the terminated primers.

7. All are correct.

8. a, b, d. Answer c is not correct, because the blocking groups must be removed differentially; for example, the dimethoxytrityl group must be removed from the 5′-hydroxyl (so that the next condensation with an incoming nucleotide can occur) without removing the blocking groups on the exocyclic amines of the bases. Answer e is not correct because the synthesis can be carried out manually, albeit more slowly and laboriously.

Recombinant DNA Construction and Cloning

9. Answers a, b, c, and d are desired characteristics. Autonomous replication allows amplification of the vector in the absence of extensive cell growth. Unique restriction sites allow the cutting of vectors at single, specific sites for the insertion of the foreign DNA. Antibiotic resistance allows for the selection of those bacteria that carry the vector or for insertional inactivation. Small size allows the insertion of long pieces of foreign DNA without interfering with the introduction of the recombinant molecule into the host bacterium. Answer e is not a desired characteristic, because vectors do not have to be circular to function effectively; an example is λ phage.

10. c, d, e, f. The oligonucleotide must have a 5′-phosphate group to serve as a substrate for DNA ligase. Therefore, ATP and polynucleotide kinase would be used, as would DNA ligase and ATP, to join the duplex form of the palindromic oligonucleotide to the blunt-ended fragment. Finally, the fragment with the linker covalently joined to it would be cut with EcoRI endonuclease to produce cohesive ends that match those of the cut vector. It is assumed that the fragment itself lacks EcoRI sites because, if one or more internal EcoRI sites were present, the fragment would be cut when it is treated with the enzyme to generate the cohesive ends.

11. b, c, d. Insertion of the DNA fragment into the gene disrupts the gene's production of the enzyme that confers drug resistance. A cell containing the altered vector is therefore sensitive to the antibiotic. When the

vector contains a gene that confers resistance to a second drug, insertional inactivation can be incorporated into a selection scheme for isolating cells that contain vectors having the inserted fragment. Cells that remain resistant to the second antibiotic while sensitive to the first probably contain the vector with the foreign DNA.

12. A genomic library is composed of a collection of clones, each of which contains a fragment of DNA from the target organism. The entire collection should contain all the sequences present in the genome of the target organism. A cDNA library is composed of a collection of clones that contain the sequences present in the mRNA of the target organism from which the mRNA was isolated. A cDNA library contains far fewer sequences than does a genomic library because only a small fraction of the genome is being transcribed into mRNA at any given time. The content of a cDNA library depends on the type of cell, the period of development, and the environmental influences on the cells from which the mRNA was isolated.

13. b. This is the better choice for reverse translation into a DNA sequence because it contains fewer amino acid residues having multiple codons; Trp and Met have one codon each and Cys has two. Thus, for the (b) sequence, Met-Trp-Cys-Trp, there are $1 \times 1 \times 2 \times 1 = 2$ different dodecameric oligonucleotide coding sequences. In contrast, Leu and Arg each have six codons, so for the (a) sequence, Met-Leu-Arg-Leu, there are $1 \times 6 \times 6 \times 6 = 216$ different coding sequences. Therefore, the probe for (b) would be simpler to construct and would hybridize to fewer incorrect clones in the cDNA library.

14. b, d. Statements a and e are incorrect because very large DNA fragments do not fit into a λ capsid and are too big to be separated by polyacrylamide gel electrophoresis. Instead, large DNA fragments can be separated by pulsed field electrophoresis. Answer c is incorrect because linkers can be used in cloning.

15. (a) 2 (b) 1 (c) 3

16. The sequence of steps is 3, 2, and 1. Heating to 95°C completely separates all double-stranded DNA molecules. The subsequent rapid cooling to 54°C causes the excess primers to hybridize to the complementary parent DNA strands. Then at 72°C the Taq DNA polymerase (which remains active through the 95°C step) carries out DNA synthesis using the four dNTPs. Amplification of DNA is achieved by repeating these steps many times.

17. a, c, d, e

Expression of Recombinant DNA

18. In eukaryotes the introns are removed from the primary transcript by processing, to produce the mRNA that is translated. Prokaryotes lack the machinery to perform this processing; consequently, the translation product of the primary transcript would not be functional, because it would contain amino acid sequences specified by the intron sequences. The problem can be circumvented by using cDNA prepared from the mRNA from the gene encoding the protein; the cDNA will contain only the sequences present in the processed RNA; that is, the intron sequences will have been removed.

19. e. An immunochemical screen could be performed by adding the radioactive antibody to lysed bacterial colonies and then examining the population by autoradiography to see which colonies contain the antigen of the cloned gene—that is, the product of the cloned gene.

20. Omitting such an obvious explanation as the destruction of the polypeptide during the bioassay, it is possible that the polypeptide might not have undergone some posttranslational modification that is required for it to function. For example, the polypeptide might need to be acetylated, methylated, or trimmed at the N terminus, or it might need to have a carbohydrate or lipid group attached to it. The bacterium in which it was produced would be unlikely to contain the enzymatic machinery necessary to carry out these modifications, or if it did, it might lack the ability to recognize the eukaryotic signals that direct these modifications. It is also possible that the bacterium might have contained a peptidase or protease that inactivated the polypeptide without destroying its antigenic properties.

Protein Engineering and Site-Specific Mutagenesis

21. b, d, e. If the sequence of the oligonucleotide were completely different from the sequence of the target gene, it could not hybridize to the target gene and serve as a primer for DNA polymerase even under low-stringency hybridization conditions. Functional domains could be better identified by deletion mutagenesis, in which relatively large regions of the gene would be systematically removed and the resulting functional consequences tested. Although oligonucleotide-directed mutagenesis can be used to make deletions, it is not the method of choice for an initial survey to find functional domains, since only a single, precisely defined deletion is produced with each oligonucleotide. For exploratory deletion analysis, nucleases are used to generate populations of deleted sequences for functional testing. Oligonucleotide-directed mutagenesis is better suited for changing specific regions when one wishes to test a specific model or hypothesis regarding the function of one or a few amino acids. At the correct conditions of hybridization stringency, the mutagenizing oligonucleotide will form a more stable hybrid with the newly produced mutant sequence than with the original unmodified sequence because it will form a perfect complement. Thus, it can be used to differentiate the mutant and the original sequences from each other.

22. You would need to know the sequence of the gene you wish to synthesize. This could be derived from the amino acid sequence of the protein the gene encodes by reverse translation using the genetic code. You would also need to know what restriction sites you wish to build into the synthetic sequence for cloning the synthetic product. You would have to decide on the individual sequences of the oligonucleotides that compose both strands of the gene. These sequences would be based on the final desired sequence, the individual lengths (30 to 80 nucleotides long) that can be easily synthesized and purified, and the requirement for overlapping ends that will be necessary to allow unique joinings of the cohesive ends of the partially duplex segments. Self-complementary oligomers would be mixed together to form duplex fragments with cohesive ends. DNA ligase and ATP would be added to join these together to form the whole duplex. If appropriate ends have been designed into the synthesis, the product can then be ligated into a vector for cloning.

1. You are studying a newly isolated bacterial restriction enzyme that cleaves double-stranded circles of pBR322 once to yield unit-length, linear, double-stranded molecules. After these molecules are denatured and are allowed to reanneal, all the double-stranded molecules are unit-length linears. In another experiment, an enzyme that cleaves double-stranded DNA at random sites is used at low concentration to cleave intact pBR322 molecules approximately once per molecule, again yielding unit-length, double-stranded linears. Denaturation and renaturation yield a high percentage of double-stranded circles with a single, randomly located nick in each strand. How do these experiments show that the new restriction enzyme cleaves pBR322 DNA at a single specific site?

2. Before the development of methods for the analysis and manipulation of genes, many attempts were made to transform both prokaryotic and eukaryotic cells with DNA. Although the experiments with the pneumococcus bacterium described in Chapter 4 were successful, most others were not. Suggest why these early efforts to transform cells were largely unsuccessful.

3. Pseudogenes are composed of nonfunctional DNA sequences that are related by sequence similarity to actively expressed genes. Some researchers have proposed that pseudogenes are copies of functional genes that have been inactivated during genome evolution. Suggest several ways that such genes could have become nonfunctional. Suppose you clone a number of closely related sequences, any of which may code for a particular protein. How can you tell which of the sequences is the functional gene, that is, which of the sequences codes for the protein?

4. Both the Maxam-Gilbert and Sanger dideoxy methods for determining DNA sequence are limited in that a stretch of only 500 or fewer bases can be analyzed at one time. Suppose you wish to sequence a mutant of the simian virus SV40, which contains 5000 base pairs. You decide to use restriction fragments of the DNA for sequencing, and you find an enzyme that makes a significant number of cuts to give fragments of 275 nucleotides or less. Why might it be a good idea also to sequence another set of fragments cleaved by another restriction enzyme?

5. The denaturation and reassociation of complementary DNA strands can be used as a tool for genetic analysis. Heating double-stranded DNA in a dilute solution of sodium chloride or increasing the pH of the solution above 11 causes dissociation of the complementary strands; when the solution of single-stranded molecules is cooled or when the pH is lowered, the complementary strands will reanneal as complementary base pairs reform. What causes base pairs to dissociate at pH 11 or higher?

Both double- and single-stranded DNA molecules can be visualized using electron microscopy in a technique called *heteroduplex analysis*. Suppose that two types of double-stranded molecules, one type containing the sequence for a single gene and the other type containing the same sequence as well as an insertion of nonhomologous DNA, are mixed and used in a reannealing experiment. If the two types of molecules undergo denaturation and reannealing, what types of molecules would you expect to see?

6. Bacterial chromosome deletions of more than 50 base pairs can be detected by electron microscopy, using heteroduplex analysis as described in problem 5. When a heteroduplex is formed between a single-stranded DNA molecule from a deletion strain and a single-stranded molecule from a nondeletion, or wild-type, strain, a single-stranded loop will be visible at the location of the deletion. Suppose you are studying a bacterial mutation, which appears to be a deletion of about 200 base pairs, located at a unique site on the bacterial chromosome, which contains over 3000 genes. Why would it be a good idea to clone DNA containing the site of the deletion, as well as the corresponding site in the wild-type strain, in order to study the deletion using heteroduplex analysis?

7. You wish to clone a yeast gene in λ phage. Why is it desirable to cleave both the yeast DNA and the λ phage DNA with the same restriction enzyme?

8. Suppose you are studying the structure of a protein that contains a proline residue, and you wish to determine whether the substitution of a glycine residue will change the conformation of the polypeptide. You have cloned the gene for the protein, and you know the sequence of the protein. Using site-specific mutagenesis, what alterations would you make in the gene sequence in order to replace proline with glycine?

9. Cleavage of a double-stranded DNA fragment that contains 500 bases with restriction enzyme A yields two unique fragments, one 100 bases and the other 400 bases in length. Cleavage of the DNA fragment with restriction enzyme B yields three fragments, two containing 150 nucleotides and one containing 200 nucleotides. When the 500-base fragment is incubated with both enzymes (this is called the *double-digest technique*), two fragments 100 bases in length and two 150 bases in length are found. Diagram the 500-base fragment, showing the cleavage sites of both enzymes. Now suppose you also have a double-stranded DNA fragment that is identical with the original fragment, except that the first 75 base pairs at the left end are deleted. How can this fragment help you construct a cleavage map for the two enzymes?

10. Problem 2 of Chapter 6 in Stryer's text refers to the expression of a eukaryotic gene, chicken ovalbumin, in *E. coli*. To avoid transcribing and translating intron sequences, you should use cDNA for protein expression. However, if you introduce *only* the chicken ovalbumin cDNA into bacteria, the level of expression of functional protein will be rather low. What other sequences should you attach to the cDNA in order to ensure optimal expression?

11. Because PCR can amplify DNA templates by a millionfold or more, no contaminating DNA must be present in the sample to be used for amplification. To see why, consider a PCR procedure that begins with 1 μg DNA (about 10^6 templates) in a reaction mixture of 100 μL. This sample can be easily amplified about one millionfold in 20 cycles (an amplification of 2^{20}). Suppose that 0.1 μL of DNA from the amplified cycle is inadvertently introduced into another reaction mixture containing 1 μg of a different DNA. Could the contaminant cause problems with PCR analysis of the second sample? Why?

12. Analysis of evidence from cases of sexual assault often includes HLA type analysis using PCR. Samples collected from a victim may contain not

only sperm but also epithelial cells from the victim. Such samples are first incubated in a protease-detergent mixture. The epithelial cells are lysed, while the sperm heads are not. The sperm heads are collected by centrifugation and then washed several times. They are then lysed in presence of a reducing agent like dithiothreitol, which makes sperm heads sensitive to the protease-detergent mixture. Lysis products are then used for PCR analysis. Why is it necessary to carry out separation of sperm and epithelial cells? Why are cells and sperm heads lysed before PCR analysis? Suppose that blood and hair samples are also found as evidence at the scene of the alleged crime. Why should precautions be taken to keep these samples isolated from each other?

13. Unlike DNA polymerase I from *E. coli*, *T. aquaticus* DNA has no proofreading activity and is therefore unable to remove mismatched bases that are randomly incorporated into newly synthesized DNA strands. Under standard conditions used for the polymerase chain reaction, misincorporation of nucleotides occurs at a rate of 1 per 900 nucleotide residues in DNA.

(a) Suppose that you are using PCR to detect copies of an oncogene in a tissue sample. You will challenge the amplified sample with a radioactive probe containing the oncogene sequence, using Southern blot analysis. Will low-frequency, random misincorporation of bases during the amplification procedure interfere with your analysis?

(b) Suppose you are using PCR with a mutant primer to introduce a deliberate alteration in a eukaryotic gene. You plan to use the amplified mutant gene for cloning and expression in *E. coli* to determine how the directed change affects the expressed protein. How might misincorporation level in the amplification procedure interfere with your cloning and expression experiments? What could you do to solve the problem?

14. *E. coli* DNA polymerase I has a 5′ → 3′ polymerase activity and also two other catalytic activities. One is a 3′ → 5′ exonuclease activity whose function is to remove from the growing 3′end of the chain those mismatched bases occasionally incorporated by the polymerase activity. The other catalytic activity is a 5′ → 3′ exonuclease, which acts on double-stranded DNA, removing mispaired DNA stretches ahead of the polymerase. The 5′ → 3′ exonuclease activity can also remove as many as 10 nucleotides from a nick (a single-strand break in DNA), beginning at the 3′-OH end and moving in the 5′ direction. If the exonuclease activity is followed by polymerase, new nucleotides are incorporated and the nick is essentially "translated" in the 5′ direction. This nick-translating ability of DNA polymerase I has been exploited to create radioactive DNA probes for use in Southern blots and other techniques. Describe how nick translation could be used for such purposes.

15. Site-directed mutagenesis allows one to introduce virtually any desired mutation in a specific gene. One very useful application of the technique involves modifying a particular protein and evaluating the effect on biological or chemical activity. In the past, two methods have been used for producing modified proteins. One is to use chemical agents or ultraviolet light to induce mutations that then cause changes in the amino acid sequence of proteins. The other is to modify certain residues in an isolated protein by treatment with chemical reagents; an example is the

inactivation of a reactive serine in the active site of proteolytic enzymes like chymotrypsin, using diisopropylfluorophosphate (see page 224 of Stryer's text for more discussion).

(a) Why is site-specific mutagenesis superior to the two older procedures described above?
(b) In order to carry out the modification of a protein using site-specific mutagenesis in the most efficient way, what sort of information should you have about the protein you wish to modify?

16. Patients with a particular form of hemophilia (a deficiency in blood clotting) have a deletion of an EcoRI restriction site within the gene for a coagulation factor protein. In one family with an affected son, PCR analysis was carried out on 200 μL blood samples from a male fetus and from several family members, to determine whether the fetus also carried the mutation. Using appropriate primer oligonucleotides, DNA fragments 150 bp in length and spanning the EcoRI polymorphic site in intron 10 were synthesized. These fragments were incubated with EcoRI, and the resulting cleavage fragments were then run on an electrophoretic gel and stained with ethidium bromide. A diagram of the gel is shown here, along with the source of the blood sample for each lane. Note that any 150-bp fragment that contains the EcoRI site will be cut by the enzyme into two fragments, 100 bp and 50 bp in length.

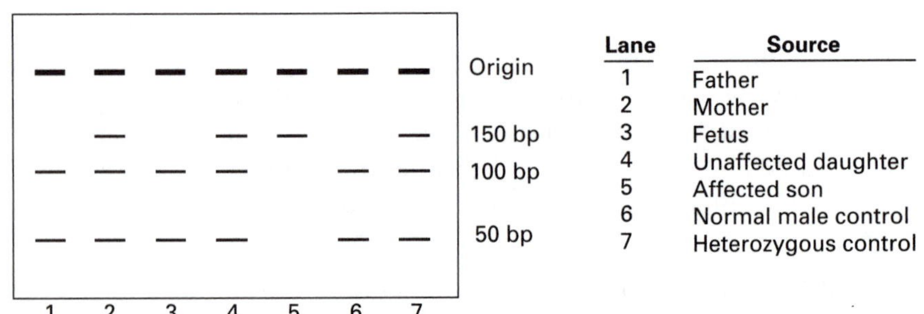

Lane	Source
1	Father
2	Mother
3	Fetus
4	Unaffected daughter
5	Affected son
6	Normal male control
7	Heterozygous control

(a) Specify the genotype for each member of the family and for the controls. Does the fetus carry the mutation?
(b) This analysis could also be done by using Southern blotting to detect single-copy sequences in genomic DNA. Why is the PCR analysis preferable?

17. Base pairing by hydrogen bond formation between complementary bases is a fundamental feature of many processes described in Chapters 4, 5, and 6 of Stryer's text. Discuss the role of base pairing in the context of each of the following:

(a) The fidelity of messenger RNA synthesis
(b) Synthesis of cDNA by reverse transcriptase
(c) The use of primers in the polymerase chain reaction (PCR)
(d) Identifying a desired clone using a radioactive DNA probe
(e) Measuring the relatedness of two DNA species without sequencing them

18. Indicate what labeled products would be obtained from a single strand of DNA sequence TCAGTACG labeled with ^{32}P at its 5′ end (^{32}P-TCAGTACG) after subjecting it to chemical reactions that cleave it once on the 5′ side of

(a) G or A.
(b) A.
(c) C or T.
(d) C.

19. The plasmid pBR322, a double-stranded circular DNA molecule containing 4.4 kilobase pairs, is commonly used in cloning experiments. A technician in a molecular biology laboratory needs to prepare a large quantity of pBR322 by growing a liter culture of *E. coli* containing the plasmid and then isolating the pBR322 DNA.

(a) How many milligrams of plasmid DNA can be prepared from a liter of bacterial cells growing at a density of 10^8 cells per ml? Assume that each cell contains 100 plasmid molecules and that the molecular weight of a base pair in the plasmid is 660.
(b) If the technician decides to use a nanogram of pBr322 as the template in a PCR experiment, how many templates will be present in the reaction mixture?

ANSWERS TO PROBLEMS

1. Cleavage of a circular molecule at one specific site, followed by denaturation, will yield single-stranded DNA molecules with a specific end-to-end base sequence; that is, the molecules have base sequences that are perfectly complementary. Such molecules will anneal to form double-stranded linears, rather than circles. Random cleavage of the original intact molecules yields double-stranded linears with a variety of end-to-end (or permuted) sequences. Denaturation and renaturation allow the random association of these linears, which results in the formation of double-stranded linears with overlapping, complementary ends. Such molecules then form circles as their overlapping ends anneal.

2. During the early years of such experiments, few ways were available to determine what happened to the DNA during transformation attempts. Among the reasons that these transformation attempts were not successful were the failure of the cells to take up the DNA, the rapid degradation of the DNA inside the cell (restriction enzymes are a good example of a cause of this particular problem), the lack of accurate transcription or translation, and the inability of the host cells to replicate and maintain the foreign DNA as they divided.

3. Among the ways that a gene could be inactivated are the insertion of a stop codon in the sequence, which would prevent the complete translation of the protein; a mutation in the promoter region of the gene, which would prevent proper transcription; and other mutations that could prevent proper splicing or processing. To distinguish a functional gene from a pseudogene, you would have to determine the sequence of the protein

and then compare it with the coding sequence for each of the gene sequences. These types of analyses remind us that protein sequencing remains a very necessary tool in molecular biology.

4. Whenever one attempts to locate all the sites cleaved by a particular enzyme, there is always a risk that very small fragments generated by the cleavages may not be detected. Determining the sequences of fragments that extend across the junctions of the original set of fragments serves as a check on the overall assignment of sequence.

5. At high pH, protons dissociate from some of the bases, making them unable to participate in base pairing. One example is guanine, for which the p*K* of the proton on N-1 is 9.2. Removal of the hydrogen at this location disrupts the ability of guanine to pair with cytosine.

If you mix the two types of double-stranded molecules, you would expect to see linear molecules that are double-stranded all along their length as well as some molecules that are only partially double-stranded. These partially double-stranded molecules will contain a single-strand loop that denotes the position of the insertion; they are formed between one strand of the molecule containing the normal gene and one strand of the molecule containing the insertion.

6. Even if you were able to isolate intact, unbroken bacterial chromosomes, formation of intact heteroduplex molecules between the deletion and wild-type DNAs is difficult because the very long single strands become entangled as they pair with each other, making them impossible to analyze by electron microscopy. In addition, the time required for complete reassociation of the strands is very long. Generating shorter, randomly cleaved DNA fragments for heteroduplex analysis permits faster reassociation and easier analysis, but since the deletion is located at a single unique site in the chromosome, the probability of finding the desired molecule among the mixture of many heteroduplex molecules is rather low. Cloning DNA molecules containing the deletion or its corresponding wild-type sequence allows you to carry out reannealing experiments that yield a high concentration of heteroduplex molecules with the loop characteristic of deletion mutations.

7. In order to insert the yeast gene into the λ phage vector, you must have complementary base pairs on the ends of each duplex in order for them to join. Because each restriction enzyme cleaves at a unique and constant site, the yeast and λ phage molecules will have complementary ends if both have been cleaved with the same enzyme. Of course, you must also make sure that the sites of cleavage are in appropriate places so that the gene is intact when it is inserted into the vector.

8. The RNA codon for proline is 5′-CC*X*-3′, and the codon for glycine is 5′-GG*X*-3′, where *X* is any ribonucleotide. Suppose you determine that the proper codon for your protein is 5′-CCC-3′. Using the scheme outlined in Figure 6-37 of the text, you would prepare an oligonucleotide primer that is complementary to the region of the gene that specifies the proline residue, except that it would contain the DNA sequence 5′-CCC-3′ instead of 5′-GGG-3′. Elongation of the primer using DNA polymerase, followed by closure and replication, will yield some progeny plasmids that will express a protein with a glycine substitution.

9. The 500-base fragment has one site that is cleaved by enzyme A. This yields two fragments with two possible patterns:

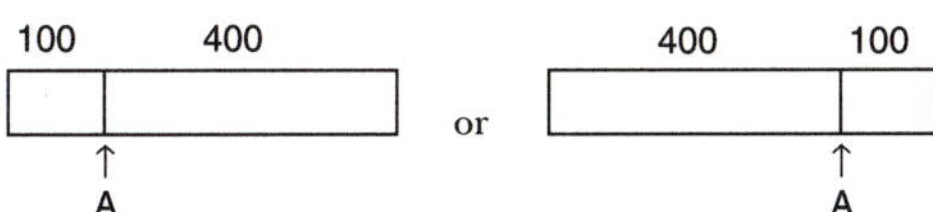

Enzyme B cleaves the 500-base molecule twice, so there are three possible cleavage patterns:

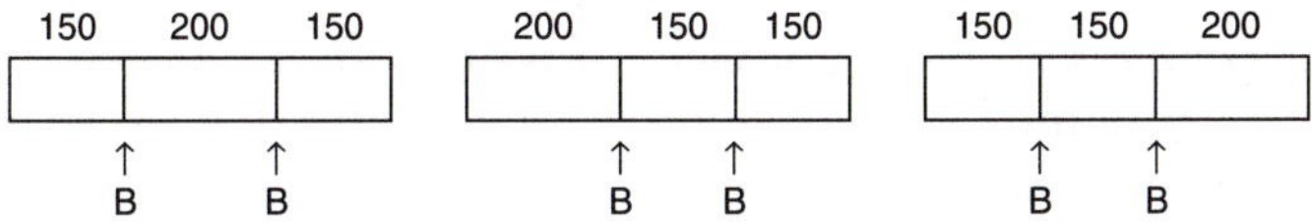

Since we cannot distinguish between ends of the molecule by this type of analysis, let us arbitrarily assume that enzyme A cuts the molecule of 100 nucleotides from the left end. We can then superimpose the possible cleavage patterns for enzyme B:

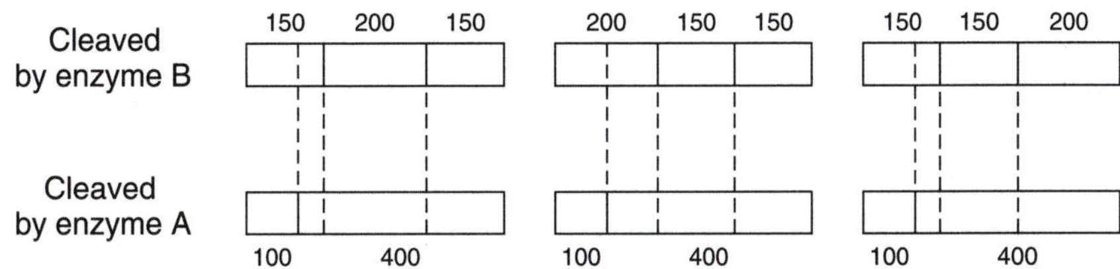

Only one of the patterns for enzyme B, that in which a cut occurs 200 bases from the left end, yields the results obtained when the fragment is incubated with both enzymes. The other patterns would yield at least one 50-base fragment. The correct pattern is therefore:

100 100 150 150

A B B

Because we still cannot distinguish between the right- and left-hand ends of the molecule, an alternative cleavage pattern can also be constructed from the analysis outlined above:

150 150 100 100

B B A

The deletion fragment serves as a marker for the left-hand end of the molecule, and using both enzymes in the double-digest technique allows us to establish which cleavage pattern is correct. For example, if the cleavage pattern shown below on the left is correct, the cleaved deletion molecule will yield four fragments, including one only 25 nucleotides in length. Alternatively, the pattern shown below on the right means that double digestion of the deletion fragment will again yield four fragments, but the smallest will have a length of 75 nucleotides.

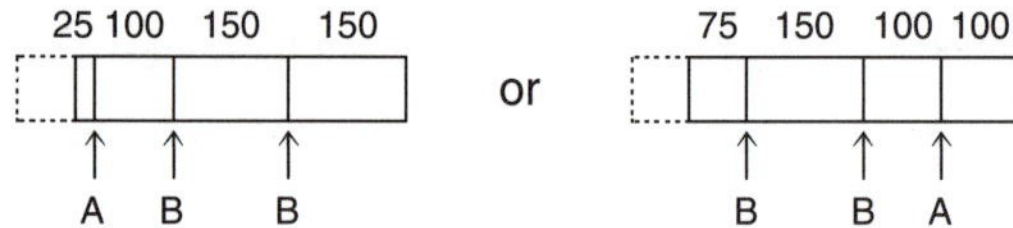

The complexity of cleavage patterns (also known as *restriction maps*) increases greatly when additional cleavages are involved. Often the best ways to use the double-digest technique is to isolate the fragments generated by one enzyme and then digest each of them with the other. This allows you to determine the location of different cleavage sites within a particular fragment. In restriction mapping, as in genetic mapping, it is important to remember that the sum of the fragment lengths generated by one enzyme must equal the sum of the fragment lengths generated by the other.

10. In order to obtain optimal expression in *E. coli,* you should attach prokaryotic DNA sequences that include the appropriate transcriptional and translational punctuation elements. For example, in Chapter 5 of Stryer's text, promoter sites that determine where transcription begins are mentioned; these include the *Pribnow box* and the *–35 region,* both of which might be required to initiate transcription of your cDNA clone in bacteria. You may also need to attach the stem-loop and GC-rich terminator sequence at the 5′-end of your cDNA, in order to cause nascent messenger RNA to dissociate from RNA polymerase when transcription is complete. In addition, you should see that proper start and stop signals for translation are also included in your cDNA, so that the mRNA code is read in the proper frame and that proper termination occurs. These signals ensure that the expressed protein has the proper amino acid sequence and is the correct length.

11. It is indeed possible that the contaminant could complicate the analysis of the second sample. If the 10^6 templates in the original sample are amplified a millionfold, then the concentration of templates at the completion of 20 cycles is $10^{12}/100\ \mu L$, or 10^{10} templates per μL. A contaminating volume of 0.1 μL will therefore contain 10^9 templates, compared with about 10^6 templates in the second sample. Such contamination could mask the identity of the DNA in that sample. Laboratory personnel take a number of precautions to avoid introduction of foreign DNA into PCR reaction mixtures. These include use of sterile containers and reaction solutions; disposable gloves; laminar flow hoods; and separate work areas for preparing reaction mixtures, pipetting template samples, and analyzing the products. For the PCR reaction itself, it is important to run reactions that contain no DNA in order to check for contamination with DNA from other sources.

12. In forensic PCR analysis, it is common to use cells from the victim as well as sperm from the alleged rapist to generate DNA templates for amplification and analysis. The differential lysis procedure allows the two to be efficiently separated, to avoid cross-contamination of DNA samples. Lysis is necessary so that the DNA templates will be accessible to DNA polymerase and substrates for amplification. During the gathering of evi-

dence from crime scenes, it is necessary to keep samples that could contain large amounts of DNA, such as blood stains, separate from those that contain little DNA, like a single shed hair. As noted in problem 11 above, the presence of large amounts of contaminating DNA can confound PCR analyses.

13. (a) There should be little interference if you utilize amplified sequences with some misincorporated bases in Southern blot experiments. The results of the experiments depend on the complementarity of relatively long sequences of DNA, and occasional mismatches should not interfere with the ability of the amplified templates and the probe to anneal with each other. Remember that the primers, which are present in excess during the initial cycles of amplification, continue to initiate the synthesis of new DNA strands. If the incorporation of mismatches is random, the chances are that any particular strand will have few of them and that it could still form complementary base pairs with the radioactive probe.

(b) In these experiments, the low frequency of base misincorporation could interfere with your analysis, because each of your clones to be used for expression experiments will be derived from a single DNA molecule obtained in the amplification process. Cloning therefore amplifies any error introduced by the polymerase in the chain reaction, and random errors could alter the expressed protein in a noncontrolled way. There are two ways to deal with this problem. You can use a DNA polymerase that has a proofreading activity; one such useful polymerase is that from phage T4 (whose polymerase molecule lacks an exonuclease activity found in *E. coli* that interferes with the PCR procedure). On the other hand, you can also prepare a number of clones and sequence them. Such a procedure will ensure that the clone you want to examine for altered expression has the sequence you wanted to generate during the amplification process.

14. Nick translation can be used with labeled nucleoside triphosphates to generate highly radioactive probes for use in Southern blot analysis as well as for other techniques that require such labeled DNA samples. The usual procedure includes isolating the DNA you wish to use for the probe, treating it with an enzyme that will create a small number of nicks or single-strand breaks per molecule, and then incubating it with DNA polymerase I and α-^{32}P-labeled nucleoside triphosphates (the α-phosphate is labeled because it is incorporated along with the base when the DNA chain is extended by DNA polymerase). The DNA molecules then will contain stretches of 10 or so radioactive sequences, and they can then be used as for autoradiography in Southern or Northern blot experiments.

15 (a) Treatment of an organism with a mutagen does not allow you to make changes in a particular region of a gene and its encoded protein. Much work would be required to find a mutant organism that has an alteration in a protein and to characterize that alteration. Treating a purified protein with a chemical agent may modify other amino acid residues in addition to the one or more specific residues of interest. Both the older approaches are quite nonspecific, compared with the ability of site-specific mutagenesis to target a specific region of a particular gene.

(b) One should have at a minimum the primary amino acid sequence for the protein and the exact DNA coding sequence for it. More information about the protein of interest would allow for more selective mutagenesis. Examples include the location of residues involved with the active site, with allosteric interactions, or with membrane association or those involved in protein–lipid or protein–nucleic acid interactions.

16. (a) The father and the normal male control have the EcoRI site, so that their amplified DNAs are cut into two fragments. The affected son has lost the site, and his DNA is not cut. The mother, daughter, and a heterozygous control have one copy of the normal gene (giving two fragments) and one copy of the mutant gene (giving one larger fragment). The amplified fetal DNA is cleaved by the enzyme, which shows that the male fetus is normal.

(b) Southern blotting, used to detect single-copy sequences in genomic DNA, requires relatively large amounts of material from which DNA is isolated, and it requires a highly radioactive probe to detect a particular sequence. Autoradiography, required to detect annealing of the probe to the genomic sequence, takes days or weeks for development. Analysis by PCR requires only nanogram amounts of DNA from very small amounts of tissue or blood, with minimal sample preparation. In addition, one can dispense altogether with radioactive probes, instead using a DNA staining reagent such as ethidium bromide to detect amplified sequences.

17. (a) Guided by a DNA template, hydrogen bonding mediates proper insertion of mononucleotides into the growing RNA chain, in the reaction catalyzed by RNA polymerase. Complementary base paring is the guiding principle in determining the order of ribonucleotide assembly.

(b) Guided by an RNA template, hydrogen bonding mediates proper insertion of mononucleotides into the growing DNA chain, in the reaction catalyzed by reverse transcriptase. Complementary base pairing guides the order of ribonucleotide assembly. Reverse transcriptase can also use cDNA as a template for the creation of a duplex DNA molecule; again, hydrogen bonding of complementary base pairs establishes the order of deoxynucleotide addition.

(c) The PCR technique requires that DNA be denatured (breaking hydrogen bonds) and then annealed to a pair of primers, whose sequences are complementary to those flanking the target sequence in DNA. DNA polymerase then carries out chain extension by adding deoxynucleotides to those primers. After denaturation of the newly formed duplexes, the strands are reannealed with primers. The process is carried out as many as 20 times. For each reaction sequence, specificity of the chain reaction for a particular DNA segment is mediated by specific base pairing between primers and templates.

(d) Radioactively labeled DNA probes are denatured and allowed to anneal to a mixture of DNA molecules. The extent of hydrogen bonding determines the homology between probe and DNA.

Southern blotting uses this hybridization technique for DNAs resolved by gel electrophoresis. Location of a specific DNA is visualized by autoradiography, which locates the band in the gel.

(e) The two species are denatured and reannealed together. The more duplex DNA formed upon annealing, the greater the degree of relatedness. The extent of duplex DNA formation is determined by measuring the melting temperature of the DNA. This will depend on the number of hydrogen bonds between strands.

18. In the chemical reactions (the Maxam-Gilbert method), the deoxynucleotide at the site of the cleavage is eliminated by modification and removal of the base from the sugar. Thus the sequences observed through chemical cleavage are

(a) TC, TCA, TCAGT, TCAGTAC
(b) TC, TCAGT
(c) T, TCAG, TCAGTA
(d) T, TCAGTA

19. (a) First determine how many plasmid molecules are present in the bacterial cell culture: 10^8 cells/ml = 10^{11} cells/L, and 10^{11} cell/L $\times$ 100 plasmids = 10^{13} plasmid molecules in the culture. Then use the molecular weight of a base in the plasmid and Avogadro's number to determine the weight of 10^{13} plasmid molecules: 10^{13} molecules is equivalent to 4.4×10^{13} kilobases.

$$\frac{4.4 \times 10^{13}\ \text{kb} \times 6.6 \times 10^{5}\ \text{g/mol}}{6.023 \times 10^{23}\ \text{kb/mol}} = 4.8 \times 10^{-5}\ \text{g, or } 0.048\ \text{mg pBR322 DNA}$$

(b) A nanogram of DNA equals 10^{-9} g. Divide this quantity by the molecular weight of a kilobase pair to obtain the number of moles of DNA, and then multiply by Avogadro's number to determine how many molecules are present in the reaction mixture.

$$10^{-9}\ \text{g}/6.6 \times 10^{5}\ \text{g/mol} = 1.51 \times 10^{-15}\ \text{mol DNA}$$

$$1.51 \times 10^{-15}\ \text{mol} \times 6.023 \times 10^{23}\ \text{molecules/mol} = 9.1 \times 10^{8}\ \text{molecules, in the reaction mixture}$$

EXPANDED SOLUTIONS TO TEXT PROBLEMS

1. (a) The direction of movement on the gel is from top to bottom, with the smallest fragment, in this case G, moving most rapidly. Since the 5′ end carries the ^{32}P label, the 5′ $\rightarrow$ 3′ sequence is read from bottom to top, opposite the direction of movement. The sequence is 5′-GGCATAC-3′. It should be noted that the fastest-moving spot in the autoradiogram is the radioactive inorganic phosphate resulting from the destruction of the guanine at the 5′ terminus. In the example shown in the text (Figure 6-5, Stryer, p. 122), the P_i spot is not shown.

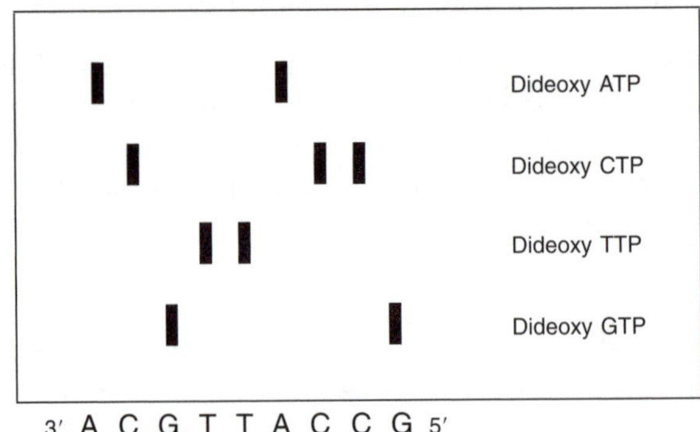

(b) Note that in the Sanger dideoxy method (Figure 6-7, Stryer, p. 123), the new DNA strands that are subjected to electrophoresis are elongated 5′ → 3′. Since the larger molecules move more slowly, the results shown above are obtained. The DNA strand serving as template in the Sanger method is the complement of the strand shown here in the figure.

2. Since *E. coli* lack the machinery to excise introns and splice exons, they would make a meaningless mRNA if presented with ovalbumin genomic DNA. Therefore, if you wish to express the ovalbumin gene in *E. coli,* you must use its cDNA, which contains the information in the eight exons, but no introns.

3. (a) No, because most human genes are much longer than 4 kb. One would obtain fragments containing only a small part of a complete gene.

(b) No, chromosome walking depends on having *overlapping* fragments. Exhaustive digestion with a restriction enzyme produces nonoverlapping, short fragments. For more details, see Stryer's discussion of Figure 6-22 (p. 132).

4. Southern blotting of an MstII digest would distinguish between the normal and mutant genes. The loss of a restriction site would lead to the replacement of two fragments on the Southern blot by a single longer fragment (see Stryer, p. 173). Such a finding would not prove that GTG replaced GAG; other sequence changes at the restriction site could yield the same result.

5. Cech replicated the recombinant DNA plasmid in *E. coli,* and then transcribed the DNA in vitro using bacterial RNA polymerase. He found that this RNA underwent self-splicing in vitro in the complete absence of any proteins from *Tetrahymena.* This observation led to the discovery that certain RNA molecules can act as catalysts. These are termed *ribozymes* to distinguish them from protein enzymes.

6. A few years ago this would have been very difficult, if not impossible. However, the availability of automated solid-phase chemical methods for synthesizing DNA has made the impossible fairly easy. A simple strategy for generating many mutants is to synthesize a group of oligonucleotides that differ only in the sequence of bases in one triplet, or codon. For

example, with the 30-mer described in this question, if a mixture of all 4 nucleotides is used in the *first and second rounds* of synthesis, the resulting oligonucleotides will begin with the sequence XYT (where X and Y denote A, C, G, or T). This will provide 16 different versions of the first triplet (codon) of the 30-mer, which will encode proteins containing either Phe, Leu, Ile, Val, Ser, Pro, Thr, Ala, Tyr, His, Asn, Asp, Cys, Arg, Ser, or Gly at the first position (see Table 5-4). In similar fashion, one can synthesize oligonucleotides in which two or more codons are simultaneously varied.

7. For PCR to amplify a DNA duplex, the polymerase must be active (primed) in both directions if each strand is to be replicated. If the DNA is linear, known sequences are needed on both sides of the portion to be amplified. To explore DNA on both sides of a single known sequence, one can digest genomic DNA with a restriction enzyme and circularize the fragments. Then, by using a pair of primers that hybridize specifically with portions of the known sequence, one can use PCR to amplify only the fragments containing the known sequence. Note that in a circular DNA a single known sequence can be used to prime polymerase action in *both directions* away from the known sequence, resulting in complete replication of the duplex.

8. These results suggest that the DNA is composed of four repeating units. If these were transcribed into mRNA and the latter translated into protein, one would expect a protein molecule whose linear sequence was composed of four repeating peptides.

9. Sequence-tagged sites (STSs) can serve as a common framework for establishing the relation between clones. An STS is a sequence 200 to 500 bp long that occurs only once in the entire genome. This sequence and those of a pair of PCR primers that can generate the STS are stored in a database. Laboratory 1 states its YAC contains STS-34, STS-102, and STS-860. Laboratory 2 can then synthesize the corresponding PCR primers and learn whether their YAC contains any of these STSs. For an illuminating discussion of this strategy, see J.D. Watson, M. Gilman, J. Witkowski, and M. Zoller, *Recombinant DNA*, 2nd ed. (W.H. Freeman, 1992), pp. 610–612.

CHAPTER

7

Portrait of an Allosteric Protein

The principles of protein structure and the methods used to investigate proteins were presented in Chapters 2 and 3. In this chapter, Stryer uses this information to describe the structure-function relationships of two very important, well-studied oxygen-carrying proteins: myoglobin and hemoglobin. First, Stryer describes the structures of myoglobin and hemoglobin; he then turns to the oxygen binding properties of hemoglobin in detail at the level of mechanism. The interaction of hemoglobin with its ligands exemplifies the allosteric proteins, which are modulated in their activity through binding specific effectors. Finally, Stryer describes how mutations in or near the hemoglobin gene can produce defective or missing hemoglobin and thereby cause molecular disease.

LEARNING OBJECTIVES

When you have mastered this chapter, you should be able to complete the following objectives:

Myoglobin Structure and Heme Properties (Stryer pages 148–154)

1. Explain the physiologic roles of *myoglobin* and *hemoglobin* in vertebrates.

2. List the major features of the structure of myoglobin as revealed by *x-ray crystallography.*

3. Describe the structure of the *heme prosthetic group* and its properties when free and when bound to the *globins.*

4. Relate the structure of the *myoglobin gene* to the functional domains of myoglobin.

Hemoglobin Structure (Stryer pages 154–157)

5. Compare the *subunit* compositions of *adult* and *fetal hemoglobins.*

6. Compare the three-dimensional structures of myoglobin and hemoglobin.

7. Describe the significance of the *invariant amino acid residues* in the amino acid sequences of hemoglobins from different species.

Oxygen Binding and Allosteric Properties of Hemoglobin (Stryer pages 157–168)

8. Contrast the *oxygen binding* properties of myoglobin and hemoglobin. Define the *cooperative binding* of oxygen by hemoglobin.

9. Explain the significance of the differences in *oxygen dissociation curves,* in which the *fractional saturation* (Y) of the oxygen-binding sites is plotted as a function of the *partial pressure of oxygen* (pO_2), for myoglobin and hemoglobin.

10. Starting with the equilibrium expression for oxygen binding by myoglobin, derive the equation

$$Y = \frac{pO_2}{pO_2 + P_{50}}$$

11. Relate the empirical expression for the fractional saturation of hemoglobin

$$\frac{(pO_2)^n}{(pO_2)^n + (P_{50})^n}$$

with the equation

$$\frac{Y}{1 - Y} = \frac{(pO_2)^n}{P_{50}}$$

and with the *Hill plot.* Explain the significance of the *Hill coefficient.*

12. Summarize how the cooperative binding of oxygen by hemoglobin makes it a better oxygen transporter.

13. Describe the *Bohr effect,* that is, the effects of CO_2 and H^+ on the binding of oxygen by hemoglobin.

14. Explain the effect of *2,3-bisphosphoglycerate* (*BPG*), also known as *2,3-diphosphoglycerate* (*DPG*), on the affinity of hemoglobin for oxygen.

15. Describe the major structural differences between the *oxygenated* and *deoxygenated* forms of the hemoglobin molecule.

16. Describe the structural bases for the effects of H^+, CO_2, and BPG on the binding of oxygen by hemoglobin. Explain the consequences of the metabolic production of CO_2 and H^+ on the oxygen affinity of hemoglobin.

17. Rationalize the existence of *fetal hemoglobin.*

18. Summarize the general properties of an *allosteric protein,* as exemplified by hemoglobin. Distinguish between *homotropic* and *heterotropic* cooperative effects.

19. Distinguish between the *sequential model* and the *concerted model* for cooperative interactions between the subunits of proteins.

Sickle-Cell Anemia and Other Molecular Diseases (Stryer pages 168–175)

20. Describe *sickle-cell anemia* as a genetically transmitted *molecular disease.*

21. Contrast the biochemical and structural properties of deoxygenated *sickle-cell hemoglobin* (*Hb S*) with those of *hemoglobin A* (*Hb A*).

22. Correlate the clinical observations on patients with sickle-cell anemia and the *sickle-cell trait* to their hemoglobins.

23. Explain how deoxyhemoglobin S forms *fibrous precipitates.*

24. Correlate the genetics of sickle-cell anemia with the geographical distribution of the disease.

25. Describe the detection of the sickle-cell gene by *endonuclease digestion* of DNA and *Southern blotting.*

26. Outline the possible functional consequences of *amino acid substitutions* in *mutant hemoglobins.*

27. Define *thalassemias,* and indicate the steps in hemoglobin synthesis where defects can occur.

28. Describe the significance of the discovery of *molecular diseases.*

SELF-TEST

Myoglobin Structure and Heme Properties

1. Match the forms of myoglobin in the left column with their corresponding properties from the right column.

(a) Ferrimyoglobin ______

(b) Oxymyoglobin ______

(c) Deoxymyoglobin ______

(1) Iron in the +2 oxidation state

(2) Iron in the +3 oxidation state

(3) Oxygen bound to the sixth coordination position of iron

(4) Empty sixth coordination position

(5) Water bound to the sixth coordination position

(6) Histidine at the fifth coordination position

2. Which of the following statements about heme structure is true?

(a) Heme contains a tetrapyrrole ring with four methyl, four vinyl, and four propionate side chains.
(b) The iron atom in heme may be present in the ferrous or the ferric state.
(c) The iron in free heme is mostly present in the ferrous state.
(d) The iron atom is coplanar with the tetrapyrrole ring in deoxymyoglobin.
(e) The axial coordination positions of heme are occupied by tyrosine residues in myoglobin.

3. Which of the following are structural features of myoglobin?

(a) Seventy-five percent of the amino acids are in α helices.
(b) All the prolines are found within α helices.
(c) The interior of the molecule consists mainly of nonpolar residues.
(d) The interior of the molecule is hollow so that it can accommodate many molecules of water.
(e) The heme group is bound on the surface of the protein.

4. Which of the following are properties of the heme binding site in myoglobin?

(a) The proximal histidine is involved in binding iron.
(b) Oxygenation involves an insertion of O_2 in a coordination complex with the iron of heme.
(c) Sequestering the heme in a nonpolar environment is important in preventing the oxidation of ferrous iron.
(d) The heme binds CO with the Fe, C, and O atoms in a linear array.
(e) The distal histidine is involved in binding iron.

5. The heme prosthetic group found in hemoglobin and myoglobin is also found in cytochromes *b* and *c*, which are proteins that transfer elec-

trons in the electron transport chain. How can the same prosthetic group serve such different functions as oxygen binding and electron transport?

---

6. Which of the following answers complete the sentence correctly? The role of the distal histidine in myoglobin is to

(a) bind O_2.
(b) prevent tight CO_2 binding.
(c) prevent tight CO binding. ✓
(d) bind to the fifth coordination position of Fe.
(e) help keep Mb in the ferrous state.

Hemoglobin Structure

7. Hemoglobin is a tetrameric protein consisting of two α and two β polypeptide subunits. The structures of the α and β subunits are remarkably similar to that of myoglobin. However, at a number of positions, hydrophilic residues in myoglobin have been replaced by hydrophobic residues in hemoglobin.

(a) How can this observation be reconciled with the generalization that hydrophobic residues fold into the interior of proteins?

(b) In this regard, what can you say about the nature of the interactions that determine the quaternary structure of hemoglobin?

---

8. Much of the information we have concerning hemoglobin structure comes from the study of crystals. Which of the following lines of evidence support the notion that crystallized hemoglobin has a structure similar to that of hemoglobin in solution?

(a) The amino acid sequence of crystallized hemoglobin is the same as that found in hemoglobin in solution.
(b) The visible absorption spectrum of crystallized hemoglobin is virtually the same as that of hemoglobin in solution.
(c) Crystallized hemoglobin is functionally active.
(d) The α helix content of crystallized hemoglobin is similar to that found in hemoglobin in solution.

9. What are the meanings of the codes that are used to designate the positions of the invariant amino acid residues in hemoglobins, as in glycine B6 or tyrosine HC2, for example?

---

10. There are nine positions in the amino acid sequences of the hemoglobins of diverse species that have invariant amino acid residues. On the

basis of the properties of amino acids, which you learned in Chapter 2, match the invariant amino acids in the left column with the appropriate role from the right column.

(a) Glycine B6 ________
(b) Histidine F8 ________
(c) Leucine F4 ________
(d) Phenylalanine CD1 ________
(e) Proline C2 ________
(f) Tyrosine HC2 ________

(1) Terminates α helix
(2) Contacts the heme (forms part of the nonpolar heme pocket)
(3) Allows the close approach of the B and E helices
(4) Cross-links two helices (H and F) with a hydrogen bond
(5) Binds to the heme iron

Oxygen Binding and the Allosteric Properties of Hemoglobin

11. Which of the following answers completes the sentence correctly? Hemoglobin differs from myoglobin in that

(a) hemoglobin is multimeric whereas myoglobin is monomeric.
(b) hemoglobin binds O_2 more tightly than does myoglobin at any given O_2 concentration.
(c) hemoglobin binds CO_2 more effectively than does myoglobin.
(d) the Hill coefficient for O_2 binding is smaller for hemoglobin than it is for myoglobin.
(e) the binding of O_2 by hemoglobin depends on the concentrations of CO_2, H^+, and BPG, whereas the binding of O_2 by myoglobin does not.

12. Which of the following statements are false?

(a) The oxygen dissociation curve of myoglobin is sigmoidal, whereas that of hemoglobin is hyperbolic.
(b) The affinity of hemoglobin for O_2 is regulated by organic phosphates, whereas the affinity of myoglobin for O_2 is not.
(c) Hemoglobin has a higher affinity for O_2 than does myoglobin.
(d) The affinity of both myoglobin and hemoglobin for O_2 is independent of pH.

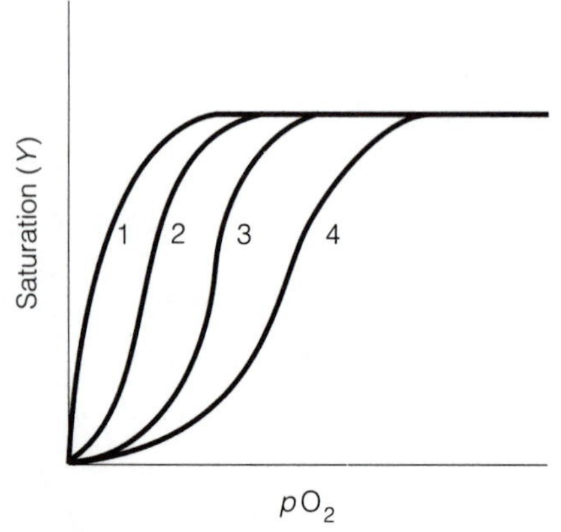

Figure 7-1
Oxygen dissociation curves.

13. Several oxygen dissociation curves are shown in Figure 7-1. Assuming that curve 3 corresponds to isolated hemoglobin placed in a solution containing physiologic concentrations of CO_2 and BPG at a pH of 7.0, indicate which of the curves reflects the following changes in conditions:

(a) Decreased CO_2 concentration ________
(b) Increased BPG concentration ________
(c) Increased pH ________
(d) Dissociation of hemoglobin into subunits ________

14. Which of the following statements concerning the Bohr effect are true?

(a) Lowering the pH shifts the oxygen dissociation curve of hemoglobin to the right.

(b) The acidic environment of an exercising muscle allows hemoglobin to bind O_2 more strongly.
(c) The affinity of hemoglobin for O_2 is diminished by high concentrations of CO_2.
(d) In the lung, the presence of higher concentrations of H^+ and CO_2 allows hemoglobin to become oxygenated.
(e) In the lung, the presence of higher concentrations of O_2 promotes the release of CO_2 and H^+.

15. Match the parameters in the left column with the appropriate definitions from the right column.

(a) Hill coefficient (n) ____________
(b) $(pO_2/P_{50})^n$ ____________
(c) P_{50} ____________
(d) Y ____________

(1) The fractional occupancy of the oxygen-binding sites
(2) The pO_2 at half-saturation of the hemes with oxygen
(3) The ratio of oxyheme to deoxyheme
(4) The cooperativity of oxygen binding

16. BPG (2,3-bisphosphoglycerate) plays a role in high-altitude adaptation. What is the effect of an increase in the level of BPG on the amount of oxygen transported to muscle in a person living at 10,000 feet as compared with a person living at sea level? Assume that the increase in BPG level has shifted P_{50} from 26 torr to 35 torr and that alveolar pO_2 is 67 torr and pO_2 in muscle capillaries is 20 torr at 10,000 feet. At sea level the alveolar pO_2 is 100 torr and pO_2 in muscle capillaries is still 20 torr. Assume that the Hill coefficient is 2.8.

____

____

17. Explain why fetal hemoglobin has a higher affinity for oxygen than does maternal hemoglobin, and why this is a necessary adaptation.

____

____

18. The oxygen dissociation curve for hemoglobin reflects allosteric effects that result from the interaction of hemoglobin with O_2, CO_2, H^+, and BPG. Which of the following structural changes occur in the hemoglobin molecule when O_2, CO_2, H^+, or BPG bind?

(a) The binding of O_2 pulls the iron into the plane of the heme and causes a change in the interaction of all four globin subunits, mediated through His F8.
(b) BPG binds at a single site between the four globin subunits in deoxyhemoglobin and stabilizes the deoxyhemoglobin form by cross-linking the β subunits.
(c) The deoxy form of hemoglobin has a greater affinity for H^+ because the molecular environment of His and the α-NH_2 groups of the α chains changes, rendering these groups less acidic when O_2 is released.
(d) The binding of CO_2 stabilizes the oxy form of hemoglobin.

19. The structure of deoxyhemoglobin is stabilized by each of the following interactions *except* for

(a) BPG binding.
(b) salt bridges between acidic and basic side chains.
(c) coordination of the hemes with the distal histidine.
(d) hydrophobic interactions.
(e) salt bridges involving N-terminal carbamates.

20. In the transition of hemoglobin from the oxy to the deoxy form, an aspartate residue is brought to the vicinity of His 146. This increases the affinity of this histidine for protons. Explain why.

____

____

Sickle-Cell Anemia and Other Molecular Diseases

21. Which of the following observations indicates that hemoglobin S differs from hemoglobin A by the substitution of Val for Glu in position 6 of the β chain?

(a) Porphyrin isolated from hemoglobin S has one more valine group than does porphyrin isolated from hemoglobin A.
(b) All the tryptic peptides of hemoglobin S are identical with those of hemoglobin A.
(c) One altered peptide of hemoglobin S migrates toward the cathode (−) more than does the corresponding peptide of hemoglobin A.
(d) The isoelectric point of hemoglobin S is slightly lower than that of hemoglobin A.

22. The N-terminal tryptic peptides from the β chains of Hb A and Hb S are as follows:

Hb A: Val-His-Leu-Thr-Pro-Glu-Glu-Lys
Hb S: Val-His-Leu-Thr-Pro-Val-Glu-Lys

(a) Would these peptides separate from each other in an electric field at pH 7.0?

____

____

(b) What is the approximate net charge on each peptide at pH 7.0?

____

____

23. Which of the following statements are true? Hemoglobin S forms fibrous precipitates

(a) because the valine at position 6 of the β chain forms a sticky hydrophobic patch on the surface of the protein.
(b) that are reversible upon oxygenation.
(c) that distort the shape of red cells.
(d) only in the deoxy form.
(e) only in homozygotes.

24. The first step in detecting the genetic mutation in sickle-cell hemoglobin is to cleave the mutant DNA with a specific restriction endonuclease. Which of the following would be appropriate subsequent steps in the analysis?

(a) The DNA fragments are separated by gel electrophoresis and are identified with a specific radiolabeled DNA probe.
(b) The DNA fragments are separated by gel filtration and are detected by their light absorption.
(c) The DNA fragments are separated by gel electrophoresis, and the abnormal fragment is identified by its altered mobility with respect to that of the normal fragment.
(d) The DNA fragments are separated by gel electrophoresis and are identified with a fluorescent antibody.

25. A mutant hemoglobin gene has been isolated and shown to encode an abnormal α chain by endonuclease digestion. The electrophoretic mobility of the mutant hemoglobin and its affinity for O_2 are essentially the same as for hemoglobin A, however. With regard to these observations, which of the following statements are correct?

(a) The mutation is most likely at the active site.
(b) There has been a substitution of a basic amino acid residue for an acidic one.
(c) There has been a substitution of an acidic amino acid residue for a neutral one.
(d) There has been a substitution that involves a homologous residue in terms of charge.
(e) The mutation is most likely on the surface of the hemoglobin.

26. The molecular defects in thalassemias and in abnormal hemoglobins, such as Hb S, can be similar in which of the following ways?

(a) The molecular defect occurs outside of the DNA encoding the protein.
(b) The molecular defect results in a stop codon that produces a short fragment of the protein.
(c) The molecular defect results in the substitution of one amino acid for another.
(d) The molecular defect occurs within the structural genes.
(e) The molecular defect results in a frameshift mutation that produces a large block of altered amino acids in the protein.
(f) The molecular defect results from the deletion of the gene encoding the protein.

ANSWERS TO SELF-TEST

Myoglobin Structure and Heme Properties

1. (a) 2, 5, 6 (b) 1, 3, 6 (c) 1, 4, 6
2. b
3. a, c
4. a, b, c
5. The oxidation state and the binding properties of heme vary markedly with its environment; therefore, the different environments caused by

the different amino acids in these two classes of proteins change the function of the heme.

6. c, e

Hemoglobin Structure

7. (a) Hydrophobic patches occur on the surface of the hemoglobin subunits where the α and β chains fit together. As a result, these patches are not on the surface of the multimeric protein.

   (b) Hydrophobic interactions play an important role in stabilizing the tetrameric subunit structure of hemoglobin.

8. b, c, d. Regarding answer b, the visible absorption spectrum of heme varies with its oxidation state and its environment. The fact that both crystallized hemoglobin and hemoglobin in solution have the same spectra is indicative of a similar structure of the heme pocket.

9. Glycine B6 means that the glycine residue is in the sixth position of the B helix; tyrosine HC2 designates the second position of the carboxyl-terminal segment. In general, the helical segments are designated A, B, C, D, E, F, G, H, and the interhelical regions, AB, BC, etc.; NA denotes the region between the amino-terminal residue and the first helix (A), and HC denotes the segment between the last helix (H) and the carboxyl-terminal residue.

10. (a) 3 (b) 5 (c) 2 (d) 2 (e) 1 (f) 4

Oxygen Binding and the Allosteric Properties of Hemoglobin

11. a, c, e

12. a, c, d

13. (a) 2 (b) 4 (c) 2 (d) 1

14. a, c, e

15. (a) 4 (b) 3 (c) 2 (d) 1

16. Using the equation

$$Y = \frac{(pO_2)^n}{(pO_2)^n + (P_{50})^n}$$

the increased level of BPG at 10,000 feet changes P_{50} and yields a ΔY (the difference in the fractional saturation of hemoglobin with oxygen in the lungs versus that in the tissues) of 0.69. Compare this with a ΔY of 0.65 at sea level. Thus, the increase in BPG results in a similar delivery of oxygen to the tissues even when alveolar pO_2 is significantly decreased. In general, the larger the value of ΔY, the higher the delivery of O_2 to the tissues.

17. Fetal hemoglobin is composed of different subunits than adult hemoglobin and binds BPG less strongly. As a result, the affinity of fetal hemoglobin for oxygen is higher, and the fetus can extract the O_2 that is transported in maternal blood.

18. a, b, c

19. c

20. The p*K* values of ionizable groups are sensitive to their environment. The change in the environment of His 146 in deoxyhemoglobin increases its affinity for protons as a result of the electrostatic attraction between the negative charge of the aspartate and the proton.

Sickle-Cell Anemia and Other Molecular Diseases

21. c

22. (a) Yes, the peptides will separate.

(b) The net charge for Hb S is between −0.9 and −1, and that for Hb A is between 0 and +0.1. Since the p*K* of His is approximately 6.0, this residue will be partially ionized with a charge of less than +0.1 at pH 7.0, which leads to the nonintegral answers.

23. a, b, c, d

24. a, c

25. d, e

26. d. A mutation within the structural gene could change one amino acid to give an abnormal hemoglobin or, alternatively, might disrupt splicing of the mRNA leading to no protein or to one that is rapidly degraded. Thus a change within the gene could give rise to a nonfunctional hemoglobin. Answers a, b, e, and f only apply only to thalassemias, whereas answer c applies only to abnormal hemoglobins.

PROBLEMS

1. An effective respiratory carrier must be able to pick up oxygen from the lungs and deliver it to peripheral tissues. Oxygen dissociation curves for substances A and B are shown in Figure 7-2. What would be the disadvantage of each of these substances as a respiratory carrier? Where would the curve for an effective carrier appear in the figure?

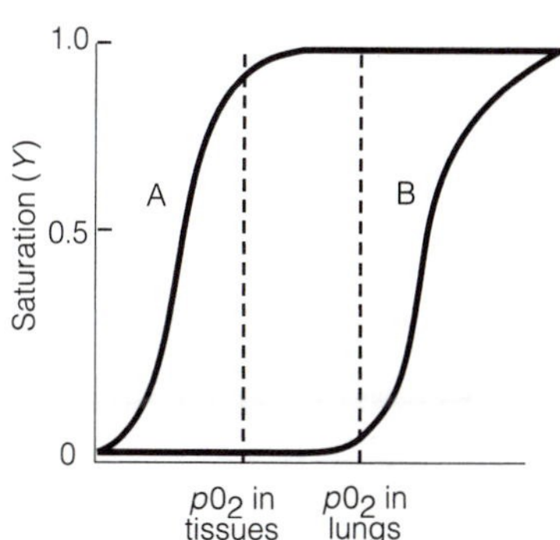

Figure 7-2
Oxygen dissociation curves for substances A and B.

2. Glutathione is a tripeptide consisting of glutamic acid, cysteine, and glycine; it is abundant in human erythrocytes as well as in many types of tissue. Reduced glutathione (GSH) acts as a reducing agent in tissues because its side-chain —SH group can be readily oxidized to form disulfide bonds. Unless glutathione is maintained substantially in its reduced form, as opposed to its oxidized form, in erythrocytes, they lose their ability to transport oxygen effectively. Why do you think this is so?

3. Although one expects proline to occur primarily in nonhelical regions of proteins, three of the four proline residues (C2, F3, and G1) of sperm whale myoglobin occur within regions designated as helical. How can this be the case? (Refer to Stryer, Figure 7-4, p. 149.)

4. The interhelical amino acid residues of sperm whale myoglobin are mostly charged and polar. How does this contribute to the stability of the protein?

5. One molecule of 2,3-bisphosphoglycerate binds to one molecule of hemoglobin in a central cavity of the hemoglobin molecule. Is the inter-

action between BPG and hemoglobin stronger or weaker than it would be if BPG bound to the surface of the protein instead? Explain your answer.

6. The partial pressure of oxygen in the capillaries of active muscles is approximately 20 torrs.

(a) Calculate the fractional saturation of myoglobin and hemoglobin under these conditions.
(b) How do the values you have calculated in part (a) relate to the physiologic role of myoglobin in muscle?

7. The partial pressure of oxygen in the venous blood of a human at rest at sea level is approximately 40 torrs.

(a) Calculate the fractional saturation of hemoglobin under these conditions.
(b) What percentage of the oxygen originally bound to hemoglobin in the alveoli remains unreleased?
(c) Is the residual oxygen bound to hemoglobin under these conditions of any benefit? Explain.

8. Patients suffering from pneumonia have a portion of their lungs filled with fluid, and therefore have reduced lung surface area available for oxygen exchange. Standard hospital treatment of these patients involves placing them on a ventilating machine set to deliver enough oxygen to keep their hemoglobin approximately 92% saturated. Why is this value selected rather than one lower or higher?

9. What major differences exist between the sequential and concerted models for allostery in accounting for hemoglobin that is partially saturated with oxygen.

10. Assuming that the concerted model for allostery applies, calculate the number of oxygen molecules that must, on the average, be bound to hemoglobin to give a [T]/[R] ratio of 1.0.

11. Predict whether each of the following manipulations will increase or decrease the tendency of Hb S molecules to polymerize in vitro. Give a brief rationale for each answer.

(a) An increase in temperature
(b) An increase in the partial pressure of oxygen
(c) Stripping the Hb S molecules of BPG
(d) An increase in the pH

12. One avenue of approach to therapy for sickle-cell anemia involves finding a way of turning on the synthesis of Hb F in afflicted adults. Briefly explain why such a manipulation might be beneficial.

13. One approach to the management of sickle-cell anemia involves a search for osmotically active agents that would expand the volume of erythrocytes.

(a) Give a brief rationale for this approach.
(b) Suppose that the volume of an erythrocyte is increased by 10%. Calculate the rate of Hb S polymer formation in the enlarged cell as a fraction of its rate in a normal-sized cell.

14. An abnormal hemoglobin, Hb Hammersmith, is characterized by the substitution of Ser for Phe at position 42 of the β chain.

(a) Can Hb Hammersmith be separated from Hb A by electrophoresis at pH 6.0? Why or why not?
(b) Give the nucleotide change or changes (on RNA) that most likely produced this mutant. (See the genetic code, Stryer, p. 109.)

15. Prenatal diagnosis of sickle-cell anemia can be carried out by the treatment of fetal DNA with the restriction endonuclease MstII, which recognizes the sequence CCTNAGG, where N is any nucleotide. (See Stryer, p. 173.) Digestion of the β^S gene with MstII yields a 1.3-kb fragment, whereas digestion of the β^A gene yields a 1.1-kb fragment.

(a) Using your knowledge of the amino acid substitution that occurs in Hb S, give the identity of nucleotide N on both the β^A and β^S genes. Also identify the mutational change at the DNA level that leads to the amino acid substitution. Explain your answer. (Refer to the genetic code, Stryer, p. 109.)
(b) Give the identity of the amino acid at position 5 in the β chain using the information provided here and the genetic code (Stryer, p. 109).

16. Explain how a mutation in the promoter region for the β chain of hemoglobin could lead to thalassemia. What analytical technique discussed by Stryer in Chapter 6 would be of use in ascertaining that this was the case?

17. The iron in hemoglobin must be in the ferrous (+2) state in order to bind oxygen. If the iron of hemoglobin becomes oxidized to the ferric (+3) state, the corresponding hemoglobin, called *methemoglobin,* cannot bind oxygen. To maintain iron in the ferrous state, red blood cells contain a reducing system to convert any methemoglobin back to hemoglobin. Some variant forms of hemoglobin, such as hemoglobin M, cannot be reduced by that system. What would be the expected clinical presentation of a patient having Hb M?

ANSWERS TO PROBLEMS

1. Substance A would never unload oxygen to peripheral tissues. Substance B would never load oxygen in the lungs. An effective carrier would have an oxygen dissociation curve between those depicted for substance A and substance B. It would be relatively saturated with oxygen in the lungs and relatively unsaturated in the peripheral tissues.

2. Reduced glutathione helps keep the iron of hemoglobin in the ferrous (+2) valence state. When the iron of heme is oxidized to the ferric (+3) state to form methemoglobin, it can no longer combine reversibly with oxygen.

3. The α helix has 3.6 residues per turn. The three proline residues that occur within helical regions of sperm whale myoglobin are all in the first three positions of a helical run, where the helix is not maximally stabilized by hydrogen bonding. Therefore, some local distortion in bond angles can be accommodated.

4. In the tertiary structure of sperm whale myoglobin, the charged amino acid residues between the helical runs are on the exterior of the protein, where they are stabilized by hydration.

5. BPG binds to hemoglobin by electrostatic interactions. These interactions between the negatively charged phosphates of BPG and the positively charged residues of hemoglobin are much stronger in the interior, hydrophobic environment than they would be on the surface, where water would compete and weaken the interaction by binding both to BPG and to the positively charged residues. Remember that the force of electrostatic interactions, given by Coulomb's law (Stryer, p. 7), is inversely proportional to the dielectric constant of the medium. The dielectric constant in the interior of a protein may be as low as 2. Hence, electrostatic interactions there are much more stable than those on the surface, where the dielectric constant is approximately 80.

6. (a) The applicable relationship for determining the fractional saturation of myoglobin is equation 5 on page 158 of Stryer:

$$Y_{\mathrm{Mb}} = \frac{pO_2}{pO_2 + P_{50}} = \frac{20}{20 + 1} = 0.952$$

For the fractional saturation of hemoglobin, equation 7 on page 158 of Stryer applies:

$$Y_{\mathrm{Hb}} = \frac{(pO_2)^n}{(pO_2)^n + (P_{50})^n} = \frac{(20)^{2.8}}{(20)^{2.8} + (26)^{2.8}} = 0.324$$

(b) Since hemoglobin has a smaller fractional saturation than myoglobin does under these conditions, hemoglobin will unload oxygen to myoglobin, as it must if myoglobin is to serve as an oxygen store in muscle.

7. (a) The applicable relationship for calculating the fractional saturation of hemoglobin is equation 7 on page 158 of Stryer (see the answer to problem 6):

$$Y_{\mathrm{Hb}} = \frac{(40)^{2.8}}{(40)^{2.8} + (26)^{2.8}} = 0.770$$

(b) The percentage of oxygen in the alveoli that remains unreleased is $(0.77/0.98) \times 100 = 79\%$.

(c) The remaining oxygen constitutes a reserve; it can be released to extremely active tissues where the partial pressure of oxygen is very low.

8. Look at the oxygen saturation curve for hemoglobin given in Figure 7-22 of Stryer. Administering enough oxygen to give saturation levels greater than approximately 92% would be wasteful of oxygen, because one reaches the point of diminishing returns. Administering oxygen in amounts less than that required for 92% saturation runs the risk of compromising oxygen delivery to the tissues.

9. According to the concerted model, hemoglobin partially saturated with oxygen is composed of a mixture of fully oxygenated molecules with all subunits in the R form and fully deoxygenated molecules, with all subunits in the T form. According to the sequential model, individual molecules would have some subunits that are oxygenated (in the R form) and some that are deoxygenated (in the T form).

10. The applicable relationship is equation 12 on page 167 of Stryer:

$$\frac{[T_j]}{[R_j]} = c^j L_0$$

Substituting the values of 9000 for L_0 and 0.014 for c given on page 168 of Stryer, we have

$$1 = (0.014)^j\ (9000)$$

$$(0.014)^j = 1/9000$$

$$j \log 0.014 = \log\ (1/9000)$$

$$j = \frac{\log\ (1/9000)}{\log 0.014}$$

$$j = \frac{-3.954}{-1.853872}$$

$$j = 2.13$$

11. (a) An increase in temperature will favor the polymerization of Hb S. The interaction between Hb molecules in polymer formation is hydrophobic in nature. Hydrophobic interactions have negative temperature coefficients; that is, they become more stable with increasing temperature.

(b) An increase in the partial pressure of oxygen will inhibit polymerization because only deoxyhemoglobin S polymerizes.

(c) BPG stabilizes deoxyhemoglobin S. Since only the deoxy form polymerizes, the removal of BPG from Hb S would inhibit polymer formation.

(d) Increasing the pH (decreasing the acidity) stabilizes oxyhemoglobin S. Since only the deoxy form polymerizes, this would inhibit polymer formation.

12. Hb F is devoid of β chains, having γ chains instead, so it does not polymerize. If adults with sickle-cell anemia could synthesize Hb F, each erythrocyte would contain a mixture of Hb S and Hb F, which would reduce the degree of the polymerization of Hb S.

13. (a) The rate of polymerization of Hb S is proportional to the tenth power of its concentration. Therefore, increasing the cell volume would decrease the Hb S concentration, which would slow the rate of Hb S polymerization.

(b) If the cell has been expanded by 10%, the Hb S concentration has been decreased to $\frac{10}{11}$, or 90.9%, of its original value. The rate of polymerization under these conditions is $(0.909)^{10} = 0.386$ of the rate in an unexpanded cell.

14. (a) No, electrophoresis at pH 6.0 will not separate Hb Hammersmith from Hb A. Both Ser and Phe have uncharged side chains. Thus, there is no difference in net charge of the two hemoglobins at pH 6.0.

(b) A C was substituted for a U. Either UUU was changed to UCU or UUC was changed to UCC. Other possibilities would involve simultaneous changes in two positions and are therefore less likely.

15. (a) Nucleotide N must be C on both the β^A and β^S genes. In Hb A, Glu is present at position 6; it is encoded by GAG (on mRNA). Therefore, the sequence CTC must be present on the informational strand of β^A DNA. The mutation that leads to the substitution of Val for Glu at position 6 is $T \rightarrow A$.

(b) The amino acid at position 6 in the β chain is Pro, which is encoded by CCU on RNA (or AGG on DNA).

16. A mutation in the promoter region could lead to decreased binding of RNA polymerase and, thus, to decreased production of mRNA for Hb β chains. This in turn could lead to a decrease in the amount of protein synthesized. A valuable technique for revealing this decreased production of mRNA is *Northern blotting,* discussed by Stryer on page 122.

17. An individual with Hb M would have a bluish appearance, a condition called *cyanosis,* because of the lower concentration of ferrous hemoglobin.

EXPANDED SOLUTIONS TO TEXT PROBLEMS

1. (a) $87\ \mu m^3 = 87 \times 10^{-12}\ cm^3 = 8.7 \times 10^{-11}$ ml. Since each milliliter contains 0.34 g of Hb, $8.7 \times 10^{-11} \times 0.34 = 2.96 \times 10^{-11}$ g of Hb per cell.

(b) The molecular weight of Hb is 66,000. Thus,

$$\frac{2.96 \times 10^{-11}}{66{,}000} = 4.49 \times 10^{-16}\ \text{mol Hb}$$

Moles of Hb × Avogadro's number = 2.71×10^8 molecules of Hb per cell.

(c) Some geometric knowledge is necessary to solve this problem. For a cube $v = s^3$ (where v = volume and s = side length). It follows that $s = v^{1/3}$ (the cube root of v). A cube whose volume is 8.7×10^{-11} cm^3 [see part (a)], therefore, has a side length of 4.44×10^{-4} cm. If the Hb molecules are 65 Å (6.5×10^{-7} cm) across, 6.84×10^2 (4.44×10^{-4} cm/6.5×10^{-7} cm) Hb molecules can be placed in a row on one side of the cube. Cubing this number $(6.84 \times 10^2)^3$ gives 3.22×10^8 molecules of Hb that can be packed into a cube of the size given in part (a). Since the actual number [see part (b)] is 2.71×10^8, the packing density is about 84% of the theoretical maximum.

2. $$70 \times 70 \times \frac{16}{100} \div 66{,}000 = 1.19 \times 10^{-2} \text{ mol Hb}$$

Moles of Hb × 4(4 Fe/Hb) × 55/85 (at wt. of Fe)

$$= 2.65 \text{ g of Fe per adult}$$

3. To answer this question we need to remember that the molecular weight of Mb is 17,800 and that each Mb binds one O_2.

(a) (8/17,800) × 32 (mol. wt. of O_2) = 1.44×10^{-2} g O_2 per kg of muscle. Therefore, in whales, 1.44×10^{-1} g O_2 per kg of muscle.

(b) $$\frac{80}{17{,}800} = 4.49 \times 10^{-3} \text{ mol/kg (or mol/liter)} = 4.49 \times 10^{-3} \text{ M}$$

$$\frac{4.49 \times 10^{-3} \text{ M}}{3.5 \times 10^{-5} \text{ M}} = 128/1$$

4. (a) Equilibrium constants express relationships between the concentrations of reactants and products. They also express the ratio of the rate constant of the forward (off) reaction to the rate constant of the reverse (on) reaction. Therefore,

$$\frac{k_{off}}{k_{on}} = \frac{k_{off}}{2 \times 10^7 \text{ M}^{-1} \text{ s}^{-1}} = 10^{-6} \text{ M}$$

Solving for k_{off} leads to the answer 20 s^{-1}.

(b) If oxygen is released from Mb 20 times a second, the mean duration of the MbO_2 complex is 0.05 s (the reciprocal of k_{off}).

5. (a) Increasing [H^+] decreases the oxygen affinity (Bohr effect). Since increasing pH decreases [H^+], the oxygen affinity is *increased.*

(b) Increasing [CO^2] *decreases* oxygen affinity (modified Bohr effect).

(c) BPG changes the oxygen saturation curves of Hb from hyperbolic to sigmoidal, thus decreasing oxygen affinity. Therefore, increasing BPG decreases oxygen affinity.

(d) The monomer units of Hb act essentially like Mb; hence, conversion to the monomer *increases* oxygen affinity.

6. Inositol (hexahydroxycyclohexane) hexaphosphate has numerous negative charges on it and phosphate groups whose distance apart approximates that found in BPG. Its charge, size, and shape are sufficiently like those in BPG to mimic its reaction with hemoglobin.

7. Consider yourself a proton (H^+) trying to escape from a carboxyl (–COOH) or to protonate its anion ($-COO^-$). (Remember that the stronger the acid, the lower the p*K* and vice versa.) In (a), the positively charged lysine side chain will help repel you (the proton) from the carboxyl, thus making it a stronger acid with a lowered p*K*. In (b), the presence of another COO^- group will make it more difficult for you to escape from the carboxyl; hence, it will weaken the acid and raise the p*K*. In (c), the presence of H_2O enhances the ionization of acids by forming H_3O^+. Therefore, if the side chain is placed at a nonpolar site, it is less likely to ionize. The result is a weaker acid with a higher p*K*.

8. (a) The dissociation constant for the conversion of P to PAB is constant regardless of pathway (P → PB → PAB or P → PA → PAB). Therefore, $K_A \times K_{AB} = K_B \times K_{BA}$. Substituting into this equation, we get $5 \times 10^{-4} \times K_{AB} = 10^{-3} \times 10^{-5}$. Solving for K_{AB}, yields 2×10^{-5} M.

(b) [A] enhances the binding of B because K_B is numerically larger than K_{AB}. It follows that [B] enhances the binding of A because K_A is larger than K_{BA}.

9. Carbon monoxide bound to one heme alters the oxygen affinity of the other hemes in the same hemoglobin molecule. Specifically, CO increases the oxygen affinity of hemoglobin and thereby decreases the amount of O_2 released in actively metabolizing tissues. Carbon monoxide stabilizes the quaternary structure characteristic of oxyhemoglobin. In other words, CO mimics O_2 as an allosteric effector.

10. (a) For maximal transport, $K = 10^{-5}$ M. In general, maximal transport is achieved when $K = ([L_A][L_B])^{0.5}$.

(b) If maximal transport is achieved when $K = ([L_A][L_B])^{0.5}$, one can make the following calculations:

$$([100/760 \times 22.4 \text{ mol/L}]\ [20/760 \times 22.4 \text{ mol/L}])^{0.5} = 1.32 \text{ M}$$

$$\frac{x \text{ (torrs)}}{1.32 \text{ M}} = 760/22.4 \text{ M}$$

$$x = 44.7 \text{ torrs}$$

Note that the answer, 44.7 torrs, is considerably higher than the physiologic value of 26 torrs. However, it is important to realize that this calculation does not take into account the Bohr effect and cooperativity in O_2 binding by hemoglobin.

11. (a) Knowledge of the specificity of trypsin leads to the conclusion that, if a hexapeptide is obtained, position 6 must be either Arg or Lys.

(b) The single base change GAG → AAG will change position 6 from Glu to Lys. No Arg codon can be formed from GAG by a single base change.

(c) The mutant hemoglobin moves more slowly toward the anode than do Hb A and Hb S because it is less negatively charged. The isoelectric points of Hb A and Hb S are approximately 6.9 and 7.1, respectively (Stryer, p. 170). Since it has an extra plus charge, this abnormal Hb will have a slightly higher isoelectric point (approximately 7.3). However, at pH 8 it will still have a negative charge and will move toward the anode, but at a slower rate than Hb A and Hb S, which have slightly larger negative charges.

12. To solve this problem we use equations 10, 11, and 12 on page 167 in Stryer. In the absence of ligand, $T_0/R_0 = 10^5$ (given) and $R_0/T_0 = 10^{-5}$. When four ligands are bound, essentially all the molecules will be in the R form.

The factor $c = K_R/K_t = (5 \times 10^{-6})/(2 \times 10^{-3}) = 2.5 \times 10^{-3}$. If one ligand is bound, the new $T/R = T_0/R_0 \times c = 10^5 \times 2.5 \times 10^{-3} = 2.5 \times 10^2$. $R/T = 1/25 \times 10^2 = 4 \times 10^{-3}$.

When two ligands are bound, $T/R = T_0/R_0 \times c^2 = 10^5 \times (2.5 \times 10^{-3})^2 = 6.25 \times 10^{-1} = 0.625$. Therefore, $R/T = 1/0.625 = 1.6/1$. The *fraction* of molecules in the R state is $1.6/2.6 = 0.615$.

For three ligands $T/R = T_0/R_0 \times c^3 = 10^5 \times (2.5 \times 10^{-3})^3 = 1.565 \times 10^{-3}$. $R/T = 1/1.565 \times 10^{-3} = 6.4 \times 10^2$, and the *fraction* of molecules in the R form is $640/641 = 0.998$.

13. Given that the subunit structures of hemoglobin A, A_2, and F are $\alpha_2\beta_2$, $\alpha_2\delta_2$ and $\alpha_2\gamma_2$, it follows that mutations in the α gene effect all three hemoglobins whereas mutations in the β, δ, or γ genes effect only one of them.

14. Deoxy Hb A contains a complementary site, and so it can add on to a fiber of deoxy Hb S. The fiber cannot then grow further, because the terminal deoxy Hb A molecule lacks a sticky patch.

15. The reaction of CO_2 with α-amino groups to form carbamates *lowers* the oxygen affinity of hemoglobin because of the negative charge on the carbamate (Stryer, p. 165). The reaction with cyanate results in an uncharged, urealike N terminus, which *increases* oxygen affinity, thus preventing sickling.

16. At very low fractional saturation, O_2 binds to hemoglobin molecules that are almost entirely in the T state. At very high fractional saturation, O_2 binds to molecules that are almost entirely in the R state. The slope is ~1 (nearly noncooperative) when binding is not accompanied by a significant switch from T to R.

CHAPTER

8

Enzymes: Basic Concepts and Kinetics

Enzymes catalyze almost all biochemical reactions, and they are intimately involved in the transformations of one form of energy into another. Stryer begins this chapter with an introduction to the catalytic power, specificity, and regulation of enzymes; he then reviews some fundamental thermodynamic principles that are essential for the understanding of whether or not chemical reactions can occur. In the second part of the chapter, enzyme catalysis, enzyme kinetics, and inhibitors of enzymatic reactions are discussed. This chapter draws on your knowledge of protein structure (Chapter 2) and the interactions between biomolecules (Chapters 1 and 7). It sets the stage for the majority of the remaining chapters of this text that deal with biochemical reactions.

LEARNING OBJECTIVES

When you have mastered this chapter, you should be able to complete the following objectives:

Introduction (Stryer pages 181–184)

1. Explain why *enzymes* are the major and most versatile *biological catalysts.*

2. List the different ways in which the *catalytic activity* of enzymes may be *regulated.*

Thermodynamic Principles (Stryer pages 185–187)

3. State the *first* and *second laws of thermodynamics.* Define the *entropy* (S), *enthalpy* (H), and *free energy* (G) of a *system,* and give their mathematical relationship.

4. Describe how ΔG can be used to predict whether a reaction can occur spontaneously.

5. Write the equation for the ΔG of a chemical reaction. Define the *standard free-energy change* (ΔG°); define $\Delta G'$ and $\Delta G^{\circ\prime}$.

6. Derive the relationship between $\Delta G^{\circ\prime}$ and the *equilibrium constant* (K'_{eq}) of a reaction. Relate each tenfold change in K'_{eq} to the change in $\Delta G^{\circ\prime}$ in kilocalories per mole (kcal/mol) or kJ/mol.

Enzyme Catalysis (Stryer pages 188–191)

7. Explain why enzymes do not alter the *equilibrium* of chemical reactions but change only the *rates* of chemical reactions.

8. Define the *transition state* and the *free energy of activation* ($\Delta G^\ddagger$), and describe the effect of enzymes on $\Delta G^\ddagger$.

9. Describe the formation of *enzyme-substrate (ES) complexes* and discuss their properties.

10. Summarize the key features of the *active sites* of enzymes.

Enzyme Kinetics and Enzyme Inhibitors (Stryer pages 192–203)

11. Outline the *Michaelis-Menten model of enzyme kinetics* and define all the terms.

12. Reproduce the derivation of the *Michaelis-Menten equation* in the text. Relate the Michaelis-Menten equation to experimentally derived plots of *velocity* (V) versus *substrate concentration* [S].

13. Define V_{max} and K_M, and explain how these parameters can be obtained from a plot of V versus [S] or a plot of $1/V$ versus $1/[S]$ (a *Lineweaver-Burk plot*).

14. Explain the significance of V_{max}, K_M, k_3, k_{cat} and k_{cat}/K_M.

15. Describe the different kinds of *enzyme inhibitors.*

16. Describe the effects of *competitive* and *noncompetitive inhibitors* on the kinetics of enzyme reactions. Contrast reversible and irreversible inhibitors.

17. Contrast the *kinetics of allosteric enzymes* with simple Michaelis-Menten kinetics; note their similarity to the oxygen dissociation kinetics of hemoglobin and myoglobin (Chapter 7).

18. Distinguish between the *sequential* and *concerted models* of allosteric enzymes. Describe the effects of *allosteric inhibitors* and *activators* on the *T* and *R forms* of allosteric enzymes.

19. Contrast the properties of substrates and *transition state analogs* of enzymes.

20. Describe the formation of *catalytic antibodies.*

21. Outline the mechanism of action of the antibiotic *penicillin.*

SELF-TEST

Introduction

1. Which of the following is *not* a general property of enzymes?

(a) Enzymes are almost exclusively proteins.
(b) Enzymes have great catalytic power.
(c) Enzymes bind substrates specifically.
(d) Enzymes use only hydrophobic interactions in binding substrates.
(e) The catalytic activity of enzymes can be regulated.

2. Which of the following are mechanisms for the regulation of the catalytic activity of enzymes?

(a) The binding of regulatory proteins
(b) The covalent modification of tryptophan residues
(c) The proteolytic cleavage of an inactive enzyme precursor
(d) The binding of regulatory peptides via disulfide bonds

3. List some of the molecular mechanisms by which the catalytic activity of enzymes is controlled. What is the common feature of these mechanisms?

____

____

Thermodynamic Principles

4. Which of the following statements is correct? The entropy of a reaction refers to

(a) the heat given off by the reaction.
(b) the tendency of the system to move toward maximal randomness.
(c) the energy of the transition state.
(d) the effect of temperature on the rate of the reaction.

5. Explain why the thermodynamic parameters ΔE and ΔS cannot be used to predict in which direction a reaction will proceed.

____

____

6. If the standard free-energy change ($\Delta G°$) for a reaction is zero, which of the following statements about the reaction are true?

(a) The entropy ($\Delta S°$) of the reaction is zero.
(b) The enthalpy ($\Delta H°$) of the reaction is zero.
(c) The equilibrium constant for the reaction is 1.0.
(d) The reaction is at equilibrium.
(e) The concentrations of the reactants and products are all 1 M at equilibrium.

7. The enzyme triose phosphate isomerase catalyzes the following reaction:

$$\text{Dihydroxyacetone phosphate} \underset{k_2}{\overset{k_1}{\rightleftharpoons}} \text{glyceraldehyde 3-phosphate}$$

The $\Delta G°'$ for this reaction is 1.83 kcal/mol. In light of this information, which of the following statements are correct?

(a) The reaction would proceed spontaneously from left to right under standard conditions.
(b) The rate of the reaction in the reverse direction is higher than the rate in the forward direction at equilibrium.
(c) The equilibrium constant under standard conditions favors the synthesis of the compound on the left, dihydroxyacetone phosphate.
(d) The data given are sufficient to calculate the equilibrium constant of the reaction.
(e) The data given are sufficient to calculate the left-to-right rate constant (k_1).

8. Phosphorylase catalyzes the reaction:

$$\text{Glycogen}_n + \text{phosphate} \rightleftharpoons \text{glucose 1-phosphate} + \text{glycogen}_{n-1}$$

$$K'_{eq} = \frac{[\text{glucose 1-phosphate}]\,[\text{glycogen}_{n-1}]}{[\text{phosphate}]\,[\text{glycogen}_n]} = 0.088$$

Based on these data, which of the following statements are correct?

(a) Because phosphorylase *degrades* glycogen in cellular metabolism, there is a paradox in that the equilibrium constant favors synthesis.
(b) The $\Delta G°'$ for this reaction at 25°C is 1.44 kcal/mol.
(c) The phosphorolytic cleavage of glycogen requires an input of energy.
(d) If the ratio of phosphate to glucose 1-phosphate in cells is high enough, phosphorylase can degrade glycogen.

9. The hydrolysis of glucose 6-phosphate to give glucose and phosphate has a $\Delta G°' = -3.3$ kcal/mol. The reaction takes place at 25°C. Initially, the concentration of glucose 6-phosphate is 10^{-5} M, that of glucose is 10^{-1} M, and that of phosphate is 10^{-1} M. Which of the following statements pertaining to this reaction are correct?

(a) The equilibrium constant for the reaction is 267.
(b) The equilibrium constant cannot be calculated because standard conditions do not prevail initially.
(c) The $\Delta G'$ for this reaction under the initial conditions is -0.78 kcal/mol.

(d) Under the initial conditions, the synthesis of glucose 6-phosphate will take place rather than hydrolysis.
(e) Under standard conditions, the hydrolysis of glucose 6-phosphate will proceed spontaneously.

Enzyme Catalysis

10. Which of the following statements is true? Enzyme catalysis of a chemical reaction

(a) decreases $\Delta G'$ so that the reaction can proceed spontaneously.
(b) increases the energy of the transition state.
(c) does not change $\Delta G^{\circ\prime}$, but it changes the ratio of products to reactants.
(d) decreases the entropy of the reaction.
(e) increases the forward and reverse reaction rates.

11. Which of the following statements regarding an enzyme-substrate complex is *not* true?

(a) The heat stability of an enzyme frequently changes upon the binding of a substrate.
(b) At sufficiently high concentrations of substrate, the catalytic sites of the enzyme become filled and the reaction rate reaches a maximum.
(c) An enzyme-substrate complex cannot be isolated because of the constant turnover of the substrate.
(d) Enzyme-substrate complexes can be visualized by x-ray crystallography or electron microscopy.
(e) Spectroscopic changes in the substrate or the enzyme can be used to detect the formation of an enzyme-substrate complex.

12. Why is there a high degree of stereospecificity in the interaction of enzymes with their substrates?

____

____

13. Explain why the forces that bind a substrate at the active site of an enzyme are usually weak.

____

____

Enzyme Kinetics and Enzyme Inhibitors

14. Which of the following statements regarding simple Michaelis-Menten enzyme kinetics are correct?

(a) The maximal velocity V_{max} is related to the maximal number of substrate molecules that can be "turned over" in unit time by a molecule of enzyme.
(b) K_M is expressed in terms of a reaction velocity (e.g., mol s^{-1}).
(c) K_M is the dissociation constant of the enzyme-substrate complex.
(d) K_M is the concentration of substrate required to achieve half of V_{max}.
(e) K_M is the concentration of substrate required to convert half the total enzyme into the enzyme-substrate complex.

15. Explain the relationship between K_M and the dissociation constant of the enzyme-substrate complex K_{ES}.

____

____

16. Note the similarity between the Michaelis-Menten equation

$$\frac{V}{V_{max}} = \frac{[S]}{[S] + K_M}$$

and the oxygen dissociation equation for myoglobin

$$Y = \frac{pO_2}{pO_2 + P_{50}}$$

Explain the relationship between the two equations.

____

____

17. From the plot of velocity versus substrate concentration shown in Figure 8-1, obtain the following parameters. (The amount of enzyme in the reaction mixture is 10^{-3} μmol.)

(a) K_M ______

(b) V_{max} ______

(c) k_3/K_M ______

(d) Turnover number ______

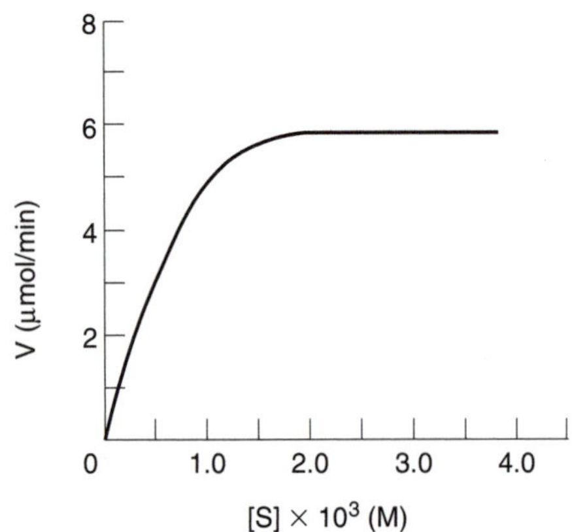

Figure 8-1
Plot of reaction velocity versus substrate concentration.

18. What is the significance of k_3/K_M or k_{cat}/K_M?

____

____

19. Which of the following statements is correct? The turnover number for chymotrypsin is 100 s^{-1}, and for DNA polymerase it is 15 s^{-1}. This means that

(a) chymotrypsin binds its substrate with higher affinity than does DNA polymerase.
(b) the velocity of the chymotrypsin reaction is always greater than that of the DNA polymerase reaction.
(c) the velocity of the chymotrypsin reaction at a particular enzyme concentration and saturating substrate levels is lower than that of the DNA polymerase reaction under the same conditions.
(d) the velocities of the reactions catalyzed by both enzymes at saturating substrate levels could be made equal if 6.7 times more DNA polymerase than chymotrypsin were used.

20. Which of the following statements about the different types of enzyme inhibition are correct?

(a) Competitive inhibition is seen when a substrate competes with an enzyme for binding to an inhibitor protein.

(b) Competitive inhibition is seen when the substrate and the inhibitor compete for the active site on the enzyme.
(c) Noncompetitive inhibition of an enzyme cannot be overcome by adding large amounts of substrate.
(d) Competitive inhibitors are often similar in chemical structure to the substrates of the inhibited enzyme.
(e) Noncompetitive inhibitors often bind to the enzyme irreversibly.

21. If the K_M of an enzyme for its substrate remains constant as the concentration of the inhibitor increases, what can be said about the mode of inhibition?

____

____

22. The kinetic data for an enzymatic reaction in the presence and absence of inhibitors are plotted in Figure 8-2. Identify the curve that corresponds to each of the following:

(a) No inhibitor ________

(b) Noncompetitive inhibitor ________

(c) Competitive inhibitor ________

(d) Mixed inhibitor ________

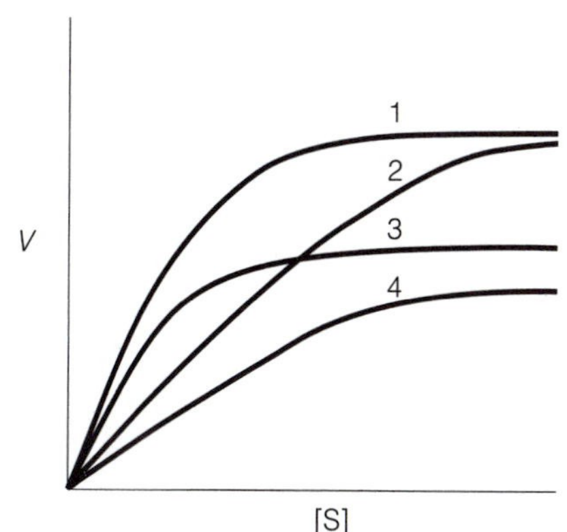

Figure 8-2
Effects of inhibitors on a plot of V versus [S].

23. Draw approximate Lineweaver-Burk plots for each of the inhibitor types in question 22.

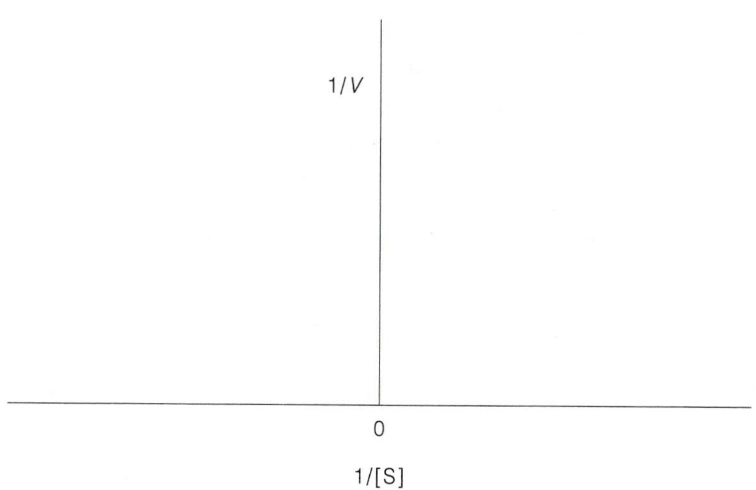

24. If T (tense) and R (relaxed) are the two possible conformational states of the subunits of an allosteric enzyme in the concerted model, which of the following statements are true?

(a) Enzyme that is saturated with substrate is in the TT form.
(b) Allosteric activators preferentially bind to subunits in the R state.
(c) Subunits in the R and T states can coexist in a single molecule at low substrate concentrations.
(d) Allosteric inhibitors increase the proportion of enzyme molecules in the TT form relative to the RR form.

25. Match the assumptions or predictions in the right column with the appropriate model for allosteric interactions in the left column.

(a) Concerted model ________

(b) Sequential model ________

(1) The R form is induced by substrate binding.
(2) The R and T forms are in equilibrium.
(3) Only RR and TT states exist.
(4) RR, RT, and TT states exist.
(5) The binding of substrate by one subunit always increases the binding affinity of the other.
(6) The binding of substrate by one subunit may increase or decrease the binding affinity of the other.

26. Which statement is not true about a transition state analog?

(a) It fits better in the active site than the substrate.
(b) It increases the rate of product formation.
(c) It can be used as a hapten to produce catalytic antibodies.
(d) It is usually a distorted or strained molecule.
(e) It is a potent inhibitor of the enzyme.

27. The inhibition of bacterial cell wall synthesis by penicillin is a classic example of a medically significant inhibition of an enzymatic reaction. Which of the following statements about the inhibition of glycopeptide transpeptidase by penicillin is true?

(a) The inhibition is noncompetitive.
(b) Penicillin binds irreversibly to an allosteric site of the enzyme.
(c) Penicillin inhibits bacterial cell wall synthesis by incorrectly cross-linking the peptides of the proteoglycan.
(d) The penicilloyl-enzyme intermediate may be dissociated by high concentrations of D-alanine.
(e) Penicillin resembles acyl-D-Ala-D-Ala, one of the substrates of the transpeptidase.

ANSWERS TO SELF-TEST

Introduction

1. d

2. a, c

3. Molecular mechanisms that control the catalytic activity of enzymes include feedback inhibition, the binding of regulatory proteins, covalent modification, and proteolytic cleavage. These mechanisms of enzyme regulation depend on conformational changes that occur at the site of modulation and are transmitted to the active site of the enzyme. Consequently, both enzyme-substrate interactions and catalytic activity are affected.

4. b

5. The thermodynamic parameters ΔE and ΔS for a chemical reaction are not easily measured. Also, both depend on changes that occur not only in the system under study but also in the surroundings. Intrinsically unfavorable reactions can take place if a change in the surroundings compensates for an increase in the energy (positive ΔE) or a decrease in the entropy (negative ΔS) of the system.

6. c, e

7. c, d

8. All the statements are correct.

   (a) Resolution of the paradox comes from the fact that the ratio of phosphate to glucose 1-phosphate in cells is usually high enough that phosphorolysis of glycogen is favored.

   (b) Using K'_{eq}, one can calculate the $\Delta G^{\circ\prime}$ for the phosphorylase reaction:

$$\Delta G^{\circ\prime} = -2.303\ RT \log_{10} K'_{eq}$$

$$= -2.303 \times 1.98\ \frac{\text{cal}}{\text{mol}\cdot{}^{\circ}\text{K}} \times 298^{\circ}\text{K} \times \log_{10}(0.088)$$

$$= -1360\ \text{cal/mol} \times -1.055$$

$$= 1440\ \text{cal/mol} = 1.44\ \text{kcal/mol}$$

   (c) In part (b) the $\Delta G^{\circ\prime}$ for the phosphorylase reaction of +1.44 kcal/mol was calculated; therefore, energy is consumed rather than released by this reaction.

   (d) In cells, the ratio of phosphate to glucose 1-phosphate is so large that phosphorylase is mainly involved with glycogen degradation.

9. a, d, e

   (a) Correct.

$$\Delta G^{\circ\prime} = -2.303\ RT \log_{10} K'_{eq}$$

$$-3.3\ \text{kcal/mol} = -1.36\ \text{kcal/mol} \times \log_{10} K'_{eq}$$

$$\log_{10} K'_{eq} = 2.426$$

$$K'_{eq} = 267$$

   (b) Incorrect. K'_{eq} is a constant; it is independent of the initial concentrations.

   (c) Incorrect.

$$\Delta G' = \Delta G^{\circ\prime} + 2.303\ RT \log_{10} \frac{[\text{glucose}]\,[\text{phosphate}]}{[\text{glucose 6-phosphate}]}$$

$$= -3.3\ \text{kcal/mol} + \left(1.36\ \text{kcal/mol} \times \log_{10} \frac{10^{-1} \times 10^{-1}}{10^{-5}}\right)$$

$$= -3.3\ \text{kcal/mol} + \left(1.36\ \text{kcal/mol} \times \log_{10} \frac{10^{-2}}{10^{-5}}\right)$$

$$= -3.3\ \text{kcal/mol} + \left(1.36\ \text{kcal/mol} \times 3\right)$$

$$= +0.78\ \text{kcal/mol}$$

(d) Correct. Under the initial conditions, $\Delta G'$ is positive; therefore, the reaction will proceed toward the formation of glucose 6-phosphate.
(e) Correct. The negative $\Delta G^{\circ\prime}$ value (at standard conditions) indicates that the reaction will proceed spontaneously toward the hydrolysis of glucose 6-phosphate.

Enzyme Catalysis

10. e

11. c. In reactions requiring two substrates, an enzyme-substrate complex of one of the substrates can be isolated if the complex is very stable.

12. The formation of an enzyme-substrate complex involves a close, complementary fitting of the atoms of the amino acid residues that make up the active site of the enzyme with the atoms of the substrate. Since stereoisomers have different spatial arrangements of their atoms, only a single stereoisomer of the substrate usually fits in the active site.

13. The enzyme-substrate and enzyme-product complexes have to be reversible so that catalysis may proceed; therefore, weak forces are involved in the binding of substrates to enzymes.

Enzyme Kinetics and Enzyme Inhibitors

14. a, d, e

15. K_M can be equal to K_{ES} when the rate constant $k_3 \ll k_2$. Since $K_M = (k_2 + k_3)/k_1$, when k_3 is negligible relative to k_2, K_M becomes equal to k_2/k_1, which is the dissociation constant of the enzyme-substrate complex.

16. These equations are related because they express the occupancy of saturable binding sites as a function of either O_2 or substrate concentration. The fraction of active sites filled, V/V_{max}, is analogous to Y, the degree of myoglobin saturation with oxygen; [S] and pO_2 are the concentrations of substrate and O_2, respectively; and K_M and P_{50} are substrate or O_2 concentrations at half-maximal saturation.

17. (a) $K_M = 5 \times 10^{-4}$ M. K_M is obtained from Figure 8-1; it is equal to [S] at $\frac{1}{2}V_{max}$. Note that the units of [S] are mM. The factor 10^3 is used to multiply the actual concentrations. For example,

$$[S] \times 10^3 = 2.0 \text{ M}$$
$$[S] = 2.0 \times 10^{-3} \text{ M}$$

(b) $V_{max} = 6\ \mu\text{mol/min}$. V_{max} is obtained from Figure 8-1; it is the maximum velocity.

(c) $k_3/K_M = 2 \times 10^5\ \text{s}^{-1}\ \text{M}^{-1}$. In order to calculate this ratio, k_3 must be known. Since $V_{max} = k_3[E_T]$, $k_3 = V_{max}/[E_T]$. Thus

$$k_3 = \frac{6\ \mu\text{mol/min}}{10^{-3}\ \mu\text{mol}}$$
$$= 6 \times 10^3\ \text{min}^{-1}$$
$$= 100\ \text{s}^{-1}$$

Using K_M from part (a),

$$\frac{k_3}{K_M} = \frac{100\ \text{s}^{-1}}{5 \times 10^{-4} \text{M}}$$

(d) The turnover number is 100 s^{-1}, equal to k_3, which was calculated in part (c).

18. The ratios k_3/K_M and k_{cat}/K_M allow one to estimate the catalytic efficiency of an enzyme. The upper limit for k_3/K_M, 10^8 to 10^9 M^{-1} s^{-1}, is set by the rate of diffusion of the substrate in the solution, which limits the rate at which it encounters the enzyme. If an enzyme has a k_3/K_M in this range, its catalytic velocity is restricted only by the rate at which the substrate can reach the enzyme, which means that the enzymatic catalysis has attained kinetic perfection.

19. d. $V_{max} = k_3[E_T]$; thus, if 6.7 times more DNA polymerase than chymotrypsin is used, V_{max} for both enzymes is the same:

$$100 \text{ s}^{-1} = 6.7 \times 15 \text{ s}^{-1}$$

Answer (a) is incorrect because the affinity of substrate for the enzyme is given by $K_{ES} = k_2/k_1$. Answer (b) is incorrect because the velocity of the enzymatic reactions is a function of K_M, V_{max}, and substrate concentration. Answer (c) is incorrect because for the same enzyme concentration, $V_{max} = k_3[E_T]$ is greater for chymotrypsin than for DNA polymerase.

20. b, c, d

21. The inhibition is noncompetitive, since the proportion of bound substrate remains the same as the concentration of the inhibitor increases.

22. (a) 1 (b) 3 (c) 2 (d) 4

23. See Figure 8-3. Plots 1 and 2 have the same $1/V$ intercept; plots 1 and 3 have the same $1/[S]$ intercept; and plots 1 and 4 have different $1/V$ and $1/[S]$ intercepts.

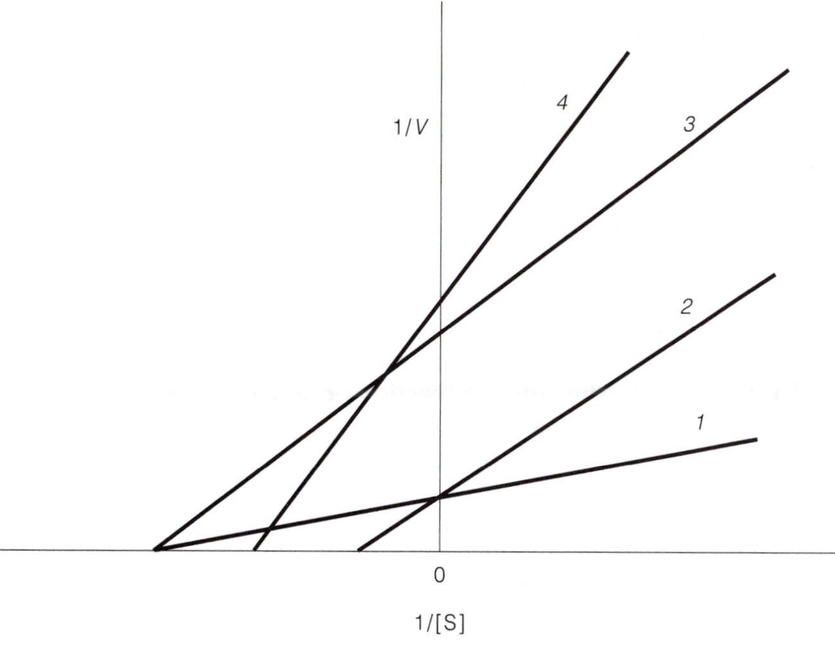

Figure 8-3
Lineweaver-Burk plots for competitive (2), noncompetitive (3), and mixed (4) inhibition, relative to the enzymatic reaction in the absence of inhibitors (1).

24. b, d

25. (a) 2, 3, 5 (b) 1, 4, 6

26. b. This answer is incorrect because transition state analogs are inhibitors of the corresponding enzymes. Therefore, they decrease rather than increase enzyme reaction rates.

27. e

PROBLEMS

1. The change in internal energy ΔE for the transition from ice to water at 0°C is approximately +1.1 kcal/mol. Yet the transition occurs spontaneously. Briefly explain why.

2. Calculate the values for $\Delta G^{\circ\prime}$ that correspond to the following values of K'_{eq}. Assume that the temperature is 25°C.

(a) 1.5×10^4
(b) 1.5
(c) 0.15
(d) 1.5×10^{-4}

3. Calculate the values for K'_{eq} that correspond to the following values of $\Delta G^{\circ\prime}$. Assume that the temperature is 25°C.

(a) −10 kcal/mol
(b) −1 kcal/mol
(c) +1 kcal/mol
(d) +10 kcal/mol

4. The enzyme hexokinase catalyzes the following reaction:

$$\text{Glucose} + \text{ATP} \rightleftharpoons \text{glucose 6-phosphate} + \text{ADP}$$

For this reaction, $\Delta G^{\circ\prime} = -4.0$ kcal/mol.

(a) Calculate the change in free energy $\Delta G'$ for this reaction under typical intracellular conditions using the following concentrations: glucose, 55 mM; ATP, 5.0 mM; ADP, 1.0 mM; and glucose 6-phosphate, 0.1 mM. Assume that the temperature is 25°C.
(b) In the typical cell, is the reaction catalyzed by hexokinase close to equilibrium or far from equilibrium? Explain.

5. The enzyme aldolase catalyzes the following reaction:

$$\text{Fructose 1,6-bisphosphate} \rightleftharpoons \text{dihydroxyacetone phosphate} + \text{glyceraldehyde 3-phosphate}$$

For this reaction, $\Delta G^{\circ\prime} = +5.7$ kcal/mol.

(a) Calculate the change in free energy $\Delta G'$ for this reaction under typical intracellular conditions using the following concentrations: fructose 1,6-bisphosphate, 0.15 mM; dihydroxyacetone phosphate, 4.3×10^{-6} M; and glyceraldehyde 3-phosphate, 9.6×10^{-5} M. Assume that the temperature is 25°C.
(b) Explain why the aldolase reaction occurs in cells in the direction written despite the fact that it has a positive free-energy change under standard conditions.

6. Stryer makes the statement (p. 188) that a decrease of 1.36 kcal/mol in the free energy of activation of an enzyme-catalyzed reaction has the effect of increasing the rate of conversion of substrate to product by a factor of 10. What effect should a decrease of 1.36 kcal/mol in the free energy of activation have on the reverse reaction, the conversion of product to substrate? Explain.

7. What is the ratio of [S] to K_M when the velocity of an enzyme-catalyzed reaction is 80% of V_{max}?

8. The simple Michaelis-Menten model (equation 14, Stryer, p. 192) applies only to the initial velocity of an enzyme-catalyzed reaction, that is, to

the velocity when no appreciable amount of product has accumulated. What feature of the model is consistent with this constraint? Explain.

9. Two first-order rate constants, k_2 and k_3, and one second-order rate constant, k_1, define K_M by the relationship

$$K_M = \frac{k_2 + k_3}{k_1}$$

By substituting the appropriate units for the rate constants in this expression, show that K_M must be expressed in terms of concentration.

10. Suppose that two tissues, tissue A and tissue B, are assayed for the activity of enzyme X. The activity of enzyme X, expressed as the number of moles of substrate converted to product per gram of tissue, is found to be five times greater in tissue A than in tissue B under a variety of circumstances. What is the simplest explanation for this observation?

11. Sketch the appropriate plots on the following axes. Assume that simple Michaelis-Menten kinetics apply. (See Stryer, pp. 192–195.)

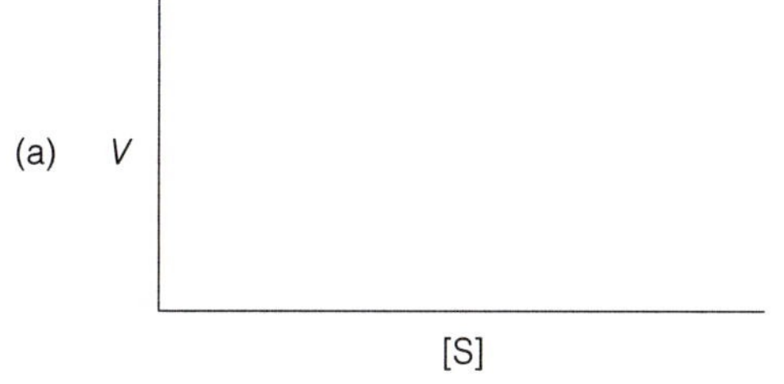

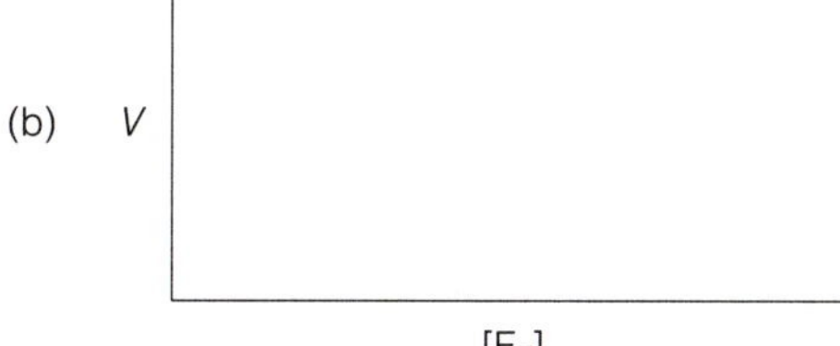

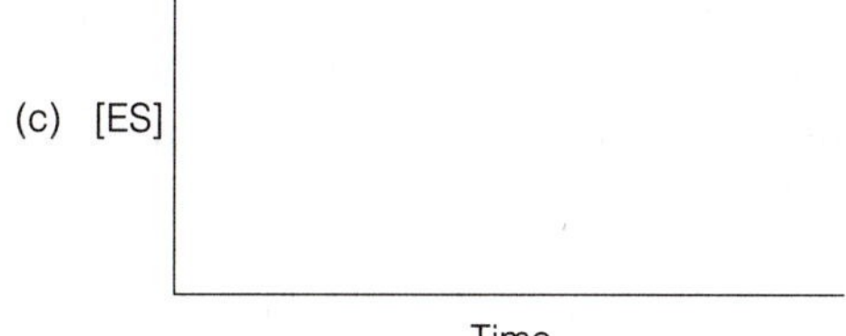

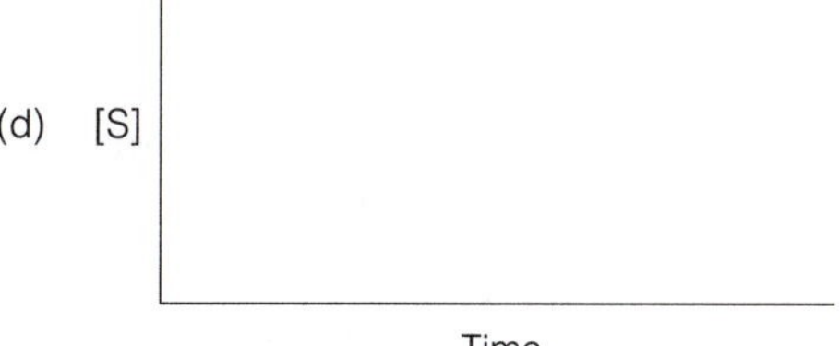

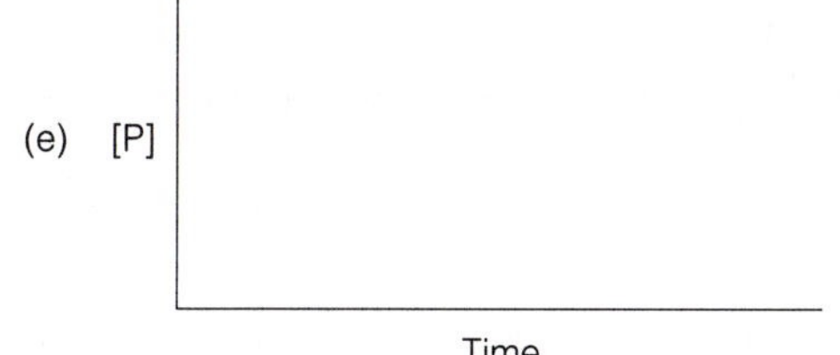

12. Suppose that the data shown in the margin are obtained for an enzyme-catalyzed reaction.

(a) From a double-reciprocal plot of the data, determine K_M and V_{max}.

(b) Assuming that the enzyme present in the system had a concentration of 10^{-6} M, calculate its turnover number.

[S] (mM)	V (mmol ml^{-1} min^{-1})
0.1	3.33
0.2	5.00
0.5	7.14
0.8	8.00
1.0	8.33
2.0	9.09

[S] (mM)	V (mmol ml^{-1} min^{-1})	
	Without X	With X
0.2	5.0	3.0
0.4	7.5	5.0
0.8	10.0	7.5
1.0	10.7	8.3
2.0	12.5	10.7
4.0	13.6	12.5

13. Suppose that the data shown in the margin are obtained for an enzyme-catalyzed reaction in the presence and absence of inhibitor X.

(a) Using double-reciprocal plots of the data, determine the type of inhibition that has occurred.
(b) Does inhibitor X combine with E, with ES, or with both? Explain.
(c) Calculate the inhibitor constant K_i for substance X, assuming that the final concentration of X in the reaction mixture was 0.2 mM.

[S] (mM)	V (mmol ml^{-1} min^{-1})	
	Without Y	With Y
0.2	5.0	2.0
0.4	7.5	3.0
0.8	10.0	4.0
1.0	10.7	4.3
2.0	12.5	5.0
4.0	13.6	5.5

14. Suppose that the data shown in the margin are obtained for an enzyme-catalyzed reaction in the presence and absence of inhibitor Y.

(a) Using double-reciprocal plots of the data, determine the type of inhibition that has occurred.
(b) Does inhibitor Y combine with E, with ES, or with both? Explain.
(c) Calculate the inhibitor constant K_i for substance Y, assuming that the final concentration of Y in the reaction mixture was 0.3 mM.

15. Although the double-reciprocal plot is the most widely used plotting form for enzyme kinetic data, it suffers from a major disadvantage. If linear increments of substrate concentration are used, thereby minimizing measurement errors in the laboratory, data points will be obtained that cluster near the vertical axis. Thus the intercept on the ordinate can be determined with great accuracy, but the slope of the line will be subject to considerable error, because the least reliable data points, those obtained at low substrate concentrations, have greater weight in establishing the slope. (Remember that many enzymes are protected against denaturation by the presence of their substrates at high concentrations.)

Because of the limitation of double-reciprocal plots described above, other linear plotting forms have been devised. One of these, the Eadie plot, is described in problem 6 on page 206 of Stryer. Another, the Hanes-Woolf plot, ([S]/V versus [S]) is perhaps the most useful in minimizing the difficulties of the double-reciprocal plot.

(a) Rearrange the Michaelis-Menten equation to give [S]/V as a function of [S].
(b) What is the significance of the slope, the vertical intercept, and the horizontal intercept in a plot of [S]/V versus [S]?
(c) Data shown in the margin were obtained for the hydrolysis of *o*-nitrophenyl-β-D-galactoside (ONPG) by *E. coli* β-galactosidase. Use both double-reciprocal and Hanes-Woolf plots to analyze these data, and calculate values for K_M and V_{max} from both plots. (We suggest that you use a graphing program to generate a scatter plot, and then fit the data using a linear curve-fitting algorithm.)

[S] (mM)	V (μmol 1^{-1} min^{-1})
0.5	8.93
1.0	14.29
1.5	16.52
2.0	19.20
2.5	19.64

(d) Make a sketch of a plot of [S]/V versus [S] in the absence of an inhibitor and in the presence of a competitive inhibitor and in the presence of a noncompetitive inhibitor.

16. Suppose that a modifier Q is added to an enzyme-catalyzed reaction with the results depicted in Figure 8-4. What role does Q have? Does it combine with E, with ES, or with both E and ES?

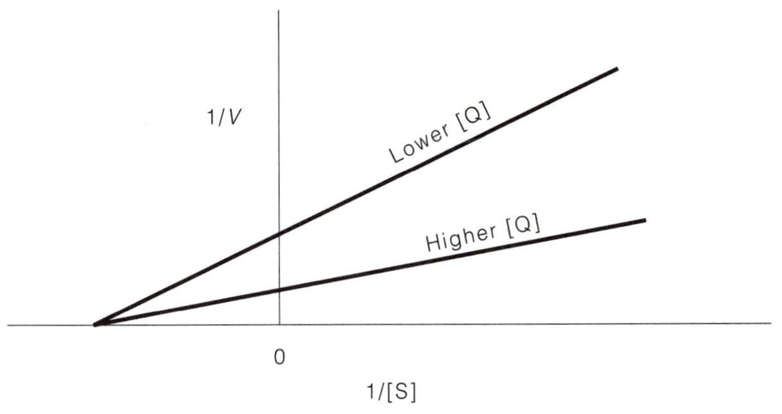

Figure 8-4
Effects of modifier Q on a plot of $1/V$ versus $1/[S]$.

ANSWERS TO PROBLEMS

1. The increase in entropy must be sufficient to offset the increase in internal energy and thereby lead to a negative change in the free energy. We may envision this as follows: as ice melts, it loses its ordered structure and its entropy therefore increases.

2. The values for $\Delta G^{\circ\prime}$ are found by substituting the values for K'_{eq} into equation 11 on page 187 of Stryer.

(a) $\Delta G^{\circ\prime} = -2.303\ RT \log_{10} K'_{e}$

$= -2.303 \times 1.98 \times 10^{-3} \times 298 \times \log_{10} (1.5 \times 10^{4})$

$= -5.7$ kcal/mol

(b) -0.24 kcal/mol

(c) $+1.1$ kcal/mol

(d) $+5.2$ kcal/mol

3. Equation 13 on page 187 of Stryer is used to find the answers.

(a) $K'_{eq} = 10^{-\Delta G^{\circ\prime}/1.36}$

$= 10^{-(-10/1.36)}$

$= 2.3 \times 10^{7}$

(b) 5.4

(c) 0.18

(d) 4.4×10^{-8}

4. (a) The applicable relationship is equation 6 on page 186 of Stryer:

$$\Delta G' = \Delta G^{\circ\prime} + RT\log_e \frac{[\text{C}][\text{D}]}{[\text{A}][\text{B}]}$$

$$= \Delta G^{\circ\prime} + RT\log_e \frac{[\text{glucose 6-phosphate}][\text{ADP}]}{[\text{glucose}][\text{ATP}]}$$

$$= -4.0 \text{ kcal/mol}$$

$$+ \left(1.98 \times 10^{-3} \times 298 \times \log_e \frac{[0.1 \times 10^{-3}][1.0 \times 10^{-3}]}{[55 \times 10^{-3}][5 \times 10^{-3}]}\right)$$

$$= -4.0 \text{ kcal/mol} - 4.7 \text{ kcal/mol}$$

$$= -8.7 \text{ kcal/mol}$$

(b) The large negative value for $\Delta G'$ means that the reaction is far from equilibrium in the typical cell and thus has a strong thermodynamic drive to go in the direction of product formation. (Remember that at equilibrium, $\Delta G' = 0$.)

5. (a) The applicable relationship is again equation 6 on page 186 of Stryer:

$$\Delta G' = \Delta G^{\circ\prime} + RT\log_e \frac{[\text{C}][\text{D}]}{[\text{A}][\text{B}]}$$

$$= \Delta G^{\circ\prime} + RT\log_e \frac{[\text{DHAP}][\text{G3P}]}{[\text{FBP}]}$$

$$= +5.7 \text{ kcal/mol} + (1.98 \times 10^{-3} \times 298)$$

$$\times \log_e \frac{(4.3 \times 10^{-6}) \times (9.6 \times 10^{-5})}{0.15 \times 10^{-3}}$$

$$= +5.7 \text{ kcal/mol} - 7.6 \text{ kcal/mol}$$

$$= -1.9 \text{ kcal/mol}$$

(b) The reaction occurs in the direction written because of the effects of the concentrations on the free-energy change. The concentration term in the equation is much smaller than 1.0, which is its value under standard conditions.

6. The rate of the reverse reaction must also increase by a factor of 10. Enzymes do not alter the equilibria of processes; they affect the rate at which equilibrium is attained. Since the equilibrium constant K_{eq} is the quotient of the rate constants for the forward and reverse reactions, both rate constants must be altered by the same factor. If the rate of the for-

ward reaction is increased by a factor of 10, the rate of the reverse reaction must increase by the same factor.

7. Start with the Michaelis-Menten equation, equation 28 on page 193 of Stryer:

$$V = V_{\max} \frac{[S]}{[S] + K_M}$$

Substituting 0.8 $V_{\max}$ for V yields

$$\begin{aligned} 0.8\, V_{\max} &= V_{\max} \frac{[S]}{[S] + K_M} \\ 0.8[S] + 0.8K_M &= [S] \\ 0.8K_M &= 0.2[S] \\ [S] &= 4K_M \\ \frac{[S]}{K_M} &= 4 \end{aligned}$$

Thus a substrate concentration four times as great as the Michaelis constant yields a velocity that is 80% of maximal velocity.

8. Equation 14 on page 192 of Stryer shows the k_3 step as being irreversible. This is true initially because P and E cannot recombine to give ES at an appreciable rate if little P is present. Note that this says nothing about the relative magnitudes of k_3 and the reverse rate constant k_4:

$$E + S \underset{k_2}{\overset{k_1}{\rightleftharpoons}} ES \underset{k_4}{\overset{k_3}{\rightleftharpoons}} E + P$$

The reverse constant k_4 may actually be quite large compared with k_3, but the reverse reaction still will not occur when little product is present, since the rate of the k_4 step depends on the concentrations of P and E as well as on the magnitude of k_4.

9. The first-order rate constants have the dimensions t^{-1}, whereas the second-order constant has the dimension $\text{conc}^{-1}\, t^{-1}$. Thus, we can carry out the following dimensional analysis:

$$\begin{aligned} K_M &= \frac{k_2 + k_3}{k_1} \\ &= \frac{t^{-1} + t^{-1}}{\text{conc}^{-1}\, t^{-1}} \\ &= \text{conc} \end{aligned}$$

10. For the activity of enzyme X to be five times greater in tissue A than in tissue B, tissue A must have five times the amount of enzyme X as does tissue B. Enzyme activity is directly proportional to enzyme concentration.

11. The sketches should resemble the following:

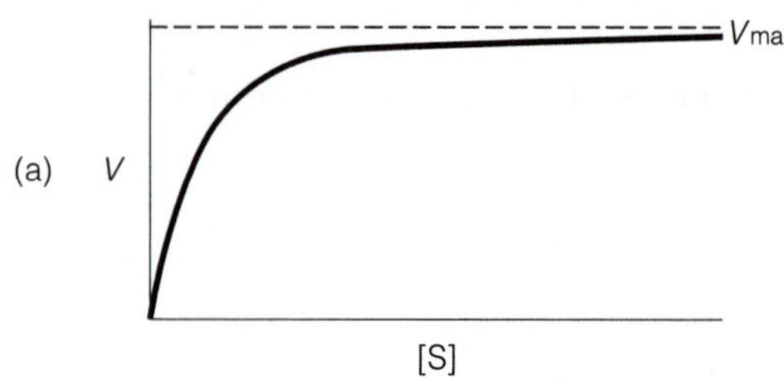

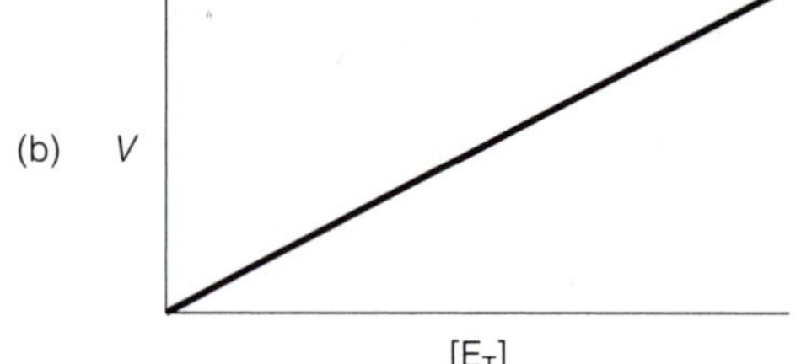

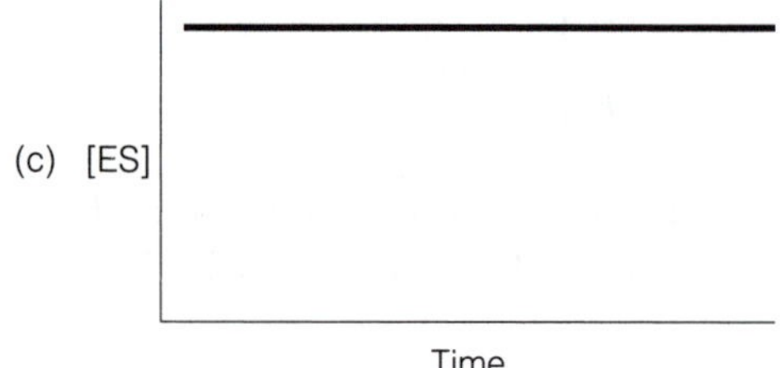

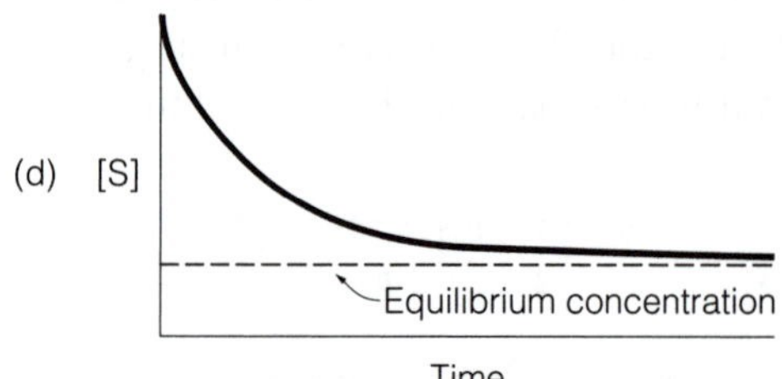

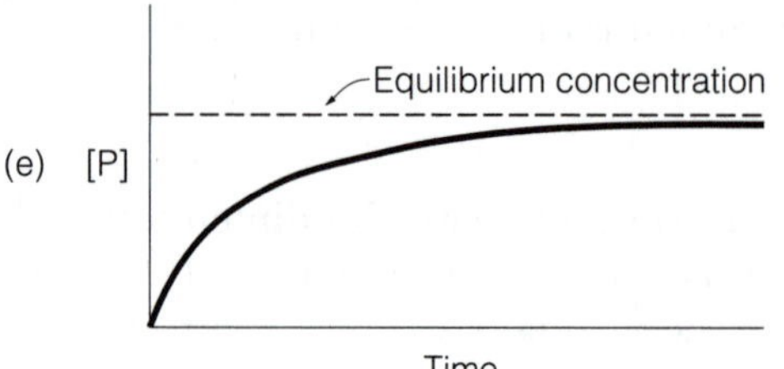

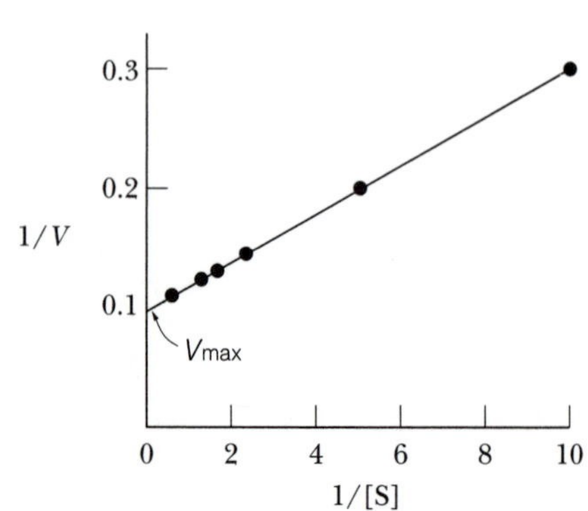

Figure 8-5
A double-reciprocal plot of data for problem 12.

12. (a) See the graph at the left, Figure 8-5. $V_{max} = 1/0.1 = 10$ mmol ml^{-1} min^{-1}.

$$\text{Slope} = \frac{0.3 - 0.1}{10} = 0.02$$

$$\text{Slope} = \frac{K_M}{V_{max}}$$

$$K_M = 0.02 \times 10 = 0.2 \text{ mM}$$

(b) The turnover number is equal to the rate constant k_3 in equation 33 on page 195 of Stryer. Rearrangement of the equation gives

$$k_3 = \frac{V_{max}}{[E_T]}$$

$$= \frac{10 \text{ mmol ml}^{-1} \text{ min}^{-1}}{10^{-6} \text{ mol liter}^{-1}}$$

$$= \frac{10 \text{ mol liter}^{-1} \text{ min}^{-1}}{10^{-6} \text{ mol liter}^{-1}}$$

$$= 10^7 \text{ min}^{-1} \quad \text{or} \quad 1.7 \times 10^5 \text{ s}^{-1}$$

13. (a) See Figure 8-6. The double-reciprocal plots intersect on the y-axis, so the inhibition is competitive.

(b) The inhibitor combines only with E, the free enzyme. A competitive inhibitor cannot combine with ES because the inhibitor and the substrate compete for the same binding site on the enzyme.

(c) An inhibitor increases the slope of a double-reciprocal plot by a factor of $1 + [I]/K_i$:

$$\text{Slope}_{\text{inhib}} = \text{slope}_{\text{uninhib}} \left(1 + \frac{[I]}{K_i}\right)$$

The slope with X is

$$\text{Slope}_{\text{inhib}} = \frac{0.333 - 0.067}{5} = 0.0532$$

The slope without X is

$$\text{Slope}_{\text{uninhib}} = \frac{0.200 - 0.067}{5} = 0.0266$$

Substituting in these values yields

$$0.0532 = 0.0266 \left(1 + \frac{0.2\text{mM}}{K_i}\right)$$

$$K_i = 0.2 \text{ mM}$$

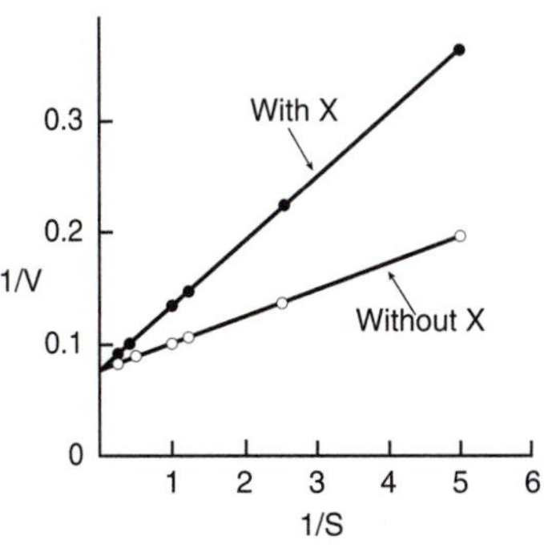

Figure 8-6
A double-reciprocal plot of data for problem 13 showing the effects of an inhibitor X.

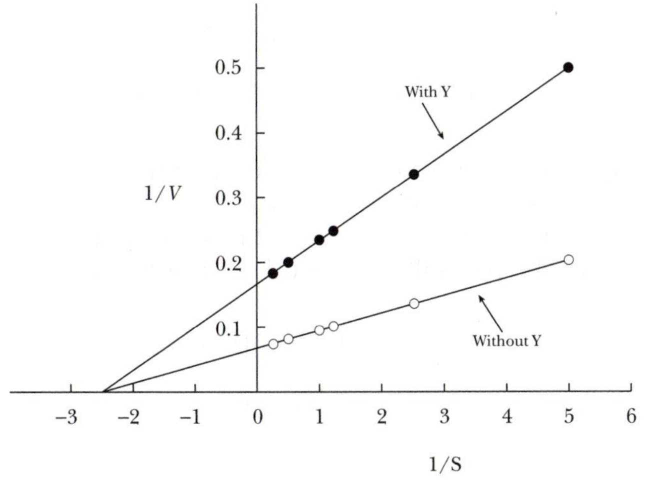

Figure 8-7
A double-reciprocal plot of data for problem 14 showing the effects of an inhibitor Y.

14. (a) See Figure 8-7. The inhibition was noncompetitive, as indicated by the fact that the double-reciprocal plots intersect to the left of the y-axis.

(b) A noncompetitive inhibitor combines at a site other than the substrate binding site. Thus, it may combine with both E and ES. In the case illustrated, the inhibitor has equal affinity for E and ES, which is shown by the fact that the plots intersect on the x-axis.

(c) Again, the slope increases by a factor of $1 + [I]/K_i$ in the presence of an inhibitor.

$$\text{Slope}_{\text{inhib}} = \text{Slope}_{\text{uninhib}}\left(1 + \frac{[I]}{K_i}\right)$$

The slope with Y is

$$\text{Slope}_{\text{inhib}} = \frac{0.500}{5.0 - (-2.5)} = 0.0667$$

The slope without Y is

$$\text{Slope}_{\text{uninhib}} = \frac{0.200}{5.0 - (-2.5)} = 0.0267$$

Substituting in these values yields

$$0.0667 = 0.0267\left(1 + \frac{0.3\ \text{mM}}{K_i}\right)$$

$$K_i = 0.2\ \text{mM}$$

15. (a) We start with the Michaelis-Menten equation:

$$V = V_{\max}[S]/(K_M + [S])$$

Cross multiplying yields

$$V(K_M + [S]) = V_{\max}[S]$$

Division of both sides by $V\ V_{\max}$ gives

$$[S]/V = (K_M + [S])/V_{\max}$$

$$[S]/V = K_M/V_{\max} + [S]/V_{\max}$$

$$[S]/V = (1/V_{\max})[S] + K_M/V_{\max}$$

(b) The equation above is in slope intercept form, $y = mx + b$, where m is the slope, and b the y-intercept. Therefore, the slope of a Hanes-Woolf plot is $(1/V_{\max})$, and the intercept on the y-axis is $K_M/V_{\max}$. The plot will intercept the x-axis when $[S]/V$ is zero. Then

$$0 = (1/V_{\max})[S] + K_M/V_{\max}$$

$$-K_M/V_{\max} = (1/V_{\max})[S]$$

$$[S] = -K_M$$

(c) See Figure 8-8. The y-intercept of the double-reciprocal plot is $1/V_{max}$. Therefore $V_{max} = 1/0.034 = 29.4\ \mu\text{mol l}^{-1}\ \text{min}^{-1}$. The slope of the double-reciprocal plot is K_M/V_{max}. Therefore,

$$0.039 = K_M/29.4$$
$$K_M = 1.15\ \text{mM}$$

See Figure 8-9. The slope of the Hanes-Woolf plot is $1/V_{max}$. Therefore $V_{max} = 1/0.035 = 28.6\ \mu\text{mol l}^{-1}\ \text{min}^{-1}$. The y-intercept of the Hanes-Woolf plot is K_M/V_{max}. Therefore,

$$0.037 = K_M/28.6$$
$$K_M = 1.06\ \text{mM}.$$

In this instance, both plots give good fits of the data, and the values derived from each for K_M and V_{max} do not differ significantly. We can conclude that the measurements at low substrate concentration are reliable.

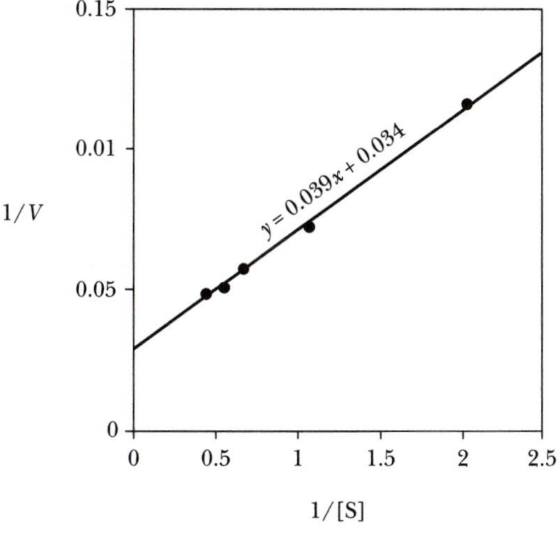

Figure 8-8
A double-reciprocal plot of data for problem 15.

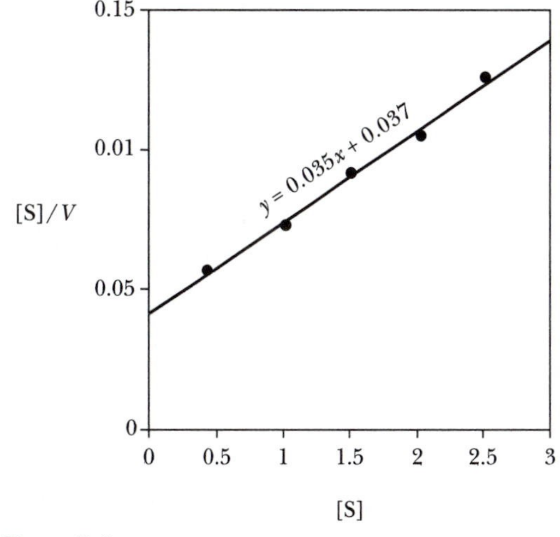

Figure 8-9
Hanes-Woolf plot of data for problem 15.

(d) See Figure 8-10.

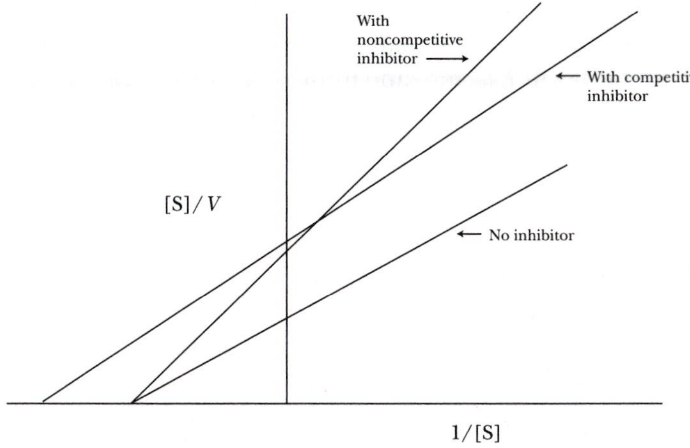

Figure 8-10
Hanes-Woolf plots depicting effects of competitive and noncompetitive inhibitors.

16. Q increases the rate of reaction, so it is an activator, or perhaps a second substrate. It combines with both E and ES.

EXPANDED SOLUTIONS TO TEXT PROBLEMS

1. (a) Note that a unit of activity is 10^{-5} mol/15 min and when the substrate concentration is much greater than K_M the enzyme operates at V_{max}. Then (2800 units $\times$ 10^{-5} mole)/(15 min $\times$ 60s) = 31.1 $\times$ 10^{-6} mol/s.

   (b) Remember grams/gram molecular weight = moles, thus 10^{-3} g/ (20 $\times$ 10^3 g/mol subunit) = 5 $\times$ 10^{-8} mol active site.

   (c) By definition, the turnover number of an enzyme is the number of molecules of substrate converted to product by one molecule of enzyme per unit time, in this case, seconds. Hence, the turnover number = 31.1 $\times$ 10^{-6} mol S s^{-1}/5 $\times$ 10^{-8} mol E = 622 molecules S per second per molecule E. *Turnover number* is discussed on page 195 of Stryer.

2. For (a) and (b), proper graphing of the data given will provide the correct answers:

$$K_M = 5.2 \times 10^{-6}\ \text{M};\ V_{max} = 6.84 \times 10^{-10}\ \text{mol/min}$$

   (c) Turnover = mol S s^{-1}/mol E = (6.84 $\times$ 10^{-10})/[(60 $\times$ 10^{-9})/ 29,600] = 337 s^{-1}

3. Penicillinase, like glycopeptide transpeptidase, forms an acyl-enzyme intermediate with its substrate but transfers it to water rather than to the terminal glycine of the pentaglycine bridge.

4. For (a) and (b), proper graphing of the data given will provide the correct answers:

   (a) In the absence of inhibitor, V_{max} is 47.6 μmol/min and K_M is 1.1 $\times$ 10^{-5} M. In the presence of inhibitor, V_{max} is the same, and the apparent K_M is 3.1 $\times$ 10^{-5} M.

   (b) Since V_{max} is not altered by the inhibitor, this is *competitive inhibition.*

   (c) Since this is competitive inhibition, equation 37 (Stryer, p. 198) applies. The only difference between equation 37 and equation 29 (Stryer, p. 193) is the factor (1 + [I]/K_i). Since in competitive inhibition V_{max} does not change, this factor describes the relationship between K_M and apparent K_M. Hence, the apparent K_M = K_M (1 + [I]/K_i). Therefore, using the data in (a) and (b), 3.1 $\times$ 10^{-5} M = (1.1 $\times$ 10^{-5} M)(1 + 2 $\times$ 10^{-3} M/K_i) and K_i = 1.1 $\times$ 10^{-3} M.

   (d) The [S]/(K_M + [S]) term in the Michaelis-Menten equation tells us the fraction of enzyme molecules bound to substrate. Thus, 1 $\times$ 10^{-5}/(1 + 3.1)10^{-5} = f_{ES} = 0.243. Since K_i = [E] [I]/[EI], [EI]/[E] = [I]/K_i = 2 $\times$ 10^{-3}/1.1 $\times$ 10^{-3} = 1.82. However, the sum of [EI] + [E] is only 0.757 of the total enzyme because the remaining 0.243 is bound to substrate. Therefore, 1.82 = f_{EI}/ (0.757 – f_{EI}). Solving this equation gives f_{EI} = 0.488.

   (e) Using [S]/(K_M + [S]), 3 $\times$ 10^{-5}/1.1 + 3) $\times$ 10^{-5} = 0.73 = f_{ES} in the absence of inhibitor. 3 $\times$ 10^{-5}/(3.1 + 3) $\times$ 10^{-5} = 0.49 = f_{ES} in the *presence* of inhibitor. This ratio, 0.73/0.49 and 33.8/22.6 (the velocity ratio) are equal.

5. For (a) and (b), proper graphing of the data given will provide the correct answers:

(a) V_{max} is 9.5 μmol/min. K_M is 1.1×10^{-5} M, the same as without inhibitor.
(b) Since K_M does not change, this is noncompetitive inhibition.
(c) To answer this question you need to obtain a value for V_{max} (47.6 μmol/min) from the graphs used in question 4(a) above. Because this is noncompetitive inhibition, use equation 39 (Stryer, p. 198) as follows: 9.5 μmol/min = 47.6 μmol/min/(1 + 10^{-4} M/K_i). Solving for K_i one obtains the answer 2.5×10^{-5} M.
(d) Since an inhibitor does not affect K_M, the fraction of enzyme molecules binding substrate = [S]/(K_M + [S]), with or without inhibitor. For solution, see 4(e).

6. (a)

$$V = \frac{V_{max}\,[S]}{(K_M + [S])}$$

$$\frac{V(K_M + [S])}{[S]} = V_{max}$$

$$V + \frac{VK_M}{[S]} = V_{max}$$

$$V = V_{max} - \frac{VK_M}{[S]}$$

$$V = V_{max} - \frac{K_M V}{[S]}$$

(b) The slope of a straight line is the constant that the x-coordinate is multiplied by in the equation for the straight line. Thus, in the Lineweaver-Burk plot, K_M/V_{max} is the slope (Stryer, p. 194); in the Eadie-Hofstee plot the slope is $-K_M$ because $V/[S]$ is plotted on the x axis; see (a). By inspection, the y-intercept is V_{max}. The x-intercept is V_{max}/K_M because one is extrapolating to [S] = 0.

(c) Note that with a competitive inhibitor V_{max} (y-intercept) stays the same but K_M increases (the slope of 2 is greater than the slope of 1). In contrast, with a noncompetitive inhibitor, K_M does not change; 1 and 3 have the same slope (while V_{max} decreases).

7. Potential hydrogen-bond donors at pH 7 are the side chains of the following residues: arginine, asparagine, glutamine, histidine, lysine, serine, threonine, tryptophan and tyrosine. For a more detailed discussion of hydrogen-bond donors and acceptors, see problem 2-8.

8. The rates of utilization of A and B are given by equation 34 (Stryer, p. 195):

$$V_A = \left(\frac{k_3}{K_M}\right)_A [E]\,[A]$$

and

$$V_B = \left(\frac{k_3}{K_M}\right)_B [E]\,[B]$$

Hence, the ratio of these rates is

$$V_A / V_B = \left(\frac{k_3}{K_M}\right)_A [A] \Big/ \left(\frac{k_3}{K_M}\right)_B [B]$$

Thus, an enzyme discriminates between competing substrates on the basis of their values of k_3/K_M rather than of K_M alone. Note that the velocity is dependent on the constants (k_3/K_M) *and* the concentrations of enzyme and substrate.

9. As discussed by Stryer on p. 187, a ten-fold change in the equilibrium constant corresponds to a standard free-energy change ($\Delta G°'$) of 1.36 kcal/mol (1.36 comes from 2.303 RT). If a mutant enzyme binds a substrate, S, 100-fold as tightly as does the native enzyme, more Gibbs free energy of activation ($\Delta G^{\ddagger}$) is needed to convert S to $S^{\ddagger}$ (transition state). In fact, the $\Delta G^{\ddagger}$ is increased by 2.72 kcal/mol (2.303 RT log 100) and the velocity of the reaction will be slowed down by a factor of 100.

CHAPTER

9

Catalytic Strategies

In the previous chapter, you learned that the catalytic action of enzymes is based on their ability to stabilize the transition states of chemical reactions. In this chapter, Stryer describes in detail the structures, active-site configurations, binding of substrates, and catalytic mechanisms of four well-understood enzymes whose three-dimensional structures are known: lysozyme, ribonuclease, carboxypeptidase A, and chymotrypsin. Using these specific examples, fundamental principles of enzyme catalysis are illustrated: specific binding of substrates, induced fit of the enzyme-substrate complex, general acid-base catalysis by active-site residues, formation and stabilization of transition states, formation of covalent intermediates, and reversibility of the catalytic steps. Since the interactions of enzymes with substrates depend on the chemical properties of amino acid residues and on protein structure in general, a review of Chapter 2 may be helpful.

LEARNING OBJECTIVES

When you have mastered this chapter, you should be able to complete the following objectives:

Lysozyme (Stryer pages 207–216)

1. Describe the *polysaccharide* and *oligosaccharide substrates* of *lysozyme.* (The nomenclature, structures, and properties of carbohydrates are described in Chapter 18 of Stryer.)

2. Explain how *x-ray crystallography* may be used to find the *active site* of an enzyme.

3. Summarize the predictions made about the *binding site* and the *catalytic groups* of lysozyme on the basis of x-ray data.

4. Outline the *catalytic mechanism* of lysozyme. Explain the roles of *general acid catalysis* and the formation and stabilization of a *carbonium ion intermediate* in the process.

5. List the experiments that support the mode of substrate binding and the mechanism of catalysis proposed for lysozyme. Note the use of *transition state analogs* in the elucidation of the mechanism.

Ribonuclease A (Stryer pages 216–218)

6. Describe the *substrate specificity* of *ribonuclease A.*

7. Summarize the chemical studies of ribonuclease A and the x-ray crystallographic evidence that implicate *histidine* residues in catalysis.

8. Describe the catalytic mechanism proposed for ribonuclease A. Explain the general acid-base catalysis by the histidine residues and the formation and stabilization of a *pentacovalent transition state* of phosphorus during catalysis by ribonuclease A.

Carboxypeptidase A (Stryer pages 218–222)

9. State the substrate specificity of *carboxypeptidase A.*

10. Describe the x-ray crystallographic information about the *zinc ion* binding site and the substrate binding site of carboxypeptidase A.

11. Describe the structural changes that occur at the active site of carboxypeptidase A upon the binding of substrate (*induced fit*).

12. Summarize the involvement of the zinc ion, glutamate 270, and arginine 127 in the catalytic mechanism of carboxypeptidase A (*activation of water*). Note how these groups stabilize the negatively charged *transition state* of the reaction.

Chymotrypsin and Other Proteolytic Enzymes (Stryer pages 222–230)

13. Describe the substrate specificity and the physiologic role of *chymotrypsin.*

14. Outline the catalytic mechanism of chymotrypsin and explain the role of the *acyl-enzyme intermediate.* Explain the relevance of *rapid burst kinetics* followed by slower *steady-state kinetics* to the mechanism of chymotrypsin.

15. Give the experimental evidence that implicates reactive serine and histidine residues in the catalytic mechanism of chymotrypsin. Describe the role of the serine-histidine-aspartate *catalytic triad.* Note the involvement of a *covalent intermediate* in the mechanism of catalysis.

16. Describe the formation and stabilization of the transient *tetrahedral intermediate* produced from the planar amide group of the substrate.

17. Compare other *serine proteases,* for example *trypsin* and *elastase,* with chymotrypsin in terms of their substrate specificities, structures, catalytic mechanisms, and *evolution* (amino acid sequence similarities).

18. Describe the roles of the key aspartate residues in the *acid protease pepsin.*

19. List other acid proteases and describe how they can be inhibited.

20. Outline the principles of *drug design* as employed in obtaining a specific inhibitor of the *HIV-1 protease.* Note how *site-specific mutagenesis* and *chemical modification* of specific amino acids are used to identify residues critical for enzyme function.

21. Name the four different classes of proteolytic enzymes and summarize their properties.

RNA Enzymes (Stryer pages 230–232)

22. Discuss the discovery that RNA molecules can act as enzymes.

23. List the properties of *ribozymes* that define them as *RNA catalysts.*

24. Relate the concepts of covalent intermediate and *transesterification* to the catalytic mechanism of *L-19 RNA.*

25. Contrast the catalytic potentials of proteins and RNA in terms of their compositions.

SELF-TEST

Lysozyme

1. Which of the following answers complete the sentence correctly? X-ray crystallographic determination of the structure of water-soluble enzymes, such as lysozyme and chymotrypsin,

(a) shows compact, almost globular structures.
(b) immediately reveals the catalytic groups.
(c) indicates variable amounts of secondary structure.
(d) can be readily performed with the substrate bound to the active site.
(e) can reveal evolutionary relationships with other enzymes.

2. Given the oligosaccharide NAM-NAG-NAM-NAG-NAM-NAG and the sugar binding sites A, B, C, D, E, and F on lysozyme, identify which of the following statements is true:

(a) Any segments of this oligosaccharide having more than three sugars would bind to lysozyme with the same affinity as the hexamer.
(b) The hexamer NAG-NAM-NAG-NAM-NAG-NAM would have the same affinity for lysozyme as the given hexamer.
(c) The hexamer in (b) would bind to lysozyme with a higher affinity than the given hexamer.
(d) The given hexamer would be cleaved between sugars 4 and 5 from the NAM end.

3. A variety of chemical experiments support the catalytic mechanism of lysozyme proposed on the basis of x-ray crystallographic data. Among them are the following:

(1) The use of ^{18}O enriched H_2O in the hydrolytic reaction
(2) The dependence of the catalytic rate on pH, with the optimum rate occurring at pH 5
(3) The cleavage pattern of hexa-NAG into tetra-NAG and di-NAG
(4) The selective chemical modification of aspartic and glutamic amino acid residues
(5) The binding of a lactone analog of tetra-NAG with high affinity

Indicate which of these experiments led to the following conclusions:

(a) Hydrolysis of the glycosidic bond occurs between sugar residues D and E. ______
(b) Acidic amino acid residues participate in catalysis. ______
(c) The oxygen of the cleaved glycosidic bond remains with sugar residue E. ______
(d) Distortion of the D sugar residue is important in catalysis. ______

4. Which of the following is a catalytic amino acid residue of lysozyme?

(a) Tryptophan 62
(b) Serine 57
(c) Histidine 24
(d) Glutamic acid 35
(e) Lysine 74

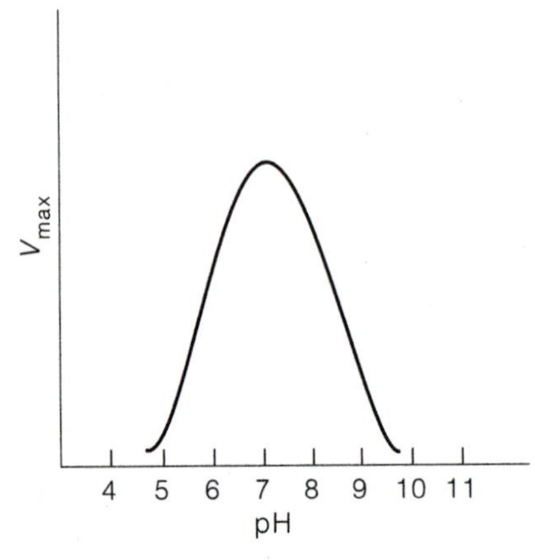

Figure 9-1
pH dependence of V_{max}.

5. A pH–enzyme activity curve (not of lysozyme) is shown in Figure 9-1. Which of the following pairs of amino acids would be likely candidates as catalytic groups: (See Stryer, p. 43, for the pK_a values of amino acid residues.)

(a) Glutamic acid and lysine
(b) Aspartic acid and histidine
(c) Histidine and cysteine
(d) Histidine and histidine
(e) Histidine and lysine

6. The proton association-dissociation equilibrium of which amino acid residues gives rise to the optimal activity of lysozyme appearing at about pH 5?

(a) Glutamic acid
(b) Histidine
(c) Cysteine
(d) Aspartic acid
(e) Tyrosine

Ribonuclease A

7. Which of the following are true statements regarding the ribonuclease reaction?

(a) Ribonuclease cleavage of RNA produces new 3′-OH and 5′-phosphate ends.
(b) Single-stranded DNA can bind to ribonuclease.
(c) A 3′,5′-cyclic phosphate intermediate is formed during the reaction.
(d) A pocket adjacent to the catalytic groups of ribonuclease specifically binds pyrimidine bases.
(e) Several main-chain NH and CO groups of ribonuclease form hydrogen bonds with the phosphate backbone of RNA.

8. Give a reason why genetic information is encoded in DNA rather than in RNA.

____

____

9. In the transition state during RNA hydrolysis by ribonuclease, the reactive phosphorus atom adopts a pentacovalent trigonal bipyramid geometry. The oxygen atoms at the apices of the pyramid correspond to

(a) H_2O and 3′-O.
(b) 2′-O and 5′-O.
(c) H_2O and 5′-O.
(d) 2′-O and 3′-O.
(e) OH^- and 2′-O.

What is a consequence of this transition state geometry?

____

____

10. Why is histidine a particularly versatile amino acid residue in terms of its involvement in enzymatic reaction mechanisms?

____

____

Carboxypeptidase A

11. Which of the following are important mechanisms by which carboxypeptidase A promotes the proteolysis of its substrate?

(a) Rearrangement of its structure to place critical amino acid residues near the scissile bond
(b) Formation of a C-terminal cyclic peptide as a covalent intermediate
(c) Deprotonation of an active site tyrosine residue to form a nucleophile
(d) Activation of water by a bound zinc ion

12. Why does carboxypeptidase A have a positive zinc ion in its active site instead of a positively charged amino acid residue such as arginine or lysine?

____

____

13. Upon the binding of substrate to carboxypeptidase A, the active site undergoes a major structural rearrangement. Which of the following is a consequence of this induced-fit rearrangement of carboxypeptidase A?

(a) The zinc ion, which was in solution, now binds to the active site.
(b) Water molecules are displaced from the active site.
(c) The electrostatic and hydrogen bond interactions between the enzyme and substrate become weaker.
(d) The active-site cavity can now accommodate an internal peptide bond of a protein substrate.

Chymotrypsin and Other Proteolytic Enzymes

14. Which of the following experimental observations provide evidence of the formation of an acyl-enzyme intermediate during the chymotrypsin reaction?

(a) There is a biphasic release of *p*-nitrophenol during the hydrolysis of *p*-nitrophenyl acetate.
(b) The active serine can be specifically labeled with organic fluorophosphates.
(c) The pH dependence of the catalytic rate is bell-shaped, when activity is plotted as a function of pH, with a maximum at pH 8.
(d) An acyl-enzyme covalent complex can be isolated at low pH values.

15. Which of the following enzymes can be irreversibly inactivated with diisopropylphosphofluoridate (DIPF)?

(a) Carboxypeptidase A
(b) Trypsin
(c) Lysozyme
(d) Elastase
(e) Pepsin

16. Three essential amino acid residues in the active site of chymotrypsin form a catalytic triad. Which of the following are roles of these residues in catalysis?

(a) The histidine residue facilitates the reaction by acting as an acid-base catalyst.
(b) The aspartate residue orients the substrate properly for reaction.
(c) The serine residue acts as a nucleophile during the reaction.
(d) The serine residue acts as an electrophile during the reaction.
(e) The aspartate residue initiates the deacylation step by a nucleophilic attack on the carbonyl carbon of the acyl intermediate.

17. Although chymotrypsin is a proteolytic enzyme, it is quite resistant to self-hydrolysis. Explain why.

____

____

18. Which of the following answers complete the sentence correctly? The three enzymes trypsin, elastase, and chymotrypsin

(a) evolved from a common ancestor.
(b) have major similarities in their amino acid sequences and three-dimensional structures.
(c) catalyze the same reaction: the cleavage of a peptide bond.

(d) catalyze reactions that proceed through a covalent intermediate.
(e) have structural differences at their active sites.

19. Match the enzymes and the catalytic or functional groups in the right column with the main proteolytic enzyme classes in the left column.

(a) Zinc proteases ____________
(b) Serine proteases ____________
(c) Thiol proteases ____________
(d) Acid proteases ____________

(1) Papain
(2) Pepsin
(3) Elastase
(4) Thermolysin
(5) Histidine
(6) Cysteine
(7) Aspartate
(8) Arginine

20. Suppose that you want to prepare an antibody with the catalytic activity of lysozyme. What type of antigen would you use?

____

____

21. For each enzyme in the left column, indicate the appropriate transition state or chemical entity in the right column that has a postulated involvement in its catalytic mechanism.

(a) Lysozyme ____________
(b) Ribonuclease ____________
(c) Carboxypeptidase A ____________
(d) Chymotrypsin ____________
(e) Pepsin ____________

(1) Zinc ion–activated water
(2) Oxyanion
(3) Pentacovalent phosphorus
(4) Carbonium ion
(5) Tetrahedral peptide bond
(6) Aspartate–activated water

22. Propose an experiment to determine whether a particular amino acid residue is essential for the activity of an enzyme.

____

____

RNA Enzymes

23. Which of the following properties of L19 RNA indicate that it is a ribozyme?

(a) It contains amino acids.
(b) It acts catalytically; i.e., each molecule converts more than one substrate molecule.
(c) It displays Michaelis-Menten kinetics.
(d) It is destroyed by proteases.
(e) It shows high selectivity for its substrate.

24. Which of the following is *not* a common property of enzymes (proteins) and ribozymes (RNS enzymes)?

(a) Their active sites may contain large hydrophobic pockets.
(b) They usually bind their substrates through multiple weak interactions.
(c) Both are subject to competitive inhibition.
(d) They exhibit saturation kinetics at high substrate concentrations.
(e) Both have a high degree of substrate specificity.

25. What are two main structural features of an enzyme (or ribozyme) that determine its substrate specificity?

____

____

ANSWERS TO SELF-TEST

Lysozyme

1. a, c, e

2. c. Because site C cannot bind NAM, the given oligosaccharide would bind to sites A–E or B–F with the approximate affinity of a pentamer.

3. (a) 3 (b) 2, 4 (c) 1 (d) 5

4. d

5. c

6. a, d

Ribonuclease A

7. b, d

8. RNA can be cleaved by hydroxyl ions even in mildly basic solutions. Such spontaneous hydrolysis of RNA proceeds via a 2′,3′-cyclic phosphate intermediate. Because DNA lacks a 2′-hydroxyl group, it is much more stable to hydrolysis than is RNA, so it is a safer repository for genetic information.

9. b. This geometry is compatible with an in-line mechanism of nucleophilic attack yielding the observed products.

10. Histidine residues may act as nucleophiles as well as general acids or basis. Because their pK values are approximately 6 to 7, histidine residues can be protonated or deprotonated at physiologic pH; they are therefore able either to donate or to accept protons.

Carboxypeptidase A

11. a, d. Induced fit and activation by water are defining features of carboxypeptidase A activity.

12. Carboxypeptidase A accelerates catalysis, in part, by inducing electronic "strain" in its substrate that renders the peptide bond more susceptible to nucleophilic attack by a water molecule. For accomplishing this task, a small, highly charged ion is more effective than a bulky, basic amino acid side chain. In addition, the zinc ion directly activates the water molecule that participates in the cleavage of the peptide bond. The enzyme enhances its catalytic properties by recruiting a zinc ion prosthetic group.

13. b

Chymotrypsin and Other Proteolytic Enzymes

14. a, d. Reactivity with DIPF (b) provides only indirect evidence, since it indicates only a highly reactive serine.

15. b, d. These enzymes have reactive serine residues in catalytic triads involved in their catalytic mechanisms.

16. a, c

17. Chymotrypsin specifically cleaves peptide bonds that are adjacent to aromatic amino acid residues or bulky, hydrophobic residues such as methionine. Since these types of residues are usually buried in the interior of proteins, including chymotrypsin, the self-hydrolysis of native, folded chymotrypsin is very inefficient. In fact, during digestion, chymotrypsin acts most effectively on partially degraded and denatured proteins.

18. a, b, c, d, e

19. (a) 4, 8 (b) 3, 5, 7 (c) 1, 5, 6 (d) 2, 7

20. Enzymes catalyze reactions by binding to and facilitating the formation of the transition state of the reaction. An antibody that binds a transition state analog might similarly bind the "true" transition state structure and thereby catalyze the reaction. To prepare an antibody with the catalytic activity of lysozyme, a transition state analog of its substrate should be used as the antigen. The D-ring lactone analog of tetra-NAG would be a likely choice.

21. (a) 4 (b) 3 (c) 1 (d) 2, 5 (e) 6

22. Site-specific mutagenesis could be used to change the particular amino acid residue into a different amino acid residue. Alternatively, specific chemical modification of the residue could be attempted. If either of these treatments make the enzyme inactive, the particular amino acid residue could be essential for the activity of the enzyme. However, it is also possible that the replacement of the amino acid or its chemical modification could inactivate the enzyme by altering its structure. Additional experiments must be performed to establish that the structure of the enzyme has not been changed.

RNA Enzymes

23. b, c, e

24. a

25. The catalytically active macromolecule must have binding groups that can interact with the substrate to position it properly for a productive reaction, namely, tight binding of the transition state of the reaction. In addition, the catalyst might have catalytic residues, for example, acid-base or nucleophilic residues, that react specifically with the substrate. Substrate and transition state binding, as well as properly placed catalytic residues, can contribute to the specificity of the reaction.

PROBLEMS

1. The peptide hydrogen of residue 59 and the carbonyl oxygen of residue 107 of the polypeptide backbone of lysozyme form hydrogen bonds with residue C of tri-*N*-acetylglucosamine (tri-NAG) (see Stryer, p. 210). Would you expect these two residues to be found in helical regions? Explain.

2. What limitations does the use of tri-NAG have for mapping the interactions that bind substrate to lysozyme? What advantage does its use have despite these limitations? Explain.

3. Glucose 1-phosphate (shown in the margin) may be cleaved to glucose and inorganic phosphate by the enzyme alkaline phosphatase:

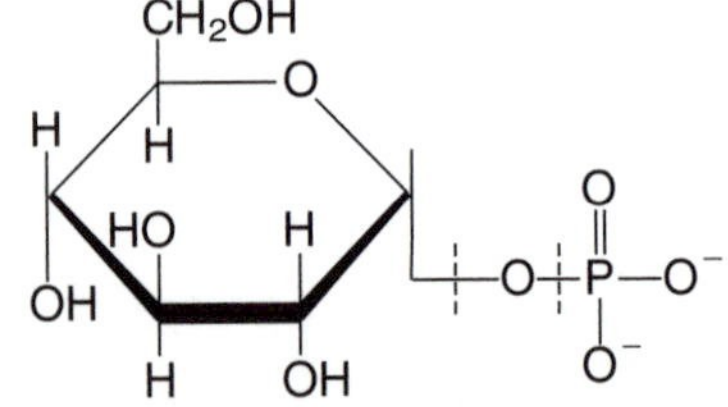

Glucose 1-phosphate
(Dashed lines indicate possible sites of cleavage)

$$\text{Glucose 1-phosphate} + H_2O \xrightleftharpoons{\text{Alkaline phosphatase}} \text{glucose} + \text{phosphate}$$

In principle, the phosphoryl group could be removed by scission of either a P–O or a C–O bond. These two possibilities may be distinguished experimentally by the use of H_2O labeled with ^{18}O as a substrate.

(a) Which product would be labeled with ^{18}O if the P–O bond is broken? Explain.
(b) Which product would be labeled with ^{18}O if the C–O bond is broken? Explain.
(c) The ^{18}O is found to be incorporated exclusively into inorganic phosphate. Which bond must therefore have been broken?

4. The proposed catalytic mechanism for lysozyme presented by Stryer on page 213 requires that the side-chain group of aspartic acid 52 be in the ionized $-COO^-$ form and that of glutamic acid 35 be in the un-ionized –COOH form.

(a) What percentage of Asp side-chain groups would be in the ionized $-COO^-$ form at pH 5.0, the pH optimum for the hydrolysis of chitin by lysozyme? Use the Henderson-Hasselbalch equation (see Stryer, p. 42). Assume that the p*K* of the side-chain carboxyl group is 3.9.
(b) Using the same method as in part (a), calculate the percentage of Glu side-chain groups that would be in the protonated –COOH form at pH 5.0. Assume a p*K* of 4.3.
(c) In view of your answer to part (b), explain how it is possible for Glu 35 in lysozyme to be present in the un-ionized form.

5. When a polysaccharide chain is cleaved by lysozyme, a C–O bond at the C-1 of sugar residue D of the substrate is broken, as indicated by the isotopic evidence presented by Stryer on page 212. Briefly explain how the cleavage of this bond rather than the C–O bond at the C-4 of sugar residue E is consistent with the notion that a carbonium ion intermediate is involved in the mechanism. See Figure 9-2.

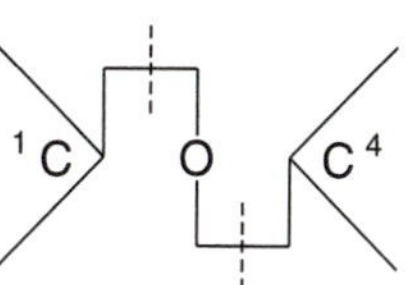

Figure 9-2
Possible cleavage sites for lysozyme.

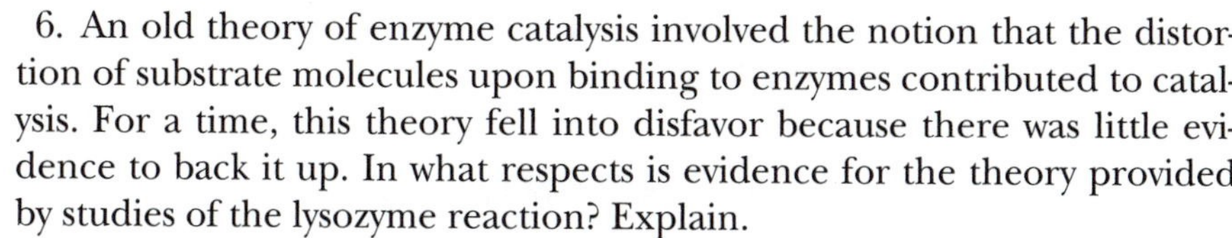

6. An old theory of enzyme catalysis involved the notion that the distortion of substrate molecules upon binding to enzymes contributed to catalysis. For a time, this theory fell into disfavor because there was little evidence to back it up. In what respects is evidence for the theory provided by studies of the lysozyme reaction? Explain.

7. Predict the cleavage products that would be obtained when the following nucleotide is cleaved with pancreatic ribonuclease:

pApUpGpUpCpGpGp

(See Stryer, pp. 216, 217, for the mechanism of ribonuclease and p. 77 for the conventions used in abbreviating the structure of nucleotides.)

8. Suppose that the nucleotide UpGp were cleaved with pancreatic ribonuclease in the presence of H_2O labeled with ^{18}O. Give the products that would be formed, and indicate which product would contain ^{18}O. Explain your answer.

9. Trypsin, chymotrypsin, and carboxypeptidase A fail to cleave peptide bonds involving proline. Trypsin, for example, will not cleave at Lys-Pro. Why do you think this is the case?

10. Place slash marks at the sites where you would expect chymotrypsin to cleave the following peptide:

Lys-Gly-Phe-Thr-Tyr-Pro-Asn-Trp-Ser-Tyr-Phe

11. A competitive inhibitor of chymotrypsin, β-phenylpropionate, protects histidine 57 against alkylation by the affinity-labeling reagent TPCK (see Stryer, p. 224). Noncompetitive inhibitors do not provide such protection. Explain.

12. Many enzymes can be protected against thermal denaturation during purification procedures by the addition of substrate. Propose an explanation for this phenomenon.

13. Transmission at many synapses in the central nervous system, including those between motor nerves and muscles, is mediated by the neurotransmitter acetylcholine. Acetylcholine is released from vesicles at the nerve endings; it diffuses across synapses to the postsynaptic membrane, where it triggers a response and is then cleaved to acetate and choline by the enzyme acetylcholinesterase:

$$CH_3\overset{O}{\overset{\|}{C}}OCH_2CH_2{-}\overset{+}{N}(CH_3)_3$$

Acetylcholine

$$\text{Acetylcholine} + H_2O \xrightleftharpoons{\text{Acetylcholinesterase}} \text{acetate} + \text{choline}$$

(a) Acetylcholinesterase is inactivated by DIPF (see Stryer, p. 224) on a mole-for-mole basis, just as chymotrypsin is. (Because of its

potent effects on acetylcholinesterase, DIPF is also called *nerve gas.*) What deduction may be made about the mechanism of acetylcholinesterase?
(b) Give a two-step reaction mechanism for acetylcholinesterase, involving an acyl-enzyme intermediate.

14. The enzyme rhodanese is abundant in mammalian liver and kidney tissue. It can catalyze the transfer of a sulfur atom from thiosulfate to cyanide to yield sulfite and thiocyanate:

$$S_2O_3{}^{2-} + CN^- \rightleftharpoons SO_3{}^- + SCN^-$$

Although this reaction is valuable for studying the mechanism of rhodanese catalysis in the laboratory, it probably does not occur to a significant extent in tissues. (Cyanide, after all, is a deadly poison.) Rather, some sulfur acceptor other than cyanide serves as the major substrate in the cell.

(a) The given reaction involves an interaction between ions with like charges. One effect that an enzyme could have would be to divide the reaction into steps so that no such unfavorable interaction would occur. It was found experimentally that when enzyme was incubated with thiosulfate labeled with ^{35}S, a stable enzyme-sulfur compound could be isolated. Give a two-step reaction mechanism for rhodanese that involves such an enzyme-sulfur intermediate.
(b) One might expect the negatively charged thiosulfate to bind to the enzyme electrostatically. In an effort to demonstrate such an interaction, neutral salts like NaCl were added to the enzyme assay system to determine whether they had any effect on thiosulfate binding. It was found that the addition of such salts decreased the ability of thiosulfate to bind? Does this support the hypothesis of electrostatic interaction? Explain.

15. Most drugs used in the management of disease, such as those that inhibit HIV-1 protease (see Stryer p. 230), are members of what class of enzyme inhibitors? Give the hallmark of this type of inhibition, and think of one advantage and one disadvantage to the use of such compounds as therapeutic agents.

ANSWERS TO PROBLEMS

1. No. If these residues were in helical regions of lysozyme, the peptide hydrogen and the carbonyl oxygen would be involved in the hydrogen bonds that maintain the α helix. They would therefore be unavailable for hydrogen bonding with the substrate.

2. Cell wall polysaccharides are alternating copolymers of *N*-acetylglucosamine (NAG) and *N*-acetylmuramate (NAM). One would therefore expect that groups on NAM are also important in binding to lysozyme. In addition, residues D, E, and F of the substrate contribute to binding, with D making a negative contribution, and E and F positive contributions (see Stryer, Figure 9-49, p. 235). The major advantage of using tri-NAG for

mapping the interactions that bind substrate to lysozyme is that it is not cleaved at an appreciable rate. Tri-NAG therefore remains stably bound in the cleft of lysozyme, where it can be detected by the difference Fourier method (see Stryer, p. 210).

3. (a) If the P–O bond is broken, the label would be incorporated into inorganic phosphate.

   (b) If the C–O bond is broken, the label would be incorporated into glucose.

   (c) Since the label is incorporated exclusively into inorganic phosphate, the P–O bond must have been broken. Such cleavage is typical of reactions involving phosphatases and is therefore called *phosphatase-type cleavage.*

4. (a) For Asp,

$$\mathrm{pH} = \mathrm{p}K + \log\frac{[\mathrm{A^-}]}{[\mathrm{HA}]}$$

$$5.0 = 3.9 + \log\frac{[\mathrm{COO^-}]}{[\mathrm{COOH}]}$$

$$1.1 = \log\frac{[\mathrm{COO^-}]}{[\mathrm{COOH}]}$$

$$12.6 = \frac{[\mathrm{COO^-}]}{[\mathrm{COOH}]}$$

   The percentage of groups in the $-\mathrm{COO^-}$ form is

$$\frac{12.6}{13.6} \times 100 = 92.6\%$$

   (b) For Glu,

$$\mathrm{pH} = \mathrm{p}K + \log\frac{[\mathrm{A^-}]}{[\mathrm{HA}]}$$

$$5.0 = 4.3 + \log\frac{[\mathrm{COO^-}]}{[\mathrm{COOH}]}$$

$$0.7 = \log\frac{[\mathrm{COO^-}]}{[\mathrm{COOH}]}$$

$$5 = \frac{[\mathrm{COO^-}]}{[\mathrm{COOH}]}$$

   The percentage of groups in the –COOH form is

$$\tfrac{1}{6} \times 100 = 16.7\%$$

(c) Because Glu 35 is present in a hydrophobic region of lysozyme, the attraction between H^+ and $-COO^-$ will be significantly greater than when Glu is free in solution, and the pK for the group will be considerably greater than 4.3. As a result, the percentage of $-COOH$ groups will be greater than calculated in part (b) of this problem. (See Stryer, p. 7, for a discussion of the effects of dielectric constants on ionic equilibria.)

5. Carbonium ions are planar, having sp^2 hybridization. A carbonium ion can be formed only if the C–O bond involving the C-1 of residue D is broken. If the C–O bond involving the C-4 of residue E were cleaved, a planar intermediate could not be involved.

6. Evidence for the theory comes from the experimental observation that hexa-NAG fits best into the active-site cleft of lysozyme when sugar residue D is distorted into a half-chair conformation. The half-chair conformation facilitates the formation of a planar carbonium ion involving the C-1 of sugar residue D. (See Stryer, p. 214, and problem 5 here.) This is substantiated by experiments showing that sugar residue D contributes negatively to the standard free energy of binding of the substrate. (See Stryer, Figure 9-49, p. 235.)

7. The points of cleavage are indicated by the arrows:

pApUpGpUpCpGpGp
↑ ↑ ↑

The products would be pApUp, GpUp, Cp, and GpGp.

8. The products would be Up and Gp. The ^{18}O would appear in Up. The bond that is broken is that between P and the 5′-O atom of G. See Figure 9-3.

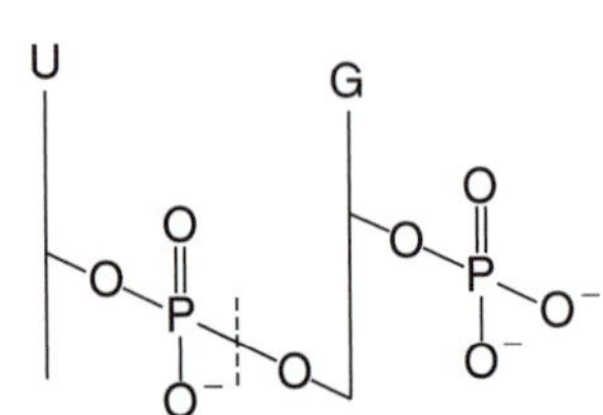

Figure 9-3
Bond cleaved by pancreatic ribonuclease.

9. Because of its ring structure, proline cannot be accommodated in the substrate binding sites of trypsin, chymotrypsin, and carboxypeptidase A. Therefore, these proteases do not cleave peptide bonds involving proline.

10. Chymotrypsin would produce the following four fragments:

Lys-Gly-Phe Thr-Tyr-Pro-Asn-Trp Ser-Tyr Phe

The Tyr-Pro bond is not cleaved. (See problem 9.)

11. The competitive inhibitor β-phenylpropionate and the affinity-labeling reagent TPCK both bind at the same site on the enzyme, namely, the substrate binding site, so the binding of the competitive inhibitor would prevent the binding of TPCK. Noncompetitive inhibitors combine with chymotrypsin at a site on the enzyme away from the substrate binding site. Therefore, noncompetitive inhibitors do not prevent the binding of TPCK to chymotrypsin.

12. When the substrate occupies a cleft in the enzyme, the weak bonds that it forms with groups on the enzyme help to stabilize the tertiary structure of the enzyme. This helps to protect the enzyme against thermal denaturation.

13. (a) Ser is involved. DIFP specifically combines with serine.

(b) Step 1:

$$\text{E-Ser-OH} + \text{acetylcholine} \rightleftharpoons \text{E-Ser-O}\overset{\displaystyle O}{\overset{\|}{C}}\text{CH}_3 + \text{choline}$$

Step 2:

$$\text{E-Ser-O}\overset{\displaystyle O}{\overset{\|}{C}}\text{CH}_3 + H_2O \rightleftharpoons \text{E-Ser-OH} + \text{acetate}$$

14. (a) Step 1:

$$S_2O_3^{2-} + E \rightleftharpoons \text{E-S} + SO_3^{2-}$$

Step 2:

$$\text{E-S} + CN^- \rightleftharpoons E + SCN^-$$

(b) Yes, it does support the hypothesis of electrostatic interaction. If an interaction between a negatively charged thiosulfate and a positively charged functional group on the enzyme occurred, the addition of a neutral salt would weaken that interaction by competition.

15. Virtually all drugs used in the management of disease are competitive inhibitors of enzymatic reactions. (The HIV-1 protease inhibitors described by Stryer are synthetic transition state analogs of the regions of HIV polyprotein cleaved by HIV-1 protease. These synthetic analogs bind tightly to the substrate binding site of HIV-1 protease and therefore preclude the binding of the native substrate, HIV polyprotein.) A characteristic of competitive inhibition is that it may be overcome by an increase in the [S]/[I] ratio (see Stryer, p. 197). Thus competitive inhibition of cellular processes is temporary and will be relieved as the concentration of the "native" substrate increases. The fact that drug therapy is temporary has obvious advantages in some instances, e.g., in the administration of local anaesthetics by a dentist, or in the mistaken administration of a drug. The disadvantage is that all drugs are somewhat less efficient than they would be if they were not competitive inhibitors.

EXPANDED SOLUTIONS TO TEXT PROBLEMS

1. Lysozyme hydrolyzes the glycosidic bond between the C-1 of NAM (M) and the C-4 of NAG (G). Since cleavage occurs between the D and E sugar binding sites (see Stryer, p. 212), G-M-G-M-G-M [(b)] should be expected to be hydrolyzed rapidly to G-M-G-M and G-M. M-G-M-G-M-G [(c)] will have more difficulty fitting into the active site, so it will be hydrolyzed more slowly. M-M-M-M-M-M [(a)], which contains only the bulkier NAM, will have the least access to the active site, so it will have the lowest rate of hydrolysis.

2. (a) G-G will prefer sites B-C because they have binding affinities that have the largest negative $\Delta G^{\circ\prime}$ (see Stryer, p. 235).

(b) Since NAM does not fit site C and since site D has a negative binding affinity, G-M will prefer sites A-B or E-F.

(c) G-G-G-G will prefer sites A-B-C. One sugar residue does not interact with the enzyme, thus avoiding site D, which is energetically unfavorable.

3. During hydrolysis by lysozyme, the oxygen of water appears in the OH group on the anomeric carbon of NAM (see Stryer, p. 212). Therefore, the glycosidic O (the ^{18}O) is in the OH on the C-4 of NAG after hydrolysis.

4. This analog lacks a bulky substituent at C-5, and so it can probably bind to site D without being strained. Consequently, the binding of residue D of this analog is likely to be energetically favorable, whereas the binding of residue D of tetra-NAG costs free energy. See P. van Eikeren and D.M. Chipman, *J. Amer. Chem. Soc.* 94(1972):4788.

5. (a) In oxymyoglobin, Fe is bonded to five nitrogens and one oxygen. In carboxypeptidase A, Zn is bonded to two nitrogens and two oxygens.

(b) In oxymyoglobin, one of the nitrogens bonded to Fe comes from the proximal histidine residue, whereas the other four come from the heme. The oxygen atom linked to Fe is that of O_2. In carboxypeptidase A, the two nitrogen atoms coordinated to Zn come from histidine residues. One of the oxygen ligands is from a glutamate side chain, the other from a water molecule.

(c) Aspartate, cysteine, and methionine. The aspartate side chain is very similar to that of glutamate; hence, it can bind Zn^{2+} or Fe^{2+}. Since sulfur atoms have an affinity for heavy metal ions, the side chains of cysteine and methionine can also bind these metal ions.

6. If tyrosine 248 were essential for catalysis, a mutant of this enzyme containing phenylalanine in its place would be inactive. The tyrosine 248 codon (TAT) was converted to that of phenylalanine (TTT) by oligonucleotide-directed mutagenesis (p. 139). The recombinant plasmid containing this gene was inserted into yeast and expressed. The striking finding was that the mutant enzyme had the same k_{cat} value as did the native enzyme but a K_M value that was sixfold higher. This experiment showed that tyrosine 248 participates in the binding of substrate but is not essential for catalysis. Moreover, it graphically illustrated the power of site-specific mutagenesis in delineating the function of a particular residue in a protein. See S.J. Gardell, C.S. Craik, D. Hilvert, M.S. Urdea, and W.J. Rutter. *Nature* 317(1985):551–555.

7. The essential sulfhydryl in its active site should be highly susceptible to chemical modification. The enzyme should be inactivated by reacting it with iodoacetamide, *N*-ethylmaleimide, or another sulfhydryl-specific reagent. Furthermore, bound substrate is likely to protect the key sulfhydryl from alkylation.

8. (a) Yes, each of the mechanisms involves the transfer of a proton from an enzyme to substrate.

(b) Examination of the appropriate figures in Stryer provides the information that glutamic acid 35 is the proton donor in lysozyme (Figure 9-11), histidine 119 is the donor in ribonuclease A (Figure 9-21), a zinc-bound water molecule is the donor in carboxypeptidase A (Figure 9-27), and histidine 57 is the donor in chymotrypsin (Figures 9-34, 37).

9. (a) Remember that chymotrypsin hydrolyzes peptide chains on the carboxyl side of amino acids having aromatic side chains (e.g., phenylalanine), whereas trypsin hydrolyzes peptide bonds on the carboxyl side of lysine and arginine residues. As shown in Figure 9-32, the affinity labeling molecule used to study chymotrypsin is TPCK, a derivative of phenylalanine. Therefore, it would be logical to use a similar derivative of lysine (tosyl-L-lysine chloromethyl ketone) for similar studies with trypsin.

(b) If the lysine derivative (TLCK) specifically binds at the active site of trypsin, substrates should protect trypsin from inactivation by TLCK and the D isomer of TLCK should not inactivate trypsin.

10. Aldehydes can react with one molecule of an alcohol to form a hemiacetal (see p. 466, Chapter 18, Stryer). Since the catalytic site of elastase contains an active serine hydroxyl group, it is reasonable to suppose that an aldehyde derivative of one of the substrates of elastase would react with the serine hydroxyl to form a hemiacetal and thus inhibit the enzyme.

11. A dimer of identical subunits is advantageous for two reasons. First, less RNA is needed to encode the protease. A small genome can be more readily replicated and packaged than a large one. Second, the protease does not become active until a critical concentration is attained. Premature excision of viral subunits from the polyprotein is thereby prevented. (See Stryer, p. 229.)

12. The monomer of HIV-1 protease is inactive; that is, a dimer is *required* for activity. Therefore, if one can prevent dimerization, one can effectively block protease action.

13. Its sulfhydryl group probably binds to the zinc ion at the active site, and its carboxylate group very likely occupies the recognition site for the C terminus of the substrate.

14. Precise positioning of catalytic residues and substrates, geometrical strain (distortion of the substrate), electronic polarization, and desolvation of the substrate.

CHAPTER

10

Regulatory Strategies

The theme of Chapter 10 is the regulation of enzyme activity. Four of the major types of regulatory mechanisms are discussed in detail: allosteric control, reversible covalent modification, proteolytic activation, and stimulation or inhibition by control proteins. Stryer uses specific examples to illustrate the general structure-function relationships involved in these control mechanisms. To illuminate allosteric control, Stryer discusses *E. coli* aspartate transcarbamoylase, a well-understood enzyme. You were introduced to allosteric interactions in Chapters 7 and 8, where oxygen binding by hemoglobin and its regulation by pH, CO_2, and 2,3-bisphosphoglycerate were described, and two theoretical models of allosteric regulation were presented. After the section on allosteric control, Stryer illustrates the regulation of enzymes by reversible phosphorylation with the example of protein kinase A. Next, Stryer turns to the activation of enzymes by proteolytic cleavage. He describes the proteolytic steps and conformational rearrangements that produce the active forms of chymotrypsin, trypsin, and pepsin from their inactive zymogens. The mechanisms of action of the digestive enzymes were presented in Chapter 9. Stryer concludes Chapter 10 with a discussion of the blood clotting cascade—the series of proteolytic activations of clotting factors that lead to the formation of fibrin clots. Several specific stimulating and inhibiting proteins are described in connection with the proteolytic enzymes.

LEARNING OBJECTIVES

When you have mastered this chapter, you should be able to complete the following objectives:

Introduction (Stryer pages 237–238)

1. List the four major regulatory mechanisms that control enzyme activity and give examples of each.

Allosteric Interactions in Aspartate Transcarbamoylase (Stryer pages 238–244)

2. Describe the reaction catalyzed by *aspartate transcarbamoylase* (*ATCase*), the regulation of ATCase by *CTP* and *ATP,* and the biological significance of this regulation.

3. Describe the subunit composition of ATCase. Explain the effects of subunit dissociation on the *allosteric behavior* of the enzyme.

4. Describe the three-dimensional structure of ATCase in terms of the subunit arrangement and the major features of its active site as revealed by the binding of N-(*phosphonacetyl*)-L-*aspartate* (*PALA*).

5. Outline the structural effects of the binding of PALA, CTP, or ATP on ATCase.

6. Describe the experimental evidence for a concerted allosteric transition during the binding of substrate analogs to ATCase.

7. Outline the effects of *heterotropic* and *homotropic* allosteric interactions on the equilibrium between the T and R forms of ATCase.

Regulation of Enzymes by Phosphorylation (Stryer pages 244–247)

8. Write the basic reactions catalyzed by *protein kinases* and *protein phosphatases.*

9. List the reasons why phosphorylation is such an effective control mechanism.

10. Describe the activation of *protein kinase A* (*PKA*) by *cyclic AMP* (*cAMP*) and the mode of interaction of PKA with its *pseudosubstrate.*

Proteolytic Activation of Enzymes (Stryer pages 247–252)

11. Define *zymogen* (*proenzyme*) and *proprotein.* Give examples of enzymes and proteins that are derived from zymogens and proproteins.

12. Outline the pathway for the secretion of *pancreatic zymogens.*

13. Describe the activation of *chymotrypsinogen* and the general structural changes that lead to the formation of the active site.

14. Describe the activation of *trypsinogen* and *pepsinogen;* compare the conformational changes that occur during the activation of the three major pancreatic zymogens.

15. Summarize the enzymes and conditions required for the activation of all the *digestive enzymes.*

16. Explain how *trypsin* is inhibited by the *pancreatic trypsin inhibitor.*

17. Describe the role of α_1-*antitrypsin* in the inactivation of *elastase.* Relate it to the clinical condition called *emphysema.*

The Cascade of Zymogen Activations in Blood Clotting
(Stryer pages 252–257)

18. Explain the biological significance of the *enzymatic cascade* in *clotting.*

19. Contrast the structure and properties of *fibrinogen* with those of *fibrin.*

20. Describe the stabilization of the *fibrin clot* by *Factor* XIII_a, an active *transamidase.* Explain the role of *thrombin* in the activation of fibrinogen and *Factor XIII.*

21. Compare the properties of thrombin with those of the *pancreatic serine proteases.*

22. Discuss the requirement for *vitamin K* in the synthesis of *prothrombin.* Outline the mechanism of prothrombin activation.

23. Describe the genetic defect in *hemophilia.* Explain how recombinant DNA technology has been used to produce human *Factor VIII* (antihemophilic factor).

Inhibition of Clotting Factors (Stryer pages 257–258)

24. Describe the general mechanisms for the control of clotting.

25. Explain the specific role of *antithrombin III* in blocking the serine proteases in the clotting cascade. Note the effect of *heparin* on antithrombin III.

26. Describe the *lysis* of fibrin clots by *plasmin* and the activation of *plasminogen* by *tissue-type plasminogen activator* (*TPA*). Note the modular structure of TPA.

SELF-TEST

Allosteric Interactions in Aspartate Transcarbamoylase

1. The dependence of the reaction velocity on the substrate concentration for an allosteric enzyme is shown in Figure 10-1 as curve A. A shift to curve B could be caused by the

(a) addition of an irreversible inhibitor.
(b) addition of an allosteric activator.
(c) addition of an allosteric inhibitor.
(d) dissociation of the enzyme into subunits.

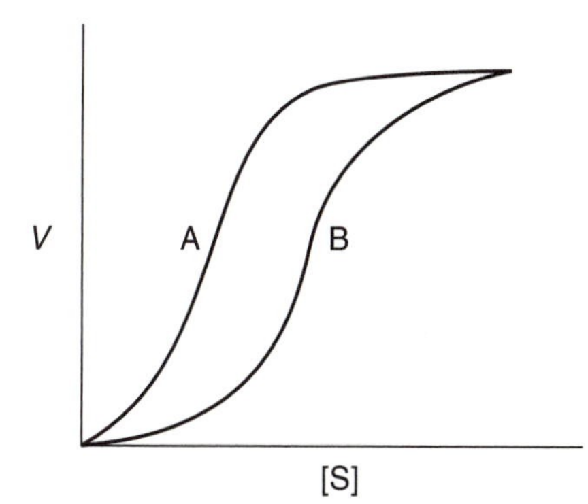

Figure 10-1
Allosteric enzyme kinetics.

2. In *E. coli,* ATCase is inhibited by CTP and is activated by ATP. Explain the biological significance of these effects.

____

____

3. Which of the following statements regarding the structure of ATCase in *E. coli* are incorrect?

(a) ATCase consists of two kinds of subunits and a total of twelve polypeptide chains.
(b) Reaction with mercurials dissociates each ATCase into three r_2 and two c_3 subunits.
(c) ATCase has a threefold axis of symmetry and a large inner cavity.
(d) The active sites of ATCase are located at the interface between c and r subunits.
(e) The separate subunits r_2 and c_3 retain their respective ligand-binding capacities.

4. Which of the following methods can provide information about the subunit dissociation of ATCase or the structural changes that occur when ATCase binds a substrate analog?

(a) X-ray crystallography
(b) Western blotting
(c) Sedimentation-velocity ultracentrifugation
(d) SDS-polyacrylamide gel electrophoresis
(e) Gel-filtration chromatography

5. Explain why the reagent *N*-(phosphonacetyl)-L-aspartate (PALA) has been especially useful in the investigation of the properties of ATCase.

____

____

6. The allosteric effect of CTP on ATCase is called

(a) homotropic activation.
(b) homotropic inhibition.
(c) heterotropic activation.
(d) heterotropic inhibition.

7. Explain how PALA can act as both an activator and an inhibitor of ATCase.

____

____

8. Match the possible experimental observation in the left column with the binding of the appropriate molecule to ATCase.

(a) Increase in the sedimentation coefficient ____________
(b) Decrease in absorbance at 430 nm of catalytic subunits containing nitrotyrosine ____________
(c) Increase in absorbance at 430 nm of catalytic subunits containing nitrotyrosine ____________

(1) ATP
(2) CTP
(3) PALA
(4) Succinate

9. Protein kinases

(a) transfer a phosphoryl group from one protein to another.
(b) use AMP as a substrate.
(c) use Thr, Ser, or Tyr as the acceptor groups for phosphoryl transfer.
(d) transfer the α phosphorus atom of ATP.
(e) are located on the external surface of cells.

10. Explain how a phosphoryl group can change the conformation of a protein.

___

___

11. Select the amino acid sequences of target proteins that may be phosphorylated by protein kinase A.

(a) RRGAI
(b) RAGSR
(c) RRGSI
(d) KRGTI
(e) RRGGI

12. Protein kinase A

(a) is activated by ATP.
(b) consists of two catalytic (C) and two regulatory (R) subunits in the absence of activator.
(c) upon binding the activator dissociates into one C_2 and two R subunits.
(d) contains a pseudosubstrate sequence in the C subunits.

Proteolytic Activation of Enzymes

13. Which of the following proteins are synthesized and secreted as zymogens or proproteins?

(a) Pepsin
(b) Lysozyme
(c) Collagen
(d) Enteropeptidase
(e) Thrombin

14. The pancreas is the source of the proteolytic enzyme trypsin. Which of the following are reasons trypsin does not digest the tissue in which it is produced?

(a) It is synthesized in the form of an inactive precursor that requires activation.
(b) It is stored in zymogen granules that are enclosed by a membrane.
(c) It is active only at the pH of the intestine, not at the pH of the pancreatic cells.
(d) It requires a specific noncatalytic modifier protein in order to become active.

15. Activation of chymotrypsinogen requires

(a) the cleavage of at least two peptide bonds by trypsin.
(b) structural rearrangements that complete the formation of the substrate cavity and the oxyanion hole.

(c) major structural rearrangements of the entire protein molecule.
(d) the concerted proteolytic action of trypsin and pepsin to give α-chymotrypsin.

16. Match the zymogens in the left column with the enzymes that participate directly in their activation or with their mechanism of activation, which are listed in the right column.

(a) Chymotrypsinogen ____________
(b) Pepsinogen ____________
(c) Trypsinogen ____________
(d) Proelastase ____________
(e) Procarboxypeptidase ____________

(1) Intramolecular hydrolysis
(2) Trypsin
(3) Enteropeptidase
(4) Carboxypeptidase

17. Explain why the new carboxyl-terminal residues of the polypeptide chains produced during the activation of pancreatic zymogens are usually Arg or Lys.

____

____

18. The inactivation of trypsin by pancreatic trypsin inhibitor involves

(a) an allosteric inhibition.
(b) the covalent binding of a phosphate to the active site serine.
(c) the facilitated self-digestion of the enzyme.
(d) denaturation at the alkaline pH of the duodenum.
(e) the nearly irreversible binding of the protein inhibitor at the active site.

19. Explain the effects of an α_1-antitrypsin deficiency on the lungs.

____

____

The Cascade of Zymogen Activations in Blood Clotting

20. Match fibrinogen and fibrin with the appropriate properties in the right column.

(a) Fibrinogen ____________
(b) Fibrin ____________

(1) Is soluble in blood
(2) Is insoluble in blood
(3) Forms ordered fibrous arrays
(4) Contains α-helical coiled coils
(5) Contains tryosine-*O*-sulfate groups
(6) Has the lower pI of the two
(7) May be cross-linked by transamidase

21. Which of the following amino acid residues are involved in the transamidase reaction catalyzed by Factor XIII_a?

(a) Asparagine
(b) Glutamine
(c) Glutamate
(d) Lysine
(e) Arginine

22. Thrombin and trypsin are both serine proteases that are capable of cleaving the peptide bond on the carboxyl side of arginine; thrombin, however, is specific for Arg-Gly bonds. Describe briefly the similarities and differences in the active sites of these two enzymes.

____

____

23. Which of the following statements about prothrombin are incorrect?

(a) It requires vitamin K for its synthesis.
(b) It can be converted to thrombin by the decarboxylation of γ-carboxyglutamate residues.
(c) It is activated by Factor IX_a and Factor VIII.
(d) It is anchored to platelet phospholipid membranes through Ca^{2+} bridges.
(e) It is part of the common pathway of clotting.

24. Explain the role of the γ-carboxyglutamate residues found in clotting factors.

____

____

Inhibition of Clotting Factors

25. Which of the following mechanisms is not involved in the control of the clotting process?

(a) The specific inhibition of fibrin formation by antielastase
(b) The degradation of Factors V_a and VIII_a by protein C, which is in turn switched on by thrombin
(c) The dilution of clotting factors in the blood and their removal by the liver
(d) The specific inhibition of thrombin by antithrombin III

26. Explain the effects of each of the following substances on blood coagulation or clot dissolution:

(a) Heparin: ____

____

(b) Dicumarol: ____

____

(c) Tissue-type plasminogen activator: ____

____

27. Which of the following statements about plasmin are true?

    (a) It is a serine protease.
    (b) It diffuses into clots.
    (c) It cleaves fibrin at connector rod regions.
    (d) It is inactivated by α_1-antitrypsin.
    (e) It contains a "kringle" region in its structure for binding to clots.

28. The antihemophilic factor (Factor VIII) is a

    (a) zymogen.
    (b) protease.
    (c) calcium-binding protein.
    (d) activating protein.
    (e) protein kinase.

29. Which of the following is not a structural domain found in TPA?

    (a) α-helical coiled-coil domain
    (b) Kringle domain
    (c) Serine protease domain
    (d) Growth factor domain
    (e) Finger domain

ANSWERS TO SELF-TEST

Allosteric Interactions in Aspartate Transcarbamoylase

1. c

2. The activation of ATCase by ATP occurs when metabolic energy is available for DNA replication and the synthesis of pyrimidine nucleotides. Feedback inhibition by CTP prevents the overproduction of pyrimidine nucleotides and the waste of precursors.

3. d

4. a, c, e

5. PALA is a bisubstrate analog; that is, it resembles a combination of both substrates, and it is a transition state analog for the carbamoyl phosphate–aspartate complex during catalysis by ATCase. X-ray diffraction analysis of ATCase with bound PALA has revealed the location of the active site and interactions that occur within it. In addition, comparisons of structures with and without PALA have indicated the large structural changes that ATCase undergoes upon binding substrates.

6. d

7. PALA is an analog of both substrates of ATCase; therefore, it binds to the active site and acts as a potent inhibitor. At low concentrations, however, binding of PALA shifts the distribution of ATCase molecules to the R conformation. This increases binding of substrates and enzymatic activity.

8. (a) 1, 3, 4 (b) 2 (c) 1, 3, 4

9. c

10. A phosphoryl group introduces two negative charges that can affect the electrostatic interactions within the protein. In addition, a phosphoryl group can form three highly directional hydrogen bonds to adjacent H-bond partners in the protein. These local effects can be transmitted to more distant parts of the protein in a manner similar to allosteric effects.

11. c, d. Lysine (K) can substitute for one of the R residues in the consensus phosphorylation sequences recognized by PKA.

12. b

Proteolytic Activation of Enzymes

13. a, c, e

14. a, b

15. b

16. (a) 2 (b) 1 (c) 3, 2 (d) 2 (e) 2

17. Because trypsin is the common activator of the pancreatic zymogens, its specificity for Arg-X and Lys-X peptide bonds will produce Arg and Lys carboxyl-terminal residues.

18. e

19. α_1-antitrypsin specifically binds to and inactivates elastase. When the activity of elastase is not effectively controlled, it can destroy the elastic proteins of the alveoli of the lungs, leading to emphysema.

The Cascade of Zymogen Activations in Blood Clotting

20. (a) 1, 4, 5, 6 (b) 2, 3, 4, 7. Fibrinogen has a lower pI than fibrin because fibrinogen contains many more negatively charged amino acid residues than does fibrin.

21. b, d

22. Because both thrombin and trypsin are serine proteases, they both have an oxyanion hole and a catalytic triad at the active site. Also, the substrate-specificity sites of both have a similar, negatively charged pocket capable of binding Arg. However, thrombin probably has just enough space to accommodate a Gly residue next to the Arg binding site in contrast to trypsin, which has no restrictions as to the amino acid residue that can be accommodated at the corresponding position.

23. b, c

24. The γ-carboxyglutamate residues are effective chelators of Ca^{2+}. This Ca^{2+} is the electrostatic anchor that binds the protein to a phospholipid membrane, thereby bringing interdependent clotting factors into close proximity.

Inhibition of Clotting Factors

25. a

26. (a) Heparin enhances the inhibitory action of antithrombin III.

    (b) Dicumarol is a vitamin K analog, and as such, it interferes with the synthesis of the factors that contain γ-carboxyglutamate residues.

    (c) Tissue-type plasminogen activator facilitates clot dissolution by converting plasminogen into plasmin directly on the clot.

27. a, b, c

28. d

29. a

PROBLEMS

1. What would be the kinetic consequences if a substrate were to have exactly equal affinities for the R form and the T form of an allosteric enzyme? (See the discussion of the Monod-Wyman-Changeux model on pp. 166–167, and the section on the kinetics of allosteric enzymes on pp. 198–199 of Stryer.)

2. In a spectroscopic study designed to elucidate the mechanism of the allosteric transition in ATCase, hybrid enzyme molecules were formed containing (in addition to regulatory subunits) three native catalytic subunits, and three modified catalytic subunits. The modified catalytic subunits contained the nitrotyrosine reporter group and a modified lysine that precluded substrate binding. Why was this modified lysine necessary to the experiment?

3. Predict whether each of the following peptide sequences is likely to be phosphorylated by protein kinase A. Briefly explain your answers, and indicate which residue would be phosphorylated.

    (a) Ala-Arg-Arg-Ala-Ser-Leu
    (b) Ala-Arg-Arg-Ala-His-Leu
    (c) Val-Arg-Arg-Trp-Thr-Leu
    (d) Ala-Arg-Arg-Gly-Ser-Asp
    (e) Gly-Arg-Arg-Ala-Thr-Ile

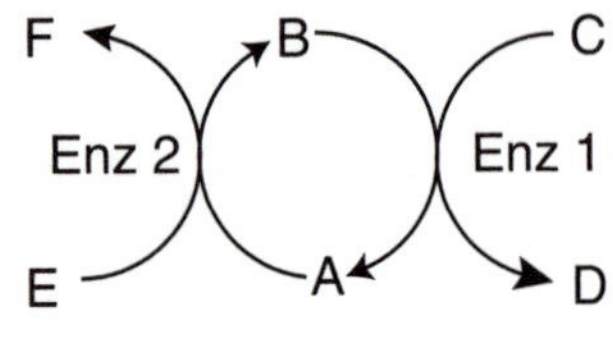

Figure 10-2

4. Consider the hypothetical metabolic sequence shown in Figure 10-2. Suppose it is known that protein kinase A phosphorylates both enzyme 1 and enzyme 2, and that an increase in intracellular cAMP levels increases the steady-state [B]/[A] ratio. In order for a given increment in cAMP concentration to result in the largest change in the steady-state [B]/[A] ratio, what should be the effect of phosphorylation on the activities of enzymes 1 and 2?

5. Figure 10-3 shows time courses for the activation of two zymogens, I and II. Which of the time courses most resembles that of the activation of trypsinogen, and which corresponds to the activation of chymotrypsinogen? Explain.

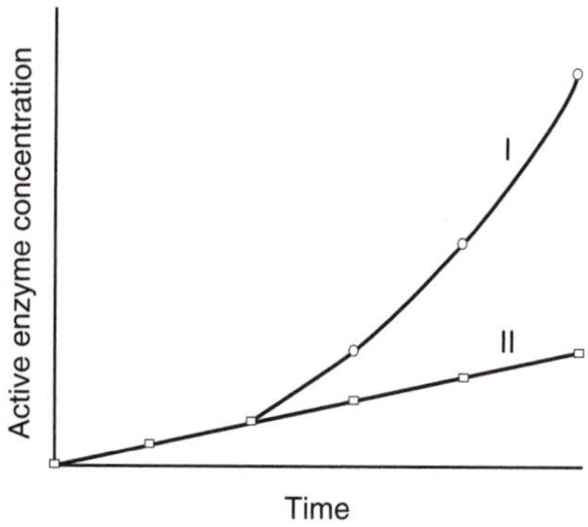

Figure 10-3
Time courses for the activation of two zymogens, I and II.

6. What is one possible significance of the fact that amino acid residues 2, 3, 4, and 5 of trypsinogen are all aspartate?

7. Trypsin has 13 lysine and 2 arginine residues in its primary structure. Why does trypsin not cleave itself into 16 smaller peptides?

8. Although thrombin has many properties in common with trypsin, the conversion of prothrombin to thrombin is not autocatalytic whereas the conversion of trypsinogen to trypsin is autocatalytic. Why is the conversion of prothrombin to thrombin not autocatalytic?

9. A patient suffering from a very rare blood disorder has shown a propensity for thrombosis since late childhood. Normal levels of plasminogen were detected in the plasma of the patient by immunoreactive techniques. Give two possible mechanisms that might account for the defect in this patient.

10. Because many clotting factors are present in blood in small concentrations, direct chemical measurements often cannot be used to determine whether the factors are within normal concentration ranges or are deficient. Once a deficiency has been established, however, plasma from the affected person can be used to screen for the presence of the deficiency in other people. A rare deficiency in Factor XII leads to a prolongation of clotting time. Assuming that you have plasma from someone in which this deficiency has been established, design a test that might help determine whether another person has a Factor XII deficiency.

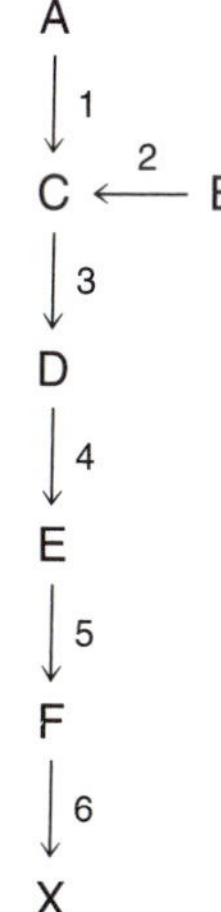

Figure 10-4
A hypothetical metabolic pathway.

11. Aspartate transcarbamoylase catalyzes the first step in the biosynthetic pathway leading to the synthesis of cytidine triphosphate (CTP). (See Stryer, p. 238.) CTP serves as an allosteric inhibitor of aspartate transcarbamoylase that shuts off the biosynthetic pathway when the cell has an ample supply of CTP. Although the first step in a pathway may often be the principal regulatory step, such is not always the case. Figure 10-4 shows a hypothetical degradative metabolic pathway in which step 3 is the principal regulatory step. In this pathway, what advantage does regulation at step 3 have over regulation at steps 1 or 2?

12. In general, regulatory enzymes catalyze reactions that are irreversible in cells, that is, reactions that are far from equilibrium. Why must this be the case?

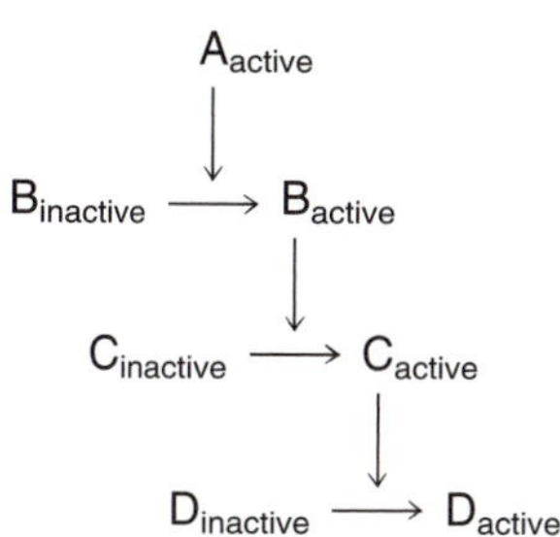

Figure 10-5
A hypothetical regulatory cascade.

13. Amplification cascades, such as the one involved in blood clotting, are important in a number of regulatory processes. Figure 10-5 shows a hypothetical cascade involving conversions between inactive and active forms of enzymes. Active enzyme A serves as a catalyst for the activation of enzyme B. Active B in turn activates C, and so forth. Assume that each enzyme in the pathway has a turnover number of 10^3. How many molecules of enzyme D will be activated per unit time when one molecule of active enzyme A is produced per unit time?

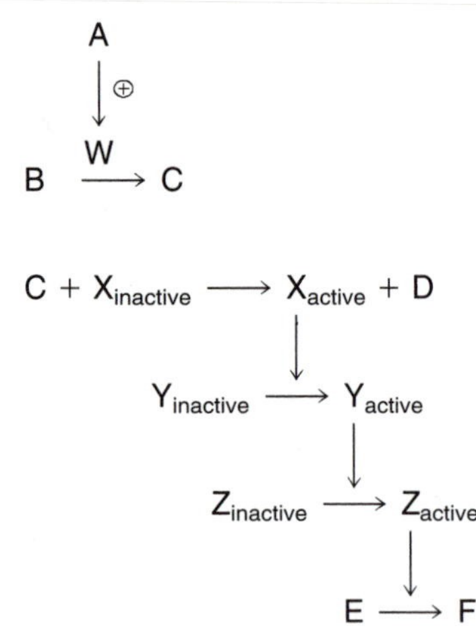

Figure 10-6
A hypothetical regulatory cascade.

14. Another hypothetical scheme for an amplification cascade is shown in Figure 10-6. The letters W, X, Y, and Z represent enzymes, whereas the letters A, B, C, D, E, and F represent metabolites. Compound A is an allosteric activator of enzyme W, which converts B to C. Compound C in turn combines stoichiometrically with the inactive form of enzyme X to give active X and compound D. Active enzyme X activates enzyme Y, which in turn activates Z. Enzyme Z catalyzes the conversion of compound E to compound F. Assuming that each enzyme in the scheme has a turnover number of 10^3, calculate the number of molecules of E that are converted to F per unit time when one molecule of A is added to the system.

15. In retroviral replication, a polyprotein containing a protease and a glycoprotein as well as reverse transcriptase is synthesized. This polyprotein cleaves itself into three separate proteins in an autocatalytic reaction. In principle, the protease could act in either an intra- or an intermolecular fashion. One way to determine which of these options occurs is to study the rate of protease cleavage as a function of the protein concentration. What would the results of such an experiment show if the cleavage mechanism were intra- rather than intermolecular?

ANSWERS TO PROBLEMS

1. If a substrate were to have equal affinities for the R and T forms, the forms would be indistinguishable kinetically and the system would behave as if all the enzyme were present in a single form. Thus, Michaelis-Menten kinetics would apply, and a plot of the reaction velocity versus the substrate concentration would be hyperbolic.

2. The point of the experiment was to show that the subunits containing the reporter group undergo conformational change because substrate is bound to a neighboring subunit of the enzyme. Thus binding of substrate to the subunit containing the reporter group had to be precluded if the experiment was to give meaningful results.

3. The consensus motif recognized by protein kinase A is Arg-Arg-X-Ser-Z, where X is a small residue and Z is a large hydrophobic residue. The site of phosphorylation is either Ser or Thr. (See Stryer, p. 245.)

(a) Ser would be phosphorylated.
(b) There would be no phosphorylation because neither Ser nor Thr is present.
(c) There would be no phosphorylation because Trp is a bulky group and residue X must be small.
(d) There would be no phosphorylation because Asp is polar and charged and residue Z must be hydrophobic.
(e) Thr would be phosphorylated.

4. Phosphorylation should increase the activity of enzyme 2 and decrease the activity of enzyme 1. That being the case, an increase in intracellular cAMP levels could greatly increase the steady-state ratio of [B]/[A]. Such coordinated, reciprocal control of opposing metabolic sequences is observed frequently in cells.

5. Curve I corresponds to the activation of trypsinogen, a process that is autocatalytic. As the process occurs, trypsin is produced, which can then cleave yet more trypsinogen. Curve II corresponds to the activation of chymotrypsinogen. The activation of chymotrypsinogen is not autocatalytic. Rather, tryspin catalyzes the conversion of chymotrypsinogen to active π-chymotrypsin. Therefore, its time course is initially linear.

6. The conversion of trypsinogen to trypsin involves the removal of an N-terminal hexapeptide (see Stryer, p. 250). The run of polar aspartate residues at positions 2 through 5 of trypsinogen ensures that the N terminus of the polypeptide is at the surface of the zymogen in a nonhelical region where it may be cleaved by either enteropeptidase or trypsin.

7. The lysine and arginine residues must be partially buried and inaccessible to the active site of trypsin.

8. Thrombin specifically cleaves Arg-Gly bonds. The two bonds that are broken when prothrombin is converted to thrombin are Arg-Thr and Arg-Ile. Therefore, the conversion cannot be autocatalytic.

9. Because the patient has a tendency to form thrombi (blood clots), a defect in fibrinolysis is likely. The fact that normal levels of plasminogen are present suggests either (1) that there is a defect in the conversion of plasminogen to plasmin or (2) that the plasmin that is produced is enzymatically inactive and fails to lyse fibrinogen. In the actual case cited, the second possibility turned out to be the correct one; however, the data presented in the problem are not sufficient for discrimination between the two alternatives.

10. Prepare two samples of blood from the person to be tested. Add normal plasma to one sample and Factor XII–deficient plasma to another. If clotting time is restored to normal in both samples, Factor XII deficiency is probably not involved. If the addition of normal plasma restores normal clotting time but the addition of Factor XII–deficient plasma does not, then a Factor XII deficiency must be suspected.

11. The pathway shown in Figure 10-4 is branched. If regulation were to occur at step 1 only, there would be no control over the production of X from B. If only step 2 were regulated, there would be no regulation over the production of X from A. Regulation at step 3 provides control of the amount of X produced from both A and B. In branched pathways, the principal regulatory step is usually after the branch point.

12. Suppose that a reaction is at equilibrium. If the enzyme catalyzing that reaction were made more active, nothing would happen. The reaction would still be at equilibrium. If, on the other hand, the reaction is displaced far from equilibrium and the enzyme catalyzing the reaction is made more active, more product will be produced. Thus, a regulatory enzyme must catalyze an irreversible step if it is to increase the flux rate through a pathway when it is allosterically activated.

13. One molecule of active A will lead to the activation of 10^9 molecules of enzyme D per unit time. Active A will produce 10^3 molecules of active B. Each of the 10^3 molecules of active B will activate 10^3 molecules of C per unit time. Since there are 10^3 molecules of B, this gives a total of 10^6

molecules of active C. Similar reasoning leads to the answer of 10^9 molecules of active D.

14. One molecule of A leads to the production of 10^{12} molecules of F per unit time. There are 10^3 molecules of C produced when one molecule of A is added to the system. For each molecule of C, there is one molecule of active enzyme X produced, since the interaction of C with X is stoichiometric and not catalytic. Since there are 10^3 molecules of C, there will be 10^3 molecules of active X, leading to 10^6 molecules of active Y, 10^9 molecules of active Z, and 10^{12} molecules of F.

15. If the protease were acting via an intramolecular mechanism, then the rate of the cleavage reaction would be independent of protein concentration. If the reaction were intermolecular, then the rate would vary as some function of protein concentration. Results of such experiments are consistent with an intramolecular mechanism of cleavage.

EXPANDED SOLUTIONS TO TEXT PROBLEMS

1. Since histidine 134 is thought to stabilize the negative charge on the carbonyl oxygen in the transition state (Figure 10-7), the protonated imidazole ring (which carries a positive charge) must be the active species. That being the case, the enzyme velocity V should be half of V_{max} at a pH of 6.5 (the pK of an unperturbed histidine side chain in a protein). Raising the pH above 6.5 will remove protons from the imidazole ring, thus causing a decrease in V; lowering the pH below 6.5 will have a reverse effect.

2. (a) Using the same logic as that used in problem 8 of Chapter 7 in Stryer's text, one can show that the change in [R]/[T] is the same as the ratio of the substrate affinities of the two forms. For example, the mathematical constant for the conversion of R to T_S is the same whether one proceeds $R \rightarrow T \rightarrow T_S \rightarrow$ or $R \rightarrow R_S \rightarrow T_S$. Let us assume that the constant for the conversion of R to T and R to R_S is 10^3. Since the affinity of R for S is 100 times that of T for S, it follows that the constant for the conversion of T to T_S is 10. The constant for the conversion of R_S to T_S is therefore equal to $10^3 \times 10/10^3$, or 10. Note that the binding of substrate with a 100-fold tighter binding to R changes the R to T ratio from 1/1000 to 1/10.

   (b) Since the binding of *one* substrate molecule changes the [R]/[T] by a factor of 100, the binding of *four* substrate molecules will change the [R]/[T] by a factor of $100^4 = 10^8$. If [R]/[T] in the absence of substrate = 10^{-7}, the ratio in the fully liganded molecule will be $10^8 \times 10^{-7} = 10$.

3. The concerted model, in contrast with the induced-fit (sequential) model, cannot account for negative cooperativity, because according to this model, the binding of substrate promotes a conformational transition of all subunits to the high-affinity R state. Hence, homotropic allosteric interactions must be cooperative if the concerted model holds. As ex-

plained in Stryer, Chapter 7 (p. 166 and Figure 7-37), in the sequential (induced-fit) model, the binding of ligand changes the conformation of the subunit to which it is bound but not that of its neighbors. This conformational change in one subunit can *increase* or *decrease* the binding affinity of other subunits in the same molecule and can, therefore, account for negative cooperativity.

4. The binding of PALA switches ATCase from the T to the R state because it acts as a substrate analog. An enzyme molecule containing bound PALA has fewer free catalytic sites than does an unoccupied enzyme molecule. However, the PALA-containing enzyme will be in the R state and hence have higher affinity for the substrates. The dependence of the degree of activation on the concentration of PALA is a complex function of the allosteric constant L_0, and of the binding affinities of the R and T states for the analog and the substrates. For an account of this experiment, see J. Foote and H.K. Schachman, *J. Mol. Biol.* 186(1985):175.

5. The activation of zymogens involves the cleavage of one or more peptide bonds. In the case of pepsinogen (see Stryer, pp. 250–251), when the catalytic site is exposed by lowering the pH, it hydrolyzes the peptide bond between the percursor and pepsin moities. Note that this activation is *autocatalytic.* Therefore, the time required for activation of half the pepsinogen molecules is independent of the total number of the molecules present.

6. If both patients have a Factor VIII deficiency, a mixture of the two bloods will not clot. However, if the second patient's bleeding disorder is due to the deficiency of another factor, a mixture of the two bloods should clot. This type of assay is called a *complementation test.*

7. The function of Factor X is to convert prothrombin to thrombin on phospholipid membranes derived from blood platelets. This proteolytic activation removes the amino-terminal fragment of prothrombin, which contains Ca^{+2}-binding sites, and releases thrombin to activate fibrinogen. Meanwhile, Factor X remains bound to the platelet membrane, where it can activate other prothrombin molecules, because during activation it retains the Ca^{+2}-binding γ-carboxyglutamate residues.

8. Apparently antithrombin III is a very poor substrate for thrombin. Remember, many enzyme inhibitors have high affinity for active sites. Thrombin, not prothrombin, can react with antithrombin III because it has available a fully formed active site.

9. One needs to remember α-helical coiled coils, introduced by Stryer in Chapter 2 (p. 30). Examination of Figure 2-34 suggests that near the axis of the superhelix some amino acid residues are located in the interior (hydrophobic) portion of the molecule. Since this is a long molecule made up of repeating units, one would expect to have hydrophobic side chains at regular intervals in the molecule.

10. Methionine 358 has a side chain that not only is essential for the binding of elastase by α_1-antitrypsin but also is most susceptible to oxidation by cigarette smoke. What is needed is a side chain resistant to oxidation yet having a strong binding affinity for elastase. A likely choice would be leucine, the side chain of which is much more stable than that of methionine but which has nearly the same volume and is very hydrophobic.

CHAPTER

11

Membrane Structure and Dynamics

In this chapter, Stryer describes the composition, structural organization, and general functions of biological membranes. After listing the common features of these membranes, Stryer introduces a new class of biomolecules, the lipids, in their roles as membrane components. The amphipathic structures of membrane lipids and their organization into bilayers in water are described first. He describes the selective permeability of a lipid bilayer and how it can be overcome by ion-carrying and channel-forming antibiotics. Next, Stryer turns to membrane proteins, the major functional constituents of biological membranes. He describes the arrangement of proteins and lipids in membranes and stresses the asymmetric, fluid nature of membranes. Stryer then discusses the erythrocyte membrane and its relationship to the cytoskeleton. He concludes the chapter by commenting on the high-resolution analyses of the structures of selected membrane proteins. The most useful background material for this chapter can be found in the first three chapters: intermolecular forces in Chapter 1, protein structure in Chapter 2, and the methods used in the investigation of proteins in Chapter 3. More detail on the structures, properties, and nomenclature of the fatty acids is found on pages 603–605.

LEARNING OBJECTIVES

When you have mastered this chapter, you should be able to complete the following objectives:

Introduction (Stryer pages 263–264)

1. List the functions of *biological membranes.*

2. Describe the common features of membranes.

Membrane Lipids (Stryer pages 264–268)

3. Define *lipid* and list the major kinds of *membrane lipids.*

4. Recognize the structures and the constituent parts of *phospholipids* (*phosphoglycerides* and *sphingomyelin*), *glycolipids,* and *cholesterol.*

5. Describe the general properties of the *fatty acid chains* found in phospholipids and glycolipids.

6. Draw the general chemical formula of a phosphoglyceride, and recognize the most common *alcohol moieties* of phosphoglycerides (e.g., *choline, ethanolamine,* and *glycerol*).

Lipid Bilayers—Structure and Properties (Stryer pages 268–274)

7. Describe the properties of an *amphipathic molecule* and the symbol used to represent it.

8. Distinguish between *oriented monolayers, micelles,* and *lipid bilayers.*

9. Describe the *self-assembly process* for the formation of lipid bilayers. Note the stabilizing intermolecular forces.

10. Outline the methods used to prepare *lipid vesicles* (*liposomes*) and *planar bilayer membranes.* Point out some applications of these systems.

11. Explain the relationship between the *permeability coefficients* of small molecules and ions and their *solubility* in a nonpolar solvent relative to their solubility in water.

12. Contrast the properties of *carrier* and *channel-forming transport antibiotics.*

13. Discuss the general properties and uses of the *transport antibiotics valinomycin* and *gramicidin A.*

14. Describe the structural basis for the binding and translocation of ions by the *carrier antibiotics.*

15. Describe the structural basis for the membrane permeability–inducing activity of the channel-forming antibiotic gramicidin A. Note that membrane transport proteins form *transmembrane channels.*

Composition, Organization, and Dynamics of Membranes (Stryer pages 275–282))

16. Give examples of the compositional and functional varieties of biological membranes.

17. Distinguish between *peripheral* and *integral membrane proteins.* Describe the use of *freeze-fracture electron microscopy* in visualizing integral membrane proteins.

18. Describe the evidence for the *lateral diffusion* of membrane lipids and proteins. Contrast the rates for *lateral diffusion* with those for *transverse diffusion.*

19. Describe the features of the *fluid mosaic model* of biological membranes.

20. Discuss the origin and the significance of *membrane asymmetry.*

21. Explain the roles of the fatty acid chains of membrane lipids and cholesterol in controlling the *fluidity of membranes.*

22. Note the roles and location of the *carbohydrate components* of membranes.

23. Describe the *reconstitution* of functional membrane systems from purified components and explain the role of *detergents* in this process.

The Red-Cell Membrane Proteins and Other Membrane Proteins
(Stryer pages 282–287)

24. Discuss the advantages of using *erythrocytes* in studies of membranes. Give examples of the membrane proteins found in erythrocytes.

25. Describe the structure of *glycophorin A* and the properties of the *anion channel.* Explain how *transmembrane* α *helices* can be predicted from *hydropathy plots.*

26. Explain the role of *spectrin* in the formation of the *membrane skeleton* of an erythrocyte.

27. Discuss the use of electron-microscopic and x-ray analyses in obtaining high-resolution images of the structures of *bacteriorhodopsin* and the *photosynthetic reaction center.*

SELF-TEST

Introduction

1. Which of the following statements about biological membranes are true?

(a) They constitute selectively permeable boundaries between cells and their environment and between intracellular compartments.
(b) They are formed primarily of lipid and carbohydrate.
(c) They sometimes generate signals.
(d) They carry out the important biological energy conversion processes of photosynthesis and oxidative phosphorylation.
(e) They sometimes contain receptors for specific stimuli.

2. Identify the functions in the right column as being associated either with lipids or with proteins, the major components of cell membranes.

(a) Lipids ______________

(b) Proteins ______________

(1) Are specific receptors for external stimuli
(2) Form a permeability barrier
(3) Act as pumps and channels
(4) Determine the fluidity of membranes
(5) Carry out concerted enzymatic reactions

3. Which of the following statements about biological membranes is not true?

(a) They contain carbohydrates that are covalently bound to proteins and lipids.
(b) They are very large, sheetlike structures with closed boundaries.
(c) They are symmetric because of the symmetric nature of lipid bilayers.
(d) They can be regarded as two-dimensional solutions of oriented proteins and lipids.
(e) They contain specific proteins that mediate their distinctive functions.

Membrane Lipids

4. Which of the following are membrane lipids?

(a) Cholesterol
(b) Glycerol
(c) Phosphoglycerides
(d) Choline
(e) Cerebrosides

5. After examining the structural formulas of the four lipids in Figure 11-1, answer the following questions:

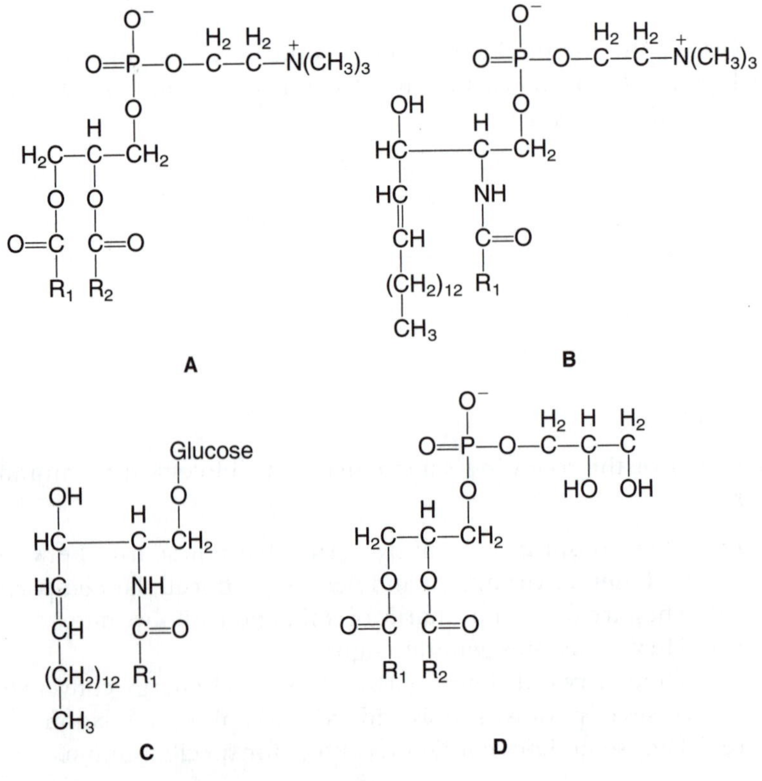

Figure 11-1
Membrane lipids R_1 and R_2 represent hydrocarbon chains.

(a) Which are phosphoglycerides? ____________

(b) Which is a glycolipid? ____________

(c) Which contain sphingosine? ____________

(d) Which contain choline? ____________

(e) Which contain glycerol? ____________

(f) Name the lipids:
A ________________
B ________________
C ________________
D ________________

(g) Draw the structures of the components—that is, the possible hydrolysis products—of lipid D. Name the components and indicate how many moles of each are present per mole of lipid.

6. Calculate the molecular weight of a phosphatidyl choline with a palmitic acid esterified to the C-1 and an oleic acid esterified to the C-2 carbon of the glycerol backbone. Calculate the percentage of the molecular weight of the lipid that is due to the fatty acid chains.

Lipid Bilayers—Structure and Properties

7. The phosphoinositol portion of the phosphatidyl inositol molecule is called which of the following?

(a) The amphipathic moiety
(b) The hydrophobic moiety
(c) The hydrophilic moiety
(d) The micelle
(e) The polar head group

8. Which of the following is not a feature that a micelle and a lipid bilayer have in common?

(a) Both assemble spontaneously in water.
(b) Both are made up of amphipathic molecules.
(c) Both are very large, sheetlike structures.
(d) Both have the thickness of two constituent molecules in one of their dimensions.
(e) Both are stabilized by hydrophobic interactions, van der Waals forces, hydrogen bonds, and electrostatic interactions.

9. A triglyceride (triacylglycerol) is a glycerol derivative that is similar to a phosphoglyceride, except that all three of its glycerol hydroxyl groups are esterified to fatty acid chains. Would you expect a triglyceride to form a lipid bilayer? Explain.

____

____

10. What is the volume of the inner water compartment of a liposome that has a diameter of 500 Å and a bilayer that is 40 Å thick?

(a) 5.6×10^5 Å^3
(b) 7.3×10^6 Å^3
(c) 3.9×10^7 Å^3
(d) 7.3×10^7 Å^3
(e) 3.9×10^8 Å^3

11. What features of liposomes make them potentially useful as a delivery system for transporting water-soluble drugs to target cells? Suggest how one could prepare a liposome that is specific for a particular type of cell.

____

____

12. Arrange the following in the order of decreasing permeability through a lipid bilayer:

(a) Isoleucine ________
(b) Tyrosine ________
(c) H_2O ________
(d) Na^+ ________
(e) O_2 ________

13. Match each of the transport antibiotics in the left column with the appropriate properties from the right column.

(a) Valinomycin ________
(b) Gramicidin A ________

(1) Is a channel former
(2) Is an ion carrier
(3) Is a cyclic peptide
(4) Is a linear peptide
(5) Has alternating L and D amino acid residues
(6) Is active in transport as a dimer
(7) Has a transmembrane diffusion constant for ions of around 10^7 s^{-1}
(8) Has a transmembrane diffusion constant for ions of around 10^3 s^{-1}

14. Explain the consequences of the alternating D and L amino acid residues on the structure of gramicidin A.

____

____

15. Show which of the properties listed on the right are characteristics of peripheral membrane proteins and which are characteristics of integral membrane proteins.

(a) Peripheral ____________

(b) Integral ____________

(1) Require detergents or organic solvent treatment for dissociation from the membrane
(2) Require mild salt or pH treatment for dissociation from the membrane
(3) Appear in freeze-fracture planes as globular particles
(4) Bind to the surface of membranes
(5) Have transmembrane domains

16. Which of the following statements about the diffusion of lipids and proteins in membranes is not true?

(a) Many membrane proteins can diffuse rapidly in the plane of the membrane.
(b) In general, lipids show a faster lateral diffusion than do proteins.
(c) Membrane proteins do not diffuse across membranes at measurable rates.
(d) Lipids diffuse across and in the plane of the membrane at equal rates.

17. Why do membrane proteins not diffuse, that is, flip-flop, across membranes?

____

____

18. In a membrane, an integral membrane protein diffuses laterally an average distance of 4×10^{-6} m in 1 minute, whereas a phospholipid molecule diffuses an average distance of 2 μm in 1 second.

(a) Calculate the ratio of the diffusion rates for these two membrane components.

____

(b) Provide reasons for the difference between the two rates.

____

19. Which of the following statements about the asymmetry of membranes are true?

(a) It is absolute for glycoproteins.
(b) It is absolute for phospholipids but is only partial for glycolipids.
(c) It arises during biosynthesis.
(d) It is structural but not functional.

20. If phosphoglyceride A has a higher T_m than phosphoglyceride B, which of the following differences between A and B may exist? (In each

case only one parameter—either chain length or double bonds—is compared.)

(a) A has shorter fatty acid chains than B.
(b) A has longer fatty acid chains than B.
(c) A has more unsaturated fatty acid chains than B.
(d) A has more saturated fatty acid chains than B.
(e) A has trans unsaturated fatty acid chains whereas B has cis unsaturated fatty acid chains.

21. Explain the role of cholesterol in cell membranes.

____

____

22. Which of the following statements about the carbohydrate components of eukaryotic cell membranes are true?

(a) They make up approximately one-third of the weight of the membrane.
(b) They are exposed to the interior of the cell.
(c) They can be recognized by lectins.
(d) They bind to membrane proteins and lipids via hydrogen bonds.
(e) They contribute to and help maintain the asymmetry of the membranes.

23. During the solubilization of membranes, the purification of integral membrane proteins, and the reconstitution of membranes, gentler detergents, such as octyl glucoside, are used in preference to sodium dodecyl sulfate (SDS). Explain why.

____

____

24. Explain how the mobility of a protein in a membrane might be restricted far beyond what a simple consideration of its native molecular weight would lead you to conclude.

____

____

The Red-Cell Membrane Proteins and Other Membrane Proteins

25. Several properties of glycophorin A are listed in the left column. Match these properties with the methods listed in the right column that can be used to demonstrate them.

(a) It is an integral membrane protein. ____________
(b) It has one transmembrane α-helical domain of 19 amino acids. ____________
(c) It has a cytoplasmic C terminus. ____________
(d) It is a glycoprotein. ____________
(e) It has a molecular weight lower than that of spectrin. ____________

(1) SDS-polyacrylamide gel electrophoresis
(2) Lectin binding
(3) Hydropathy plot
(4) Extraction with detergent
(5) PAS staining
(6) Binding of ferritin-labeled specific antibodies

26. Which of the following statements about the anion channel (the band 3 protein) is incorrect?

 (a) It facilitates the exchange of HCO_3^- for Cl^-.
 (b) It comprises about one-third of the total membrane protein.
 (c) Its C-terminal half contains the single transmembrane helical domain.
 (d) It has the peripheral protein ankyrin associated with its cytosolic domain.

27. Why is an α helix the preferred structure for transmembrane protein segments?

____

____

28. Which of the following answers complete the sentence correctly? Spectrin

 (a) is an integral membrane protein.
 (b) is a glycoprotein.
 (c) helps to maintain the shape and integrity of erythrocytes.
 (d) binds to plasma membrane components through ankyrin and the band 4.1 protein.
 (e) consists of two polypeptide chains, each of which is organized into multiple, triple-stranded, α-helical coiled coils.

29. Which of the following characteristics do bacteriorhodopsin and the photosynthetic reaction center have in common?

 (a) Compact, essentially globular shapes
 (b) The same number of transmembrane segments
 (c) Similar lengths for the transmembrane helices
 (d) Similar sizes of the hydrophilic domains
 (e) Similar orientations of the transmembrane helices with respect to the plane of the membrane

ANSWERS TO SELF-TEST

Introduction

1. a, c, d, e

2. (a) 2, 4 (b) 1, 3, 5

3. c

Membrane Lipids

4. a, c, e

5. (a) A, D (b) C (c) B, C (d) A, B (e) A, D (f) A is phosphatidyl choline, B is sphingomyelin, C is cerebroside, and D is phosphatidyl glycerol

 (g) $R_1{-}C({=}O){-}O^-$ and $R_2{-}C({=}O){-}O^-$ — **Fatty acids** (2 moles); $H_2C{-}OH$ / $HC{-}OH$ / $H_2C{-}OH$ — **Glycerol** (2 moles); $^-O{-}P({=}O)(OH){-}O^-$ — **Phosphate** (1 mole)

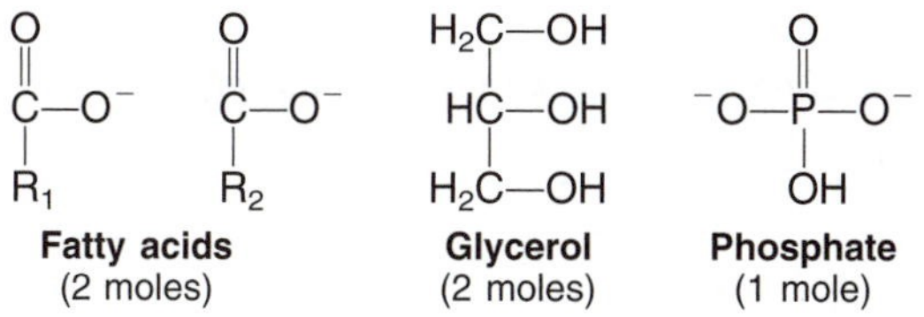

6. The total molecular weight of the molecule is 759 daltons; the fatty acid chains account for 71% of the molecular weight. The molecule contains a total of 42 carbon atoms, 8 oxygen atoms, 82 hydrogen atoms, 1 phosphorus atom, and 1 nitrogen atom, giving a molecular weight of 759. The oleoyl chain has a molecular weight of 281 daltons, and the palmitoyl chain has a molecular weight of 255 daltons. The percentage of the molecular weight of the phospholipid due to the fatty acids is therefore

$$\frac{281 + 255}{759} \times 100\% = 70.6\%$$

Lipid Bilayers—Structure and Properties

7. c, e

8. c

9. No. Although a triglyceride has hydrophobic fatty acyl chains attached to a glycerol backbone, it lacks a polar head group; therefore, it is not an amphipathic molecule and is incapable of forming a bilayer.

10. c. The radius of the water compartment is

$$\frac{500\ \text{Å} - (2 \times 40\ \text{Å})}{2} = 210\ \text{Å}$$

and the volume of a sphere is $\frac{4}{3}\pi r^3$

11. Liposomes are essentially impermeable to water-soluble molecules. Therefore, water-soluble drugs could be trapped inside the liposomes and then be delivered into the target cells by fusing the liposomes with the cell membrane. To make a liposome specific for a particular type of cell, antibodies that have been prepared against a surface protein of the target cell could be attached to the liposome via a covalent bond with a bilayer lipid, for example, phosphatidyl ethanolamine. This would enable the liposome to recognize the target cells. Of course, strategies would also have to be devised to prevent the premature, nonspecific fusion of the liposome with other cells.

12. e, c, a, b, d. Though not discussed in the text, a lipid bilayer is even more permeable for O_2 than for water because O_2 lacks a significant dipole and is a smaller molecule. Also, recall that Ile is more hydrophobic than Tyr (see Stryer, p. 21 and Table 11-2).

13. (a) 2, 3, 8 (b) 1, 4, 5, 6, 7

14. The alternating sequence of D and L amino acid residues in gramicidin A allows the formation of a helical channel with all the nonpolar side chains facing out toward the hydrophobic interior of the membrane. The polar peptide groups face the interior of the channel and transiently bind the ions as they move through it.

Composition, Organization, and Dynamics of Membranes

15. (a) 2, 4 (b) 1, 3, 5

16. d

17. Membrane proteins are very bulky molecules that contain numerous charged amino acid residues and polar sugar groups (in the case of glycoproteins) that are highly hydrated. Such molecules do not diffuse through the hydrophobic interior of the lipid bilayer.

18. (a) Rate of protein diffusion:

$$4 \times 10^{-6}\ \text{m/min} = \frac{1\ \text{min}}{60\ \text{s}} = 6.7 \times 10^{-8}\ \text{m/s}$$

Rate of phospholipid diffusion:

$$2\ \mu\text{m/s} = 2 \times 10^{-6}\ \text{m/s}$$

Ratio of phospholipid diffusion rate to protein diffusion rate:

$$\frac{2 \times 10^{-6}\ \text{m/s}}{6.7 \times 10^{-8}\ \text{m/s}} = 30$$

(b) The difference in diffusion rates is due primarily to the difference in mass between phospholipids, which have a molecular weight of approximately 800, and proteins, which have a molecular weight greater than 10,000. In addition, integral membrane proteins may associate with peripheral proteins, which would further decrease their lateral diffusion. See also question 21.

19. a, c

20. b, d, e. Trans unsaturated fatty acid chains have a straighter conformation than do cis unsaturated chains; the packing of trans chains in bilayers is therefore more highly ordered, so they require higher temperatures to melt.

21. Cholesterol modulates the fluidity of membranes. By inserting itself between the fatty acid chains, cholesterol prevents their "crystallization" at temperatures below T_m and sterically blocks large motions of the fatty acid chains at temperatures above T_m. In fact, high concentrations of cholesterol abolish phase transitions of bilayers. This modulating effect of cholesterol maintains the fluidity of membranes in the range required for biological function.

22. c, e

23. Although sodium dodecyl sulfate (SDS) is a very effective detergent for solubilizing membrane components, the strong electrostatic interactions of its polar head groups with charged groups on the membrane proteins disrupt protein structure. A detergent such as octyl glucoside, which has an uncharged head group, allows the proteins to retain their three-dimensional structures while it interacts with their hydrophobic domains. Recall the use of SDS as a protein denaturant in the determination of protein molecular weights (see Chapter 3) and in the separation of membrane proteins (see Figure 11-25 in Stryer).

24. The protein could be free to move in the membrane; alternatively, it could be anchored to the underlying cytoskeleton, rendering it virtually fixed in place. Also, see question 18.

The Red-Cell Membrane Proteins and Other Membrane Proteins

25. (a) 4 (b) 3 (c) 6 (d) 2, 5 (e) 1

26. c

27. Transmembrane protein segments usually consist of nonpolar amino acids. The main-chain peptide CO and NH groups, however, are polar and tend to form hydrogen bonds with water. In an α helix, these groups hydrogen-bond to each other, thereby decreasing their overall polarity and facilitating the insertion of the protein segment into the lipid bilayer.

28. c, d, e

29. a, c, e

PROBLEMS

1. The ability of bacteria, yeasts, and fungi to convert aliphatic hydrocarbons to carbon dioxide and water has been studied intensively over the past decade because of concerns about the effects of crude oil spills on the environment. Microorganisms cannot survive when they are placed in high concentrations of crude oil or any of its components. However, they can utilize hydrocarbons very efficiently when they are placed in a medium in which an extensive lipid-water interface is created by agitation and aeration. Why?

2. Phytol, a long-chain alcohol, appears as an ester in plant chlorophyll. When consumed as part of the diet, phytol is converted to phytanic acid.

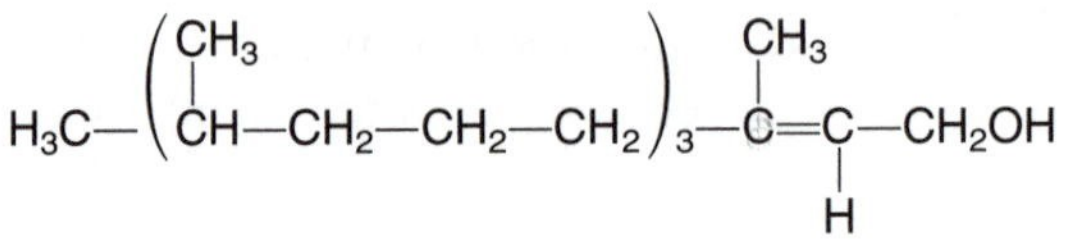

$$H_3C-\left(\overset{CH_3}{\overset{|}{CH}}-CH_2-CH_2-CH_2\right)_3-\overset{CH_3}{\overset{|}{C}}=\underset{\underset{H}{|}}{C}-CH_2OH$$

Phytol

$$H_3C-\left(\overset{CH_3}{\overset{|}{CH}}-CH_2-CH_2-CH_2\right)_3-\overset{CH_3}{\overset{|}{CH}}-CH_2-C\begin{smallmatrix}\nearrow O\\ \searrow O^-\end{smallmatrix}$$

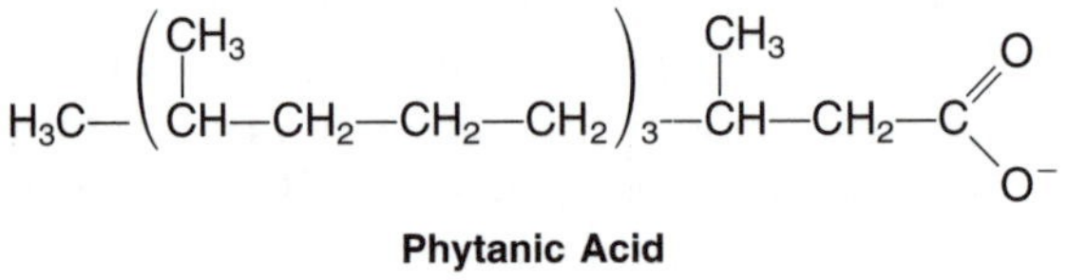

Phytanic Acid

People who cannot oxidize phytanic acid suffer from a number of neurological disorders that together are known as *Refsum's disease.* The symptoms may be related to the fact that phytanic acid accumulates in the membranes of nerve cells. What general effects of phytanic acid on these membranes would be observed?

3. Bacterial mutants that are unable to synthesize fatty acids will incorporate them into their membranes when fatty acids are supplied in their growth medium. Suppose that each of two cultures contains a mixture of several types of straight-chain fatty acids, some saturated and some unsaturated, ranging in chain length from 10 to 20 carbon atoms. If one culture is maintained at 18°C and the other is maintained at 40°C over several generations, what differences in the composition of the cell membranes of the two cultures would you expect to observe?

4. The formation of lipid bilayers in a mixture of water and phospholipids is a rapid and spontaneous process of self-assembly. For such a process to be spontaneous, the free-energy change must be negative ($\Delta G < 0$). Given that

$$\Delta G = \Delta H - T\,\Delta S$$

explain why the free-energy change for membrane assembly is negative.

5. Given two bilayer systems, one composed of phospholipids having saturated acyl chains 20 carbons in length and the other having acyl chains of the same length but with cis double bonds at C-5, C-8, C-11, and C-14, compare the effect of the acyl chains on T_m for each system.

6. Hopanoids are pentacyclic molecules that are found in bacteria and in some plants. A typical bacterial hopanoid, bacteriohopanetetrol, is shown in the margin. Compare the structure of this compound with that of cholesterol. What effect would you expect a hopanoid to have on a bacterial membrane?

H_3C H_3C CH_3 CH_3 H_3C H_3C
H_3C—CH
CH_2
CH_2
HC—OH
HO—CH
HC—OH
CH_2
CH_2OH

Bacteriohopanetetrol

7. As early as 1972, it was known that many biological membranes are asymmetric with respect to distribution of phospholipids between the inner and outer leaflets of the bilayer. Once such asymmetry is established, what factors act to preserve it?

8. As discussed on page 279 of Stryer's text, the length and degree of saturation of the fatty acyl chains in membrane bilayers can affect the melting temperature T_m.

(a) The value of T_m for a pure sample of phosphatidyl choline that contains two 12-carbon fatty acyl chains is −1°C. Values for phosphatidyl choline species with longer acyl chains increase by about 20°C for each two-carbon unit added. Why?
(b) Suppose you have a phosphatidyl choline species that has one palmitoyl group esterified to C-1 of the glycerol moiety, as well as an oleoyl group esterified at C-2 of glycerol (see page 266 of Stryer's text). How would T_m for this species compare with that of dipalmitoylphosphatidyl choline, which contains two esterified palmitoyl groups?
(c) Suppose you have a sample of sphingomyelin that has palmitate esterified to the sphingosine backbone. Compare the T_m for this phospholipid with that of dipalmitoylphosphatidyl choline.
(d) The transition temperature for dipalmitoylphosphatidyl ethanolamine is 63°C. Suppose you have a sample of this phospholipid in excess water at 50°C, and you add cholesterol until it constitutes about 50% of the total lipid, by weight, in the sample. What would you expect when you attempt to determine the transition temperature for the mixture?

9. At least two segments of the polypeptide chain of a particular glycoprotein span the membrane of an erythrocyte. All the sugars in the glycoprotein are *O*-linked.

(a) Which amino acids might be found in the portion of the chain that is buried in the lipid bilayer, and how many amino acids could those segments contain?

(b) Why would you expect to find serine or threonine residues in the glycoprotein?

(c) Describe a procedure that would enable you to determine on which side of the erythrocyte membrane the sugar residues are located.

10. (a) Many integral membrane proteins are composed of a number of membrane-spanning segments, which form bundles of α helices packed closely together, often forming a membrane channel or pore. Each membrane-spanning sequence of most integral membrane proteins is an α helix composed of 18 to 20 amino acids. What is the width of the hydrocarbon core of the membrane?

(b) The sequence of one of the α helices in a particular integral membrane protein is shown below, and the 19 residues in this helix are plotted in a helical wheel plot. Such a plot projects the side chains of the amino acid residues along the axis of the α helix (z-axis) onto an x-y plane. In an α helix, a full turn occurs every 3.6 residues, so that each successive residue is 100° apart on the helix wheel. Compare the location of hydrophobic side chains on the helix surface with those that are polar or hydrophilic. Where are the hydrophobic side chains, and how are they accommodated in the membrane? Where are the polar side chains? How are they accommodated in the protein-membrane complex?

Helix sequence:

Ser Val Tyr Asp Ile Leu Glu Arg Phe Asn Glu Thr Met Asn His Ala Val Ser Gly

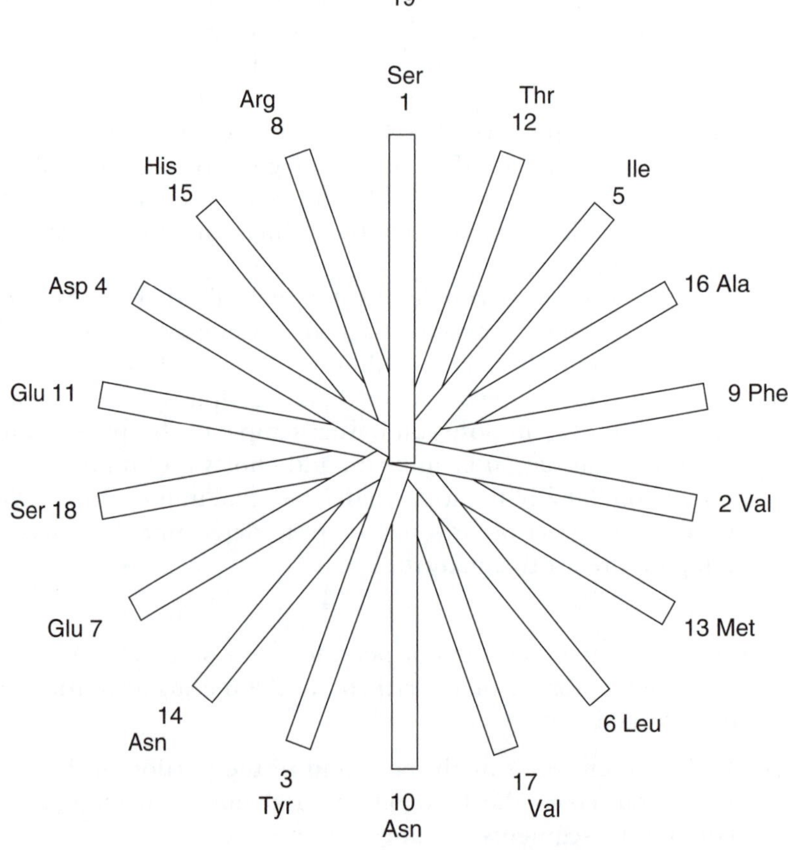

11. A series of experiments that shed some light on the movement of lipids in membranes were conducted by Rothman and Kennedy, using a gram-positive bacterium. They used 2,4,6-trinitrobenzenesulfonic acid (TNBS), which reacts with amino groups in phosphoethanolamine residues. Note that TNBS, shown below, is charged at physiologic pH and cannot penetrate intact membrane vesicles. Incubation of TNBS with intact bacterial cells and with disrupted cells revealed that about two-thirds of the phosphoethanolamine molecules are located on the outside of the membrane, with the remaining residues on the inside. Rothman and Kennedy then incubated growing cells with a pulse of radioactive inorganic phosphate, to label newly synthesized phosphoethanolamine molecules in the membrane. Using TNBS once again to distinguish between residues on the two sides of the membrane, they determined that immediately after the radioactive pulse, all newly synthesized phosphoethanolamine residues were located on the inner face of the membrane. After 30 minutes, however, the original distribution of phosphoethanolamine residues on the inner and outer faces of the bacterial cell membrane was restored. What do these observations suggest about the movement of phospholipids in membranes?

NO_2, O_2N, NO_2, SO_3^- + $H_2N—R$ ⟶ NO_2, O_2N, NO_2, NH—R + SO_3^{2-} + H^+

2,4,6-Trinitrobenzenesulfonic acid (TNBS)

12. Mycoplasma cells can be grown under conditions so that their plasma membrane contains one type of glycolipid, such as mono- or diglucosylated sphingosine molecules.

(a) Membranes prepared from mycoplasma cells undergo a phase transition when heated. Suppose that sample A is isolated from cells whose glycolipids contain a very high percentage of unsaturated fatty acyl chains, while sample B is isolated from cells whose glycolipids contain a high percentage of saturated fatty acyl chains of the same length. When heated, which sample will exhibit a higher melting temperature? Why?

(b) Glycolipids from samples A and B are analyzed for the carbohydrate content of their polar head groups. Those from sample A have a higher percentage of diglucosyl residues than those from sample B, which have mostly monoglucosyl residues. Explain how this observation is consistent with the lipid content of the two samples.

13. In mammals, lysophosphoglycerides (1-monoacylglycerol-3-phosphates) are generated in small quantities in order to trigger physiologic responses. Hydrolysis of a fatty acyl group from the C-2 position of a glycerophospholipid yields a lysophosphoglyceride. The reaction is catalyzed by phospholipase A_2, whose activity is strictly regulated. However, large quantities of phospholipase A_2 are found in snake venom, and the active venom enzyme can generate high concentrations of lysophosphoglycerides from membranes of snakebite victims. Lysophosphoglycerides are so named because in high concentrations they can disrupt membrane structure. Why?

ANSWERS TO PROBLEMS

1. All organisms require water for many biochemical reactions, and thus organisms can live only in an aqueous environment. While a bacterial cell placed in a solution of crude oil might at least survive, it would not have enough water for growth and division. Microorganisms can best utilize crude oil or its hydrocarbon components when the microorganisms are present at a boundary layer between water and lipid. Aeration and agitation increase the effective area of such a layer. Some microorganisms that degrade hydrocarbons have a glycolipid-rich cell wall in which those compounds are soluble. After being solubilized, the compounds are transferred to the cytoplasmic membrane, where water-requiring reactions that initiate hydrocarbon degradation occur.

2. The four methyl side chains of each phytanic acid molecule interfere with the ordered association of fatty acyl chains; thus, they increase the fluidity of nerve cell membranes. This increase in fluidity could interfere with myelin function or ion transport, but the actual molecular basis for the symptoms is not yet known. Many of the symptoms of Refsum's disease can be eliminated by adopting diets that are free of phytol. The primary source of phytol in the human diet is from dairy products and other fats from ruminants. Cows, for example, consume large quantities of chlorophyll as they ingest grasses and plant materials. The symbiotic bacteria that inhabit the bovine rumen readily degrade chlorophyll, releasing free phytol, which is then converted to phytanic acid. Up to 10% of the fatty acids in bovine blood plasma are found as phytanic acid, which can then be incorporated into cell membranes and milk. For those who have Refsum's disease, it is therefore necessary to restrict consumption of beef as well as dairy products like milk and butter. Because humans do not degrade chlorophyll extensively during digestion, restriction of green plants in the diet is usually unnecessary.

3. You would expect to find that the bacteria grown at the higher temperature will have incorporated a higher number of the longer fatty acids and a greater proportion of the saturated fatty acids. The membranes of bacteria grown at 18°C will have more short-chain fatty acids and more that are unsaturated. These cells select fatty acids that will remain fluid at a lower temperature in order to prevent their membranes from becoming too rigid. The cells grown at the higher temperatures can select fatty acids that pack more closely. Cells in both cultures thus employ strategies designed to achieve optimal membrane fluidity.

4. Before the bilayer forms, water must arrange itself in a highly ordered way around the acyl chains of each phospholipid; as the phospholipids begin to assemble, the process is driven by the association of hydrophobic chains and the tendency of water to form a more disordered structure with itself. Thus, the entropy of the system increases as assembly occurs. There are, of course, many more water molecules than phospholipids in the system. Even though the completed bilayer is ordered, the overall system has undergone a large increase in entropy ($\Delta S > 0$) that exceeds the small change in enthalpy and results in a negative value for ΔG.

5. The higher the number of cis double bonds, the less ordered the bilayer structure will be and the more fluid the membrane system will be.

You would therefore expect T_m for the bilayer system containing the acyl chains with four unsaturated bonds to be much lower than that for the system containing the saturated fatty acid chains.

6. Like cholesterol, bacteriohopanetetrol is a pentacyclic molecule with a rigid, platelike, hydrophobic ring structure; it has a hydrophilic region as well, although that region is on the opposite end of the molecule when compared with cholesterol. In bacterial membranes hopanoids may have a function similar to that of cholesterol in mammalian membranes; that is, they may moderate bacterial membrane fluidity by blocking the motion of fatty acyl chains and by preventing their crystallization.

7. Phospholipids have polar head groups, so their transfer across the hydrophobic interior of the bilayer as well as their dissociation from water at the bilayer surface would require a positive change in free energy. Without the input of free energy to make the process a spontaneous one, the transfer of the polar head group is very unlikely, so the asymmetric distribution of the phospholipids is preserved.

8. (a) The longer the acyl groups, the larger the number of noncovalent interactions that can form among the hydrocarbon chains. Higher temperatures are therefore required to disrupt the interactions of phospholipid species that have longer fatty acyl groups.

(b) The cis double bond in oleate produces a bend in the hydrocarbon chain, interfering with the formation of noncovalent bonds between the acyl chains. Less heat energy is therefore required to cause a phase transition; in fact, the melting temperature for phosphatidyl choline with a palmitoyl and an oleoyl unit is –5°C, while T_m for dipalmitoylphosphatidyl choline is 41°C.

(c) As shown in Figures 11-6 and 11-7 of Stryer's text, the structures of phosphatidyl choline and of sphingomyelin are very similar to each other; both contain phosphoryl choline and both have a pair of hydrocarbon chains. Given similar chain lengths in palmitoylspingomyelin and in dipalmitoylphosphatidyl choline, you would expect that values of T_m for the two molecules are similar. Both species in fact exhibit a phase transition at 43°C.

(d) At 50°C, you should expect cholesterol to diminish or even to abolish the transition, by preventing the close packing of the fatty acyl chains that impart rigidity to the molecular assembly. At higher temperatures, cholesterol in the mixture also prevents larger motions of fatty acyl chains, making the assembly less fluid. Studies show that in mixtures containing 30 to 35 mol % cholesterol, phase transitions are extinguished.

9. (a) You should expect to find nonpolar amino acid residues in the portions of the glycoprotein chain that are buried in the membrane. These amino acids could include many of those shown on the left half of Table 11-2 on page 284 of Stryer's text. Because the core portion of the membrane is 30 Å wide, up to 20 amino acids could be included in the buried segments, assuming that the amino acids are part of an α helix, in which the translation distance for each amino acid is 1.5 Å.

(b) The glycoprotein has *O*-linked carbohydrate residues, and in most such proteins the sugars are attached to the side chains of serine or threonine residues. Were the sugars *N*-linked instead, you would expect to find one or more asparagine residues in the glycoprotein.

(c) First, you would use a lectin that binds to the carbohydrate residues of the intact erythrocyte, and then you would determine the number of bound lectins per intact cell. In another experiment, you would lyse the cells and repeat the lectin binding. If the carbohydrate residues are located only on the exterior surface of the cell, as is usually the case, then the number of bound lectins per cell will be the same in each of the two experiments.

10. (a) In an α helix, each amino acid residue extends 1.5 Å (1.5×10^{-1} nm) along the helix axis. Therefore a span of 20 amino acids in an α helix will be about 30 Å in length, corresponding to the width of the hydrophobic core of the membrane.

(b) The plot clearly shows that hydrophobic amino acids are concentrated along one side of the surface of the helix. The side chains of those residues are likely to face the hydrophobic core of the membrane. Polar side chains are located on the opposite side of the helical surface; they are likely to be on the side of the chain that faces other α-helical bundles. They could form hydrogen or ionic bonds with polar residues in other bundles.

11. In model membrane systems, the transfer of phospholipid head groups from one side of the bilayer to the other is very slow, presumably because of the energy required to move the polar head group through the hydrophobic bilayer. The experiments carried out by Rothman and Kennedy indicate that a process which mediates the flip-flop of membrane lipids is operating in bacterial cells. Phospholipid synthesis takes place on the cytosolic face (the inner leaflet) of the membrane, and some of the newly synthesized lipids are moved through the bilayer to the outside surface of the membrane bilayer. While aminophospholipid translocases that can move polar lipids across membranes have been found in eukaryotes, it is not yet known how such a process occurs in bacteria.

12. (a) You would expect sample B, with a higher percentage of saturated fatty acids, to have a higher melting point. The saturated chains will aggregate more closely with each other, requiring more thermal energy to disrupt that aggregation. The acyl chains of unsaturated fatty acids are kinked, and therefore cannot aggregate in regular arrays like saturated acyl chains of the same length. They are therefore disrupted at a lower temperature.

(b) In the membrane, the cross-sectional area occupied by unsaturated fatty acids is larger than that occupied by saturated chains, because of kinks in the hydrocarbon chains due to double bonds. A diglucosyl head group is larger (i.e., has a larger cross-section size) than that of a monoglucosyl derivative, so that the larger head group would match the increase in cross-sectional area in the interior of the bilayer composed of unsaturated fatty acyl chains.

13. Glycerophospholipids contain two fatty acyl groups esterified to glycerol, to which a polar head group is also attached at the C-3 carbon. When phospholipase A_2 removes one of the fatty acyl chains, the polar head group is too large in relation to the single hydrocarbon chain to allow optimal packing in the bilayer. The regular association of the hydrocarbon tails is disrupted, and the plasma membrane dissolves.

EXPANDED SOLUTIONS TO TEXT PROBLEMS

1. $1\ \mu m^2 = (10^{-6}\ m)^2 = 10^{-12}\ m^2$. $70\ Å^2 = 70\ (10^{-10}\ m)^2 = 70 \times 10^{-20}\ m^2$. Since the bilayer has two sides, $2 \times 10^{-12}/(70 \times 10^{-20}) = 2.86 \times 10^6$ molecules.

2. Using the diffusion coefficient equation, $s = (4Dt)^{1/2}$, one gets $s = (4 \times 10^{-8} \times 10^{-6})^{1/2}$ or $(4 \times 10^{-8} \times 10^{-3})^{1/2}$ or $(4 \times 10^{-8} \times 1)^{1/2}$. Solving for s gives 2×10^{-7} cm, 6.32×10^{-6} cm, and 2×10^{-4} cm, respectively.

3. The gram molecular weight of the protein divided by Avogadro's number $= 10^5\ g/(6.02 \times 10^{23}) = 1.66 \times 10^{-19}$ g/molecule. 1.66×10^{-19} g/1.35 (density) $= 1.23 \times 10^{-19}\ cm^3$/molecule. The volume of a sphere equals $\frac{4}{3}\pi r^3 = 1.23 \times 10^{-19}\ cm^3$. Solving for r, one gets 3.08×10^{-7} cm. By substituting this value into the equation given,

$$D = \frac{1.38 \times 10^{-16} \times 310}{6 \times 3.14 \times 1 \times 3.08 \times 10^{-7}}$$

$$= 7.37 \times 10^{-9}\ cm^2/s$$

By substituting this value for D and the times given in the problem into the equation shown in the answer to problem 2, one obtains the distances traversed: 1.72×10^{-7} cm in 1 μs, 5.42×10^{-6}cm in 1 ms, and 1.72×10^{-4} cm in 1 s.

4. As its name implies, a *carrier* antibiotic must move from side to side when it shuttles ions across a membrane. By contrast, *channel formers* allow ions to pass through their pores much like water through a pipe. Lowering the temperature caused a phase transition from a fluid to a nearly frozen membrane. In the nearly frozen state the *carrier* is immobilized, whereas the pore of the *channel former* remains intact, allowing ions to pass through it.

5. The initial decrease in the amplitude of the paramagnetic resonance spectrum results from the reduction of spin-labeled phosphatidyl cholines in the outer leaflet of the bilayer. Ascorbate does not traverse the membrane under these experimental conditions, and so it does not reduce the phospholipids in the inner leaflet. The slow decay of the residual spectrum is due to the reduction of phospholipids that have flipped over to the outer leaflet of the bilayer.

CHAPTER

12

Membrane Channels and Pumps

This chapter describes various transmembrane proteins that serve as channels or pumps for the transport of ions or polar molecules across cell membranes. First, Stryer describes ion channels that are regulated by ligand binding or by voltage and explains their importance in nerve function. Next, Stryer turns to pumps, which as their name implies, require an input of energy in the form of ATP or ion gradients to drive the transport of ions and molecules across membranes, in a direction opposite to their thermodynamic gradient. Finally, Stryer describes rhodopsin, a proton pump that is driven by light energy, as an example of the biological conversion of one form of energy into another. Chapters 21 and 26 on oxidative phosphorylation and photosynthesis will present detailed views of other key energy transduction processes. Looking back, Chapter 2 on protein structure and function, Chapter 3 on the methods used in the purification of proteins, Chapter 6 on the use of recombinant DNA technology in the elucidation of protein sequences, Chapter 8 on the basic concept of free energy in governing the direction of spontaneous chemical processes, and Chapter 11 on the structure of membrane proteins and the permeability of membranes contain important background information for this chapter.

LEARNING OBJECTIVES

When you master this chapter, you should be able to complete the following objectives:

Introduction (Stryer pages 291–292)

1. Distinguish between *channels* and *pumps*. List the forms of energy that can drive *active transport*.

Acetylcholine Receptor Channel (Stryer pages 292–297)

2. Explain the function of *neurotransmitters* and *ligand-gated channels* in the transmission of nerve impulses across *synapses*. Outline the effects of *acetylcholine* on the *postsynaptic membrane*.

3. Describe the subunit structure, ligand binding sites, and channel architecture of the *acetylcholine receptor* from *Torpedo*.

4. Outline the *patch-clamp technique* and note its use in the electrical measurements of membranes.

5. Explain how the *Xenopus* oocyte expression system was used to learn about the properties of the acetylcholine receptor.

6. Describe the activation and inactivation mechanisms of the acetylcholine receptor.

Voltage-Gated Channels (Stryer pages 297–309)

7. Define *action potential* and explain its mechanism in terms of the transient changes in Na^+ and K^+ permeability of the plasma membrane of a neuron.

8. Describe the role of *tetrodotoxin* in the inactivation and purification scheme of the *sodium channel*.

9. Outline the cDNA cloning and sequencing, as well as the predicted structure, of the sodium channel.

10. Explain the ion selectivity of the sodium channel.

11. Use the expression for the voltage dependence of the equilibrium constant (K) between closed and open states of ion channels to obtain gating charge z, V_0, or K values.

12. Outline the possible role of the *S4 segments* of sodium channels as *voltage sensors*. List the sequence of steps in the cycling of sodium channels during an action potential.

13. Describe the structure of the *potassium channel* as revealed by studies of the *Shaker* cDNA sequence.

14. Compare the structure, ion selectivity, and inactivation mechanism of the potassium channel with the sodium channel.

15. Contrast *gap junctions* with other membrane channels. Describe the functions of gap junctions, their overall structure, and mechanism of closure by Ca^{2+}.

16. List the common structural and mechanistic motifs of membrane channels.

Active Transport Driven by ATP (Stryer pages 309–316)

17. Distinguish between *active* and *passive transport.* Use the concepts of free-energy change (ΔG) and electrochemical potential to predict active or passive transport.

18. Describe the functions of the *Na^+-K^+ ATPase* or *Na^+-K^+ pump.*

19. Describe the structure and the functional sites of the Na^+-K^+ pump. Discuss its inhibition by *cardiotonic steroids.*

20. Outline the reaction cycle of the Na^+-K pump.

21. Note the similarities between the sarcoplasmic *Ca^{2+}-ATPase* and the Na^+-K^+ pump in terms of functional sites and reaction cycles.

Transport Driven by Ionic Gradients (Stryer pages 316–318)

22. Define *symporter* and *antiporter.*

23. Compare the rates of active transport with rates of transport through channels.

24. Describe the role of the *sodium-calcium exchanger* and compare its capacity to extrude Ca^{2+} with the Ca^{2+}-ATPase.

25. Using lactose permease as the example, explain how a *proton gradient* or *proton-motive force* promotes the accumulation of lactose in bacteria.

Active Transport Driven by Light (Stryer pages 318–320)

26. Describe *bacteriorhodopsin* as a light-driven proton pump. Recognize the structure of *retinal.*

27. Outline the cycle of conformational transitions of the *Schiff base* of retinal initiated by the absorption of light by bacteriorhodopsin.

28. Discuss the molecular events that result in the formation of the proton gradient.

SELF-TEST

Introduction

1. Which properties are common to membrane channels and pumps?

   (a) Both are integral, transmembrane proteins.
   (b) Both contain phosphorylation sites.
   (c) Both contain multiple subunits or domains.
   (d) Both have polar amino acid residues within their pores.
   (e) Both allow bidirectional flux of the transported molecule.

Acetylcholine Receptor Channel

2. List the events in the transmission of nerve impulses in synapses in their proper sequence.

(a) Binding of acetylcholine to acetylcholine receptor

______

(b) Depolarization of the postsynaptic membrane

______

(c) Release of acetylcholine from synaptic vesicles into the synaptic cleft

______

(d) Increase in postsynaptic membrane permeability to Na^+ and K^+

______

(e) Increase of acetylcholine concentration in the synaptic cleft from ~10 nM to 500 μM

______

3. Acetylcholine opens a single kind of cation channel that has a very similar permeability to Na^+ and K^+, yet the influx of Na^+ is much larger than the efflux of K^+. Explain this fact.

______

______

4. Explain the use of cobratoxin in the purification of acetylcholine receptor.

______

______

5. Which of the following is not a characteristic of the structure of the acetylcholine receptor of *Torpedo?*

(a) It has five subunits of four different kinds.
(b) It contains two binding sites for acetylcholine.
(c) It has a uniform channel 20 Å in diameter.
(d) It can be partly active when the γ or δ chains are missing.
(e) It has pentagonal symmetry.

6. Which of the following statements are correct? The patch-clamp technique

(a) can be used only on giant squid axons.
(b) allows the measurement of ion flux through a single channel.
(c) involves the microinjection of mRNA into a cell with a micropipet.
(d) can be used to monitor the expression of active channel proteins in *Xenopus* oocytes.
(e) has a time resolution of seconds.

7. The equilibrium scheme for the multiple conformational states of the acetylcholine receptor is

$$\begin{array}{ccccccc} C & \overset{1}{\rightleftharpoons} & AC & \overset{2}{\rightleftharpoons} & A_2C & \overset{3}{\rightleftharpoons} & A_2O \\ & & {\scriptstyle 4}\Big\updownarrow & & {\scriptstyle 5}\Big\updownarrow & & \\ & & AI & \rightleftharpoons & A_2I & & \end{array}$$

Answer the following questions about these states and their transformations:

(a) Which states are open to cation flow?

____

(b) Which states are closed to cation flow?

____

(c) Which steps are rapid?

____

(d) Which steps are slow?

____

(e) Which steps are responsible for desensitization?

____

8. Which of the statements about cholinesterase are correct? Acetylcholinesterase

(a) has a low turnover number.
(b) exists in free form in the synaptic cleft.
(c) forms an acetyl-enzyme intermediate during catalysis.
(d) is inhibited by organic fluorophosphates.
(e) allows synapses to transmit action potentials at high frequency.

9. Explain how the acetylcholine receptor excludes anion flow through the channel.

____

____

Voltage-Gated Channels

10. Which of the following statements about the plasma membrane of a neuron are correct?

(a) In the resting state, the membrane is more permeable to Na^+ than to K^+.
(b) In the resting state, the membrane potential is approximately +30 mV.
(c) The Na^+ and K^+ gradients across the membrane are maintained by the Na^+-K^+ pump.
(d) The equilibrium potential for K^+ across the membrane is near −75 mV.
(e) During the action potential, the membrane potential varies between the limits of +30 mV and −75 mV.

11. Place the following events of the action potential in their correct sequence:

(a) The spontaneous closing of sodium channels _______
(b) A membrane potential of −75 mV _______
(c) The depolarization of the plasma membrane to approximately −40 mV _______
(d) The opening of the potassium channels _______
(e) The opening of sodium channels _______
(f) A membrane potential of +30 mV _______
(g) A membrane potential of −60 mV _______

12. Tetrodotoxin and saxitoxin are very specific competitive inhibitors of the sodium channel. Explain why.

____

____

13. Which of the following statements about the sodium channel, purified and reconstituted in lipid bilayers, are correct?

(a) It is about 10 times more permeable to Na^+ than to K^+.
(b) It is sensitive to voltage.
(c) It is inhibited by cobratoxin.
(d) It becomes inactivated spontaneously.
(e) It consists of seven hydrophobic transmembrane segments.

14. Match the sodium (eel electric organ) and potassium (Shaker) channels with the corresponding properties listed in the right column.

(a) Sodium channel _______________

(b) Potassium channel _______________

(1) Four 70-kd subunits
(2) Single 260-kd polypeptide chain
(3) Fourfold symmetry
(4) Tetrodotoxin binding site
(5) Positively charged, voltage-sensing S4 helical segment
(6) Ball-and-chain inactivation mechanism
(7) 3-Å–diameter channel

15. Explain each of the following characteristics of the sodium channel in terms of its structure:

(a) Na^+ selectivity _______________

____

(b) Voltage gating _______________

____

(c) Inactivation _______________

____

16. What will be the free-energy change generated by five gating charges moving in an action potential ($\Delta V = 0.1$ V)?

____

____

17. The K^+ channel is over 100 times more permeable to K^+ than to Na^+. Explain the molecular mechanism for this selectivity.

____

____

18. Which of the following are not characteristics of gap junctions between cells?

(a) They allow the exchange of ions and metabolites between cells.
(b) They allow the exchange of cytoplasmic proteins.

(c) They are essential for the nourishment of cells that are distant from blood vessels.
(d) They are made up of 10 molecules of connexin.
(e) They are controlled by Ca^{2+} and H^+ concentrations in cells.

19. Explain how relatively small conformational changes in the structure of connexin in response to Ca^{2+} can close a 20-Å channel?

____

____

Active Transport Driven by ATP

20. An uncharged molecule is transported from side 1 to side 2 of a membrane.

(a) If its concentration is 10^{-3} M on side 1 and 10^{-6} M on side 2, will the transport be an active or a passive process? Explain your answer.

____

____

(b) If the concentration is 10^{-1} M on side 1 and 10^{-4} M on side 2, how will the free-energy change compare with that in question (a)? Explain.

____

____

(c) How will the rate of transport in (a) and (b) compare? Explain your answer.

____

____

21. The orientation of the Na^+-K^+ pump in cell membranes determines the side of the membrane where the various processes involved in the transport of Na^+ and K^+ will take place. Assign each of the steps or processes in the right column to the intracellular or extracellular side of the pump.

(a) Intracellular side ________________	(1) Binding of vanadate
	(2) Binding of cardiotonic steroids
(b) Extracellular side ________________	(3) Hydrolysis of ATP and phosphorylation of the pump
	(4) Binding of K^+
	(5) Binding of Na^+

22. Explain why an electric current is generated during the transport of Na^+ and K^+ by the Na^+-K^+ pump.

____

____

23. Which of the following statements about the subunit structure of the Na^+-K^+ pump are correct?

(a) The pump contains α and β subunits, associated as $\alpha_2\beta_2$ tetramers.
(b) The β subunits contain the ATPase activity.

(c) The α subunits contain the binding sites of cardiotonic steroids.
(d) The α subunits have large extracellular domains and contain oligosaccharide chains.
(e) The transmembrane channel is made by the interface between α and β subunits.

24. The proposed model for the mechanism of the Na^+-K^+ pump is based on the existence of four conformational states of this enzyme. Match each conformational state in the left column with the appropriate descriptions from the right column.

(a) E_1 ____________

(b) E_1-P ____________

(c) E_2 ____________

(d) E_2-P ____________

(1) High affinity for K^+
(2) High affinity for Na^+
(3) Phosphorylated by ATP
(4) Ion-binding sites face the cytosol
(5) Ion-binding sites face the extracellular space
(6) Dephosphorylated upon the binding of K^+
(7) Stabilized by cardiotonic steroids
(8) Stabilized by vanadate

25. Which of the following statements describe properties that are common to the Ca^{2+}-ATPase of the sarcoplasmic reticulum and the Na^+-K^+ pump?

(a) Both are very abundant membrane proteins in the sarcoplasmic reticulum.
(b) Both have homologous α subunits containing numerous transmembrane helices.
(c) Both contain an aspartate residue that is phosphorylated by ATP.
(d) Both translocate the same number of ions per transport cycle.
(e) Both probably have four major conformational states.

26. Explain how the Ca^{2+}-ATPase can be made to synthesize ATP in vitro.

____

____

____

____

Transport Driven by Ionic Gradients

27. Which of the following statements about the sodium-calcium exchanger is incorrect? The sodium-calcium exchanger

(a) is a symporter for sodium and calcium transport.
(b) is an antiporter for sodium and calcium transport.
(c) is driven by the Na^+ gradient generated by the Na^+-K^+ pump.
(d) has a lower affinity for Ca^{2+} than the Ca^{2+}-ATPase.
(e) has a higher transport rate for Ca^{2+} than the Ca^{2+}-ATPase.

28. Ascribe the characteristics in the right column either to active transport or to transport of ions or molecules through channels.

(a) Active transport ____________
(b) Transport through channels____________

(1) Flux ~ 10^7 s^{-1}
(2) Flux 3×10 to 2×10^3 s^{-1}
(3) Ions can flow from either side of the membrane
(4) Flux in a specific direction

29. Which of the following statements about the lactose permease of *E. coli* under physiologic conditions are correct?

(a) It derives energy for transport from a Na^+ gradient.
(b) It derives energy for transport from a H^+ gradient.
(c) It derives energy for transport from a Ca^{2+} gradient.
(d) It is an antiport for lactose and H^+.
(e) It is a symport for lactose and Na^+.

Active Transport Driven by Light

30. Which of the following statements about the retinal prosthetic group of bacteriorhodopsin are correct?

(a) It is covalently bound to the protein via a Schiff base.
(b) It absorbs light with a peak at 350 nm.
(c) It becomes isomerized from an all-*trans* to an all-*cis* retinal upon the absorption of light.
(d) It dissociates from the protein upon the absorption of light.
(e) It undergoes a series of isomerizations in the time scale from picoseconds to milliseconds.

31. Place the following events in the excitation of bacteriorhodopsin by light in the proper order:

(a) Eversion of deprotonated Schiff base to face the cytosol ________
(b) Transfer of the proton from Arg 82 to the external medium ________
(c) Isomerization of retinal to all-trans form ________
(d) Conformational transition of the protein at M_{412} ________
(e) Photoisomerization of all-*trans*- to 13-*cis*-retinal ________
(f) Transfer of proton from Schiff base to Asp 85 ________
(g) Restoration of protein conformation to BR_{568} state ________
(h) Reprotonation of Schiff base from Asp 96 ________

ANSWERS TO SELF-TEST

Introduction

1. a, c, d

Acetylcholine Receptor Channel

2. c, e, a, d, b

3. The electrochemical gradient for Na^+ influx is steeper than that for K^+ efflux. The concentration gradients across the membrane are similar

for both ions, but Na^+ moves from the positive to the negative side of the membrane, whereas K^+ moves from the negative to the positive side; that is, the membrane potential favors Na^+ influx.

4. Cobratoxin binds specifically and with very high affinity to the acetylcholine receptor; therefore, a column with covalently attached cobratoxin can be used in the affinity purification of the receptor from a mixture of macromolecules in the postsynaptic membrane that has been solubilized by adding nonionic detergents.

5. c

6. b, d

7. (a) A_20 (b) C, AC, A_2C, AI, A_2I (c) 1, 2, 3 (d) 4, 5 (e) 4, 5

8. c, d, e

9. The interior of the pore has three rings of negatively charged amino acid residues that exclude anion flow through the channel.

Voltage-Gated Channels

10. c, d, e

11. g, c, e, f, a, d, b, g

12. Both these toxins contain a guanido group. This positively charged group interacts with a carboxylate group at the entrance of the Na^+ channel on the extracellular side of the membrane to obstruct the conductance pathway.

13. a, b, d

14. (a) 2, 3, 4, 5, 6 (b) 1, 3, 5, 6, 7

15. (a) The sodium channel contains a negatively charged group within the pore and has a diameter of about 5 Å. This permits hydrated Na^+ ions to pass through much more easily than the larger K^+ ions.

    (b) The S4 transmembrane segments of each repeating unit of the sodium channel contain from four to eight positively charged residues (arginine or lysine) paired with negatively charged residues on other transmembrane segments. Upon depolarization, the S4 segments are thought to move in a spiral motion, displacing the positively charged groups and opening the channel.

    (c) Inactivation of sodium channels, similarly to potassium channels, appears to occur by a cytoplasmic domain of the protein attached to a flexible segment. When the channel is open, the cytoplasmic domain binds to a complementary site in the pore and occludes it. In the closed form of the channel, the cytoplasmic domain moves freely.

16. 11.6 kcal/mol.

$$\Delta G = ZF\Delta V = 5 \times 23.1 \frac{\text{kcal}}{\text{V} \bullet \text{mol}} \times 0.1 \text{ V} = 11.6 \frac{\text{kcal}}{\text{mol}}$$

17. At the entrance of the pore of the potassium channel there is a glutamate residue that binds cations. The pore is about 3 Å in diameter at its narrowest point so that only dehydrated small cations can fit; however, the energy required to dehydrate Na^+ and smaller cations is too large and is not compensated by favorable polar interactions that occur in the case of K^+.

18. b, d

19. After binding Ca^{2+}, the connexin subunits in each hemichannel rotate and slide relative to one another to adopt a more perpendicular orientation relative to the plane of the membrane. This effectively closes the channel. Because six subunits participate cooperatively in this process, small displacements in each subunit lead to the closure of the large channel.

Active Transport Driven by ATP

20. (a) The transport will be a passive process. Because the concentration on side 1 is higher than that on side 2, the molecule will move spontaneously down its concentration gradient. One can use the expression

$$\Delta G = 2.303\ RT \log \frac{c_2}{c_1}$$

$$= 2.303\ RT \log \frac{10^{-6}}{10^{-3}}$$

to show that ΔG has a negative value. Assuming a temperature of 25°C

$$\Delta G = 2.303 \times 1.98 \frac{\text{cal}}{\text{mol} \bullet \text{K}} \times 298 \text{ K} \times (-3)$$

$$= -4.07 \frac{\text{kcal}}{\text{mol}}$$

A negative ΔG value in the direction of movement is the hallmark of passive transport.

(b) Since the ratios of the concentrations in (a) and (b) are equal, the free-energy change for the transport is the same in both cases.

$$\Delta G = 2.303 \times 1.98 \frac{\text{cal}}{\text{mol} \bullet \text{K}} \times 298 \text{ K} \times \log \frac{10^{-4}}{10^{-1}}$$

$$= -4.07 \frac{\text{kcal}}{\text{mol}}$$

(c) The rate of a chemical process is always equal to a rate constant multiplied by the concentration of the chemical species undergoing the change (see Stryer, p. 192). Thus, the rate of transport will be equal to $k(c_1)$. Since the concentration of c_1 is greater in (b) than in (a), the rate of transport will be greater in (b).

21. (a) 1, 3, 5 (b) 2, 4

22. Since three Na^+ ions are transported out for every two K^+ ions that are transported in, there is a net efflux of one positively charged ion. The net movement of ions sets up an electric current.

23. a, c

24. (a) 2, 3, 4, 8 (b) 2, 4 (c) 1, 5 (d) 1, 5, 6, 7

25. b, c, e

26. Ca^{2+}-ATPase can be reconstituted into phospholipid vesicles such that there is a high concentration of Ca^{2+} inside the vesicles and the phosphorylation site of the Ca^{2+}-ATPase is exposed to the outside of the vesicles. If P_i and ADP are added to the vesicle suspension, passive transport of Ca^{2+} from the inside to the outside of the vesicles will cause the phosphorylation of the Ca^{2+}-ATPase and synthesis of ATP.

Transport Driven by Ionic Gradients

27. a. The sodium-calcium exchanger transports these cations in the opposite direction; therefore, it is an antiporter, not a symporter.

28. (a) 2, 4 (b) 1, 3

29. b

Active Transport Driven by Light

30. a, e. Answer (c) is incorrect because all-*trans*-retinal isomerizes to 13-*cis*-retinal rather than to all-*cis*-retinal.

31. e, f, b, d, a, h, c, g

PROBLEMS

1. In addition to the Na^+-K^+ ATPase, eukaryotic cells contain other ATP-driven pumps. One such pump is the H^+-K^+ ATPase, in which a hydrogen ion is extruded from the cytoplasm in exchange for a potassium ion at the expense of ATP hydrolysis. Given that the interior of most animal cells is electrically negative with respect to the exterior, explain why the Na^+-K^+ ATPase can contribute to the membrane potential but the H^+-K^+ ATPase cannot.

2. In experiments to investigate the mechanism of transport of two substances, X and Y, across cell membranes, cells were incubated in media containing various concentrations of X and Y, and the initial rate of transport of each of the substances into the cell was determined. The results

that were obtained are depicted in Figure 12-1. What conclusion is suggested by the results? Explain. It may be helpful to refer to the discussion of enzyme kinetics beginning on page 192 of Stryer.

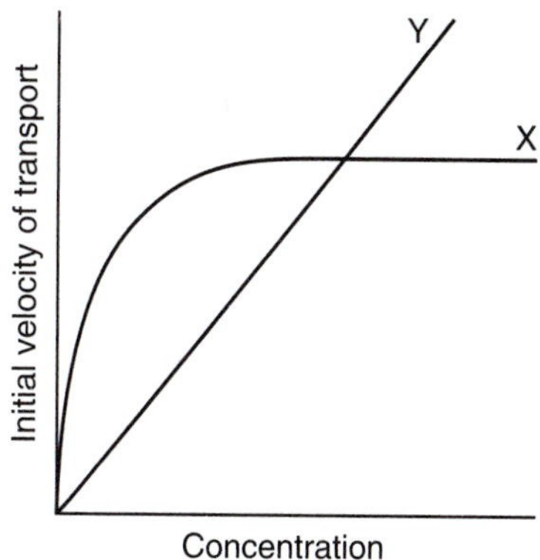

Figure 12-1
Initial velocity of transport versus concentration for substances X and Y.

3. Calculate the free-energy change for the transport of an uncharged species from a concentration of 5 mM outside a cell to a concentration of 150 mM inside. Assume that the temperature is 25°C.

4. Now repeat the calculation of problem 3 for an ion with a charge of +1 that is crossing a membrane with a potential of −60 mV, with the interior negative with respect to the exterior.

5. In dog skeletal muscle, the extracellular and intracellular concentrations of Na^+ are 150 mM and 12 mM, and those of K^+ are 2.7 mM and 140 mM, respectively.

(a) Calculate the free-energy change as three Na^+ are transported out and two K^+ are transported in by the Na^+-K^+ pump. Assume that the temperature is 25°C and that the membrane potential is −60 mV.

(b) Does the hydrolysis of a single ATP provide sufficient energy for the process in part (a)? Explain.

6. Design an experiment using ATP labeled with ^{32}P in the γ position that would suggest that the Na^+-K^+ ATPase reaction involves a stable enzyme-phosphate intermediate. (See Stryer, p. 311.)

7. Suggest an experiment involving a membrane preparation, ATP labeled in the β position with ^{32}P, and unlabeled ADP that would lend additional support to the finding that the Na^+-K^+ ATPase reaction is a double displacement. (Hint: Your experiment should involve the reisolation of ADP from the reaction mixture.)

8. Predict the effect of digitoxigenin on the force of contraction of an isolated muscle fiber when it is (a) added to a medium bathing the muscle fiber or (b) injected directly into the muscle fiber. Explain your answers.

9. When erythrocyte ghosts are equilibrated in a medium containing ADP, phosphate, and K^+ and are then transferred to a medium containing high Na^+, ATP is formed within the erythrocytes. What is the significance of this observation?

10. Figure 12-2 depicts a typical action potential that might be measured in an isolated axon, such as the giant axon of a squid. Give the events that are responsible for (a) the rising phase of the action potential and (b) the falling phase. Specify in each case whether ion flow occurs with or against concentration gradients, electrical gradients, or both.

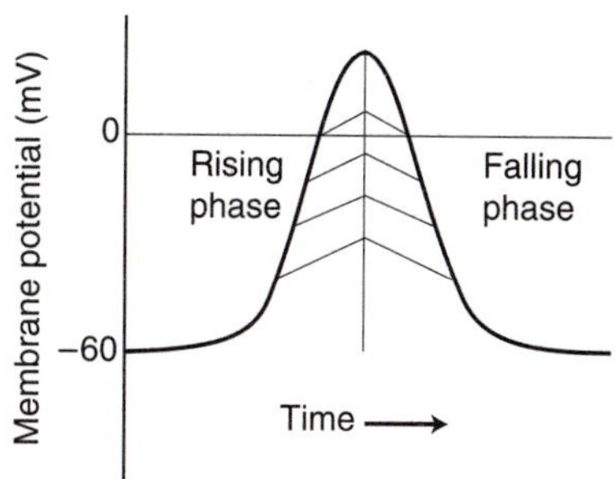

Figure 12-2
An action potential.

11. When the sciatic nerve is removed from a frog, placed in an isotonic salt solution, and stimulated electrically, it will generate action potentials that can be measured by an electrode placed at some distance from the site of stimulation. When metabolic poisons are added to the preparation, the nerve retains the capability of generating action potentials even though the supply of ATP to drive its Na^+-K^+ pump has been depleted and it is thus incapable of carrying out active transport. Explain how this can be the case.

12. Suppose that you want to find a drug that would act at the neuromuscular junction to induce muscle relaxation.

    (a) Give three possible approaches you could take, explain the rationale for each, and give a specific example of a known drug that would have the desired effect.
    (b) Is there a major potential hazard in the use of any of the three types of agents you have described in part (a)? What is it?

13. Surgeons opening the abdomen frequently paint the abdominal musculature with succinylcholine to prevent the patient from eviscerating. Explain why this measure works.

14. Suppose that a Glu residue is present in the narrow region of the sodium channel. A mutant protein is found in which that Glu is replaced by Val.

    (a) Compare the Na^+ conductance of the mutant as opposed to the normal channel. Explain.
    (b) Compare the sodium permeability as a function of pH in each case.
    (c) Compare the sensitivity of the normal and the mutant channels to tetrodotoxin.
    (d) Compare the magnitude of an action potential in nerves containing sodium channels of the mutant type, as opposed to the normal type.

15. How could you produce a synthetic vesicle in which the uptake of lactose from the medium into the vesicle against a concentration gradient could be driven by light?

ANSWERS TO PROBLEMS

1. The difference between the two ATPase systems lies in the stoichiometry of their exchange of ions. In the case of the Na^+-K^+ ATPase, three Na^+ ions are extruded and two K^+ ions are taken up during each pump cycle, making the interior of the cell more negative (less positive) for each pump cycle. The H^+-K^+ ATPase, in contrast, extrudes one H^+ ion for each K^+ taken up, so its operation is electrically neutral.

2. The curve for X shows saturation, which would be expected if some membrane carrier is involved in the transport of substance X. The curve for Y shows no saturation, which is consistent with the notion that substance Y diffuses through the membrane without a carrier. Such behavior is shown by lipid soluble substances, which dissolve in the hydrophobic tails of membrane phospholipids and can thus enter cells without a carrier.

3. We use the following equation from page 309 of Stryer:

$$\Delta G = RT \log_e \frac{c_2}{c_1}$$

$$= 1.99 \times 298 \times \log_e \frac{150}{5}$$

$$= +2.0 \text{ kcal/mol}$$

4. The relevant equation is given on page 309 of Stryer:

$$\Delta G = RT \log_e \frac{c_2}{c_1} + ZF\Delta V$$

$$= 1.99 \times 298 \times \log_e \frac{150}{5} + [(+1) \times 23.062 \times (-0.060)]$$

$$= +2.0 - 1.4$$

$$= +0.6 \text{ kcal/mol}$$

Note that the membrane potential favors the entry of a positively charged ion.

5. (a) To solve this problem, we first calculate the free-energy change for transporting three Na^+ and then that for two K^+. The total free-energy change will be the sum of the two.

Again we use the following equation from page 309 of Stryer:

$$\Delta G = RT \log_e \frac{c_2}{c_1} + ZF\Delta V$$

Substituting in the values for Na^+ yields

$$\Delta G_{Na^+} = 1.99 \times 298 \times \log_e \frac{150}{12} + [(+1) \times 23.062 \times 0.060]$$

$$= +1.5 + 1.4$$

$$= +2.9 \text{ kcal/mol}$$

Note that when Na^+ is transported out of the cell, work must be done against both a concentration gradient and an electrical gradient.

Now we carry out the corresponding calculation for the K^+ ion.

$$\Delta G_{K^+} = 1.99 \times 298 \times \log_e \frac{140}{2.7} + [(+1) \times 23.062 \times (-0.060)]$$

$$= +2.3 - 1.4$$

$$= +0.9 \text{ kcal/mol}$$

Note that potassium ion is being transported against a concentration gradient but with an electrical gradient. Accordingly, the sign for the electrical term in the equation is negative.

To get the total energy expenditure, we must account for the stoichiometry of transport by summing the energy required for the transport of Na^+ and that required for the transport of K^+.

$$\Delta G = 3\,\Delta G_{Na^+} + 2\,\Delta G_{K^+}$$

$$= (3 \times 2.9) + (2 \times 0.9)$$

$$= 10.5 \text{ kcal/mol}$$

(b) Although the free-energy change for the hydrolysis of ATP under standard conditions is −7.3 kcal/mol, the free-energy change for the ATP concentrations that exist in typical cells is approximately −12 kcal/mol. Thus, the energy furnished by the hydrolysis of a single ATP is sufficient.

6. The Na^+-K^+ ATPase reaction is a two-step process involving two independent half-reactions:

$$E + ATP \xrightleftharpoons{Na^+, Mg^{2+}} E-P + ADP \qquad (1)$$

$$E-P + H_2O \xrightarrow{K^+} E + P_i \qquad (2)$$

In the first half-reaction, ATP and unmodified enzyme interact in the presence of Na^+ and Mg^{2+} to give phosphorylated enzyme and the first product of the overall reaction, ADP. In the second half-reaction, the phosphorylated enzyme is hydrolyzed by water in the presence of K^+ to give unmodified enzyme and inorganic phosphate. An overall reaction that consists of two independent half-reactions is known as a *double displacement.*

The following experiment would suggest that such an overall reaction occurs. Incubate a fragmented membrane preparation with K^+ ion to hydrolyze any phosphate that might be bound to the enzyme. Then wash the membrane preparation to remove all K^+, and transfer the membrane to a medium containing γ-labeled ATP, Na^+, and Mg^{2+}. After a suitable incubation period, wash the membrane preparation to remove any unreacted labeled ATP. Then carry out scintillation counting on the membrane preparation to detect the presence of labeled phosphate. The presence of radioactivity in the membrane fraction would suggest that a stable enzyme-phosphate intermediate had been formed. In fact, a covalent aspartyl phosphate derivative is formed at the active site of the ATPase. (See p. 311 of Stryer.)

7. You could incubate the β-labeled ATP and the unlabeled ADP with the membrane preparation in the presence of Na^+ and Mg^{2+} but in the absence of K^+. After a suitable period, reisolate the ADP from the reaction mixture and assay it for radioactivity. If an E-P intermediate has been produced in this half-reaction (as it is), the ADP should be radioactive. A characteristic of double-displacement reactions is that the half-reactions will occur separately.

8. (a) Extracellularly, digitoxigenin acts to inhibit the K^+-dependent dephosphorylation of the E-P intermediate of the Na^+-K^+ ATPase (step 2 in the solution to problem 6). Therefore, it increases the intracellular concentration of Na^+, which diminishes the gradient that drives the extrusion of Ca^{2+} from the intracellular fluid. Thus, the intracellular concentration of Ca^{2+} increases as well, which increases the contractile force of muscle.

(b) Direct intracellular injection of the digitoxigenin should have little or no effect on the force of contraction, because the K^+ hydrolysis of the E-P intermediate takes place on the exterior surface of the plasma membrane.

9. The Na^+-K^+ pump is apparently reversible. The ionic gradient established across the membrane provides sufficient free energy to drive the synthesis of ATP, much as an ionic gradient across the inner membrane of mitochondria drives the synthesis of ATP in oxidative phosphorylation.

10. Nerve cells, like most animal cells, have a higher concentration of K^+ inside than outside and a higher concentration of Na^+ outside than inside. In addition, there is a membrane potential; that is, the inside of the cell is negative (in this case −60 mV) with respect to the outside.

(a) The rising phase of the action potential is due to the influx of Na^+ ions down a concentration gradient and an electrical gradient.
(b) The falling phase is due to the efflux of K^+ ions down a concentration gradient but against an electrical gradient.

11. For each action potential that is generated in an axon, only a very few Na^+ ions enter and a very few K^+ ions depart the cell. Thus, in a poisoned nerve cell, many tens of thousands of impulses may be conducted before ionic equilibrium across the membrane is achieved. The active transport of Na^+ and K^+ across the membrane may be best viewed as necessary in the long run but not in the short run.

12. (a) Signal transmission at the neuromuscular junction involves the acetylcholine system. To be effective as a relaxant, a substance must prevent the repetitive transmission of signals by acetylcholine. One approach would be to find a drug that binds to the acetylcholine receptor but does not induce the depolarization of the motor end plate (and hence cause muscle contraction). Furthermore, the drug should not be cleaved by acetylcholinesterase so that its effect is (relatively) long-lasting. An example of such an agent is curare.

A second approach would be to find an agent, such as succinylcholine or decamethonium, that is cleaved very slowly or not at all by acetylcholinesterase and binds to the acetylcholine receptor to produce the persistent depolarization of the motor end plate. A muscle treated with such an agent would go through an initial contraction cycle, but it would be refractory to further stimuli because the receptor is blocked and the depolarization of the motor end plate is persistent.

A third approach would be to find an inhibitor of acetylcholinesterase. As a result of adding such an inhibitor, acetylcholine would be maintained at the receptor site and the persistent depolarization of the motor end plate would result. Physostigmine and neostigmine are examples of this third type of agent.

(b) The major hazard in the use of any of the three types of muscle relaxants is that they will also relax the muscle groups used for respiration.

13. Succinylcholine binds to the acetylcholine receptor and induces persistent depolarization of the motor end plate. (See the answer to the previous question.) Because the receptor is blocked, the muscle will not contract in response to further stimuli.

14. (a) The mutant sodium channel would have decreased Na^+ conductance. Charge attraction between Na^+ ions and carboxylate anions is important in drawing Na^+ into the channel. In the mutant channel, Val, with its uncharged side chain, will not attract Na^+ ions.

(b) In the normal channel, Na^+ conductance decreases as pH is lowered below 5.4 and the negatively charged carboxylate side chains are titrated to uncharged carboxyl groups. The mutant channel would show no such sensitivity to pH, because no charged groups are present in the side chain of Val.

(c) The normal channel would be sensitive to tetrodotoxin, whereas the mutant channel would less sensitive. Tetrodotoxin contains a positively charged guanido group (Stryer, p. 299) that presumably interacts electrostatically with a negatively charged group in the sodium channel.

(d) The magnitude of the action potential would be reduced in nerves containing the mutant channel. The action potential is generated by sodium flowing from the outside of the cell to the inside. Decreased sodium conductance in the mutant would reduce the amount of sodium influx, and hence reduce the magnitude of the action potential.

15. Form reconstituted vesicles containing lactose permease and bacteriorhodopsin, with each oriented in the membrane so that its cytoplasmic face is toward the inside of the vesicle. Illumination of such vesicles will cause hydrogen ions to be extruded by bacteriorhodopsin. The resulting gradient of hydrogen ions will then drive the entry of lactose by its permease.

EXPANDED SOLUTIONS TO TEXT PROBLEMS

1. To answer this problem, refer to problem 7-12 in Stryer, which is essentially identical with this one, and to Stryer's discussion on pp. 167–168. From the data given, we get $L_0 = [T_0]/[R_0] = 10^5$ and $c = 1/20 = 5 \times 10^{-2}$. The ratio of closed to open channels = [T]/[R]. When no ligands are bound, $[T]/[R] = 10^5$. For one ligand, $[T]/[R] = L_0 \times c = 10^5 \times 5 \times 10^{-2} = 5 \times 10^3$.

For two ligands, [T]/[R] is $10^5 \times (5 \times 10^{-2})^2 = 2.5 \times 10^2$. For three ligands, [T]/[R] is $10^5 \times (5 \times 10^{-2})^3 = 1.25 \times 10$. For four ligands, it is $10^5 \times (5 \times 10^{-2})^4 = 0.625$.

From these ratios of closed/open, one can calculate that the fraction of open channels is respectively 10^{-5}, 2×10^{-4}, 3.98×10^{-3}, 7.41×10^{-2}, and 0.615.

2. All three of these molecules contain highly reactive phosphoryl groups that readily react with the active-site serine of acetylcholinesterase

to form a stable derivative. Without active acetylcholinesterase, synaptic transmission at the cholinergic synapses is impossible, resulting in respiratory paralysis.

3. (a) The binding of the first acetylcholine increases the open/closed channels by a factor of $1.2 \times 10^{-3}/(5 \times 10^{-6}) = 240$, whereas the binding of the second acetylcholine increases this ratio by a factor of $14/(1.2 \times 10^{-3}) = 11.7 \times 10^3$.

   (b) For the free-energy calculation, refer to Table 8-1 and the accompanying discussion by Stryer (pp. 186–187). Note that for a tenfold change in the equilibrium constant, there is a standard free-energy ($\Delta G^{\circ\prime}$) change of 1.36 kcal/mol. Also note that the $\Delta G^{\circ\prime}$ varies with the log of K'_{eq}. Therefore, the free-energy change during the binding of the first acetylcholine is the log $(240) \times 1.36 = 3.24$ kcal/mol. For the second binding, $\Delta G^{\circ\prime} = \log(11.7 \times 10^3) \times 1.36 = 5.54$ kcal/mol.

   (c) No. The MWC model predicts that the binding of each ligand will have the same effect on the closed/open ratio. The acetylcholine receptor channel is not perfectly symmetric. The two α chains are not in identical environments. Also, the presence of desensitized states in addition to the open and closed ones indicates that a more complex model is required.

4. (a) Since 37% of the channels are open at −25 mV and 59% are open at −20 mV, 50% of the channels will be open at a voltage somewhere between −20 and −25 mV. Then we can say that

$$\frac{5\text{ mV}}{0.22\ (59\% - 37\%)} = \frac{x}{0.09\ (59\% - 50\%)}$$

Solving this equation gives $x = 2$ mV, where x is the *difference* in the voltage causing 59% of the channels to open and the voltage causing 50% of the channels to open. Therefore the voltage required to open half the channels is −20 mV − 2 mV = −22 mV.

   (b) $$K = \frac{f_c}{f_o} = \frac{(1-f_o)}{f_o} = e^{-zF(V-V_0)/RT},$$

where f_c = fraction closed and f_o = fraction open (see Stryer, pp. 302–303).

Then,

$$\log_e \frac{(1-f_o)}{f_o} = -zF(V-V_0)/RT$$

$$\log_e \frac{f_o}{(1-f_o)} = zF(V-V_0)/RT$$

From the information given we calculate the information shown in the table in the margin.

Volts	f_o	$1-f_o$	$\log_e(f_o/(1-f_o))$
−0.045	0.02	0.98	−3.89
−0.04	0.04	0.96	−3.18
−0.035	0.09	0.91	−2.31
−0.03	0.19	0.81	−1.45
−0.025	0.37	0.63	−0.53
−0.02	0.59	0.41	0.36
−0.015	0.78	0.22	1.27
−0.01	0.89	0.11	2.09
−0.005	0.95	0.05	2.94
0	0.98	0.02	3.89
0.005	0.99	0.01	4.60

If we plot these data, with or without the aid of a computer program, we obtain the following graph:

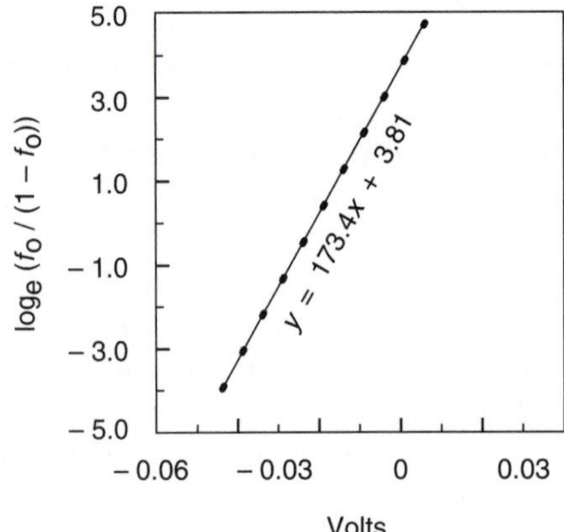

Since the slope of the line $= 173.4$, and $= zF/RT$

Slope $= 173.4 = zF/RT$

$$z = \text{slope} \times \frac{RT}{F} = 173.4 \times \frac{1.987 \times 298}{23060} = 4.45$$

The answer to (a) can be obtained from the equation for this straight line because when $f_0 = 0.5$, $y = 0$. Then, $0 = 173.4x + 3.81$ and $x = -3.81/173.4\text{ V} = -22\text{ mV}$.

(c) The free-energy change in transporting a charged species across a membrane is given by $\Delta G = RT \log_e c_2/c_1 + ZF\Delta V$, where Z is the electrical charge of the transported species, V is the potential in volts, and F is the faraday. To calculate the ΔG contributed by the movement of the gating charge, we need only consider the $ZF\Delta V$ portion of this equation. Substituting, we get 4.5×23 kcal V^{-1} $\text{mol}^{-1} \times 0.05\text{ V} = 5.2$ kcal/mol.

5. Normally the *open state* of sodium channels lasts for only about 1 ms because it spontaneously converts to an *inactive state.* Its return to a closed but activatable state requires repolarization (see Figure 12-22 in Stryer). Since BTX keeps the sodium channels open after depolarization, it apparently blocks the transition from the open to the inactivated state.

6. (a) Open *channels* enable ions to flow rapidly through membranes in a thermodynamically downhill direction, that is, from higher to lower concentration. Therefore, chloride ions will flow into the cell.

(b) This flow of chloride ions *increases* the membrane polarization. Since depolarization triggers an action potential, the chloride flux is inhibitory.

(c) If the GABA_A receptor resembles the acetylcholine receptor, its channel must consist of 5 subunits.

7. Since $\Delta G = 2.3\,RT\log c_2/c_1 + ZF\Delta V$, we can substitute and get the following:

$$\Delta G = 2.3\,RT\log \frac{1.5\times 10^{-3}}{4\times 10^{-7}} + 2 \times 23 \times 6 \times 10^{-2}$$

$$= 4.86 \text{ kcal/mol (chemical work)} + 2.76 \text{ kcal/mol (electrical work)}$$

8. (a)

$$\text{Conductance} = g = \frac{i}{V - E_r}$$

$$= \frac{5\text{ pA}}{5 \times 10^{-2}\text{ V} - 0\text{V}} = 100 \text{ pS (picosiemens)}$$

(See Stryer, p. 295.)

(b) Since a picoampere is 10^{-12} amperes, it is the flow of 6.24×10^{6} charges per second. Therefore, $5 \times 6.24 \times 10^{6} \times 10^{-3}\text{ s} = 3.12 \times 10^{4}$ charges per ms.

(c) Since 3.12×10^{4} molecules flow through the channel in 1 ms,

$$\frac{1\text{ ms}}{3.12 \times 10^{4}\text{ molecules}} = \frac{t\text{ (time)}}{1\text{ molecule}}$$

Solving this equation gives $t = 32 \times 10^{-6}$ ms = 32 ns (nanoseconds).

9. Membrane vesicles containing a high concentration of lactose in their inner volume could be formed. The binding of lactose to the inner face of the permease would be followed by the binding of a proton. Both sides would then evert. Because the lactose concentration on the outside is low, lactose and the proton will dissociate from the permease. The downhill flux of lactose will drive the uphill flux of protons in this in vitro system.

Note that this situation is the opposite of that depicted in Figure 12-44 in Stryer. Instead of a proton gradient driving the uptake of lactose, a lactose gradient is driving the release of protons.

CHAPTER

13

Signal Transduction Cascades

In Chapter 11 you learned how biological membranes serve as semipermeable boundaries that isolate the cell from its surroundings and separate intracellular compartments from one another. Chapter 12 described how selective, controlled breaching of this barrier can generate changing ion gradients across the bilayer and thereby produce electrical signals. In this chapter Stryer shows how external molecules bind to integral membrane protein receptors to initiate specific responses within the cell. He describes how these binding and transmission mechanisms lead to an amplification of the initial signal and to specific effects that adapt the cell to its environment through effects on intracellular enzymes and regulatory proteins.

First, Stryer discusses bacterial chemotaxis, which is the directed movement of bacteria toward chemical substances in their surroundings that are attractants and away from those that are repellents. Next, Stryer describes the excitable intracellular retinal receptor of vision that is activated by light, rather than by a molecule. This process introduces you to the ubiquitous G proteins that recur as members of many hormone-mediated amplification cascades. Stryer then explains how the discovery of cyclic AMP (cAMP) revealed the details of a general pathway common to many hormones. Cyclic AMP serves as an intracellular second messenger, the hormone itself being regarded as the first messenger, that delivers the stimulus arising from the bound hormone to the targeted enzymes. The hormone-initiated activation of adenylate cyclase, an integral membrane enzyme that forms cAMP, exemplifies how an external signal from a hormone leads to a change in the level of an intracellular allosteric effector.

Stryer describes the structure of the seven-helix transmembrane β-adrenergic receptor and indicates how it transmits to the intracellular side of the plasma membrane a signal arising from binding the hormone epinephrine on the extracellular surface of the cell. The description of the information-transmission pathway from hormone stimulus to G proteins to adenylate cyclase is completed by a discussion of how cAMP activates specific protein kinases, primarily protein kinase A, to modulate the activities of the target proteins. The cAMP binds to the regulatory subunits of the inactive protein kinase tetramer, which leads to the release of the active catalytic subunits. The catalytic subunits then phosphorylate serine and threonine residues in the target enzymes to activate or inactivate them. A small number of molecules of a hormone result in an amplified response because each activated enzyme in the triggered cascade forms numerous phosphorylated products.

Stryer next describes an analogous hormone-stimulated system—the phosphoionositide cascade. In this system, the hormone activates, by means of G proteins, a specific phospholipase (phospholipase C) that cleaves a plasma membrane phospholipid, phosphatidyl inositol 4,5-bisphosphate (PIP_2), to form two second messengers. The inositol phosphate derivative, inositol 1,4,5-trisphosphate (IP_3), which is short-lived, triggers the opening of ion channels so that the Ca^{+2} concentration in the cytosol is increased. The remainder of the PIP_2 molecule, diacylglycerol (DAG), is another second messenger that activates protein kinase C. The increased Ca^{+2} levels and the activated protein kinase C affect a variety of biochemical reactions. The common features of the seven-helix receptors (serpentine receptors) and G proteins are presented next. Stryer then describes the structure of Ca^{+2}-binding proteins, focusing on calmodulin, and explains how the binding of the ion is highly specific and leads to a large conformational change in the protein—qualities desirable in molecules serving as Ca^{2+} sensors and signal transducers.

Stryer then introduces another class of receptors, the transmembrane receptor tyrosine kinases that are often activated by a ligand-induced dimerization. The activated dimers phosphorylate some of their own tyrosine residues to provide docking sites for effector proteins on the cytosolic side of the membrane. Once bound, these effector enzymes are themselves phosphorylated and thereby activated by the tyrosine receptor kinase. A description of the susceptibility of signal transduction pathways to malfunctions that produce cancer follows, and the roles of oncogenes and their normal cellular counterparts (protooncogenes) in cell growth and differentiation close the chapter.

When you have mastered this chapter, you should be able to complete the following objectives:

Introduction (Stryer pages 325–326)

1. List the components of *signal transduction cascades.*

Bacterial Chemotaxis (Stryer pages 326–332)

2. Define *chemotaxis* and outline the sequence of events constituting it.

3. Describe the *flagella* and the *motors* of the chemotactic apparatus of *E. coli.* Relate their actions to the smooth swimming and tumbling motions of a bacterium in response to a *temporal gradient* of *attractant* or *repellent* substances.

4. Describe the general structures of *methyl-accepting chemotaxis proteins* (*MCPs*) and their role as *transducers* of chemotactic signals.

5. Explain the use of the che *gene mutants* of *E. coli* in the analysis of the *central processing system* of chemotaxis.

6. Describe the effect of *phosphorylation* of the *che* gene product *CheY* on flagellar rotation.

7. List the essential constituents of *two-component regulatory systems.*

8. Discuss the roles of reversible *methylations* of the *MCPs* and the reversible phosphorylations of the *che* proteins in the process of *adaptation.* Explain the necessity for adaptation during chemotaxis.

Retinal Photoreceptors (Stryer pages 332–339)

9. Distinguish between *cone* and *rod photoreceptor cells.*

10. Describe the regions of the rod cell and list their components and functions.

11. Describe the properties of *retinal* and outline its synthesis from *vitamin A.*

12. Describe the overall structure of *rhodopsin,* its functional regions, and its orientation in the *disc* membrane. Relate the effect of light absorption on the *cation-specific channels* of the rod cell.

13. List the sequence of events from the absorption of light by *11-*cis-*retinal* to the release of *all-*trans-*retinal* from opsin and the regeneration of 11-*cis*-retinal.

14. Discuss the light-induced *hyperpolarization* of the plasma membrane as a result of the closing of *cation-specific channels.* Relate this hyperpolariza-

tion to the decrease in the concentration of *guanosine 3′,5′-monophosphate* (*cyclic GMP, cGMP*) in the *outer segment* of the rod cell.

15. Outline the *enzymatic cascade* leading to the hydrolysis of cGMP. Explain the roles of *photoexcited rhodopsin; transducin,* a *G protein;* and *activated phosphodiesterase.*

16. Compare the properties of transducin with those of other members of the *G protein family* (see Stryer, pages 346–347).

17. Describe the restoration of the dark state by the *GTPase activity* of transducin, *rhodopsin kinase,* the binding of *arrestin* to phosphorylated rhodopsin, and the synthesis of cGMP by *guanylate cyclase.* Explain the role of Ca^{2+} in increasing cGMP concentration.

18. Describe the functions of the three kinds of cones in color vision and appreciate how the absorption spectrum of retinal can be modified by its protein environment.

Cyclic AMP—Its Formation and Its Role as a Second Messenger

(Stryer pages 340–341)

19. Recount the history of the discovery of *adenosine 3′,5′-monophosphate* (*cyclic AMP,* or *cAMP*), noting the observations of enzyme regulation by *covalent modification* and hormone action in a cell-free extract.

20. Write the reactions catalyzed by *adenylate cyclase* and *cAMP phosphodiesterase,* and give the standard free-energy changes for each reaction. Note the analogies between these reactions and those involving cGMP.

21. List the salient features of the *second-messenger concept.* Appreciate that a hormone need not enter a cell to affect intracellular reactions.

22. Cite the evidence that implicates cAMP as a second messenger.

23. Note the multiplicity of hormones that use cAMP as their intracellular effector and the diversities of the physiologic responses elicited by changing cAMP levels.

Guanyl Nucleotide–Binding Proteins and Adenylate Cyclase

(Stryer pages 341–343)

24. Describe the roles of the *guanyl nucleotide–binding proteins* (*G proteins*) in coupling a *hormone-receptor complex* to adenylate cyclase and in amplifying the stimulus.

25. Describe the structure of the *β-adrenergic receptor,* which binds *epinephrine,* and explain the phenomenon of *desensitization,* or *adaptation,* in hormone function.

26. Locate G proteins in the cell and describe their structures, catalytic characteristics, and molecular mechanisms of activation and inactivation. Describe the role of *GTP* in G protein activity.

27. Describe the mechanism by which cAMP modulates the activities of *protein kinases* in eukaryotes. Relate *protein kinase A (PKA)* to cAMP signaling and note its diverse actions.

28. List three steps in the *cAMP cascade* that contribute to the amplification of the hormonal stimulus, and explain how the amplified response is achieved.

29. Appreciate that there are families of G proteins that enable diverse hormones to effect a variety of physiologic functions.

The Phosphoinositide Cascade and the Modulation of Ca^{2+} Concentrations
(Stryer pages 343–350)

30. Draw the structure of *phosphatidyl inositol 4,5-bisphosphate* (*PIP_2*). Write the reaction catalyzed by *phospholipase C* to produce the second messengers *inositol 1,4,5-trisphosphate* (*IP_3*) and *diacylglycerol* (*DAG*). Note that there are several forms of mammalian phospholipase C.

31. Outline the *phosphoinositide cascade,* and note the diversity of the elicited physiologic responses.

32. Describe the biochemical fates of the second messengers produced from PIP_2. Note that the phosphoinositide cascade often produces *arachidonate*—a precursor of the prostaglandin family of hormones.

33. Describe the effects of IP_3 on Ca^{2+} released from endoplasmic reticulum of smooth muscle, and list some biochemical processes affected by intracellular Ca^{2+} concentration.

34. Describe the effects of DAG on *protein kinase C* and the mechanistic and functional relationships between IP_3, Ca^{2+}, and DAG. Describe the biochemical and physiologic effects of *phorbol esters.*

35. Summarize the functions and interrelationships of the *serpentine receptors* and the G proteins.

36. Outline the features that suit Ca^{2+} in its role as a eukaryotic signaling ion.

37. Describe the structure of *calmodulin* and its biochemical function. Relate calmodulin to the *calmodulin-dependent protein kinase II and the Ca^{2+}-ATP ion* pump. Note the value of calcium ionophores, calcium buffers, and fluorescent indicators in studying the functions of Ca^{2+} in cells.

38. Describe the *EF hand structural motif* of *calcium-binding proteins,* and explain how it binds Ca^{2+}.

The Receptor Tyrosine Kinases (Stryer pages 350–354)

39. Describe the general structures of the *receptor tyrosine kinases* and outline the process that converts them from inactive proteins to active enzymes.

40. Note the essential role of the receptor tyrosine kinases in controlling cell growth and differentiation.

41. List some of the hormones that activate tyrosine kinases.

42. Describe the *epidermal growth factor* and *epidermal growth factor receptor.* Outline the effect of their formation of a hormone-receptor complex on the tyrosine kinase activity of the complex.

43. Note the role of autophosphorylation of the kinase in the signal transduction process.

44. Define *oncogene* and relate oncogene function to tyrosine kinases.

45. Explain the role of *SH2 domains* in tyrosine kinase function. Appreciate that *protooncogenes,* which provide normal, essential functions in cell growth and proliferation, can give rise to cancer upon mutation to oncogenes.

Signal Transduction and Cancer (Stryer pages 354–356)

46. Describe the effect of Rous sarcoma virus gene product (*v-src*) on cell growth and note the relationship of *v-src* to its cellular counterpart, *c-src.* Distinguish the oncogene and *protooncogene.*

47. Appreciate that the diversity of oncogene-directed molecular pathologies are often mediated by their effects on signal transduction pathways.

48. Outline the biochemical mechanism of *v-ras protein*–induced cancer and note the role of the normal (noncarcinogenic) *c-ras* protein in cellular differentiation. Appreciate that a diminished GTPase activity, in this case, leads to cancer.

SELF-TEST

Introduction

1. Signal transduction cascades are produced by molecular assemblies of which of the following components?
   (a) Enzymes
   (b) Regulatory proteins
   (c) Receptors
   (d) Transmembrane channels
   (e) Nuclear pores

Bacterial Chemotaxis

2. Match the major components of the chemotaxis system of *E. coli* in the left column with the appropriate descriptions from the right column.

(a) Chemoreceptor ________	(1) Contains four rings, a hook, and a rod
(b) Processing system ________	(2) Has binding sites for attractants or repellents
(c) Motor ________	(3) Includes cytosolic peripheral membrane proteins
(d) Flagellum ________	(4) Contains flagellin and adopts a helical configuration

3. For a bacterium moving toward an increasing concentration of an attractant, which of the following statements are correct?

(a) Tumbling will be less frequent.
(b) Tumbling will be more frequent.
(c) The counterclockwise rotation of flagella will occur more frequently.
(d) The clockwise rotation of flagella will occur more frequently.

4. The experiment of Koshland and Macnab that demonstrated that bacteria sense temporal gradients of attractants consisted of rapidly mixing a suspension of bacteria devoid of an attractant with a solution containing an attractant. The observed result was that tumbling was suppressed within a second after mixing, indicating that the bacteria detected the change in the concentration of the attractant over time. Explain what result would have been expected if bacteria detected differences in concentration over space.

____

____

5. Which of the following are properties of methyl-accepting chemotaxis proteins?

(a) MCPs may bind attractants or repellents directly.
(b) MCPs may interact with soluble, periplasmic chemosensors.
(c) MCPs contain a motor binding site in the cytosolic region.
(d) MCPs are reversibly methylated in the cytosolic domain.
(e) MCPs include four proteins having highly homologous periplasmic domains.

6. Explain the role of the reversible methylation of MCPs in chemotaxis by *E. coli.*

____

____

7. The proposed model for the transduction of chemotactic signals via CheY, the tumble regulator, includes which of the following?

(a) Phosphorylated CheY promotes the counterclockwise rotation of the motor.
(b) MCPs with bound repellents activate CheY.
(c) MCPs with bound attractants activate CheY.
(d) The methylation of MCPs increases the probability of activation of CheY.
(e) The activation of CheY requires ATP.

8. Which of the following answers complete the sentence correctly? Two-component regulatory systems in bacteria

(a) are usually activated by a methylation of one of the components.
(b) consist of a pair of sensor and effector proteins.
(c) allow the organism to monitor its environment and internal state.
(d) use an autophosphorylating kinase as one of their components.
(e) usually use a methylesterase as one of their components.

SIGNAL TRANSDUCTION CASCADES

Retinal Photoreceptors

9. Match the two main regions of the rod cell in the left column with the appropriate functions or properties from the right column.

(a) Outer segment ________
(b) Inner segment ________

(1) Contains the synaptic terminal
(2) Contains discs
(3) Carries out protein synthesis and the generation of ATP
(4) Contains the photoreceptors that absorb light
(5) Contains the majority of Na^+-K^+ ATPase pumps
(6) Contains cation-specific channels that are controlled by cyclic GMP

10. Which of the following statements about retinal, the chromophore of rhodopsin, is incorrect?

(a) It is derived from vitamin A in several steps including a dehydrogenation (oxidation) and an isomerization.
(b) In the dark, it is covalently bound to opsin through a Schiff base linkage.
(c) In the dark, it is present as the 11-*cis*-retinal isomer.
(d) When bound to rhodopsin, it absorbs light maximally at 500 nm.
(e) It becomes the all-cis isomer after absorbing light.

11. With its seven transmembrane helices, rhodopsin has a structure similar to that of which of the following integral membrane proteins?

(a) Nicotinic acetylcholine receptor channel
(b) Na^+-K^+ pump
(c) Sarcoplasmic Ca^{2+}-ATPase
(d) Chemoreceptor MCPs
(e) Bacteriorhodopsin light-driven pump
(f) Photoreceptors of retinal cones

12. Place the following events in the excitation of rhodopsin by light in their correct sequence:

(a) Deprotonation of the Schiff base ________
(b) Conversion of all-*trans*-retinal to 11-*cis*-retinal ________
(c) Hydrolysis of the Schiff base ________
(d) Conversion of 11-*cis*-retinal to all-*trans*-retinal ________
(e) Triggering of the enzyme cascade ________
(f) Series of conformational changes of the rhodopsin ________

13. Explain briefly the major roles of the following participants in the enzymatic cascade that is triggered by the photoexcitation of rhodopsin:

(a) Transducin ________________________________

____

(b) Cyclic GMP (cGMP)__________

(c) Activated phosphodiesterase __________

14. How does the photoexcited system return to the dark state?

15. Which of the following statements about color vision are correct?
    (a) It is mediated by three different chromophores.
    (b) It is mediated by three different photoreceptors.
    (c) It involves only the cone cells.
    (d) It involves only the rod cells.
    (e) It involves serpentine, seven-helix transmembrane proteins.

Cyclic AMP—Its Formation and Its Role as a Second Messenger

16. Which of the following statements about adenosine 3′,5′-monophosphate (cyclic AMP, or cAMP) are correct?
    (a) ATP is converted to cAMP by the enzyme adenylate cyclase in one step.
    (b) Cyclic AMP contains a phosphorous atom in a phosphodiester bond.
    (c) ATP reacts with adenosine to form cAMP and ADP in a reaction catalyzed by adenylate kinase.
    (d) Cyclic AMP is converted to 5′-AMP by a phosphodiesterase-catalyzed reaction with H_2O.
    (e) Cyclic AMP is susceptible to hydrolysis to P_i and the ribonucleoside adenosine by phosphomonoesterases.

17. Which of the following statements about cAMP and the second-messenger mechanism of hormone function are correct?
    (a) The hormonal stimulus leads to increased amounts of adenylate cyclase.
    (b) The formation of a hormone-receptor complex leads to the activation of adenylate cyclase.
    (c) Cyclic AMP acts as an allosteric modulator to affect the activities of specific protein kinases.
    (d) Cyclic AMP interacts with a hormone-receptor complex to dissociate the hormone.
    (e) The hormone-receptor complex enters the cell and affects the activities of target enzymes.

18. Why do thyroid cells respond to thyroid stimulatory hormone (TSH) whereas muscle cells do not?

19. Would you expect drinking coffee to have an effect on blood glucose levels, given that caffeine is an inhibitor of cAMP phosphodiesterase? Explain.

20. Suppose a patient is suffering from a disorder in which adenylate cyclase is impaired and, as a result, cAMP levels aren't readily increased by hormones. Explain why the infusion of cAMP probably won't remedy the problem.

Guanyl Nucleotide–Binding Proteins and Adenylate Cyclase

21. Which of the following statements about GTP and its role in the cAMP-mediated hormone response system are correct?

(a) GTP is associated with the α subunit of a guanyl nucleotide–binding protein (Gs protein).
(b) GTP reduces the magnitude of the hormone response because it is converted to cGMP—a compound that antagonizes the effects of cAMP.
(c) GTP maintains the steady-state level of cAMP by rephosphorylating AMP to ATP in a nucleotide kinase–catalyzed reaction.
(d) GTP activates Gs protein so that its Gs_{α}–GTP subunit interacts with adenylate cyclase.
(e) GTP couples the stimulus from a hormone-receptor complex or an activated receptor to a system that produces an allosteric effector.
(f) The effect of GTP on hormone response is antagonized by the GTPase activity of the Gs_{α} subunit of the Gs protein.
(g) A single GTP binding event with a stimulatory G protein leads to the formation of one cAMP molecule.

22. Which of the following are correct statements about Gs proteins and their functioning in cAMP-mediated hormonal systems?

(a) Gs proteins bind hormones.
(b) Gs proteins are integral membrane proteins.
(c) Gs proteins are heterotrimers.
(d) Gs proteins bind adenylate cyclase.
(e) In their GDP form and in the absence of hormone, Gs proteins bind to hormone receptors and are converted to their GTP forms.
(f) When Gs protein in the GDP form binds to a hormone-receptor complex, GTP exchanges with GDP.
(g) The α subunit of G proteins is a GTPase.

23. If cells with β-adrenergic receptors are exposed for extended times to epinephrine, a hormone that activates adenylate cyclase, the G protein fails to carry out efficiently the GDP-GTP exchange reaction and adeny-

late cyclase is no longer activated. What is this phenomenon called, what is its biological function, and how does it occur?

24. Which of the following statements about cAMP and its functioning in hormone action are correct?

(a) Most effects of cAMP in eukaryotic cells are exerted through the activation of protein kinase A (PKA).
(b) Cyclic AMP binds the catalytic subunits of PKA and activates the enzyme allosterically.
(c) Cyclic AMP binds the regulatory subunits of PKA and activates the enzyme by releasing the catalytic subunits.
(d) Cyclic AMP is bound by the activated hormone receptor and PKA simultaneously to convey the hormonal signal in order to activate the kinase.

25. During cAMP-mediated hormone activation, what are the three steps at which amplification occurs?

The Phosphoinositide Cascade and the Modulation of Ca^{2+} Concentrations

26. Which of the following statements about the phosphoinositide cascade are correct?

(a) The phosphoinositide cascade depends on the hydrolysis of a phospholipid component of the plasma membrane.
(b) A polypeptide hormone interacts with a G_{M1} ganglioside on the cell surface to trigger the phosphoinositide cascade.
(c) In some cases, the G protein system acts to transduce the stimulus from the receptor to the phosphoinositidase.
(d) At least four kinds of phospholipase C play a crucial role in the phosphoinositide cascade.
(e) The phosphoinositide cascade directly produces two different second messengers.

27. Which of the following are the second messengers that are produced by the phosphoinositide cascade?

(a) Phosphatidyl inositol 4,5-bisphosphate (PIP_2)
(b) Inositol 1,4,5-trisphosphate (IP_3)
(c) Inositol 4-phosphate
(d) Inositol 1,3,4,5-tetrakisphosphate
(e) Inositol 1,3,4-trisphosphate
(f) Diacylglycerol (DAG)

28. Which of the following statements about inositol 1,4,5-trisphosphate (IP_3) are correct?

(a) IP_3 leads to the uptake of Ca^{2+} by the endoplasmic reticulum and the sarcoplasmic reticulum.
(b) IP_3 may be rapidly inactivated by either a phosphatase or a kinase.

(c) IP_3 opens calcium ion channels in the membranes of the endoplasmic reticulum and the sarcoplasmic reticulum.
(d) IP_3 reacts with CTP to form CDP–inositol phosphate, a precursor of PIP_2.
(e) IP_3 acts by altering the intracellular-to-extracellular Na^+-to-K^+ ratio, thereby altering the transmembrane potential.

29. Which of the following statements about the actions or targets of the second messengers of the phosphoinositide cascade are correct?

(a) Diacylglycerol (DAG) activates protein kinase C (PKC).
(b) Most of the effects of IP_3 and DAG are antagonistic.
(c) DAG increases the affinity of PKC for Ca^{2+}.
(d) PKC requires Ca^{2+} for its activity.
(e) The phorbol esters activate PKC because they mimic DAG.

30. How is a pseudosubstrate involved in the regulation of PKC?

____

____

31. Which of the following statements about G proteins are correct?

(a) G proteins are activated by seven-helix receptors only.
(b) G proteins make up a large family of proteins that are involved in regulating enzymes, chemotaxis, visual excitation, and ion channels.
(c) G proteins cycle between a GTP form and an ATP form by means of a self-catalyzed exchange reaction.
(d) Activated seven-helix proteins switch them on by causing the exchange of GTP for GDP.
(e) Certain G proteins can participate in a reaction with NAD^+ that leads to their covalent modification, for example, reactions catalyzed by cholera and pertussis toxins.

32. Which of the following statements about Ca^{2+} and its roles in the regulation of cellular metabolism are correct?

(a) The solubility product of calcium phosphate is small; therefore, low Ca^{2+} levels must be maintained in the cell to avoid its precipitation.
(b) Intracellular Ca^{2+} is maintained at concentrations that are several orders of magnitude less than the extracellular concentration by ATP-dependent Ca^{2+} pumps.
(c) The transient opening of ion channels in the plasma membrane or endoplasmic reticulum can rapidly raise cytosolic Ca^{2+} levels.
(d) The binding of Ca^{2+} by a protein can induce a large conformational change because the ion simultaneously coordinates to several groups on the protein.
(e) Ca^{2+} is bound by a family of regulatory proteins that have a characteristic EF hand, helix-loop-helix structure.
(f) When calmodulin binds Ca^{2+} at its low-affinity site, it undergoes a conformational change that allows it to interact with target proteins.

33. Explain how a Ca^{2+} ionophore could mimic the effects of a hormone.

____

____

34. If it were incubated with cells in vitro, why would EGTA prevent either a Ca^{2+}-triggering hormone or a Ca^{2+} ionophore from acting?

____

____

35. Which of the following answers complete the sentence correctly? Calmodulin

(a) is a member of the EF hand family of calcium-binding proteins.
(b) activates target molecules by recognizing negatively charged β sheets.
(c) serves as a calcium sensor in most eukaryotic cells.
(d) is activated when intracellular Ca^{2+} concentrations rise above 0.5 μM.
(e) activates CAM kinase II, which then phosphorylates many different proteins.
(f) undergoes a large conformational change upon binding three or four Ca^{2+} ions.

The Receptor Tyrosine Kinases

36. Which of the following answers complete the sentence correctly? Receptor tyrosine kinases

(a) are seven-helix transmembrane receptors.
(b) are integral membrane enzymes.
(c) activate their targets via the G protein cascade.
(d) are often activated by ligand-induced dimerization.
(e) can phosphorylate themselves on their cytoplasmic domains when activated.
(f) that have been activated by hormone binding are recognized by target proteins having SH2 (src protein homology region 2) sequences.
(g) are so named because they contain extraordinarily high amounts of tyrosine.

Signal Transduction and Cancer

37. Which of the following statements about the tyrosine kinases or hormones that affect them are correct?

(a) Nerve growth factor (NGF) and epidermal growth factor (EGF) stimulate target cells to divide.
(b) NGF and EGF are protein kinases that phosphorylate tyrosine residues.
(c) EGF and insulin share a common mechanism of signal transduction across the plasma membrane.
(d) Receptors for EGF and insulin are integral membrane proteins.
(e) Some oncogenes encode tyrosine kinases.
(f) One oncogene, v-*erbB*, produces a truncated form of an EGF receptor that behaves as though it were an activated EGF receptor–hormone complex.

38. Match each compound in the left column with its characteristic from the right column.

(a) Cyclic AMP ____________
(b) GTP ____________
(c) G proteins ____________
(d) Adenylate cyclase ____________
(e) PIP_2 ____________
(f) Ca^{2+} ____________
(g) IP_3 ____________
(h) DAG ____________
(i) A specific phosphodiesterase ____________
(j) Insulin receptor ____________
(k) Glucagon receptor ____________
(l) β-Adrenergic receptor ____________
(m) Arachidonic acid ____________

(1) Cleaved by phospholipase C
(2) Binds the regulatory subunits of specific protein kinases
(3) Converts cAMP to AMP
(4) Exchanges with GDP on G_α subunits
(5) A product of PIP_2 catabolism
(6) Binds epinephrine
(7) A second messenger that raises intracellular Ca^{2+} levels
(8) Leads to increased cAMP levels when complexed to a polypeptide hormone
(9) Transduces hormone stimulus to adenylate cyclase
(10) Has inducible tyrosine kinase activity
(11) Activated by $G_{S\alpha}$–GTP
(12) Intracellular concentration increased by IP_3
(13) Activates protein kinase C

39. Which of the following answers complete the sentence correctly? A mammalian protein, ras,

(a) has a viral counterpart, v-ras, that is oncogenic.
(b) is a protooncogene.
(c) is a component of a signaling pathway for cell growth and differentiation.
(d) can be converted to an oncogene by a point mutation.
(e) is similar to the G proteins in its GTPase activity.

40. Which of the following statements about hormones in mammals are correct?

(a) Hormones are enzymes.
(b) Hormones are synthesized in specific tissues.
(c) Hormones are secreted into the blood.
(d) Hormones alter one or more activities in the cells to which they are targeted.
(e) Hormones display specificity toward the tissues with which they interact.
(f) Hormones are involved in biochemical amplification systems.

Introduction

1. a, b, c, d

Bacterial Chemotaxis

2. (a) 2 (b) 3 (c) 1 (d) 4

3. a, c

4. After the rapid mixing of the bacteria with a solution containing an attractant, the concentration of the attractant around each bacterium would be uniform, so there would be no spatial gradient. If bacteria sensed spatial rather than temporal gradients, the tumbling rate of the bacteria would not be expected to change as rapidly as it did.

5. a, b, d. Answer (e) is incorrect, because it is the cytoplasmic domains that are highly homologous because they interact with a common central processing system.

6. The reversible methylation of MCPs allows the bacteria to adapt to different basal levels of attractants and repellents and to respond to changes in the relative concentrations of attractants and repellents rather than to their absolute levels.

7. b, d, e. Answer (a) is incorrect because phosphorylated CheY induces clockwise rotation of the flagella, leading to tumbling.

8. b, c, d. Two-component regulators comprise a transmitter that senses the environment and a response regulator that effects the intracellular response. In chemotaxis in *E. coli,* CheA, an autophosphorylating kinase is the transmitter that activates CheY, the response regulator, to convey the external signal from the receptor to the flagellar motor. Methylation and methyl ester hydrolysis are accessory elements that vary the basic two-component system in order to modulate chemotaxis.

Retinal Photoreceptors

9. (a) 2, 4, 6 (b) 1, 3, 5

10. e

11. a, f

12. d, f, a, e, c, b

13. (a) Photoexcited rhodopsin binds inactive transducin (T–GDP), catalyzes the exchange of GTP for GDP, and releases T_α–GTP. This form of transducin then activates cGMP phosphodiesterase. Hydrolysis of GTP bound to T_α deactivates phosphodiesterase and allows the binding of $T_{\beta\gamma}$, regenerating T–GDP.

(b) In the dark, cGMP keeps the cation-specific channels open. Activated phosphodiesterase hydrolyzes cGMP to 5′-GMP, leading to the closing of the channels.

(c) When phosphodiesterase is activated by transducin, it hydrolyzes cGMP, which leads to the closing of the cation-specific channels and to hyperpolarization of the plasma membrane.

14. The return to the dark state requires the deactivation of both cGMP phosphodiesterase and photoexcited rhodopsin and the formation of cGMP. Phosphodiesterase is deactivated by the hydrolysis of GTP bound to T_α to return transducin to the inactive state. Photoexcited rhodopsin is deactivated by rhodopsin kinase, which catalyzes the phosphorylation of photoexcited rhodopsin at multiple sites. The phosphorylated rhodopsin binds arrestin, which blocks the binding of transducin. Guanylate cyclase catalyzes the synthesis of cGMP from GTP. Guanylate cyclase is inhibited by high Ca^{2+} levels, and therefore light-induced lowering of Ca^{2+} reactivates cGMP formation.

15. b, c, e

Cyclic AMP—Its Formation and Its Role as a Second Messenger

16. a, b, d. Answer (e) is incorrect because a phosphomonoesterase cannot cleave a phosphodiester bond.

17. b, c. Answer (a) is incorrect because the hormone leads to an increase in the activity of adenylate cyclase, not an increase in the amount of the enzyme. Answer (e) is incorrect because the hormone need not enter the cell to carry out its action.

18. The simplest explanation for the tissue specificity of hormones is the presence or absence of receptors for particular hormones on the extracellular surfaces of the tissues. In this case, the targeted thyroid cells have a receptor for TSH, whereas muscle cells do not.

19. Yes. Caffeine would be expected to increase the steady-state level of cAMP. In the liver, increasing intracellular cAMP levels would stimulate glycogenolysis and thereby increase blood glucose levels (Stryer, pp. 594–595).

20. Aside from the likelihood that serum phosphodiesterases might destroy it, cAMP is a polar molecule that does not readily traverse the plasma membrane. Even if a more hydrophobic derivative, such as dibutyryl-cAMP, were used to overcome the permeability problem, there would be no tissue specificity, and all cells would have increased cAMP levels, leading to a massive, nonspecific response.

Guanyl Nucleotide–Binding Proteins and Adenylate Cyclase

21. a, d, e, f. Answer (g) is incorrect because the GTP form of the G_S protein activates adenylate cyclase such that it forms many cAMP molecules; that is, an amplification occurs.

22. c, d, f, g. Answers (a) and (b) are incorrect because G_S proteins are peripheral membrane proteins inside cells. They don't bind the hormone but rather the activated hormone-receptor complex, and they carry the signal to adenylate cyclase. Answer (e) is incorrect because the hormone

receptor must have the hormone bound to it or it must have been activated by hormone binding before the G protein will bind.

23. The phenomenon is called *desensitization* or *adaptation.* It allows the system to adapt to a given level of hormone so that it can respond to changes in hormone concentrations rather than to absolute amounts. Desensitization is effected by phosphorylation by β-adrenergic receptor kinase at multiple sites on the carboxyl-terminal region of the β-adrenergic receptor when it has epinephrine bound to it. These covalent modifications of the hormone-receptor complex allow β-arrestin to bind it and further inhibit, but not completely prevent, the GDP-GTP exchange. These events thereby decrease the activation of adenylate cyclase. However, the desensitized receptor can still respond to an increase in epinephrine concentrations. Ultimately, a phosphatase reverses the effects of the modification and resensitizes the receptor.

24. a, c

25. A single hormone molecule combines with a single receptor to form several stimulatory $G_{S\alpha}$–GTP molecules. Each of these stimulatory molecules activates an adenylate cyclase molecule to form many cAMP molecules. The cAMP molecules activate protein kinases, mainly PKA molecules, each of which can phosphorylate many target enzymes.

The Phosphoinositide Cascade and the Modulation of Ca^{2+} Concentrations

26. a, c, d, e

27. b, f. Answer (a) is incorrect because PIP_2 is the precursor of the second messengers. The other incorrect choices are all downstream products of IP_3 metabolism.

28. b, c. Answer (a) is incorrect because IP_3 causes the release, not the uptake, of Ca^{2+}. Answer (b) is correct because not only does a phosphatase act on IP_3 but a specific kinase phosphorylates it to form the inactive tetrakisphosphate derivative. Answer (d) is incorrect because free inositol reacts with CDP–diacylglycerol to form phosphatidyl inositol (see p. 687 in Stryer), which is then phosphorylated to form PIP_2.

29. a, c, d, e. Answer (b) is incorrect because most of the effects of IP_3 and Ca^{2+} are synergistic, not antagonistic.

30. The N-terminal domain of PKC contains an amino acid sequence like that of its substrates which interacts with the active site formed by a C-terminal domain. This pseudosubstrate sequence lacks the critical target serine residue, and, consequently, although it binds to the active site, it cannot be phosphorylated. It thus occludes the active site until DAG binds and displaces it, thereby freeing the active site to react with the true protein substrates.

31. a, b, d, e

32. a, b, c, d, e, f

33. The ionophore allows Ca^{2+} to enter cells by effectively rendering the membrane permeable to the ion. Since the extracellular Ca^{2+} concentration is higher than the intracellular concentration, the ion enters the cell

and the cytosolic level increases. Because some hormones act to raise intracellular Ca^{2+} levels in order to carry out their physiological roles, the ionophore could lead to the same response.

34. EGTA is a specific Ca^{2+} chelator. It would bind tightly to the ion and markedly lower its concentration in the extracellular medium. Consequently, when a hormone or a Ca^{2+} ionophore acted to allow Ca^{2+} influx, none could occur because the concentration gradient of Ca^{2+} would be insufficient.

35. a, c, d, e, f. Answer (b) is incorrect because positively charged amphipathic α helices are recognized.

The Receptor Tyrosine Kinases

36. b, d, e, f. Answer (g) is incorrect because the name arises from the amino acid that they phosphorylate in their target proteins.

Signal Transduction and Cancer

37. a, c, d, e, f. Answer (b) is incorrect because the hormones themselves do not have tyrosine kinase activities; only their activated receptors are active.

38. (a) 2 (b) 4 (c) 9 (d) 11 (e) 1 (f) 12 (g) 7 (h) 13 (i) 3 (j) 10 (k) 8 (l) 6 (m) 5

39. a, b, c, d, e

40. b, c, d, e, f

PROBLEMS

1. *E. coli* cells that are grown in a medium that is deficient in methionine show impaired chemotactic response. Why do you think this is the case?

2. Figure 13-1 depicts the response of bacteria to two chemotactic agents, X and Y. Which agent is likely to be an attractant, and which is likely to be a repellent? Give the reasons for your choices.

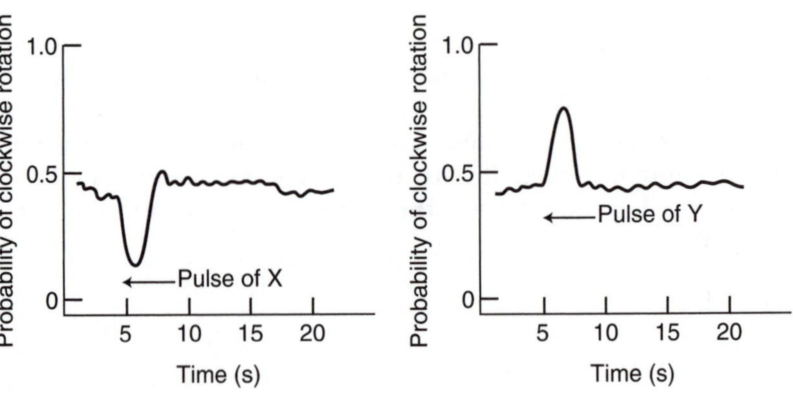

Figure 13-1
Bacterial response to two chemotactic agents.

3. Which of the compounds in Figure 13-2 would be expected to have the smallest absorbance per mole in the visible light range? Give the reason for your choice.

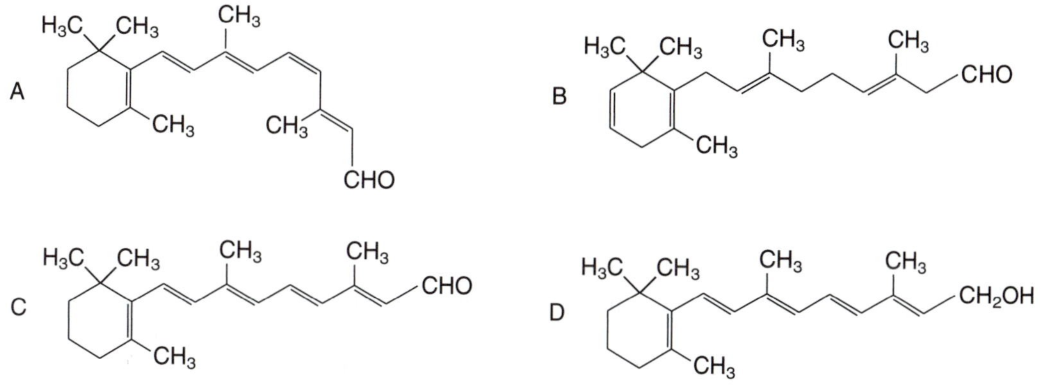

Figure 13-2
Four light-absorbing compounds.

4. Figure 13-3 shows the absorption spectrum of a light-absorbing pigment that is involved in color vision.

(a) What color light is absorbed by the pigment?
(b) When the pigment is extracted into an organic solvent, what is the color of the resulting solution?

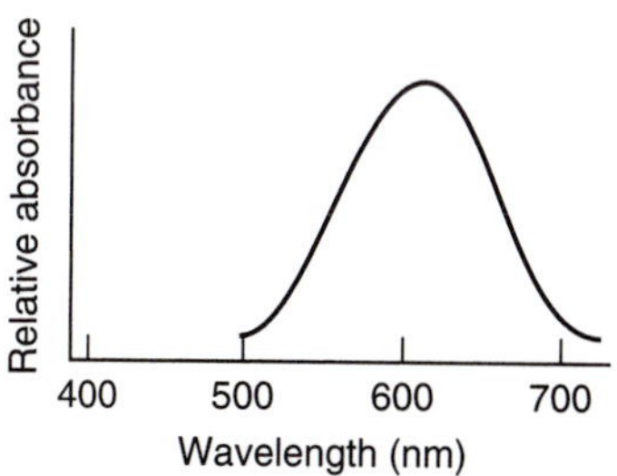

Figure 13-3
Absorption spectrum of a color vision pigment.

5. 11-*cis*-Retinal is covalently linked to a lysine side chain of opsin by the formation of a Schiff base:

$$\text{R—CHO} + \text{H}_2\text{N—(CH}_2)_4\text{—Opsin} \overset{\text{H}^+}{\rightleftharpoons} \text{R—CH=N}^{\oplus}\text{H—(CH}_2)_4\text{—Opsin} + \text{H}_2\text{O}$$

Such linkages can typically be stabilized in the laboratory by reduction with borohydride:

$$\text{R—CH=N}^{\oplus}\text{H—(CH}_2)_4\text{—Opsin} \xrightarrow{\text{BH}_4^-} \text{R—CH}_2\text{—NH—(CH}_2)_4\text{—Opsin}$$

Would you expect a rod cell preparation that has been treated with borohydride to be active in the cycle of visual excitation? Why or why not?

6. A color vision pigment that absorbs red light is chemically cleaved to separate the retinal from the protein. The same is done for a pigment that absorbs blue light. Then a new pigment is constituted using the retinal from the red-absorbing pigment and the opsin from the blue-absorbing pigment. What color of light will be absorbed by the new pigment?

7. Suppose that an experiment on a light-excitable cell involving a patch electrode is carried out as described by Stryer on page 337 and the following experimental results are obtained:

[*cGMP*] (μM)	Y
7.0	0.2
9.0	0.4
10.0	0.5
11.1	0.6
14.1	0.8

Calculate the minimum number of cyclic GMP molecules that are required to open each cation-specific channel of the cell. You should refer to the discussion of Hill plots on page 158 of Stryer.

8. Give the major criteria that must be fulfilled by a substance for it to be classified as a hormone.

9. Bee venom is particularly rich in phospholipase A_2, an enzyme that hydrolytically removes the fatty acyl residue at position 2 of phospholipids. The action of phospholipase A_2 on phosphatidyl choline is shown in Figure 13-4. One of the mediators of the inflammatory response following a bee sting (swelling, redness, pain, heat, and loss of function) is lysophosphatidyl choline, the remainder of the phospholipid following the hydrolysis of the fatty acyl residue at position 2. Lysophosphatidyl choline stimulates mast cells to release histamine, which triggers the inflammatory response.

$CH_2-O-C(=O)-R_1$

$R_2C(=O)-O-C-H$ (cleavage site A_2)

$CH_2-O-P(=O)(O^-)-O-CH_2-CH_2-N^+(CH_3)_3$

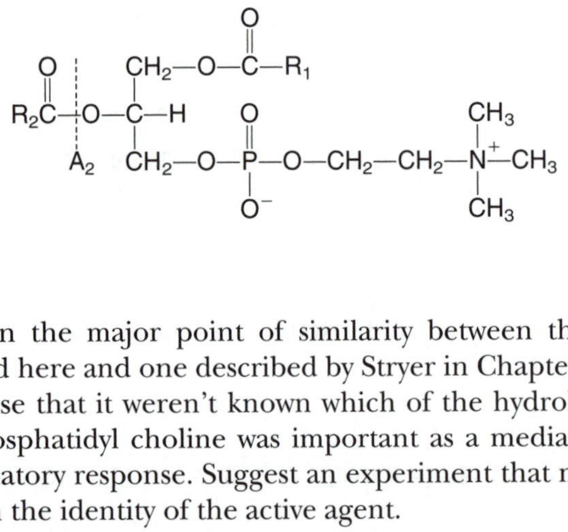

Figure 13-4
Action of phospholipase A_2 on phosphatidyl choline.

(a) Explain the major point of similarity between the system described here and one described by Stryer in Chapter 13.
(b) Suppose that it weren't known which of the hydrolysis products of phosphatidyl choline was important as a mediator of the inflammatory response. Suggest an experiment that might help establish the identity of the active agent.

10. Suppose that epinephrine stimulates the conversion of A to B in liver cells via a regulatory cascade involving G_S protein (Stryer, p. 342), cAMP, protein kinase A, and enzymes E_1 and E_2 as shown in Figure 13-5. Assume that each catalytically active enzyme subunit in the regulatory cascade has a turnover number of 1000 s^{-1}. Assume further that 10 G_S–GTP are formed for each molecule of epinephrine bound to receptor. Calculate the theoretical number of molecules of A that would be coverted to molecules of B per second as a result of the interaction of one molecule of epinephrine with its receptor on a liver cell membrane. (See Stryer, pp. 340–343.)

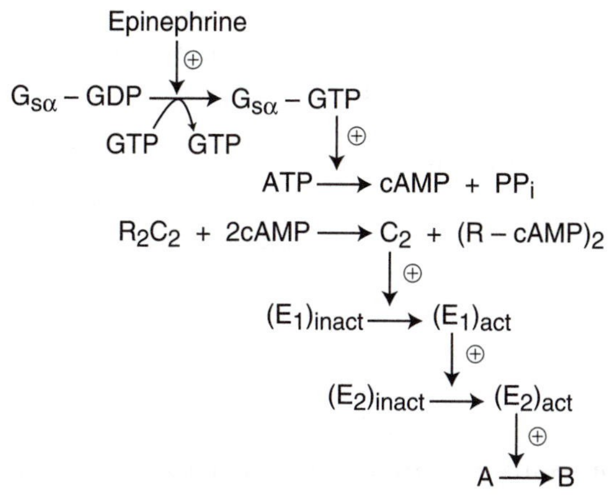

Epinephrine

$$G_{s\alpha} - GDP \xrightarrow[GTP \quad GTP]{\oplus} G_{s\alpha} - GTP$$

$$\downarrow \oplus$$

$$ATP \longrightarrow cAMP + PP_i$$

$$R_2C_2 + 2cAMP \longrightarrow C_2 + (R - cAMP)_2$$

$$\downarrow \oplus$$

$$(E_1)_{inact} \longrightarrow (E_1)_{act}$$

$$\downarrow \oplus$$

$$(E_2)_{inact} \longrightarrow (E_2)_{act}$$

$$\downarrow \oplus$$

$$A \longrightarrow B$$

Figure 13-5
Hypothetical regulatory cascade for problem 10.

11. In the early days of research on insulin action, it was not known whether insulin might enter cells and directly mediate intracellular effects or whether it might act through a second messenger. In a classic experiment, Pedro Cuatrecasas attached insulin covalently to sepharose beads many times the size of fat cells and showed that the addition of the insulin–sepharose complexes to isolated fat cells gave the same stimulation of glucose oxidation as did addition of insulin alone.

(a) What conclusion might follow from this experiment? Explain.
(b) What assumption have you made about the effects of adding sepharose alone to fat cells?

12. Suppose that you are trying to isolate a receptor for a polypeptide hormone from liver cells. Suggest an effective means of purification involving specialized-column chromatography and highly purified polypeptide hormone. Also explain how you would get the receptor off the column.

13. In kinetic studies on the interaction of human growth hormone with its receptor, it was found that although the Hill coefficient (Stryer, p. 158) was 1.0, receptors needed to dimerize in order to transduce the signal. Explain how this might occur.

14. To be effective, intracellular signal substances must be readily inactivated when their effects are no longer needed. Give a method of inactivation for each of the following classes of intracellular messengers:

(a) G proteins
(b) Cyclic nucleotides
(c) Phosphoproteins
(d) Calcium ion
(e) Inositol 1,4,5-trisphosphate (IP_3)
(f) Diacylglycerol

15. A tissue is known to increase cyclic AMP production upon stimulation by a certain hormone. Addition of an analog of GTP in which the terminal phosphate group is replaced by a sulfate to a homogenate of the tissue results in sustained production of cyclic AMP. Propose an explanation for this observation.

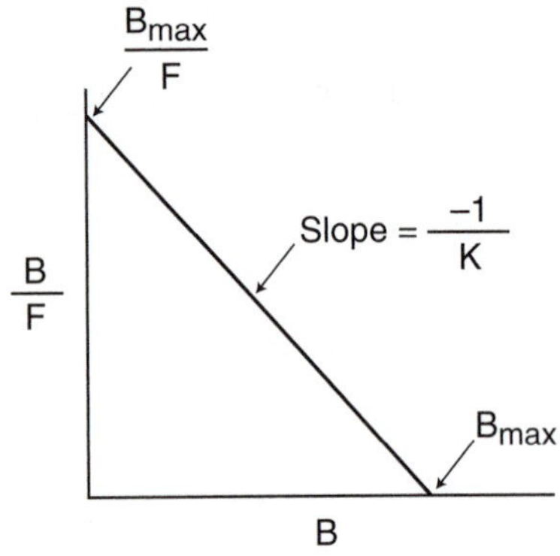

Figure 13-6
Scatchard plot for a ligand and highly purified receptor.

16. Just as steady-state kinetic studies of enzymes can yield valuable information about reaction mechanisms, so can equilibrium binding studies on the interaction of hormonal ligands with receptors. Typically, in such studies, radioactively labeled hormones are incubated with a receptor preparation long enough for equilibrium to be achieved. The amounts of free and bound hormone may be readily measured because insoluble hormone-receptor complexes may be separated from free hormone by filtration. A typical linear plotting form for such data is the Scatchard plot (Figure 13-6), in which the bound/free ratio is plotted against the amount bound. The slope of such a line is equal to -1/K, where K is the dissociation constant for the ligand-receptor complex; the intercept on the B/F axis is B_{max}/F; and the intercept on the B axis is B_{max}, a value that can be used to estimate the number of receptors in the system. (A Scatchard plot is the equilibrium binding analog of the Eadie plot for steady-state enzyme kinetic data, presented in problem 6 on page 206 of Stryer. B is the analog of *V* of the Eadie plot, B_{max} is the analog of V_{max}, and F is the analog of S.) When a system consists of a ligand and a highly purified receptor, the expected linear plot is usually obtained. On the other hand, when insulin is incubated with a tissue homogenate known to contain insulin receptors, the resulting plot is a curve that is concave upward (Figure 13-7). Propose two reasons that might account for the more complex observations in the latter system. (Hint: Think about the fact that tissue homogenates contain many molecular components. Also, think about some mechanisms used to explain allosterism.)

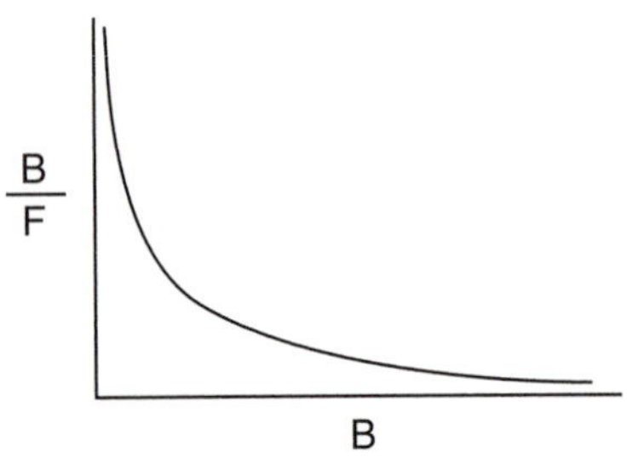

Figure 13-7
Scatchard plot for insulin and liver cell homogenate.

ANSWERS TO PROBLEMS

1. The chemotactic response depends on the rapid and reversible methylation of several glutamate side chains of the methyl-accepting chemotaxis proteins (MCPs). *S*-adenosylmethionine is the methyl-group donor in these methylation reactions. A methionine deficiency would result in a decreased pool of *S*-adenosylmethionine and, hence, in an impaired chemotactic response. (See p. 331 of Stryer.)

2. X is likely to be an attractant and Y is likely to be a repellent. The addition of X causes a decrease in clockwise rotation (an increase in counterclockwise rotation), which leads to smooth swimming. The addition of Y causes an increase in clockwise rotation (a decrease in counterclockwise rotation), which leads to tumbling.

3. Compounds that absorb visible light significantly have long sequences of alternating single and double bonds; that is, they are conjugated. Compound B is unconjugated, so it would have negligible absorbance in the visible range of the spectrum. Compound A is 11-*cis*-retinal, C is all-*trans*-retinal, and D is all-*trans*-retinol.

4. (a) Red light is absorbed by the pigment.

   (b) A solution of the pigment would be blue or blue-green. The pigment absorbs red light, but it transmits light at the blue end of the spectrum.

5. The photoexcitation of rhodopsin leads initially to the isomerization of 11-*cis*-retinal to all-*trans*-retinal and ultimately to the cleavage of all-*trans*-retinal from the protein. In a rod cell preparation that has been treated with borohydride, the Schiff base would be stabilized by reduction, so the removal of the all-*trans*-retinal would be impaired. As a result, the cycle of visual excitation could not occur. (See p. 334 of Stryer for a discussion of these events.)

6. The new pigment will absorb blue light. The retinal is the same in all color vision pigments. The protein component, however, varies among the pigments, giving absorption maxima at different wavelengths.

7. We start with the Hill equation from page 158 of Stryer:

$$\log \frac{Y}{1-Y} = n \log [\text{cGMP}] - n \log [\text{cGMP}]_{50}$$

The second term on the right side of the equation represents the concentration of cGMP that gives 50 percent opening. From inspection of the data obtained, that value is 10 μM. We then get the values in Table 13-1. The resulting Hill plot is shown in Figure 13-8. The Hill coefficient n is equal to the slope of the plot and is given by

$$n = \frac{0.8 - (-0.7)}{0.4} = 3.8$$

Since $n = 3.8$, we estimate that four molecules of cGMP must be hydrolyzed to open each channel.

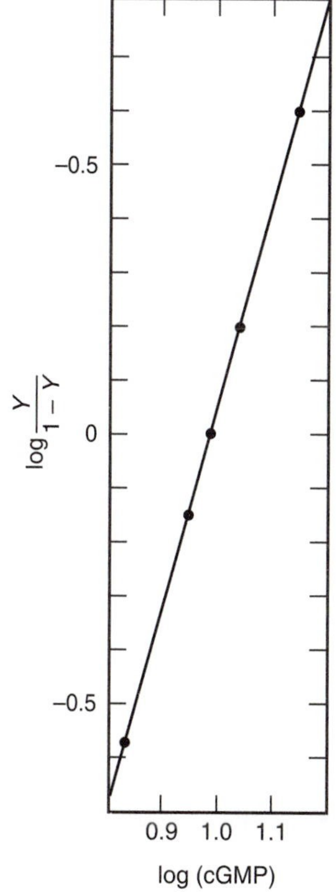

Figure 13-8
Hill plot for problem 7.

Table 13-1
Values for Hill Plot

Y	$1-Y$	$\frac{Y}{1-Y}$	$\log \frac{Y}{1-Y}$	[cGMP]	log [cGMP]
0.2	0.8	0.25	−0.60	7.0	0.85
0.4	0.6	0.67	−0.17	9.0	0.95
0.5	0.5	1.0	0	10.0	1.0
0.6	0.4	1.5	0.18	11.1	1.05
0.8	0.2	4.0	0.60	14.1	1.15

8. The major criteria for classifying a substance as a hormone are as follows:

   (1) It should be produced by one type of cell and have effects on another type of cell.
   (2) Its effects should involve the chemical amplification of the original signal.
   (3) It should be produced in response to a stimulus, and its production should cease upon cessation of the stimulus.

(4) The substance should be selectively destroyed following cessation of the stimulus.
(5) The addition of the purified substance to tissues should mimic physiologic stimuli.
(6) Specific inhibitors of the physiologic response should also abolish the response elicited by the addition of the purified substance to tissues.
(7) Specific receptors for the hormone should exist and should be more abundant in tissues that are more sensitive to the hormone.

9. (a) The bee venom system resembles the phosphoinositide cascade discussed by Stryer on page 345. In that system, a membrane phospholipid is also converted into an active mediator of the response of several hormones.

(b) One could inject each of the hydrolysis products—lysophosphatidyl choline and the fatty acid—into tissues separately to see which elicits the inflammatory response.

10. The theoretical number of molecules of A converted to B per second would be 10^{13}. One molecule of epinephrine would result in the production of 10 G_S–GTP. Each activated α subunit would stimulate adenylate cyclase to produce 1000 cAMP molecules for a total of 10,000 molecules of cAMP. Each of these cAMP molecules would activate one catalytic subunit of protein kinase. (Remember that a molecule of protein kinase exists as an R_2C_2 complex. Two cAMP molecules combine with two R subunits to give two catalytically active C subunits.) Each of the 10,000 active C subunits would result in the production of 1000 molecules of active E_1, for a total of 10^7 molecules of active E_1. Each molecule of active E_1 would in turn activate 1000 molecules of E_2 for a total of 10^{10} molecules of active E_2. Since each molecule of active E_2 would convert 1000 molecules of A to B per second, the total would be $1000 \times 10^{10} = 10^{13}$ per second.(Note: This is a greatly oversimplified example, but it illustrates the profound chemical amplification that can occur in systems under hormonal control.)

11. (a) The best conclusion is that insulin need not enter the cell to have an effect. The results are consistent with the notion that insulin affects cells by combining with a membrane receptor and releasing some second messenger that mediates its intracellular effects. Note that the experiment doesn't prove that insulin doesn't enter cells.

(b) You very likely assumed that the addition of sepharose alone gave no stimulation of glucose oxidation.

12. The most useful technique would be affinity chromatography. Covalently attach the purified hormone to sepharose beads, and use the complex to fill a column (see problem 11). Homogenize liver cells, and add the homogenate to the column. Receptors will stick on the column because they are complementary in shape to the covalently bound hormone. Elute the receptors from the column as hormone-receptor complexes by adding free hormone to compete with the hormone that is bound to the column.

13. A given molecule of growth hormone could contain two domains, each of which could bind receptor. Thus a single molecule of growth hormone could tie two receptors together.

14. (a) G proteins are active while GTP is bound, but become inactive as GTP is hydrolyzed to GDP and phosphate.

(b) Cyclic nucleotides are converted by phosphodiesterases to 5′-mononucleotides.

(c) Phosphates are cleaved from phosphoproteins by phosphatases.

(d) Calcium ions are pumped from the cell interior into the extracellular fluid.

(e) Inositol 1,4,5-trisphosphate can be degraded to inositol and inorganic phosphate by the sequential action of phosphatases.

(f) Diacylglycerol may be converted to phosphatidate, or hydrolyzed to glycerol and fatty acids.

15. The observation could be explained if the sulfate-containing analog of GTP bound to G protein, stimulating cyclic AMP production, but failed to be hydrolyzed to GDP and sulfate by the GTPase activity of G_S. Thus the production of cyclic AMP would persist.

16. A homogenate of a tissue containing insulin receptors contains other molecules that may bind to insulin less specifically than do insulin receptors. Thus we could have a whole population of ligands, each with a different affinity for insulin. The result would be a Scatchard plot that is concave upward. The second reason for such behavior has its analog in allosteric enzyme behavior. Remember that binding of substrate to one subunit of an enzyme may increase or decrease the binding affinity of other subunits for substrate. (You may wish to review the discussion of the sequential and concerted models for allosterism on p. 166 of Stryer.) In the case of the interaction of insulin with its receptor, we may postulate that binding of insulin to some receptors on a cell may inhibit binding of insulin to other receptor molecules, an instance of so-called negative cooperativity. This would also result in a Scatchard plot that is concave upward.

EXPANDED SOLUTIONS TO TEXT PROBLEMS

1. Phosphorylation mediates the excitatory response in chemotaxis and receptor tyrosine kinase cascades, whereas it participates in recovery and adaptation in vision. Autophosphorylation of cheA is essential in chemotaxis. Likewise, autophosphorylation of receptor tyrosine kinases is crucial in growth control cascades. By contrast, phosphorylation of photoexcited rhodopsin leads to its deactivation.

2. The binding of an attractant rapidly leads to the formation of less phosphorylated cheY, and hence the probability of tumbling decreases. On a slower time scale, the binding of attractant leads to the formation of less phosphorylated cheB (the more potent form of the methylesterase). Also, the rate of methylation is increased. Consequently, the methylation level of the receptor increases, which speeds the production of phosphorylated cheY to bring the tumbling probability back to the initial value. Hence, excitation is followed by adaptation.

3. (a) Loss of *cheB* (the gene for the methylesterase) would give receptors that are fully methylated. The bacterium would persistently tumble, always convinced that it is heading in the wrong direction.

   (b) Loss of *cheR* (the gene for methyltransferase) would give receptors that are completely demethylated. The bacterium would swim smoothly all the time, confident that it is heading for greener pastures.

4. (a) The photoresponse would be prolonged because deactivation of R* would be inhibited.

   (b) The ampl10itude of the photoresponse would be decreased because less cGMP would be hydrolyzed.

   (c) EGTA would lower the calcium level, which would increase guanylate cyclase activity. The photoresponse would be diminished by the higher cGMP level.

   (d) The photoresponse would be markedly prolonged because GTPγS bound to transducin cannot be hydrolyzed. The phosphodiesterase would be persistently activated.

5. The absorption maximum is likely to be intermediate between that of the red and the green receptors. The capacity to distinguish hues between red and green depends on having receptors with well-separated absorption maxima. The separation between the hybrid and green receptors is less than in the normal case. Hence, color vision is impaired.

6. In the dark the catalytic subunits (α and β) of cGMP phosphodiesterase (PDE) are inhibited by a pair of γ subunits. T_α–GTP, the active form of transducin, activates PDE by prying away its γ subunits.

The R_2C_2 complex of protein kinase A (PKA) is inactive. The binding of cAMP to the R (regulatory) chains releases catalytically active C chains.

The regulatory domain of protein kinase C (PKC) inhibits the catalytic domain by occupying the substrate binding site. The binding of diacylglycerol in the presence of Ca^{2+} disrupts this interaction and enables a protein substrate to enter.

7. ADP-ribosylation by cholera toxin leads to persistent activation of adenylate cyclase because G_S–GTP is very slowly hydrolyzed to G_S–GDP. The abnormally high cAMP level leads to complete activation of protein kinase A, which then phosphorylates chloride channels. The proportion of open channels is much higher than normal, which leads to excessive loss of salt and water.

8. The Ca^{2+}-ATPase pump contains a region (D) that binds this internal calmodulinlike sequence (C). The catalytic site is blocked by the formation of the $C{\cdot}D$ complex. Ca^{2+}–calmodulin (C') disrupts this internal complex and activates the pump by binding D. Thus, *calcium–calmodulin reverses the effect of an autoinhibitory region of its target.* Once again we see that stimulation of one protein by another can be mediated by relief of a built-in inhibitory constraint.

9. Most receptor tyrosine kinases need to be phosphorylated to transmit signals. This phosphorylation is achieved by autophosphorylation catalyzed by the receptor dimer. Since dimerization is required for activation, a search for compounds that block dimerization might be fruitful.

CHAPTER
14

Antibodies and T-Cell Receptors

Chapter 14 deals with the proteins of the immune system. The cells and proteins of this system cooperate to detect and inactivate foreign (nonself) molecules, microorganisms, and viruses. The humoral immune response acts through soluble antibodies, secreted into the circulatory system, that bind antigens with high specificity and affinity. The cellular immune response acts through receptor proteins similar to antibodies on the surface of specialized cells (T lymphocytes) that bind to peptides presented on another kind of cell-surface protein, the MHC proteins.

After defining essential immunological terms, Stryer describes the kinetics of the appearance in serum of immunoglobulins following the introduction of an antigen into an animal. He then describes the clonal selection theory that explains the origin of antibody specificity, outlines the production of monoclonal antibodies, and discusses the structure of immunoglobulin G, including the molecular details of antibody-antigen recognition. Next Stryer relates the variable and constant regions of immunoglobulins to the organization of the genes that encode them. He explains how somatic recombination of a large variety of *V*-, *D*-, and *J*-segment genes with a few *C* genes, plus imprecise joining, and addition of extra nucleotides can generate the enormous diversity of antibody molecules. The five classes of immunoglobulins with their characteristic polypeptide compositions and functions are introduced, and the complement cascade, which lyses infected cells and microorganisms, is described. Gene rearrangements also give rise to the five different classes of immunoglobulins, and alternative mRNA splicing leads to the formation of membrane-bound or soluble immunoglobulins. Finally Stryer turns to the cellular immune response. He describes the functions of killer T cells and helper T cells, and discusses the diversity, polypeptide composition, and structures of major histocompatibility complex (MHC) proteins and T-cell receptors. Stryer concludes this chapter with a discussion of the HIV infection of helper T cells.

Preceding chapters in Stryer's text that deal with protein structure (Chapters 2 and 7), molecular recognition (Chapters 8 and 10), and flow of genetic information (Chapters 5 and 6) constitute important background for this chapter.

LEARNING OBJECTIVES

When you have mastered this chapter, you should be able to complete the following objectives:

Introduction and the Clonal Selection Theory (Stryer pages 361–364)

1. Define or describe the following immunological terms: *antibody* (*immunoglobulin*), *plasma cell, B lymphocyte* (*B cell*), *antigen* (*immunogen*), *antigenic determinant* (*epitope*), and *hapten.*

2. Describe the *humoral immune response* in terms of the immunoglobulins involved and the order of their appearance after *immunization.*

3. Explain how a small foreign molecule can give rise to an immune response.

4. Distinguish between the *instructive theory* and the *selective* (*clonal selection*) *theory* of antibody formation. Describe an important experiment that supports the clonal selection hypothesis.

5. List the salient features of the clonal selection theory.

Immunoglobulin Structures, Monoclonal Antibodies, and Antibody-Antigen Interactions (Stryer pages 364–372)

6. Relate the intact structure of an *immunoglobulin G* (*IgG*) molecule to the F_{ab} and F_c fragments produced by proteolysis. Describe the functions performed by the different regions of IgG.

7. Note the similarities and differences between naturally occurring antibodies and enzymes.

8. Describe *multiple myeloma* and the nature of the antibodies produced by *myeloma cells.*

9. Outline the procedure for forming a *hybridoma cell* by fusing a myeloma cell with a lymphocyte. Describe *monoclonal antibodies,* list some of their uses, and explain how they are produced.

10. Sketch the polypeptide chains of an IgG molecule, and relate the *heavy chain (H)–light chain (L) subunit composition* (H_2L_2) to the F_{ab} and F_c fragments of the molecule. Understand the function of the *hinge region* of IgG.

11. Describe the features of the *constant* (*C*), *variable* (*V*), and *hypervariable* amino acid sequences of the L and H chains of IgG molecules.

12. Relate the hypervariable sites of the H and L chains of IgG to the *complementarity-determining regions* (*CDRs*) of the immunoglobulin. Describe the function of the constant regions of the H and L chains of IgG.

13. Describe the *domain structure* of IgG as revealed by crystallography, and note the presence of the *immunoglobulin fold* as a common structural feature. Locate the *antigen-binding sites.*

14. Summarize the types of bonds that form complexes between immunoglobulins and haptens or antigens. Note the similarities of *combining sites* of immunoglobulins with the active sites of enzymes.

Immunoglobulin Gene Structure and Expression (Stryer pages 372–378)

15. Compare the relative spatial relationships of the *immunoglobulin genes* that encode the V and C regions of the L chains in *embryonic* (*germ-line*) *cells* and *antibody-producing* (*plasma*) *cells.*

16. Describe the contributions of the number of immunoglobulin genes (the *germ-line repertoire*), *somatic recombination,* and *somatic mutation* in the generation of antibody diversity. Describe the roles of the *V, C, D,* and *J segment genes, DNA recombination in alternative reading frames,* and *terminal deoxyribonucleotidyl transferase* in these processes.

17. Calculate the diversity of immunoglobulin structures that arise from the *combinatorial association* of different genes and from somatic mutation.

18. List the different classes of immunoglobulins (*isotypes*) and give their functions. Note the common occurrence of (κ) and (λ) L chains in all classes and the (α, μ, δ, γ, or ε) H chains that provide the structural bases for the function of each class.

19. Outline the components and the mechanism of action of the *complement system* in the lysis of microorganisms and infected cells.

20. Describe the phenomenon of *class switching,* and note its significance in maintaining constant recognition specificity among the immunoglobulin classes.

T-Cell Function and the Cellular Immune Response (Stryer pages 378–386)

21. Contrast the usual targets of the humoral immune response and the *cellular immune response.*

22. Describe the formation and recognition of foreign peptides displayed on cell surfaces in complexes with *major histocompatibility complex* (*MHC*) proteins.

23. Distinguish between *killer T cells* and *helper T cells* and describe their functions.

24. Describe the three classes of proteins (*class I, class II,* and *class III*) encoded by the *MHC genes.* Note the diversity of the class I and class II MHC proteins and their significance in *transplantation rejection.*

25. Sketch the subunit structures of the class I and class II MHC proteins. Indicate the peptide-binding sites.

26. Describe the interaction of a peptide with the peptide-binding site of HLA-B27.

27. Compare the structures and features of *T-cell receptors* with those of the immunoglobulins. Note the sizes and conformations of the epitopes that are recognized by each kind of protein.

28. Explain the origins of the diversity of T-cell receptors.

29. Outline a model that accounts for the recognition of a *combined epitope* by a T-cell receptor.

30. Explain the role of the *CD4* and *CD8 coreceptors* in the activation of T cells, and describe the functions of *perforin* and *interleukins.*

31. List examples of protein types that contain the immunoglobulin fold structural motif.

32. Describe the *human immunodeficiency virus* (*HIV*) and the disease it causes (*AIDS*).

33. Outline the mechanism of infection and lysis of helper T cells by HIV.

SELF-TEST

Introduction and the Clonal Selection Theory

1. Match the immunological term in the left column with its description or definition in the right column.

(a) Antigen______________

(b) Antibody______________

(c) Epitope______________

(d) Hapten ______________

(1) A particular site on an immunogen to which an antibody binds
(2) Protein synthesized in response to an immunogen
(3) Macromolecule that elicits antibody formation
(4) Small foreign molecule that elicits antibody formation

2. Small foreign molecules do not usually elicit the formation of soluble antibodies, and the cellular immune system also ordinarily responds only to macromolecules. Explain how an antibody can sometimes be directed against a small foreign molecule.

____

____

3. Which of the following events usually follow the injection of an antigen into a rabbit?

(a) IgG antibodies reach a relatively constant concentration in the serum approximately three weeks after injection.
(b) IgG antibodies appear in large quantities approximately three days after antigen injection.
(c) A second administration of the antigen a month or more after the initial injection elicits still higher levels of IgG antibodies.
(d) IgM antibodies appear approximately 10 days after antigen injection.

4. Describe a key experiment with a fragment of IgG that bound antigen which contradicted a prediction of the instructive theory of antibody formation. State the prediction of the instructive theory.

____

____

5. Which of the following statements about the clonal selection theory of antibody formation are correct?

(a) Each specific antibody is made by a single cell or its clonal descendents.
(b) All the cells producing antibodies die when the antigen stimulating those cells is absent.
(c) If the surface antibody of an immature, antibody-producing cell interacts with its antigen during fetal life, the cell is stimulated to divide and begin production of soluble antibodies.
(d) Each antibody-producing clone contains *unique* sequences of DNA that encode the polypeptide chains of its antibodies.

Immunoglobulin Structures, Monoclonal Antibodies, and Antibody-Antigen Interactions

6. Match the structure or feature listed in the right column with the appropriate IgG fragment on the left.

(a) F_{ab} ______________________

(b) F_c ______________________

(1) Contains an antigen combining site.
(2) One is formed per IgG molecule.
(3) Contains an H-chain fragment.
(4) Contains an intact L chain.
(5) Mediates complement fixation in the intact IgG.
(6) Two are formed from an IgG molecule.
(7) Forms a precipitate upon binding an antigen.

7. Explain why the antibodies produced in an animal in response to a hapten display a range of binding constants for the antigen eliciting them.

____

____

8. What is the unique feature of the immunoglobulins produced in a patient with multiple myeloma?

____

____

9. Which of the following statements about the immunoglobulins produced by a clone of a myeloma cell is correct?

(a) Myeloma antibodies generally have shorter C regions than do normal IgG immunoglobulins.
(b) Myeloma antibodies display a range of binding constants for the antigen that elicits their synthesis.
(c) Myeloma antibodies are directed against several epitopes on the surface of the antigen that elicits their synthesis.
(d) Myeloma antibodies are monoclonal antibodies.

10. Because they are nearly homogeneous populations, myeloma immunoglobulins were useful for studying antibody structure. What rendered them less than ideal for the study of antigen-antibody interactions?

____

____

11. What key discovery did Milstein and Kohler make that overcame the difficulty in studying the antibody-antigen interactions referred to in question 10?

____

____

12. Which of the following statements about monoclonal antibodies are correct?

(a) Monoclonal antibodies can be produced in large amounts by growing many mice, each of which has been immunized with the same antigen.
(b) Monoclonal antibodies that bind nearly any antigen can be produced.
(c) Monoclonal antibodies are useful for quantifying specific proteins in human blood.
(d) Immobilized monoclonal antibodies can be used to purify scarce proteins by affinity chromatography.
(e) Fluorescent-labeled monoclonal antibodies can be used in conjunction with fluorescence-activated cell sorting to isolate cells having a given cell-surface antigen.

13. Which of the following statements about the L and H chains of IgG are correct?

(a) The H chains of IgG molecules have variable and constant regions of amino acid sequences.

(b) Bence-Jones proteins provided the first information about the amino acid sequence of the L chains of IgG.
(c) The constant region of L chains exists in two allotypes (κ and λ), which have a single difference at one amino acid position.
(d) The variable region of the L chain has a counterpart of the same length and amino acid sequence in the variable region of the H chain.

14. Which of the following statements about IgG structure are correct?

(a) Each of the two antigen-combining sites on an IgG molecule can bind to a structurally distinct epitope.
(b) Both interchain and intrachain disulfide bonds stabilize IgG structure.
(c) Both the L and the H chains of IgG contain domains with similar structures.
(d) The F_c unit of IgG binds to a complement-pathway component to initiate the lysis of an infected cell.
(e) The hypervariable regions of the L chain are the sole determinants for the binding of the IgG to the specific antigen.

15. Which of the following are properties of antigen-binding sites of IgG?

(a) They are located between the two sheets of β strands of the V domains.
(b) They are made up of loops formed by the complementarity-determining regions of both the V_L and the V_H domains.
(c) They may undergo conformational changes upon binding of the hapten or antigen.
(d) They contain a specific amino acid residue that covalently binds to the antigen.
(e) They form numerous weak electrostatic, hydrogen bond, van der Waals, and hydrophobic interactions with the antigen surface.

Immunoglobulin Gene Structure and Expression

16. If the mRNA encoding the L chain of an IgG molecule were isolated, radiolabeled, and hybridized to genomic DNA that has been isolated from either the plasma cell producing the antibody or from germ-line cells from the same organism, what would you observe with respect to the relative locations of the L-chain gene sequences on the two DNAs?

____

____

17. Some antibody diversity arises from the combination of one *V* gene and one *C* gene from pools containing numerous different copies of each. Why doesn't this mechanism account completely for the observed diversity?

____

____

18. Select the mechanism that does not contribute to the diversity of the sequences of antibodies.

(a) Selection from hundreds of V_L and V_H genes in the germ line
(b) Recombination of diverse *V*, *J*, and *D* gene segments in somatic cells

(c) Recombination of V and J genes in different joining frames
(d) Insertion of nucleotides between V_H and D gene segments by deoxyribonucleotidyl transferase
(e) Trimming of complementarity-determining regions by peptidyl transferase

19. Match the following gene segments with their approximate number in the germ line.

(a) V_L ________
(b) V_H ________
(c) J_L ________
(d) J_H ________
(e) D ________

(1) 3
(2) 4
(3) 5
(4) 15
(5) 250

20. Match the immunoglobulin class listed in the left column with its property or function from the right column.

(a) IgA ________
(b) IgD ________
(c) IgE ________
(d) IgG ________
(e) IgM ________

(1) Most prevalent soluble antibodies
(2) Unknown function
(3) First soluble antibodies to appear in serum after immunization
(4) Protect against parasites
(5) Major antibodies in tears, saliva, and mucus

21. Which of the following statements about class switching are correct?

(a) RNA splicing joins the sequences that encode V_HDJ_H regions to sequences that encode different class C_H regions.
(b) Plasma cells that initially synthesize IgM switch to form IgG with the same antigen specificity.
(c) Class switching doesn't affect the variable region of the H chains.
(d) Class switching allows a given recognition specificity of an antibody to be coupled with different effector functions.

22. Order the following steps in the complement cascade in their proper sequence:

(a) Activation of C3 and C5 ________
(b) Entry of C9 into the membrane and formation of pores ________
(c) Binding of multiple F_C regions to C1q ________
(d) Cleavage of C4 and C2 ________
(e) Penetration of the membrane by the membrane attack complex (C5b, C6, C7, and C8) ________
(f) Lysis of the target cell ________
(g) Activation of C1 ________

23. Match the type of T cell with the corresponding feature listed in the right column.

(a) Helper T cell__________
(b) Killer T cell __________

(1) Detects foreign peptides presented on cell surfaces
(2) Recognizes class I MHC protein plus peptide
(3) Recognizes class II MHC protein plus peptide
(4) Expresses CD8
(5) Expresses CD4
(6) Stimulates B cells
(7) Lyses infected cells

24. Which of the following statements about the major histocompatibility complex (MHC) proteins are correct?

(a) MHC proteins play a role in the rejection of transplanted tissues.
(b) One class of MHC proteins is present on the surfaces of nearly all cells, binds fragments of antigens, and presents them to T-cell receptors.
(c) MHC proteins are encoded by multiple genes.
(d) One class of MHC proteins provides components of the complement system.
(e) The genes encoding MHC proteins produce three classes of soluble proteins.

25. Which are properties of the peptide-binding site of HLA-B27, a MHC class I protein?

(a) Consists of a deep groove with a β sheet floor and α-helical walls
(b) Can bind peptides of 7 to 11 amino acid residues with equal affinity
(c) Interacts specifically with four amino acid residues of the peptide and nonspecifically with the rest of the amino acid residues
(d) Binds peptides that retain their α-helical conformation
(e) Has a very high affinity for peptides

26. Which of the following statements about T-cell receptors are correct?

(a) T-cell receptors recognize soluble foreign molecules in the extracellular fluid.
(b) T-cell receptors recognize T cells.
(c) For a T-cell receptor to recognize a foreign molecule, the molecule must be bound by proteins encoded by the genes of the major histocompatibility complex.
(d) The T-cell receptor is structurally similar to an IgG immunoglobulin in that it has two H and two L chains.
(e) The T-cell receptor is encoded by genes that arise through the recombination of a repertoire of *V, J, D,* and *C* DNA sequences.
(f) T-cell receptors primarily recognize fragments derived from foreign macromolecules.

27. Match the receptor proteins in the left column with appropriate structural features in the right column.

(a) Class I MHC proteins ____________
(b) Class II MHC proteins ____________
(c) T-cell receptors ____________

(1) Contain an immunoglobulin fold motif
(2) Contain domains homologous to the V and C domains of immunoglobulins
(3) Contain a β-microglobulin chain
(4) Contain two transmebrane polypeptide chains
(5) Contain one transmebrane chain
(6) Are encoded by six different genes that are highly polymorphic
(7) Are encoded by *V, D,* and *J* gene segments

28. Which one of the following is not a property of the human immunodeficiency virus (HIV)?

(a) It is a retrovirus, i.e., has an RNA genome that produces viral DNA in the host cell.
(b) It contains a bilayer membrane with two kinds of glycoproteins.
(c) It interacts with the CD4 coreceptor of helper cells through the gp120 glycoprotein.
(d) It injects its RNA into the cell, and releases the coat into the medium surrounding the cell.
(e) It impairs and destroys the host cell by affecting its permeability and interfering with the expression of host-cell mRNA transcripts and recognition of peptide–class II MHC complexes on target cells.

ANSWERS TO SELF-TEST

Introduction and the Clonal Selection Theory

1. (a) 3 (b) 2 (c) 1 (d) 4

2. If the small molecule (hapten) becomes attached to a macromolecule (carrier), it can act as an immunogen and serve as an epitope (haptenic determinant) to which an antibody can be selected to bind.

3. a, c. Answer (d) is incorrect because IgM antibodies appear a few days after immunization and decrease after 10 days.

4. A fragment containing an antigen-binding site was obtained after the limited proteolysis of IgG. The fragment was denatured and tested for its ability to bind antigen. In the denatured state, it failed to associate with the antigen, but when it was renatured in the absence of the antigen, it regained the ability to form a complex with the antigen. The instructive theory predicted that the antigen serves as a template upon which the

antibody directed against it is folded. The experiment indicated that this was not the case.

5. a, d. Answer (b) is incorrect because some cells from the antibody-producing clone survive in the absence of the antigen and remain ready to begin antibody production immediately should the antigen reappear. Answer (c) is incorrect because immature antibody-producing cells are killed when their surface antibodies bind antigens. This process destroys the cells that are capable of forming antibodies against the organism's own macromolecules.

Immunoglobulin Structures, Monoclonal Antibodies, and Antibody-Antigen Interactions

6. (a) 1, 3, 4, 6 (b) 2, 3, 5. Answer (7) is inappropriate for either fragment because the F_c fragment lacks an antigen-binding site and the F_{ab} fragment contains only one. The insoluble lattice of antigen-antibody molecules forms because intact IgG molecules each have two antigen-binding sites and can therefore link several antigens together.

7. The hapten stimulates several B lymphocytes that bear different surface antibodies which recognize it to differentiate into plasma cells and secrete antibodies. Each plasma cell secretes a different kind of antibody that binds the hapten through a unique array of noncovalent interactions between the hypervariable regions of the antibody and the hapten.

8. Most of the antibodies in these patients arise from a clone of a single malignant plasma cell that has proliferated uncontrollably. These cells all produce a single molecular species of immunoglobulin.

9. d. Answers (a), (b), and (c) are incorrect because myeloma immunoglobulins have the same molecular architecture as normal IgG immunoglobulins and are directed against a single epitope on the surface of the antigen that elicits them. Since they are identical molecules, they display a single association constant for the antigen they bind.

10. The antigen eliciting the immunoglobulin in a myeloma cell is generally unknown.

11. Milstein and Kohler discovered that any particular antibody-producing spleen cell could be fused to a strain of myeloma cell to form a hybridoma. The hybridoma was immortal and could produce large amounts of a monoclonal antibody to the antigen that had stimulated the spleen cell, thereby facilitating the study of antigen-antibody interactions. Note: Although not stated in Stryer, the reason that the hybridoma cell secretes the antibody specified by the spleen cell and not the one made by the myeloma cell is that a strain of nonsecreting myeloma cells is used.

12. b, c, d, e

13. a, b, c. Answer (d) is incorrect because the variable sequences of the L and H chains in a given IgG are different from one another.

14. b, c. Answer (a) is incorrect because the two antigen-combining sites on an IgG molecule are directed toward the same epitope. Thus, an IgG can bind only one kind of antigen. Answer (d) is incorrect because it is

the F_c unit of the IgG-antigen complex, not of IgG alone, that associates with a component of the complement pathway to trigger cell lysis. Finally, answer (e) is incorrect because the hypervariable regions of both H and L chains form the antigen-combining site.

15. b, c, e

Immunoglobulin Gene Structure and Expression

16. The mRNA probe would hybridize to one region on the DNA from the plasma cell but to widely separated regions on the germ-line DNA. The gene on the plasma cell DNA encodes the intact gene sequence for the V, J, and L regions of the L chain, including introns. The genes encoding the V, J, and L regions of the L chain are in distant locations in the germ-line DNA because the DNA has not yet rearranged to bring these regions into proximity.

17. There aren't enough unique *V* and *C* genes to provide a sufficient number of different sequences when they are recombined in all the possible combinations. Joining (*J*) and diversity (*D*) genes increase the number of possible combinations. (See Stryer, pp. 373–374.)

18. e

19. (a) 5 (b) 5 (c) 2 (d) 3 (e) 4

20. (a) 5 (b) 2 (c) 4 (d) 1 (e) 3

21. b, c, d. Answer (a) is incorrect because the sequence rearrangements of class switching take place through DNA recombination.

22. c, g, d, a, e, b, f

T-Cell Function and the Cellular Immune Response

23. (a) 1, 3, 5, 6 (b) 1, 2, 4, 7

24. a, b, c, d. Answer (e) is incorrect because MHC proteins are all bound to the cell surface and are not soluble.

25. a, c, e

26. c, e, f. Answer (a) is incorrect because T-cell receptors recognize fragments of foreign macromolecules only when the antigen is bound on the surface of a cell. Answer (d) is incorrect because the T-cell receptor is composed of one α and one β chain, each having sequences that are homologous to the V regions of the chains of immunoglobulins.

27. (a) 1, 3, 5, 6 (b) 1, 4, 6 (c) 1, 2, 4, 7

28. d. This answer is incorrect because the HIV RNA and other core components enter the helper T cells as the viral membrane fuses with the cell membrane.

1. Most antigens are polyvalent; that is, they have more than one antibody-binding site. In the case of macromolecules that contain regular, repeating sequences, like polysaccharides, it is easy to understand how a molecule might have multiple binding sites. In the case of proteins with nonrepeating sequences, it is more difficult to envision how polyvalence might be accounted for. Yet proteins with single polypeptide chains are polyvalent as antigens. What feature of antibody production accounts for this behavior?

2. Assuming that antigen-antibody precipitates have latticelike structures, draw simple sketches showing possible arrangements of antigen and antibody molecules in a precipitate in which the ratio of antibodies to antigens is (a) 1.14 and (b) 2.83.

3. Suppose that dinitrophenol is attached to a protein with many potential DNP binding sites and that the resulting antigen is used to stimulate antibody production in rabbits. When serum is harvested and the immunoglobulin fraction is purified and mixed together with antigen, no precipitate forms, yet fluorescence measurements reveal the presence of antigen-antibody complexes. Explain this seeming paradox.

4. In people who are heterozygous for two hemoglobin types, say HbA and HbS, a given red blood cell contains approximately a 50 percent mixture of the two types. What happens when people are heterozygous for immunoglobulin genes?

5. When polyacrylamide gel electrophoresis (Stryer, pp. 46–47) of a monoclonal antibody preparation is conducted, a single sharp band appears. When the antibody preparation is treated with β-mercaptoethanol, two bands appear. Why is this the case?

6. Pepsin cleaves IgG molecules on the carboxyl-terminal side of the interchain disulfide bonds between heavy chains. How many physical pieces would result from the cleavage of IgG by pepsin? How many of the pieces derive from the F_c region of IgG?

7. Quantitative measures of the interactions between antigens and antibodies are frequently given as association constants, the reciprocal of dissociation constants. (See problem 1 on p. 389 of Stryer.) The association constant for the binding of a given hapten to an antibody is 10^9 M^{-1}, and the second-order rate constant for its binding is 10^8 M^{-1} s^{-1}. Calculate the rate constant for the dissociation of the hapten from the antibody.

8. The addition of a bifunctional DNP affinity-labeling reagent (one with two affinity-labeling groups) to myeloma protein produces light and heavy chains that are cross-linked through Tyr 34 and Lys 54, respectively. What conclusion is suggested by this observation?

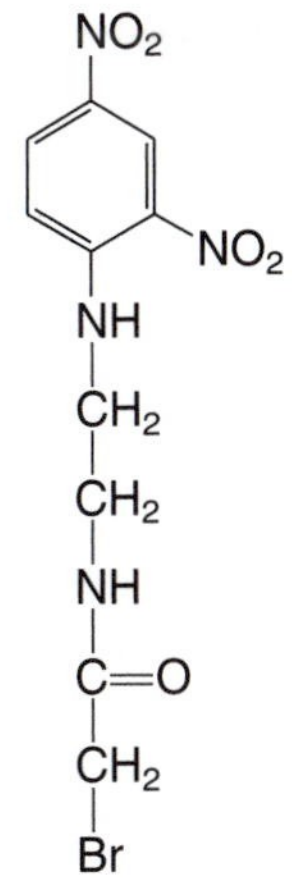

NO$_2$ (ring) NO$_2$ — NH — CH$_2$ — CH$_2$ — NH — C=O — CH$_2$ — Br

Figure 14-1
Affinity-labeling agent.

9. Dinitrophenol is covalently attached to a protein, and the DNP-labeled protein is then used to produce a monoclonal antibody. A DNP affinity-labeling agent (Figure 14-1) is then added to the antibody preparation in

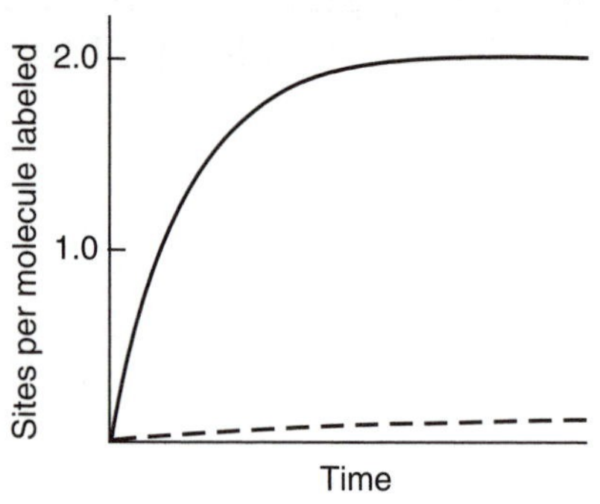

Figure 14-2
Time course for the labeling of the antibody in the presence (dashed line) and absence (solid line) of DNP–lysine.

the presence and absence of excess ε-DNP–lysine, giving the results depicted in Figure 14-2.

(a) Which group on the affinity-labeling agent is recognized by the hapten binding sites on the antibody?
(b) Which group on the affinity-labeling agent is likely to be involved in covalently labeling the antibody?
(c) What is the role of the added ε-DNP–lysine?

10. An antibody that is prepared as described in problem 9 is treated with two radiolabeled affinity-labeling agents, A and B. β-Mercaptoethanol is then added, and the mixture is poured over a molecular sieving column, giving the results depicted in Figure 14-3. Explain what conclusion likely follows from this experiment.

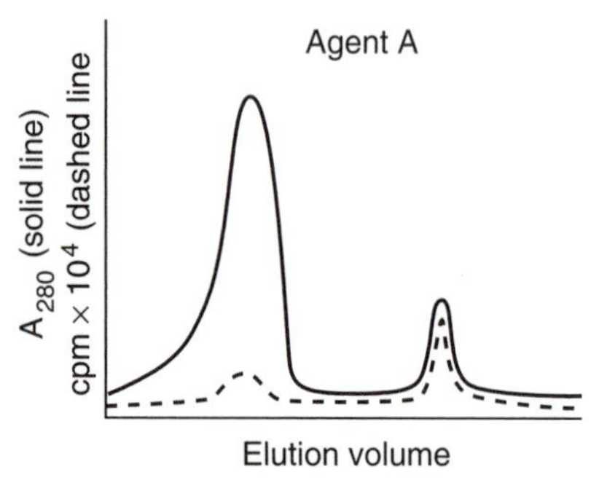

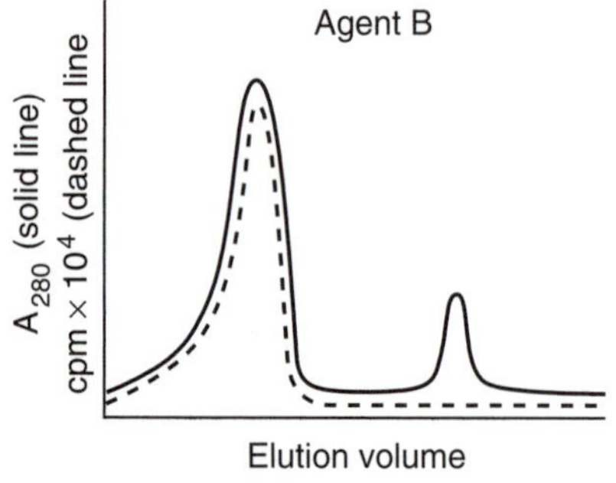

Figure 14-3
Results for problem 10.

11. Removal of the thymus gland from a young animal not only compromises the production of cytotoxic T cells, but also decreases antibody production by B cells. Why is this the case?

12. Propose a method using the technique of affinity chromatography (Stryer, p. 50) that would allow one to select lymphocytes producing antibody to one particular antigen from a heterogeneous population of immature lymphocytes. Explain the rationale behind your proposal.

13. Suppose that antibody is prepared against the extracellular portion of a particular hormone receptor known to have intracellular tyrosine kinase activity, and known to carry out autophosphorylation of its tyrosine kinase. Suppose further that addition of this antibody to target cells elicits intracellular responses similar to those obtained by addition of hormone itself. Propose a mechanism by which addition of antibody may mimic the effects of adding hormone. (You may wish to refer to problem 12 in Chapter 13 of the *Companion,* together with its answer, in tackling this question. Hint: F_{ab} fragments do not elicit the hormonelike response.)

14. Although the immune system is designed so that antibodies are not ordinarily directed toward one's own tissue components, sometimes that process goes awry, leading to so-called *autoimmune* diseases. One human disease thought to involve such an autoimmune mechanism is myasthenia gravis, a relatively common disorder in which antibodies directed toward acetylcholine receptors lead to a decrease in receptor numbers. Those suffering from myasthenia gravis show weakness and fatigability of skeletal muscle. Eventually death results from loss of function of breathing muscles. Medical therapy for people suffering from myasthenia gravis may include two types of drugs, immunosuppressive agents and inhibitors of acetylcholinesterase. Give a brief rationale for the use of each.

ANSWERS TO PROBLEMS

1. A given antigenic protein will stimulate the production of a mixed population of antibodies, with each type of antibody being specific for a different region in the tertiary structure of the antigenic protein.

2. The sketches for the two precipitates are shown in Figure 14-4. In (A) the Ab/Ag ratio is 8/7 = 1.14. In (B) the ratio is 17/6 = 2.83

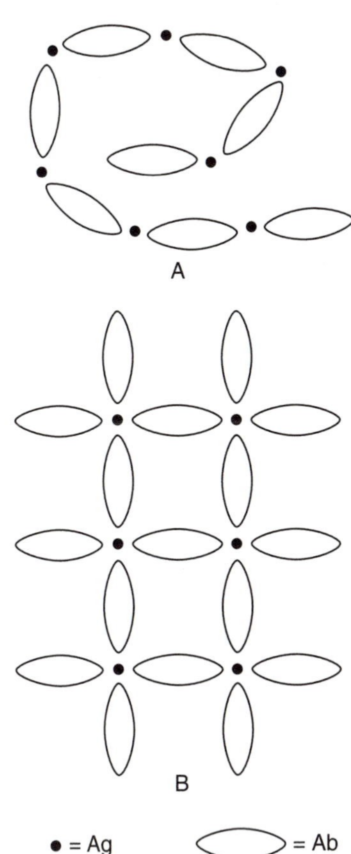

Figure 14-4
Possible lattice structures for Ab/Ag rations of (A) 1.14 and (B) 2.83.

3. The results would occur if the haptens were clustered so densely on the antigen that bivalent antibodies combined preferentially with two neighboring haptens on a given antigen molecule. Such behavior is actually found in some systems and is termed *monogamous bivalency.*

4. All antibodies produced by a single lymphocyte are identical. Therefore only one of two alleles is expressed, a process known as *allelic exclusion.*

5. Polyacrylamide gel electrophoresis separates proteins on the basis of size. β-Mercaptoethanol reduces the disulfide bonds that link the light and heavy antibody chains. Thus, the light and heavy chains are separated from one another on the gel.

6. Three pieces result from the cleavage of IgG by pepsin, as shown in Figure 14-5. One piece contains both F_{ab} units. The other two pieces derive from the bisection of the F_c region.

7. The mass action expression for an antigen-antibody reaction is

$$\text{Ag} + \text{Ab} \underset{k_{off}}{\overset{k_{on}}{\rightleftharpoons}} \text{Ag} - \text{Ab}$$

Since

$$K_a = k_{on}/k_{off}$$

substituting in the given values yields

$$10^9\text{M}^{-1} = \frac{10^8\ \text{M}^{-1}\text{s}^{-1}}{k_{off}}$$

$$k_{off} = 10^{-1}\text{s}^{-1}$$

8. The observation suggests that both Tyr 34 on the light chain and Lys 54 on the heavy chain are involved in binding of hapten.

9. (a) In the preparation of the antibody, DNP served as the hapten. The DNP portion of the affinity-labeling agent will therefore be recognized by the hapten binding site on the antibody.

(b) The reagent alkylates the side chains of cysteine present in the hapten binding site of the antibody in a reaction involving the elimination of HBr. Thus, the end of the molecule opposite the ring is involved.

(c) When DNP–Lys is present, the DNP ring binds to the hapten binding site of the antibody, thus competitively precluding the binding of the affinity agent. Because little or no affinity-labeling agent is bound to the hapten binding site of the antibody, the amino acid side chains cannot be affinity-labeled.

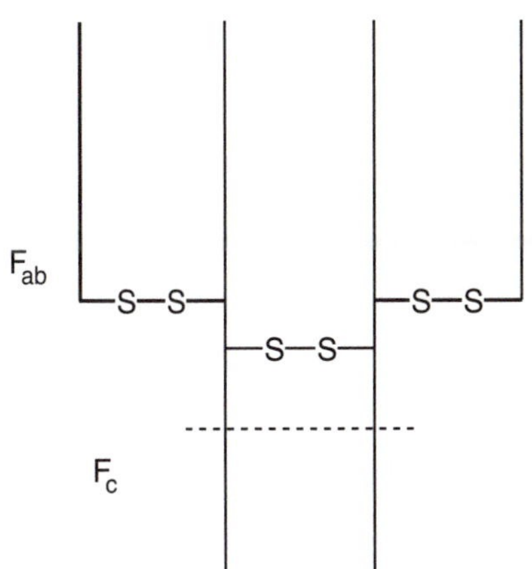

Figure 14-5
Cleavage of IgG by pepsin.

10. The β-mercaptoethanol reduces disulfide bonds in the antibody, producing a mixture of light and heavy chains, which is then resolved on the molecular sieving column. Affinity-labeling agent B apparently has labeled the heavy chains at the hapten binding site, and agent A has labeled the light chains. (Remember that larger molecules elute first from a molecular sieving column.)

11. In addition to cytotoxic T cells, helper T cells are also derived from the thymus, which stimulate B cells to produce antibody.

12. One could attach antigen to insoluble beads, and then pass the lymphocyte preparation over a column containing the beads. Lymphocytes capable of producing antibody toward the particular antigen would have a receptor complementary to that antigen on their cell surface and hence would be bound to the beads. Other lymphocytes would pass through the columnn without binding. Bound lymphocytes could then be released from the column by the addition of free antigen, which would compete with bound antigen for combination with the cell-surface receptors. The fact that such a purification method works is a consequence of the clonal selection theory, which states that many, many types of lymphocytes are produced, each having a specific receptor on the cell surface capable of reacting with some previously unencountered antigen. Combination of antigen with that cell-surface receptor causes the lymphocyte to divide and stimulates antibody production.

13. Addition of hormone to receptors in some instances causes the hormone-receptor complex to dimerize. The tyrosine kinase domains of two such approximated receptor monomers will then become capable of cross-phosphorylation, with a consequent increase in tyrosine kinase activity. The addition of bivalent antibodies would also have the effect of drawing monomeric receptor units together in such a way that they could cross-phosphorylate and thus trigger intracellular effects.

14. Use of immunosuppressive agents will retard the synthesis of antibodies against acetylcholine receptors (as well, of course, as inhibiting the synthesis of many other useful and protective antibodies). Administration of acetylcholinesterase inhibitors will lead to an increase in acetylcholine concentrations at motor end plates. This increase will lead in turn to an increased number of acetylcholine-acetylcholine receptor complexes by mass action, and will therefore compensate somewhat for the decrease in receptor concentration.

EXPANDED SOLUTIONS TO TEXT PROBLEMS

1. (a) $\Delta G^{\circ\prime} = -2.303\ RT \log_{10} K'_{\text{dis}} = -1.36 \times \log_{10}(3 \times 10^{-7}) = 8.9$ kcal/mol. Note that this *positive* value is the $\Delta G^{\circ\prime}$ for *dissociation* of the F_{ab}-hapten complex. Since binding is the reverse of dissociation, the $\Delta G^{\circ\prime}$ for binding = −8.9 kcal/mol.

(b) K_a = 1/dissociation constant = $1/3 \times 10^{-7}$ M = 3.3×10^6 M. Note [see part (a)] that the $\Delta G^{\circ\prime}$ of binding = $-2.303\, RT \log_{10} K_a$.

(c) As noted in Chapter 7, problem 4, of Stryer, an equilibrium constant is equal to the ratio of the rate constant of the forward (off) reaction to the rate constant of the reverse (on) reaction. Therefore, $k_{off}/k_{on} = 120\ s^{-1}/k_{on} = 3 \times 10^{-7}$ M. Solving this equation gives $k_{on} = 4 \times 10^8\ M^{-1}s^{-1}$. This value is close to the diffusion-controlled limit for combination of a small molecule with a protein (Stryer, p. 196). Hence, the extent of structural change is likely to be small because extensive conformational transitions take time.

2. The active site in lysozyme also accommodates six sugar residues (Stryer, p. 211). The range of sizes of the combining sites of antibodies is like that of active sites of enzymes.

3. The fluorescence enhancement and shift to the blue indicate that water is largely excluded from the combining site when the hapten is bound. Hydrophobic interactions contribute significantly to the formation of most antigen-antibody complexes.

4. (a) Using $K'_{eq} = 10^{-\Delta G^{\circ\prime}/1.36}$ (Stryer, p. 187), $K_a = 10^{7/136} = 10^{5.15}$. Since K_{dis} is the reciprocal of K_a, the log of K_{dis} = –log of K_a = –5.15 and $K_{dis} = 7.1 \times 10^{-6}$ M.

(b) $\Delta G^{\circ\prime} = 2 \times -7 + 3 = -11$ kcal/mol. Using the same logic as in (a), $K_a = 10^{11/1.36} = 10^{8.1}$ and $K_{dis} = 10^{-8.1} = 8 \times 10^{-9}$ M.

A comparison of bivalent binding with univalent binding is given by $K_a2/K_a1 = K_{dis}1/K_{dis}2 = 7.1 \times 10^{-6}/8 \times 10^{-9} = 888$, meaning that the bivalent binding is 888 times tighter than the univalent binding.

5. (a) An antibody-combining site is formed by CDRs from both the H and L chains. The V_H and V_L domains are essential. A small portion of F_{ab} fragments can be further digested to produce F_V, a fragment that contains just these two domains. C_H1 and C_L contribute to the stability of F_{ab} but not to antigen binding.

(b) A synthetic F_V analog 248 residues long was prepared by expressing a synthetic gene consisting of a V_H gene joined to a V_L gene through a linker. See J.S. Huston, et al., *Proc. Nat. Acad. Sci.* 85 (1988):5879.

6. (a) Multivalent antigens lead to the dimerization or oligomerization of transmembrane immunoglobulins, an essential step in their activation. This mode of activation is reminiscent of that of receptor tyrosine kinases (Stryer, p. 351).

(b) An antibody specific for a transmembrane immunoglobulin will activate a B cell by cross-linking these receptors. This experiment can be carried out using, for example, a goat antibody to cross-link receptors on a mouse B cell.

7. μ_s and μ_m are formed by alternative splicing of the same primary transcript (p. 115). The mRNA for μ_m encodes a C-terminal transmembrane anchor that is not present in μ_s.

8. B cells do not express T-cell receptors. Hybridization of T-cell cDNAs with B-cell mRNAs removes cDNAs that are expressed in both cells. Hence, the mixture of cDNAs following this hybridization are enriched in those encoding T-cell receptors. This procedure, called *subtractive hybridization,* is generally useful in isolating low-abundance cDNAs. Hybridization should be carried out using mRNAs from a closely related cell that does not express the gene of interest. See S.M. Hedrick, M.M. Davis, D.I. Cohen, E.A. Nielsen, and M.M. Davis, *Nature* 308(1984):149, for an interesting account of how this method was used to obtain genes for T-cell receptors.

CHAPTER

15

Molecular Motors

The transduction of chemical energy into mechanical energy—for example, the use of ATP hydrolysis to drive the contraction of muscles or to move cells, or the exploitation of transmembrane proton-motive force to rotate bacterial flagella—is the subject of Chapter 15. Stryer describes how nanometer motions of proteins can be converted into the coordinated movements of cellular organelles, bacteria, and even animals themselves. He first describes the structure of vertebrate skeletal muscle by showing how thick and thin protein filaments give rise to the striated appearance revealed in micrographs. Muscle contraction occurs by the oriented sliding of these filaments past one another. He next explains that the thick filaments are primarily myosin and the thin filaments are principally an actin polymer with associated troponin and tropomyosin molecules. He describes the ATPase and actin-binding activities of the globular domains of myosin as well as its spontaneous assembly into filaments. The subunit composition and domain structure of myosin is then presented to show where the ATPase, actin-binding, and filament-forming regions are located within the molecule. He next shows the structure of the F-actin polymer, which is composed of a linear coiled array of G-actin monomers. The association of myosin and the actin polymer to form actomyosin is described. Actomyosin threads that have been formed in vitro shorten when they are incubated with ATP, Mg^{2+}, and K^+, thus providing a hint of the molecular basis for the contraction of skeletal muscle. The opposite polarities of the thick and thin filaments within a sarcomere further indicate how coordinated molecular movement can result in the shortening of myofibrils.

The exact mechanism of the energy transduction remains obscure. However, the repeated association and dissociation of the S1 heads of myosin with actin and the conformational changes in myosin that are effected by the binding of ATP, its hydrolysis to ADP and P_i, and the release of the hydrolysis products suggest how the power stroke occurs. Reconstituted, moving assemblies of myosin-coated beads traversing along actin filaments reveal that myosin is the motor driving movement along the track. Stryer next describes how Ca^{2+} functions to regulate striated muscle contraction through its interaction with the actin-associated troponin complex and the resulting effects on the location of tropomyosin on the actin polymer. He explains that the TnC component of the troponin complex is related to the EF hand superfamily of Ca^{2+}-binding proteins. Increases in cytoplasmic Ca^{2+} resulting from neural stimulation of the sarcomere cause a conformational change in the troponin complex that results in movements of tropomyosin to enable myosin to bind to actin.

Stryer next considers the more universal roles of actin and of myosin homologs in eukaryotic cells. The structures and functions of microfilaments and intermediate filaments are mentioned, and those of the microtubules are considered in depth. Microtubules are found in nearly every cell and serve multiple structural and functional roles. They are composed primarily of α-tubulin and β-tubulin monomers that form tubular structures which have larger diameters than microfilaments or intermediate filaments. Microtubules also contribute the basic macromolecular assembly of the axoneme, which is the fundamental structural component of the cilia and flagella of eukaryotic cells. Dynein and kinesin are auxiliary proteins that interact with microtubules to bend cilia and flagella and to move vesicles along microtubules, respectively. They are ATPases that transduce chemical energy into mechanical movement as does myosin. Microtubules also participate in the separation of the chromosomes during mitosis. The rapid association of GTP–tubulin with microtubules and the rapid dissociation of GDP–tubulin from the ends of tubulin polymers in conjunction with the GTPase activity of the tubulins explains the dynamic instability of microtubules. Selection and stabilization of correctly targeted microtubules from among the many being formed and degraded give rise to the final microtubular structures that are observed. Vesicle transport in neurons is explained as their ATP-dependent, kinesin-driven movement along microtubule tracks, which is analogous to the myosin-actin interactions that slide muscle filaments past each other.

The chapter ends with a description of the flagellar motor of bacteria and a discussion of how the interplay of two kinds of protein components forming the motor give rise to directional rotation as a consequence of protons moving from outside the membrane into the cytosol. The examples given in this chapter demonstrate how energy in one form (chemical) can be converted into another form (kinetic) by the regulated activities of proteins.

LEARNING OBJECTIVES

When you have mastered this chapter, you should be able to complete the following objectives:

Introduction (Stryer page 391)

1. List the two sources of energy that power coordinated molecular movement.

Microscopic Vertebrate Muscle Structure (Stryer pages 392–393)

2. Define *sarcomere, sarcoplasm,* and *myofibril.*

3. Identify the *A band, H zone, M line, I band,* and *Z line* of a sarcomere in an electron micrograph of a myofibril.

4. Relate the locations of the *thick filaments* and *thin filaments* to the I band, the A band, and the H zone of a sarcomere.

5. Relate the structures of the thick and thin filaments to their compositions of *actin, myosin, tropomyosin,* and the *troponin* complex. Note the *cross-bridges* between the thick and thin filaments.

6. Describe muscle contraction in terms of the *sliding-filament model,* and relate contraction to changes in the sizes of the A band, I band, and H zone.

Myosin, Actin, and Actomyosin in Vertebrate Skeletal Muscle (Stryer pages 393–398)

7. List the three critical biological activities of myosin and explain why it is called a *mechanoenzyme.*

8. Describe the composition of *actomyosin,* and note the effects of adding ATP or ATP, K^+, and Mg^{2+} to a solution of actomyosin. Relate these effects to *muscle contraction.*

9. Sketch the general structure of the polypeptide backbone of a myosin molecule, and note its dimensions. Describe the subunit composition of myosin, and note its *α-helical coiled-coil structure* and its *dual globular head.*

10. Describe the fragmentation of myosin into *light meromyosin* (LMM) and *heavy meromyosin* (HMM) by proteolysis and the further fragmentation of HMM into its *S1* and *S2 subfragments.* Sketch the polypeptide structures of the fragments, and associate them with the *ATPase activity,* the *actin-binding sites,* and the *thick-filament-forming domains* of the intact molecule.

11. Describe the amino acid sequence regularities of the regions of the myosin polypeptide chains that form the α-helical coiled coil and explain how their composition contributes to the structure.

12. Note the dimensions of *G-actin,* and distinguish between G-actin and *F-actin.* Define *critical concentration* as it relates to the polymerization of actin. Contrast the roles of the ATPase activities of actin and myosin.

13. Describe the composition and polarity of a *decorated filament.*

Muscle Structure at the Biochemical Level and Muscle Contraction
(Stryer pages 398–404)

14. Note the *bipolarity* of the thick and thin filaments with respect to the Z lines of a sarcomere. Relate the opposing polarities of the molecules to the coordinate and macroscopic movement that occurs during contraction.

15. Provide an overview of a model for the mechanism of the *power stroke* during muscle contraction. Explain the roles of actin and the ATPase activity of myosin in the model. Discuss the contributions of the *hinges* of myosin to its functioning in muscle contraction.

16. Recount the experiment that displayed the unidirectional movement of myosin molecules along an actin cable. Describe the experiment indicating that the force of muscle contraction is generated by the S1 head of myosin.

17. Describe the roles of Ca^{2+}, *tropomyosin,* and the *troponin complex* in the contraction of skeletal muscle. Note the resemblance of *troponin C* to *calmodulin,* and outline the flow of information from Ca^{2+} to myosin during the stimulation of contraction.

The Filamentous Cytoskeleton (Microfilaments, Intermediate Filaments, and Microtubules) and Eukaryotic Flagellar and Ciliary Bending
(Stryer pages 405–411)

18. Describe some of the roles of actin and myosinlike proteins in the movements of nonmuscle cells. Note the diversity in myosin structures, and describe how myosin I moves vesicles along actin tracks.

19. Describe the *filamentous composition* and outline the functions of the *cytoskeleton.* Note the composition and diameters of *microfilaments, intermediate filaments,* and *microtubules.*

20. Sketch the general structure of a *microtubule,* noting its diameter and showing its alternating subunits of *α-tubulin* and *β-tubulin.*

21. Distinguish between the *cilia* and *flagella* of eukaryotic cells in terms of their relative lengths and physiologic functions. Describe the structure of the *axoneme,* and sketch the *9 + 2 array* of the microtubule doublets and singlets that form its basic ring motif.

22. Locate the *dynein* in an axoneme, and compare its biochemical activities with those of myosin. Explain the roles of *dynein cross-bridges, radial spokes,* and *nexin links* in causing the bending rather than the contraction of the axoneme.

23. List some of the causes and consequences of the *immotile-cilia syndrome.*

24. Describe the dynamics of the polymerization and depolymerization of microtubules, noting the role of the *microtubule-organizing centers* (MTOC) on the *centrisomes* and *poles* of the *mitotic spindles* and the roles of *GTP* and *GDP* in these processes.

25. Note the roles of *dynamic instability* and *selection* in target location by growing microtubules. Describe the effects of *colchicine* and *taxol* on the polymerization and depolymerization of microtubules.

26. Describe the *kinesin*-dependent movement of *vesicles* and *organelles* along microtubules. Distinguish between the plus and minus ends of a microtubule, and relate this polarity to the *anterograde* and *retrograde transport* in a cell.

27. Compare *kinesin* to dynein and myosin.

Prokaryotic Flagellar Rotation (Stryer pages 411–413)

28. Outline the structure, dimensions, and function of *bacterial flagella.* Distinguish between the mechanism by which eukaryotic and prokaryotic flagella generate motile force.

29. Recount an experiment that demonstrated the role of *proton-motive force* in flagellar rotation. Present a model that explains the production of rotary motion from the effects of a proton gradient on the transmembrane flagellar motor.

SELF-TEST

Microscopic Vertebrate Muscle Structure

1. Energy is required to drive the contraction of striated muscle, the beating of flagella or cilia, and the intracellular transport of vesicles along microtubules. Which of the following statements about energy transduction in these systems are correct?

   (a) The proton-motive force across the plasma membrane surrounding the cell provides the energy for these movements.
   (b) The binding of ATP to proteins, which then undergo a conformational change, provides the energy for these processes.
   (c) The hydrolysis of ATP is coupled to the phosphorylation of tyrosine residues on the proteins of these systems to drive them.
   (d) The hydrolysis of protein-bound ATP and the release of ADP + P_i lead to conformational transitions that complete a movement cycle.
   (e) The binding of GTP to oriented proteins at the interface of the moving assemblies drives these processes.

2. Which of the following statements about striated vertebrate muscle are correct?

   (a) The muscle cells (myocytes) are multinucleate and are surrounded by an electrically excitable plasma membrane.
   (b) Myocytes contain parallel myofibrils that are immersed in the sarcoplasm.
   (c) Each myofibril is composed primarily of thick and thin protein filaments.
   (d) The functional unit of a muscle cell that is directly responsible for contraction is called a *sarcomere.*

3. The accompanying figure is a schematic diagram of a longitudinal segment of a skeletal muscle myofibril. Label the structures indicated in the figure by matching them with the listed choices.

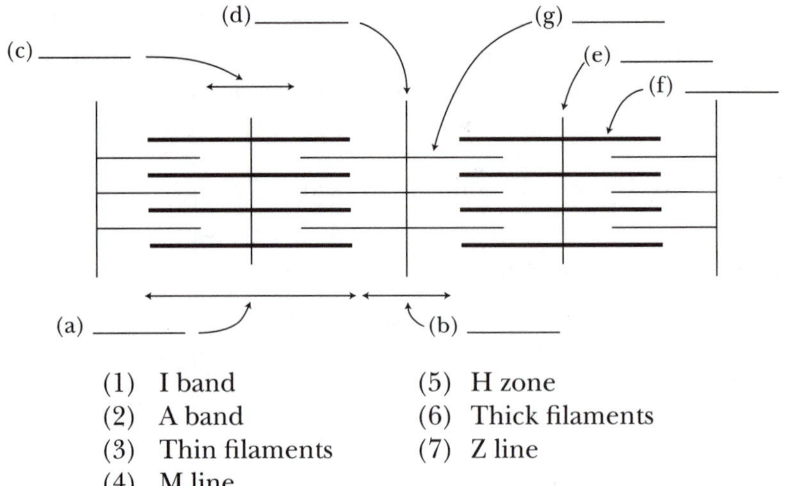

(1) I band
(2) A band
(3) Thin filaments
(4) M line
(5) H zone
(6) Thick filaments
(7) Z line

4. During contraction, reversible cross-bridges form between which of the two structures listed in question 3?

____

____

5. Which of the following structures change their dimensions or alter their positions with respect to each other during contraction? Answer with *Yes* or *No.* (Refer to question 3 for a diagram of a myofibril.)

(a) I band_______________
(b) A band _______________
(c) Thin filaments_______________
(d) M line to M line_______________
(e) H zone _______________
(f) Thick filaments _______________
(g) Z line to Z line_______________
(h) Z line to edge of adjacent H zone_______________
(i) Z line to M line _______________

Myosin, Actin, and Actomyosin in Vertebrate Skeletal Muscle

6. Which of the following statements about myosin are correct?

(a) Myosin binds to polymerized actin.
(b) In vitro, myosin assembles spontaneously into the thin filaments.
(c) Myosin is an ATPase.
(d) Myosin has domains that interact with one another to effect its physiologic functions.
(e) Myosin is composed of two polypeptide chains, one of which forms an α-helical coiled coil and the other a globular head.

7. Which of the following statements about heavy meromyosin (HMM) and light meromyosin (LMM) are correct?

(a) HMM and LMM are formed from myosin by tryptic cleavage.
(b) HMM can be cleaved by papain to yield two globular proteins that polymerize to form the thin filaments.
(c) LMM is an α-helical coiled coil composed of two polypeptide chains that can form filaments in vitro.
(d) HMM contains the two globular heads of the myosin molecule, hydrolyzes ATP, and binds actin in vitro.

8. Which of the following statements about actomyosin or its constituents are correct?

(a) The polymerization of G-actin is controlled by the binding of guanyl nucleotides.
(b) The addition of ATP to a solution of actomyosin leads to the contraction of the actomyosin threads.
(c) F-actin is composed of a helical array of two strings of G-actin monomers.
(d) Cross-bridges form between the globular heads of myosin and the F-actin polymer.
(e) Mixing solutions of F-actin and myosin leads to the formation of a viscous solution of actomyosin.

9. What features of the tails of myosin molecules contribute to their ability to interact with one another to form thick filaments?

____

____

10. Which of the following answers complete the sentence correctly? Actin

(a) is formed from a 42-kd monomer (G-actin).
(b) monomers can exist with either ATP or ADP bound to themselves.
(c) exhibits an ATPase activity that helps power muscle contraction.
(d) in its F form binds myosin in an oriented manner.
(e) is so named because it activates the ATPase activity of myosin.

Muscle Structure at the Biochemical Level and Muscle Contraction

11. Which of the following statements concerning the contraction of vertebrate skeletal muscle are correct?

(a) Coherent movement occurs during contraction because actomyosin complexes are highly ordered and oppositely oriented within the sarcomere.
(b) Cross-bridges form between adjacent myosin molecules to stabilize the thick filaments.
(c) Whereas myosin is obviously an asymmetric molecule that forms an asymmetric filament, thin filaments are formed from globular actin (G-actin) subunits and are thus nondirectional or symmetric polymers.
(d) The polarity of the thin filaments reverses at the Z line.

12. Which of the following statements concerning events related to the power stroke of muscle contraction are correct?

(a) The hydrolysis of ATP to ADP and P_i by myosin is fast relative to the release of the ADP and P_i from the protein.
(b) The binding of actin to myosin stimulates the ATPase activity of myosin by facilitating the release of ADP and P_i.
(c) Actin and myosin are joined by cross-bridges that are stabilized by the binding of ATP to the myosin head domains.
(d) Repeated cycles of ATP binding, ATP hydrolysis, and the resulting association and dissociation of cross-bridges and conformational changes in myosin contribute to the contractile process.
(e) In the region of overlapping thick and thin filaments of a sarcomere, the cross-bridges will either all be formed or all be dissociated, depending on the phase of the power stroke.

13. Considering only the power stroke of skeletal muscle contraction and the events that precede and follow it, place the following states or processes that occur in going from the resting state to the contracted state and back again in their proper order.

(a) The thick filament moves with respect to the thin filament. ________
(b) The S1 heads of myosin interact with actin, P_i is released, and myosin changes its conformation. ________
(c) ATP binds to myosin. ________
(d) The S1 heads of myosin are dissociated from the thin filament and contain bound ADP and P_i. ________
(e) ATP is hydrolyzed to ADP and P_i, and myosin undergoes a conformational change. ________
(f) ADP dissociates from the myosin to complete the power stroke. ________

14. Which of the domains of myosin is primarily responsible for generating the force of skeletal muscle contraction?

(a) The hinge between the S1 and S2 domains
(b) The hinge between the S2 and the LMM domains
(c) The LMM domain
(d) The S1 globular head
(e) The S2 domain that connects the S1 domain to the α-helical coiled coil
(f) The α-helical coiled coil

15. Assign the proteins in the right column to the appropriate myofibrillar component in the left column.

(a) Thin filament ________
(b) Thick filament ________

(1) Tropomyosin
(2) Myosin
(3) Actin
(4) Troponin complex

16. Which of the following statements about the troponin complex and tropomyosin of skeletal muscle are correct?

(a) Tropomyosin is a two-stranded, α-helical rod that winds about the α-helical coiled-coil tail of myosin.

(b) The troponin complex is composed of three subunits.
(c) The subunits of the troponin complex bind actin, tropomyosin, or Ca^{2+}.
(d) The conformational change induced in the troponin complex by the binding of Ca^{2+} is transmitted to tropomyosin.
(e) Troponin C binds Ca^{2+} and is structurally similar to calmodulin.
(f) The movement of tropomyosin allows the S1 heads of the thick filament to interact with the thin filaments.
(g) The regulation of contraction by Ca^{2+} is mediated by proteins.

17. Considering the events associated with the regulation of the contraction of skeletal muscle, place the following processes in the order in which they occur upon the stimulation of a muscle to contract:

(a) A nerve impulse leads to the release of Ca^{2+} from the sarcoplasmic reticulum. ________
(b) Tropomyosin changes position on the thin filament. ________
(c) The formation of the complex between troponin C and Ca^{2+} causes the other components of the troponin complex to move. ________
(d) An ATP-dependent pump lowers the Ca^{2+} concentration of the sarcoplasm. ________
(e) Ca^{2+} binds to low-affinity sites of troponin C in the troponin complex. ________
(f) The S1 heads of myosin become free to bind to actin. ________

18. Which of the following statements about actin and myosin in nonmuscle cells are correct?

(a) Some nonmuscle cells contain large amounts of actin.
(b) The amino acid sequences of the actins in nonmuscle cells are highly diverged from those of the actin in muscle cells.
(c) The amino acid sequences of the myosins in nonmuscle cells are highly diverged from those of the myosin of muscle cells.
(d) Myosins of at least two classes exist in nonmuscle cells.

The Filamentous Cytoskeleton (Microfilaments, Intermediate Filaments, and Microtubules) and Eukaryotic Flagellar and Ciliary Bending

19. Which of the following statements about the cytoskeleton or its components are correct?

(a) A filamentous scaffolding composed primarily of proteins forms the cytoskeleton of eukaryotic cells.
(b) The cytoskeleton of eukaryotic cells is composed of three major classes of assemblies—microfilaments, intermediate filaments, and microtubules.
(c) The microfilaments of the cytoskeleton are stable once they are formed, and a given filament exists for the life of the cell.
(d) Intermediate filaments are composed of two or three F-actin polymers that are coiled about one another.
(e) The keratins are examples of intermediate filaments.

20. Which of the following statements concerning microtubules are correct?

(a) Microtubules are filaments composed of α-helical coiled-coil polypeptide chains.

(b) Microtubules contain α-tubulin and β-tubulin protomers that are disposed in a helical array around a hollow core to form a cylindrical structure.
(c) The cilia and flagella of eukaryotic cells contain nine microtubule doublets that surround a pair of microtubule singlets.
(d) The outer microtubules in an axoneme are linked together and to an ATPase called *dynein.*
(e) The powered movements of dynein in an axoneme shorten the structure.

21. When defective dynein causes the immotile-cilia syndrome, sometimes not only chronic respiratory disorders but also infertility occurs in males. Explain.

___

___

22. Which of the following statements about microtubule assembly are correct?

(a) Nucleation centers are required to initiate the efficient formation of a new microtubule.
(b) Colchicine blocks the polymerization of microtubules and thus the movement of chromosomes.
(c) Microtubule-organizing centers serve as foci at which α-tubulin and β-tubulin polymerize to form microtubules.
(d) α-Tubulin and β-tubulin are potent ATPases.
(e) GTP–tubulin adds to growing microtubules more readily than does GDP–tubulin.
(f) GDP–tubulin rapidly dissociates from the ends of microtubules.
(g) GDP–tubulin remains in microtubules if it is in an internal position.
(h) Growing microtubules find their targets by forming along preexisting microfilament tracts.
(i) Taxol stabilizes microtubules and interferes with chromosome movements.

23. A neuron can move a vesicle approximately a meter from the central cell body to the end of an axon in a day. How are microtubules involved in this process, what provides the energy for the movement, and what protein is directly involved in the movement?

___

___

24. Which mechanoenzyme catalyzes retrograde intracellular transport in eukaryotes?

___

___

Prokaryotic Flagellar Rotation

25. Contrast the protein compositions and molecular movements of the flagella of bacteria and of eukaryotic cells.

___

___

26. How would the movements of a starved bacterium (one whose energy stores had been depleted) be affected by having the pH value of the media in which it resides lowered below that of its cytoplasm?

____

____

ANSWERS TO SELF-TEST

Microscopic Vertebrate Muscle Structure

1. b, d

2. a, b, c, d

3. (a) 2 (b) 1 (c) 5 (d) 7 (e) 4 (f) 6 (g) 3. The M lines lie halfway between the Z lines and correspond to the middle of the bare zone separating regions where oppositely oriented myosin molecules point toward each other (see Stryer Figure 15-19).

4. 3 and 6. Cross-bridges form between thick filaments and thin filaments.

5. (a) Yes (b) No (c) No (d) Yes (e) Yes (f) No (g) Yes (h) No (i) Yes

Myosin, Actin, and Actomyosin in Vertebrate Skeletal Muscle

6. a, c, d. Answer (b) is incorrect because myosin assembles to form the thick filaments. Answer (e) is incorrect because myosin is composed of six polypeptide chains. Two heavy chains intertwine their *C*-terminal portions to form the α-helical coiled coil, with their *N*-terminal segments forming two globular heads. Four more polypeptide chains of two kinds associate with the globular heads.

7. a, c, d. Answer (b) is incorrect because the two globular heads arising from papain digestion cannot polymerize, for they lack α-helical coiled-coil structures.

8. c, d, e. Answer (b) is incorrect because ATP dissociates the actomyosin complex. The contraction of the actomyosin threads requires ATP, K^+, and Mg^{2+} to allow repeated association of myosin and actin.

9. The absence of proline, which would interfere with α helix formation, and the abundance of regularly spaced leucine, alanine, and glutamate residues in seven-long repeating motifs that form long α helices with hydrophobic pockets and knobs on one face and charged residues on the opposite face allow two such helices in two myosin molecules to form a long coiled-coil rod.

10. a, b, d, e. Answer (c) is incorrect because the ATPase activity of actin is involved in the formation and disassembly of F-actin.

Muscle Structure at the Biochemical Level and Muscle Contraction

11. a, d. Answer (b) is incorrect because cross-bridges form between the S1 heads of myosin and actin. Answer (c) is incorrect because thin filaments also have polarity or directionality (see Stryer, pp. 397–398).

12. a, b, d. Answer (c) is incorrect because the binding of ATP to actomyosin dissociates the cross-bridges. Answer (e) is incorrect because, at any instant, numerous cross-bridges will be in all stages of forming and breaking because the process is dynamic and asynchronous.

13. d, b, a, f, c, e, d

14. d. Although the whole molecule is required for muscle contraction, the activities of the S1 head domains provide the biochemical activity for the power generation. The cyclic changes in the conformation of the head domains and in their affinities for actin, ATP, ADP, and P_i are the basis of the energy transduction.

15. (a) 1, 3, 4 (b) 2

16. b, c, d, e, f, g. Answer (a) is incorrect because tropomyosin is associated with actin, not myosin.

17. a, e, c, b, f, d

18. a, c, d. Answer (b) is incorrect because the actins have highly conserved amino acid sequences.

The Filamentous Cytoskeleton (Microfilaments, Intermediate Filaments, and Microtubules) and Eukaryotic Flagellar and Ciliary Bending

19. a, b, e. Answer (c) is incorrect because actin filaments are dynamic structures that are continuously being formed and degraded. Answer (d) is wrong because intermediate filaments don't contain actin.

20. b, c, d. Answer (a) is incorrect because microtubules are assembled from relatively globular tubulin subunits and lack the α-helical coiled-coil structure of myosin and the intermediate filaments. Answer (e) is incorrect because dynein movements lead to the bending not the contraction of the axoneme.

21. Defective dynein molecules not only immobilize the cilia of the respiratory tract, but also render sperm immotile.

22. a, b, c, e, f, g, i. Answer (h) is incorrect because microtubules form randomly, and only those that reach a target that stabilizes them are not rapidly depolymerized.

23. The microtubules, which form a network of fibers that traverse the cell, provide tracts along which vesicles and organelles can move. ATP hydrolysis by the protein kinesin acts as a molecular engine, in a manner analogous to the way myosin acts, to power the movements.

24. Cytoplasmic dynein moves particles toward the center of the cell along microtubule tracks.

Prokaryotic Flagellar Rotation

25. Bacterial flagella are formed principally from flagellin subunits, whereas eukaryotic flagella contain microtubules, among other components. Eukaryotic flagella bend owing to an internally generated force. Bacterial flagella are rotated by a motor attached to one of their ends.

26. The bacterium would commence swimming because protons would move from the higher concentration outside the cell through the intramembrane motors of the flagella, causing their counterclockwise rotation and thereby coordinated movement.

PROBLEMS

1. Decide whether each of the following will remain unchanged or will decrease upon muscle contraction. Assume that the sliding-filament model applies. Refer to page 393 of Stryer.

   (a) The distance between adjacent Z lines
   (b) The length of the A band
   (c) The length of the I band
   (d) The length of the H zone

2. The symmetry of thick and thin filaments in a sarcomere is such that six thin filaments ordinarily surround each thick filament in a hexagonal array. (See Figure 15-2 on p. 392 of Stryer.) In electron micrographs of cross sections of fully contracted muscle, the ratio between thin and thick filaments has been found to be double that of resting muscle.

   (a) Propose an explanation for this observation based on the sliding-filament model.
   (b) How might the explanation you have given in part (a) also account for the long-appreciated fact that a fully contracted muscle is, paradoxically, "weaker" than a resting muscle?

3. Ethylenediaminetetraacetate (EDTA), a heavy-metal chelator, acts as a muscle relaxant. Explain the basis for this effect.

4. Onion root-tip cells that have been incubated in solutions of colchicine for several days become polyploid rather than diploid. Propose an explanation for this observation.

5. The mold products vincristine and vinblastine interfere with the polymerization of microtubules as does colchicine. Vincristine and vinblastin are widely used in the treatment of rapidly growing cancers. Explain the basis for their effects.

6. Myosin, kinesin and dynein all contain a globular head which harbors ATPase activity and a tail region which can be thought of as a protein-binding region. The specificity of these protein-binding domains changes the role of each of these three proteins in tissues. What is the specificity of each of these three protein-binding domains, and how does it relate to the roles of myosin, kinesin, and dynein in tissues?

MOLECULAR MOTORS

7. Design an experiment involving ATP labeled in the γ phosphoryl group with ^{32}P that would show that actin stimulates the hydrolysis of ATP by myosin.

8. A single kinesin molecule can move a single vesicle from the nucleus of a nerve cell to the end of its axon. Each "step" in the kinesin cycle is 8 nm, and each cycle uses one ATP.

(a) How many cycles of kinesin binding and ATP hydrolysis will it take for a vesicle to reach the little toe from the upper cervical region of a tall human. Assume that the length of the nerve cell is 6 ft.
(b) Given that the rate of vesicle transport is 400 mm per day, how long will it take the vesicle to reach the little toe?
(c) How long does each step take and what is the rate of ATP hydrolysis?

[ATP] (μM)	Velocity V (μm s^{-1})
10	0.16
25	0.32
50	0.48
100	0.64
200	0.77
250	0.79

9. The dependence of kinesin motility on ATP concentration can be treated by classical Michaelis-Menten kinetics. (See Stryer, p. 192.)

(a) From a double-reciprocal plot of the data in the margin, determine K_M and V_{max}.
(b) At the start of a motility assay there is 200 μM ATP, and the velocity of kinesin-driven motility is 0.58 μm s^{-1}. What would be the velocity after half this ATP was hydrolyzed to ADP, which is a competitive inhibitor of kinesin-driven motility, with $K_i = 150\ \mu$M? The rate equation in the presence of a competitive inhibitor is

$$V = \frac{V_{max}[S]}{K_M\left(1 + \frac{[I]}{K_i}\right) + [S]}$$

(c) The maximal microtubule-activated ATPase activity of kinesin is 100 μmole ATP split per minute in an assay containing 20 nanomole of kinesin. What is the turnover number of kinesin? (See Stryer, p. 195.)
(d) What is the mechanochemical cycle time of kinesin, i.e., the time taken to hydrolyze an ATP molecule when kinesin is working at maximal activity?
(e) Given your determination of V_{max} from part (a) and the turnover number from part (c), estimate the probable "step size" for kinesin, i.e., the distance a kinesin molecule moves along a microtubule for each ATP hydrolyzed. (The authors are grateful to Professor Jonathan M. Scholey of the University of California, Davis, for his kindness in submitting problem 9.)

ANSWERS TO PROBLEMS

1. (a) The distance between adjacent Z lines will decrease.

(b) The length of the A band will remain unchanged.

(c) The length of the I band will decrease.

(d) The length of the H zone will decrease. See Figure 15-1.

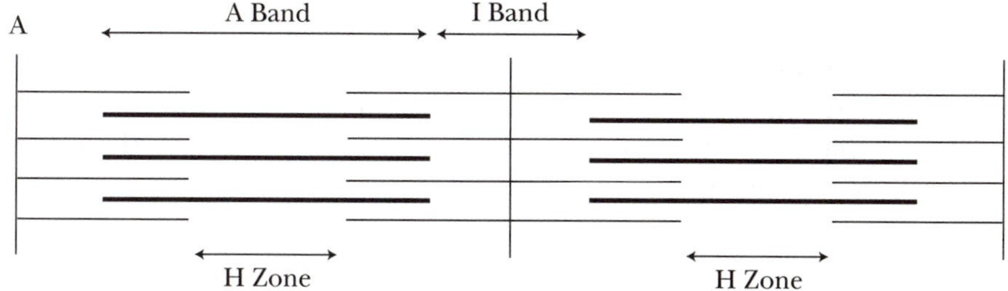

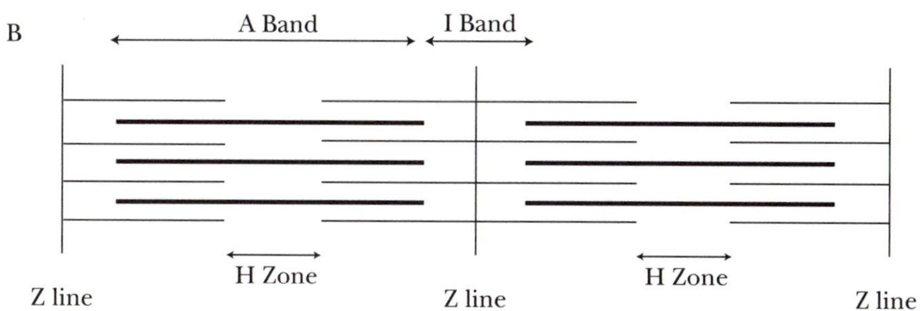

Figure 15-1
Schematic illustration of (A) uncontracted and (B) partially contracted sarcomeres.

2. (a) The number of thin filaments per thick filament doubles because the thin filaments override one another when the sarcomere is fully contracted. See Figure 15-2.

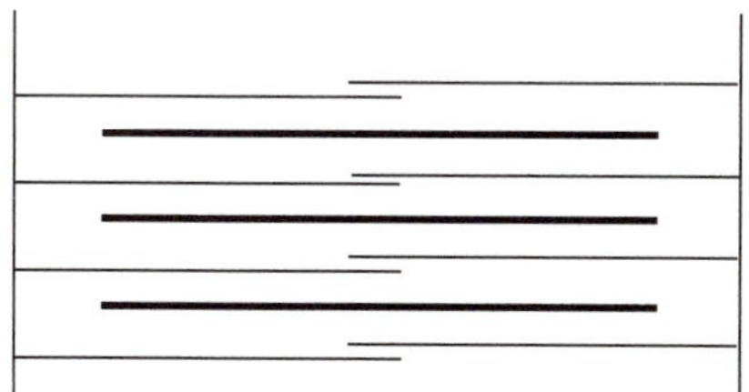

Figure 15-2
Schematic illustration of a fully contracted sarcomere.

(b) Thin filaments and thick filaments must have the same polarity to interact with one another to form cross-bridges, but the polarity of the thick filaments reverses halfway between the Z lines. (See Figure 15-20 on p. 399 of Stryer.) Thus, in the region where the thin filaments override one another in a fully contracted sarcomere, the ends of the thin filaments have the wrong polarity to interact with an adjacent thick filament. Hence, a slightly smaller number of cross-bridges are formed, so the fully contracted muscle develops less tension; that is, it is "weaker."

3. The cycle of muscle contraction is initiated by the release of calcium ion from the tubules of the sarcoplasmic reticulum. Relaxation must therefore involve the sequestration or removal of that calcium. EDTA binds calcium, thereby mimicking its removal by the sarcoplasmic reticulum.

4. Colchicine inhibits the polymerization of microtubules (see Stryer, p. 408). When colchicine is added to mitotic onion root-tip cells, the formation of spindle fibers does not occur. Chromosomes divide at the centromere, but the chromatids are not pulled to the opposite poles of the cell. Thus, the number of chromosomes per cell increases for some of the cells and decreases for other cells.

5. Cancer cells have a greater than normal rate of cell division. In the presence of vincristine or vinblastine, mitotic spindle fibers fail to form, and cell division is retarded. Since cancer cells grow more rapidly, they are more sensitive to these drugs.

6. The tail of myosin binds other myosin molecules to form the thick filaments of striated muscle. Without this self-aggregating property of

myosin, striated muscle would not form. The tail regions of kinesin or dynein bind the surface of a vesicle that is being transported. Changing their tail regions would change the vesicles or molecules that are being transported. Finally, cellular and ciliary dynein must have different tail regions so that the former will transport vesicles and the latter will polymerize with tubulin in cilia.

7. Incubate the labeled ATP with myosin and measure the amount of labeled inorganic phosphate liberated as a function of time without actin. Then add actin. The result should be a burst in the amount of labeled P_i released. (See the mechanism presented in Figure 15-7 on p. 394 of Stryer.)

8. (a)

$$\frac{6 \text{ ft} \times 12 \text{ in/ft} \times 2.54 \text{ cm/in} \times 0.01 \text{ m/cm}}{8 \times 10^{-9} \text{ m/step}} = 2.3 \times 10^{8} \text{ steps}$$

and 2.3×10^8 molecules of ATP hydrolyzed.

(b)

$$\frac{6 \text{ ft} \times 12 \text{ in/ft} \times 2.54 \text{ cm/in} \times 0.01 \text{ m/cm}}{0.4 \text{ m/day}} = 4.6 \text{ days}$$

(c)

$$\frac{4.6 \text{ days} \times 24 \text{ h/day} \times 60 \text{ min/h} \times 60 \text{ s/min}}{2.3 \times 10^{8} \text{ steps}} = 1.7 \times 10^{-3} \text{ s/steps}$$

and the inverse of 1.7×10^{-3} seconds/step gives 588 ATPs hydrolyzed per second.

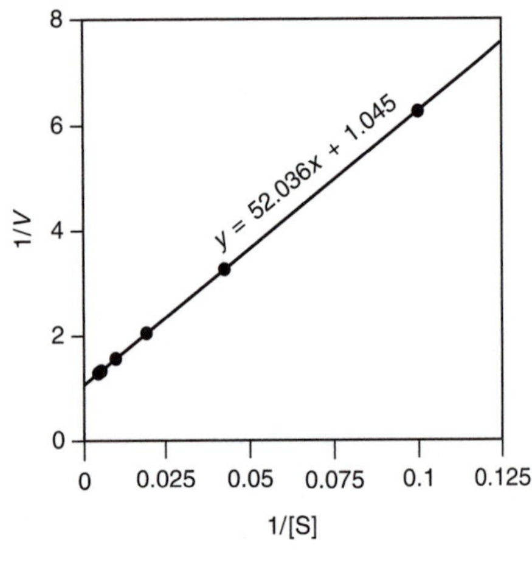

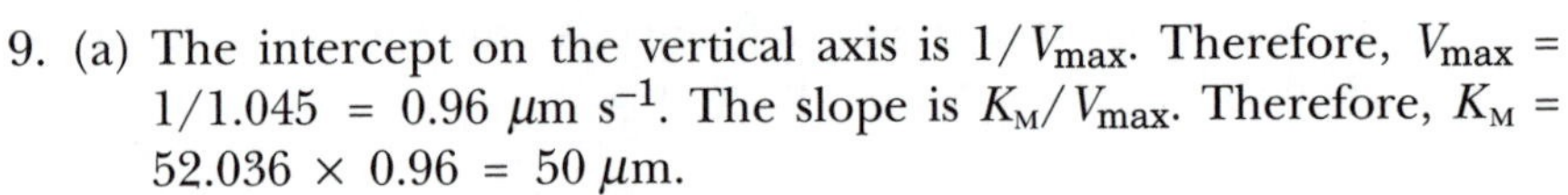

9. (a) The intercept on the vertical axis is $1/V_{max}$. Therefore, $V_{max} = 1/1.045 = 0.96\ \mu\text{m s}^{-1}$. The slope is K_M/V_{max}. Therefore, $K_M = 52.036 \times 0.96 = 50\ \mu\text{m}$.

(b) When half the ATP is hydrolyzed, the ATP concentration is 100 μM. We start with the rate equation for competitive inhibition:

$$V = \frac{V_{max}[S]}{K_M\left(1 + \frac{[I]}{K_i}\right) + [S]}$$

Substitution of values into the equation gives

$$V = \frac{0.96\ (100)}{50\left(1 + \frac{100}{150}\right) + 100}$$

$$V = 0.52\ \mu\text{m s}^{-1}$$

(c) The turnover number is equal to k_3 in equation 33 on page 195 of Stryer:

$$V_{max} = k_3[E_T]$$

$$k_3 = \frac{V_{max}}{[E_T]} = \frac{100\ \mu\text{mole min}^{-1}}{20 \times 10^{-3}\ \mu\text{mole}} = 5000\ \text{min}^{-1}$$

or $5000/60 = 83.3\ \text{s}^{-1}$.

(d) Cycle time $= 1/k_3 = 0.012$ s

(e) $V_{max} = 0.96\ \mu\text{m s}^{-1} = 960\ \text{nm s}^{-1}$

$k_3 = 83.3\ \text{s}^{-1} = 83.3$ ATP split s^{-1}

Therefore, the distance moved per ATP is $960/83.3 = 11.5$ nm.

EXPANDED SOLUTIONS TO TEXT PROBLEMS

1. (a) ATP for skeletal muscle and eukaryotic cilia, and proton-motive force for bacterial flagella.

(b) Two in each system: myosin-actin, dynein-microtubule, and motA-motB, respectively.

(c) Thick and thin filaments of muscle are located in the cytosol. The axoneme of cilia is enveloped by the plasma membrane; the contents are continuous with the cytosol. The rotatory motor that drives bacterial flagella is located in the cytoplasmic membrane of the cell; the flagella are extracellular appendages.

2. A large load resists the movement of the thin filament relative to the thick filament. ADP release is slower with a large load than a small one because the power stroke takes longer to complete.

3. The rapid decrease in the level of ATP following death has two consequences. First, the cytosolic level of calcium rises rapidly because the Ca^{2+}-ATPase pumps in the plasma membrane and sarcoplasmic reticulum membrane no longer operate. High Ca^{2+}, through troponin and tropomyosin, enables myosin to interact with actin. Second, a large proportion of S1 heads will be associated with actin. Recall that ATP is required to dissociate the actomyosin complex. In the absence of ATP, skeletal muscle is locked in the contracted (rigor) state.

4. Remember, dissociation constants express the ratio of the rate constants of the off and on reactions: $K_{dis} = k_{off}/k_{on}$ (see Chapter 7, problem 4, of Stryer). Since k_{on} is limited by diffusion (Stryer, Chapter 8, p. 196) to a value somewhat less than $10^9\ \text{M}^{-1}\ \text{s}^1$, we can do the following calculations. For calmodulin, $K_{dis} = k_{off}/k_{on}$. Substituting, $10^{-6} = k_{off}/{<}10^9$ and $k_{off} = {<}10^3 \text{M}^{-1}\ \text{s}^{-1}$. A similar calculation for troponin gives its $k_{off} = {<}10^4$

$M^{-1}\ s^{-1}$. Thus, as predicted, troponin C is a faster switch than calmodulin, which enables skeletal muscle to contract and relax rapidly. The price paid is that a higher concentration of Ca^{2+} is needed to activate troponin C, and hence more ATP must be spent to pump Ca^{2+} out of the cytosol to return it to the basal state.

5. The effect is mediated through Ca^{2+}–calmodulin, which stimulates myosin light-chain kinase (MLCK) (see Figure 13-47 on p. 350). Phosphoryl groups introduced by MLCK are removed from myosin by a Ca^{2+}-independent phosphatase.

6. In principle, phalloidin could (a) block the assembly of microfilaments by capping the plus (growing) end, (b) sever already formed microfilaments, (c) prevent the disassembly of microfilaments, and (d) interfere with the binding of microfilaments to membranes and other cellular structures. In fact, phalloidin binds to F-actin and blocks depolymerization.

7. (a) Substituting into the equation given for force (F), we get $F = 6\pi \times 10^{-2}\ g\ cm^{-1}\ s^{-1} \times 10^{-4}\ cm \times 0.5 \times 10^{-4}\ cm\ s^{-1} = 9.4 \times 10^{-10}\ g\ cm\ s^{-2} = 9.4 \times 10^{-10}$ dyne.

(b) If one erg = dyne cm and the bead moved 0.5 μm (5×10^{-5} cm), the work per second = 9.4×10^{-10} dyne $\times\ 5 \times 10^{-5}$ cm = 4.7×10^{-14} erg.

(c) To calculate the energy content of ATP in ergs, you need to know that 1 kcal = 4.19×10^{10} ergs. Then for ATP hydrolysis within the cell $\Delta G = -12$ kcal/mol = -50.2×10^{10} ergs per mole. Dividing ergs per mole by Avogadro's number (6.02×10^{23}), we get 8.3×10^{-13} ergs/molecule (the number used in Stryer's answer). The hydrolysis of 40 ATP molecules could yield 3.3×10^{-11} ergs, much more energy than the actual work performed in moving the 2-μm-diameter bead. Thus, the hydrolysis of ATP by a single kinesin motor provides more than enough free energy to power the transport of micrometer-size cargos at micrometer-per-second velocities.

CHAPTER

16

Protein Folding and Design

This chapter addresses the fundamental problem of how polypeptide chains fold into functional proteins. First Stryer discusses the allowed and preferred conformations of the main chain and of individual amino acid residues in proteins. He also introduces the major kinds of supersecondary structures found in proteins. The next sections present examples of the methods used in the study of the folding of several proteins. Stryer then turns to the catalyzed folding of proteins that occurs in cells and concludes the chapter with a discussion of current advances in the design of proteins. Chapters 2 and 3, on protein structure and methodologies, as well as several of the chapters on protein structure and function (Chapters 7, 9, 10, 14) contain pertinent information for this chapter.

LEARNING OBJECTIVES

When you have mastered this chapter, you should be able to complete the following objectives:

Introduction (Stryer pages 417–420)

1. Note that *protein folding* proceeds through stabilization of *intermediate states* rather than through a sampling of all possible conformations.

2. Define *molten globule.* Outline the experiments with α-lactalbumin that provided evidence for the existence of molten globule states.

Conformation of Polypeptide Chains (Stryer pages 420–424)

3. Discuss the significance of *Ramachandran plots.* Contrast the conformational states of Gly and Pro in proteins compared with other amino acid residues.

4. Rationalize the conformational preferences of different amino acids in proteins and polypeptides.

5. Describe the different kinds of *supersecondary structures* or *folding motifs* found in proteins: *four-helix bundle, β hairpin, Greek key motif, $\beta\alpha\beta$ motif.*

Folding of Proteins (Stryer pages 424–429)

6. Explain how proteins can be *unfolded.*

7. List examples of the methods that are currently used to study the *folding* process of proteins.

8. Outline the *rapid kinetic studies* of the folding of cytochrome *c.* Explain the use of *circular dichroism spectroscopy* in these studies, and give the time sequence of the folding steps.

9. Describe the sequence of steps in the *pulsed exchange NMR spectroscopy* experiment in the folding of lysozyme. Explain the significance of the results.

10. Discuss the role of *disulfide bonds* in the refolding of reduced *bovine pancreatic trypsin inhibitor (BPTI).*

11. Note that *subdomains* of BPTI can also fold into a functional structure.

Catalyzed Folding of Proteins (Stryer pages 429–431)

12. Describe the function and catalytic mechanism of *protein disulfide isomerase (PDI).*

13. Outline the function and catalytic mechanism of *peptidyl prolyl isomerase (PPI).*

14. Describe the role of *molecular chaperones.* Note that ATP is used during chaperone activity.

Protein Stability and Design (Stryer pages 431–437)

15. Discuss the *stabilization* of proteins by disulfide bonds and specific amino acid residues.

16. List the reasons why *marginal stability* of proteins is advantageous.

17. Note that proteins may tolerate extensive amino acid substitutions while preserving the native fold. Use the examples of globins and λ repressor to illustrate this principle.

18. Discuss the methods and advances in the prediction of three-dimensional structures of proteins.

19. List the possible motivations for the design of proteins.

20. Describe the construction and properties of a *parallel two-stranded polypeptide rod.*

21. Outline the design and synthesis of a *four-helix bundle protein.* Note its properties.

22. Describe the design, construction, and properties of a Zn^{2+}-binding protein.

SELF-TEST

Introduction

1. Explain why the folding of a protein cannot occur by the conversion of all possible conformational states into the native, energetically most favorable one.

____

____

2. Which of the following is not a property of a molten globule?

(a) It has a hydrophobic core.
(b) It contains secondary-structure elements.
(c) It has a defined three-dimensional structure.
(d) It is compact.
(e) It is more flexible than the native protein.

Conformation of Polypeptide Chains

3. Which of the following answers completes the sentence correctly? A Ramachandran plot

(a) represents the sterically allowed conformations of a polypeptide backbone.
(b) gives the frequency of occurrence of amino acids in β sheet structures.
(c) predicts α-helical structures from given amino acid sequences.
(d) shows the x-ray diffraction pattern of a protein.

4. Indicate which of the following statements are True (T) or False (F).

(a) The conformation of a main chain is completely defined if the ϕ and ψ angles are known for each amino acid residue. ______
(b) The ϕ angle is the angle between the α carbon and the carbonyl carbon of an amino acid residue. ______
(c) There are four allowed regions in the Ramachandran plot for Ala in actual proteins. ______
(d) The right- and left-handed α helices are predicted to have the same ϕ and ψ angles. ______
(e) Pro has more allowed conformations than all the other amino acids. ______
(f) Because it has no side chain, Gly can adopt a much wider range of ϕ and ψ angles than the other amino acid residues. ______

5. Match each amino acid or class of amino acid residues in the right column with the secondary structure element in the left column in which it is most likely to occur.

(a) α helix ______________	(1) Gly
	(2) Pro
(b) β sheet ______________	(3) Glu
(c) β turn ______________	(4) Amino acids with branching at the β carbon
	(5) Ala

6. Which statements are correct about the Greek key folding motif? The Greek key motif

(a) contains β hairpins.
(b) consists of four antiparallel α helices.
(c) consists of four antiparallel β strands.
(d) contains two parallel β strands connected with an α helix.
(e) contains loops that are mostly exposed to solvent.

Folding of Proteins

7. Proteins can usually be unfolded by which of the following treatments:

(a) Increase in temperature to 70°C
(b) Dilution of a concentrated urea solution from 7 to 0.1 M
(c) Addition of guanidine hydrochloride to a 5 M concentration
(d) Decrease of pH to about 2
(e) Change of the solution from H_2O to D_2O

8. Explain the kinds of different structural information that can be obtained from circular dichroism measurements of proteins in the far-ultraviolet region or in the near-ultraviolet region.

____

____

9. Match the conditions or results (a, b, c, d) of the pulsed-label NMR studies of the folding of lysozyme with the explanatory statements in the right column.

(a) Lysozyme is dissolved in a high concentration of denaturant in D_2O. ________

(b) The denaturant is diluted with D_2O, and H_2O is added at various times at high pH (~9). ________

(c) The pH of the solution is decreased to 4 and two-dimensional NMR is performed. ________

(d) Amides in helices become protected in about 60 ms from the start of folding, while the amides in the β sheet become protected in about 600 ms. ________

(1) Isotope exchange stops and the location of amide protons can be determined.
(2) Lysozyme is unfolded and all amide protons are exchanged with deuteriums.
(3) Different regions of secondary structure become stabilized at different times during the folding of lysozyme.
(4) Folding is initiated and protons exchange for exposed amide deuteriums at a high rate.

10. Which statements about the folding of reduced BPTI in the presence of oxidizing agents are correct?

(a) The major intermediates in the folding of BPTI contain only native disulfide bonds.
(b) All three disulfide bonds form simultaneously.
(c) BPTI inhibits trypsin only in the completely refolded and oxidized native state.
(d) Folding intermediates can be separated by density-gradient centrifugation.
(e) Subdomains fold into structures resembling those found in the intact protein.

Catalyzed Folding of Proteins

11. Which of the following statements about protein disulfide isomerase are incorrect?

(a) It accelerates disulfide interchange during protein folding.
(b) It binds only peptide and protein sequences that contain His-Cys pairs.
(c) It contains Cys residues in its active site.
(d) It catalyzes the cis-trans isomerization of peptide bonds adjacent to Cys residues.
(e) It forms a disulfide bond with the substrate during catalysis.

12. Which of the following are properties of peptidyl prolyl isomerase?

(a) It acts on X-Pro peptide bonds.
(b) It acts on Pro-X peptide bonds.
(c) It accelerates cis-trans isomerization by about 300-fold.
(d) It assists the formation of molten globules.
(e) It stabilizes a transition state with a planar peptide bond.

13. Explain why refolding of denatured proteins in vitro is usually performed in dilute solutions.

____

____

Protein Stability and Design

14. Which of the following changes would stabilize a protein?

(a) Introduction of disulfide bonds
(b) Substitution of a pair of amino acid residues hydrogen-bonded to each other with Gly
(c) Replacement of a small amino acid residue in the crowded hydrophobic core of a protein with Trp
(d) Decrease in the number of Pro and hydroxy-Pro residues in a collagen molecule

15. Which of the following processes is not favored by the marginal stability of proteins?

(a) Conformational changes of proteins during signal transduction
(b) Translocation of proteins through membranes
(c) Switch in the substrate specificity of enzymes
(d) Rearrangement of protein-folding intermediates to give the native state
(e) Remodeling of tissues

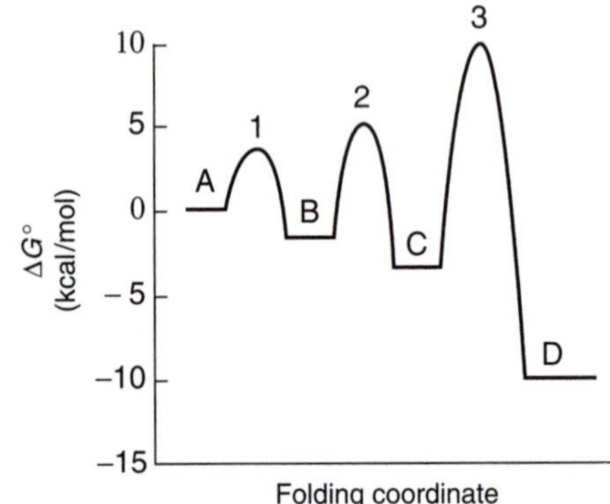

Figure 16-1
A free-energy diagram for the folding of a protein.

16. Examine the free-energy ($\Delta G°$) diagram for the folding of a protein (Figure 16-1) and identify the following:

(a) Native state ______________
(b) Folding intermediate(s) ______________
(c) Unfolded state ______________
(d) Slowest folding step ______________
(e) Most rapid folding step ______________

17. Refer to Figure 16-1. Estimate the free-energy difference between the unfolded and the native state of the protein.

____

____

18. Explain how it is possible to change numerous amino acid residues in λ repressor and still preserve its function.

____

____

19. In attempting to predict the three-dimensional structure of a protein, what information is not required?

(a) Knowledge of the amino acid sequence of the protein
(b) Crystallization and low-resolution x-ray structure of the protein

(c) Computer databases of protein folds and structures
(d) Availability of the three-dimensional structure of a homologous protein (>40% identity)

20. Match the de novo designed protein with the corresponding properties listed in the column on the right.

(a) Parallel two-stranded rod ____________
(b) Four-helix bundle ____________
(c) Zn^{2+}-binding protein ____________

(1) Contains β sheets
(2) Contains α helices
(3) Contains Leu residues in nearly every fourth position of the sequence
(4) Contains His residues in its loops
(5) Contains Gly residues in its loops
(6) Consists of two peptides, α and β
(7) Angle between pairs of its helices is about 20°
(8) Stabilized by electrostatic interactions between Lys and Glu residues

21. Explain why Leu residues are chosen for the *a* and *d* positions of the *abcdefg* repeating sequence of the parallel two-stranded rod.

____

____

22. Why is Gly selected for the connecting hairpin loops of the four-helix bundle protein?

____

____

23. What pH would you choose to study the binding of Zn^{2+} to the metal-binding protein described in Stryer p. 437? Explain your answer.

____

____

ANSWERS TO SELF-TEST

Introduction

1. Even a small protein has an enormous number of possible conformational states. The conversion of all these states into the native folded form would take on the order of 10^{27} years. Therefore a random search through all possible conformations is not a realistic mechanism, considering that proteins can fold within seconds.

2. c. This answer is incorrect because a molten globule has fluctuating interactions between secondary-structure elements, but lacks a unique three-dimensional structure and is devoid of biological activity.

Conformation of Polypeptide Chains

3. a. A Ramachandran plot gives the possible ϕ and ψ angles for the main polypeptide chain containing different amino acid residues.

4. (a) T (b) F (c) F (d) F (e) F (f) T

5. (a) 3, 5 (b) 4 (c) 1, 2 (See Table 16-1 in Stryer.)

6. a, c, e. Answer (b) corresponds to a four-helix bundle, and answer (d) describes a $\beta\alpha\beta$ folding motif.

Folding of Proteins

7. a, c, d. Answer (b) is incorrect because proteins unfolded by high concentrations of urea are refolded by dilution. Answer (e) is incorrect because, in general, the properties of D_2O are similar to those of H_2O.

8. Far-ultraviolet circular dichroism spectroscopy probes the peptide-bond region of proteins and provides information on their content of secondary structure, especially α helix. Circular dichroism spectra in the near ultraviolet region, where aromatic amino acid residues absorb light, provides information on the three-dimensional environment of these residues.

9. (a) 2 (b) 4 (c) 1 (d) 3

10. a, e. Answer (b) is incorrect because the native disulfide bonds form at different times during refolding and oxidation. Partially reoxidized BPTI can inhibit trypsin; therefore, answer (c) is incorrect. Answer (d) is incorrect because the density of BPTI intermediates is about the same; their separation is accomplished, instead, by high-pressure liquid chromatography based on charge differences between the partially oxidized forms.

Catalyzed Folding of Proteins

11. b, d. Answer (b) is incorrect because PDI preferentially binds protein or peptide sequences that contain Cys residues, but is otherwise not selective.

12. a, c. Answer (b) is incorrect because a Pro-X bond, in contrast to the X-Pro bond, is not a substrate for PPI. Answer (d) is incorrect because prolyl isomerization is the last, rate-limiting step in protein folding. Molten globules form in ms. Answer (e) is incorrect because the transition state is a twisted, nonplanar peptide bond.

13. Dilute solutions are preferred because refolding from concentrated solutions is much more likely to produce aggregates.

Protein Stability and Design

14. a. Answers (b), (c), and (d) are incorrect because favorable interactions between amino acid residues are replaced by unfavorable ones.

15. c. The recognition and binding of substrates by enzymes depends on multiple, specific interactions at the active site, which depend on a defined protein structure. In fact, binding of substrates stabilizes enzymes substantially.

16. (a) D (b) B, C (c) A (d) 3 (e) 1. The activation-energy barriers 1, 2, and 3 determine the rates of the different steps. The higher the barrier, the slower the reaction. For a review of the concepts of free energy, energy of activation, and reaction rates see pages 185–188 in Stryer.

17. The difference in $\Delta G°$ between A and D is −10 kcal/mol. This means that D is more stable than A by 10 kcal/mol.

18. If the hydrophobic and hydrophilic amino acid residue patterns are preserved, then numerous amino acid residues can be changed without affecting the structure and function of the protein. Changes in the polar and charged residues on the surface of the protein are often not critical because other similar residues can preserve the interactions with water. Changes in hydrophobic residues are also possible, provided the packing of the core is not disturbed.

19. b. If crystals of a protein are available and x-ray structural information can be obtained, then the three-dimensional structure can be determined directly. Such information is not available for most proteins; therefore, predictions and modeling are performed.

20. (a) 2, 3, 6, 8 (b) 2, 5, 7 (c) 1, 4

21. Every fourth position of each helix faces the interior of the rod. Leu residues provide hydrophobic stabilization in the interior of the rod.

22. Gly has a predilection for hairpin loops because of its conformational flexibility in proteins and peptides.

23. pH 7 or higher. Since the metal-binding site is made up of three histidines, they must be in the unprotonated, uncharged form to be able to bind the positively charged metal ion. The pK_a of His residues is approximately 6.5; therefore, a pH higher than this must be chosen to have a substantial fraction of unprotonated His.

PROBLEMS

1. Early experiments on the problem of protein folding suggested that the native three-dimensional structure of a protein was an automatic consequence of its primary structure.

   (a) Cite experimental evidence that shows that this is the case.

Later, the discovery that proteins are synthesized directionally on ribosomes, from the amino to the carboxy terminus, complicated the earlier view of protein folding.

   (b) Explain what the complicating circumstance might be.

The discovery of chaperone proteins allows both earlier views to be reconciled.

   (c) Explain how that might be the case.

2. Predict whether each of the following substitutions in the amino acid sequence of the myosin tail would significantly affect the secondary structure of the myosin tail. Briefly explain your answers. (Refer to pp. 396, 422, and 435 of Stryer.)

(a) Ile for Leu
(b) Met for Ala
(c) Asp for Glu

3. Each amino acid in a run of several amino acid residues in a polypeptide chain has ϕ values of approximately $-140°$ and ψ values of approximately $+147°$. This run of amino acids is separated from another run of amino acids with similar values for ϕ and ψ by amino acid residue X with a ϕ value of $+100°$ and a ψ value of $-40°$. What kind of structure is formed by the two runs? What is the likely identity of X?

4. As described by Stryer on pages 425–426, studies of the exchange rate between hydrogen and deuterium have been valuable in studying protein structure. (An exchangeable hydrogen is defined as one bound to an atom with an unshared pair of electrons.) A fully deuterated, largely α-helical polypeptide is found to have a number of deuteriums that readily exchange with hydrogen without perturbation. Addition of urea brings about the exchange of further deuteriums for hydrogen. Explain these observations.

5. The free-energy difference between folded and unfolded states of a typical 100-residue protein is 10 kcal/mol (Stryer, p. 418). What would be the ratio between folded and unfolded states at 25°C?

6. Trypsin and chymotrypsin are members of the well-studied class of enzymes known as the *trypsin family of serine proteases,* all of which have serine present at one active site, and most of which have a histidine at another site involved in a charge-relay system. The consensus sequence at the histidine site is [LIVM]-[ST]-A-[STAG]-H-C, and the consensus sequence at the serine site is G-D-S-G-G? What are the prospects for having α helices at the active sites of serine proteases?

7. Suppose that a polypeptide chain consists of three β pleated sheet regions separated from one another by short loops. Draw topological diagrams using fat arrows for the β pleated sheet regions to indicate all possible conformations.

8. The lens of the eye contains (in addition to other proteins) γ-crystalline, a single polypeptide chain consisting of two domains linked together by a short segment. Each of the domains, in turn, consists of two Greek key motifs joined by a short segment. Make a topological sketch for γ-crystalline.

9. Helical wheel representations of α-helical regions, such as that shown in Figure 16-30 on page 435 of Stryer, are useful in understanding some of the short-range properties of the α helix. In such a representation, each amino acid residue is separated from its neighbor in the primary structure by 100 degrees of rotation. (This is necessary because there are 360 degrees in a circle, and 3.6 residues per turn in an α helix.)

(a) Use Figure 16-2 to construct a wheel representation for the following peptide, placing the Leu at position 1. Also indicate the hydrophobic and hydrophilic faces.

Leu-Ala-Asp-Ala-Leu-Gln-Gln

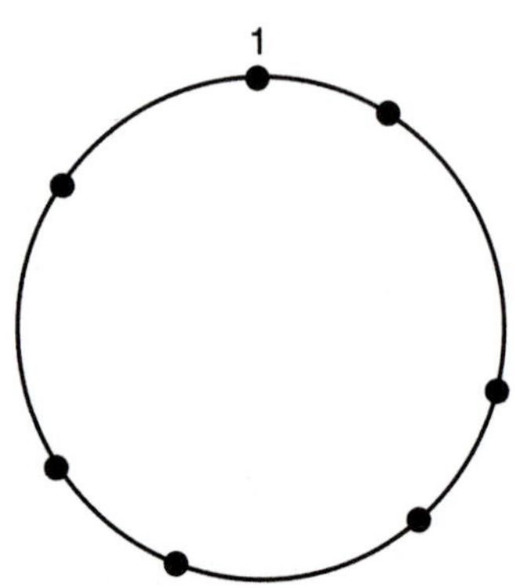

Figure 16-2
Helical wheel for problem 9.

(b) When a helical region is very long, what happens to hydrophilic and hydrophobic faces?

ANSWERS TO PROBLEMS

1. (a) The experiment by Anfinsen on ribonuclease, described by Stryer in pages 37–38, is the classic observation. When native ribonuclease is treated with mercaptoethanol to disrupt disulfide bonds and with urea as a denaturant, it unfolds, as indicated by the fact that it becomes enzymically inactive. When urea is removed by dialysis and disulfide bonds reform by oxidation, it regains enzymic activity, suggesting that its native structure has been restored.

(b) The discovery that proteins are synthesized directionally on ribosomes beginning at the amino terminus complicates matters somewhat because folding of the amino end of the polypeptide chain could begin before the carboxyl end had been synthesized. Such folding could represent the most stable conformation over a short range, but there would be no guarantee that it would be part of the energy minimum for the entire molecule.

(c) Chaperone proteins could bind to an initially synthesized polypeptide and prevent it from undergoing final folding until the entire molecule was synthesized.

2. The myosin tail is composed of a two-stranded α-helical coil. Therefore, the amino acid sequence must be compatible with the maintenance of the α helix.

(a) Whereas Leu is sterically compatible with the α helix, Ile is not. Ile branches at the β carbon atom. This would disrupt the α helix and would therefore distort the myosin tail.
(b) Both Met and Ala are compatible with the α helix, so the substitution of Met for Ala should not affect the secondary structure of the myosin tail.
(c) Although Asp and Glu both have negative charges in their side chains at physiologic pH, Asp has one less methylene group. The negative charge in its side chain is close enough to the hydrogen-bonding groups in the backbone of the α helix to interact with them and disrupt regular hydrogen bonding. Accordingly, the substitution of Asp for Glu would alter the secondary structure of the myosin tail.

3. From the Ramachandran plot on page 421 of Stryer, we see that antiparallel β conformation is accommodated by ϕ values of approximately $-140°$ and ψ values of approximately $+147°$. Thus the two runs are likely antiparallel β sheets. The residue separating them must allow a hairpin turn. The ϕ and ψ values reported for it suggest Gly as a possibility.

4. The most readily exchangeable hydrogens would be those in the side chains of amino acids, particularly those containing ionizable groups like Asp, Glu, His, Lys, and Arg. Peptide hydrogens, because they are involved

in the hydrogen bonds with carbonyl oxygens that maintain α-helical structure, would be slow to exchange. Addition of urea would open the α helix into a random coil. Because peptide hydrogens are no longer involved in hydrogen bonds with carbonyl oxygens, they now exchange more readily.

5. The equilibrium constant for a process gives the ratio of products to reactants, so this question amounts to asking for the equilibrium constant for the reaction:

$$\text{Unfolded} = \text{folded}$$

The applicable relationship is equation (13) on page 187 of Stryer:

$$K'_{eq} = 10^{-\Delta G°'/1.36}$$

Substituting in values yields

$$K'_{eq} = 10^{-(-10)/1.36} = 2.25 \times 10^{7}$$

Thus, an energy difference this large ensures that virtually all protein molecules will be in the folded state.

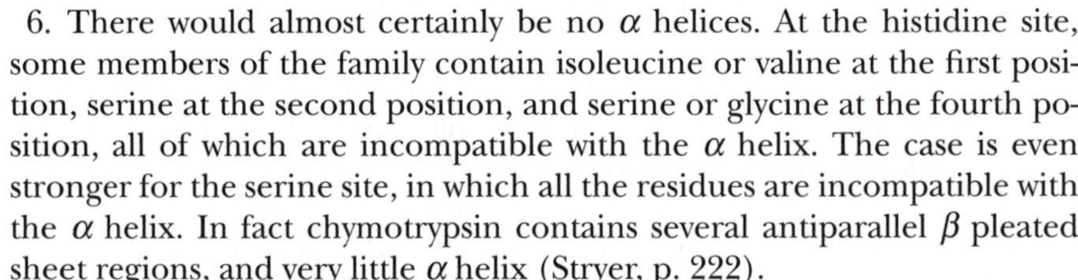

6. There would almost certainly be no α helices. At the histidine site, some members of the family contain isoleucine or valine at the first position, serine at the second position, and serine or glycine at the fourth position, all of which are incompatible with the α helix. The case is even stronger for the serine site, in which all the residues are incompatible with the α helix. In fact chymotrypsin contains several antiparallel β pleated sheet regions, and very little α helix (Stryer, p. 222).

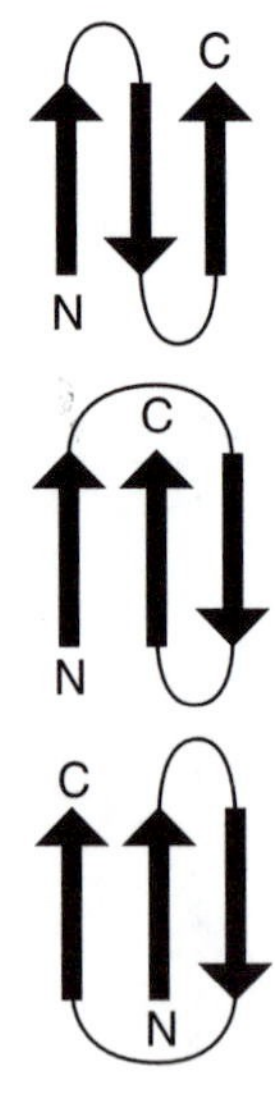

Figure 16-3
Possible conformations for three β pleated sheet regions.

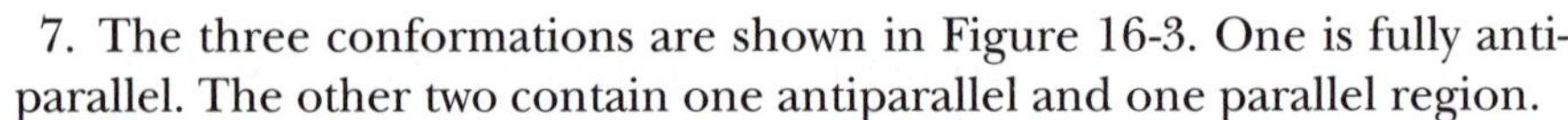

7. The three conformations are shown in Figure 16-3. One is fully antiparallel. The other two contain one antiparallel and one parallel region.

8. A possible topological sketch is shown in Figure 16-4.

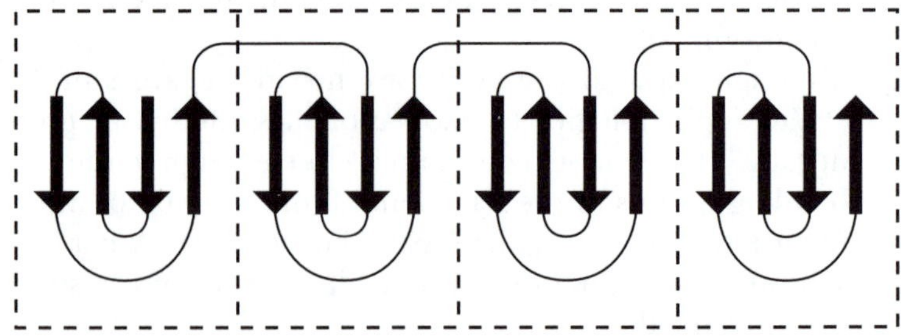

Figure 16-4
Topological diagram for γ-crystalline.

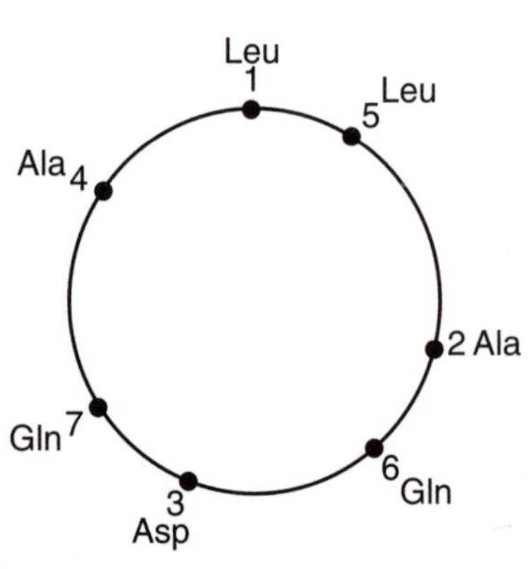

Figure 16-5
A completed helical wheel for problem 9.

9. (a) A helical wheel is shown in Figure 16-5. The Leu pairs in positions 1 and 5 define the hydrophobic face, while the Gln residues at positions 6 and 7, together with the Asp in position 3, define the hydrophilic face.

(b) In a very long α helix, the hydrophobic and hydrophilic faces will rotate as one moves along the helix. This is necessary because there are 3.6 residues per turn, not an integral number.

EXPANDED SOLUTIONS TO TEXT PROBLEMS

1. Locating calcium-binding EF hands in an amino acid sequence involves searching for a distinctive pattern of hydrophobic and oxygen-containing side chains in successive 29-residue segments (Stryer, Figure 16-28, p. 434). Although this analysis is difficult by eye, it is easily done by computer. The four hands are sequences 12–40, 48–76, 85–113, and 121–149. Note that the sequences EF, EI, EL, or EV occur at positions 1–2 and 21–22 in each EF hand. Also, the sequence DGXGXI (or V) is seen at positions 12–17.

2. (a) Since the replacement of seven amino acids with alanine had little effect on the enzymatic activity of lysozyme, there must be considerable information redundancy in the amino acid sequence of this enzyme. This part of lysozyme can tolerate substantial amino acid variation while maintaining an active molecular structure.

   (b) The side chain of alanine is much smaller than that of leucine (methyl versus isobutyl). The absence of the three carbon atoms probably creates a cavity or space in the hydrophobic core of the molecule. This means that some of the atoms are now farther apart and the strength of their van der Waals attractions is diminished. Since the bonds that stabilize the protein conformation are weaker, the melting temperature will lower.

   (c) As shown in Table 16-1, in an α helix alanine is preferred over serine. However, when exposed to solvent the hydrophilic side chain of serine may be preferred over alanine. This probably explains why the T_m is raised only 1°C.

   (d) Multiple alanine mutagenesis has the potential for rather easily finding a number of noncritical residues and thus focuses attention on positions that are critically important for folding and stability.

3. This result argues against a sequential folding mechanism in which the N terminus folds first as it comes off the ribosome. If the folding of the C-terminal domain depended on the prior folding of the N-terminal domain, the mutant in which these termini are joined would not be expected to fold correctly and the molecule would not be enzymatically active.

4. (a) The gene for each heavy chain is altered so that it encodes a coiled-coil sequence following the C_H1 domain. The F_{ab} units will then associate by forming an α-helical coiled coil.

   (b) A bispecific antibody has been used to increase the destruction of target cells by killer T cells. One of the F_{ab} units of this antibody was specific for a cell-surface protein on killer T cells, and the other was specific for an interleukin receptor on a target cell. [See S.A. Kostelny, M.S. Cole, and J.Y. Tso, *J. Immunol.* 148(1992): 1547.]

5. One possibility a priori is that cyclophilin catalyzes the cistrans isomerization of a peptide bond in CsA. The newly formed isomer would then dissociate from cyclophilin, bind to calcineurin, and block its phosphatase activity. Alternatively, calcineurin might be inhibited by CsA only when the cyclic peptide is bound to cyclophilin. In fact, the inhibitory species is a complex of CsA and cyclophilin. We see here that the simplest mechanism is not always the one utilized by nature. The research that elucidated this subtle mechanism of action of an important drug is reviewed in S.L. Schreiber, *Cell* 70(1992):365.

6. A water-soluble globular protein has been prepared by joining four amphipathic helices through short linkers (four-helix bundle, Stryer, p. 436). In this case (Figure 16-31A), the hydrophobic portion of the helix is on the inside of the bundle, and the hydrophilic portion of the helix is on the outside, making the bundle water-soluble. For a membrane ion channel the reverse is needed; the outside of the bundle must now be hydrophobic and the inside hydrophilic. Such a bundle, with helical segments of about 21 residues (to span the bilayer), might fold in a membrane and conduct ions.

CHAPTER

17

Metabolism: Basic Concepts and Design

This chapter is an introduction to the next two parts of the text, which are devoted to metabolism. Metabolism is the interconnected, integrated ensemble of chemical reactions cells use to extract energy and reducing power from their environments, synthesize the building blocks of their macromolecules, and carry out all the other processes that are required to sustain life. Basic thermodynamic postulates were presented in Chapter 8 and should now be reviewed for a better understanding of the principles of metabolism presented in this chapter. Because energy is an essential concept in understanding metabolism, Stryer begins this chapter with a review of free-energy changes in the context of a description of coupled reactions. He explains how an energetically unfavorable reaction can occur if it is coupled to one that occurs spontaneously. He presents the three ways in which separate reactions can be linked energetically.

The most important molecules for storing and carrying energy in metabolic processes are described next, including ATP, the universal currency of energy in biological systems, and the ubiquitous electron carriers NAD^+ and FAD. The relationship of these activated carrier molecules and of other enzyme cofactors to the nutritionally essential vitamins is presented. The role of creatine phosphate, the specialized energy storage molecule of vertebrate muscle, is also given. Stryer organizes a broad outline of energy metabolism by classifying it into three stages and gives an overview of its regulation. He describes nuclear magnetic resonance (NMR) spectroscopy as a noninvasive tool for investigating the metabolism of tissues, and concludes the chapter by indicating the importance of ribonucleotides in the prebiotic world as well as in present-day metabolism.

LEARNING OBJECTIVES

When you have mastered this chapter, you should be able to complete the following objectives:

Introduction (Stryer pages 443–444)

1. Define *metabolism.*

2. State the significance of the *free-energy change* (ΔG) of reactions and the relationship of ΔG to $\Delta G°$, the *equilibrium constant,* and the *concentrations of reactants* and *products* of the reaction.

3. Describe the *additivity* of ΔG values for *coupled reactions* and explain the ability of a thermodynamically favorable (exergonic) reaction to drive an energetically unfavorable (endergonic) one.

4. List the three general mechanisms by which distinct reactions can be energetically coupled.

Adenosine Triphosphate and Phosphoryl Transfer Potential (Stryer pages 445–449)

5. Give the structure of *adenosine triphosphate* (*ATP*) and draw a *phosphoester* bond and a *phosphoanhydride* bond. Describe the role of ATP as the major energy-coupling agent (*energy currency*) in metabolism.

6. Describe the *ATP-ADP cycle* of energy exchange in biological systems.

7. Describe the structural and electronic bases for the *high phosphoryl group –transfer potential* of ATP and give the free energy liberated by the hydrolysis of ATP under standard and cellular conditions.

8. Recognize those compounds that have a *high group–transfer potential,* that is, compounds that release large amounts of free energy upon hydrolysis or oxidation.

9. Describe how *creatine phosphate* serves as a "high-energy" buffer in vertebrate muscle.

10. Explain how coupling a reaction with the hydrolysis of ATP can change the equilibrium ratio of the concentrations of the products to the concentrations of the reactants by a factor of 10^8.

Electron Carriers (Stryer pages 449–451)

11. Recognize the structures of *nicotinamide adenine dinucleotide* (*NAD^+*) and *flavin adenine dinucleotide* (*FAD*), describe their reduction to *NADH* and *$FADH_2$*, and explain their roles in metabolism.

12. Contrast the metabolic roles of *NADPH* and *NADH.*

13. Note that most *high-energy compounds* and *reduced electron carriers* are kinetically stable and require enzymes for their reactions.

Coenzymes (Activated Group Carriers) and Vitamins (Stryer pages 451–455)

14. Describe the structure of *coenzyme A (CoA)* and its role as a *carrier of acetyl* or *acyl groups.*

15. List the major *activated carriers,* both of electrons and activated groups, in metabolic reactions.

16. Define *coenzyme* and list the *water-soluble vitamins* that are components of coenzymes.

17. Recognize the *lipid-soluble vitamins* and state their physiologic roles.

18. Relate the role of *ascorbate (vitamin C)* in collagen formation. Outline the *post-translational modification* of proline to hydroxyproline.

Outline of Energy Metabolism (Stryer pages 455–459)

19. Describe the three major *stages in the extraction of energy* from foodstuffs.

20. Discuss the three major mechanisms for the *regulation of metabolism.*

21. Define *energy charge* and compare it with the *phosphorylation potential.*

22. Describe how *nuclear magnetic resonance (NMR)* methods are used to study metabolism in intact tissues.

23. Explain why ribonucleotides and their derivatives probably play a key role in metabolism.

SELF-TEST

Introduction

1. Which of the following are not a function or purpose of metabolism?
   (a) To extract chemical energy from substances obtained from the external environment
   (b) To form and degrade the biomolecules of the cell
   (c) To convert exogenous foodstuffs into building blocks and precursors of macromolecules
   (d) To equilibrate extracellular substances and the biomolecules of the cell
   (e) To assemble the building-block molecules into macromolecules

2. If the $\Delta G'$ of the reaction A $\rightarrow$ B is -3.0 kcal/mol, which of the following statements are correct? (Recall that the prime symbol means that a thermodynamic parameter is measured at pH 7.0.)
   (a) The reaction will proceed spontaneously from left to right at the given conditions.
   (b) The reaction will proceed spontaneously from right to left at standard conditions.

(c) The equilibrium constant favors the formation of B over the formation of A.
(d) The equilibrium constant could be calculated if the initial concentrations of A and B were known.
(e) The value of $\Delta G^{\circ\prime}$ is also negative.

3. Glucose 1-phosphate is converted to fructose 6-phosphate in two successive reactions:

$$\text{Glucose-1-P} \longrightarrow \text{glucose-6-P} \qquad \Delta G^{\circ\prime} = -1.7 \text{ kcal/mol}$$

$$\text{Glucose-6-P} \longrightarrow \text{fructose-6-P} \qquad \Delta G^{\circ\prime} = -0.4 \text{ kcal/mol}$$

What is $\Delta G^{\circ\prime}$ for the overall reaction?

(a) −2.1 kcal/mol
(b) −1.7 kcal/mol
(c) −1.3 kcal/mol
(d) 1.3 kcal/mol
(e) 2.1 kcal/mol

4. The reaction

$$\text{Phosphoenolpyruvate} + \text{ADP} + \text{H}^+ \rightarrow \text{pyruvate} + \text{ATP}$$

has a $\Delta G^{\circ\prime} = -7.5$ kcal/mol. Calculate $\Delta G^{\circ\prime}$ for the hydrolysis of phosphoenolpyruvate.

5. Inside cells, the $\Delta G'$ value for the hydrolysis of ATP to ADP + P_i is approximately −12 kcal/mol. Calculate the approximate ratio of [ATP] to [ADP] [P_i] found in cells at 37°C.

(a) 5000/1
(b) 4000/1
(c) 2000/1
(d) 1000/1
(e) 200/1

6. Which of the following are ways by which two reactions can be coupled energetically to each other?

(a) As common intracellular components of a compartment, two reactions become automatically coupled.
(b) An ionic gradient across a membrane that is formed by one reaction can drive another reaction that uses the gradient to render it exergonic.
(c) A shared, common intermediate can couple two reactions.
(d) A protein that is activated by binding another molecule or by being covalently modified can provide energy to drive another reaction.

7. Which of the following statements about the structure of ATP are correct?

(a) It contains three phosphoanhydride bonds.
(b) It contains two phosphate ester bonds.
(c) The sugar moiety is linked to the triphosphate by a phosphate ester bond.
(d) The nitrogenous base is called *adenosine.*
(e) The active form is usually in a complex with Mg^{2+} or Mn^{2+}.

8. Which of the following factors contributes to the high phosphate group–transfer potential of ATP?

(a) The greater resonance stabilization of ADP and P_i than of ATP
(b) The increase in the electrostatic repulsion of oxygens upon hydrolysis of ATP
(c) The interaction of the terminal phosphoryl group with the ribose group in ADP
(d) The formation of a salt bridge between the base amino group and the negative charges of the phosphate oxygens in ATP

9. Which of the following are high-energy compounds?

(a) Glycerol 3-phosphate
(b) Adenosine diphosphate
(c) Glucose 1-phosphate
(d) Acetyl phosphate
(e) Fructose 6-phosphate

10. ATP falls in the middle of the list of compounds having high phosphate group–transfer potentials. Explain why this is advantageous for energy coupling during metabolism.

____

____

11. Nucleoside phosphates can be interconverted enzymatically. Which of the following reactions are incorrect?

(a) ATP + GDP $\longrightarrow$ ADP + GTP
(b) AMP + GTP $\longrightarrow$ ATP + GDP
(c) ATP + AMP $\longrightarrow$ ADP + ADP
(d) ATP + UMP $\longrightarrow$ ADP + UDP
(e) ADP + AMP $\longrightarrow$ ATP + adenosine

12. Which of the following are features of the ATP-ADP cycle in biological systems?

(a) ATP hydrolysis is used to drive reactions that require an input of free energy.
(b) The oxidation of fuel molecules forms ADP + P_i from ATP.
(c) The oxidation of fuel molecules forms ATP from ADP + P_i.
(d) Light energy drives ATP hydrolysis.
(e) A transmembrane proton-motive force drives ATP synthesis.

13. Which of the following statements about the phosphoryl transfer potential of skeletal muscle are correct?

(a) The ATP of muscle can sustain contraction for less than a second.
(b) Creatine phosphate serves as a phosphoryl reservoir that replenishes the ATP pool.
(c) Creatine phosphate can support contraction for up to four minutes.
(d) The phosphoguanidino group of creatine phosphate has a large negative standard free energy of hydrolysis.
(e) Creatine phosphate is formed by a reaction between creatine and ATP.
(f) The ATP pool of muscle can be partially replenished as a result of the activity of the enzyme adenylate kinase.

Electron Carriers

14. Which of the following answers complete the sentence correctly? NAD^+

(a) is a flavin nucleotide.
(b) is the major electron acceptor used in fuel metabolism.
(c) contains a nicotinamide ring that accepts a hydride ion during reduction.
(d) loses a plus charge upon reduction.
(e) contains ATP as a part of its structure.

15. Which of the following answers complete the sentence correctly? During the reduction of FAD,

(a) a flavin group is transferred.
(b) an equivalent of a hydride ion is transferred.
(c) the isoalloxazine ring becomes charged.
(d) two hydrogen atoms are added to the isoalloxazine ring.
(e) the adenine ring opens.

Coenzymes (Activated Group Carriers) and Vitamins

16. Match the four cofactors in the left column with the appropriate structural features and properties from the right column.

(a) ATP ______________
(b) FAD ______________
(c) NAD^+ ______________
(d) CoA ______________

(1) Nicotinamide ring
(2) Adenine group
(3) Phosphoanydride bond
(4) Sulfur atom
(5) Isoalloxazine ring
(6) Ribose group
(7) Acyl group transfer
(8) Electron transfer
(9) Phosphate transfer

17. ATP and NADH release large amounts of free energy upon the transfer of the phosphate group to H_2O and electrons to O_2, respectively. However, both molecules are relatively stable in the presence of H_2O or O_2. Explain why.

____

____

18. Which of the following correctly pairs a coenzyme with the group transferred by that coenzyme?

(a) CoA, electrons
(b) Biotin, CO_2
(c) ATP, one-carbon unit
(d) NADPH, phosphoryl group
(e) Thiamine pyrophosphate, acyl group

19. Which of the following water-soluble vitamins forms part of the structure of CoA?

(a) Pantothenate
(b) Thiamine
(c) Riboflavin
(d) Pyridoxine
(e) Folate

20. Which of the vitamins in question 19 is referred to as *vitamin* B_1?

____

____

21. Match the lipid-soluble vitamins in the left column with the appropriate biological functions they or their derivatives serve in the right column.

(a) Vitamin A __________
(b) Vitamin D __________
(c) Vitamin E __________
(d) Vitamin K __________

(1) Protection of unsaturated membrane lipids from oxidation
(2) Carboxylation of glutamate residues of clotting factors
(3) Participation in Ca^{2+} and phosphorus metabolism
(4) Precursor of retinal, the light-absorbing group in visual pigments
(5) Related to fertility in rats
(6) Activates transcription of some growth and development genes

22. The hydroxylation of proline in nascent collagen polypeptide chains does not require which of the following?

(a) O_2
(b) Dioxygenase
(c) Ascorbate
(d) Pyridoxal phosphate
(e) α-Ketoglutarate

23. Why does hydroxylation increase the stability of the collagen triple helix?

(a) It promotes hydrogen bonding with water.
(b) It increases hydrogen bonding between polypeptide chains.
(c) It expands the helix and allows the glycine residues to fit better in the interior.

(d) It decreases the melting temperature of nascent collagen.
(e) It helps neutralize the charge on lysine residues.

Outline of Energy Metabolism

24. Which of the following statements about the third of the three stages of metabolism that generate energy from foodstuffs are correct?

(a) It is common to the oxidation of all fuel molecules.
(b) It involves the breakdown of the macromolecular components of food into smaller units, such as amino acids, sugars, and fatty acids.
(c) It releases relatively little energy compared with the second stage.
(d) It involves the conversion of sugars, fatty acids, and amino acids into a few common metabolites.
(e) It produces most of the ATP and CO_2 in cells.

25. Which of the following are reasons the biochemical pathway for the catabolism of a molecule is almost never the same as the pathway for the biosynthesis of that molecule?

(a) It would be extremely difficult to regulate the pathway if it served both functions.
(b) The free-energy change would be unfavorable in one direction.
(c) The reactions never take place in the same type of cell.
(d) Enzyme-catalyzed reactions are irreversible.
(e) Biochemical systems are usually at equilibrium.

26. Which of the following are ways by which metabolism is regulated?

(a) Accessibility of substrates
(b) Pressure fluxes
(c) Amounts of enzymes
(d) Control of enzyme activities
(e) Temperature cycles

27. Which of the following statements about the energy charge are correct?

(a) It can have a value between 0 and 1.
(b) It is around 0.1 in energy-consuming cells, such as the muscle cells.
(c) It can regulate the rates of reactions in energy-consuming and energy-producing pathways.
(d) It is also called the *phosphorylation potential.*
(e) It is buffered in the sense that its value is maintained within narrow limits.

28. Explain why NMR spectroscopy is a very promising tool for the investigation of metabolic events in intact tissues.

____

____

29. Phosphorus NMR spectroscopy of muscle cells allows the intracellular measurement of which of the following?

(a) pH
(b) ADP
(c) ATP
(d) Nucleic acids
(e) Phosphocreatine

Introduction

1. d. The intracellular and extracellular concentrations of most substances are not at equilibrium, and one of the functions of metabolism is to maintain these nonequilibrium concentrations.

2. a, d. The expression for $\Delta G'$ contains two variables: $\Delta G^{\circ\prime}$ (a derivative of K'_{eq}) and the ratio of the product concentrations to the reactant concentrations. Therefore, $\Delta G'$ alone cannot provide information about $\Delta G^{\circ\prime}$ or K'_{eq}. Answer (d) is correct because $\Delta G^{\circ\prime}$ and K'_{eq} can be calculated when $\Delta G'$ and the reactant and product concentrations are known.

3. a

4. −14.8 kcal/mol. The overall reaction can be separated into two steps:

$$\begin{array}{llr} (1) & \text{Phosphoenolpyruvate} + H_2O \longrightarrow \text{pyruvate} + P_i & \Delta G^{\circ\prime} = ? \\ (2) & ADP + P_i + H^+ \longrightarrow ATP + H_2O & \Delta G^{\circ\prime} = +7.3\ \text{kcal/mol} \\ \hline & \text{Phosphoenolpyruvate} + ADP + H^+ \longrightarrow \text{pyruvate} + ATP & \Delta G^{\circ\prime} = -7.5\ \text{kcal/mol} \end{array}$$

Therefore, $\Delta G^{\circ\prime}$ for step 1 is

$$\Delta G^{\circ\prime} = -7.5\ \text{kcal/mol} - (+7.3\ \text{kcal/mol})$$

$$= -14.8\ \text{kcal/mol}$$

5. c.

$$ATP \longrightarrow ADP + P_i \qquad \Delta G^{\circ\prime} = -7.3\ \text{kcal/mol}$$

Using

$$\Delta G' = \Delta G^{\circ\prime} + 2.3\ RT \log_{10} \frac{[ADP][P_i]}{[ATP]}$$

$$-12\frac{\text{kcal}}{\text{mol}} = -7.3\frac{\text{kcal}}{\text{mol}} + 2.3 \times 1.98\frac{\text{cal}}{\text{mol} \cdot {}^{\circ}\text{K}} \times 310\ {}^{\circ}\text{K} \times \log_{10}\frac{[ADP][P_i]}{[ATP]}$$

$$\frac{[ADP][P_i]}{[ATP]} = 10^{-3.33}$$

$$\frac{[ATP]}{[ADP][P_i]} = 2140$$

6. b, c, d. Being in the same compartment does not necessarily couple two reactions.

Adenosine Triphosphate and Phosphoryl Transfer Potential

7. c, e

8. a

9. b, d

10. The intermediate phosphate group–transfer potential of ATP means that, although ATP hydrolysis can drive a very large number of thermodynamically unfavorable biochemical reactions in metabolic pathways, it can itself be regenerated by coupling with other reactions that release more free energy than −7.3 kcal/mol. Thus, ATP can act as an effective carrier, since it both accepts and donates phosphoryl groups.

11. b, e. These reactions are impossible as written because the number of "high-energy" (phosphoanhydride) bonds in the reactants and products are different.

12. a, c, e

13. a, b, d, e, f. Answer (c) is incorrect because the amount of creatine phosphate is sufficient to maintain only a few seconds of intense contraction. The ATP for the creatine kinase–catalyzed reaction given in (d) is generated by glycolysis in anaerobic muscle or by respiration in muscle with sufficient oxygen.

Electron Carriers

14. b, c, d. Answer (e) is incorrect because ADP, not ATP, forms a part of the structure of NAD^+.

15. d

Coenzymes (Activated Group Carriers) and Vitamins

16. (a) 2, 3, 6, 9 (b) 2, 3, 5, 6, 8 (c) 1, 2, 3, 6, 8 (d) 2, 3, 4, 6, 7

17. Although the transfer reactions of the cofactors ATP and NADH have large negative free-energy changes, there are high activation-energy barriers that greatly slow spontaneous reactions with H_2O or O_2, respectively. In other words, cofactors with high group transfer potentials and fuel molecules are thermodynamically unstable yet kinetically stable. Consequently, specific enzymes are required to catalyze their reactions.

18. b

19. a

20. Thiamine

21. (a) 4, 6 (b) 3 (c) 1, 5 (d) 2

22. d

23. b

Outline of Energy Metabolism

24. a, e

25. a, b

26. a, c, d

27. a, c, e

28. NMR spectroscopy is a noninvasive technique that can provide information on the amounts and types of some biological molecules that are present in living tissues. It is based on the energy absorption of naturally occurring atomic nuclei when they are placed in a magnetic field.

29. a, c, e. Although it contains phosphorus, ADP is "invisible" under normal conditions because it is at a low concentration with respect to ATP, phosphocreatine, and P_i.

PROBLEMS

1. Two reactions involving L amino acids and the value of their respective free-energy changes are as follows:

$$\text{Glutamate} + \text{pyruvate} \rightleftharpoons \alpha\text{-ketoglutarate} + \text{alanine} \qquad \Delta G^{\circ\prime} = -0.24 \text{ kcal/mol}$$

$$\text{Glutamate} + \text{oxaloacetate} \rightleftharpoons \alpha\text{-ketoglutarate} + \text{aspartate} \qquad \Delta G^{\circ\prime} = -1.15 \text{ kcal/mol}$$

(a) Write the overall reaction for production of alanine and oxaloacetate from aspartate and pyruvate.
(b) Show that under standard conditions the net formation of alanine and oxaloacetate from aspartate and pyruvate is thermodynamically unfavorable.
(c) Suppose that at 25°C the molar concentrations of reactants and products are as follows:

$$[\text{Pyruvate}] = [\text{aspartate}] = 10^{-2}\text{ M}$$

$$[\text{Alanine}] = 10^{-4}\text{ M}$$

$$[\text{Oxaloacetate}] = 10^{-5}\text{ M}$$

Is the spontaneous synthesis of alanine and oxaloacetate possible under these conditions? Why?

2. Phosphocreatine can be used as a phosphoryl donor for the synthesis of ATP in a reaction catalyzed by creatine kinase. Refer to pages 447 and 448 of Stryer's text for free energies of hydrolysis of ATP and creatine phosphate.

(a) What effect does creatine kinase have on the value of $\Delta G^{\circ\prime}$ for the reaction?
(b) From the typical concentrations of ATP, ADP, creatine phosphate, and creatine cited on page 448 of Stryer's text, calculate $\Delta G'$ for the reaction in a resting muscle cell at 25°C.

(c) Suppose that during muscle contraction the concentration of creatine phosphate drops to 1 mM, while ATP concentration drops to 3.9 mM. At these concentrations, will creatine phosphate serve as a donor of phosphoryl groups to ADP at 25°C?

3. Under standard conditions, the free energy of hydrolysis of L-glycerol phosphate is −2.2 kcal/mol, and for ATP hydrolysis it is −7.3 kcal/mol. Show that when ATP is used as a phosphoryl donor for the formation of L-glycerol phosphate, the value of the equilibrium constant is altered by a factor of over 10^5.

4. When a hexose phosphate is hydrolyzed to free hexose and inorganic phosphate, the ratio of the concentration of hexose to the concentration of hexose phosphate at equilibrium is 99 to 1. What is the free-energy change for the reaction under standard conditions?

5. The process of catabolism releases free energy, some of which is stored as ATP and some of which is lost as heat to the surroundings. Explain how these observations are consistent with the fact that catabolic pathways are essentially irreversible.

6. The equilibrium constant for the oxidative deamination of glutamate to form α-ketoglutarate is 10^{-5}.

$$\text{Glutamate} + \text{NADP}^+ \rightleftharpoons \alpha\text{-ketoglutarate} + \text{NH}_4^+ + \text{NADPH} + \text{H}^+$$

Although the equilibrium constant is very small, α-ketoglutarate is readily formed from glutamate in the cell. How is this possible?

7. Chemotrophs derive free energy from the oxidation of fuel molecules, such as glucose and fatty acids. Which compound, glucose or a saturated fatty acid containing 18 carbons, would yield more free energy per carbon atom when subjected to oxidation in the cell?

8. In a typical cell, the concentrations of pyridine nucleotides and flavins are relatively low compared with the number of substrate molecules that must be oxidized. What does this observation suggest about the rate of oxidation and reduction of these electron carriers?

9. The flow of electrons from reduced pyridine nucleotides, such as NADH, provides energy that can drive the formation of ATP. Why must the reaction

$$\text{NADH} \longrightarrow \text{NAD}^+ + \text{H}^+ + 2\ e^-$$

have a negative value for $\Delta G°'$?

10. In an erythrocyte, the concentration of glucose is 5.0 mM, and the volume of the cell is 1.5×10^{-16} mm^3. If the oxidation of glucose yields 686 kcal/mol, how much heat can be generated by the complete oxidation of the glucose in an erythrocyte?

11. Microorganisms use derivatives of the vitamin B series as cofactors for biochemical reactions. Many bacteria, including *E. coli,* can grow and divide in a medium containing glucose and simple inorganic salts, including nitrogen in the form of ammonia. What does this tell you about vitamin B metabolism in these microorganisms?

12. Why is it desirable for a cell to regulate the *first* reaction in a biosynthetic pathway?

13. Refer to problem 5 on page 462 of Stryer's text. The formation of a number of other important compounds in biosynthetic reactions involves the generation of pyrophosphate and its subsequent hydrolysis to two molecules of P_i. For example, the formation of UDP–glucose from UTP and glucose 1-phosphate yields PP_i, which is then cleaved. What does this tell you about the group transfer potential of UDP–glucose?

14. How can adenylate kinase be used to generate more ATP when it is needed in rapidly contracting muscle?

15. If the free energy of ATP hydrolysis in an erythrocyte is −12.0 kcal/mol, what is the ratio of [ATP] to [ADP][P_i] in the cell?

16. On page 453 of Stryer's text, pantothenate, part of coenzyme A, is listed as a vitamin. Coenzyme A also contains AMP and mercaptoethylamine. Why do you think these components are not listed as vitamins?

17. The coenzyme biotin acts as a carrier of carbon dioxide molecules in carboxylase enzymes. During catalytic cycles biotin undergoes successive carboxylation and decarboxylation, but the coenzyme itself is not chemically altered at the end of each cycle. Yet a small amount of biotin is required on a daily basis in the diet. Why?

18. Sucrose phosphorylase catalyzes the phosphorolytic cleavage of sucrose in certain microorganisms.

$$\text{Sucrose} + P_i \longrightarrow \text{glucose 1-phosphate} + \text{fructose}$$

(a) Use the information below to calculate the standard free-energy change for the phosphorolysis of sucrose:

$$\text{Sucrose} + H_2O \longrightarrow \text{glucose} + \text{fructose} \qquad \Delta G^{\circ\prime} = -7.0 \text{ kcal/mol}$$

$$\text{Glucose 1-phosphate} + H_2O \longrightarrow \text{glucose} + P_i \qquad \Delta G^{\circ\prime} = -5.0 \text{ kcal/mol}$$

(b) Calculate the equilibrium constant for the phosphorolysis of glucose at 25°C.

19. An infant carries a mutation that leads to a 90% decrease in the activity of proline hydroxylase.

(a) What symptoms are you likely to observe in this infant?
(b) How might this disorder be similar to that caused by a deficiency of ascorbic acid (vitamin C)?
(c) Could this genetic disorder be relieved by the administration of large doses of vitamin C?

20. Would you expect prolyl hydroxylase to hydroxylate free proline? Why or why not?

21. What role do you think that iron likely plays in the hydroxylation of proline? Explain your answer with reference to the role of iron in hemoglobin and myoglobin.

22. Suppose that you use site-specific mutagenesis to prepare four kinds of synthetic helical collagen containing the repeating triplet sequences that follow. Predict the relative thermal stability of helices containing each of the four sequences, giving the most stable first.

(a) $[\text{Gly–Pro–Y}]_n$
(b) $[\text{Gly–Pro–Hyp}]_n$
(c) $[\text{Gly–X–Y}]_n$
(d) $[\text{Gly–Pro–Pro}]_n$

23. The rate of collagen turnover in the body is frequently assessed clinically by measuring the concentration of hydroxyproline in the urine.

(a) Give two reasons why hydroxyproline is a better choice than proline for these purposes.
(b) Why might this procedure tend to underestimate the rate of collagen turnover?

ANSWERS TO PROBLEMS

1. (a) The individual reactions proceed as follows:

$$\text{Glutamate} + \text{pyruvate} \longrightarrow \alpha\text{-ketoglutarate} + \text{alanine}$$

$$\alpha\text{-Ketoglutarate} + \text{aspartate} \longrightarrow \text{glutamate} + \text{oxaloacetate}$$

The coupled reaction is obtained by combining the two reactions:

$$\text{Pyruvate} + \text{aspartate} \longrightarrow \text{alanine} + \text{oxaloacetate}.$$

(b) For the synthesis of alanine and α-ketoglutarate, $\Delta G^{\circ\prime}$ is -0.24 kcal/mol, and for the synthesis of glutamate and oxaloacetate, $\Delta G^{\circ\prime} = +1.15$ kcal/mol, under standard conditions. Summing those two values gives $\Delta G^{\circ\prime} = +0.91$ kcal/mol for the overall reaction, which is therefore not thermodynamically favorable under standard conditions.

(c) For the conditions specified,

$$\Delta G = \Delta G^{\circ\prime} + 1.364 \log_{10} \frac{[\text{alanine}][\text{oxaloacetate}]}{[\text{pyruvate}][\text{aspartate}]}$$

$$= +0.91 \text{ kcal/mol} + 1.364 \log_{10} \frac{(10^{-4}\text{ M})(10^{-5}\text{ M})}{(10^{-2}\text{ M})(10^{-2}\text{ M})} \text{ kcal/mol}$$

$$= +0.91 + 1.364(-5) = -5.91 \text{ kcal/mol}$$

The negative value for the ΔG indicates that the reaction can proceed spontaneously.

2. (a) Although the enzyme controls the rate at which equilibrium is attained, it has no effect on the equilibrium constant, K'_{eq}. Because $\Delta G'$ is a function of the equilibrium constant, the action of the enzyme has no effect on its value.

(b) Under the conditions described in Stryer's text,

$$\Delta G' = -3 \text{ kcal/mol} + 1.364 \log_{10} \frac{[4 \text{ mM}][13 \text{ mM}]}{[0.013 \text{ mM}][25\text{mM}]}$$

$$\Delta G' = -3 + 1.364 \log_{10} 160 = -3 + 1.364(2.204)$$

$$\Delta G' = 0.006, \text{ virtually zero.}$$

At these concentrations, there is no free energy released in the reaction, and there is no net formation of ATP.

(c) Under these conditions, the concentration of creatine is 37 mM, and the concentration of ADP increases to about 0.113 mM. Using the free-energy equation above,

$$\Delta G' = -3 \text{ kcal/mol} + 1.364 \log_{10} \frac{[3.9 \text{ mM}][37 \text{ mM}]}{[0.113 \text{ mM}][1 \text{ mM}]}$$

$$\Delta G' = -3 \text{ kcal/mol} + 1.364 \log_{10} 1277 = -3 + 1.364(3.105)$$

$$\Delta G' = -3 + 4.24 = 1.24 \text{ kcal/mol}$$

The positive value of $\Delta G'$ shows that the reaction will not proceed toward the net formation of ATP, so that under these conditions, creatine phosphate does not serve as a donor of phosphoryl groups to ATP. Instead, the reaction procedes toward the net formation of creatine phosphate, with phosphoryl groups donated from ATP.

3. For the synthesis of L-glycerol phosphate from glycerol and phosphate, the value of $\Delta G^{\circ\prime}$ is +2.2 kcal/mol. Under standard conditions,

$$\Delta G^{\circ\prime} = -1.364 \log_{10} K'_{eq}$$

$$\log_{10} K'_{eq} = \frac{+2.2}{-1.364}$$

$$= -1.612$$

$$K'_{eq} = \text{antilog } -1.612$$

$$= 2.45 \times 10^{-2}$$

The overall value of $\Delta G^{\circ\prime}$ for the formation of L-glycerol phosphate using ATP as a phosphoryl donor is equal to the sum of the free-energy values for the two individual reactions:

$$\Delta G^{\circ\prime} = (+2.2) + (-7.3) = -5.1 \text{ kcal/mol}$$

The equilibrium constant for the overall reaction is

$$K'_{eq} = \frac{-5.1}{-1.364}$$

$$= +3.739$$

$$K'_{eq} = \text{antilog } 3.739$$

$$= 5.48 \times 10^3$$

The ratio of the two equilibrium constants is 2.24×10^5.

4. The reaction is

$$\text{Hexose phosphate} + H_2O \rightleftharpoons \text{hexose} + P_i$$

The expression for the equilibrium constant is

$$K'_{eq} = \frac{[\text{hexose}][P_i]}{[\text{hexose phosphate}]} = \frac{99}{1}$$

Thus, $\log_{10} K'_{eq}$ is approximately equal to 2. Using the free-energy equation,

$$\Delta G^{\circ\prime} = -1.364(2)$$

$$= -2.728 \text{ kcal/mol}$$

5. The heat that is lost contributes to an increase in the entropy of the surroundings. A positive change in entropy means that the free energy for a catabolic process is more likely to be negative. Reactions with negative free-energy values are irreversible in that they require an input of energy to proceed in the opposite direction.

6. The removal of the products, α-ketoglutarate and NADPH, for use in other metabolic reactions causes the equilibrium to be shifted toward formation of more α-ketoglutarate. In addition, increased concentrations of glutamate and $NADP^+$ can displace the equilibrium toward the direction of ketoacid synthesis. In either instance, the concentrations of the reactants or products will affect the overall direction of the biochemical reaction.

7. Of the 18 carbon atoms in a saturated fatty acid, 17 are saturated (as $-CH_2-$groups) and are more reduced than the partially oxidized carbon atoms in glucose. In glucose, five of the six carbons are partially oxidized to the hydroxymethyl level, and the sixth is at the more oxidized aldehyde level. A greater number of electrons per carbon are available in the fatty acid, so more metabolic energy is available from it than from glucose.

8. Pyridine nucleotides, such as NADH, serve as acceptors and donors of electrons in many metabolic reactions, including those that generate energy for the cell. Because the absolute number of pyridine nucleotides in the cell is low, the cycle of oxidation and reduction for these compounds must occur rapidly for the oxidation of fuel molecules to proceed at a sufficient rate.

9. For the synthesis of ATP to proceed spontaneously, the overall value of $\Delta G^{\circ\prime}$ must be negative. The value of $\Delta G^{\circ\prime}$ for ATP synthesis is positive,

so a negative value would be expected for the oxidation of NADH to NAD^+.

10. The concentration of glucose in the erythrocyte is 5×10^{-3} mole per liter, or 5×10^{-6} mole per milliliter. The number of moles of glucose in an erythrocyte is

$$1.5 \times 10^{-16}\ \text{mm}^3 \times 5 \times 10^{-6}\ \text{mol/mm}^3 = 7.5 \times 10^{-22}\ \text{mol}$$

The heat released by the oxidation of glucose is

$$7.5 \times 10^{-22}\ \text{mol} \times 6.86 \times 10^5\ \text{cal/mol} = 5.2 \times 10^{-16}\ \text{cal}$$

11. The fact that these bacteria use vitamin B derivatives as cofactors means that they must be able to synthesize them from the simple compounds supplied in the growth medium.

12. Regulation of the first reaction in a biosynthetic pathway ensures that the intermediates in the pathway will be synthesized only when the ultimate product is required. In this way the cell can conserve energy as well as precursors of all intermediates. Such a regulatory scheme will be found when none of the intermediates are utilized in other pathways.

13. As Stryer points out in the answer to his problem (Stryer, p. 1023), the cleavage of pyrophosphate ensures that the coupled reactions will proceed toward the net formation of desired product; that is, the overall reaction will have a rather large negative free-energy value. The fact that PP_i is formed during the synthesis of UDP–glucose suggests that the free energy released by the coupled reactions for the formation of UDP–glucose and the hydrolysis of UTP is small. Therefore, the free energy of the hydrolysis of UDP–glucose would be similar to that of UTP. Thus, you should surmise that the group transfer potential of glucose from UDP–glucose would be high. This is the case, as UDP–glucose serves as a donor of glucose residues for the synthesis of glycogen.

14. Adenylate kinase catalyzes the interconversion of two molecules ADP to ATP and AMP. In contracting muscle, when ATP is hydrolyzed to yield ADP and P_i, an increase in ADP concentration drives the reaction toward the net formation of ATP. In this way, more ATP is made available for contraction.

15. To determine the ratio of the concentrations of products to reactants in the cell, we begin by using the known values of the free-energy changes:

$$\Delta G' = \Delta G^{\circ\prime} + 1.364 \log_{10} Q$$

where Q is the ratio of the concentrations of products to reactants, or $[\text{ADP}][\text{P}_i]/[\text{ATP}]$.

$$-12.0 = -7.3 + 1.364 \log_{10} Q$$

$$\log_{10} Q = \frac{-4.7}{1.364} = -3.446$$

$$Q = \text{antilog } -3.446$$

$$= 3.58 \times 10^{-4}$$

The ratio of the concentration of ATP to the concentration of the products is the inverse of the value for Q.

$$\frac{[\text{ATP}]}{[\text{ADP}][\text{P}_\text{i}]} = \frac{1}{Q} = 2.8 \times 10^3$$

16. AMP and mercaptoethylamine are not listed as vitamins because they can be synthesized de novo from other precursors in cells. Pantothenate is required in the diet because one or more of the biochemical steps needed to synthesize it are deficient in higher organisms. All three components of coenzyme A can be put together by a cell to form the required cofactor.

17. In cells, there is constant synthesis and degradation of enzymes in response to the need for enzymatic activity. Enzyme degradation through proteolysis will often release coenzymes like biotin. Although some biotin molecules can be incorporated into newly synthesized proteins, others are carried by the blood to the kidney, where they are then excreted. Daily excretion of coenzymes leads to a requirement for their replenishment in the diet.

18. (a) To calculate the standard free-energy change for the phosphorolysis of sucrose, add the standard free-energy changes of the two reactions written so that their sum yields the required net reaction:

$$\text{H}_2\text{O} + \text{Sucrose} \longrightarrow \text{glucose} + \text{fructose} \qquad \Delta G^{\circ\prime} = -7.0 \text{ kcal/mol}$$

$$\text{Glucose} + \text{P}_\text{i} \longrightarrow \text{glucose 1-phosphate} + \text{H}_2\text{O} \qquad \Delta G^{\circ\prime} = +5.0 \text{ kcal/mol}$$

Net:

$$\text{Sucrose} + \text{P}_\text{i} \longrightarrow \text{glucose 1-phosphate} + \text{fructose} \qquad \Delta G^{\circ\prime} = -2.0 \text{ kcal/mol}$$

(b) The equation that describes the relationship between the standard free energy and the equilibrium constant is

$$\Delta G^{\circ\prime} = -1.364 \log_{10} K'_{\text{eq}} \text{ at } 25°\text{C}$$

Solving for K'_{eq},

$$\log K'_{\text{eq}} = -\frac{\Delta G^{\circ\prime}}{1.364} = -\frac{-2.0}{1.364} = 1.47$$

$$K'_{\text{eq}} = \text{antilog } 1.47 = 30$$

19. (a) The triple helix of collagen is stabilized by hydroxylation of proline residues. Defective hydroxylation of proline will lead to forms of collagen that have a lower than normal melting temperature. The abnormal collagen cannot properly form fibers; skin lesions and blood vessel fragility are likely consequences. The clinical picture would be likely to resemble scurvy.

(b) Ascorbic acid is a potent reducing agent that maintains proline hydroxylase in an active form, probably by keeping its iron atom in the reduced ferrous state. Collagen synthesized during ascorbic acid deficiency (and therefore during a deficiency of proline hydroxylase activity) is insufficiently hydroxylated and hence has a lower melting temperature. The symptoms of scurvy are similar to those described in (a) above.

(c) It is unlikely that increasing ascorbate concentration would overcome the absence of enzyme activity. With only 10% of the available molecules able to carry out their normal hydroxylating activity, even an excess of ascorbate could not make up for the fact that many fewer enzyme molecules are available.

20. Prolyl hydroxylase hydroxylates the prolyl residues of nascent (newly synthesized) polypeptides. Free proline is not a substrate for the enzyme.

21. Iron likely binds the molecular oxygen that is used in these reactions. The clue to this role is the fact that iron must be maintained in the reduced (ferrous) state by ascorbate in order to be effective, just as the iron of hemoglobin and myoglobin must be maintained in the ferrous state to bind reversibly with oxygen (see Stryer, p. 151).

22. The order of relative stability is b > d > a > c. Hydroxyproline stabilizes the collagen helix more than proline does, and proline stabilizes it more than the amino acids X and Y do.

23. (a) Hydroxyproline is the better choice because (1) hydroxyproline is not found to any extent in any other protein, and (2) hydroxyproline is not recycled for the biosynthesis of new collagen. Rather, proline is incorporated into the polypeptides and is then converted to hydroxyproline.

(b) The method tends to underestimate the rate of collagen turnover because some hydroxyproline is metabolized to other end products rather than being directly excreted.

EXPANDED SOLUTIONS TO TEXT PROBLEMS

1. The direction of a reaction when the reactants are initially present in equimolar amounts is dependent on $\Delta G^{\circ\prime}$. Since by convention reactions are written from left to right, if $\Delta G^{\circ\prime}$ is negative, K'_{eq} is positive and the *direction* is to the right because at equilibrium the product concentrations will exceed those of the reactants. If $\Delta G^{\circ\prime}$ is positive, the reverse is true. The $\Delta G^{\circ\prime}$ values for these reactions are (a) +3 kcal/mol (left), (b) −5.1 kcal/mol (right), (c) +7.5 kcal/mol (left), and (d) −4 kcal/mol (right).

2. Consider a large rock that has been sitting on the side of a mountain for a million years. It has a large amount of potential energy but no kinetic energy—until you push it! Or consider a mixture of H_2 and O_2; it's perfectly stable until you light it! Notice that the thermodynamics of a reaction tell you little, if anything, about its kinetics.

3. (a) Note that phosphoenolpyruvate is *formed* in this reaction; hence, its contribution to $\Delta G^{\circ\prime}$ is *plus* 14.8 kcal/mol. Therefore, $\Delta G^{\circ\prime}$ for the entire reaction is $+14.8 - 7.3 = +7.5$ kcal/mol. $K'_{eq} = 10^{-7.5/1.36} = 3.06 \times 10^{-6}$.

(b) Substituting into $K'_{eq} = 3.06 \times 10^{-6}$ = [ADP]/[ATP] × [PEP]/[Pyr], we get $3.06 \times 10^{-6} = 1/10 \times$ [PEP]/[Pyr]. [PEP]/[Pyr] $= 3.06 \times 10^{-5}$, and [Pyr]/[PEP] $= 1/3.06 \times 10^{-5} = 3.28 \times 10^{4}$.

4. $$\Delta G^{\circ\prime} = +5 - 3.3 = 1.7 \text{ kcal/mol}$$

$$K'_{eq} \frac{[G-1-P]}{[G-6-P]} = 10^{-1.7/1.36} = 5.62 \times 10^{-2}$$

The reciprocal of this is $\frac{[G-6-P]}{[G-1-P]}$ or 17.8.

5. (a) $\Delta G^{\circ\prime} = +7.5 - 7.3 = +0.2$ kcal/mol.

(b) The hydrolysis of PP_i drives the reaction toward the formation of acetyl CoA by making $\Delta G^{\circ\prime}$ strongly negative ($0.2 - 8.0 = -7.8$ kcal/mol).

6. (a) By definition, $\log K = -pK$. Since $\Delta G^{\circ\prime} = -2.3\, RT \log K$, by substitution one obtains $\Delta G^{\circ\prime} = 2.3\, RT\, pK$.

(b) $\Delta G^{\circ\prime} = 1.36 \times 4.8 = 6.53$ kcal/mol at 25°C.

7. The activated form of sulfate in most organisms is 3′-phosphoadenosine 5′-phosphosulfate. See P.W. Robbins and F. Lipmann. *J. Biol. Chem.* 229(1957): 837.

8. (a) One hertz is a frequency of one cycle per second; 129 MHz = 129 $\times 10^{6}$ Hz. One *part per million* of this is 129 Hz. Therefore, 2.4 ppm is 2.4×129 Hz = 310 Hz.

(b) Both the protonation and deprotonation rates must be faster than 310 s^{-1}.

(c) The pK' for the equilibrium of $H_2PO_4^{2-}$ and HPO_4^{-} is 7.21 (see Appendix C). Hence, the dissociation constant K is 6.16×10^{-8} M.

Since $K_{dis} = k_{off}/k_{on}$, the rate constant for the association (k_{on}) = k_{off}/K_{dis}. Because k_{off} is greater than 310 s^{-1} [see part (b)], k_{on} must be greater than $310/6.16 \times 10^{-8}$, or 5×10^{9} $M^{-1}s^{-1}$.

9. Arginine phosphate, like creatine phosphate, contains a *phosphoguanidino* group; these compounds are called *phosphagens.* The transfer of the phosphoryl group of arginine phosphate to ADP is catalyzed by arginine phosphokinase in a reaction similar to the one catalyzed by creatine kinase.

To test the hypothesis that arginine phosphate is acting in a manner similar to that of creatine phosphate, you would monitor the concentrations of ATP and arginine phosphate in invertebrate muscle during contraction. If arginine phosphate is indeed serving as a reservoir of high-potential phosphoryl groups, its concentration will decrease while that of ATP will remain constant (or nearly so) during the early stages of contraction.

10. An ADP unit (or a closely related derivative, in the case of CoA).

CHAPTER

18

Carbohydrates

Carbohydrates are one of the four major classes of biomolecules; the others are proteins, nucleic acids, and lipids, all of which were discussed in earlier chapters. In Chapter 18, Stryer describes the chemical nature of carbohydrates and summarizes their principal biological roles. First, he introduces monosaccharides, the simplest carbohydrates, and describes their chemical properties. Since these sections assume familiarity with the properties of aldehydes, ketones, alcohols, and stereoisomers, students with a limited background in organic chemistry should review these topics in any standard organic chemistry text. Next, Stryer discusses simple derivatives of monosaccharides, including sugar phosphates and disaccharides. *Sugar* is the common name for monosaccharides and their derivatives. You have already seen some monosaccharide derivatives in the structures of nucleic acids, ATP, and other coenzymes in Chapter 17. Finally, Stryer discusses polysaccharides and oligosaccharides as storage and structural polymers and as components of proteoglycans and glycoproteins. Once again, these topics are not completely new; polysaccharides were mentioned in connection with lysozyme substrates in Chapter 9, and glycoproteins and glycolipids were described as components of biological membranes in Chapter 11.

LEARNING OBJECTIVES

When you have mastered this chapter, you should be able to complete the following objectives:

Introduction (Stryer pages 463–464)

1. List the main roles of *carbohydrates* in nature.

Monosaccharide Structures and Chemical Properties (Stryer pages 464–469)

2. Define *carbohydrate* and *monosaccharide* in chemical terms.

3. Relate the absolute configuration of monosaccharide D or L *stereoisomers* to those of *glyceraldehyde.*

4. Associate the following monosaccharide class names with their corresponding structures: *aldose* and *ketose; triose, tetrose, pentose, hexose,* and *heptose; pyranose* and *furanose.*

5. Distinguish among *enantiomers, diastereoisomers,* and *epimers* of monosaccharides.

6. Draw the *Fisher* (open-chain) *structures* and the most common *Haworth* (ring) *structures* of D-*glucose,* D-*fructose,* D-*galactose,* and D-*ribose.*

7. Explain how ring structures arise through the formation of *hemiacetal* and *hemiketal* bonds. Draw a ring structure, given a Fisher formula.

8. Distinguish between α and β *anomers* of monosaccharides and explain *mutarotation.*

9. Compare the *chair, boat,* and *envelope conformations* of monosaccharides.

Simple Monosaccharide Derivatives (Stryer pages 469–472)

10. Define O-*glycosidic* and N-*glycosidic* bonds in terms of *acetal* and *ketal* bonds. Draw the bonds indicated by such symbols as α-1,6 or β-1,4. Stryer's text and other sources also use $\alpha(1\rightarrow6)$ or $\beta1{:}4$ to designate the same bonds.

11. Explain the role of O-glycosidic bonds in the formation of monosaccharide derivatives, *disaccharides,* and *polysaccharides.*

12. Give examples of important phosphorylated sugars.

13. Draw the structures of *sucrose, lactose,* and *maltose.* Give the natural sources of these common disaccharides.

14. Explain the consequences of human *lactase deficiency.*

Polysaccharides (Stryer pages 472–475)

15. Describe the structures and biological roles of *glycogen, starch, amylose, amylopectin, cellulose* and *chitin.*

16. Give examples of enzymes involved in the digestion of carbohydrates in humans.

17. List the major kinds of *glycosaminoglycans* and name their sugar components.

18. Describe the structural organization of the *proteoglycan* from cartilage. Note the roles of proteoglycans in other tissues.

Glycoproteins and Their Receptors (Stryer pages 475–479)

19. Describe the types of covalent bonds that link carbohydrates and proteins.

20. Name the sugars that are frequent constituents of *glycoproteins*. Write a sample sequence for an N-*linked* oligosaccharide.

21. Explain the role of *asialoglycoprotein receptor* in the removal of glycoproteins from blood.

22. Give examples of *lectins* and outline their functions and uses.

23. Describe the properties of the *homing receptor* of lymphocytes and note the role of *selectins* in the development of the nervous system.

SELF-TEST

Introduction

1. Which are roles of carbohydrates in nature? Carbohydrates

(a) serve as energy stores in plants and animals.
(b) are major structural components of mammalian tissues.
(c) are constituents of nucleic acids.
(d) are conjugated to many proteins and lipids.
(e) are found in the structures of all the coenzymes.

2. In the human diet, carbohydrates constitute approximately half the total caloric intake, yet only 1% of tissue weight is carbohydrate. Explain this fact.

____

____

Monosaccharide Structures and Chemical Properties

3. Examine the following five sugar structures:

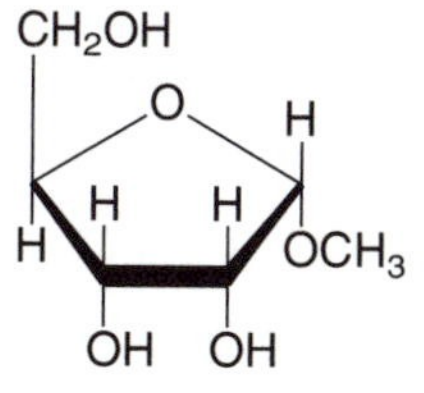

A: furanose ring with CH_2OH, O, H, OCH_3, H, H, H, OH, OH

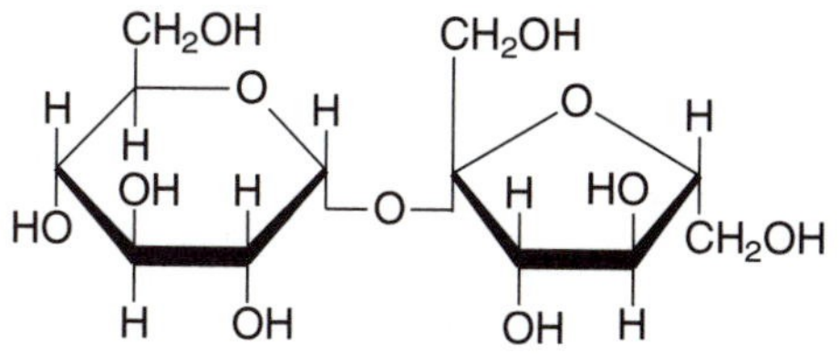

B: two pyranose rings linked by O (CH_2OH; H, O, H; H; OH H; HO; H OH) – (CH_2OH; O, H; H HO; CH_2OH; OH H)

C: furanose ring with H_2COH, O, CH_2OH, H HO, H, OH, OH H

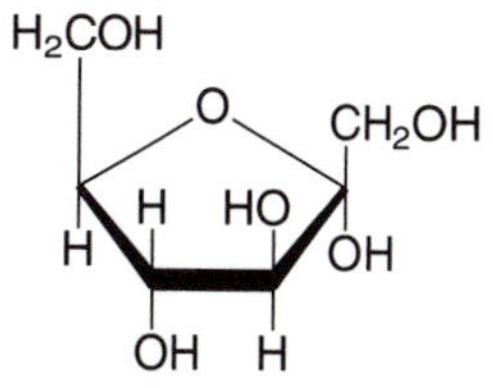

D: two pyranose rings linked by O (CH_2OH; H, O, H; H; OH H; HO; H OH) – (CH_2OH; H, O, H; H; OH H; OH; H OH)

E: pyranose ring with H OH, HO, H, OH H, H, H, O, OH, CH_2OH

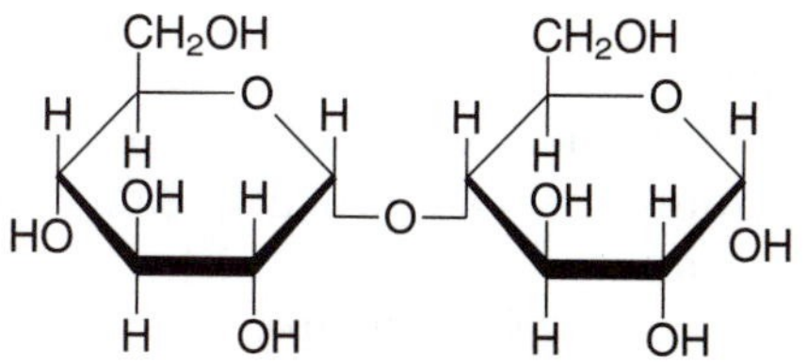

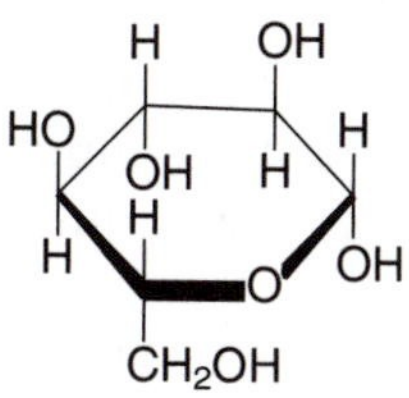

Which of these sugars

(a) contain or are pentoses?__________

(b) contain or are ketoses? __________

(c) contain the same monosaccharides? __________

Name those monosaccharides.__________

____

(d) will yield different sugars after chemical or enzymatic hydrolysis of glycosidic bonds? __________

(e) are reducing sugars? __________

(f) contain a β-anomeric carbon? __________

(g) will mutarotate?__________

(h) is sucrose? __________

(i) are released upon the digestion of starch?__________

A	B	C
CHO	CHO	CHO
HCOH	HCOH	HOCH
HOCH	HCOH	HCOH
HCOH	HCOH	HOCH
CH_2OH	CH_2OH	CH_2OH

Aldopentoses

4. Consider the aldopentoses shown in the margin.

(a) Name the types of stereoisomers represented by each pair.

A and B are __________

B and C are __________

A and C are __________

(b) Name sugar B. __________

(c) Draw the α-anomeric form of the furanose Haworth ring structure for sugar A.

5. Identify the properties common to D-glucose and D-ribose. Both monosaccharides

(a) are reducing sugars.
(b) form intramolecular hemiacetal bonds.
(c) have functional groups that can form glycosidic linkages.
(d) occur in hexose form.
(e) are major constituents of glycoproteins.

6. Referring to the structure of ATP as shown below, which of the following statements are true?

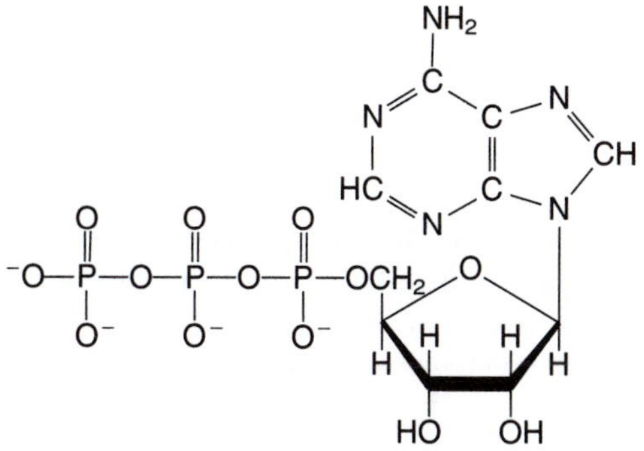

Adenosine triphosphate (ATP)

The structure of ATP

(a) contains a β-*N*-glycosidic linkage.
(b) contains a pyranose ring.
(c) exists in equilibrium with the open Fischer structure of the sugar.
(d) preferentially adopts a chair conformation.
(e) contains a ketose sugar.

7. Draw the structure of the disaccharide glucose $\alpha(1\rightarrow6)$-galactose in the β-anomeric form.

8. Lactase deficiency is characterized by the inability to hydrolyze

(a) α-1,4 glucosidic bonds.
(b) β-1,6 galactosidic bonds.
(c) β-1,4 glucosidic bonds.
(d) β-1,4 galactosidic bonds.
(e) α-1,6 glucosidic bonds.

9. Glucose readily enters all tissues from the blood, but only specialized tissues (liver and kidney) are capable of releasing glucose into the blood. Give one reason why most tissues cannot release glucose to blood.

____

____

Polysaccharides

10. If one carries out the partial mild acid hydrolysis of glycogen or starch and then isolates from the product oligosaccharides all the trisaccharides

present, how many different kinds of trisaccharides would one expect to find? Disregard α or β anomers.

(a) 1
(b) 2
(c) 3
(d) 4
(e) 5

11. A sample of bread gives a faint positive color with Nelson's reagent for reducing sugars. After an equivalent bread sample has been masticated, the test becomes markedly positive. Explain this result.

____

____

12. Why does cellulose form dense linear fibrils, whereas amylose forms open helices?

____

____

13. For the polysaccharides in the left column, indicate all the descriptions in the right column that are appropriate.

(a) Amylose ____________
(b) Cellulose ____________
(c) Dextran ____________
(d) Glycogen ____________
(e) Starch ____________

(1) Contains α-1,6 glucosidic bonds
(2) Is a storage polysaccharide in yeasts and bacteria
(3) Can be effectively digested by humans
(4) Contains β-1,4 glucosidic bonds
(5) Is a branched polysaccharide
(6) Is a storage polysaccharide in humans
(7) Is a component of starch

14. α-Amylase

(a) removes glucose residues sequentially from the reducing end of starch.
(b) breaks the internal α-1,6 glycosidic bonds of starch.
(c) breaks the internal α-1,4 glycosidic bonds of starch.
(d) cleaves the α-1,4 glycosidic bond of lactose.
(e) can hydrolyze cellulose in the presence of an isomerase.

15. Which of the following statements about glycosaminoglycans are true?

(a) They contain derivatives of either glucosamine or galactosamine.
(b) They constitute 5% of the weight of proteoglycans.
(c) They contain positively charged substituent groups.
(d) They bind substantial amounts of water and cations.
(e) They have repeating units of four sugar groups.

16. Which of the following statements about the proteoglycan from cartilage are true?

(a) It contains heparin sulfate.
(b) It contains chondroitin sulfate.
(c) It has a core protein with hyaluronate filaments bound to it.
(d) It contains two kinds of proteins: a core protein and a link protein.
(e) It has a mass of about 2×10^6 daltons.

Glycoproteins and Their Receptors

17. Glycoproteins

(a) contain oligosaccharides linked to the side chain of lysine or histidine residues.
(b) contain oligosaccharides linked to the side chain of serine or threonine residues.
(c) contain linear oligosaccharides with a terminal glucose residue.
(d) bind to liver cell-surface receptors that recognize sialic acid residues.
(e) are mostly cytoplasmic proteins.

18. Match the different kinds of carbohydrates listed in the left column with the corresponding properties in the right column.

(a) Cellulose ____________
(b) Keratan sulfate ____________
(c) Oligosaccharide in an IgG molecule ____________

(1) Contains GlcNAcβ1$\rightarrow$4Man bonds
(2) Contains Glcβ1$\rightarrow$4Glc bonds
(3) Contains SO_3^- groups
(4) Contains more than two kinds of sugars
(5) Is not bound covalently to protein
(6) Is a branched carbohydrate
(7) Has a structural role
(8) Serves a recognition role
(9) Is a component of proteoglycans

19. Which of the following statements are true? Asialoglycoprotein receptor

(a) is an integral membrane protein of liver cells.
(b) mediates cell-to-cell recognition in the liver.
(c) binds to syndecan, a cell-surface proteoglycan.
(d) activates killer T cells.
(e) removes old glycoproteins from blood.

20. Which of the following statements about lectins are correct? Lectins

(a) are produced by plants and bacteria.
(b) contain a single binding site for carbohydrate.
(c) are glycosaminoglycans.
(d) Recognize specific oligosaccharide patterns.
(e) mediate cell-to-cell recognition.

21. Which of the following statements are true? Selectins
    (a) circulate in blood as free proteins.
    (b) are cell-surface receptor proteins.
    (c) are carbohydrate-binding adhesive proteins.
    (d) recognize and bind collagen in the extracellular matrix.
    (e) mediate the binding of leukocytes to sites of injury during the inflammation process.

ANSWERS TO SELF-TEST

Introduction

1. a, c, d

2. Most of the carbohydrates in the human diet are used as fuel to supply the energy requirements of the organism. Although some carbohydrate is stored in the form of glycogen, the mass stored is relatively small compared with adipose tissue and muscle mass. The carbohydrate present in nucleic acids, glycoproteins, glycolipids, and cofactors, although functionally essential, contributes relatively little to the weight of the body.

Monosaccharide Structures and Chemical Properties

3. (a) A
   (b) B, C
   (c) B and C contain fructose; B, D, and E contain or are glucose. Note that glucose is in the α-anomer form in sugars B and D and is in the β-anomer form in sugar E.
   (d) A, B, D
   (e) C, D, E
   (f) B and E. In structure B, the fructose ring is flipped over.
   (g) C, D, E
   (h) B
   (i) D and E. Although E is in the β-anomer form, recall that in solution it can mutarotate.

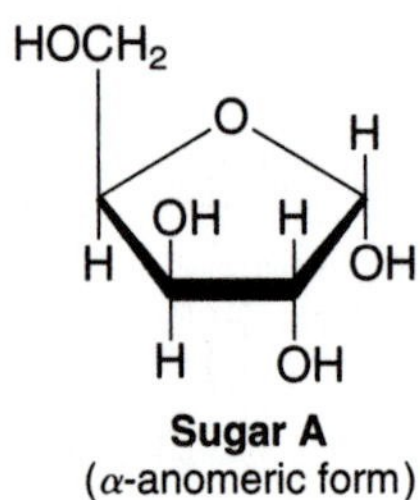

Sugar A
(α-anomeric form)

4. (a) A and B are epimers. B and C are diastereoisomers. A and C are enantiomers.
   (b) D-ribose.
   (c) See the structure in the margin.

5. a, b, c. Note that *glycosidic* refers to bonds involving any sugars; however, *glucosidic* and *galactosidic* refer specifically to bonds involving the anomeric (reducing) carbons of glucose and galactose, respectively.

Simple Monosaccharide Derivatives

6. a

7. See the structure in the margin.

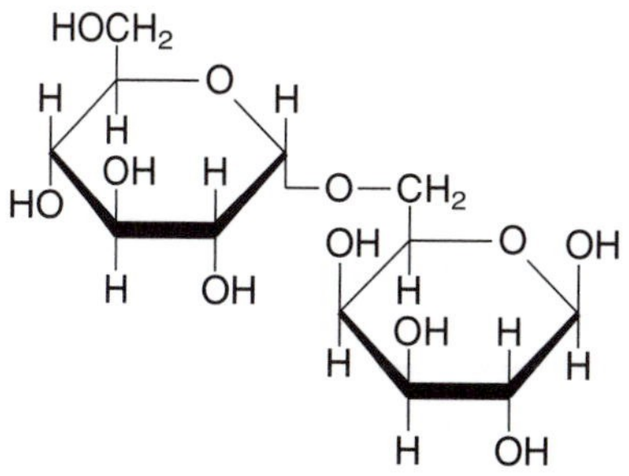

8. d

9. Inside cells, glucose is rapidly phosphorylated to glucose 6-phosphate. This anionic sugar phosphate is not transported through cell membranes.

Glucose α (1–6)-galactose
(β-anomeric form)

Polysaccharides

10. Both c and d are correct. Since there are two glucosidic bonds in each trisaccharide and each bond can be $\alpha(1\rightarrow4)$ or $\alpha(1\rightarrow6)$, the total number of possible kinds of trisaccharides is four. However, two consecutive $\alpha(1\rightarrow6)$ bonds would be very rare in glycogen or starch; therefore, one would be more likely to find three kinds.

11. The carbohydrate in bread is mostly starch, which is a polysaccharide mixture containing D-glucose residues linked by glucosidic bonds. All the aldehyde groups in each polysaccharide, except one at the free end, are involved in acetal bonds and do not react with Nelson's reagent. During mastication, α-amylase in saliva breaks many of the internal α-1,4 glucosidic bonds and exposes reactive aldehyde groups (reducing groups). Note: Nelson's reagent consists of copper sulfate in a hot alkaline solution; a reducing sugar, such as glucose, reduces the copper, which in turn reduces the arsenomolybdate in the reagent, producing a blue complex.

12. Both cellulose and amylose are linear polymers of D-glucose, but the glucosidic linkages of cellulose are β-1,4 whereas those of amylose are α-1,4. The different configuration at the anomeric carbons determines a different spatial orientation of consecutive glucose residues. Thus, cellulose is capable of forming a linear, hydrogen-bonded structure, whereas amylose forms an open helical structure (see Figure 18-14 in Stryer).

13. (a) 3, 7 (b) 4 (c) 1, 2, 5 (d) 1, 3, 5, 6 (e) 1, 3, 5

14. c

15. a, d

16. b, d, e. Answer (c) is incorrect because keratan sulfate and chondroitin sulfate chains are bound to the core protein.

Glycoproteins and their Receptors

17. b

18. (a) 2, 5, 7 (b) 3, 7, 9 (c) 1, 4, 6, 8

19. a, e

20. a, d, e

21. b, c, e

PROBLEMS

1. Glucose and other dietary monosaccharides like fructose and galactose are very soluble in water at neutral pH. For example, over 150 g of glucose can be dissolved in 100 ml water at 25°C.

   (a) What features of the chemical structure of glucose make it so soluble in water?
   (b) What features of the proteoglycans found in cartilage make them so highly hydrated and contribute to their ability to spring back after deformation?

2. Indicate whether the following pairs of molecules are enantiomers, epimers, diastereoisomers, or anomers.

   (a) D-Xylose and D-lyxose
   (b) α-D-Galactose and β-D-galactose
   (c) D-Allose and D-talose
   (d) L-Arabinose and D-arabinose

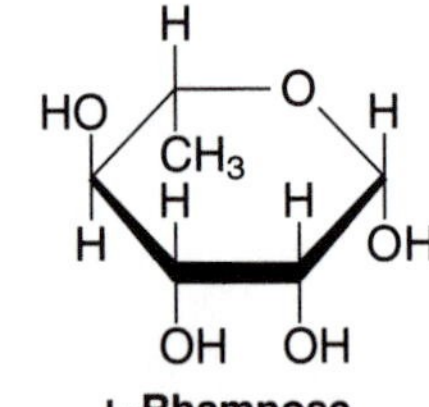

L-Rhamnose

3. L-Rhamnose is a common 6-deoxy sugar. It is derived from D-glucose. What biochemical steps would be needed for the conversion of D-glucose to L-rhamnose?

4. What is the name of the compound that is the mirror image of α-D-glucose?

5. Compound X, an aldose, is enzymatically reduced using NADPH as an electron donor, yielding D-sorbitol. This sugar alcohol is then oxidized at the C-2 position with NAD^+ as the electron acceptor; the products are NADH and a ketose, compound Y.

   (a) Name compound X and write its structure.
   (b) Will sorbitol form a furanose or pyranose ring? Why?
   (c) Name compound Y and write its structure.

CH_2OH
HCOH
HOCH
HCOH
HCOH
CH_2OH

D-Sorbitol

6. Compare the number of dimers that can be prepared from a pair of alanine molecules and from a pair of D-galactose molecules, each of which is present as a pyranose ring. For the galactose molecules, pairs may be made using the α or β anomers.

7. In 1952, Morgan and Watkins showed that *N*-acetylgalactosamine and its α-methylglucoside inhibit the agglutination of type A erythrocytes by type A–specific lectins, whereas other sugars had little effect. What did this information reveal about the structure of the glycoprotein on the surface of type A cells?

8. The steps in the conversion of glucose 6-phosphate to myoinositol 1-phosphate are shown in Figure 18-1. The enzyme inositol 1-phosphate cyclase, which catalyzes this series of steps, has as a prosthetic group a molecule of NAD^+.

   (a) Describe the role of NAD^+ in the reaction.
   (b) What type of reaction occurs when the carbocyclic ring is formed?

(c) In solution, less than 1% of glucose 6-phosphate is present in the open-chain form, yet the enzyme quantitatively converts all the hexose 6-phosphate to myoinositol 1-phosphate. Explain.
(d) Which carbon atoms undergo inversion of configuration?

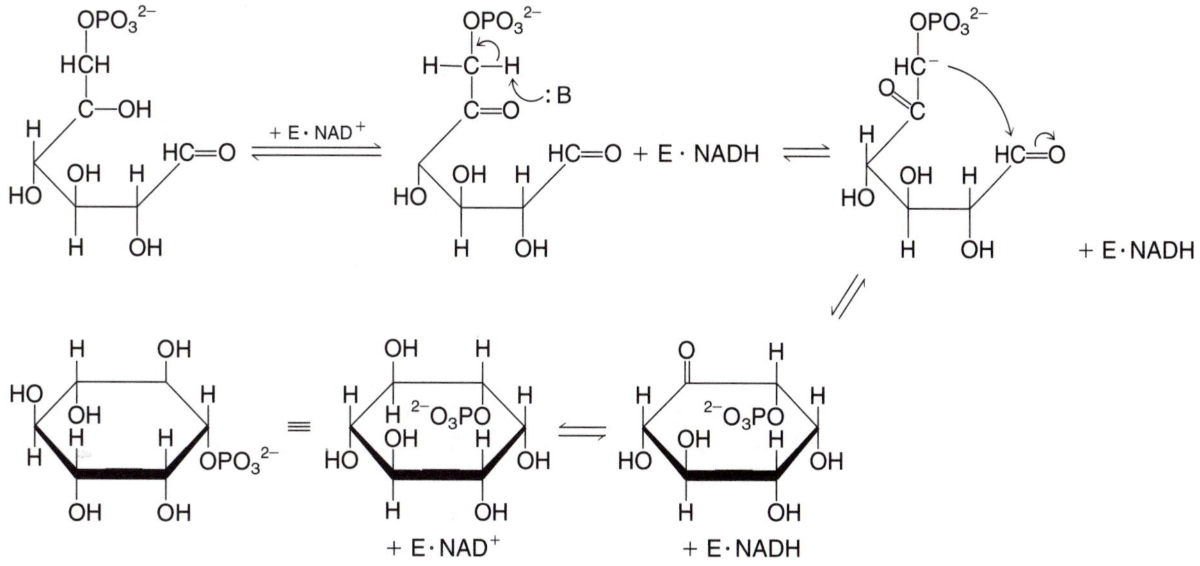

Figure 18-1
Steps in the conversion of glucose 6-phosphate to myoinositol 1-phosphate. [From C. Walsh, *Enzymatic Reaction Mechanisms,* W. H. Freeman and Company, 1979.]

9. *Penicillium* and other fungi contain an enzyme called *glucose oxidase,* which converts D-glucose to the compound shown in the margin.

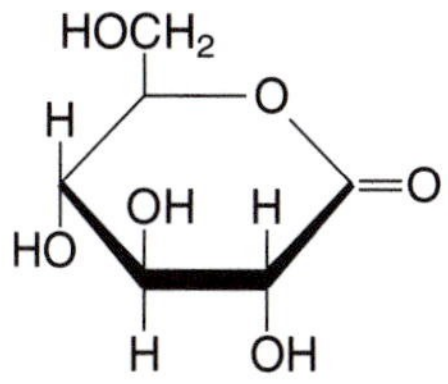

(a) What is the generic name for compound formed?
(b) Can the compound open to form a straight-chain compound? What else is required for this reaction to occur?
(c) Why would you expect that NAD^+ or FAD^+ would be required by the enzyme?

10. Storage polysaccharides like starch and glycogen often contain over a million glucose units. The energetic cost of synthesizing polysaccharides is high (see Stryer, p. 588). Suppose that in a liver cell, the glucosyl residues in large numbers of glycogen molecules were replaced with an equivalent number of molecules of free glucose. What problems would this cause for the liver cell?

11. Shown below is one example of the storage oligosaccharides that account in part for the flatulence caused by eating beans, peas, and other legumes. These oligosaccharides cannot be digested by enzymes in the small intestine, but they can be metabolized by anaerobic microorganisms in the large intestine. There, they undergo oxidation, with the production of large quantities of carbon dioxide, hydrogen sulfide, and other gases. Solutions are now on the market containing one or more enzymes that, when ingested with the offending legumes at mealtime, convert the oligosaccharides to digestible products.

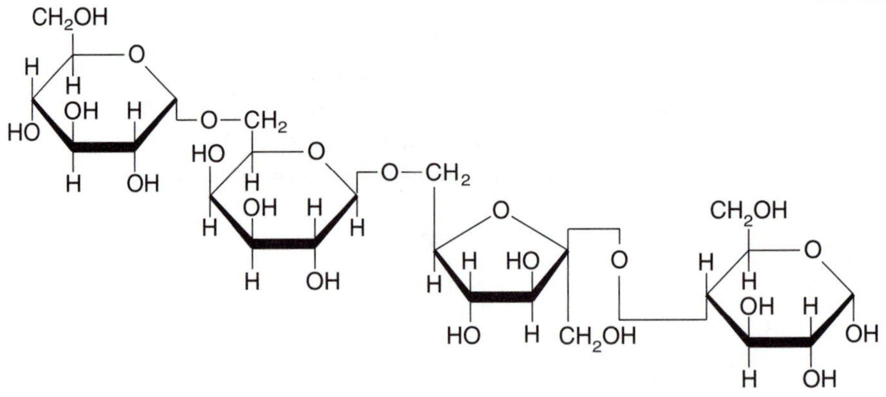

(a) Name the oligosaccharide shown above, using the notation employed in Stryer's text on page 476.
(b) Given that free hexoses can pass easily through intestinal cells into the blood, what types of enzymes do you think are included in the commercial products that aid in legume oligosaccharide digestion?
(c) The concentration of oligosaccharides in beans can be reduced by cooking or by sprouting. What happens to those oligosaccharides when cooking is employed? When the beans are sprouted before cooking or eating?
(d) When small amounts of cellulose are ingested purposely or accidentally (by pets or young children, for example), there is usually no gas production. In fact, the primary concern about paper ingestion by pets or small children is intestinal blockage. Why?

12. On page 468 of Stryer's text, reducing sugars are defined as those with a free aldehyde or keto group that can reduce cupric ion to the cuprous form. The reactive species in the reducing sugar reaction is the open-chain form of the aldose or ketose. The reaction can be used to estimate the total amount of glucose in a solution such as blood plasma. An aqueous solution of glucose contains only a small amount of the open-chain form. How can the reaction be used to provide a *quantitative* estimate of glucose concentration?

13. You have a sample of glycogen that you wish to analyze using exhaustive methylation and acid hydrolysis. Using a sample of 0.4 g, you incubate the glycogen with methyl iodide, which methylates all *free* primary or secondary alcohol groups on sugars. Then you subject the sample to acid hydrolysis, which cleaves glycosidic linkages between adjacent glucose residues. You then determine the yield of 2,3-dimethylglucose in your sample.

(a) Why is a 2,3-dimethylglucose residue produced from a branch point in glycogen?
(b) The yield of 2,3-dimethylglucose is 0.247 mmol. What fraction of the total residues in each sample are branch points? The molecular weight of a glycosyl residue in glycogen is 162.
(c) Could you use this technique to determine the anomeric nature of the glycogen branch? Why?

14. Why is the structural analysis of an oligosaccharide containing eight monosaccharide residues more complicated than a similar analysis for an octanucleotide or an octapeptide?

15. On page 473 of Stryer's text, the yearly production of cellulose is listed as 10^{15} kg. If you assume that each molecule of cellulose has 10^6 glucose residues, how many cellulose molecules are synthesized on earth per second?

ANSWERS TO PROBLEMS

1. (a) Glucose and other hexose monosaccharides have five hydroxyl groups and an oxygen in the heterocyclic ring that can all form hydrogen bonds with water. The ability to form these hydrogen bonds with water and other polar molecules enables hexoses and other carbohydrates to dissolve easily in aqueous solution.

   (b) In addition to hydrogen bonding of water to hydroxyl groups and oxygen atoms in the repeating disaccharide units of cartilaginous proteoglycans like keratan sulfate and chondroitin sulfate, these molecules also contain charged sulfate and carboxylate groups that can also interact with water. Compression of these large hydrated polyanions can drive some water out of the cavities between them, but the high degree of hydration of the molecules, as well as charge repulsion between the sulfate and carboxylate groups, contributes to the tendency of these compounds to resume their normal conformations after deformation.

2. (a) D-Xylose and D-lyxose differ in configuration at a single asymmetric center; they are epimers.

   (b) α-D-Galactose and β-D-galactose have differing configurations at the C-1, or anomeric, carbon; they are anomers.

   (c) D-Allose and D-talose are diastereoisomers because they have opposite configurations at one or more chiral centers, but they are not complete mirror images.

   (d) L-Arabinose and D-arabinose are mirror images of each other and are therefore enantiomers.

3. The conversion requires inversions of configuration at C-3, C-4, and C-5, as well as a reduction of the carbon at position 6. The sequence of reactions in vivo is rather complex and occurs while D-glucose is attached to deoxythymidine diphosphate. The steps include oxidation at C-4, dehydration and reduction at C-5 and C-6, and reduction with inversion at C-4; at some point, inversion at C-3 also occurs.

4. Although the mirror image of a D compound is an L compound, the mirror image of an α compound is an α compound. (An α compound has a 1-hydroxyl group in the α position.) Thus, α-L-glucose is the compound that is the mirror image, or enantiomer, of α-D-glucose.

5. (a) Compound X is D-glucose; it is the only aldose whose reduction will yield a hexitol with the same conformation as that of D-sorbitol.

   (b) Sorbitol cannot form a hemiacetal because it has no aldehyde or ketone group. Therefore, neither type of ring can be formed by sorbitol.

   (c) Compound Y is D-fructose, a ketose that is produced by the oxidation of sorbitol.

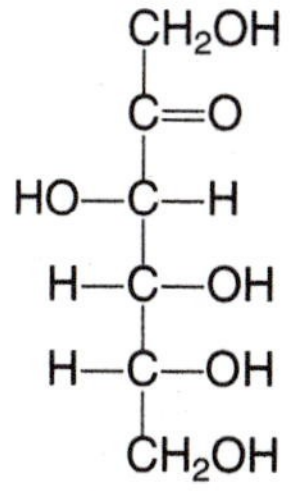

CH_2OH
C=O
HO—C—H
H—C—OH
H—C—OH
CH_2OH

D-Fructose

6. Only one dimer, alanylalanine, can be made from two alanine molecules linked via a peptide bond. However, the presence of several hydroxyl groups and the aldehydic function at the C-1 position of each D-galactose molecule provides an opportunity to make a larger number of dimers. Both the α and β forms of one molecule can form glycosidic linkages with the C-2, C-3, C-4, or C-6 hydroxyl groups of the other. Recall that the C-5 position is not available, because it participates in the formation of the pyranose ring. To these eight dimers can be added those dimers formed through glycosidic linkages involving the $\alpha\alpha$, $\alpha\beta$, or $\beta\beta$ configurations. Thus, 11 possible dimers exist. If one is allowed to use L forms, then the number of possible dimers increases greatly. This variety of linkages makes the sugars very versatile molecules and yields many different structures that may be useful in biology. However, this variety has also made the systematic study of the chemistry of polysaccharides very difficult.

7. The observations of Morgan and Watkins suggested that the sugar *N*-acetylgalactosamine in α linkage is the determinant of blood group A specificity. The galactose derivative binds to type A lectins, occupying the sites that would otherwise bind to glycoproteins, having *N*-acetylgalactosamine end groups, on the surfaces of type A cells.

8. (a) First, NAD^+ acts as an electron acceptor, oxidizing the C-5 carbon to the keto form. Later it acts as an electron donor, reducing the C-6 keto group to yield the inositol product. The reduction also regenerates the initial enzyme-NAD^+ complex and keeps the aldol [see (b)] from reopening.

(b) An intramolecular aldol condensation involving the C-6 carbanion and the C-1 aldehyde leads to the formation of the carbocyclic ring.

(c) An equilibrium exits between the open-chain form and the α and β anomers. As the open-chain form is consumed during the reaction, more of the open-chain form is generated from the anomers. Eventually all anomers are converted to the open-chain form, which is the substrate for the cyclase.

(d) None of the carbons undergo inversion if one compares the carbons in the myoinositol 1-phosphate with those in the glucose 6-phosphate. The C-1 carbon is no longer an anomeric carbon, because it has been reduced.

9. (a) The compound is a lactone, a cyclic ester formed between a carboxylate and a hydroxyl group.

(b) Formation of a straight-chain compound requires hydrolysis of the ester linkage. Esters react with water in the presence of acid or base to yield carboxylic acid and alcohol groups, both part of the open-chain form of the molecule, called *D-gluconic acid.*

(c) Because the C-1 atom is oxidized, an oxidized pyridine nucleotide is needed to act as an electron acceptor.

10. The primary consequence of a high concentration of free glucose molecules in the cell would be a dramatic and probably catastrophic increase in osmotic pressure. In aqueous solutions, colligative properties like boiling and freezing points, vapor pressure, and osmotic pressure depend primarily on the number of molecules in the solution. Thus a glycogen molecule

containing a million glucose residues exerts one millionth the osmotic pressure of a million molecules of free glucose. Osmotic pressure exerted by high glucose concentration would induce entry of water into the cell, in an attempt to equalize pressure inside and outside the cell. Unlike bacterial or plant cells, which have a rigid cell wall that can help resist high pressures, animal cells have a comparatively fragile plasma membrane, which will burst when osmotic pressures are too high.

11. (a) Gluα1$\rightarrow$6Galα1$\rightarrow$6Fruβ1$\rightarrow$4Glu.

(b) The solution must contain enzymes that hydrolyze the glycosidic linkages between the monosaccharides. For example, an activity that would be required for the oligosaccharide shown would be a type of α-1,6-glycosidase, which would cleave the $\alpha(1\rightarrow6)$ linkage between glucose and galactose. The glycosidases are needed to convert the oligosaccharides to free hexoses, which then pass easily into the circulation. The three common sugars found in the oligosaccharide shown in this problem are easily metabolized by the liver and other cells.

(c) Cooking by heating in water probably hydrolyzes some of the glycosidic linkages found in the oligosaccharides. Sprouting or germinating beans undergo a reduction in oligosaccharide concentration because hydrolase proteins induced during germination produce free hexoses, which can be used in the developing plant tissues as a source of carbon for biosynthesis.

(d) Because cellulose is an unbranched polymer of glucose residues joined by β-1,4 linkages, the molecule is resistant to hydrolysis, even by anaerobic bacteria in the human intestine. Small amounts of cellulose and other indigestible complex carbohydrates are virtually unaltered as they pass through the digestive system. Thus no gases from carbohydrate breakdown are generated in the large intestine. Intestinal blockage may result from ingestion of large quantities of cellulose because there are no enzymes available to cleave the glycosidic linkages.

12. In water, an equilibrium exists among three forms of glucose. Two-thirds is present as the β anomer, one-third as the α anomer, and less than 1% as the open-chain form. When excess cupric ion reacts with the open-chain form, glucose is oxidized to gluconic acid. Through the law of mass action, the α and β anomers of glucose are then converted to the open-chain aldose form. Continued production of gluconic acid from the open-chain form leads to the ultimate conversion of all glucopyranoses to the open-chain form, which reacts quantitatively with cupric ion. Thus the total amount of glucose in a known volume of blood plasma or other solution can be determined.

13. (a) A glycosyl residue at a branch point has three of its five carbons linked to other glucose residues; these are carbons 1, 4, and 6. Only C-2 and C-3 of a branch point residue will have alcohol or hydroxyl groups that are free and therefore available for methylation. Thus residues at a branch point are converted to 2,3-dimethylglucose after methylation and hydrolysis. Those glucosyl residues not at a branch point would be converted to 2,3 6-trimethylglucose

by the same procedure, except for the single residue at the reducing end, which could be converted to 1,2,3,6-tetramethylglucose.

(b) The original sample of 0.4 g corresponds to 0.4 g ÷ 162 g/mole, or 2.46×10^{-3} mole, or 2.46 mmol glucose residues, which is 10% of the total sample. Thus 10% of the glucosyl residues are at branch points.

(c) The analysis using methylation and acid hydrolysis does not allow determination of the anomeric linkage. Acid hydrolysis cleaves both α- and β-anomeric linkages and does not allow distinctions between them.

14. In oligosaccharides, there are a number of different types of potential glycosidic linkages that can be formed among eight residues, because each free hydroxyl group as well as the anomeric carbon on a particular monosaccharide could be linked to similar groups on adjacent residues. An octooligosaccharide could be linear or branched, and could be composed of as many as eight different monosaccharides, each of which could require additional steps to analyze completely. Analysis of an oligonucleotide is somewhat less complicated, because usually only four different bases will be found during the analysis, and the linkage between adjacent nucleotides is almost always $3' \rightarrow 5'$; in addition, the oligonucleotide molecule is not likely to be branched. Although there may be as many as eight different amino acid residues in an octapeptide, all 20 different amino acids found in most proteins are relatively easy to characterize (see Stryer, Chapter 3) and the octapeptide is unlikely to be branched.

15. One way to solve this problem is to find the number of grams of glucose polymerized into cellulose each second.

$$10^{15} \text{ kg cellulose produced each year} = 10^{18} \text{ g glucose}$$

$$\frac{10^{18} \text{ g glucose per year}}{3.15 \times 10^{7} \text{ seconds per 365-day year}} = 3.17 \times 10^{10} \text{ g/s}$$

Then calculate the number of moles of glucose produced per second and divide this number by 10^6 to calculate the number of moles of cellulose produced per second.

$$\frac{3.17 \times 10^{10} \text{ g/s}}{162 \text{ g/mole}} = 1.96 \times 10^{8} \text{ moles glucose produced per second}$$

$$\frac{1.96 \times 10^{8} \text{ moles glucose/s}}{10^{6} \text{ glucose molecules/cellulose molecule}} = 1.96 \times 10^{2} \text{ moles cellulose produced per second}$$

Then convert moles of cellulose produced per second to molecules produced, using Avogadro's number.

$$1.96 \times 10^{2} \text{ moles cellulose} \times 6.023 \times 10^{23} \text{ molecules/mole} = 11.8 \times 10^{25} \text{ molecules cellulose produced per second}$$

EXPANDED SOLUTIONS TO TEXT PROBLEMS

1. Although it can be risky business, chemists have always tried to gain some insight into molecular structure from knowledge of the empirical formula. Since the empirical formula for carbohydrates is $(CH_2O)_n$, it is not surprising that in the latter half of the nineteenth century the name *carbohydrate* was coined.

2. To answer this problem, one must know the structures of the molecules in question and a couple of definitions. By definition, *epimers* are a pair of molecules that differ from each other only in their configuration at a single asymmetric center. *Anomers* are special epimers that differ only in their configuration at a carbonyl carbon; hence, they are usually acetals or hemiacetals. An aldose-ketose pair is obvious. Inspection of Fischer representations of the molecular pairs leads to the conclusion that (a), (c), and (e) are aldose-ketose pairs; (b) and (f) are epimers; and (e) are anomers.

3. A mild oxidant, Tollens' reagent converts aldoses to aldonic acids and free silver as follows:

$$RCHO + 2\ Ag(NH_3)_2 + H_2O \rightarrow RCO_2^- + 2\ Ag^0 + 3\ NH_4^+ + NH_3$$

However, cyclic hemiacetals are oxidized directly to lactones, which are hydrolyzed to the corresponding aldonic acid under alkaline conditions. Thus, in the case of glucose, the major first reaction product is D-δ-gluconolactone. To prepare aldonic acids, Br_2 is usually used as the oxidant because it gives fewer side reactions than does Tollens' reagent.

(Problem 3)

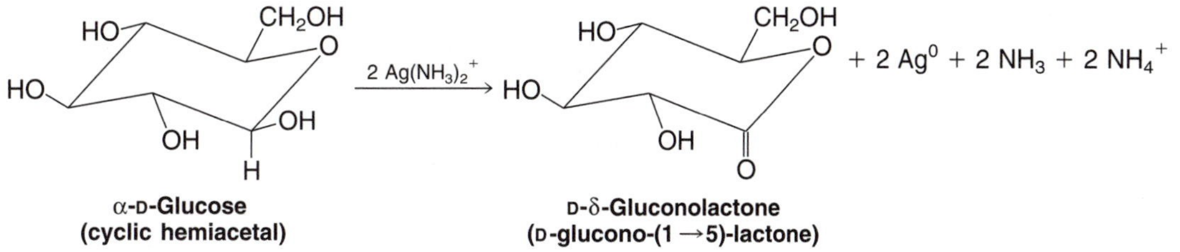

$\xrightarrow{2\ Ag(NH_3)_2^+}$ $+\ 2\ Ag^0 + 2\ NH_3 + 2\ NH_4^+$

α-D-Glucose (cyclic hemiacetal) **D-δ-Gluconolactone (D-glucono-(1→5)-lactone)**

4. The reason the specific rotation of α-D-glucopyranose changes after it is dissolved in water is that the ring form is in equilibrium with a small amount of the straight-chain form of glucose (see problem 5). The straight-chain form then converts to either α-D-glucopyranose or β-D-glucopyranose. This process, called *mutarotation,* continues until after 1–2 hours a thermodynamically stable mixture of the α and β anomers is obtained. Its specific rotation is 52.7°. The difference in the specific rotations of the two anomers is 93.3° (112° – 18.7°), and the difference between the equilibrium value and that of the β anomer is 34° (52.7° – 18.7°). Since the optical rotation of the equilibrium mixture is closer to that of the β anomer than it is to that of the α anomer, obviously more than half the equilibrium mixture is in the β configuration. The fraction present in the α configuration is 34° ÷ 93.3° = 0.36. The fraction in the β configuration is 1 – 0.36 = 0.64.

5. Glucose reacts slowly because the predominant hemiacetal ring form (which is inactive) is in equilibrium with the active straight-chain free aldehyde. The latter can react with terminal amino groups to form a Schiff base, which can then rearrange to the stable amino ketone, sometimes referred to as Hb A_{Ic}, which accounts for approximately 3% to 5% of the hemoglobin in normal adult human red cells. In the diabetic, its concentration may rise 6% to 15% owing to the elevated concentrations of glucose (see Stryer, p. 780).

(Problem 5)

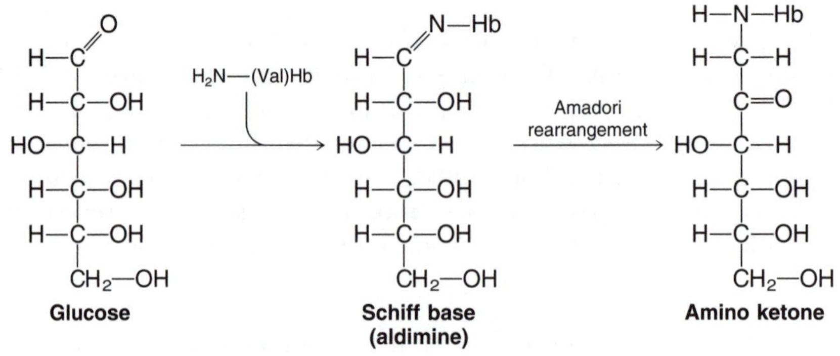

H_2N—(Val)Hb

Amadori rearrangement

Glucose **Schiff base (aldimine)** **Amino ketone**

6. Whereas pyranosides have a series of three adjacent hydroxyls, furanosides have only two. Therefore, oxidation of pyranosides uses *two* equivalents of periodate and yields *one* mole of formic acid, whereas oxidation of furanosides uses only *one* equivalent of periodate and yields *no* formic acid.

(Problem 6)

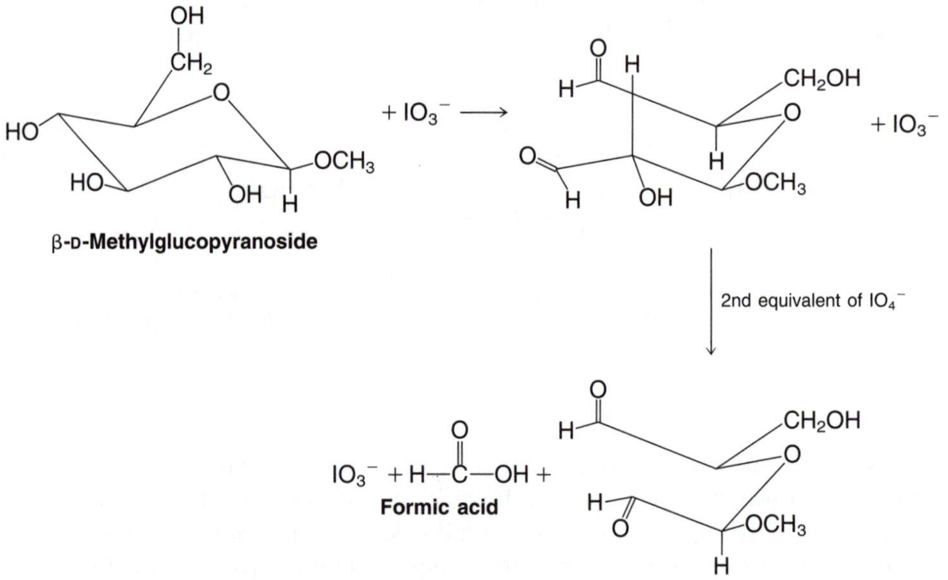

β-D-Methylglucopyranoside

2nd equivalent of IO_4^-

IO_3^- + H—C(=O)—OH +

Formic acid

(Problem 6)

β-D-Methylfructofuranoside

+ IO_4^- ⟶ + IO_3^-

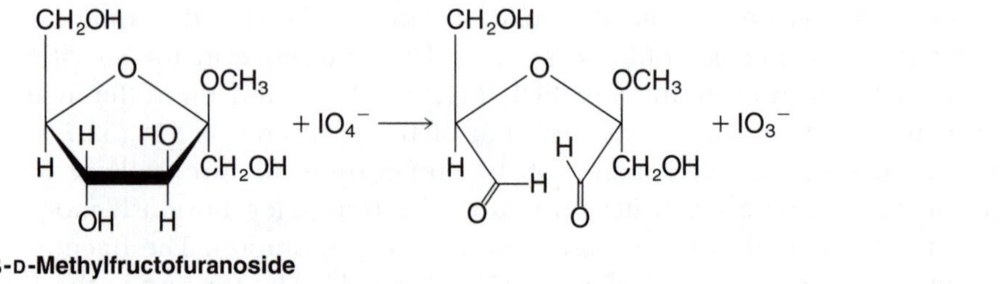

7. The formation of acetals (such as methylglucoside) is acid-catalyzed. In a mechanism similar to that of the esterification of carboxylic acids (shown in most organic chemistry texts), the anomeric hydroxyl group is replaced. The resulting carbocation is susceptible to attack by the nucleophilic oxygen of methanol, leading to the incorporation of this oxygen into the methylglucoside molecule.

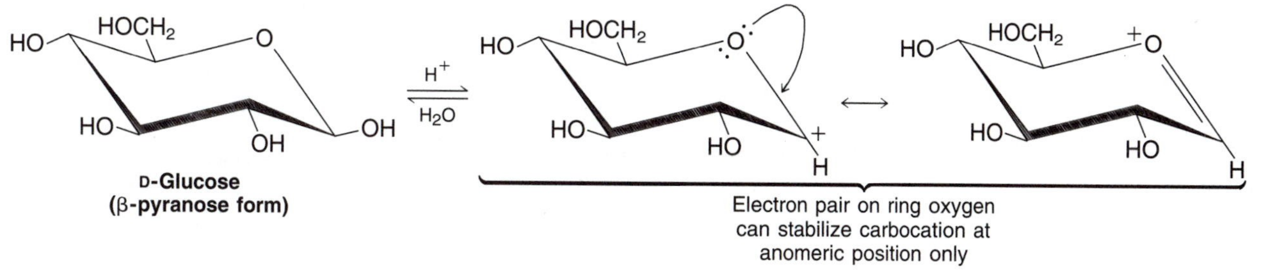

D-Glucose
(β-pyranose form)

Electron pair on ring oxygen can stabilize carbocation at anomeric position only

8. By inspection, A, B, and D are the pyranosyl forms of D-aldohexoses because the CH_2OH is above the plane of the ring. In Haworth projections, OH's above the ring are to the left in Fischer projections, and those below the ring are to the right. Therefore, A is *β*-D-mannose, B is *β*-D-galactose, and D is *β*-D-glucosamine. By similar use of the Haworth projection, C can be identified as *β*-D-fructose. All these sugars are *β* because, in Haworth projections, when the CH_2OH attached to the C-5 carbon (the carbon that determines whether the sugar is D or L) is above the ring, if the anomeric hydroxyl is also above the ring, the sugar is *β*.

9. The trisaccharide itself should be a competitive inhibitor of cell adhesion if the trisaccharide unit of the glycoprotein is critical for the interaction.

CHAPTER

19

Glycolysis

Following introductions to the general principles of metabolism (Chapter 17) and the structures and properties of carbohydrates in the preceding chapter, Stryer turns to the metabolism of carbohydrates via the glycolytic pathway. Glycolysis is the sequence of reactions that converts glucose into pyruvate with the concomitant trapping of a portion of the energy as ATP. Stryer starts with a historical account of the discovery of some of the reactions of glycolysis. He then gives an overview of the key structures and types of reactions in glycolysis. Next, he describes the sequence of the reactions of glycolysis in three stages: (1) the conversion of glucose into fructose 1,6-bisphosphate, (2) the formation of triose phosphate intermediates and the oxidation of glyceraldehyde 3-phosphate, which leads to the formation of one ATP, and (3) the conversion of 3-phosphoglycerate into pyruvate and the formation of a second ATP. A consideration of all the reactions of glycolysis shows that conversion of one glucose molecule to two of pyruvate yields two ATP molecules. Along with glucose, fructose and galactose are important dietary sugars, so their mode of entry into glycolysis is also described. The regulation of glycolysis by the enzymes that catalyze the irreversible reactions in the pathway is discussed, and the differing fates of pyruvate are outlined. Depending on the organism, cell type, and metabolic state, pyruvate is converted to ethanol, lactate, or acetyl CoA. The three-dimensional arrangement of the NAD^+-binding sites of dehydrogenases, the induced-fit rearrangement of the hexokinase structure, and the mechanisms of action of several enzymes in the glycolytic pathway are described. Triosephosphate isomerase is discussed as an example of the evolution of catalytic perfection. The formation and breakdown of 2,3-bisphosphoglycerate, an important modulator of oxygen affinity in hemoglobin (Chapter 7), is described as a side pathway of glycolysis. The chapter concludes with a description of the family of glucose transporters as examples of the ability of isoforms of proteins to perform diverse and specialized functions. In this chapter and others covering metabolism, a firm understanding of the properties of enzymes (Chapter 8), their catalytic mechanisms (Chapter 9), and their regulation (Chapter 10) is essential.

LEARNING OBJECTIVES

When you have mastered this chapter, you should be able to complete the following objectives:

Introduction (Stryer pages 483–485)

1. Define *glycolysis* and explain its role in the generation of *metabolic energy.*

2. List the alternative end points of the glycolytic degradation of *glucose.*

3. Summarize the kinds of reactions and the types of intermediates found in glycolysis.

Stages of Glycolysis (Stryer pages 485–491)

4. Outline the three stages of glycolysis.

5. Describe the steps in the conversion of glucose to *fructose 1,6-bisphosphate,* including all the intermediates and enzymes. Note the steps where *ATP* is consumed.

6. List the reactions that convert fructose 1,6-bisphosphate, a hexose, into the triose *glyceraldehyde 3-phosphate.*

7. Outline the steps of glycolysis between glyceraldehyde 3-phosphate and *pyruvate.* Recognize all the intermediates and enzymes and the cofactors that participate in the ATP-generating reactions.

8. Write the net reaction for the transformation of glucose into pyruvate and enumerate the ATP and NADH molecules formed.

Entry of Fructose and Galactose into Glycolysis (Stryer pages 491–493)

9. Describe the two pathways for the conversion of *fructose* into glyceraldehyde 3-phosphate.

10. Explain how *galactose* is transformed into *glucose 6-phosphate.* Note the role of *UDP-activated sugars* in this process.

11. Describe the biochemical defect in *galactosemia.*

Control of Glycolysis (Stryer pages 493–496)

12. List the key enzymes in the regulation of glycolysis.

13. Describe the allosteric regulation of *phosphofructokinase.*

14. Explain the role of *fructose 2,6-bisphosphate* in the regulation of phosphofructokinase. Describe the *fused-domain structure* of *phosphofructokinase 2 (PFK2)/fructose bisphosphatase 2* that forms and degrades fructose 2,6-bisphosphate.

15. Discuss the regulation of *hexokinase.* Contrast the properties and physiologic roles of hexokinase and *glucokinase.*

16. Define *isozyme.* Describe the regulation of the isozymes of *pyruvate kinase.*

The Fates of Pyruvate (Stryer pages 496–498)

17. Outline the reactions for the conversion of pyruvate into *ethanol, lactate,* or *acetyl CoA.*

18. Explain the role of *alcoholic fermentation* and lactate formation in the regeneration of NAD^+.

Structural Characteristics and Mechanisms of Action of Glycolytic Enzymes (Stryer pages 498–503)

19. Describe the structure of the NAD^+-binding region common to many *NAD^+-linked dehydrogenases.*

20. Discuss the *induced-fit rearrangements* that occur in hexokinase upon glucose binding.

21. Summarize the most important features of the catalytic mechanisms of *aldolase* (*Schiff base* formation, *electron sink*), *triosephosphate isomerase I* (*enediol intermediate, kinetic perfection, substrate channeling* by *induced fit*), *glyceraldehyde 3-phosphate dehydrogenase* (*thioester* formation, *covalent enzyme intermediate* in *energy coupling*).

Triose Bisphosphates (Regulation by 2,3-Bisphosphoglycerate and Phosphoryl Group Transfer by 1,3-Bisphosphoglycerate) (Stryer pages 503–505)

22. Describe the formation of *2,3-bisphosphoglycerate* (*2,3-BPG*) from *1,3-bisphosphoglycerate* and the conversion of 2,3-BPG into *3-phosphoglycerate.*

23. Explain the role of 2,3-BPG in the interconversion of 3-phosphoglycerate and *2-phosphoglycerate.*

24. Explain the role of the *enol → ketone conversion* in the *phosphoryl transfer* catalyzed by *pyruvate kinase.*

SELF-TEST

Introduction

CH_2OH—$C{=}O$—$CH_2OPO_3^{2-}$ $\rightleftharpoons$ $O{=}C(H)$—$H{-}C{-}OH$—$CH_2OPO_3^{2-}$

1. Examine the reaction in the margin. What type of a reaction is it?
   (a) Aldol cleavage
   (b) Dehydration
   (c) Phosphoryl transfer
   (d) Phosphoryl shift
   (e) Isomerization
   (f) Phosphorylation coupled to oxidation.

2. Draw the structures of
   (a) Glycerate

   (b) 1,3-Bisphosphoglycerate; indicate the phosphate ester bond and mixed anhydride bond

Stages of Glycolysis

3. Examine the following chemical structures of five of the intermediates of glycolysis:

A: $O{=}C(O^-)$—$H{-}C{-}OH$—$CH_2O{-}P(=O)(O^-){-}O^-$

B: $^{2-}O_3POH_2C$, O, $CH_2OPO_3^{2-}$, H, HO, H, OH, OH, H

C: $O{=}C(O^-)$—$C{-}OPO_3^{2-}$—CH_2

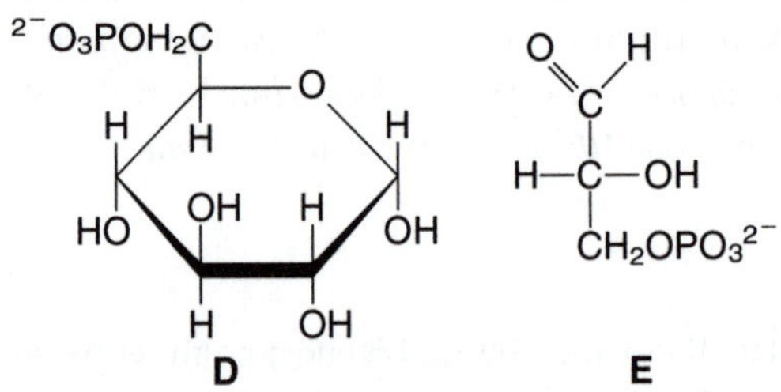

(a) Name the five intermediates:
A ________________
B ________________
C ________________
D ________________
E ________________

(b) Give the order in which these intermediates occur in glycolysis.

____

(c) Name the enzyme or enzymes that produce the following intermediates:

B ____

D ____

E ____

(d) Which of these intermediates is a "high-energy" compound, that is, has a large phosphate group–transfer potential?

____

(e) Name the immediate precursor of intermediate A.

____

4. Which of the following answers completes the sentence correctly? Hexokinase

(a) catalyzes the conversion of glucose 6-phosphate into fructose 1,6-bisphosphate.
(b) requires Ca^{2+} for activity.
(c) uses inorganic phosphate to form glucose 6-phosphate.
(d) catalyzes the transfer of a phosphoryl group to a variety of hexoses.
(e) catalyzes a phosphoryl shift reaction.

5. During the phosphoglucose isomerase reaction, the pyranose structure of glucose 6-phosphate is converted into the furanose ring structure of fructose 6-phosphate. Does this conversion require an additional enzyme? Explain.

____

____

6. The steps of glycolysis between glyceraldehyde 3-phosphate and 3-phosphoglycerate involve all the following, except

(a) ATP synthesis.
(b) the utilization of P_i.
(c) the oxidation of NADH to NAD^+.
(d) the formation of 1,3-bisphosphoglycerate.
(e) catalysis by phosphoglycerate kinase.

7. Which of the following answers complete the sentence correctly? The phosphofructokinase and the pyruvate kinase reactions are similar in that

(a) both generate ATP.
(b) both involve a "high-energy" sugar derivative.
(c) both involve three-carbon compounds.
(d) both are essentially irreversible.
(e) both enzymes undergo induced-fit rearrangements after binding of the substrate.

8. The reaction

$$\text{Phosphoenolpyruvate} + \text{ADP} + \text{H}^+ \rightarrow \text{pyruvate} + \text{ATP}$$

has a $\Delta G^{\circ\prime} = -7.5$ kcal/mol and a $\Delta G' = -4.0$ kcal/mol under physiologic conditions. Explain what these free-energy values reveal about this reaction.

____

____

9. If the C-1 carbon of glucose were labeled with ^{14}C, which of the carbon atoms in pyruvate would be labeled after glycolysis?

(a) The carboxylate carbon
(b) The carbonyl carbon
(c) The methyl carbon

10. Starting with fructose 6-phosphate and proceeding to pyruvate, what is the net yield of ATP molecules?

(a) 1
(b) 2
(c) 3
(d) 4
(e) 5

Entry of Fructose and Galactose into Glycolysis

11. In the liver, the main conversion pathway from fructose to pyruvate includes which of the following intermediates?

(a) Fructose 6-phosphate
(b) Fructose 1,6-bisphosphate
(c) Glyceraldehyde
(d) Dihydroxyacetone phosphate
(e) Phosphoenolpyruvate
(f) 1,3-Bisphosphoglycerate

12. Galactose metabolism involves the following reactions:

(1) Galactose + ATP $\longrightarrow$ galactose 1-phosphate + ADP + H^+

(2) ?

(3) UDP–galactose $\rightleftharpoons$ UDP–glucose

(a) Write the reaction for step 2. ________________

____

(b) Which step is defective in galactosemia? ____________

____

(c) Which enzymes catalyze steps 1, 2, and 3? ____________

____

(d) How is NAD^+ involved in the reaction in step 3? ________

____

13. Galactose is an important component of glycoproteins. Explain why withholding galactose from the diet of galactosemic patients has no effect on their synthesis of glycoproteins.

____

____

14. The essentially irreversible reactions that control the rate of glycolysis are catalyzed by which of the following enzymes?

(a) Pyruvate kinase
(b) Aldolase
(c) Glyceraldehyde 3-phosphate dehydrogenase
(d) Phosphofructokinase
(e) Hexokinase
(f) Phosphoglycerate kinase

15. Phosphofructokinase activity is enhanced by which of the following?

(a) Increased ATP concentration
(b) Increased fructose 2,6-bisphosphate concentration
(c) Decreased citrate concentration
(d) Decreased AMP concentration
(e) Increased H^+ concentration

16. When blood glucose levels are low, glucagon is secreted. Which of the following are the effects of increased glucagon levels on glycolysis and related reactions in liver?

(a) Phosphorylation of phosphofructokinase 2 and fructose bisphosphatase 2 occurs.
(b) Dephosphorylation of phosphofructokinase 2 and fructose bisphosphatase 2 occurs.
(c) Phosphofructokinase is activated.
(d) Phosphofructokinase is inhibited.
(e) Glycolysis is accelerated.
(f) Glycolysis is slowed down.

17. In which of the following is the enzyme correctly paired with its allosteric effector?

(a) Hexokinase: ATP
(b) Phosphofructokinase: glucose 6-phosphate
(c) Pyruvate kinase (L isozyme): alanine
(d) Phosphofructokinase: AMP
(e) Glucokinase: fructose 2,6-bisphosphate

18. Match hexokinase and glucokinase with the descriptions from the right column that are appropriate.

(a) Hexokinase __________

(b) Glucokinase __________

(1) Is found in the liver
(2) Is found in nonhepatic tissues
(3) Is specific for glucose
(4) Has a broad specificity for hexoses
(5) Requires ATP for reaction
(6) Has a high K_M for glucose.
(7) Is inhibited by glucose 6-phosphate.

The Fates of Pyruvate

19. Which of the following are metabolic products of pyruvate in higher organisms?

(a) Glycerol
(b) Lactic acid
(c) Acetone
(d) Acetyl CoA
(e) Ethanol

20. Write the complete reaction for the net conversion of glucose into lactate, including all the cofactors.

___

___

21. Since lactate is a "dead-end" product of metabolism in the sense that its sole fate is to be reconverted into pyruvate, what is the purpose of its formation?

___

___

Structural Characteristics and Mechanisms of Action of Glycolytic Enzymes

22. Which of the following answers completes the sentence correctly? The characteristic three-dimensional structure of hexokinase and several other kinases includes two lobes that come together when a substrate is bound. This substrate-induced closing of the cleft

(a) decreases the polarity of the environment at the active center.
(b) allows for the tight binding of NAD^+.
(c) facilitates the hydrolysis of ATP.
(d) exposes most of the substrate on the surface of the enzyme.
(e) rearranges the active site structure into four α helices and six parallel β pleated sheets.

23. Match the enzymes in the left column with the appropriate reactive species or steps that occur during catalysis listed in the right column.

(a) Aldolase ___
(b) Glyceraldehyde 3-phosphate dehydrogenase ___
(c) Pyruvate kinase ___

(1) Thioester intermediate
(2) Enolpyruvate
(3) Enolate anion
(4) Removal of hydride ion
(5) Protonated Schiff base

24. Which of the following statements about triosphosphate isomerase (TIM) is not correct?

(a) The mechanism of action of TIM involves a conformational change in the structure of the enzyme that prevents escape of an activated intermediate.
(b) The rate-limiting step in the reaction catalyzed by TIM is the release of the product glyceraldehyde 3-phosphate.
(c) The k_{cat}/K_M ratio for the reaction catalyzed by TIM is close to the diffusion-controlled limit for a bimolecular reaction.

(d) The isomerization of a hydrogen atom from one carbon atom to another in the TIM-catalyzed reaction is assisted by a base (the γ-carboxyl of a glutamate residue) in the enzyme.
(e) TIM catalyzes an intramolecular oxidation-reduction reaction.

25. For the glyceraldehyde 3-phosphate dehydrogenase reaction, explain how the oxidation of the aldehyde group ultimately gives rise to an acyl phosphate product.

Triose Bisphosphates (Regulation by 2,3-Bisphosphoglycerate and Phosphoryl Group Transfer by 1,3-Bisphosphoglycerate)

26. Which of the following answers completes the sentence correctly? The compound 2,3-Bisphosphoglycerate

(a) is formed from 1,3-bisphosphoglycerate.
(b) is derived directly from glyceraldehyde 3-phosphate.
(c) is formed by a kinase that phosphorylates 3-phosphoglycerate.
(d) can be readily converted in good yield to 1,3-bisphosphoglycerate.
(e) is synthesized and degraded by the enzymes of the glycolytic pathway.

27. Examine the reaction scheme in the margin.

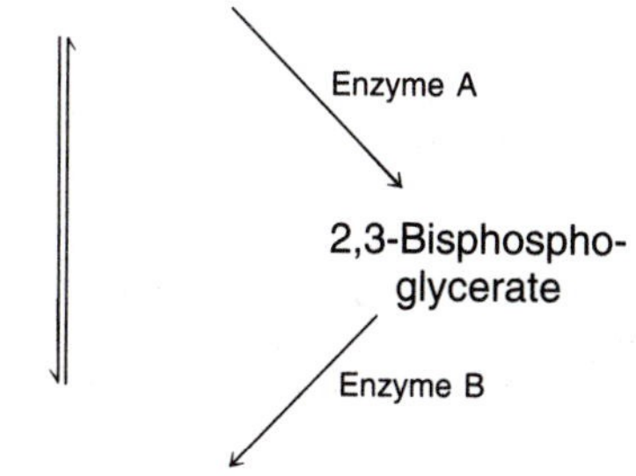

(a) Name enzymes A and B.

(b) The two phosphate groups on 1,3-BPG appear on 2,3-BPG after the reaction catalyzed by enzyme A. True or false? Explain.

(c) The 2,3-BPG participates in catalytic amounts in which reaction of glycolysis?

(d) The 2,3-BPG decreases the affinity of hemoglobin for O_2. True or false? Explain.

28. Explain the high phosphoryl group–transfer potential of phosphoenolpyruvate as it is displayed in the pyruvate kinase reaction.

29. Which of the following statements about glucose transporters is not correct?
(a) They are transmembrane proteins.
(b) They accomplish the movement of glucose across animal cell plasma membranes.

(c) Their tissue distribution and concentration can depend on the tissue type and metabolic state of the organism.
(d) Their glucose binding site is moved from one side of the membrane to the other by rotation of the entire protein.
(e) They constitute a family of five isoforms of a protein.

ANSWERS TO SELF-TEST

Introduction

1. e

2. (a)

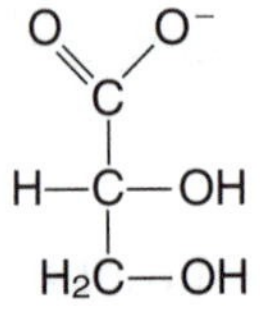

O O⁻
C
H—C—OH
H_2C—OH

Glycerate

(b)

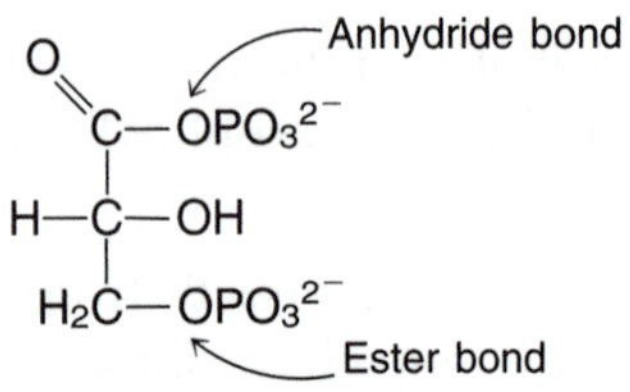

O
C—OPO_3^{2-} ← Anhydride bond
H—C—OH
H_2C—OPO_3^{2-} ← Ester bond

1,3-Bisphosphoglycerate

Stages of Glycolysis

3. (a) A is 3-phosphoglycerate; B is fructose 1,6-bisphosphate; C is phosphoenolpyruvate; D is glucose 6-phosphate; and E is glyceraldehyde 3-phosphate.

(b) D, B, E, A, C

(c) B is produced by phosphofructokinase. D is produced by hexokinase or glucokinase. E is produced by aldolase or triose phosphate isomerase; aldolase produces E directly from fructose 1,6-bisphosphate, and triose phosphate isomerase converts dihydroxyacetone phosphate into E.

(d) C

(e) 1,3-Bisphosphoglycerate

4. d

5. No. The reaction catalyzed by phosphoglucose isomerase is a simple isomerization between an aldose and a ketose and involves the open-chain structures of both sugars. Since glucose 6-phosphate and fructose 6-phosphate are both reducing sugars, their Haworth ring structures are in equilibrium with their open-chain forms. This equilibration is very rapid and does not require an additional enzyme. Note that this isomerization reaction is of the same type as that catalyzed by triosephosphate isomerase.

6. c

7. d, e

8. The large negative $\Delta G^{\circ\prime}$ value indicates that equilibrium favors product formation by a very large margin. The −4 kcal/mol value for $\Delta G'$ means that under physiologic conditions the reaction will also proceed toward product formation essentially irreversibly. The fact that $\Delta G'$ has a smaller negative value than $\Delta G^{\circ\prime}$ indicates that the concentrations of the products are considerably greater than the concentrations of the reactants under physiologic conditions.

9. c

10. c

Entry of Fructose and Galactose into Glycolysis

11. c, d, e, f. In the liver, a specific fructokinase forms fructose 1-phosphate, which is subsequently cleaved into glyceraldehyde and dihydroxyacetone phosphate.

12. (a) Galactose 1-phosphate + UDP–glucose $\rightleftharpoons$ glucose 1-phosphate + UDP–galactose

(b) Step 2 is defective in galactosemia.

(c) Galactokinase catalyzes step 1; galactose 1-phosphate uridyl transferase catalyzes step 2; and UDP–galactose 4-epimerase catalyzes step 3.

(d) NAD^+ is tightly bound to the epimerase and is involved in forming a transient keto structure at C-4 that is an intermediate in the epimerization.

13. Because their epimerase activity is normal, galactosemic patients are able to synthesize UDP–galactose from UDP–glucose. The UDP–galactose is then used in the synthesis of glycoproteins.

Control of Glycolysis

14. a, d, e

15. b, c

16. a, d, f

17. c, d

18. (a) 1, 2, 4, 5, 7 (b) 1, 3, 5, 6

The Fates of Pyruvate

19. b, d

20. Glucose + 2 ADP + 2 $P_i \longrightarrow$ 2 lactate + 2 ATP + 2 H_2O

21. The reduction of pyruvate to lactate converts NADH to NAD^+, which is required in the glyceraldehyde 3-phosphate dehydrogenase reaction. This prevents glycolysis from stopping owing to too low a concentration of NAD^+ and allows continued production of ATP.

Structural Characteristics and Mechanisms of Action of Glycolytic Enzymes

22. a

23. (a) 3, 5 (b) 1, 4 (c) 2

24. b. The rate-limiting step of the reaction, which by definition can be no faster than the rate at which the product appears, is the diffusion-controlled encounter of the substrate with the enzyme, not the release of product.

25. The oxidation of the aldehyde group by NAD^+ is an energetically favorable reaction that leads to the formation of a high-energy thioester bond between the substrate and the thiol group of a cysteine residue of the enzyme. Inorganic phosphate then attacks the thioester bond, which gives rise to an acyl phosphate product, 1,3-bisphosphoglycerate.

Triose Bisophosphates (Regulation by 2,3-Bisphosphoglycerate and Phosphoryl Group Transfer by 1,3-Bisphosphoglycerate)

26. a

27. (a) Enzyme A is bisphosphoglycerate mutase, and enzyme B is 2,3-bisphosphoglycerate phosphatase.

(b) False. The carbon atoms and the phosphate on the C-3 carbon of 2,3-BPG arise from 3-phosphoglycerate. Only the phosphate on C-2 of 2,3-BPG comes from 1,3-BPG. In fact 3-phosphoglycerate is an obligatory intermediate in the conversion of 1,3-BPG into 2,3-BPG.

(c) The interconversion of 3-phosphoglycerate and 2-phosphoglycerate is catalyzed by phosphoglyceromutase, which requires catalytic amounts of 2,3-BPG to be active.

(d) True. 2,3-BPG is present in high concentrations in erythrocytes; it is an allosteric effector of hemoglobin that stabilizes the T form (or deoxy form) of hemoglobin and decreases the affinity of hemoglobin for O_2.

28. When pyruvate kinase transfers the phosphoryl group from phosphoenolpyruvate to ADP, the remaining enediol reminant, enolpyruvate, is much more unstable than its ketone tautomer, pyruvate. This enol-ketone tautomerization drives the overall reaction toward ATP formation by removing the enolpyruvate by converting it to pyruvate. Stryer (p. 504) shows the structures involved in this conversion.

29. d

1. Inorganic phosphate labeled with ^{32}P is added with glucose to a glycogen-free extract from liver, and the mixture is then incubated in the absence of oxygen. After a short time, 1,3-bisphosphoglycerate (1,3-BPG) is isolated from the mixture. On which carbons would you expect to find radioactive phosphate? If you allow the incubation to continue for a longer period, will you find any change in the labeling pattern? Why?

2. Mannose, the 2-epimer of glucose, and mannitol, a sugar alcohol, are widely used as dietetic sweeteners. Both compounds are transported only slowly across plasma membranes, but they can be metabolized by the liver. Propose a scheme by which mannitol and mannose can be converted into intermediates of the glycolytic pathway. You may wish to take advantage of the fact that hexokinase is relatively nonspecific. Why should such sugars be brought into glycolysis as early in the sequence as possible?

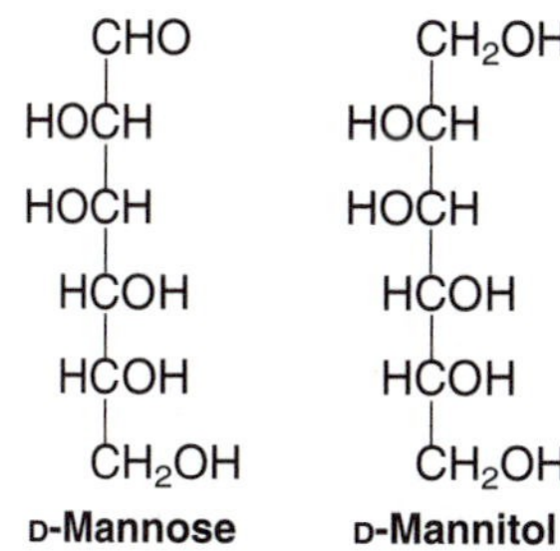

3. The value of $\Delta G^{\circ\prime}$ for the hydrolysis of sucrose to glucose and fructose is −7.30 kcal/mol. You have a solution that is 0.10 M in glucose and that contains sufficient sucrase enzyme to bring the reaction rapidly to equilibrium.

(a) What concentration of fructose would be required to yield sucrose at an equilibrium concentration of 0.01 M, at 25°C?
(b) The solubility limit for fructose is about 3.0 M. How might this limit affect your experiment?

4. Assay of hexokinase activity versus the concentration of glucose 6-phosphate yields the double-reciprocal plot shown in Figure 19-1.

(a) Explain the results shown in the figure and how the effect of glucose 6-phosphate on hexokinase is consistent with the overall control of the metabolism of glucose.
(b) Sketch the curve you would expect to see when the activity of liver glucokinase is determined at various concentrations of glucose 6-phosphate. Explain why the kinetic behavior of the enzyme is consistent with its role in glucose metabolism in the liver.

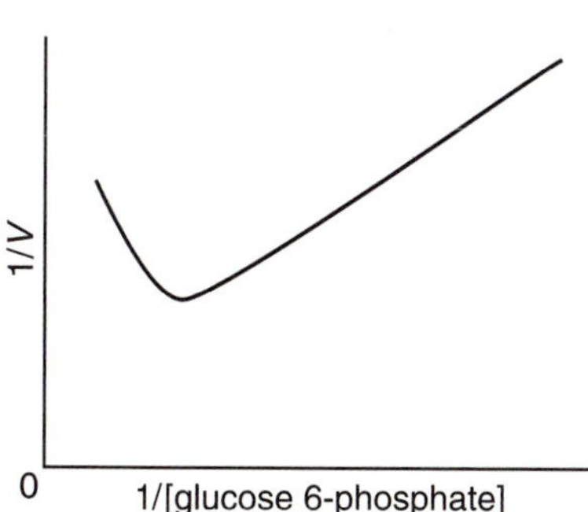

Figure 19-1
The influence of the concentration of glucose 6-phosphate on hexokinase.

5. In 1905, Harden and Young, two English chemists, studied the fermentation of glucose using cell-free extracts of yeast. They monitored the conversion of glucose to ethanol by measuring the evolution of carbon dioxide from the reaction vessel. In one set of experiments, Harden and Young observed the evolution of CO_2 when inorganic phosphate (P_i) was added to a yeast extract containing glucose. In the graph in Figure 19-2, curve A shows what happens when no P_i is added. Curve B shows the effect of adding P_i in a separate experiment. As the evolution of CO_2 slows with time, more P_i is added to stimulate the reactions; this is shown in curve C.

(a) Why is glucose fermentation dependent on P_i?
(b) During fermentation, what is the ratio of P_i consumed to CO_2 evolved?
(c) How does the formation of ethanol ensure that the fermentation process is in redox balance?

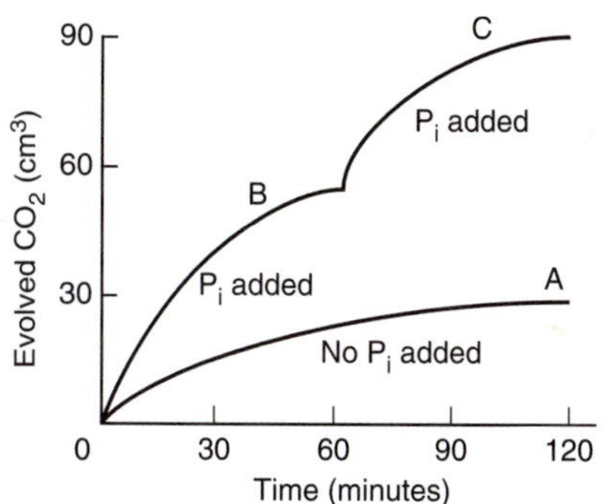

Figure 19-2
Evolution of CO_2 in the Harden-Young experiment.

(d) Harden and Young found that they could recover phosphate from the reaction mixture, but it was not precipitable by magnesium citrate, as is P_i. Name at least three organic compounds that would be phosphorylated when P_i is added to the fermenting mixture.

(e) As the rate of CO_2 evolution decreased, Harden and Young found that an unusual compound accumulated in the reaction mixture. In 1907, Young identified the compound as a hexose bisphosphate. Name the compound and explain why it might accumulate when P_i becomes limiting.

(f) Later, Meyerhof showed that the addition of adenosine triphosphatase (ATPase, an enzyme that hydrolyzes ATP to yield ADP and P_i) to the reaction mixture stimulates the evolution of CO_2. Explain this result.

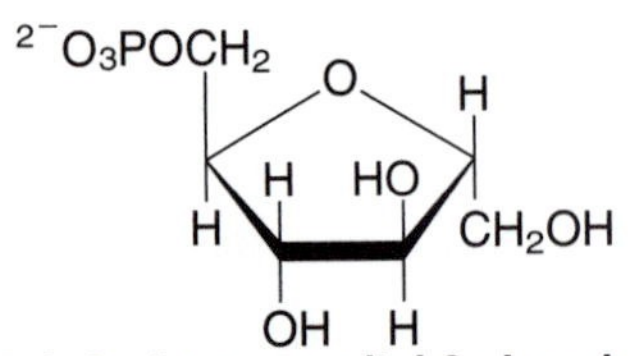

2,5-Anhydro-D-mannitol 6-phosphate
(in β configuration)

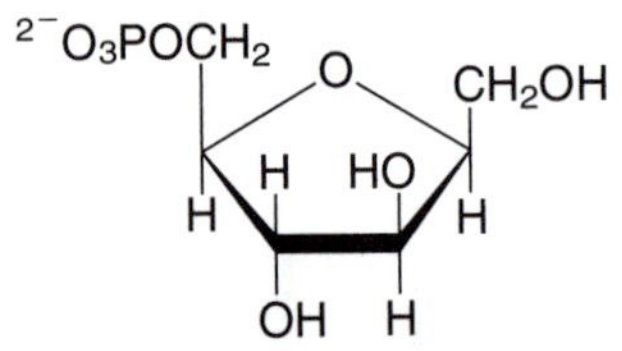

2,5-Anhydro-D-glucitol 6-phosphate
(in α configuration)

6. In solution, 80% of the fructose 6-phosphate is in the β-anomeric form and 20% is in the α-anomeric form, with the half-time for anomerization being about 1.5 seconds. To determine which of the two anomers is a substrate for phosphofructokinase (PFK), Voll and his colleagues employed two model substrates (shown in the margin) that have C-2 configurations corresponding to the anomeric forms of fructose 6-phosphate.

(a) Why will neither of the substrates in the margin undergo mutarotation?

When the mannitol derivative is incubated with PFK and ATP, its rate of phosphorylation is about 80% of that for fructose 6-phosphate; K_M for fructose 6-phosphate is 0.04 mM, whereas K_M for the mannitol derivative is 0.40 mM. The glucitol derivative binds to PFK with an affinity almost equal to that of the mannitol derivative, but it is not phosphorylated by PFK; it is a competitive inhibitor of fructose 6-phosphate, with a K_i of 0.35 mM.

(b) On the basis of these observations, which anomer of fructose 6-phosphate is a substrate for PFK?

7. In Table 19-3, on page 504 of the text, Stryer provides the concentrations of many of the intermediates of the glycolytic pathway in erythrocytes. Given these concentrations, how much free energy is liberated when pyruvate and ATP are produced through the action of pyruvate kinase?

8. The absence of UDP–galactose 4-epimerase in the erythrocytes of a few people has been described; the tissues of these subjects exhibit relatively normal activity. What effect would the absence of epimerase activity have on galactose assimilation in the erythrocytes of these people? Would you expect to see the accumulation of galactitol in the lens? Why?

9. One step in the conversion of galactose to glucose 6-phosphate involves the epimerization of galactose to glucose while attached to UDP. The inversion of the hydroxyl group at C-4 is catalyzed by UDP–galactose 4-epimerase, which contains a tightly bound NAD^+. Fluorescence studies have shown that this NAD^+ is transiently reduced to NADH during catalysis.

(a) Write a series of steps that shows how NAD^+ is involved in the epimerization of galactose and glucose.

(b) When the mechanism of action of UDP–galactose 4-epimerase was studied, investigators considered a 1,1 hydrogen shift or dehydration and rehydration as possible mechanisms for epimerization at C-4. However, when the enzyme was incubated with ^{3}H-H_2O and either UDP–galactose or UDP–glocuse, no tritium was incorporated into either of the products. In addition, when one or the other of the substrates was incubated with ^{18}O-H_2O, there was no incorporation of ^{18}O into either of the products. What do both sets of observations tell you about the mechanism of action of the epimerase?

10. Ahlfors and Mansour studied the activity of purified sheep phosphofructokinase (PFK) as a function of the concentration of ATP, in experiments that were carried out at a constant concentration of fructose 6-phosphate. Typical results are shown in Figure 19-3. Explain these results, and relate them to the role of PFK in the glycolytic pathway.

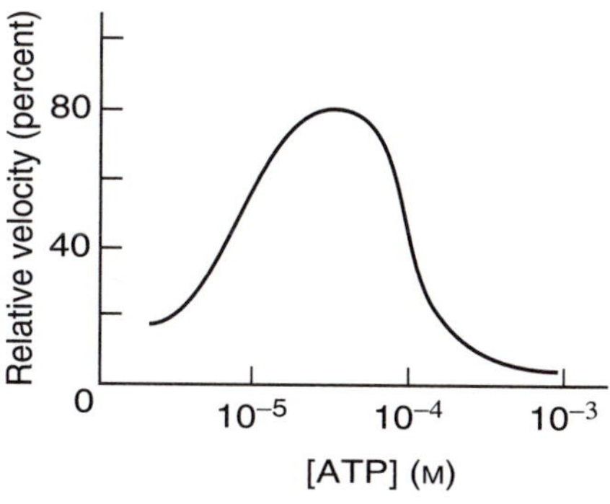

Figure 19-3
The effects of ATP concentration on sheep PFK.

11. Purified glyceraldehyde 3-phosphate dehydrogenase rapidly and irreversibly loses its catalytic activity when it is incubated with iodoacetate. Structural analysis of the acetylated protein reveals that a single cysteine residue at position 149 in the primary sequence is converted to the carboxymethyl form. In the presence of glyceraldehyde 3-phosphate, the enzyme is protected against inactivation by iodoacetate. Can you suggest a role for the cysteine residue at position 149 in the enzyme? Write out the reaction that occurs when iodoacetate reacts with the cysteine residue. What effect would you expect iodoacetate to have on glycolysis in living cells?

12. Hexokinase catalyzes the formation of glucose 6-phosphate at a maximum velocity of 2×10^{-5} mol/min, whereas V_{max} for the formation of fructose 6-phosphate is 3×10^{-5} mol/min. The value of K_M for glucose is 10^{-5} M, whereas K_M for fructose is 10^{-3} M. Suppose that in a particular cell the observed rates of phosphorylation are 1.0×10^{-8} mol/min for glucose and 1.5×10^{-5} mol/min for fructose.

(a) Estimate the concentrations of glucose and fructose in the cell.
(b) Which of these hexoses is more important in generating energy for this cell?

13. What is the effect of increasing the concentration of each of the following metabolites on the net rate of glycolysis? Briefly explain each answer.

(a) Glucose 6-phosphate
(b) Fructose 1,6-bisphosphate
(c) Citrate
(d) Fructose 2,6-bisphosphate

14. The following series of experiments are conducted with aldolase:

(a) The enzyme is incubated with sodium borohydride ($NaBH_4$, a powerful reducing agent) and then is dialyzed to remove the reducing agent. The enzyme is then found to be catalytically active.
(b) When incubated with $NaBH_4$ and ^{14}C-labeled glyceraldehyde 3-phosphate, and then dialyzed, the enzyme remains catalytically active.
(c) When aldolase is incubated with $NaBH_4$ in the presence of ^{14}C-labeled dihydroxyacetone phosphate and then dialyzed, it is inactivated.

Samples from each of the three experiments described above are hydrolyzed in 1 N HCl at 100°C for 24 hours. The neutralized hydrolysis products are chromatographed, and a radioactive spot is found only with the sample from experiment (c). Analysis of the spot reveals that it is ε–*N*–*β*-(1,3-dihydroxypropyl)lysine, shown below:

$$\begin{array}{l} H_2C-OH \\ \;\;| \\ HC-NH-(CH_2)_4-\underset{\displaystyle NH_3^+}{\underset{|}{CH}}-COO^- \\ \;\;| \\ H_2C-OH \end{array}$$

ε – N – β - (1,3-Dihydroxypropyl)lysine

Draw the intermediate formed between the substrate and the lysine residue in experiment (c), before the reduction and hydrolysis are carried out, and describe the role of that intermediate in the overall reaction catalyzed by the enzyme. Then provide a concise explanation for the results of the three experiments described above, including the reason that aldolase is inactivated only in experiment (c).

15. Base-catalyzed isomerizations can cause interconversions among glucose, fructose, and mannose in aqueous solution, and such reactions are nonstereospecific. How could you determine whether the reaction catalyzed by phosphoglucose isomerase is stereospecific?

16. You and your laboratory partner are studying oxamate, which is said to inhibit glycolysis. A cell-free muscle extract is incubated in a 100% argon atmosphere with oxamate and glucose. After 30 minutes, samples are assayed for levels of various glycolytic intermediates. Compared with a control experiment carried out in the absence of oxamate, you note an accumulation of glyceraldehyde 3-phosphate, dihydroxyacetone phosphate, and fructose 1,6-bisphosphate. Levels of phosphoenolpyruvate and pyruvate are relatively low, and there is virtually no lactate. Your partner concludes that glyceraldehyde 3-phosphate dehydrogenase is inhibited by oxamate. However, you suspect that another enzyme could be affected. You repeat the experiment, but this time you include an excess of NAD^+. After 30 minutes and compared with the control, there is no accumulation of any hexose or triose phosphates, and only pyruvate appears to accumulate in the extract. Your partner immediately agrees that another enzyme is inhibited by oxamate. Which enzyme is it? Briefly explain your answer.

$$H_3\overset{+}{N}-\overset{\displaystyle O}{\overset{\|}{C}}-C\begin{matrix}\nearrow O\\ \searrow O^-\end{matrix}$$

Oxamate

17. Glycerol can enter the glycolytic pathway through phosphorylation to glycerol 3-phosphate, catalyzed by glycerol kinase, and then by oxidation to dihydroxyacetone phosphate, catalyzed by glycerol 3-phosphate dehydrogenase. Glyceraldehyde is funneled into the glycolytic pathway through phosphorylation to glyceraldehyde 3-phosphate, catalyzed by triose kinase. Glycerate enters the glycolytic pathway when it is phosphorylated to 3-phosphoglycerate by glycerate kinase. Lactate-forming bacteria can metabolize glycerol, glyceraldehyde, or glycerate in the presence of oxygen, but only one of these substrates can be converted to lactate under anaerobic conditions. Which one, and why?

18. Suppose that in erythrocytes the formation of 2,3-bisphosphoglycerate and its conversion to 3-phosphoglycerate were unregulated. What implications would this have for ATP production during glycolysis?

1. After a short incubation time, labeled phosphate will be found on C-1 of 1,3-BPG. Inorganic phosphate enters the glycolytic pathway at the step catalyzed by glyceraldehyde 3-phosphate dehydrogenase:

$$\text{Glyceraldehyde 3-phosphate } (\text{H–C(=O)–CH(OH)–}CH_2OPO_3^{2-}) + NAD^+ + P_i \xrightleftharpoons{\text{Glyceraldehyde 3-phosphate dehydrogenase}} \text{1,3-Bisphosphoglycerate (1,3-BPG) } (^{2-}O_3PO\text{–C(=O)–CH(OH)–}CH_2OPO_3^{2-}) + NADH + H^+$$

Glyceraldehyde 3-phosphate **1,3-Bisphosphoglycerate (1,3-BPG)**

However, after a longer incubation time, the radioactive label will be found on both C-1 and C-3 of 1,3-BPG because the step subsequent to the formation of 1,3-BPG involves the phosphorylation of ADP to form ATP, which will be radioactively labeled in the γ phosphoryl group:

$$\text{1,3-Bisphosphoglycerate } (\text{O=C(–}OPO_3^{2-}\text{)–CH(OH)–}CH_2OPO_3^{2-}) + ADP \xrightleftharpoons{\text{Phosphoglycerate kinase}} \text{3-Phosphoglycerate } (\text{O=C(–}O^-\text{)–CH(OH)–}CH_2OPO_3^{2-}) + \mathbf{ATP}$$

1,3-Bisphosphoglycerate **3-Phosphoglycerate**

In other glycolytic reactions, the radioactively labeled ATP can phosphorylate at C-1 of fructose 6-phosphate and C-6 of glucose, both of which are equivalent to C-3 in 1,3-BPG. Thus, after prolonged incubation, both labeled inorganic phosphate *and* labeled ATP will be present in the mixture, and 1,3-BPG with a radioactive label at both C-1 and C-3 will be present in the extract. One must assume that a small amount of unlabeled ATP is available at the start to initiate hexose phosphorylation.

2. The first step is the conversion of mannitol to mannose. This requires oxidation at the C-1 of mannitol, using a dehydrogenase enzyme with NAD^+ or $NADP^+$ as an electron acceptor. One could then propose a number of schemes, using nucleotide derivatives with isomerase or epimerase activities in combination with one or more phosphorylated intermediates. An established pathway uses hexokinase and ATP for the synthesis of mannose 6-phosphate; this is then converted by mannose phosphate isomerase to form fructose 6-phosphate, an intermediate of the glycolytic pathway. Bringing such sugars into the glycolytic pathway as soon as possible means that already existing enzymes can be used to process the intermediates derived from each of a number of different sugars. Otherwise, a separate battery of enzymes would be needed to obtain energy from each of the sugars found in the diet.

3. (a) The reaction you are concerned with is

$$\text{Glucose} + \text{fructose} \longrightarrow \text{sucrose}$$

For the reaction in this direction, $\Delta G°' = +7.30$ kcal/mol. The equilibrium constant K'_{eq} is

$$K'_{eq} = \frac{[\text{sucrose}]}{[\text{glucose}][\text{fructose}]}$$

At the start of the reaction, [sucrose] = 0, [glucose] = 0.10 M, and [fructose] = x. At equilibrium, [sucrose] = 0.01 M, [glucose] = 0.09 M, and [fructose] = x – 0.01 M. First, calculate K'_{eq} for the reaction at equilibrium:

$$\Delta G°' = -1.364 \log_{10} K'_{eq}$$

$$7.3 = -1.364 \log_{10} K'_{eq}$$

$$\log_{10} K'_{eq} = \frac{7.3}{-1.364} = -5.35$$

$$K'_{eq} = \text{antilog } (-5.35) = 4.46 \times 10^{-6} \text{ M}$$

Then find the concentration of fructose that satisfies the conditions at equilibrium. Assume that the unknown concentration of fructose at equilibrium (x – 0.01 M) is approximately equal to x.

$$K'_{eq} = 4.46 \times 10^{-6} \text{ M} = \frac{10^{-2} \text{ M}}{(0.09 \text{ M})(x)}$$

$$x = 2.49 \times 10^{4} \text{ M}$$

(b) The concentration of fructose required to generate sucrose at a concentration of 0.01 M exceeds the solubility limit for fructose; it is therefore impossible to establish such conditions in solution.

4. (a) The double-reciprocal plot shows that at relatively low concentrations of glucose 6-phosphate, velocity increases in proportion to substrate concentration. At higher concentrations of glucose 6-phosphate, the velocity of the hexokinase reaction decreases as the concentration of glucose 6-phosphate increases. In the cell, the concentration of glucose 6-phosphate rises when the levels of other intermediates in the glycolytic pathway increase. For example, when the energy requirements of the cell are being met and the levels of ATP and citrate are relatively high, the activity of phosphofructokinase decreases and the concentration of fructose 6-phosphate increases. A subsequent rise in the concentration of glucose 6-phosphate occurs because it is in equilibrium with fructose 6-phosphate, and the activity of hexokinase decreases. This causes the rate of formation of glucose 6-phosphate to decrease, which allows the cell to stop utilizing glucose until it is needed for the generation of energy, NADPH, or glycogen. The reaction catalyzed by hexokinase is not the first *committed* step of the glycolytic pathway (that is, it is not the first irreversible step that is unique to glycolysis), because there are a number of possible fates of glucose 6-phosphate pathway in the cell; for example, it can be oxidized in the pentose phosphate pathway, or it can be used as a source of glucosyl units in the synthesis of glycogen. It would therefore not be appropriate for glycolytic intermediates to regulate the synthesis of glucose 6-phosphate, because the demands of other pathways would not be met by such a control system.

(b) Glucokinase is not inhibited by glucose 6-phosphate (Stryer, p. 495). Thus you would expect the graph of reciprocal values of enzyme velocity versus concentration of glucose 6-phosphate to be a straight line (see the accompanying figure). Glucokinase functions as an enzyme designed to convert glucose to glucose 6-phosphate when blood glucose levels are high, as they would be after a meal. The enzyme is designed to secure glucose for use in glycogen synthesis; there is no need for the enzyme to be inhibited by its substrate. When blood glucose concentration drops below 5mM, the velocity of the glucokinase reaction slows, allowing other tissues like brain and muscle to take up glucose.

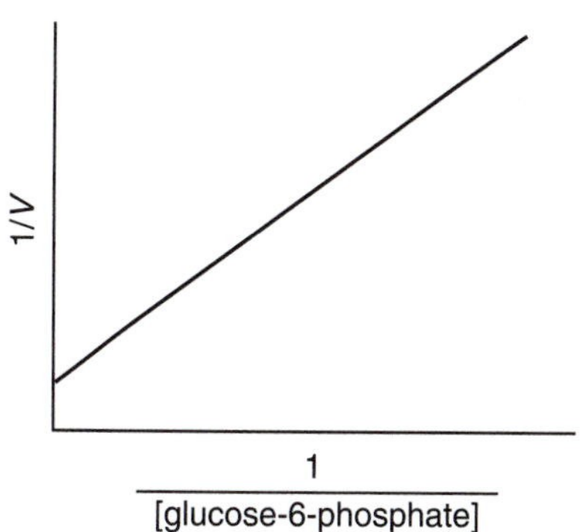

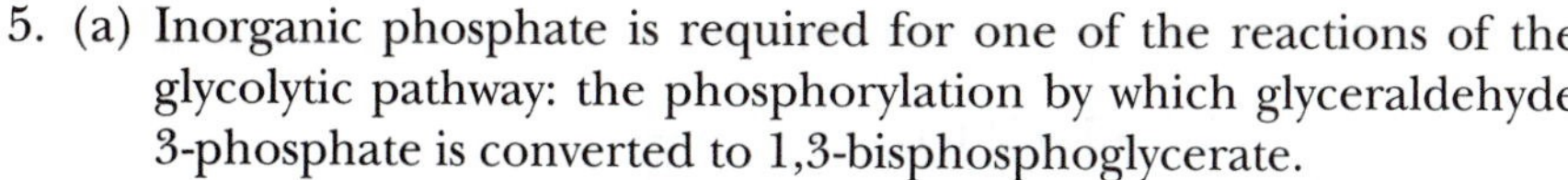

5. (a) Inorganic phosphate is required for one of the reactions of the glycolytic pathway: the phosphorylation by which glyceraldehyde 3-phosphate is converted to 1,3-bisphosphoglycerate.

(b) The P_i/CO_2 ratio is 1.0, with one P_i being consumed for each pyruvate that undergoes decarboxylation.

(c) In most cells, the absolute concentration of NAD^+ and NADH is low. Successive and continuous reduction and oxidation of NAD^+ and NADH, respectively, is necessary for them to continue to serve as donors and acceptors of electrons. In this case, NAD^+ must be constantly available for the continued activity of glyceraldehyde 3-phosphate dehydrogenase in the glycolytic pathway. The NADH generated during the oxidation of glyceraldehyde 3-phosphate is reoxidized to NAD^+ when acetaldehyde is reduced to ethanol.

(d) Initially, glyceraldehyde 3-phosphate is phosphorylated when P_i is added to the fermenting mixture. ADP is phosphorylated when 1,3-bisphosphoglycerate donates a phosphoryl group to the nucleotide and is itself converted to 3-phosphoglycerate. The ATP formed in this reaction can be used in two earlier reactions of glycolysis, the phosphorylations of glucose and of fructose 6-phosphate.

(e) The hexose bisphosphate is fructose 1,6-bisphosphate, the only such intermediate in the glycolytic pathway. This compound accumulates when glycolytic flux is blocked at the glyceraldehyde 3-phosphate dehydrogenase step by limited P_i availability. As the phosphorylation of glucose continues, intermediates from the steps preceding the formation of 1,3-bisphosphoglycerate build up.

(f) The hydrolysis of ATP to ADP and P_i makes more P_i available for the phosphorylation of glyceraldehyde 3-phosphate. Under such conditions, glycolytic activity and CO_2 production are stimulated.

6. (a) Each of the model substrates lacks a C-2 hydroxyl group, which is required for the opening of the ring and mutarotation to the other anomeric form.

(b) The observations show that both substrates bind to phosphofructokinase (PFK) with almost equal affinities, but only the mannitol derivative, which has a β conformation at C-2, is phosphorylated. Thus, it is likely that the β anomer of fructose 6-phosphate is the substrate for PFK.

7. From Table 19-2, page 491 of the text, the value of $\Delta G^{\circ\prime}$ for the reaction catalyzed by pyruvate kinase is –7.5 kcal/mol. We want to calculate $\Delta G'$, the free energy released for the concentrations found in erythrocytes:

$$\Delta G' = \Delta G^{\circ\prime} + 1.364 \log_{10} Q$$

where Q is the ratio of the products of the concentrations of products and substrates:

$$\Delta G' = -7.5 + 1.364 \log_{10} \frac{[\text{pyruvate}][\text{ATP}]}{[\text{phosphoenolpyruvate}][\text{ADP}]}$$

$$= -7.5 + 1.364 \log_{10} \frac{(51)(1850)}{(23)(138)}$$

$$= -7.5 + 1.364 \log_{10} 29.73$$

$$= -7.5 + (1.364)(1.472)$$

$$= -5.5 \text{ kcal/mol}$$

8. The absence of the epimerase activity in the erythrocytes means that they are unable to convert UDP–galactose to UDP–glucose. An accumulation of UDP–galactose and galactose 1-phosphate would be observed in the erythrocytes because galactose 1-phosphate uridyl transferase and galactose kinase are still active. Glucose 1-phosphate, the other product of transferase activity, can be isomerized to glucose 6-phosphate by phosphoglucomutase. Although one might expect to see some deterioration of erythrocyte function as a result of the absence of epimerase activity, this does not occur, perhaps because other tissues in the body can metabolize galactose. The ability of lens tissue to utilize galactose and its metabolites in the normal manner means that galactitol concentrations in the lens are not elevated, in contrast to patients who lack galactose 1-phosphate uridyl transferase.

9. (a) NAD^+ probably accepts the hydrogen atom attached to C-4 of galactose, and a 4-keto sugar is formed. NADH then transfers a hydrogen to the other side of the carbon atom to form the corresponding epimer.

HO–C(–H) with NAD^+ ⇌ NADH + H^+, O=C ⇌ NAD^+, H–C(–OH)

C-4 of UDP – galactose | 4-Keto intermediate | C-4 of UDP – glucose

(b) The two observations suggest that the transfer of hydrogen is direct, from one epimer to NAD^+ and then to the corresponding epimer. Direct transfer of a hydrogen atom provides no opportunity for tritium exchange with water. Direct hydrogen transfer also rules out hydration and dehydration, so that there is no exchange of oxygen when the reaction is carried out in ^{18}O-H_2O.

10. The rate of the reaction catalyzed by PFK initially increases with the ATP concentration because ATP is a substrate for the reaction; it binds at the active site of PFK with fructose 6-phosphate and serves as a phosphoryl donor. At higher concentrations, ATP binds not only at the active site but also at the allosteric site; this alters the conformation of the enzyme and decreases the level of its activity. The effects of ATP on PFK are consistent with the role of PFK as a control element for the glycolytic pathway. When concentrations of ATP are relatively low, the activity of PFK is stimulated so that additional fructose 1,6-bisphosphate is made available for subsequent energy-generating reactions; when concentrations of ATP are higher and the demand of the cell for energy is lower, ATP inhibits PFK activity, thereby allowing glucose and other substrates to be utilized in other pathways. In many cells, ATP concentration is maintained at relatively high and constant levels, so that PFK is always subject to inhibition by ATP. Inhibition can be relieved by fructose 2,6-bisphosphate, which is synthesized when glucose is readily available. This allows cells to carry out glycolysis even when ATP levels are high, permitting the synthesis of building blocks from glucose.

11. The rapid reaction of glyceraldehyde 3-phosphate dehydrogenase with iodoacetate and the protection against inactivation provided by glyceraldehyde 3-phosphate suggest that the cysteine residue at position 149 is unusually reactive and that it may be located near the active site. Indeed, other evidence has confirmed that the cysteine residue at position 149 is the one referred to in Stryer's description of the action of glyceraldehyde 3-phosphate dehydrogenase (p. 502 of the text). Normally the aldehyde substrate reacts with this cysteine to form a hemithioacetal. Iodoacetate reacts with the sulfhydryl group of the residue to form a carboxymethyl derivative, thereby inactivating the enzyme. This reaction is as follows:

$$\text{E—Cys—S}^- + \text{I—CH}_2\text{—COO}^- \longrightarrow \text{E—Cys—S—CH}_2\text{—COO}^- + \text{I}^-$$

Carboxymethyl enzyme
(inactive)

Glyceraldehyde 3-phosphate protects against inactivation by blocking the access of iodoacetate to the active site. Early experiments showed that iodoacetate can completely inhibit glycolysis in living cells; the irreversible inactivation of glyceraldehyde 3-phosphate dehydrogenase probably accounts for these observations.

12. (a) The values of V_{max}, K_M, and V, the measured velocity, are given for glucose and fructose. You can use the Michaelis-Menten equation by solving for [S], the substrate concentration:

$$V = V_{max}\frac{[S]}{[S] + K_M}$$

$$V_{max}[S] = V[S] + VK_M$$

$$[S](V_{max} - V) = VK_M$$

$$[S] = \frac{VK_M}{V_{max} - V}$$

Then you can use the values provided to calculate the concentrations of the two sugars in the cell. For glucose, [S] = 5×10^{-9} M; whereas for fructose, [S] = 1×10^{-3} M.

(b) The phosphorylation of a hexose such as glucose or fructose is the initial step in the oxidation of the sugar, a process in which energy in the form of ATP is generated. Fructose is more important in the provision of energy for the cell because *V*, the observed rate of formation, is 1500 times faster for fructose 6-phosphate than it is for glucose 6-phosphate.

13. (a) Initial increases in the concentration of glucose 6-phosphate increase the pace of glycolysis by raising the level of substrate for glucose 6-phosphate isomerase and all subsequent reactions of the glycolytic pathway. However, glucose 6-phosphate is also an allosteric inhibitor of hexokinase. Thus, high concentrations of glucose 6-phosphate could ultimately inhibit glycolysis by decreasing the entry of glucose into the pathway.

(b) Fructose 1,6-bisphosphate is a product of the reaction catalyzed by phosphofructokinase 1, which is the primary regulator of glycolysis. Increasing the level of this intermediate increases the rate of glycolysis by raising the level of substrates for all subsequent reactions of the glycolytic pathway.

(c) Citrate, an intermediate of the citric acid cycle, is a feedback inhibitor of phosphofructokinase 1. Thus, increasing the concentration of citrate decreases the pace of glycolysis.

(d) Fructose 2,6-bisphosphate, which is formed from fructose 6-phosphate in a reaction catalyzed by phosphofructokinase 2, increases the rate of glycolysis because it activates phosphofructokinase 1, increasing its affinity for fructose 6-phosphate and decreasing its sensitivity to ATP.

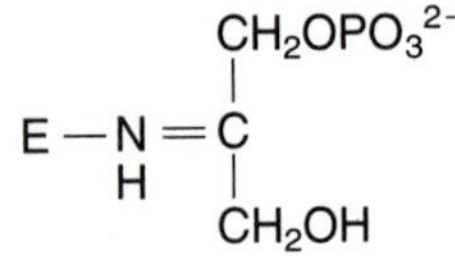

14. Your drawing should illustrate the Schiff base linkage formed between dihydroxyacetone phosphate (DHAP) and the amino group in the side chain of a specific lysine residue of the aldolase protein. This is the first step in the aldol condensation of DHAP and glyceraldehyde 3-phosphate. Normally the Schiff base intermediate is a transient one, but when it is reduced by $NABH_4$, the bond between lysine and DHAP is no longer easily cleaved (it is in fact stable in hot acid) and the enzyme is inactivated. The derivative recovered after acid proteolysis has no phosphate group because it is removed by acid. Because the enzyme alone has no Schiff base and because it cannot form a Schiff base with glyceraldehyde 3-phosphate, aldolase is unaffected by $NABH_4$ under the conditions in experiments (a) and (b). One can conclude that aldolase can form a Schiff base with DHAP but not with glyceraldehyde 3-phosphate.

15. Run the reaction backwards using the enzyme and fructose 6-phosphate, which has a nonchiral carbon atom at position 2, to produce aldose phosphate isomers. If the reaction is stereospecific, only glucose 6-phos-

phate, but not its C-2 epimer, mannose 6-phosphate, should be produced. Experiments like this have confirmed that the phosphoglucose isomerase does indeed catalyze the isomerization of glucose 6-phosphate and fructose 6-phosphate with absolute specificity. Another enzyme, mannose 6-phosphate isomerase, catalyzes the isomerization of mannose 6-phosphate to its epimer, glucose 6-phosphate.

16. Lactate dehydrogenase, which reduces pyruvate to lactate and regenerates NAD^+, is inhibited by oxamate. The key observation is that an excess of NAD^+, when incubated with glucose and oxamate, leads to an accumulation of pyruvate with no accumulation of other glycolytic intermediates. Excess NAD^+ allows glyceraldehyde 3-phosphate dehydrogenase to continue operating, to compensate for the failure of lactate dehydrogenase to regenerate NAD^+ from NADH. Extracts incubated only with glucose in the presence of oxamate will begin to accumulate pyruvate, but will also accumulate intermediates preceding glyceraldehyde 3-phosphate, owing to the lack of NAD^+ required for continued synthesis of 1,3-BPG.

17. Only glyceraldehyde can be converted to lactate under anaerobic conditions. The pathway for glyceraldehyde to lactate produces net formation of one ATP with no net oxidation per molecule metabolized.

Every glycerol molecule converted to lactate under anaerobic conditions generates 2 NADH, one produced during the conversion of glycerol 3-phosphate into DHAP and another during the formation of 1,3-BPG from glyceraldehyde 3-phosphate. Because there is only one step, catalyzed by lactate dehydrogenase, that regenerates an NAD^+ molecule for every glyceraldehyde molecule metabolized, NADH accumulates. The glycolytic pathway is interrupted because there is no NAD^+ available to accept electrons from glycerol 3-phosphate or glyceraldehyde 3-phosphate.

Glycerate cannot be metabolized under anaerobic conditions, because during its conversion to lactate there is no net formation of ATP. In addition, the pathway from glycerate to lactate has no pathway for generation of NADH, which would be required to balance the generation of NAD^+ during the reduction of pyruvate to form lactate.

18. Unregulated formation of 2,3-bisphosphoglycerate and its conversion to 3-phosphoglycerate in erythrocytes would cause fewer ATP molecules to be produced per glucose metabolized in glycolysis. Recall that 2,3-bisphosphoglycerate is synthesized from 1,3-bisphosphoglycerate, which is also a substrate for phosphoglycerate kinase. Diversion of 1,3-BPG into another pathway means that the ATP normally formed in the conversion of 1,3-BPG into 3-phosphoglycerate is lost.

The net reaction catalyzed by phosphoglycerate kinase is

$$\text{1,3-BPG} + \text{ADP} \longrightarrow \text{3-phosphoglycerate} + \text{ATP}$$

while the net reaction catalyzed by bisphosphoglycerate mutase and 2,3-BPG phosphatase is

$$\text{1,3-BPG} + H_2O \longrightarrow \text{3-phosphoglycerate} + P_i + 2\,H^+$$

EXPANDED SOLUTIONS TO TEXT PROBLEMS

1. D-glucopyranose is a cyclic hemiacetal which is an equilibrium with its open-chain form which contains an active aldehyde group. In contrast, the anomeric carbon atoms of glucose and fructose are joined in an α-glycosidic linkage in sucrose. Hence sucrose is not in equilibrium with an active aldehyde or ketone form.

2. (a) The key is the aldolase reaction. Note that the carbons attached to the phosphate in glyceraldehyde 3-phosphate and dihydroxyacetone phosphate are interconverted by triose phosphate isomerase and *both* become the terminal carbon of the glyceric acids. Hence, the label is in the methyl carbon of pyruvate.

   (b) By definition, specific activity is radioactivity/mol (or mmol). In this case the specific activity is halved (to 5 mCi/mmol) because the number of moles of product (pyruvate) is twice that of the labeled substrate (glucose).

3. Glucose + 2 P_i + 2 ADP $\longrightarrow$ 2 lactate + 2 ATP

   (a) To obtain the answer, −29.5 kcal/mol, you add the $\Delta G^{\circ\prime}$ values given in Table 15-2 and the value given for the reduction of pyruvate to lactate. Remember that the values for the three carbon molecules must be doubled, since each hexose yields two trioses.

   (b)
$$\Delta G' = -29.5 + 1.36 \log \frac{(5 \times 10^{-5})^2 \times (2 \times 10^{-3})^2}{(5 \times 10^{-3})(10^{-3})^2(2 \times 10^{-4})^2}$$
$$= -29.5 + 1.36 \log 50$$
$$= -27.2 \text{ kcal/mol}$$

   The concentrations of ATP, ADP, P_i, and lactate are squared because, in reactions such as A $\longrightarrow$ 2 B, the $K_{eq} = [B]^2/[A]$.

4.
$$-7.5 = -1.36 \log \frac{[Pyr]\,[ATP]}{[PEP]\,[ADP]}$$
$$5.515 = \log \frac{[Pyr]}{[PEP]} + \log 10$$
$$\frac{[Pyr]}{[PEP]} = 10^{4.515}$$
$$\frac{[PEP]}{[Pyr]} = 10^{-4.515} = 3.06 \times 10^{-5}$$

   Note that this problem is a variation of problem 17-3 in Stryer.

5. Since $\Delta G^{\circ\prime}$ for the aldolase reaction is +5.7 kcal/mol (Stryer, p. 491), $K_{eq} = 10^{-5.7/1.36} = 6.5 \times 10^{-5}$. If we let the concentration of each of the trioses (DHAP and G-3P) formed during the reaction be *X*, then the concentration of F-1,6-BP is 10^{-3} M − *X* at equilibrium, since we started with millimolar F-1,6-BP. Then,

$$\frac{X^2}{10^{-3} - X} = 6.5 \times 10^{-5}$$

Solving this quadratic equation leads to the answer of 2.24×10^{-4} M for X (DHAP and G-3P). Subtracting this from 10^{-3} gives the F-1,6-BP concentration of 7.76×10^{-4} M.

6. The 3-phosphoglycerate labeled with ^{14}C accepts the phosphate attached to C-1 of 1,3-BPG (see Stryer, p. 504). Therefore, the resulting 2,3-BPG is ^{14}C-labeled in all three carbon atoms and is ^{32}P-labeled in the phosphorus atom attached to the C-2 hydroxyl.

7. Hexokinase has a low ATPase activity in the absence of a sugar because it is in a catalytically inactive conformation (Stryer, p. 499). The addition of xylose closes the cleft between the two lobes of the enzyme. However, the xylose hydroxymethyl group (at C-5) cannot be phosphorylated. Instead, a water molecule at the site normally occupied by the C-6 hydroxymethyl group of glucose acts as the phosphoryl acceptor from ATP.

8. (a) 2,3-Bisphosphoglycerate (BPG) lowers the oxygen affinity of hemoglobin (Stryer, p. 160). The rate of synthesis of 2,3-BPG is controlled by the level of 1,3-BPG, a glycolytic intermediate.

(b) The lowered level of glycolytic intermediates leads to less 2,3-BPG and, hence, a higher oxygen affinity.

(c) Glycolytic intermediates are present at a higher than normal level. The level of 2,3-BPG is increased, which makes the oxygen affinity lower than normal.

9. (a) The fructose 1-phosphate pathway (Stryer, p. 491) forms glyceraldehyde 3-phosphate. Phosphofructokinase, a key control enzyme, is bypassed. Furthermore, fructose 1-phosphate stimulates pyruvate kinase.

(b) The rapid, unregulated production of lactate can lead to metabolic acidosis.

10. The metal ion serves as an electron sink, as does the protonated Schiff base in animal aldolases.

11. EDTA removes the metal ion from the catalytic site of prokaryotic aldolases, whereas sodium borohydride reduces the Schiff base intermediate in catalysis by animal aldolases.

12. (a) Glycolysis increases because ATP can no longer inhibit PFK, phosphofructokinase.

(b) Glycolysis increases because citrate can no longer inhibit PFK.

(c) Glycolysis will increase because in the absence of fructose 2,6-bisphosphatase, the level of fructose 2,6-bisphosphate will increase, resulting in activation of PFK.

(d) Glycolysis will decrease because fructose 1,6-bisphosphate can no longer activate pyruvate kinase.

CHAPTER

20

Citric Acid Cycle

The citric acid cycle, also known as the *tricarboxylic acid cycle* or the *Krebs cycle,* is the final oxidative pathway for carbohydrates, lipids, and amino acids. It is also a source of precursors for biosynthesis. Stryer begins Chapter 20 with a general description of the reactions of the citric acid cycle. He then discusses in detail the reaction mechanisms of the pyruvate dehydrogenase complex and citrate synthase, as well as the stereospecificity of some of the reactions. In the following sections, Stryer summarizes the biosynthetic roles of the citric acid cycle and describes its relationship to the glyoxylate cycle found in bacteria and plants. He concludes the chapter with a discussion of control mechanisms. The chapters on enzymes (Chapters 8 through 10), the introduction to metabolism (Chapter 17), and the chapter on glycolysis (Chapter 19) contain essential background material for this chapter.

LEARNING OBJECTIVES

When you have mastered this chapter, you should be able to complete the following objectives:

Introduction (Stryer page 509)

1. Outline the role of the *citric acid cycle* in aerobic metabolism.

2. Locate the enzymes of the cycle in eukaryotic cells.

Reactions of the Citric Acid Cycle (Stryer pages 509–514)

3. Name all the *intermediates* of the citric acid cycle and draw their structures.

4. List the enzymatic reactions of the citric acid cycle in their appropriate sequence. Name all the enzymes.

5. Give examples of *condensation, dehydration, hydration, decarboxylation, oxidation,* and *substrate-level phosphorylation* reactions.

6. Indicate the steps of the cycle that yield CO_2, *NADH*, $FADH_2$, and *GTP*. Note the biological roles of GTP.

7. Calculate the *yield of ATP* from the complete oxidation of pyruvate or of acetyl CoA.

The Pyruvate Dehydrogenase Complex, Citrate Synthase, and Stereospecific Enzymatic Reactions (Stryer pages 514–522, 526–527)

8. Describe *pyruvate dehydrogenase* as a *multienzyme complex.*

9. List the *cofactors* that participate in the pyruvate dehydrogenase complex reactions and discuss the roles they play in the overall reaction.

10. Compare the pyruvate dehydrogenase complex and the *α-ketoglutarate dehydrogenase* complex.

11. Describe the consequences and the biochemical basis of *thiamine deficiency.*

12. Outline the enzymatic mechanism of *citrate synthase.*

13. Explain the importance of the *induced-fit* structural rearrangements in citrate synthase during catalysis.

14. Describe the *asymmetric reaction* of citrate. Follow the fate of each carbon atom of citrate through the cycle.

15. Define *chiral* and *prochiral* molecules. Familiarize yourself with the *RS designation of chirality.*

16. Describe the stereospecific transfer of hydrogen by NAD^+ *dehydrogenases.*

The Citric Acid Cycle as a Source of Biosynthetic Precursors and the Glyoxylate Cycle (Stryer pages 522–524)

17. Indicate the citric acid cycle intermediates that may be used as *biosynthetic precursors.*

18. Describe the role of *anaplerotic reactions* and discuss the *pyruvate carboxylase* reaction.

19. Compare the reactions of the *glyoxylate cycle* and those of the citric acid cycle. List the reactions that are unique to the glyoxylate cycle.

20. Describe the control of *isocitrate dehydrogenase* in bacteria and plants.

Control of the Pyruvate Dehydrogenase Complex and the Citric Acid Cycle (Stryer pages 524–525)

21. Summarize the *regulation* of the pyruvate dehydrogenase complex through reversible *phosphorylation.* List the major *activators* and *inhibitors* of the kinase and phosphatase.

22. Indicate the *control points* of the citric acid cycle and note the activators and inhibitors.

SELF-TEST

Introduction

1. Which of the following statements are correct? The citric acid cycle
   (a) does not exist as such in plants and bacteria, because its functions are performed by the glyoxylate cycle.
   (b) oxidizes acetyl CoA derived from fatty acid degradation.
   (c) produces most of the CO_2 in anaerobic organisms.
   (d) provides succinyl CoA for the synthesis of carbohydrates.
   (e) provides precursors for the synthesis of glutamic and aspartic acids.

2. If a eukaryotic cell were broken open and the subcellular organelles were separated by zonal ultracentrifugation on a sucrose gradient, in which of the following would the citric acid cycle enzymes be found?
   (a) Nucleus
   (b) Lysosomes
   (c) Golgi complex
   (d) Mitochondria
   (e) Endoplasmic reticulum

Reactions of the Citric Acid Cycle

3. Given the biochemical intermediates of the pyruvate dehydrogenase reaction and the citric acid cycle (see Figure 20-1), answer the following questions:

CITRIC ACID CYCLE

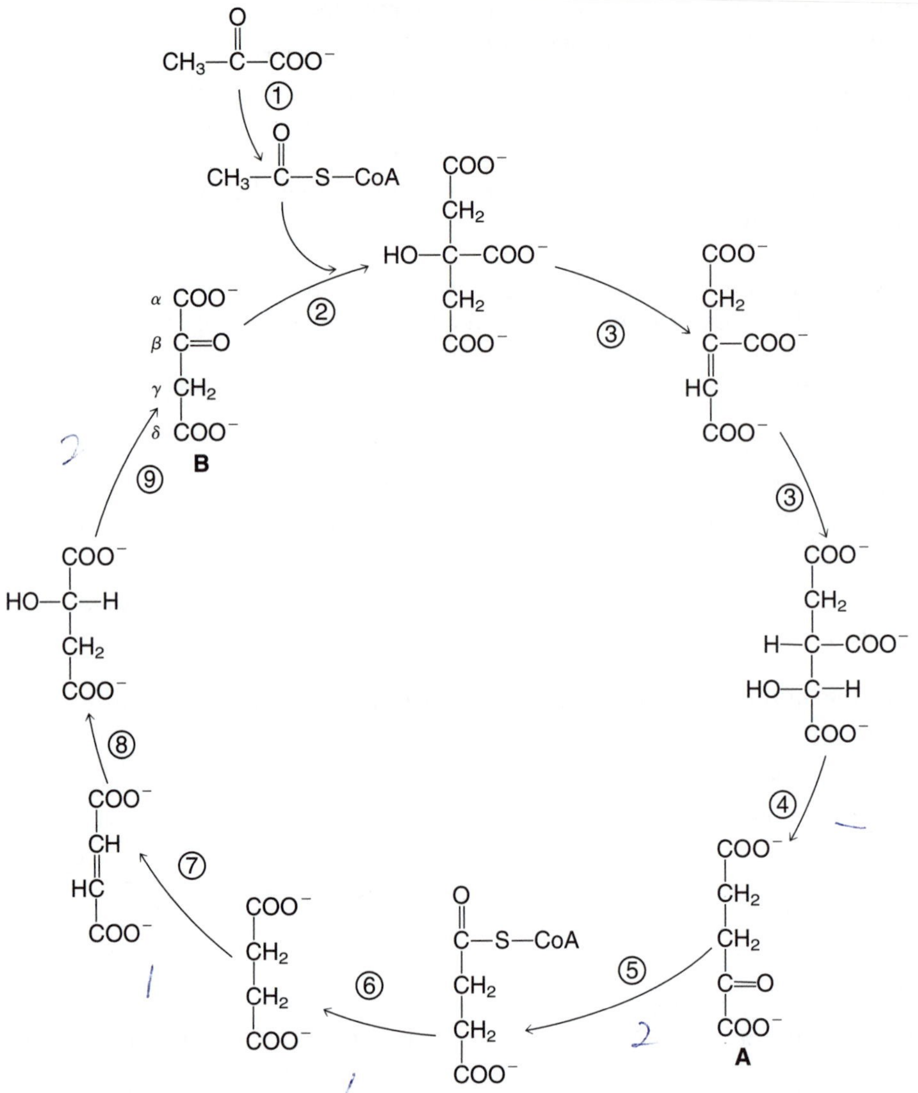

$CH_3-CO-COO^-$ (pyruvate)

① → $CH_3-CO-S-CoA$

② $CH_3-CO-S-CoA$ + B → $^-OOC-CH_2-C(OH)(COO^-)-CH_2-COO^-$

③ → $^-OOC-CH_2-C(COO^-){=}CH-COO^-$

③ → $^-OOC-CH_2-CH(COO^-)-CH(OH)-COO^-$

④ → A: $^-OOC-CH_2-CH_2-CO-COO^-$

⑤ → $CoA-S-CO-CH_2-CH_2-COO^-$

⑥ → $^-OOC-CH_2-CH_2-COO^-$

⑦ → $^-OOC-CH{=}CH-COO^-$

⑧ → $^-OOC-CH(OH)-CH_2-COO^-$

⑨ → B: α COO^-, β $C{=}O$, γ CH_2, δ COO^-

Figure 20-1
Citric acid cycle and the pyruvate dehydrogenase reaction.

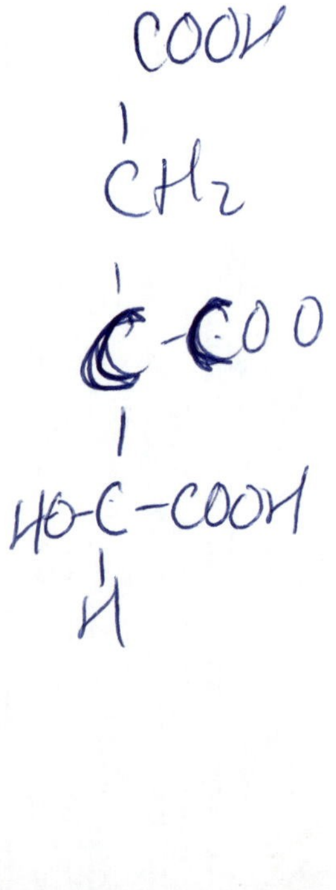

(a) Name the intermediates:

A ______________________

B ______________________

(b) Draw the structure of isocitrate and show those atoms that come from acetyl CoA in bold letters.

(c) Which reaction is catalyzed by α-ketoglutarate dehydrogenase?

____

(d) Which enzyme catalyzes step 2?

____

(e) Which reactions are oxidations? Name the enzyme catalyzing each of them.

____

(f) At which reaction does a substrate-level phosphorylation occur? Name the enzyme and the products of this reaction.

____

____

(g) Which of the reactions require an FAD cofactor? Name the enzymes.

____

____

(h) Indicate the decarboxylation reactions and name the enzymes.

____

____

4. Which of the following answers complete the sentence correctly? Succinate dehydrogenase

(a) is an iron-sulfur protein like aconitase. ✓
(b) contains FAD and NAD^+ cofactors like pyruvate dehydrogenase.
(c) is an integral membrane protein unlike the other enzymes of the citric acid cycle. ✓
(d) carries out an oxidative decarboxylation like isocitrate dehydrogenase.

5. The conversion of malate to oxaloacetate has a $\Delta G^{\circ\prime} = +7.1$ kcal/mol, yet in the citric acid cycle the reaction proceeds from malate to oxaloacetate. Explain how this is possible.

____

____

6. Considering the citric acid cycle steps between α-ketoglutarate and malate, how many high-energy phosphate bonds, or net ATP molecules, can be generated?

(a) 4
(b) 5
(c) 7
(d) 10
(e) 12

7. The standard free-energy change (in terms of net ATP production) when glucose is converted to 6 CO_2 and 6 H_2O is about how many times as great as the free-energy change when glucose is converted to 2 lactate molecules?

(a) 2
(b) 7
(c) 14
(d) 28

The Pyruvate Dehydrogenase Complex, Citrate Synthase, and Stereospecific Enzymatic Reactions

8. Match the cofactors of the pyruvate dehydrogenase complex in the left column with their corresponding enzyme components and with their roles in the enzymatic steps that are listed in the right column.

(a) Coenzyme A ________
(b) NAD^+ ________
(c) Thiamine pyrophosphate ________
(d) FAD ________
(e) Lipoamide ________

(1) Pyruvate dehydrogenase component
(2) Dihydrolipoyl dehydrogenase
(3) Dihydrolipoyl transacetylase
(4) Oxidizes the hydroxyethyl group
(5) Decarboxylates pyruvate
(6) Oxidizes dihydrolipoamide
(7) Accepts the acetyl group from acetyllipoamide
(8) Provides a long, flexible arm that conveys intermediates to different enzyme components
(9) Oxidizes $FADH_2$

9. Which of the following enzymes have impaired activity in vitamin B_1 deficiency?

(a) Succinate dehydrogenase
(b) Pyruvate dehydrogenase
(c) Isocitrate dehydrogenase
(d) α-Ketoglutarate dehydrogenase
(e) Dihydrolipoyl transacetylase
(f) Transketolase

10. What are the potential advantages of a multienzyme complex with respect to the isolated enzyme components? Explain.

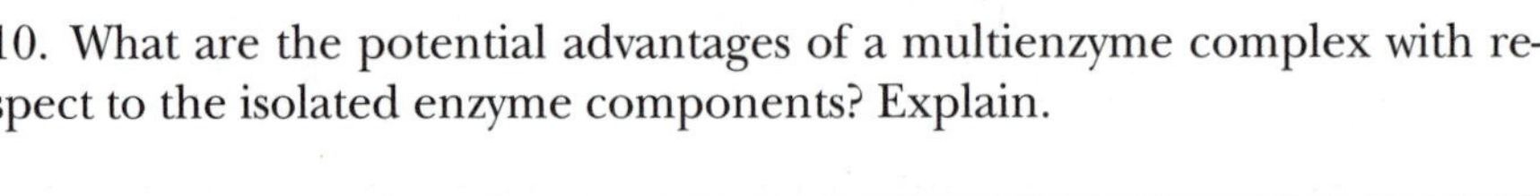

11. Which of the following statements concerning the enzymatic mechanism of citrate synthase is correct?

(a) Citrate synthase uses an NAD^+ cofactor.
(b) Acetyl CoA binds to citrate synthase before oxaloacetate.
(c) The histidine residues at the active site of citrate synthase participate in the hydrolysis of acetyl CoA.
(d) After citryl CoA is formed, additional structural changes occur in the enzyme.
(e) Each of the citrate synthase subunits binds one of the substrates and brings the substrates into close proximity to each other.

12. Citrate synthase binds acetyl CoA, condenses it with oxaloacetate to form citryl CoA, and then hydrolyzes the thioester bond of this intermediate. Why doesn't citrate synthase hydrolyze acetyl CoA?

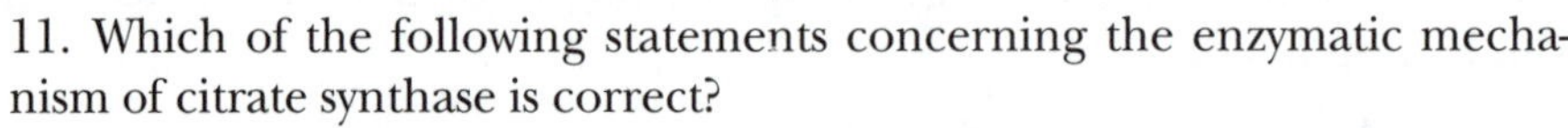

13. If the methyl carbon atom of pyruvate is labeled with ^{14}C, which of the carbon atoms of oxaloacetate would be labeled after one turn of the

citric acid cycle? (See the lettering scheme for oxaloacetate in Figure 20-1.)

(a) None. The label will be lost in CO_2.
(b) α
(c) β
(d) γ
(e) δ

14. Draw the structures of succinate, malate, and malonate and indicate which of the molecules are chiral or prochiral.

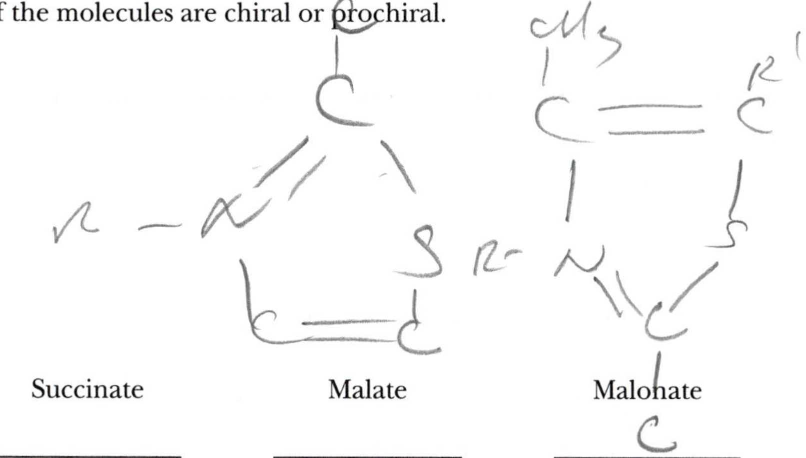

Succinate ______ Malate ______ Malonate ______

15. It has been shown that there are two kinds of NAD^+ (or $NADP^+$) dehydrogenases: A-stereospecific and B-stereospecific. Which of the following statements about these dehydrogenases is correct?

(a) The A-stereospecific dehydrogenases react with the R configuration, and the B-stereospecific dehydrogenases react with the S configuration of the nicotinamide ring.
(b) The A- and B-stereospecific dehydrogenases have distinct active-site pockets: one for the binding of NAD^+ and another for the binding of $NADP^+$.
(c) The A-stereospecific dehydrogenases transfer two hydrogen atoms, and the B-stereospecific dehydrogenases transfer a hydride ion.
(d) Both types of enzymes bind the same cofactors and carry out the same kinds of reactions, but they bind the cofactors with a different spatial orientation for the nicotinamide ring relative to the ribose ring.

The Citric Acid Cycle as a Source of Biosynthetic Precursors and the Glyoxylate Cycle

16. Match the intermediates of the citric acid cycle in the left column with their biosynthetic products in mammals, listed in the right column.

(a) Isocitrate ______
(b) α-Ketoglutarate ______
(c) Succinyl CoA ______
(d) *cis*-Aconitate ______
(e) Oxaloacetate ______

(1) Aspartic acid
(2) Glutamic acid
(3) Cholesterol
(4) Porphyrins
(5) None

17. Which of the following answers complete the sentence correctly? Anaplerotic reactions

(a) are necessary because the biosynthesis of certain amino acids requires citric acid cycle intermediates as precursors.
(b) can convert acetyl CoA to oxaloacetate in mammals.
(c) can convert pyruvate into oxaloacetate in mammals.
(d) are not required in mammals, because mammals have an active glyoxylate cycle.
(e) include the pyruvate dehydrogenase reaction operating in reverse.

18. Which of the following answers complete the sentence correctly? Pyruvate carboxylase

(a) catalyzes the reversible decarboxylation of oxaloacetate.
(b) requires thiamine pyrophosphate as a cofactor.
(c) is allosterically activated by NADH.
(d) requires ATP.
(e) is found in the cytoplasm of eukaryotic cells.

19. Malate synthase, an enzyme of the glyoxylate cycle, catalyzes the condensation of glyoxylate with acetyl CoA. Which enzyme of the citric acid cycle carries out a similar reaction? Would you expect the binding of glyoxylate and acetyl CoA to malate synthase to be sequential? Why?

____

____

20. All organisms require three- and four-carbon precursor molecules for biosynthesis, yet bacteria can grow on acetate whereas mammals cannot. Explain why this is so.

____

____

21. Starting with acetyl CoA, what is the approximate yield of high-energy phosphate bonds (net ATP formed) via the glyoxylate cycle?

(a) 3
(b) 6
(c) 9
(d) 12
(e) 15

22. If a ^{14}C label were introduced in the carbonyl carbon of oxaloacetate, where would the label appear after one round of the glyoxylate cycle?

(a) In the methylene carbons of succinate
(b) In the carboxylate carbons of succinate
(c) In the carbonyl carbon of glyoxylate
(d) In the carboxylate carbon of glyoxylate

Control of the Pyruvate Dehydrogenase Complex and the Citric Acid Cycle

23. Although O_2 does not participate directly in the reactions of the citric acid cycle, the cycle operates only under aerobic conditions. Explain this fact.

____

____

24. Which of the following answers complete the sentence correctly? The pyruvate dehydrogenase complex is activated by

(a) phosphorylation of the pyruvate dehydrogenase component (E_1).
(b) stimulation of a specific phosphatase by Ca^{2+}.
(c) inhibition of a specific kinase by pyruvate.
(d) decrease of the $NADH/NAD^+$ ratio.
(e) decreased levels of insulin.

25. First select the enzymes in the left column that regulate the citric acid cycle. Then match those enzymes with the appropriate control mechanisms in the right column.

(a) Citrate synthase ____________
(b) Aconitase ____________
(c) Isocitrate dehydrogenase ____________
(d) α-Ketoglutarate dehydrogenase ____________
(e) Succinyl CoA synthetase ____________
(f) Succinate dehydrogenase ____________
(g) Fumarase ____________
(h) Malate dehydrogenase ____________

(1) Feedback-inhibited by succinyl CoA
(2) Allosterically activated by ADP
(3) Inhibited by NADH
(4) Regulated by the availability of acetyl CoA and oxaloacetate
(5) Inhibited by ATP

26. Although the ATP/ADP ratio and the availability of substrates and cycle intermediates are very important factors affecting the rate of the citric acid cycle, the $NADH/NAD^+$ ratio is of paramount importance. Explain why.

____

____

ANSWERS TO SELF-TEST

Introduction

1. b, e

2. d

Reactions of the Citric Acid Cycle

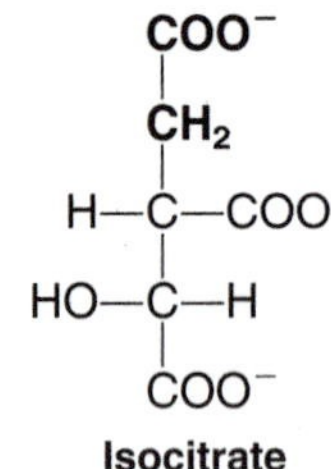

Isocitrate

3. (a) A: α-ketoglutarate; B: oxaloacetate

(b) See the structure of isocitrate in the margin.

(c) Reaction 5

(d) Citrate synthase

(e) Step 1, pyruvate dehydrogenase; step 4, isocitrate dehydrogenase; step 5, α-ketoglutarate dehydrogenase; step 7, succinate dehydrogenase; step 9, malate dehydrogenase

(f) Step 6; the enzyme is succinyl CoA synthetase; the products of the reaction are succinate, CoA, and GTP.

(g) Step 1, dihydrolipoyl dehydrogenase component of the pyruvate dehydrogenase complex; step 5, dihydrolipoyl dehydrogenase component of the α-ketoglutarate dehydrogenase complex; step 7, succinate dehydrogenase.

(h) Step 1, pyruvate dehydrogenase; step 4, isocitrate dehydrogenase; step 5, α-ketoglutarate dehydrogenase.

4. a, c

5. Although this step is energetically unfavorable at standard conditions, in mitochondria the concentrations of malate and NAD^+ are relatively high and those of the products, oxaloacetate and NADH, are quite low, so the overall $\Delta G'$ for this reaction is negative.

6. b

7. c. From glucose to lactate 2 ATP are formed; from glucose to CO_2 and H_2O about 27 ATP are formed.

The Pyruvate Dehydrogenase Complex, Citrate Synthase, and Stereospecific Enzymatic Reactions

8. (a) 3, 7 (b) 2, 9 (c) 1, 5 (d) 2, 6 (e) 3, 4, 8

9. b, d, f

10. A multienzyme complex can carry out the coordinated catalysis of a complex reaction. The intermediates in the reaction remain bound to the complex and are passed from one enzyme component to the next, which increases the overall reaction rate and minimizes side reactions. In the case of isolated enzymes, the reaction intermediates would have to diffuse randomly between enzymes.

11. d

12. Citrate synthase binds acetyl CoA only after oxaloacetate has been bound and the enzyme structure has rearranged to create a binding site for acetyl CoA. After citryl CoA is formed, there are further structural changes that bring an aspartate residue and a water molecule into the vicinity of the thioester bond for the hydrolysis step. Thus, acetyl CoA is protected from hydrolysis.

13. c and d. Both of the middle carbons of oxaloacetate will be labeled because succinate is a symmetrical molecule.

14. See the structures in the margin. Succinate is a prochiral molecule; the prochiral centers are the methylene carbons. Malate is a chiral and prochiral molecule; the carbon atom bearing the OH group is a chiral center, and the methylene carbon is a prochiral center. Malonate is neither chiral nor prochiral.

COO^-–CH_2–CH_2–COO^-
Succinate

COO^-–HO—C—H–CH_2–COO^-
Malate

COO^-–CH_2–COO^-
Malonate

15. d

The Citric Acid Cycle as a Source of Biosynthetic Precursors and the Glyoxylate Cycle

16. (a) 5 (b) 2 (c) 4 (d) 5 (e) 1

17. a, c

18. a, d

19. The condensation of glyoxylate and acetyl CoA carried out by malate synthase in the glyoxylate cycle is similar to the condensation of oxaloacetate and acetyl CoA carried out by citrate synthase in the citric acid cycle. The initial binding of glyoxylate, which induces structural changes in the enzyme that allow the subsequent binding of acetyl CoA, would be expected in order to prevent the premature hydrolysis of acetyl CoA. See question 12.

20. Bacteria are capable, via the glyoxylate cycle, of synthesizing four-carbon precursor molecules for biosynthesis (e.g., malate) from acetate or acetyl CoA. Mammals do not have an analogous mechanism; in the citric acid cycle, the carbon atoms from acetyl CoA are released as CO_2, and there is no net synthesis of four-carbon molecules.

21. a. One NADH is formed that can yield approximately 2.5 molecules of ATP.

22. a

Control of the Pyruvate Dehydrogenase Complex and the Citric Acid Cycle

23. The citric acid cycle requires the oxidized cofactors NAD^+ and FAD for its oxidation-reduction reactions. The oxidized cofactors are regenerated by transfer of electrons through the electron transport chain to O_2 to give H_2O (see Chapter 21).

24. b, c, d

25. a, c, d (a) 4, 5 (c) 2, 3, 5 (d) 1, 3, 5

The regulation of citrate synthase is explained differently in different texts. Although ATP is given as an allosteric inhibitor of citrate synthase by Stryer, other texts do not take this view. There is general agreement, however, that citrate synthase is quite sensitive to the levels of available oxaloacetate and acetyl CoA.

26. The oxidized cofactors NAD^+ and FAD are absolutely required as electron acceptors in the various dehydrogenation reactions of the citric acid cycle. When these oxidized cofactors are not available, as when their reoxidation stops in the absence of O_2 or respiration, the citric acid cycle also stops.

PROBLEMS

1. A cell is deficient in pyruvate dehydrogenase phosphate phosphatase. How would such a deficiency affect cellular metabolism?

2. ATP is an important source of energy for muscle contraction. Pyruvate dehydrogenase phosphate phosphatase is activated by calcium ion, which increases greatly in concentration during exercise. Why is activation of the phosphatase consistent with the metabolic requirements of muscle during contraction?

3. The oxidation of a fatty acid with an even number of carbon atoms yields a number of molecules of acetyl CoA, whereas the oxidation of an odd-numbered fatty acid yields molecules of not only acetyl CoA but also propionyl CoA, which then gives rise to succinyl CoA. Why does only the oxidation of odd-numbered fatty acids lead to the *net* synthesis of oxaloacetate?

4. In addition to its role in the action of pyruvate dehydrogenase, thiamine pyrophosphate (TPP) serves as a cofactor for other enzymes, such as pyruvate decarboxylase, which catalyzes the *nonoxidative* decarboxylation of pyruvate. Propose a mechanism for the reaction catalyzed by pyruvate decarboxylase. What product would you expect? Why, in contrast to pyruvate dehydrogenase, are lipoamide and FAD not needed as cofactors for pyruvate decarboxylase?

5. The flavins FAD and FMN (flavin mononucleotide) are both bright-yellow compounds; in fact, the name *flavin* was taken from *flavus,* the Latin word for "yellow." The corresponding reduced compounds, $FADH_2$ and $FMNH_2$, are nearly colorless. What portion of a flavin molecule accounts for these color changes? The enzyme glucose oxidase, found in fungi, catalyzes the conversion of free glucose to gluconic acid. Glucose oxidase utilizes two molecules of FAD as cofactors. How could you use the light-absorbing properties of flavins as a means of monitoring glucose oxidase activity?

6. Some microorganisms can grow using ethanol as their sole carbon source. Propose a pathway for the utilization of this two-carbon compound;

the pathway should convert ethanol into one or more molecules that can be used for energy generation and as biosynthetic precursors.

7. In the early 1900s, Thunberg proposed a cyclic pathway for the oxidation of acetate. In his scheme, two molecules of acetate are condensed, with reduction, to form succinate, which in turn is oxidized to yield oxaloacetate. The decarboxylation of oxaloacetate to pyruvate followed by the oxidative decarboxylation of pyruvate to acetate complete the cycle. Assuming that electron carriers like NAD^+ and FAD would be part of the scheme, compare the energy liberated by the Thunberg scheme with that liberated by the now-established citric acid cycle. Which of the steps in Thunberg's scheme was not found in subsequent studies?

8. When an isolated rat heart is perfused with sodium fluoroacetate, the rate of glycolysis decreases and hexose monophosphates accumulate. In cardiac cells, fluoroacetate is condensed with oxaloacetate to give fluorocitrate. Under these conditions, cellular citrate concentrations increase, while the levels of other citric acid cycle components decrease. What enzyme is inhibited by fluorocitrate? How can you account for the decrease in glycolysis and the buildup of hexose monophosphates?

9. Transient phosphorylation of succinyl CoA synthetase in vitro can be achieved by incubating the enzyme with labeled GTP or with labeled inorganic phosphate and succinyl CoA. One can also observe the reversible formation of succinyl phosphate during the reaction. Propose a pathway for the steps catalyzed by the enzyme. How is the reaction catalyzed by succinyl CoA synthetase similar to that carried out by glyceraldehyde 3-phosphate dehydrogenase?

10. Recent studies suggest that succinate dehydrogenase activity is affected by oxaloacetate. Would you expect the enzyme activity to be enhanced or inhibited by oxaloacetate?

11. In addition to the carboxylation of pyruvate, there are other anaplerotic reactions that help to maintain appropriate levels of oxaloacetate. For example, the respective amino groups of glutamate and aspartate can be removed to yield the corresponding α-keto acids. How can these α-keto acids be used to replenish oxaloacetate levels?

12. Lipoic acid and FAD serve as prosthetic groups in the enzyme isocitrate dehydrogenase. Describe their possible roles in the reaction catalyzed by the enzyme.

13. Malonate anion is a potent competitive inhibitor of succinate dehydrogenase, which catalyzes the conversion of succinate to malate.

(a) Why is malonate unreactive?
(b) In work that led to the elucidation of the citric acid cycle, Hans Krebs employed malonate as an inhibitor of succinate dehydrogenase. Earlier studies by Martius and Knoop had shown that in animal tissues there is a pathway from citrate to succinate. Krebs had also noticed that citrate catalytically enhances respiration in minced muscle tissues. Knowing that malonate reduces the rate of respiration in animal cells, he then added citrate to malonate-poisoned muscle. In another experiment, Krebs added fumarate to malonate-poisoned muscle. What changes in succinate concentra-

$$\begin{array}{c} COO^- \\ | \\ CH_2 \\ | \\ CH_2 \\ | \\ COO^- \end{array}$$
Succinate

$$\begin{array}{c} COO^- \\ | \\ CH_2 \\ | \\ COO^- \end{array}$$
Malonate

tion did Krebs observe in each of the experiments with malonate-treated muscle, and what was the significance of each finding?

(c) Krebs carried out a final set of crucial studies by showing that citrate can be formed in muscle suspensions if oxaloacetate is added. What is the significance of this experiment, and how did it provide a coherent scheme for terminal oxidation of carbon atoms?

14. The conversion of citrate to isocitrate in the citric acid cycle actually occurs by a dehydration-rehydration reaction with aconitate as an isolatable intermediate. A single enzyme, aconitase, catalyzes the conversion of citrate to aconitate and aconitate to isocitrate. An equilibrium mixture of citrate, aconitate and isocitrate contains about 90, 4, and 6 percent of the three acids, respectively.

(a) Why must citrate be converted to isocitrate before oxidation takes place in the citric acid cycle?

(b) What are the respective equilibrium constants and standard free-energy changes for each of the two steps (citrate $\rightarrow$ aconitate; aconitate $\rightarrow$ isocitrate)? for the overall process at 25°C?

(c) Could the citric acid cycle proceed under standard conditions? Why or why not?

(d) Given the thermodynamic data you have gathered about the reactions catalyzed by aconitase, how can the citric acid cycle proceed under cellular conditions?

15. Bacteria called *methanophiles* (or *methanotrophs*) can use methane as a fuel. Propose an energy-conserving reaction sequence for converting methane into CO_2. What is the likely yield of ATP of this pathway?

ANSWERS TO PROBLEMS

1. Pyruvate dehydrogenase phosphate phosphatase removes a phosphoryl group from pyruvate dehydrogenase, activating the enzyme complex and accelerating the rate of synthesis of acetyl CoA. Cells deficient in phosphatase activity cannot activate pyruvate dehydrogenase, so that the rate of entry of acetyl groups into the citric acid cycle will decrease, as will aerobic production of ATP. Under such conditions, stimulation of glycolytic activity and a subsequent increase in lactate production would be expected, as the cell responds to a continued requirement for ATP synthesis.

2. As discussed in problem 1, the phosphatase activates pyruvate dehydrogenase, stimulating the rate of both glycolysis and the citric acid cycle. Calcium-mediated activation of pyruvate dehydrogenase therefore promotes increased production of ATP, which is then available for muscle contraction.

3. The entry of acetyl groups from acetyl CoA into the citric acid cycle does not contribute to the net synthesis of oxaloacetate, because two carbons are lost as CO_2 in the pathway from citrate to oxaloacetate. Only the entry of compounds with three or more carbons, like succinate, can in-

crease the relative number of carbon atoms in the pathway. Thus while odd-numbered fatty acids contribute to the net synthesis of oxaloacetate, those compounds with an even number of fatty acids do not.

4. The mechanism is similar to that shown on page 515 of Stryer, in which the C-2 carbanion of TPP attacks the α-keto group of pyruvate. The subsequent decarboxylation of pyruvate is enhanced by the delocalization of electrons in the ring nitrogen of TPP. The initial product is hydroxyethyl-TPP, which is cleaved upon protonation to yield acetaldehyde and TPP. In contrast to the reaction catalyzed by pyruvate dehydrogenase, no net oxidation occurs, so lipoamide and FAD, which serve as electron acceptors, are not needed.

5. The conjugated π system in the isoalloxazine ring of oxidized flavins like FAD and FMN accounts for their intense yellow color. The reduced molecules are partially saturated, and their remaining double bonds are not conjugated, making them nearly colorless. To monitor glucose oxidase activity, one could use spectrophotometry to determine the wavelength of maximum absorption for FAD and then monitor the oxidation of the flavin in the enzyme by observing changes in absorption. As glucose is oxidized and FAD is reduced to $FADH_2$, one would observe a corresponding decrease in absorption.

6. The microorganism first converts ethanol to acetic acid, or acetate, by carrying out two successive oxidations, with acetaldehyde as an intermediate. Two molecules of a reduced electron carrier such as NADH will also be produced. Next, acetate is activated through the action of acetyl CoA synthetase to form acetyl CoA, and then the acetyl group is transferred to oxaloacetate to form citrate. After citrate is converted to isocitrate, two enzymes of the glycolytic pathway, isocitrate lyase and malate synthase, assist in the net formation of oxaloacetate from isocitrate and another molecule of acetyl CoA, as discussed in Stryer's text on page 523. Oxaloacetate can then be used for generation of energy as well as production of biosynthetic intermediates. Note that a small amount of oxaloacetate and other intermediates of the citric acid cycle must be present initially in order for acetate to enter the pathway.

7. As shown in Figure 20-2, there are at least four steps that generate reduced electron carriers. For each acetate group consumed, 3 NADH and 1 $FADH_2$ are generated, and their subsequent reoxidation in the electron transport chain provides energy for the generation of 9 molecules of ATP. The same number of reduced electron carriers is generated through the action of pyruvate dehydrogenase and the enzymes of the citric acid cycle, so that the energy liberated by both schemes is the same. Each of the reactions shown in Thunberg's scheme is known to occur, except for the condensation of two acetyl groups to form succinate.

8. The accumulation of citrate and the decrease in the levels of other citric acid cycle intermediates suggest that aconitase is inhibited by fluorocitrate. Excess citrate inhibits phophofructokinase, causing a decrease in the rate of glycolysis and an accumulation of hexose monophosphates such as glucose 6-phosphate and fructose 1,6-bisphosphate.

9. The enzyme carries out three reversible steps leading to the formation of succinate and GTP. The first is the hydrolysis of succinyl CoA; the free

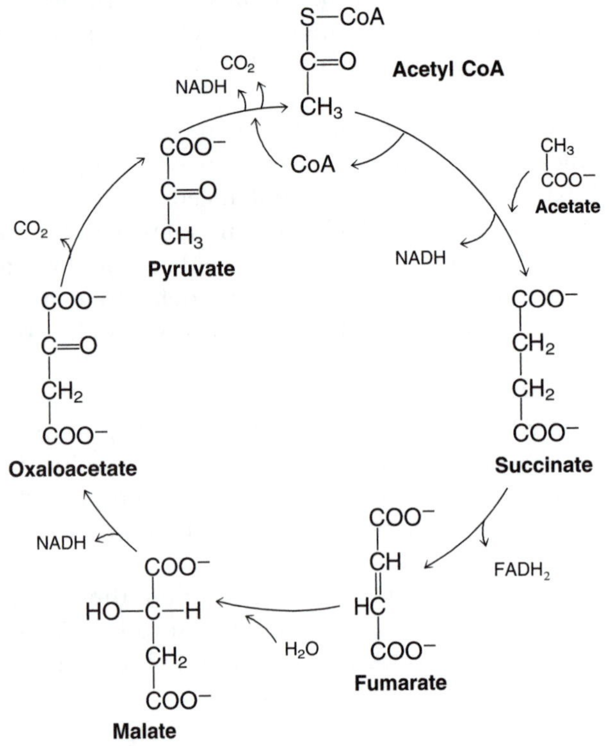

CITRIC ACID CYCLE

Figure 20-2
Thunberg's cycle.

energy of hydrolysis is used in the second step to form succinyl phosphate, which is noncovalently bound to the enzyme. Finally, the phosphate group is then transferred to an amino acid side chain in the enzyme. (Studies have shown that a specific histidine residue is phosphorylated.) The phosphate group is then transferred to GDP, yielding GTP. The overall reaction is another example of a substrate-level phosphorylation, in which the energy for the transfer of the phosphate is derived from a chemical reaction, rather than electron transfer. A similar reaction is catalyzed by glyceraldehyde 3-phosphate dehydrogenase.

10. Oxaloacetate is derived from succinate; it is a product of the reaction catalyzed by succinate dehydrogenase. When levels of oxaloacetate are high, one would expect the activity of the enzyme to be reduced. Low levels of oxaloacetate would call for an increase in succinate production.

11. Examination of the structures of the α-keto acid analogs of glutamate and aspartate shows that they are in fact both citric acid cycle intermediates, α-ketoglutarate and oxaloacetate. Glutamate, when it is deaminated, thus contributes directly to the insertion of additional molecules of oxaloacetate. α-Ketoglutarate, as a component of the citric acid cycle, is a precursor of oxaloacetate.

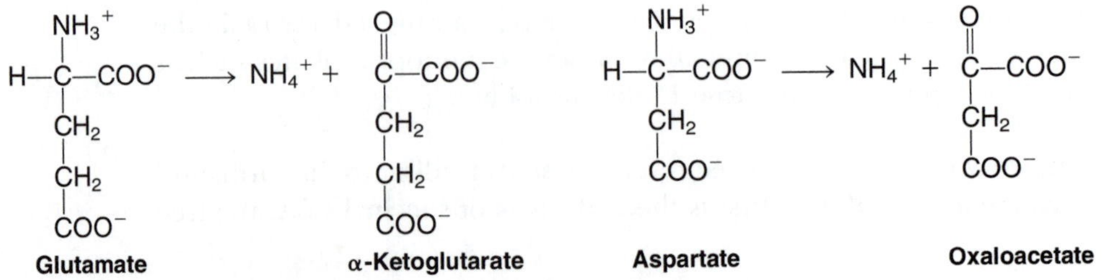

Glutamate → NH_4^+ + α-Ketoglutarate

Aspartate → NH_4^+ + Oxaloacetate

12. Lipoic acid contains a sulfhydryl group that could act as an acceptor for electrons from isocitrate. Those electrons could then be transferred to NAD^+ via FAD. The roles of the prosthetic groups would be similar to those they play in the reaction catalyzed by pyruvate dehydrogenase.

13. (a) Succinate, which has two methylene groups, loses two hydrogens during its oxidation by succinate dehydrogenase. Malonate, which has only one methylene group, cannot be dehydrogenated and is therefore unreactive.

(b) In both experiments, Krebs observed an increase in the concentration of succinate. We know now that both citrate and fumarate can be viewed as precursors of succinate or other components of the citric acid cycle. A malonate-induced block in the conversion of succinate to fumarate would cause an increase in succinate concentration. The first experiment, in which citrate addition caused an increase in succinate concentration, showed that the pathway from citrate to succinate is physiologically significant and is related to the process of respiration using carbohydrates as a fuel. The second experiment with fumarate suggested that a pathway from fumarate to succinate exists which is separate from the reaction catalyzed by succinate dehydrogenase. Krebs realized that a cyclic pathway could account for all these observations. The piece of the puzzle that was left was to learn how citrate might be generated from pyruvate or acetate. The results from those experiments are described in (c).

(c) The generation of citrate from oxaloacetate enabled Krebs to devise a scheme that incorporated two-carbon molecules from acetate or pyruvate into citrate, with oxaloacetate serving as an acceptor of those carbon atoms. He then was able to use the results of his experiments and those of others to show how a cyclic pathway could function to carry out oxidation of carbon molecules while regenerating oxaloacetate. Krebs was prepared for the development of the cyclic scheme because he had shown earlier that the ornithine cycle, which is used for urea synthesis, is also a cyclic metabolic pathway.

14. (a) The oxidation of isocitrate involves oxidation of a secondary alcohol. Citrate has an alcohol function, but it is a tertiary alcohol, which is much more difficult to oxidize. Isomerization of citrate to isocitrate provides an easier route to oxidative decarboxylation.

(b) For the citrate-aconitate pair, the equilibrium constant is equal to the ratio of product and substrate concentration.

$$K_{eq} = \frac{[\text{aconitate}]}{[\text{citrate}]} = \frac{4}{90} = 4.44 \times 10^{-2}$$

The value of $\Delta G^{\circ\prime}$ is $-1.36 \log_{10} (4.44 \times 10^{-2}) =$

$$-1.36(0.647 - 2) = +1.84 \text{ kcal/mol.}$$

Similar calculations for the aconitate-isocitrate pair give

$$K_{eq} = 6/4 = 1.5 \quad \text{and} \quad \Delta G^{\circ\prime} = -0.24 \text{ kcal/mol}$$

The overall standard-free energy value for the conversion of citrate to isocitrate is the sum of the two values for the individual reactions.

$$\Delta G^{\circ\prime} = 1.84 + (-0.24) = +1.6 \text{ kcal/mol}$$

(c) Under standard conditions, the citric acid cycle could not proceed, because the positive free-energy value for the reaction indicates that it would proceed toward net formation of citrate.
(d) The net conversion of citrate to isocitrate can occur in the mitochondrion if the isocitrate produced is then converted to α-ketoglutarate. This would lower the concentration of isocitrate, pulling the reaction toward net formation of that molecule. Concentrations of citrate could also be increased, driving the reaction once again toward the formation of isocitrate. Although accurate concentrations of metabolites in mitochondria are difficult to establish, it appears that both mechanisms may operate to ensure net isocitrate synthesis.

15. Methane is first oxidized by a monooxygenase to methanol; NADH, the reductant, is converted to NAD^+ while methanol and water are generated. Methanol is then oxidized to formaldehyde; PQQ, a novel quinone, is the electron acceptor in this step. Formaldehyde is oxidized to formic acid, which in turn is oxidized to CO_2. NADH is formed in each of these two steps. About 4 ATP are formed (2.5 ATP from NADH and 1.5 from PQQH2). See G. Gottschalk, *Bacterial Metabolism,* 2nd ed. (Springer-Verlag, 1986), pp. 155 and 163.

EXPANDED SOLUTIONS TO TEXT PROBLEMS

1. To answer this problem one must follow carbon atoms around the citric acid cycle as shown by Stryer on pages 513 and 520. Remember that the randomization of carbon occurs at succinate, a truly symmetrical molecule. Also, this problem (and the answers given) assumes that all pyruvate goes to acetyl CoA. In fact, this is not necessarily true, since pyruvate can also enter the cycle at oxaloacetate (Stryer, Chapter 22).

(a) After one round of the citric acid cycle, the label emerges in C-2 and C-3 of oxaloacetate.
(b) After one round of the citric acid cycle, the label emerges in C-1 and C-4 of oxaloacetate.
(c) The label emerges in CO_2 in the formation of acetyl CoA from pyruvate.
(d) The fate is the same as in (a).
(e) C-1 of G-6-P becomes the methyl carbon of pyruvate [see Chapter 19, problem 2(a)] and hence has the same fate as in (a).

2. (a) Isocitrate lyase and malate synthase are required in addition to the enzymes of the citric acid cycle.

(b) $2 \text{ Acetyl CoA} + 2\, NAD^+ + FAD + 3\, H^2O \rightarrow$

$\text{oxaloacetate} + 2 \text{ CoA} + 2\, NADH + FADH_2 + 3H^+$

(c) No, because they lack these two enzymes and hence cannot carry out the glyoxylate cycle.

3. Addition of the $\Delta G^{\circ\prime}$ values in Table 20-1 gives the answer −9.8 kcal/mol.

4. The coenzyme stereospecificity of glyceraldehyde 3-phosphate dehydrogenase is the opposite of that of alcohol dehydrogenase (type B versus type A, respectively). The type A and B dehydrogenases are discussed in detail by Stryer on pp. 521–522.

5. Thiamine thiazolone pyrophosphate is a transition state analog. The sulfur-containing ring of this analog is uncharged, and so it closely resembles the transition state of the normal coenzyme in thiamine-catalyzed reactions (e.g., the uncharged resonance form of hydroxyethyl-TPP, Stryer, p. 515). See J.A. Gutowski and G.E. Lienhard, *J. Biol. Chem.* 251 (1976): 2863, for a discussion of this analog.

6. (a) The steady-state concentrations of the products (OAA +NADH) are much lower than those of the substrates (Mal $+NAD^+$). Hence, the reaction is pushed "uphill" by the overwhelming mass action caused by the differences in concentration.

(b)

$$\frac{[\text{OAA}][\text{NADH}]}{[\text{Mal}][\text{NAD}^+]} = 10^{7/-1.36} = 7.08 \times 10^{-6}$$

Since $[\text{NADH}]/[\text{NAD}^+] = 1/8$,

$$\frac{[\text{OAA}]}{[\text{Mal}]} = 7.08 \times 10^{-6} \times 8 = 5.67 \times 10^{-5}$$

The reciprocal of this is 1.75×10^4, the smallest [Mal]/[OAA] ratio permitting net OAA formation.

7. The enolate anion of acetyl CoA attacks the carbonyl carbon atom of glyoxylate to form a C–C bond. This reaction is like the condensation of oxaloacetate with the enolate anion of acetyl CoA (Stryer, p. 519). Glyoxylate contains a hydrogen atom in place of the $-CH_2COO^-$ of oxaloacetate; the reactions are otherwise nearly identical.

8. The phosphoryl group on Ser 113 is transferred to bound ADP and then to water, rather than directly to water. See D.C. Laporte, *J. Cell Biochem.* 51(1993):14. Note the similarity (though in opposite directions) between the kinase and phosphatase of this enzyme, whose active site catalyzes both reactions. During kinase activity, ATP $\rightarrow$ ADP + a phosphoryl group whereas during phosphatase activity phosphoryl + ADP $\rightarrow$ ADP-P.

9. The kinase that synthesizes fructose 2,6-bisphosphate (F-2,6-BP) and the phosphatase that hydrolyzes F-2,6-BP to fructose 6-phosphate are present on a single polypeptide chain (p. 494). However, in contrast with the bifunctional enzyme that modifies bacterial isocitrate dehydrogenase, the kinase and phosphatase active sites are on different domains.

CHAPTER

21

Oxidative Phosphorylation

The reduced coenzymes NADH and $FADH_2$ that are formed during glycolysis (Chapter 19) and as a result of the functioning of the citric acid cycle (Chapter 20) ultimately transfer their electrons via a series of carriers to oxygen and release a large amount of useful energy in the process. The energy from this electron transfer is coupled to the formation of ATP via a transmembrane proton-motive force composed of a pH gradient and an electric potential. This process, known as *oxidative phosphorylation,* produces the bulk of the ATP in aerobic organisms. Stryer begins with an introduction to the concept of reduction potential as another measure of the free energy of chemical reactions (see Chapters 8 and 17) involving electron transfers. He then lists the three proton pumps and the electron carriers they bear, as well as two interpump, mobile electron carriers. The location of these electron transfer chain constituents in the mitochondria of eukaryotes, and the sequence of electron transfer reactions that occur between the reduced cofactors and the final electron acceptor, O_2, are also outlined. Next, Stryer discusses the generation of a proton gradient as a consequence of the electron flow and the use of the resulting proton-motive force by ATP synthase to form ATP from ADP and P_i. Because ATP synthesis occurs inside mitochondria whereas most of the reactions that utilize ATP take place in the cytosol, the membrane shuttle systems for ATP-ADP and other cofactors and biomolecules are described. Stryer then discusses the reasons for the variation in the estimation of the number of ATP molecules formed per glucose molecule oxidized. The control of oxidative phosphorylation, the mechanisms that uncouple electron transfer and oxidative phosphorylation, and the enzyme systems that degrade toxic, partially reduced products of oxygen are discussed. Stryer ends the chapter by stressing the central role of proton gradients in interconverting free energy in cells.

LEARNING OBJECTIVES

When you have mastered this chapter, you should be able to complete the following objectives:

Introduction (Stryer pages 529–531)

1. Define *oxidative phosphorylation* and *respiration.*

2. Realize that *coupling* of *oxidation* to *phosphorylation* by a *proton gradient* forms *ATP.*

3. Describe the *compartments* and *membranes of mitochondria* and locate the *respiratory assemblies* and the *N* and *P sides* of the inner membrane.

Redox Potentials and Free-Energy Changes (Stryer pages 531–533)

4. Relate the concepts of *redox potential* and *free-energy change.*

5. Describe the meaning and the measurement of the redox potential (E_0) for a *redox couple* relative to the *standard reference half-cell.*

6. Explain the meaning of E'_0.

7. Calculate $\Delta G^{\circ\prime}$ for *oxidation-reduction reactions* from the redox potentials for the individual redox couples.

The Respiratory Chain (Stryer pages 534–544)

8. List the components of the *respiratory chain* and the *electron-carrying molecules.*

9. Describe the entry of electrons from *NADH* into *NADH-Q reductase* (*Complex I*) and their path through this *proton pump.* State the roles of *flavin mononucleotide* (*FMN*), *iron-sulfur clusters,* and *coenzyme Q.*

10. Relate disorders in energy generation in mitochondria to human disease.

11. Discuss the role of *ubiquinone* (coenzyme Q, or Q) as a mobile electron carrier between NADH-Q reductase and *cytochrome reductase* (*Complex III*).

12. Describe the entry of electrons into the respiratory chain at the *succinate-Q reductase complex* (*Complex II*) from *flavoproteins* such as *succinate dehydrogenase* (a component of Complex II), *glycerol phosphate dehydrogenase,* and *fatty acyl CoA dehydrogenase* via $FADH_2$. Appreciate that Complex II is not a proton pump.

13. Describe the structures, properties, and functions of the *cytochromes* and contrast the features of cytochromes b, c_1, c, a, and a_3.

14. List the components of the *cytochrome reductase complex,* and explain how *ubiquinol* transfers its electrons to cytochromes c_1 and b (b_K–566 and b_H–560) and ultimately to cytochrome c. Appreciate the roles of *heme* in these processes.

15. Explain how a two-electron carrier, ubiquinol, can interact with a one-electron carrier, the *Fe-S cluster.*

16. Describe the *cytochrome oxidase proton pump* (*Complex IV*) and its electron-carrying groups.

17. Outline the mechanism for the reduction of O_2 to H_2O on cytochrome oxidase. Describe the path of the electrons from the *heme* a–Cu_A cluster to the *heme* a_3–Cu_B cluster and note the changes in the oxidation states of Fe and O. Note the formation of a *superoxide anion* intermediate.

18. Describe the salient features of the three-dimensional structure of cytochrome *c* and relate them to its interaction with cytochrome reductase and cytochrome oxidase. Appreciate that electrons can be transferred between groups not in contact with one another.

19. Discuss the functional implications of the conservation of the structure of cytochrome *c* in evolution.

20. List some of the compounds that specifically block electron transport and locate their sites of action.

21. Provide an estimate of the number of protons pumped per pair of electrons transferred by Complexes I, III, and IV.

Coupling of Oxidation and Phosphorylation to Form ATP by a Proton-Motive Force (Stryer pages 544–548)

22. Describe the *chemiosmotic model* of oxidative phosphorylation and give the evidence that supports it.

23. Describe the mitochondrial location, the subunit structure, and the function of eukaryotic *ATP synthase.*

24. Outline the proposed mechanism of ATP synthesis by ATP synthase during proton flow. Relate *catalytic cooperativity* to the *binding-change mechanism.*

Mitochondrial Transport Systems (Stryer pages 548–551)

25. Explain the roles of the *glycerol phosphate* and *malate-aspartate shuttles* in carrying the electrons of cytoplasmic NADH into the mitochondrion. Estimate the ATP yields in each case.

26. Describe the *ATP-ADP translocase* mechanism.

27. List the most important *mitochondrial transport systems* for ions and metabolites and appreciate the common feature of their structures.

ATP Yield and Control of Oxidative Phosphorylation, Uncouplers, Scavengers of Toxic Derivatives of O_2, and Power Transmission by Proton Gradients (Stryer pages 551–555)

28. Estimate the net yield of ATP from the complete oxidation of glucose, taking into account the different shuttles for the cytoplasmic reducing equivalents of NADH. Discuss the sources of uncertainty in the estimation.

29. Explain the difficulty in relating the stoichiometry of protons transported through the enzyme assembly to ATP molecules formed.

30. Describe *respiratory control* and relate it to *energy charge.*

31. Explain the effect of *uncouplers* on oxidative phosphorylation. Note how regulated uncoupling can be used for *thermogenesis.*

32. Explain why oxygen is a potentially toxic substance. Summarize the reactions and the biological roles of *superoxide dismutase, catalase,* and the *peroxidases.*

33. List examples of *energy conversions* by proton gradients and appreciate their central roles in free-energy interconversions.

SELF-TEST

Introduction

1. Explain the difference between oxidative phosphorylation and substrate level phosphorylation as seen in glycolysis and the citric acid cycle.

____

____

2. Which of the following statements regarding mitochondria and their components are correct?

(a) Mitochondria are approximately 20 μm in diameter.
(b) The matrix compartment contains the enzymes of glycolysis.
(c) Mitochondria have two membrane systems: an inner membrane and an outer membrane.
(d) The inner membrane contains pores and is readily permeable to most small metabolites.
(e) The inner membrane has a large surface area because it is highly folded.

Redox Potentials and Free-Energy Changes

3. Which of the following statements about the redox potential for a reaction are correct?

(a) It is used to describe phosphate group transfers.
(b) It is unrelated to the free energy of the reaction.
(c) It can be used to predict when a given compound will reduce another, given a suitable catalyst.
(d) It can be used to predict whether a given oxidation will provide sufficient energy for the formation of ATP from ADP + P_i.
(e) It can be used to predict the rate of O_2 uptake at the expense of a given substrate.

4. The equation for the reduction of cytochrome *a* by cytochrome *c* is as follows:

$$\text{Cyt } a\ (+3) + \text{cyt } c\ (+2) \longrightarrow \text{cyt } a\ (+2) + \text{cyt } c\ (+3)$$

where

Cyt *a* (+3):cyt *a* (+2)	$E'_0 = 0.27$ V
Cyt *c* (+3):cyt *c* (+2)	$E'_0 = 0.22$ V

Which of the following answers completes the sentence correctly? Under standard conditions—that is, 1 M concentrations of reactants and products—this reaction

(a) proceeds spontaneously.
(b) yields sufficient energy for ATP synthesis.
(c) does not alter the absorption spectra of the cytochromes.
(d) involves the transfer of two electrons.

5. Calculate $\Delta G^{\circ\prime}$ for the reaction

$$\text{Succinate} + \text{FAD} \longrightarrow \text{fumarate} + \text{FADH}_2$$

Fumarate:succinate	$E'_0 = 0.03$ V
FAD:$FADH_2$	$E'_0 = 0$ V

(a) −1.38 kcal/mol
(b) −0.69 kcal/mol
(c) 0.14 kcal/mol
(d) 0.69 kcal/mol
(e) 1.38 kcal/mol

6. The parameters $\Delta G^{\circ\prime}$ and $\Delta E'_0$ can be used to predict the direction of chemical reactions at standard conditions. On the other hand, $\Delta G'$ can be used for any concentrations of reactants and products to predict whether a chemical reaction will proceed. Using the expressions

$$\Delta G' = \Delta G^{\circ\prime} + 2.3\, RT \log_{10} \frac{[\text{products}]}{[\text{reactants}]}$$

and

$$\Delta G^{\circ\prime} = -nF\,\Delta E'_0$$

derive an expression for $\Delta E'$. Explain the significance of this redox potential.

7. Obtain $\Delta G'$ and $\Delta E'$ for the reaction given in question 5 when the succinate concentration is 2×10^{-3} M, the fumarate concentration is 0.5×10^{-3} M, the FAD concentration is 2×10^{-3} M, the $FADH_2$ concentration is 0.2×10^{-3} M, and the temperature is 37°C. (R = 1.98 cal/mol K)

The Respiratory Chain

8. Place the following respiratory-chain components in their proper sequence. Also, indicate which are mobile carriers of electrons.

(a) Cytochrome c ______

(b) NADH-Q reductase ______

(c) Cytochrome oxidase ______

(d) Ubiquinone ______

(e) Cytochrome reductase ______

9. Match the enzyme complexes of the respiratory chain in the left column with the appropriate electron carrying groups from the right column.

(a) Cytochrome oxidase ______	(1) Heme c_1
	(2) FAD
	(3) Heme a_3
(b) Cytochrome reductase ______	(4) Heme b-560
	(5) Iron-sulfur complexes
(c) NADH-Q reductase ______	(6) Cu_A and Cu_B
	(7) FMN
	(8) Heme a
(d) Succinate-Q reductase ______	(9) Heme b-566

10. Which of the following statements about the enzyme complexes of the electron transport system are correct?

(a) They are located in the mitochondrial matrix.
(b) They cannot be isolated from one another in functional form.
(c) They have very similar visible spectra.
(d) They are integral membrane proteins located in the inner mitochondrial membrane.
(e) They interact with one another via mobile carriers of electrons.

11. Which of the following statements about ubiquinol are correct?

(a) It is the mobile electron carrier between cytochrome reductase and cytochrome oxidase.
(b) It is an integral membrane protein.
(c) Its oxidation involves the simultaneous transfer of two electrons to the Fe-S center of cytochrome reductase.
(d) It is oxidized to ubiquinone via a semiquinone intermediate.
(e) It is a lipid-soluble molecule.

12. Which cytochrome has a protoporphyrin IX heme that is not covalently bound to protein?

(a) Cytochrome a
(b) Cytochrome a_3
(c) Cytochrome b
(d) Cytochrome c
(e) Cytochrome c_1

13. Explain the roles of cytochrome c_1 and the b cytochromes (b-560 and b-566) in the oxidation of ubiquinol to ubiquinone.

____

____

14. In the reduction of O_2 to H_2O by cytochrome oxidase, four electrons and four protons are used. How can this occur when a single electron at a time is transferred by heme iron and by copper?

____

____

15. Which of the following statements about the structure and properties of cytochrome c are incorrect?

(a) It has negatively charged patches on its surface that permit interactions with cytochrome reductase and cytochrome oxidase.
(b) It contains very little α helix or β pleated sheet secondary structure.
(c) It contains a heme with an Fe bonded to a histidine and a methionine residue.
(d) It has retained a highly conserved conformation throughout evolution.
(e) It is soluble in water.

Coupling of Oxidation and Phosphorylation to Form ATP by a Proton-Motive Force

16. Which of the following experimental observations provide evidence supporting the chemiosmotic model of oxidative phosphorylation?

(a) A closed membrane or vesicle compartment is required for oxidative phosphorylation.
(b) A system of bacteriorhodopsin and ATPase can produce ATP in synthetic vesicles when light causes proton pumping.
(c) A proton gradient is generated across the inner membrane of mitochondria during electron transport.

(d) Submitochondrial particles show a symmetric distribution of the respiratory chain components and ATP synthase across the inner mitochondrial membrane.

(e) ATP is synthesized when a proton gradient is imposed on mitochondria.

17. Which of the following statements about an aerated, functional mitochondrial preparation in which the reduced substrate is succinate are correct?

(a) Approximately 1.5 ATP molecules will be formed per succinate oxidized to fumarate.

(b) Approximately two protons will be pumped by the succinate-*Q* reductase complex.

(c) The addition of CN^- will result in the synthesis of only one ATP per succinate.

(d) Reduction of NADH-Q reductase will take place.

(e) Reduction of cytochrome reductase will occur.

18. Match each inhibitor in the left column with its *primary effect* from the right column.

(a) Azide ________

(b) Atractyloside ________

(c) Rotenone ________

(d) Dinitrophenol ________

(e) Carbon monoxide ________

(f) Oligomycin ________

(g) Antimycin A ________

(1) Inhibition of electron transport

(2) Uncoupling of electron transport and oxidative phosphorylation

(3) Inhibition of ADP-ATP translocation

(4) Inhibition of ATP synthase

19. Predict the oxidation-reduction states of NAD^+, NADH-Q reductase, ubiquinone, cytochrome c_1, cytochrome *c*, and cytochrome *a* in liver mitochondria that are amply supplied with isocitrate as substrate, P_i, ADP, and oxygen but are inhibited by

(a) Rotenone: ________

(b) Antimycin A: ________

(c) Cyanide: ________

20. The reduction potential for methylene blue is similar to that of oxygen, and it can be reduced by components of the electron transport chain. Explain why massive doses of methylene blue are sometimes given for cyanide poisoning.

____

____

21. Explain why one cannot precisely predict the sites in the electron transport chain where the coupling of oxidation to phosphorylation occurs on the basis of the redox potentials of the electron transport chain components.

22. Which of the following statements about the mitochondrial ATP-synthesizing complex are correct?

(a) It contains more than 10 subunits.
(b) It is located in the intermembrane space of mitochondria.
(c) It contains a subassembly that constitutes the proton channel.
(d) It is sensitive to oligomycin inhibition.
(e) It translocates ATP through the mitochondrial membranes.

23. Match the major units of the ATP-synthesizing system in the left column with the appropriate components and functions from the right column.

(a) F_0 ______
(b) F_1 ______
(c) Stalk between F_0 and F_1 ______

(1) Regulates proton flow and ATP synthesis
(2) Contains the proton channel
(3) Contains the catalytic sites for ATP synthesis
(4) Contains F_1 inhibitor
(5) Contains α, β, γ, δ, and ε subunits
(6) Contains the protein that confers oligomycin sensitivity

24. Which of the following statements about the proposed mechanism for ATP synthesis by ATP synthase are correct?

(a) ATP synthase forms ATP only when protons flow through the complex.
(b) ATP synthase contains sites that change in their affinity for ATP as protons flow through the complex.
(c) ATP synthase binds ATP more tightly when protons flow through the complex.
(d) ATP synthase has two active sites per complex.
(e) ATP synthase has active sites that may not be functionally equivalent at a given time.

Mitochondrial Transport Systems

25. The inner mitochondrial membrane contains translocases—that is, specific transport proteins—for which pairs of substances?

(a) NAD^+ and NADH
(b) Glycerol 3-phosphate and dihydroxyacetone phosphate

(c) AMP and ADP
(d) Citrate and pyruvate
(e) Glutamate and aspartate ✓

26. Explain why the rate of eversion of the binding site from the matrix to the cytosolic side is more rapid for ATP than for ADP when the ATP-ADP translocase functions in the presence of a proton gradient.

____

____

ATP Yield and Control of Oxidative Phosphorylation, Uncouplers, Scavengers of Toxic Derivatives of O_2 and Power Transmission by Proton Gradients

27. Approximately how many ATP are formed for each extramitochondrial NADH that is oxidized to NAD^+ by O_2 via the electron transport chain. Assume that the glycerol phosphate shuttle is operating.

(a) 1.0
(b) 1.5 ✓
(c) 2.5
(d) 3.0

28. How many ATP molecules are generated during the complete oxidative degradation of each of the following to CO_2 and H_2O? Assume the glycerol phosphate shuttle is operating.

(a) Acetyl CoA 10
(b) Phosphoenolpyruvate 13.5
(c) glyceraldehyde 3-phosphate 16

29. What is meant by the term *respiratory control?*

____

____

30. Which of the following answers completes the sentence correctly? The rate of flow of electrons through the electron transport chain is most directly regulated by

(a) the ATP:ADP ratio. ✓
(b) the concentration of acetyl CoA.
(c) the rate of oxidative phosphorylation.
(d) feedback inhibition by H_2O.
(e) the catalytic rate of cytochrome oxidase.

31. Which of the following answers completes the sentence correctly? Uncouplers, such as dinitrophenol (DNP) or thermogenin, uncouple electron transport and phosphorylation by

(a) inhibiting cytochrome reductase.
(b) dissociating the F_0 and F_1 units of ATP synthase.
(c) blocking electron transport.

(d) dissipating the proton gradient.
(e) blocking the ATP-ADP translocase.

32. Which of the following are the products of the reaction of superoxide dismutase?
(a) $O_2^-\cdot$
(b) H_2O
(c) H_2O_2
(d) O_2
(e) H_3O^+

ANSWERS TO SELF-TEST

Introduction

1. Oxidative phosphorylation is the process in which ATP is formed as electrons are transferred from NADH or $FADH_2$ to O_2. The energy coupling occurs through a proton gradient. In substrate level phosphorylation, the syntheses of ATP or GTP are coupled by shared intermediates to other chemical reactions that release sufficient energy to drive the formation of ATP or GTP from ADP and GDP, respectively. Substrate level phosphorylations have defined stoichiometries because they use a shared intermediate to couple reactions. Oxidative phosphorylation has a less precise stoichiometry because its driving force is a delocalized energy gradient composed of a proton concentration and electric charge difference across a membrane.

2. c, e

Redox Potentials and Free-Energy Changes

3. c, d

4. a. The $\Delta E'_0$ for the reaction is 0.05 V. Calculating $\Delta G^{\circ\prime}$:

$$\Delta G^{\circ\prime} = -nF\Delta E'_0$$

where n, the number of electrons transferred, is 1, and F, the energy per volt and per mole, is 23.06 kcal/V mol.

$$\Delta G^{\circ\prime} = -1 \times 23.06 \text{ kcal/V mol} \times 0.05 \text{ V}$$

$$= -1.15 \text{ kcal/mol}$$

Therefore, the reaction will proceed spontaneously; but, when considered by itself, it is insufficiently exergonic to drive ATP synthesis, which requires −7.3 kcal/mol under standard conditions. In the cell, this comparison is relatively meaningless because ATP is not synthesized during oxidative phosphorylation by direct chemical coupling of redox reactions to ATP formation.

5. e. The $\Delta E'_0$ for this reaction is –0.03 V. Calculating $\Delta G^{\circ\prime}$:

$$\begin{aligned}\Delta G^{\circ\prime} &= -nF\,\Delta E'_0\\ &= -2 \times 23.06\text{ kcal/V mol} \times (-0.03\text{ V})\\ &= 1.38\text{ kcal/mol}\end{aligned}$$

6. The same proportionality constants that relate $\Delta G^{\circ\prime}$ and $\Delta E'_0$—that is, $\Delta G^{\circ\prime} = -nF\,\Delta E'_0$—can be used to relate $\Delta G'$ and $\Delta E'$. F is only the caloric equivalent of a volt, and n is the number of electrons participating in the redox reaction. Therefore,

$$\Delta G' = -nF\,\Delta E'$$

Substituting this in the expression for $\Delta G'$

$$\Delta G' = \Delta G^{\circ\prime} + 2.3\,RT\log_{10}\frac{[\text{products}]}{[\text{reactants}]}$$

$$-nF\,\Delta E' = -nF\,\Delta E'_0 + 2.3\,RT\log_{10}\frac{[\text{products}]}{[\text{reactants}]}$$

$$\Delta E' = \Delta E'_0 - \frac{2.3\,RT}{nF}\log_{10}\frac{[\text{products}]}{[\text{reactants}]}$$

$\Delta E'$ is a measure of the direction in which an oxidation-reduction reaction will proceed for any given concentrations of reactants and products.

7. Substituting into the equation for $\Delta G'$,

$$\begin{aligned}\Delta G' &= \Delta G^{\circ\prime} + 2.3\,RT\log_{10}\frac{[\text{products}]}{[\text{reactants}]}\\ &= 1.38\text{ kcal/mol} + 2.3 \times 1.98\text{ cal/mol K} \times\\ &\qquad 310\text{ K}\log_{10}\frac{(0.5\times10^{-3})(0.2\times10^{-3})}{(2\times10^{-3})(2\times10^{-3})}\\ &= 1.38\text{ kcal/mol} - 2.26\text{ kcal/mol}\\ &= -0.88\text{ kcal/mol}\end{aligned}$$

Since $\Delta G' = -nF\,\Delta E'$,

$$\begin{aligned}\Delta E' &= \frac{-\Delta G'}{nF}\\ &= -\frac{(-0.88\text{ kcal/mol})}{(2)(23\text{ kcal/mol V})}\\ &= 0.019\text{ V}\end{aligned}$$

Alternatively, $\Delta E'$ could be calculated first using the expression derived in question 6, and $\Delta G'$ could be subsequently determined from $\Delta G' = -nF\,\Delta E'$.

8. The proper sequence is b, d, e, a, and c. The mobile carriers are (a) and (d).

9. (a) 3, 6, 8 (b) 1, 4, 5, 9 (c) 5, 7 (d) 2, 5

10. d, e

11. d, e

12. c

13. See Figure 21-13 in Stryer. Ubiquinol transfers one electron to cytochrome c_1 via an Fe-S cluster on cytochrome reductase. The semiquinone derived from Q in this process donates an electron to cytochrome *b*-566, giving rise to ubiquinone. In turn, the electron from cytochrome *b*-566 is transferred to cytochrome *b*-560, which then reduces another semiquinone to ubiquinol. Thus, the *b* cytochromes act as a recycling device that allows ubiquinol, a two-electron carrier, to transfer its electrons, one at a time, to the Fe-S cluster of cytochrome reductase. Cytochrome c_1 accepts the electrons from the Fe-S cluster and transfers them to cytochrome *c* via the *b* cytochromes.

14. See Figure 21-15 in Stryer. Molecular O_2 is bound between the Fe^{2+} and Cu^+ ions of the a_3-Cu_B center of cytochrome oxidase. The oxygen remains bound while four electrons and four protons are sequentially added to its various intermediates, resulting in the net release of 2 H_2O. The a-Cu_A center supplies the electrons for this process. Although four electrons are used in the reduction of O_2 to 2 H_2O, the individual steps of the reaction cycle involve single electron transfers.

15. a

Coupling of Oxidation and Phosphorylation to Form ATP by a Proton-Motive Force

16. a, b, c, e

17. a, e

18. (a) 1 (b) 3 (c) 1 (d) 2 (e) 1 (f) 4 (g) 1

19. Upstream from the inhibitor, reduced respiratory-chain components will accumulate; downstream, oxidized components will be present. The point of inhibition is the crossover point.

(a) Rotenone inhibits the step at NADH-Q reductase; therefore, NADH and NADH-Q reductase will be more reduced, and ubiquinone, cytochrome c_1, cytochrome *c*, and cytochrome *a* will be more oxidized.

(b) Antimycin A blocks electron flow between cytochromes *b* and c_1; therefore, NADH, NADH-Q reductase, and ubiquinol will be more reduced, and cytochrome c_1, cytochrome *c*, and cytochrome *a* will be more oxidized.

(c) Cyanide inhibits the transfer of electrons from cytochrome oxidase to O_2, so all the components of the respiratory chain will be more reduced.

20. Cyanide blocks the transfer of electrons from cytochrome oxidase to O_2. Therefore, all the respiratory-chain components become reduced and electron transport ceases; consequently, phosphorylation stops. An artificial electron acceptor with a redox potential similar to that of O_2, such as methylene blue, can reoxidize various components of the respiratory chain, reestablish a proton gradient, and thereby restore ATP synthesis.

21. The free-energy change ($\Delta G^{\circ\prime}$) for each electron transfer step of the respiratory chain can be calculated from the redox-potential change ($\Delta E'_0$) for that step, using the equation $\Delta G^{\circ\prime} = -nF\ \Delta E'_0$. Since $\Delta G^{\circ\prime}$ for the synthesis of ATP from ADP is +7.3 kcal/mol, $\Delta G^{\circ\prime}$ for an electron transfer reaction in which the coupling of oxidation to phosphorylation could occur must be more negative than −7.3 kcal/mol. However, we cannot conclude that there is a direct, quantitative relationship between a given redox reaction and the phosphorylation of ADP in the cell because the free-energy coupling is via a spatially delocalized proton-motive force.

22. a, c, d

23. (a) 2 (b) 3, 5 (c) 1, 4, 6

24. b, e

Mitochondrial Transport Systems

25. b, e. Answer (a) is incorrect because only the electrons of NADH are transported, not the entire molecule. Answers (c) and (d) are incorrect because these pairs of compounds are not transported by the same translocase.

26. The proton gradient and the membrane potential make the cytosolic side of the inner mitochondrial membrane more positive than the matrix side; therefore, ATP, which has one more negative charge than ADP, is more attracted to the cytosolic side than is ADP.

ATP Yield and Control of Oxidative Phosphorylation, Uncouplers, Scavengers of Toxic Derivatives of O_2, and Power Transmission by Proton Gradients

27. b. The oxidation of NADH by the electron transport chain leads to the synthesis of approximately 2.5 ATP. However, when the reducing equivalents of an extramitochondrial NADH enter the mitochondrial matrix via the glycerol phosphate shuttle, they give rise to $FADH_2$, which yields 1.5 ATP.

28. (a) About 10 ATP: citric acid cycle (3 NADH × 2.5, 1 $FADH_2$ × 1.5, 1 GTP)

(b) About 13.5 ATP: citric acid cycle (10 ATP), pyruvate dehydrogenase reaction (1 NADH, intramitochondrial × 2.5), pyruvate kinase reaction (1 ATP)

(c) About 16 ATP: citric acid cycle (10 ATP), pyruvate dehydrogenase reaction (1 NADH × 2.5), pyruvate kinase reaction (1 ATP), phosphoglycerate kinase reaction (1 ATP), glyceraldehyde 3-phosphate dehydrogenase reaction (1 NADH, extramitochondrial, which yields 1.5 ATP by the glycerol phosphate shuttle)

29. The regulation of the rate of oxidative phosphorylation by the availability of ADP is referred to as *respiratory control.*

30. a

31. d

32. c, d

PROBLEMS

1. Nitrite (NO_2^-) is toxic to many microorganisms. It is therefore often used as a preservative in processed foods. However, members of the genus *Nitrobacter* oxidize nitrite to nitrate (NO_3^-), using the energy released by the transfer of electrons to oxygen to drive ATP synthesis. Given the E'_0 values below, calculate the maximum ATP yield per mole of nitrate oxidized.

$$NO_3^- + 2\,H^+ + 2\,e^- \longrightarrow NO_2^- + H_2O \qquad E'_0 = +0.42\text{ V}$$

$$\tfrac{1}{2}O_2 + 2\,H^+ + 2\,e^- \longrightarrow H_2O \qquad E'_0 = +0.82\text{ V}$$

2. A newly discovered compound called *coenzyme U* is isolated from mitochondria.

(a) Several lines of evidence are presented in advancing the claim that coenzyme U is a previously unrecognized carrier in the electron transport chain:
(1) When added to a mitochondrial suspension, coenzyme U is readily taken up by mitochondria.
(2) Removal of coenzyme U from mitochondria results in a decreased rate of oxygen consumption.
(3) Alternate oxidation and reduction of coenzyme U when it is bound to the mitochondrial membrane can be easily demonstrated.
(4) The rate of oxidation and reduction of coenzyme U in mitochondria is the same as the overall rate of electron transport.
Which of the lines of evidence do you find the most convincing? Which are the least convincing? Why?

(b) In addition to the evidence cited in (a), the following observations were recorded when coenzyme U was incubated with a suspension of submitochondrial particles:
(1) The addition of NADH caused a rapid reduction of coenzyme U.
(2) Reduced coenzyme U caused a rapid reduction of added cytochrome *c*.

(3) In the presence of antimycin A, the reduction of coenzyme U by added NADH took place as rapidly as in the absence of antimycin A. However, the reduction of cytochrome *c* by reduced coenzyme U was blocked in the presence of the inhibitor.

(4) The addition of succinate caused a rapid reduction of coenzyme U.

Assign a tentative position for coenzyme U in the electron transport chain.

3. Coenzyme Q can be selectively removed from mitochondria using lipid solvents. If these mitochondria are then incubated in the presence of oxygen with an electron donor that is capable of reducing NAD^+, what will be the redox state of each of the carriers in the electron transport chain?

4. The treatment of submitochondrial particles with urea removes F_1 subunits. When these treated particles are incubated in air with an oxidizable substrate and calcium ion, the concentration of calcium inside the particles increases.

(a) What do these observations tell you about the source of free energy required for accumulation of calcium ion?
(b) Would you expect the accumulation of calcium ion to be more sensitive to DNP or to oligomycin? Why?

5. In plants, the synthesis of succinate from acetyl CoA in the glyoxylate cycle generates NADH. Glyoxysomes do not have an electron transport chain. Suggest a way in which NADH could be reoxidized in a plant cell so that the glyoxylate cycle can continue.

6. Analysis of the electron transport pathway in a pathogenic gram-negative bacterium reveals the presence of five electron transport molecules with the redox potentials listed in Table 21-1.

Table 21-1

Reduction potentials for pathogenic gram-negative bacterium

Oxidant	*Reductant*	*Electrons transferred*	E'_0 *(V)*
NAD^+	NADH	2	−0.32
Flavoprotein *b* (oxidized)	Flavoprotein *b* (reduced)	2	−0.62
Cytochrome *c* (+3)	Cytochrome *c* (+2)	1	+0.22
Ferroprotein (oxidized)	Ferroprotein (reduced)	2	+0.85
Flavoprotein *a* (oxidized)	Flavoprotein *a* (reduced)	2	+0.77

(a) Predict the sequence of the carriers in the electron transport chain.
(b) How many molecules of ATP can be generated under standard conditions when a pair of electrons is transported along the pathway?
(c) Why is it unlikely that oxygen is the terminal electron acceptor?

7. Calculate the minimum value of $\Delta E'_0$ that must be generated by a pair of electron carriers to provide sufficient energy for ATP synthesis. Assume that a pair of electrons is transferred.

8. The value of E'_0 for the reduction of $NADP^+$ is −0.32 V.

(a) Calculate the equilibrium constant for the reaction catalyzed by NADPH dehydrogenase:

$$NADP^+ + NADH \longrightarrow NADPH + NAD^+$$

(b) What function could NADPH dehydrogenase serve in the cell?

9. Mitchell and Moyle carried out an elegant series of experiments using a sensitive pH meter to measure changes in hydrogen ion concentration in an anaerobic suspension of rat mitochondria. Explain each of the following observations:

(a) The anaerobic suspension was incubated with β-hydroxybutyrate, an oxidizable substrate. When a small amount of oxygen was then injected into the suspension, an immediate decrease in pH was observed, followed by a gradual return to the original hydrogen ion concentration.
(b) When the experiment described in (a) was carried out in the presence of Triton X, a detergent that disrupts membranes, no decrease in pH was observed when oxygen was injected into the suspension.
(c) The anaerobic suspension was incubated in the absence of an oxidizable substrate. When a small amount of ATP was injected into the suspension, an immediate decrease in pH was observed followed by a gradual return to the original pH.

10. Why is it important for the value of $\Delta E'_0$ for the NAD^+:NADH redox couple to be less negative than those for the redox couples of oxidizable compounds that are components of the glycolytic pathway and the citric acid cycle?

11. Arsenate, AsO_4^{3-}, is an uncoupling reagent for oxidative phosphorylation, but unlike DNP it does not transport protons across the inner mitochondrial membrane. How might arsenate function as an uncoupler?

12. A newly isolated soil bacterium grows without oxygen but requires ferric ion in the growth medium. Succinate suffices as a carbon source, but neither hexoses nor pyruvate can be utilized. The bacteria require riboflavin as a growth supplement. Neither niacin nor thiamin is required, and neither substance nor compounds derived from them can be found in the cells. The electron carriers found in the bacteria are cytochrome *b*, cytochrome *c*, FAD, and coenzyme Q.

(a) Propose a reasonable electron transport chain that takes these observations, including the requirement for ferric ion, into account.
(b) Which of the other observations help to explain why the bacterium cannot utilize pyruvate or hexoses?
(c) Why is riboflavin required for the growth of the bacterium?

13. Yeast can grow both aerobically and anaerobically on glucose. Explain why the rate of glucose consumption decreases when yeast cells that have been maintained under anaerobic conditions are exposed to oxygen.

14. The addition of the drug dicyclohexylcarbodiimide (DCCD) to mitochondria markedly decreases both the rate of electron transfer from NADH to O_2 and the rate of ATP formation. The subsequent addition of 2,4-dinitrophenol leads to an increase in the rate of electron transfer without changing the rate of ATP formation. What does DCCD inhibit?

15. Mitochondria isolated from the liver of a particular patient will oxidize NADH at a relatively high rate even if ADP is absent. The P:O ratio for oxidative phosphorylation by these mitochondria is less than normal. Predict the likely symptoms of this disorder.

16. A 24-year-old male suffers from a muscle disorder, dementia, and several other neurological problems. A skeletal muscle biopsy is obtained, the intact mitochondria isolated, and respiratory activities measured under the following conditions:

		Number atoms O consumed/min/mg		
Protein	*Substrate*	*No addition*	*ADP + P_i*	*DNP*
Normal	Pyruvate + malate	20	100	100
Patient	Pyruvate + malate	20	40	100

(a) For the normal control, explain the basis for the increase in oxygen consumption following the addition of ADP + P_i. Then provide an explanation for increases in O_2 consumption when DNP (2,4-dinitrophenol) is added.
(b) Explain the differences observed in the patient's mitochondria. What possible defects would explain these results?
(c) Suppose that fragments of sonicated mitochondria from normal subjects and from the patient are used with pyruvate and malate as substrates. How would you expect the results of the experiments to differ from those using intact mitochondria? Provide a brief explanation for your answer.

OH

NO_2

NO_2

2, 4-Dinitrophenol (DNP)

17. Acidic aromatic compounds like 2,4-dinitrophenol (DNP) act as uncouplers of electron transport and oxidative phosphorylation because they carry protons across the inner mitochondrial membrane, disrupting the proton gradient. The structure of the neutral, protonated form of DNP is shown in the margin.

(a) Although you might expect that 2,4-dinitrophenylate anion (pK_a = 4.0) would be unable to cross the inner mitochondrial membrane, the deprotonated form of DNP is membrane-soluble. Explain why, by drawing resonance forms of 2,4-dinitrophenolate anion, showing how negative charge is distributed over the phenyl ring structure.
(b) Suppose you were studying the effects of DNP on proton transport in an artificial phospholipid membrane system. You observe that the rate of proton transport by DNP increases at temperatures above the transition temperature for the membrane. Why?

18. Earlier experiments with the cytochrome reductase system showed that antimycin A causes a crossover (see Stryer's problem 12, p. 558) be-

tween cytochrome *b* and the *c* cytochromes. This work and other experiments led to the proposed pathway for electron flow in cytochrome reductase:

$$\text{Q} \longrightarrow \text{cytochrome } b \longrightarrow \text{cytochrome } c_1 \longrightarrow \text{cytochrome } c$$

However, further experiments placed the accuracy of this pathway in doubt. When mitochondria under low oxygen concentrations were pulsed with oxygen, the *b* cytochromes became transiently reduced at the same time that the *c* cytochromes were oxidized. The reduction of *b* cytochromes under such conditions was enhanced when antimycin A was used.

(a) To understand why the reduction of cytochrome *b* was surprising, consider what would happen when oxygen is suddenly available to the reduced electron carriers in the proposed pathway. In what sequence would oxidized carriers appear?

(b) Based on your understanding of the current view of the flow of electrons from Q to cytochrome *c* through cytochrome *c* reductase, describe how an oxygen pulse could cause the transient reduction of *b* cytochromes in mitochondrian under low oxygen tension.

ANSWERS TO PROBLEMS

1. The two reactions that generate nitrate are

$$NO_2^- + H_2O \longrightarrow NO_3^- + 2\,H^+ + 2\,e^- \qquad E'_0 = -0.42\text{V}$$

$$\tfrac{1}{2}O_2 + 2\,H^+ + 2\,e^- \longrightarrow H_2O \qquad E'_0 = +0.82\text{ V}$$

The net reaction is

$$NO_2^- + \tfrac{1}{2}O_2 \longrightarrow NO_3^- \qquad E'_0 = +0.40\text{ V}$$

The Nernst equation is used to calculate the free energy liberated by the oxidation of nitrite under standard conditions:

$$\Delta G^{\circ\prime} = -nF\,\Delta E'_0$$

$$= -2(23.06\text{ kcal/V mol})\,(0.40\text{ V})$$

$$= -18.45\text{ kcal/mol}$$

Therefore, each mole of nitrite oxidized yields energy sufficient to drive the formation of 2.52 moles of ATP under standard conditions ($\Delta G^{\circ\prime} = -7.3$ kcal/mol).

2. (a) If coenzyme U is a component of the electron transport chain, it should undergo successive reduction and oxidation in the mitochondrion, and its rate of electron transfer should be close to the overall rate. Observations 3 and 4 are therefore the most convincing. In addition to electron carriers, other compounds can be taken up by mitochondria, and some of them, such as pyruvate, can affect the rate of oxygen consumption because they are substrates that donate electrons to carriers. Therefore, observations 1 and 2 are less convincing.

(b) The first two observations show that coenzyme U lies along the electron transport chain between NADH, which can reduce it, and cytochrome *c*, which is reduced by it. The fact that antimycin A blocks cytochrome *c* reduction by coenzyme U suggests that the carrier lies before cytochrome reductase. Succinate, which can transfer electrons to Q, can also transfer electrons to coenzyme U, so the position of coenzyme U in the chain is similar to that of Q.

3. The removal of ubiquinone from the electron transport chain means that no electrons can be transferred beyond Q in the pathway. You would therefore expect all carriers preceding Q to be more reduced and those beyond Q to be more oxidized. The removal of Q gives results similar to those observed by Chance in his crossover experiments (see Stryer, problem 12, p. 558).

4. (a) The treated submitochondrial particles can carry out electron transport and can establish a proton gradient, but because the F_1 subunits have been removed, they cannot synthesize ATP. The source of free energy for calcium accumulation must therefore be the proton-motive force generated by electron transport.

(b) DNP disrupts the proton-motive force by carrying protons across the mitochondrial membrane, thereby dissipating the free energy required for calcium accumulation. Oligomycin inhibits ATP synthase in mitochondria. Because the treated particles do not depend on ATP as a source of energy for calcium accumulation, oligomycin will have little effect.

5. Plant cells could use shuttle mechanisms to carry electrons from the glyoxysome to the mitochondrion. For example, oxaloacetate could be used as an acceptor of electrons from NADH, yielding malate, which could then be transferred to mitochondria, where the oxaloacetate would be regenerated. The transamination reaction discussed by Stryer on page 549 of the text could be used to return the four-carbon unit to the glyoxysome. The actual mechanism for generation of NAD^+ in the glyoxysome is unknown.

6. (a) The carrier with the most negative reduction potential has the weakest affinity for electrons and so transfers them most easily to an acceptor. The carrier with the most positive reduction potential will be the strongest oxidizing substance and will have the greatest affinity for electrons. A carrier should be able to pass electrons to any carrier having a more positive reduction potential. Thus, the probable order of the carriers in the chain is flavoprotein *b*, NADH, cytochrome *c*, flavoprotein *a*, and ferroprotein.

(b) $\Delta E'_0 = +0.85\text{ V} - (-0.62\text{ V}) = 1.47\text{ V}$

The total amount of free energy released by the transfer of two electrons is

$$\begin{aligned}\Delta G^{\circ\prime} &= -nF\,\Delta E'_0 \\ &= -(2)(23.06\text{ kcal/V mol})(1.47\text{ V}) \\ &= -67.80\text{ kcal/mol}\end{aligned}$$

Because +7.3 kcal/mol is required to drive ATP synthesis under standard conditions, the number of molecules of ATP synthesized per pair of electrons is 67.80/7.3 = 9.3.

(c) It is unlikely that oxygen is the terminal electron acceptor, because the reduction potential for the ferroprotein is slightly more positive than that of oxygen, so under standard conditions, the ferroprotein could not transfer electrons to oxygen.

7. The minimum amount of free energy that is needed to drive ATP synthesis under standard conditions is −7.3 kcal/mol. The value of $\Delta E'_0$ needed to generate this amount of free energy can be determined using the equation

$$\Delta G^{\circ\prime} = -nF\,\Delta E'_0$$

$$\Delta E'_0 = -\frac{\Delta G^{\circ\prime}}{nF}$$

If a pair of electrons is transferred, then

$$\Delta E'_0 = -\frac{-7.3 \text{ kcal/mol}}{2(23.06 \text{ kcal/V mol})}$$

$$= +0.159 \text{ V}$$

8. (a) The transfer of a pair of electrons from NADH to $NADP^+$ occurs with no release of free energy:

$$NAD^+ + H^+ + 2\,e^- \longrightarrow NADH \qquad E'_0 = -0.32 \text{ V}$$

$$NADP^+ + H^+ + 2\,e^- \longrightarrow NADPH \qquad E'_0 = -0.32 \text{ V}$$

For the overall reaction, $\Delta E'_0 = 0.00$ V. Because

$$\Delta G^{\circ\prime} = -nF\,\Delta E'_0$$

then

$$\Delta G^{\circ\prime} = 0 = -2.303\,RT \log_{10} K'_{eq}$$

$$\log_{10} K'_{eq} = 0$$

$$K'_{eq} = 1$$

This means that the concentrations of the products equals the concentrations of the reactants; the reaction is at equilibrium.

(b) In the cell, NADPH dehydrogenase serves to replenish NADPH when the reduced cofactor is needed for biosynthetic reactions. On the other hand, metabolites such as isocitrate and glucose 6-phosphate are substrates for $NADP^+$-linked dehydrogenases. NAD^+ can accept reducing equivalents generated through the action of these enzymes.

9. (a) The injection of oxygen allowed electrons to be transferred from the oxidizable substrate to oxygen via the electron transport chain in the inner mitochondrial membrane. Protons were pumped

into the medium surrounding the organelles, resulting in a drop in pH. As the protons moved back into the mitochondria through ATP synthase or other routes, the hydrogen ion concentration returned to its original level.

(b) The detergent disrupted the integrity of the closed membrane compartment in the mitochondria, making a proton gradient impossible to establish.

(c) ATP is a substrate for ATP synthase, which can also catalyze the hydrolysis of ATP (see Stryer's text, p. 546). During ATP hydrolysis, protons were pumped back through the proton channel of the synthase complex from the inner mitochondrial membrane into the medium, resulting in an elevation of the hydrogen ion concentration.

10. NADH is a primary source of electrons for the respiratory chain. Oxidizable substrates must have a more negative reduction potential to donate electrons to NAD^+. A more negative redox potential for the NAD^+: NADH couple would make it unsuitable as an electron acceptor.

11. Arsenate chemically resembles inorganic phosphate; therefore, it can enter into many of the same biochemical reactions as P_i (see p. 503 of Stryer's text). Arsenate can replace phosphate during oxidative phosphorylation, presumably forming an arsenate anhydride with ADP. Such compounds are unstable and are rapidly hydrolyzed, effectively causing the uncoupling of electron transport and oxidative phosphorylation.

12. (a) Because neither niacin nor NAD^+ is found in these cells, the transfer of electrons must procede from succinate to FAD and then to those carriers that have successively more positive redox potentials. In order, these are coenzyme Q, cytochrome *b*, cytochrome *c*, and ferric ion, which serves as the terminal electron acceptor in the absence of oxygen.

(b) In order to utilize hexoses in the glycolytic pathway, NAD^+ is required as a cofactor for glyceraldehyde 3-phosphate dehydrogenase. Because the bacterium has no NAD^+, the glycolytic pathway is not operating. Similarly, the fact that thiamin is neither required nor found in the cells means that pyruvate dehydrogenase cannot be used to convert pyruvate to acetyl CoA.

(c) Riboflavin is a precursor of FAD, which is required as an acceptor of electrons from succinate.

13. The decrease in glucose consumption when oxygen is introduced is known as the *Pasteur effect.* Under anaerobic conditions, glucose cannot be oxidized completely to CO_2 and H_2O, because NADH and QH_2 generated in the citric acid cycle cannot be reoxidized in the absence of oxygen. In order to regenerate NAD^+, needed for continued operation of the glycolytic pathway, pyruvate is converted to ethanol and CO_2. There is a net production of only two molecules of ATP from each glucose molecule metabolized by the glycolytic pathway. This means that the pace of glycolysis must be relatively high to generate sufficient amounts of ATP for cell maintenance.

When oxygen is introduced, reduced cofactors in the citric acid cycle can be reoxidized in the electron transport chain and oxidative phosphorylation occurs. Under these conditions the yeast cell can utilize glucose much more efficiently, producing 38 molecules of ATP for each glucose molecule oxidized completely to CO_2 and H_2O. The rate of glucose consumption is greatly reduced under aerobic conditions because less glucose is needed to provide the amount of ATP needed to maintain the cell. Other factors that decelerate the pace of glycolysis include increases in concentrations of citrate and ATP under aerobic conditions. Both molecules are key regulators of phosphofructokinase 1, a rate-limiting glycolytic enzyme (see Stryer, p. 767).

14. The fact that the rate of electron transport increases without a concomitant increase in ATP synthesis suggests that DCCD inhibits ATP synthase. Experiments show that indeed DCCD blocks proton flow through the c subunit of F_0, inhibiting ATP synthase activity. When DCCD is added to respiring mitochondria, protons cannot move back into the mitochondrial matrix and the rate of ATP synthesis decreases. The flow of electrons through the respiratory chain slows as the need to maintain the proton gradient decreases. The metabolic uncoupler 2,4-dinitrophenol is an effective ion carrier that dissipates the proton-motive force generated by electron transfer by allowing protons to freely cross the inner mitochondrial membrane. Although the rate of electron transport increases in response to the dissipation of the proton gradient, the rate of phosphorylation of ADP through oxidative phosphorylation does not increase, because ATP synthase remains inhibited by DCCD.

15. From these observations it appears that electron transfer and ATP synthesis are uncoupled, so that there is no way to control electron transport by limiting ADP availability. One would expect the patient to have a very high rate of metabolism, along with a possibly elevated temperature. The fact that much of the energy available from electron transport is not utilized means that the energy is released as heat rather than being utilized for the formation of ATP.

16. (a) Pyruvate and malate are in this case sources of electrons for the electron transport chain. ADP and P_i are substrates for ATP formation. When they are added to the mitochondria, electron transfer *and* oxidative phosphorylation can proceed at a maximal rate, with maximum consumption of oxygen. DNP is an uncoupler of oxidative phosphorylation and electron transport; it allows electron transport to proceed maximally, whether or not ADP and P_i are available.

(b) The patient is unable to carry out oxidative phosphorylation at maximum rates even when ADP and P_i are present; however, electron transport proceeds at a high rate when the uncoupler DNP is used. Thus the patient's mitochondria must be partially defective with respect to ATP synthesis; the defect could be partial uncoupling, a deficiency in the ATPase (or ATP synthase), or failure to transport ATP from the mitochondria to the cytosol.

(c) In order for ATP synthesis to occur, a proton gradient across the inner mitochondrial membrane must be generated during electron transport. Such a gradient requires an intact compartment outside so that an excess of protons can accumulate. Fragments of mitochondria do not permit formation of such a gradient. Thus whether you use mitochondrial fragments from normal or from affected subjects, you will see a certain level of electron transport (or a certain level of oxygen consumed per minute per milligram of protein) that is not affected by the addition of ATP and P_i or of DNP.

17. (a) See the accompanying figure for structures. Resonance structures which include the two nitro groups show that the negative charge can be distributed among a number of forms. Therefore the dinitrophenylate anion is soluble in the membrane bilayer. This explains why DNP can rapidly carry protons across the inner mitochondrial membrane.

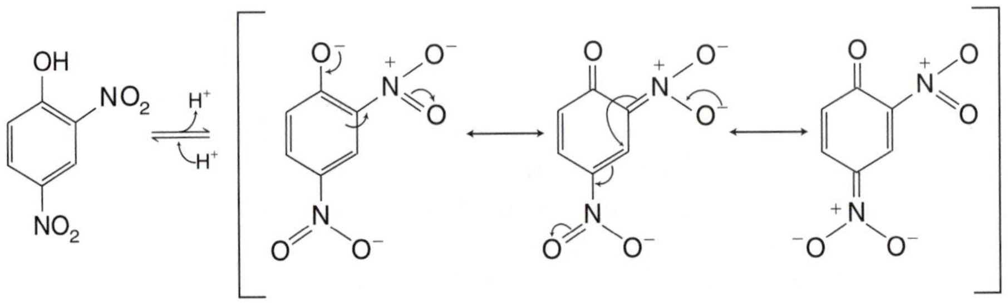

(b) DNP is a mobile proton carrier that is soluble in the membrane bilayer. Below the transition temperature for the phospholipid, the bilayer is in the gel state, where molecules like DNP may not be able to diffuse rapidly from one side of the bilayer to the other. At temperatures above the transition temperature, the bilayer is in the fluid state, in which DNP is more mobile, more rapidly transporting protons across the bilayer.

18. (a) Oxygen is the terminal electron acceptor in the respiratory chain, accepting electrons from cytochrome *c*. You would expect oxidized cytochrome *c* to appear first, followed by the oxidized carriers c_1, hemes in *b* cytochromes, and Q.

(b) The availability of oxygen will allow the *c* cytochromes and the 2Fe-2S complex to be oxidized, but as QH_2 is oxidized, it is converted to $Q\cdot^-$, which donates an electron to heme b_L, which then reduces heme b_H during the first part of the Q cycle. Although the two hemes are then oxidized in the second phase of the Q cycle, they will be reduced initially during an oxygen pulse. Antimycin A is now known to block electron transfer between heme b_H and $Q\cdot^-$ or Q, which means that during an oxygen pulse, electrons can go no farther than hemes b_H and b_L. Thus, the movement of electrons through the Q cycle explains the transient reduction of both hemes in cytochrome *b* when oxygen is introduced to the mitochondria.

1. (a) 12.5 (b) 15 (c) 32 (d) 13.5 (e) 30 (f) 16

These answers are readily obtained if one remembers the ATP yields from the various parts of glycolysis and remembers that (1) cytoplasmic NADH yields only 1.5 ATP, (2) pyruvate → acetyl CoA + 1 NADH yields 2.5 ATPs, (3) each acetyl CoA traversing the citric acid cycle yields 10 ATP (7.5 from 3 NADH, 1.5 from 1 $FADH_2$, and 1 from GTP), and (4) galactose requires 1 ATP for activation, just as glucose does.

2. Biochemists use E'_0, the value at pH 7, whereas chemists use E_0, the value in 1 N H^+. The prime denotes that pH 7 is the standard state. Since the pH of 1 N H^+ is 0, the difference in the [H^+] between the two systems is 10^7. Multiplying the log of this difference, 7, by 0.059 V (2.3 RT/faraday) = 0.041 V, the difference of the two systems.

3. (a) Blocks electron transport and proton pumping at site 3.

(b) Blocks electron transport and ATP synthesis by inhibiting the exchange of ATP and ADP across the inner mitochondrial membrane.

(c) Blocks electron transport and proton pumping at site 1.

(d) Blocks ATP synthesis without inhibiting electron transport by dissipating the proton gradient.

(e) Blocks electron transport and proton pumping at site 3.

(f) Blocks electron transport and proton pumping at site 2.

4. (a) Since P:O is the number of high-energy phosphate bonds formed per atom of oxygen and oxygen accepts two electrons, P/O = ~P/O = ATP/O = ATP/$2e^-$. The amount of ATP synthesized is directly proportional to the number of protons pumped (H^+_{pu}) and inversely proportional to the number of protons required for ATP synthesis (H^+_{re}). Hence ATP synthesized = H^+_{pu}/H^+_{re}.

(b) Since NADH donates two electrons to one atom of oxygen, the P/O = ATP formed = $10H^+_{pu}/4H^+_{re}$ = 2.5. For succinate ($FADH_2$) oxidation the P/O = $6H^+_{pu}/4H^+_{re}$ = 1.5.

5. For oxidation by NAD^+, $\Delta E'_0 = -0.32 - 0.03 = -0.35$ V, and $\Delta G^{\circ\prime} = -(-0.35) \times 2 \times 23 = +16.1$ kcal/mol. For oxidation by FAD, $\Delta E'_0 = 0 - 0.03 = -0.03$ V, and $\Delta G^{\circ\prime} = -(-0.03) \times 2 \times 23 = +1.38$ kcal/mol.

As the data above indicate, the oxidation of succinate by NAD^+ is not thermodynamically feasible because it would require a very large *input* of energy (16.1 kcal/mol).

6. Cyanide can be lethal because it binds to the ferric form of cytochrome oxidase and thereby inhibits oxidative phosphorylation. Nitrite converts ferrohemoglobin to ferrihemoglobin, which also binds cyanide.

Thus, ferrihemoglobin competes with cytochrome oxidase for cyanide. This competition is therapeutically effective because the amount of ferrihemoglobin that can be formed without impairing oxygen transport is much greater than the amount of cytochrome oxidase.

7. The absolute configuration of thiophosphate is opposite to that of ATP in the reaction catalyzed by ATP synthase. This result is consistent with an in-line phosphoryl transfer reaction occurring in a single step. The retention of configuration in the Ca^{2+}-ATPase reaction points to two phosphoryl transfer reactions—inversion by the first, and a return to the starting configuration by the second. The Ca^{2+}-ATPase reaction proceeds by a phosphorylated enzyme intermediate. See M.R. Webb, C. Grubmeyer, H.S. Penefsky, and D.R. Trentham. *J. Biol. Chem.* 255(1980):255.

8. $\Delta G^{\circ\prime} = -0.2 \times 2 \text{ (or 3, or 4)} \times 23 = -9.23, -13.8, \text{ or } -18.5 \text{ kcal/mol}$

These become positive values (energy input) for the $\Delta G'$ of ATP synthesis. Therefore,

$$\Delta G' = \Delta G^{\circ\prime} + 1.36 \log \frac{[\text{ATP}]}{[\text{ADP}][\text{P}_i]}$$

and for two protons,

$$9.23 = 7.3 + 1.36 \log \frac{[\text{ATP}]}{[\text{ADP}][\text{P}_i]}$$

$$\frac{[\text{ATP}]}{[\text{ADP}][\text{P}_i]} = 10^{1.93/1.36} = 26.2$$

Similar calculations with three and four protons gives ratios of 6.1×10^4 and 1.7×10^8, respectively.

Suspensions of isolated mitochondria synthesize ATP until this ratio is greater than 10^4, which shows that the number of protons translocated per ATP synthesized is at least three.

9. Such a defect (called *Luft's syndrome*) was found in a 38-year-old woman who was incapable of performing prolonged physical work. Her basal metabolic rate was more than twice normal, but her thyroid function was normal. A muscle biopsy showed that her mitochondria were highly variable and atypical in structure. Biochemical studies then revealed that oxidation and phosphorylation were not tightly coupled in these mitochondria. In this patient, much of the energy of fuel molecules was converted into heat rather than ATP. The development of mitochondrial medicine is lucidly reviewed in R. Luft, *Proc. Nat. Acad. Sci.* 91(1994):8731.

10. Dicylohexylcarbodiimide reacts readily with carboxyl groups, as was discussed earlier in regard to its use in peptide synthesis (Stryer, p. 69). Hence, the most likely targets are aspartate and glutamate side chains. In fact, aspartate 61 of subunit c of *E. coli* F_0 is specifically modified by this reagent. Conversion of this aspartate to an asparagine by site-specific mutagenesis also eliminates proton conduction. See A.E. Senior. *Biochim. Biophys. Acta* 726(1983):81.

11. Recall that during the cleavage of fructose 1,6-bisphosphate, glyceraldehyde 3-phosphate and dihydroxyacetone phosphate are formed. Only glyceraldehyde 3-phosphate can proceed through glycolysis to pyruvate.

Although it can be converted to glycerol by some cells, dihydroxyacetone phosphate is a dead-end in glycolysis; to proceed further it must be converted to glyceraldehyde 3-phosphate by triose phosphate isomerase.

12. Inhibitors of electron transport cause the electron carriers between the source of electrons (e.g., NADH, $FADH_2$) and the point of inhibition to become more reduced while the electron carriers between the point of inhibition and O_2 become more oxidized. Thus there is a "crossover point" from reduced carriers to oxidized carriers. Therefore, from the information given we conclude that the inhibitor acts somewhere between QH_2 and cytochrome *c*; it prevents the reduction of cytochrome *c* by QH_2.

CHAPTER

22

Pentose Phosphate Pathway and Gluconeogenesis

Chapters 19, 20, and 21 introduced you to the metabolic pathways that produce the bulk of ATP in aerobic organisms from glucose and other fuel molecules. In this chapter, Stryer describes two additional metabolic pathways that involve glucose: the pentose phosphate pathway and gluconeogenesis. The role of the pentose phosphate pathway is to produce reduced nicotinamide adenine dinucleotide phosphate (NADPH), which is the currency of reducing power utilized for most reductive biosyntheses. In addition, this pathway generates ribose 5-phosphate needed for DNA synthesis and produces various three-, four-, five-, six-, and seven-carbon sugars from glucose. The pentose phosphate pathway is linked to glycolysis (Chapter 19) by the common intermediates glucose 6-phosphate, fructose 6-phosphate, and glyceraldehyde 3-phosphate.

Gluconeogenesis is a biosynthetic pathway that generates glucose from noncarbohydrate precursors. All the reversible steps of glycolysis also occur in gluconeogenesis; only the essentially irreversible enzymatic steps of glycolysis are bypassed by different reactions in gluconeogenesis. A few of these reactions take place in the mitochondrial matrix and involve oxaloacetate and malate, which are also intermediates of the citric acid cycle (Chapter 20).

LEARNING OBJECTIVES

When you have mastered this chapter, you should be able to complete the following objectives:

The Pentose Phosphate Pathway (Stryer pages 559–565)

1. Distinguish the structures of *NADH, NADPH,* and *ATP,* and contrast their functions in metabolism.

2. List the major physiologic functions of the *pentose phosphate pathway,* which is also known as the *pentose shunt,* the *hexose monophosphate pathway,* and the *phosphogluconate oxidative pathway.*

3. Describe the reactions of the *oxidative branch* of the pentose phosphate pathway and the regulation of *glucose 6-phosphate dehydrogenase* by *NADP*$^+$ levels.

4. Explain how the pentose phosphate pathway and the glycolytic pathway are linked through reactions catalyzed by *transaldolase* and *transketolase.*

5. Outline the sugar interconversions of the *nonoxidative branch* of the pentose phosphate pathway.

6. Describe the different product stoichiometries obtained from the pentose phosphate pathway under conditions in which (1) more *ribose 5-phosphate* than NADPH is needed; (2) there is a balanced requirement for both; and (3) more NADPH than ribose 5-phosphate is needed, and the carbon atoms give rise to either CO_2 or pyruvate.

7. Explain the use of *radiolabeled glucose* in the elucidation of the relative activities of glycolysis and the pentose phosphate pathway in various tissues.

Enzymatic Mechanisms of Transketolase and Transaldolase, Glucose 6-Phosphate Dehydrogenase, and the Formation of Reduced Glutathione (Stryer pages 566–569)

8. Compare the role of *thiamine pyrophosphate* (*TPP*) in transketolase with its role in *pyruvate dehydrogenase* and *α-ketoglutarate dehydrogenase.* Outline the enzymatic mechanisms of transketolase and transaldolase.

9. Discuss the effects of *glucose 6-phosphate dehydrogenase deficiency* on red cells in drug-induced hemolytic anemia, and relate them to the biological roles of *glutathione.*

10. Describe the reduction of glutathione by *glutathione reductase.*

Gluconeogenesis (Stryer pages 569–574)

11. Describe the physiologic significance of *gluconeogenesis.* List the primary precursors of gluconeogenesis.

12. List the irreversible reactions of glycolysis and the enzymatic steps of gluconeogenesis that bypass them.

13. Name the major organs that carry out gluconeogenesis. Locate the various enzymes of gluconeogenesis in cell compartments.

14. Describe the enzymatic steps in the conversion of *pyruvate* to *phosphoenolpyruvate.* Name the enzymes, intermediates, and cofactors involved in these reactions.

15. Explain the role of *biotin* as a carrier for *activated* CO_2 in the *pyruvate carboxylase* reaction. Describe the control of pyruvate carboxylase by *acetyl CoA* and its role in maintaining the level of citric acid cycle intermediates.

16. Calculate the number of high-energy phosphate bonds consumed during gluconeogenesis and compare it with the number formed during glycolysis.

Regulation of Gluconeogenesis and the Cori Cycle (Stryer pages 574–578)

17. Describe the coordinated control of glycolysis and gluconeogenesis at the level of the *phosphofructokinase* and *fructose 1,6-bisphosphatase* reactions and the *pyruvate kinase* and *pyruvate carboxylase* as well as *phosphoenolpyruvate carboxykinase* reactions. Contrast the role of *fructose 2,6-bisphosphate* in the regulation of glycolysis and gluconeogenesis.

18. Discuss the effects of the hormones *insulin* and *glucagon* on the enzymes of glycolysis and gluconeogenesis.

19. Explain how *substrate cycles* may *amplify metabolic signals* or *produce heat.*

20. Outline the *Cori cycle* and explain its biological significance.

21. Contrast the properties and roles of the *H* and *M isozymes* of *lactate dehydrogenase.*

SELF-TEST

The Pentose Phosphate Pathway

1. Which of the following properties are common to NADH and NADPH?

   (a) They can be used interchangeably as cofactors for dehydrogenases.
   (b) Both have the same net charge at pH 7.0.
   (c) Both contain two ribose moieties, a nicotinamide ring, an adenine base, and a phosphate anhydride bond.
   (d) They have equivalent oxidation-reduction potentials.
   (e) Both are readily translocated through mitochondrial membranes.

2. Which of the following compounds is not a product of the pentose phosphate pathway?

   (a) NADPH
   (b) Glycerate 3-phosphate

(c) CO_2
(d) Ribulose 5-phosphate
(e) Sedoheptulose 7-phosphate

3. Figure 22-1 shows the first four reactions of the pentose phosphate pathway. Use it to answer the following questions:

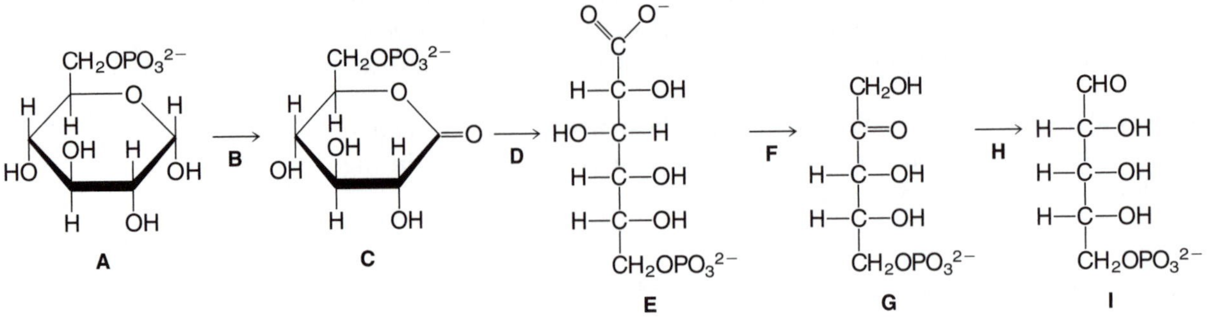

Figure 22-1
Reactions of the pentose phosphate pathway.

(a) Which reactions produce NADPH?

(b) Which reaction produces CO_2?

(c) Which compound is ribose 5-phosphate?

(d) Which compound is 6-phosphoglucono δ-lactone?

(e) Which compound is 6-phosphogluconate?

(f) Which reaction is catalyzed by phosphopentose isomerase?

(g) Which enzyme is deficient in drug-induced hemolytic anemia?

(h) Which compound can be a group acceptor in the transketolase reaction?

4. Which of the following statements about glucose 6-phosphate dehydrogenase are correct?

(a) It catalyzes the committed step in the pentose phosphate pathway.
(b) It is regulated by the availability of NAD^+.
(c) One of its products is 6-phosphogluconate.
(d) It contains thiamine pyrophosphate as a cofactor.
(e) It is important in the metabolism of glutathione in erythrocytes.

5. The nonoxidative branch of the pentose phosphate pathway *does not* include which of the following reactions?

(a) Ribulose 5-P $\rightleftharpoons$ ribose 5-P
(b) Xylulose 5-P + ribose 5-P $\rightleftharpoons$ sedoheptulose 7-P + glyceraldehyde 3-P
(c) Ribulose 5-P + glyceraldehyde 3-P $\rightleftharpoons$ sedoheptulose 7-P

(d) Sedoheptulose 7-P + glyceraldehyde 3-P $\rightleftharpoons$ fructose 6-P + erythrose 4-P
(e) Ribulose 5-P $\rightleftharpoons$ xylulose 5-P

6. Even if glucose 6-phosphate dehydrogenase is deficient, the synthesis of ribose 5-phosphate from glucose 6-phosphate can proceed normally. Explain how this is possible.

___

___

7. Which of the following conversions take place in a metabolic situation that requires much more NADPH than ribose 5-phosphate, as well as complete oxidation of glucose 6-phosphate to CO_2? The arrows represent one or more enzymatic steps.

(a) Glucose 6-phosphate $\longrightarrow\longrightarrow$ ribulose 5-phosphate ✓
(b) Fructose 6-phosphate + glyceraldehyde 3-phosphate $\longrightarrow\longrightarrow$ ribose 5-phosphate
(c) Ribose 5-phosphate $\longrightarrow\longrightarrow$ fructose 6-phosphate + glyceraldehyde 3-phosphate ✓
(d) Glyceraldehyde 3-phosphate $\longrightarrow\longrightarrow$ pyruvate
(e) Fructose 6-phosphate $\longrightarrow\longrightarrow$ glucose 6-phosphate ✓

8. If glucose is labeled on the C-2 carbon with ^{14}C, will the glyceraldehyde 3-phosphate resulting from conversions occurring in the glycolytic and the pentose phosphate pathways be labeled in the same position? Explain your answer.

___

___

9. Liver synthesizes fatty acids and lipids for export to other tissues. Would you expect the pentose phosphate pathway to have a low or a high activity in this organ? Explain your answer.

___

___

Enzymatic Mechanisms of Transketolase and Transaldolase, Glucose 6-Phosphate Dehydrogenase, and the Formation of Reduced Glutathione

10. Transaldolase and transketolase have which of the following similarities?

(a) Both require thiamine pyrophosphate.
(b) Both form a Schiff base with the substrate.
(c) Both use an aldose as a group donor.
(d) Both use a ketose as a group donor. ✓
(e) Both form a covalent addition compound with the donor substrate. ✓

11. Match the cofactors in the left column with the appropriate enzymes from the right column.

(a) Thiamine pyrophosphate ______
(b) $NAPD^+$ ______
(c) Biotin ______
(d) GTP ______

(1) Glucose 6-P dehydrogenase
(2) 6-Phosphogluconate dehydrogenase
(3) Pyruvate dehydrogenase
(4) Pyruvate carboxylase
(5) Transketolase
(6) Phosphoenolpyruvate carboxykinase

12. Which of the following statements regarding reduced glutathione is incorrect?

(a) It contains one γ-carboxyglutamate, one cysteine, and one glycine residue.
(b) It keeps the cysteine residues of proteins in their reduced states.
(c) It is regenerated from oxidized glutathione by glutathione reductase.
(d) It reacts with hydrogen peroxide and organic peroxides.
(e) It is decreased relative to oxidized glutathione in glucose 6-phosphate dehydrogenase deficiency.

13. Suggest reasons why glucose 6-phosphate dehydrogenase deficiency may be manifested in red blood cells but not in adipocytes, which also require NADPH for their metabolism.

______

______

Gluconeogenesis

14. Which of the following statements about gluconeogenesis are correct?

(a) It occurs actively in the muscle during periods of exercise.
(b) It occurs actively in the liver during periods of exercise or fasting.
(c) It occurs actively in adipose tissue during feeding.
(d) It occurs actively in the kidney during periods of fasting.
(e) It occurs actively in the brain during periods of fasting.

15. Glucose can be synthesized from which of the following noncarbohydrate precursors?

(a) Adenine
(b) Alanine
(c) Lactate
(d) Palmitic acid
(e) Glycerol

16. Which statement about glucose 6-phosphatase is correct? Glucose 6-phosphatase

(a) is bound to the inner mitochondrial membrane.
(b) is stabilized by an associated Ca^{2+}-binding protein.

(c) is directly associated with a glucose transporter.
(d) produces glucose and phosphate in a reaction that consumes energy.
(e) has an identical active site with hexokinase.

17. The following is the sequence of reactions of gluconeogenesis from pyruvate to phosphoenolpyruvate:

$$\text{Pyruvate} \underset{A}{\rightleftharpoons} \text{oxaloacetate} \underset{B}{\rightleftharpoons} \text{malate} \underset{C}{\rightleftharpoons} \text{oxaloacetate} \underset{D}{\rightleftharpoons} \text{phosphoenolpyruvate}$$

Match the capital letters indicating the reactions of the gluconeogenic pathway with the following statements:

(a) Occur in the mitochondria ______
(b) Occur in the cytosol ______
(c) Produces CO_2 ______
(d) Consumes CO_2 ______
(e) Requires ATP ______
(f) Requires GTP ______
(g) Is regulated by acetyl CoA ______
(h) Requires a biotin cofactor ______
(i) Is also a reaction of the citric acid cycle ______
(j) Is an anaplerotic reaction ______

18. Explain why a CO_2 is added to pyruvate in the pyruvate carboxylase reaction only to be subsequently removed by the phosphoenolpyruvate carboxykinase reaction. Identify the high-energy intermediate in the carboxylation reaction.

______

______

19. Glycogen is the storage form of glucose in the liver and muscles. In terms of absolute amounts, muscles contain more total glycogen than does liver, yet muscle glycogen is not a direct precursor of blood glucose. Explain this fact.

______

______

20. How many "high-energy" bonds are required to convert oxaloacetate to glucose?

(a) 2
(b) 3
(c) 4
(d) 5
(e) 6

Regulation of Gluconeogenesis and the Cori Cycle

21. Which of the following statements correctly describe what happens when acetyl CoA is abundant?

(a) Pyruvate carboxylase is activated.
(b) Phosphoenolpyruvate carboxykinase is activated.

(c) Phosphofructokinase is activated.
(d) If ATP levels are high, oxaloacetate is diverted to gluconeogenesis.
(e) If ATP levels are low, oxaloacetate is diverted to gluconeogenesis.

22. In the coordinated control of phosphofructokinase (PFK) and fructose 1,6-bisphosphatase (F-1,6-BPase),

(a) citrate inhibits PFK and stimulates F-1,6-BPase.
(b) fructose 2,6-bisphosphate inhibits PFK and stimulates F-1,6-BPase.
(c) acetyl CoA inhibits PFK and stimulates F-1,6-BPase.
(d) AMP inhibits PFK and stimulates F-1,6-BPase.
(e) NADPH inhibits PFK and stimulates F-1,6-BPase.

23. Indicate which of the conditions listed in the right column *increase* the activity of the glycolysis or gluconeogenesis pathways.

(a) Glycolysis ________
(b) Gluconeogenesis ________

(1) Increase in ATP
(2) Increase in AMP
(3) Increase in F-2,6-BP
(4) Increase in citrate
(5) Increase in acetyl CoA
(6) Increase in insulin
(7) Increase in glucagon
(8) Starvation
(9) Fed state

24. Explain how the levels of phosphoenolpyruvate carboxykinase are regulated by diverse hormones.

____

____

25. Which of the following statements about the Cori cycle and its physiologic consequences are correct?

(a) It involves the synthesis of glucose in muscle.
(b) It involves the release of lactate by muscle.
(c) It involves lactate synthesis in the liver.
(d) It involves ATP synthesis in muscle.
(e) It involves the release of glucose by the liver.

26. Explain how a substrate cycle can generate heat.

____

____

27. In muscle, lactate dehydrogenase produces lactate from pyruvate, whereas in the heart it preferentially synthesizes pyruvate from lactate. Explain how this is possible.

____

____

The Pentose Phosphate Pathway

1. c, d

2. b

3. (a) B, F (b) F (c) I (d) C (e) E (f) H (g) B (h) I

4. a, e

5. c

6. Ribose 5-phosphate can be synthesized from fructose 6-phosphate and glyceraldehyde 3-phosphate, both of which are glycolytic products of glucose 6-phosphate. These reactions are carried out by transketolase and transaldolase in a reversal of the nonoxidative branch of the pentose phosphate pathway and do not involve glucose 6-phosphate dehydrogenase.

7. a, c, e. Glucose 6-phosphate is converted to ribulose 5-phosphate, producing CO_2 and NADPH in the process. Then ribulose 5-phosphate, via ribose 5-phosphate, is transformed into fructose 6-phosphate and glyceraldehyde 3-phosphate. These two glycolytic intermediates are converted back to glucose 6-phosphate, and the cycle is repeated until the equivalent of six carbon atoms from glucose 6-phosphate are converted to CO_2.

8. No, the glyceraldehyde 3-phosphates resulting from the different pathways will be labeled on different carbons. For conversion via the glycolytic pathway, the ^{14}C label will appear on the C-2 of glyceraldehyde 3-phosphate. Conversion via the pentose phosphate pathway will result in ribose 5-phosphate and xylulose 5-phosphate that are labeled on their C-1 carbons. These two sugars will give rise to sedoheptulose 7-phosphate and fructose 6-phosphate, which will be labeled on both their C-1 and C-3 carbons. Upon the conversion of the fructose 6-phosphate to glyceraldehyde 3-phosphate the labels will be on C-1 and C-3.

9. The activity of the pentose phosphate pathway in the liver is high. The biosynthesis of fatty acids and lipids requires reducing equivalents in the form of NADPH. In all organs that carry out reductive biosyntheses, the pentose phosphate pathway supplies a large proportion of the required NADPH. In Chapter 24 of Stryer (pp. 620–621), you will learn that a reaction catalyzed by $NADP^+$-linked malate enzyme produces part of the NADPH required for fatty acid synthesis.

Enzymatic Mechanisms of Transketolase and Transaldolase, Glucose 6-Phosphate Dehydrogenase, and the Formation of Reduced Glutathione

10. d, e

11. (a) 3, 5 (b) 1, 2 (c) 4 (d) 6. (For a review of the cofactors of pyruvate dehydrogenase, see Stryer, pp. 514–517.)

12. a. Answer (a) is incorrect because the glutamate residue in glutathione is not γ-carboxyglutamate; rather, the glutamate in glutathione forms a peptide bond with the adjacent cysteine residue via its γ-carboxyl group. Recall the structure and role of γ-carboxyglutamate residues in clotting factors (Stryer, p. 255).

13. The glucose 6-phosphate dehydrogenase in erythrocytes and that in adipocytes are specified by distinct genes; they have the same function but different structures—that is, they are isozymes. Furthermore, NADPH synthesis by the pentose phosphate pathway may not be as critical in the cells of other tissues as it is in erythrocytes because other tissues have other sources of NADPH (see the answer to question 9).

Gluconeogenesis

14. b, d

15. b, c, e

16. b. The other answers are incorrect because glucose 6-phosphatase is bound to the luminal side of the endoplasmic reticulum membrane. It is associated with the glucose 6-phosphate and phosphate transporters, but not with the glucose transporter (see Figure 22-7 in Stryer). The hydrolysis of glucose 6-phosphate is an exergonic reaction. The active site of hexokinase is distinct from that of the phosphatase, since the hexokinase binds ATP.

17. (a) A, B (b) C, D (c) D (d) A (e) A (f) D (g) A (h) A (i) B (j) A. For (i), answer C is incorrect because this reaction occurs in the cytosol and not in the mitochondria, where the enzymes of the citric acid cycle are located.

18. The carboxylation reaction produces an activated carboxyl group in the form of a high-energy *carboxybiotin intermediate.* The cleavage of this bond and release of CO_2 in the phosphoenolpyruvate carboxykinase reaction or the transfer of the CO_2 to acceptors in other reactions in which biotin participates allows endergonic reactions to proceed. Thus, the formation of phosphoenolpyruvate from oxaloacetate is driven by the release of CO_2 ($\Delta G^{\circ\prime} = -4.7$ kcal/mol) and the hydrolysis of GTP ($\Delta G^{\circ\prime} = -7.3$ kcal/mol).

19. Muscle lacks glucose 6-phosphatase; therefore, glucose 6-phosphate cannot be dephosphorylated in muscle for export into the blood. Muscle stores glycogen for its own use as a source of readily available fuel during periods of activity.

20. c. The two steps in gluconeogenesis that consume GTP or ATP are

$$\text{Oxaloacetate} + \text{GTP} \longrightarrow \text{phosphoenolpyruvate} + \text{GDP} + CO_2$$

$$\text{3-Phosphoglycerate} + \text{ATP} \longrightarrow \text{1,3-bisphosphoglycerate} + \text{ADP}$$

Since two oxaloacetate molecules are required to synthesize one glucose molecule, a total of four "high-energy" bonds are required.

Regulation of Gluconeogenesis and the Cori Cycle

21. a, d

22. a

23. (a) 2, 3, 6, 9 (b) 1, 4, 5, 7, 8

24. The promoter of the carboxykinase gene contains sequences that respond to insulin, glucagon, glucocorticoids, and thyroid hormone. Binding of specific transcription factors that recognize hormones or cAMP to the regulatory DNA sequences controls the expression of the gene and the levels of the enzyme.

25. b, d, e

26. A substrate cycle, such as the one established by phosphofructokinase and fructose 1,6-bisphosphatase, causes the net hydrolysis of ATP. The energy from the hydrolysis of ATP into ADP and P_i is released as heat.

27. Muscle and heart have distinct lactate dehydrogenase isozymes. Heart lactate dehydrogenase contains mostly H-type subunits. This enzyme has higher affinity for substrates and is inhibited by high concentrations of pyruvate; that is, it is designed to form pyruvate from lactate. In contrast, muscle lactate dehydrogenase, which consists of M-type subunits, is more effective for forming lactate from pyruvate.

PROBLEMS

1. The conversion of glucose 6-phosphate to ribose 5-phosphate via the enzymes of the pentose phosphate pathway and glycolysis can be summarized as follows:

$$\text{Glucose 6-phosphate} + \text{ATP} \longrightarrow 6 \text{ ribose 5-phosphate} + \text{ADP} + \text{H}^+$$

Which enzyme uses the molecule of ATP shown in the equation?

2. Aminotransferases are enzymes that catalyze the removal of amino groups from amino acids to yield α-keto acids. How could the action of such enzymes contribute to gluconeogenesis? Consider the utilization of alanine, aspartate, and glutamate in your answer.

3. Citrate inhibits phosphofructokinase and stimulates fructose 1,6-bisphosphatase. Why are these effects consistent with cellular strategies for regulating glucose synthesis and degradation?

4. Several disorders of biotin metabolism are known, and their symptoms include deficiencies in the levels of a number of different carboxylase enzymes, each requiring biotin. One of these disorders leads to the accumulation of the biotin derivative biocytin (shown in the figure below) in the blood and to its excretion in the urine. Normally, biocytin is virtually undetectable in blood or urine. Deficiencies in carboxylase activity and accumulation of biocytin in blood of affected people can be relieved by daily administration of biotin; however, biotin supplements do not lower blood biocytin concentrations. The disorder that causes biocytin accumulation results from a deficiency in a particular enzyme. Describe the reaction catalyzed by that enzyme and relate the reaction to biotin metabolism in the cell.

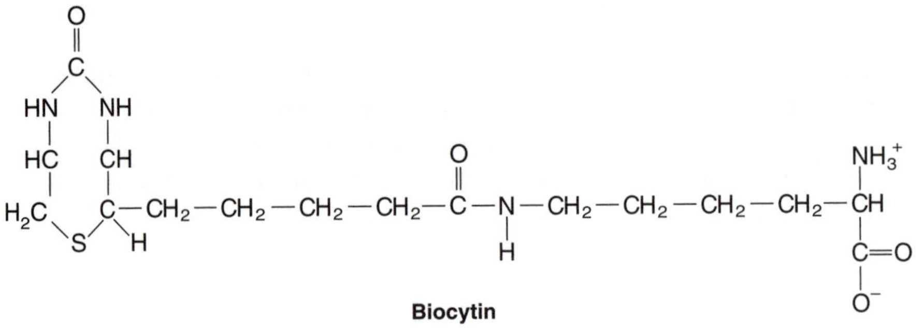

Biocytin

5. Liver and other organ tissues contain relatively large quantities of nucleic acids. During digestion, nucleases hydrolyze RNA and DNA, and among the products is ribose 5-phosphate.

(a) How can this molecule be used as a metabolic fuel?
(b) Another product formed by the degradation of nucleic acids is 2-deoxyribose 5-phosphate. Can this molecule be converted to glycolytic intermediates through the action of the pentose phosphate pathway? Explain your answer.

6. Bacteria can carry out gluconeogenesis using components of the citric acid cycle or amino acids, but unlike mammals, they can generate glucose from acetate via the glyoxylate pathway. Because bacteria have no organelles, the enzymes of the citric acid cycle and the glyoxylate pathway are not compartmentalized.

(a) Why is regulation of the glyoxylate pathway necessary in bacteria?
(b) Studies show that bacterial isocitrate lyase is allosterically inhibited by high concentrations of phosphoenolpyruvate. Would you expect to see the inhibition of isocitrate lyase when bacteria are utilizing glucose as a sole carbon source? Why?
(c) Would you expect the glyoxylate pathway to be more active than the citric acid cycle when bacteria are growing on acetate? Why?
(d) Would you expect to find glucose 6-phosphatase in bacteria?

7. You have glucose that is radioactively labeled with ^{14}C at C-1, and you have an extract that contains the enzymes that catalyze the reactions of the glycolytic and the pentose phosphate pathways, along with all the intermediates of the pathways.

(a) If the enzymes of the *oxidative* branch of the pentose phosphate pathway are *not* active in your extract, is it possible to obtain labeled sedoheptulose 7-phosphate using glucose labeled with ^{14}C at C-1? Explain.
(b) Suppose that in a second experiment *all* the enzymes of both the oxidative branch and the nonoxidative branch of the pentose phosphate pathway are active. Will the labeling pattern of sedoheptulose 7-phosphate be different? Explain.
(c) Can sedoheptulose 7-phosphate form a heterocyclic ring?

8. Why is the pentose phosphate pathway more active in cells that are dividing than in cells that are not?

9. Even if the concentration of lactate and other precursors were high, why is it unlikely that liver cells would be carrying out gluconeogenesis under anaerobic conditions?

10. A bacterium isolated from a soil culture can utilize ribose as a sole source of carbon when grown anaerobically. Experiments show that in the anaerobic pathways leading to ATP production, three molecules of ribose are converted to five molecules of CO_2 and five molecules of ethanol. These organisms also use ribose for the production of NADPH. The assimilation of ribose begins with its conversion to ribose 5-phosphate, with ATP serving as a phosphoryl donor.

(a) Explain how ribose can be converted to CO_2 and ethanol under anaerobic conditions. Write the overall reaction, showing how much ATP can be produced per pentose utilized.
(b) Write an equation for the generation of NADPH using ribose as a sole source of carbon.

11. Mature erythrocytes, which lack mitochondria, metabolize glucose at a high rate. In response to the increased availability of glucose, erythrocytes generate lactate and also evolve carbon dioxide.

(a) Why is generation of lactate necessary to ensure the continued utilization of glucose?
(b) In erythrocytes, what pathway is likely to be used for the generation of carbon dioxide from glucose? Can glucose be completely oxidized to CO_2 in erythrocytes? Explain.

12. In liver, V_{max} for fructose bisphosphatase is three to four times higher than V_{max} for phosphofructokinase, whereas in muscle it is only about 10 percent of that of phosphofructokinase. Explain this difference.

13. In gluconeogenesis, four enzymes are required to bypass the irreversible steps of glycolysis. Liver and kidney tissue express all four of these enzyme activities, while not all of them are found in other tissues. Muscle cells as well as erythrocytes are likely to have at least one of the enzymes used for gluconeogenesis. Which one, and why?

14. The two-step conversion of pyruvate to phosphoenolpyruvate by the enzymes pyruvate carboxylase and phosphoenolpyruvate carboxykinase is greatly stimulated by very small amounts of CO_2. Why?

15. Von Gierke's disease, a deficiency in liver glucose 6-phosphatase activity, can lead to a number of metabolic problems, including an inability to maintain proper blood glucose levels and to metabolize glycogen properly (see p. 598 of Stryer). When the enzyme is assayed using cell extracts from one group of affected subjects, glucose 6-phosphatase activity is absent from the extracts under a number of conditions, including freezing and thawing the extract before the assay. Extracts taken from another group of affected subjects have glucose 6-phosphatase activity in freeze-thawed samples, but the enzyme activity is absent in untreated extracts. Explain the differences in the two groups of cell extracts.

16. Avidin, a 70-kd protein that inhibits biotin enzymes, is found in egg white. The protein interferes with gluconeogenesis by binding biotin.

(a) What do you think is the function of avidin in the egg? Would gluconeogenesis be required for avian development? If so, how might inhibitory effects of avidin be avoided?
(b) People who eat raw eggs may suffer from biotin insufficiency, but those who eat cooked eggs do not. Why?

17. A biochemist needs to determine whether a particular tissue homogenate has a high level of pentose phosphate pathway activity. She incubates one sample with ^{14}C-1 glucose, and another with ^{14}C-6 glucose. Then she measures the specific activity of radioactive CO_2 generated by each samples. Her measurements show that the specific activity of CO_2 from the experiment using glucose labeled at C-1 is much higher than that from the sample in which glucose labeled at C-6 was employed. What is her conclusion?

ANSWERS TO PROBLEMS

1. Phosphofructokinase uses ATP to convert fructose 6-phosphate to fructose 1,6-bisphosphate, which is then cleaved by aldolase to yield dihydroxyacetone phosphate (DHAP) and glyceraldehyde 3-phosphate. The conversion of DHAP to a second molecule of glyceraldehyde 3-phosphate provides the molecules that are needed for the synthesis of ribose 5-phosphate.

2. Examination of the structures of the α-keto acid analogs of alanine, aspartate, and glutamate shows that each can be used for gluconeogenesis. Specific transaminases convert alanine to pyruvate, aspartate to oxaloacetate, and glutamate to α-ketoglutarate. These amino acids, along with others whose carbon skeletons can be used for the synthesis of glucose, are termed *glucogenic amino acids.*

3. When citrate levels are high, the citric acid cycle has ample fuel molecules that can be oxidized for the generation of ATP. Under such conditions, gluconeogenesis is stimulated and the rate of glycolysis is decreased. Citrate helps achieve these goals by inhibiting phosphofructokinase, which catalyzes a rate-determining step in glycolysis, and by increasing the activity of fructose 1,6-bisphosphatase, a key enzyme in gluconeogenesis.

4. The enzyme, known as *biotinidase,* cleaves the amide linkage in biocytin, converting it into lysine and free biotin. Biotinidase activity is essential for the final step in the cellular proteolysis of carboxylases during enzyme turnover. The free biotin generated by the action of biotinidase can be incorporated into newly synthesized carboxylase enzymes. The turnover of biotin in cells keeps the daily requirement for biotin relatively low; the requirement increases when biocytin cannot be cleaved. Biotin supplements relieve deficiencies in carboxylase activity, but biocytin accumulation and excretion continue because biotinidase remains deficient.

5. (a) The most direct route for the oxidative degradation of ribose 5-phosphate is its conversion to glycolytic intermediates by the nonoxidative enzymes of the pentose phosphate pathway, as described by Stryer on pages 561 to 562 of his text. The overall reaction is

$$\text{3 Ribose 5-phosphate} \longrightarrow \text{2 fructose 6-phosphate} + \text{glyceraldehyde 3-phosphate}$$

(b) The formation of glycolytic intermediates from 2-deoxyribose 5-phosphate is not possible, because unlike ribose 5-phosphate,

2-deoxyribose 5-phosphate lacks a hydroxyl group at C-2. It is therefore not a substrate for phosphopentose isomerase, whose action is required to convert ketopentose phosphates to substrates that can be utilized by other enzymes of the pentose phosphate pathway. Most deoxyribose phosphate molecules are used in salvage pathways to form deoxynucleotides (Stryer, p. 755).

6. (a) When ATP levels are low in the bacterial cell, the flux of carbon atoms through the citric acid cycle must increase to generate the reduced cofactors needed for the generation of ATP. Because the glyoxylate cycle bypasses the steps that generate NADH and $FADH_2$, it must be suppressed when ATP synthesis is needed. On the other hand, high levels of ATP and an abundance of components of the citric acid cycle cause the glyoxylate cycle to accelerate, leading to the synthesis of succinate and other biosynthetic intermediates.

(b) Glucose is converted to biosynthetic intermediates via the glycolytic and pentose phosphate pathways as well as the citric acid cycle. When ample glucose is available, isocitrate lyase activity can be inhibited because the glyoxylate pathway is not needed to generate such intermediates. High concentrations of PEP generated by glycolysis inhibit isocitrate lyase activity. You would expect the glyoxylate cycle to be less active.

(c) When acetate is the sole carbon source, acetyl CoA is formed. The net synthesis of succinate and other intermediates of the citric acid cycle cannot be accomplished from acetyl CoA. Instead, succinate is generated by the glyoxylate cycle, which is activated because low levels of PEP do not inhibit isocitrate lyase.

(d) No. Glucose 6-phosphatase activity is found in cells that export glucose to other cells. Bacteria usually utilize glucose as a fuel and as a source of carbon atoms and therefore have no need for an enzyme that allows glucose to leave the bacterial cell.

7. (a) Yes. The most direct route would be the conversion of glucose to fructose 6-phosphate, followed by the condensation of fructose 6-phosphate with erythrose 4-phosphate to form sedoheptulose 7-phosphate and glyceraldehyde 3-phosphate. The labeled carbon of glucose becomes the C-1 of fructose 6-phosphate and C-1 of sedoheptulose 7-phosphate.

(b) The labeling pattern will be the same, although the amount of labeled carbon incorporated into the heptose will be reduced. In the oxidative branch of the pentose phosphate pathway, the labeled glucose is converted to glucose 6-phosphate with the ^{14}C label on C-1. Glucose 6-phosphate then undergoes successive oxidations and decarboxylation to form ribulose 5-phosphate. The label is lost when the C-1 carbon is removed during decarboxylation.

(c) Sedoheptulose 7-phosphate is a ketose and can form a heterocyclic ring through a hemiketal linkage. The most likely link would be between the keto group at C-2 and the hydroxyl group at C-6.

8. Cells have a high rate of nucleic acid biosynthesis when they grow and divide. Among the precursors needed is ribose 5-phosphate, which is synthesized through the action of the enzymes of the glycolytic and the pentose phosphate pathways. Biosynthetic reactions requiring NADPH occur at a high rate in growing and dividing cells. For these reasons, the enzymes of the pentose phosphate pathway will be extremely active in dividing cells.

9. Gluconeogenesis requires six high-energy phosphate bonds for every two molecules of pyruvate converted to glucose. These phosphate molecules come from ATP, most of which is generated in liver by oxidative phosphorylation in the presence of oxygen. Under anaerobic conditions, the only source of ATP is glycolysis, but only two molecules of ATP are produced per glucose converted to pyruvate. The extra price of generating glucose from pyruvate would lead to a deficit in the supply of ATP. Stryer also points out on pages 574 and 575 that the balance between gluconeogenesis and glycolysis is stringently controlled; therefore, it seems unlikely that ATP generation via glycolysis would occur if cellular conditions favored gluconeogenesis.

10. (a) To generate ATP, ethanol, and CO_2, ribose must first be converted to ribose 5-phosphate, with ATP serving as a phosphate donor. Then, in the nonoxidative branch of the pentose phosphate pathway, three molecules of ribose 5-phosphate are converted to two molecules of fructose 6-phosphate and one molecule of glyceraldehyde 3-phosphate. Two molecules of ATP are required for the production of fructose 1,6-bisphosphate from fructose 6-phosphate. The formation of a total of five molecules of glyceraldehyde 3-phosphate is achieved through the action of aldolase and triose phosphate isomerase. These five molecules are converted to five molecules of pyruvate, yielding ten ATP molecules and five NADH molecules. To keep the anaerobic cell in redox balance, the pyruvate molecules are converted to five molecules of ethanol, with the production of five CO_2 molecules and five NAD^+. The overall reaction is

$$3 \text{ Ribose} + 5 \text{ ADP} + P_i \longrightarrow 5 \text{ ethanol} + 5\, CO_2 + 5 \text{ ADP}$$

(b) Ribose 5-phosphate molecules must first be converted to glucose 6-phosphate for the oxidative enzymes of the pentose pathway to generate NADPH. The stoichiometry of the reactions is

$$6 \text{ Ribose 5-phosphate} \longrightarrow 4 \text{ fructose 6-phosphate} + 2 \text{ glyceraldehyde 3-phosphate}$$

$$4 \text{ Fructose 6-phosphate} \longrightarrow 4 \text{ glucose 6-phosphate}$$

$$2 \text{ Glyceraldehyde 3-phosphate} \longrightarrow \text{glucose 6-phosphate} + P_i$$

$$5 \text{ Glucose 6-phosphate} + 10\, NADP^+ + 5\, H_2O \longrightarrow 5 \text{ ribose 5-phosphate} + 10\, NADPH + 10\, H^+ + 5\, CO_2$$

The net reaction is

$$\text{Ribose 5-phosphate} + 10\, NADP^+ + 5\, H_2O \longrightarrow 10\, NADPH + 10\, H^+ + 5\, CO_2 + P_i$$

11. (a) Because erythrocytes lack mitochondria, they cannot use the citric acid cycle to regenerate the NAD^+ needed to sustain glycolysis. Instead, they regenerate NAD^+ by reducing pyruvate through the action of lactate dehydrogenase; NAD^+ is then reduced in the reaction catalyzed by glyceraldehyde 3-phosphate dehydrogenase during glycolysis. Failure to oxidize the NADH generated in the glycolytic pathway will cause a reduction in the rate of glucose breakdown.

(b) In erythrocytes, the pentose phosphate pathway is the only route available to yield CO_2 from glucose. Glucose can be completely oxidized by first entering the oxidative branch of the pathway, generating NADPH and ribose 5-phosphate. Transaldolase and transketolase then convert the pentose phosphates to fructose 6-phosphate and glyceraldehyde 3-phosphate. Part of the gluconeogenic pathway is used to convert both the products to glucose 6-phosphate. The net reaction is

$$\text{Glucose 6-P} + 12\ NADP^+ + 7\ H_2O \longrightarrow 6\ CO_2 + 12\ NADPH + 12\ H^+ + P_i$$

12. In contrast to muscle tissue, which oxidizes glucose to yield energy, liver tissue generates glucose primarily for export to other tissues. Thus, one would expect the rate of gluconeogenesis in the liver to be greater than the rate of glycolysis. Therefore, the relative catalytic capacity (as measured by V_{max}) of fructose bisphosphatase, a key enzyme in gluconeogenesis, should be expected to exceed that of phosphofructokinase, which is a regulatory enzyme of the glycolytic pathway.

13. Both muscle tissue and erythrocytes are most likely to require fructose 1,6-bisphosphatase, which is needed to help generate glucose 6-phosphate from glycolytic intermediates produced by the pentose phosphate pathway. Red cells use the oxidative portion of the pathway to produce NADPH and ribose 5-phosphate. Ribose 5-phosphate molecules are converted into glyceraldehyde 3-phosphate and fructose 6-phosphate by transaldolase and transketolase. The pathway from glyceraldehyde 3-phosphate to glucose 6-phosphate includes the conversion of fructose 1,6-bisphosphate to fructose 6-phosphate, catalyzed by fructose 1,6-bisphosphatase. Muscle cells use the enzyme in converting glycolytic intermediates to ribose 5-phosphate for the synthesis of nucleotides and cofactors like NAD^+.

14. In the two reactions CO_2 is essentially a catalytic molecule. While it is a substrate for the conversion of pyruvate into oxaloacetate, it is regenerated when oxaloacetate reacts with GTP to form phosphoenolpyruvate. Only a small amount of CO_2 is therefore required to initiate the reactions.

15. In liver cells, glucose 6-phosphatase protein is located on the lumenal side of the endoplasmic reticulum, where it cleaves glucose 6-phosphate transported into the lumen by the transport protein T1 (see Figure 22-7 in Stryer). If in an affected person glucose 6-phosphate is inactive, its activity will not be observed in both untreated and in freeze-thawed cells. However, consider people in which the T1 transport protein is defective; in this case glucose 6-phosphate will be unable to enter the lumen of the endoplasmic reticulum where it can be hydrolyzed. Freezing and thawing of cell extracts ruptures membranes of the endoplasmic reticulum, exposing

glucose 6-phosphatase to its substrate, glucose 6-phosphate. Therefore extracts from subjects with T1 deficiencies will exhibit no enzyme activity in untreated extracts but will have glucose 6-phosphatase activity in freeze-thawed samples. There is now good evidence that at least two and possibly more different types of deficiencies in glucose 6-phosphatase function can be explained on the basis of the organization of the transport and hydrolysis system in the endoplasmic reticulum.

16. (a) One possible function of avidin in egg white may be to prevent growth of bacteria after accidental infection. The binding of biotin by avidin could prevent the conversion of pyruvate to oxaloacetate, interfering with both the citric acid cycle and the synthesis of glucose 6-phosphate in bacteria. It seems likely that embryonic development in the fertilized egg would require biotin enzymes. Even if the rate of gluconeogenesis were low, pyruvate carboxylase would be needed to replenish oxaloacetate in the citric acid cycle. As you will see in Chapter 24, another biotin enzyme, acetyl CoA carboxylase, is needed for the synthesis of fatty acyl chains, which can be used for phospholipid synthesis. Dividing cells have a high rate of fatty acid and phospholipid synthesis to support the production of membranes. One way to avoid the inhibitory effects of avidin would be to prevent the entry of the protein into the yolk or other areas of the egg where the embryo is growing. The avidin protein could also be hydrolyzed early in development, at which time other proteins, like ovalbumin, are degraded to provide amino acids for new protein synthesis and for oxidation.

(b) Avidin is a small protein that is denatured by heat. Therefore, those who eat cooked eggs are ingesting inactive avidin, which cannot bind biotin.

17. The experiments show that the activity of the pentose phosphate pathway is high. In the pentose phosphate pathway, glucose labeled at C-1 is decarboxylated, while glucose labeled at C-6 is not. On the other hand, both C-1- and C-6-labeled glucose are decarboxylated to the same extent by the combined action of the glycolytic pathway and the citric acid cycle. Because in these experiments, the specific activity (ratio of labeled CO_2 to total CO_2) is higher for C-1-labeled glucose, much of the glucose in the experiment must be moving through the pentose phosphate pathway.

EXPANDED SOLUTIONS TO TEXT PROBLEMS

1. Since the C-1 of glucose is lost during the conversion to pentose, carbon atoms 2 through 6 of glucose become carbon atoms 1 through 5 of the pentose. That is, each pentose carbon is numerically 1 less than its counterpart in glucose.

2. Note that in the oxidative decarboxylation of 6-phosphogluconate, oxidation occurs at the carbon β to the carboxyl group. A similar β-oxidation occurs during the decarboxylation of isocitrate in the citric acid cycle. In both cases a β-keto acid intermediate is formed. Since β-keto acids are relatively unstable, they are easily decarboxylated.

3. Ribose 5-P is first converted to xylulose 5-P (labeled in C-1) via ribulose 5-P. Transketolase can then catalyze the conversion of ribose 5-P + xylulose 5-P to sedoheptulose 7-P (labeled in C-1 and C-3) and glyceraldehyde 3-P. Transaldolase then transfers carbons 1 through 3 of sedoheptulose 7-P to glyceraldehyde 3-P, forming erythrose 4-P, which is unlabeled (from carbons 4 through 7 of sedoheptulose), and fructose 6-P, which is labeled in C-1 and C-3 (from C-1 and C-3 of sedoheptulose).

4. (a) To make six pentoses, four glucose 6-phosphates must be converted to fructose 6-phosphate (no ATP required), and one glucose 6-phosphate must be converted to two molecules of glyceraldehyde 3-phosphate (this requires one ATP). These are converted to pentoses by the following reactions (Stryer, p. 563):

2 Fructose 6-phosphate + *2 glyceraldehyde 3-phosphate* ⟶
2 erythrose 4-phosphate + *2 xylulose 5-phosphate*

2 Fructose 6-phosphate + 2 erythrose 4-phosphate ⟶
2 glyceraldehyde 3-phosphate + 2 sedoheptulose 7-phosphate

2 Glyceraldehyde 3-phosphate + 2 sedoheptulose 7-phosphate ⟶
2 xylulose 5-phosphate + *2 ribose 5-phosphate*

(b) What really happens is that six molecules of glucose 6-phosphate are converted to six CO_2 + six ribulose 5-phosphates + 12 NADPH + 12 H^+ (see Table 22-1). The ribulose phosphates are then converted back to five molecules of glucose 6-phosphate by the action of transketolase and transaldolase (see Stryer, p. 562). By these reactions three pentoses are converted to two hexoses and one triose. Thus six pentoses can be converted to four hexoses plus two trioses, and the latter can be converted to the fifth hexose.

5. The lactate level in the maternal circulation, and hence in the fetal circulation, increases during pregnancy because the mother is carrying a growing fetus with its own metabolic demands. The shift to H_4 in the fetal heart enables the fetus to use lactate as a fuel. Consequently, there is less need for gluconeogenesis by the liver and kidneys of the mother.

6. The double bond in Schiff bases can be reduced to stable molecules with sodium borohydride ($NaBH_4$). Since transaldolase forms a Schiff base with a ketose substrate (Stryer, p. 567), this enzyme-substrate complex can be reduced with tritiated $NaBH_4$ to yield a stable radioactive derivative of the active-site lysine. *Fingerprinting* the labeled enzyme (Stryer, p. 171) will identify the lysine at the active site.

7. The $\Delta E'_0$ for the reduction of glutathione by NADPH is +0.09 V. Then $\Delta G^{\circ\prime} = -nF\Delta E'_0 = -2 \times 23.06 \times 0.09 = -4.15$ kcal/mol. Also, $K'_{eq} = 10^{-\Delta G^{\circ\prime}/1.36} = 10^{4.15/1.36} = 1.126 \times 10^3$. Thus,

$$K_{eq} = \frac{[GSH]^2[NADP^+]}{[GSSG][NADPH]} = 1126$$

After substituting the given concentrations for GSH and GSSG,

$$\frac{[0.01\text{M}]^2[\text{NADP}^+]}{[0.001\text{M}][\text{NADPH}]} = 1.126 \times 10^3$$

Therefore,

$$\frac{[\text{NADP}^+]}{[\text{NADPH}]} = 1.126 \times 10^4$$

and

$$\frac{[\text{NADPH}]}{[\text{NADP}^+]} = \frac{1}{1.126 \times 10^4} = \frac{8.9 \times 10^5}{1}$$

8. Under normal circumstances when glucose is abundant, the resulting high liver concentration of fructose 2,6-bisphosphate inhibits gluconeogenesis by blocking fructose 1,6-bisphosphatase. In this genetic disorder the phosphatase is active irrespective of the glucose level. When phosphofructokinase is also active, substrate cycling (Stryer, p. 576) between fructose 6-phosphate and fructose 1,6-bisphosphate is increased, thus generating more heat. Because of this *substrate cycling,* the concentration of fructose 1,6-bisphosphate is lower than normal. It follows that less pyruvate is formed, resulting in less acetyl CoA. In turn, less ATP is formed by the citric acid cycle and oxidative phosphorylation.

9. The lipoic acid is covalently attached to a lysine side chain in the dihydrolipoyl transacetylase component (E_2) of pyruvate dehydrogenase (p. 516). This flexible unit carries acetyl groups from bound thiamine pyrophosphate to coenzyme A.

10. Pyruvate carboxylase has a covalently bound biotin cofactor and is therefore inhibited by avidin. Since oxaloacetate is an intermediate in the conversion of pyruvate to glucose, reactions (b) and (e) are inhibited.

CHAPTER

23

Glycogen Metabolism

The topic of carbohydrate metabolism presented in Chapters 19 and 21 is further developed in this chapter with a detailed discussion of the metabolism of glycogen, the storage form of glucose. Glycogen is important in the metabolism of higher animals because its glucose residues can be easily mobilized by the liver to maintain blood glucose levels and by muscle to satisfy the energy needs of contraction during bursts of activity. Stryer first reviews the structure and the physiologic roles of glycogen. You were introduced to the structure of glycogen, a polymer of glucose, in the chapter on carbohydrates (Chapter 18). With glucose as the ending and starting points, Stryer next presents the enzymatic reactions of glycogen degradation and synthesis. This is followed by a discussion of the control of these reactions by allosteric mechanisms and the phosphorylation and dephosphorylation of the key enzymes in response to hormonal signals. AMP, ATP, glucose, and glucose 6-phosphate act as allosteric effectors, and the hormones insulin, glucagon, and epinephrine function as signals in transduction pathways that control critical enzyme phosphorylations and dephosphorylations. Stryer describes the structures and the control mechanisms for phosphorylase, phosphorylase kinase, glycogen synthase, and protein phosphatase 1. The differences in glycogen metabolism in muscle and liver are related to the differing physiologic functions these tissues perform. He concludes the chapter with a discussion of the biochemical bases of several glycogen storage diseases.

LEARNING OBJECTIVES

When you have mastered this chapter, you should be able to complete the following objectives:

Introduction (Stryer pages 581–582)

1. Describe the structure of *glycogen* and its role in the *liver* and *muscle.*

2. Discuss the properties of *glycogen granules* and their location within cells.

Reactions of Glycogen Degradation and Synthesis (Stryer pages 582–589)

3. Explain the advantage of the *phosphorolytic cleavage* of glycogen over its *hydrolytic cleavage.*

4. Outline the steps in the degradation of glycogen, and relate them to the action of phosphorylase, *transferase,* and *α-1,6-glucosidase* (which is also known as the *debranching enzyme*).

5. Compare the reaction mechanisms of *phosphoglucomutase* and *phosphoglyceromutase.*

6. Explain the importance of *glucose 6-phosphatase* in the release of glucose by the liver. Note the absence of this enzyme in the brain and muscle.

7. Contrast the synthesis of glycogen with its degradation. Justify the existence of distinct pathways for the biosynthesis and degradation of glycogen.

8. Explain the roles of *UDP–glucose* and *inorganic pyrophosphatase* in the synthesis of glycogen.

9. Outline the steps in the synthesis of glycogen, name the pertinent enzymes, and note the requirement for a *primer.* Describe the actions of *glycogenin* and its effect on *glycogen synthase.*

10. Discuss the efficiency of glycogen as a storage form of glucose.

11. Describe the roles of *pyridoxal 5′-phosphate, general acid-base catalysis,* and the *carbonium ion intermediate* in the mechanism of action of *glycogen phosphorylase.*

Control of Glycogen Metabolism, and Structural Features of the Enzymes of Glycogen Metabolism (Stryer pages 590–594)

12. Describe the phosphorylation of phosphorylase by *phosphorylase kinase* and it dephosphorylation by *protein phosphatase 1* (*PP1*).

13. Explain the relationships between *phosphorylase* a and *phosphorylase* b, the *T* (*tense*) and *R* (*relaxed*) forms of each, and the allosteric effectors that mediate their interconversions in skeletal muscle.

14. Contrast the regulation of liver phosphorylase and muscle phosphorylase.

15. Contrast the important structural features of phosphorylase *a* and phosphorylase *b.* Note the variety of binding sites, their functional roles, and the critical location of the phosphorylation and AMP binding sites near the subunit interface.

16. Describe the major compositional features of phosphorylase kinase and its activation by *protein kinase A* (*PKA*). Explain the effects of *calmodulin* and Ca^{2+} on glycogen metabolism in muscle and liver.

17. Distinguish between the active and inactive forms of glycogen synthase in terms of their states of phosphorylation and contrast the effects of phosphorylation on glycogen synthase and glycogen phosphorylase.

Reaction Cascades in the Phosphorylation and Dephosphorylation of Glycogen Synthase and Phosphorylase (Stryer pages 594–597)

18. Compare the effects of *glucagon* and *epinephrine* on glycogen metabolism in the liver and muscle.

19. List the sequence of events from the binding of hormones by their receptors to the phosphorylation of glycogen synthase and phosphorylase. Explain the roles of *cAMP* and PKA in these processes.

20. Explain the roles of protein phosphatase 1, its *G subunit*, and *inhibitor 1* in the control of the activities of glycogen phosphorylase and synthase.

21. Outline the effects of *insulin* on glycogen synthase.

Regulation of Blood Glucose Levels, and Glycogen-Storage Diseases (Stryer pages 597–599)

22. Describe the events that lead to the inactivation of phosphorylase and the activation of glycogen synthase by glucose in the liver. Note the role of phosphorylase *a* as the glucose sensor in liver cells and the participation of PP1 in glucose sensing.

23. Give examples of glycogen storage diseases, and relate the biochemical defects with the clinical observations.

SELF-TEST

Introduction

1. Answer the following questions about the glycogen fragment in Figure 23-1:

Figure 23-1
Fragment of glycogen. (*R* represents the rest of the glycogen molecule.)

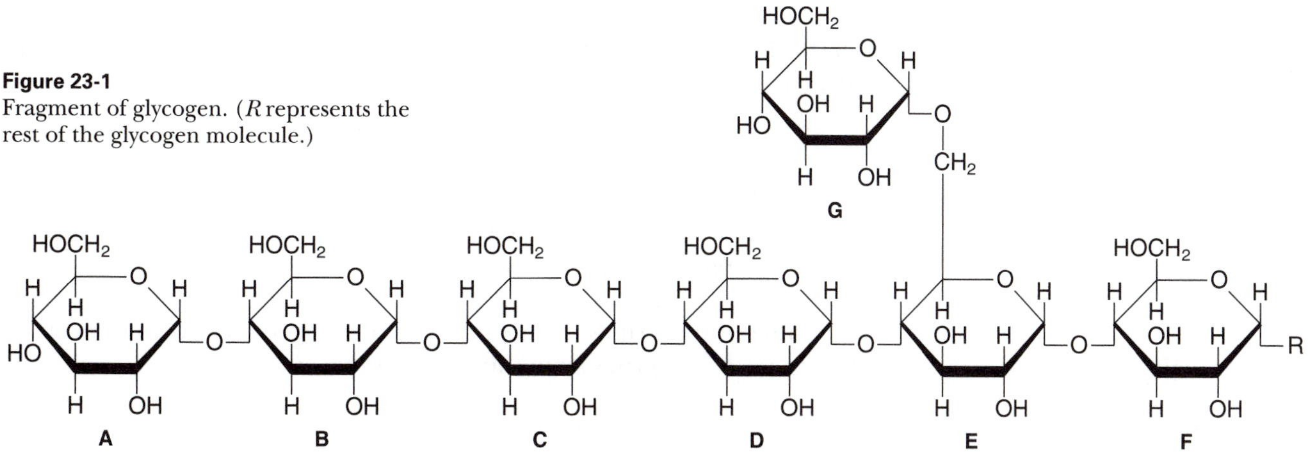

(a) Which residues are at nonreducing ends?

(b) An α-1,6 glycosidic linkage occurs between which residues?

(c) An α-1,4 glycosidic linkage occurs between which residues?

(d) Is the glycogen fragment a substrate for phosphorylase *a*? Explain:

(e) Is the glycogen fragment a substrate for the debranching enzyme? Explain:

(f) Is the glycogen fragment a substrate for the branching enzyme? Explain:

2. Which of the following statements about glycogen storage are incorrect?

(a) Glycogen is stored in muscles and liver.
(b) Glycogen is a major source of stored energy in brain.
(c) Glycogen reserves can exceed the equivalent of 10,000 kcal in the average human.
(d) Glycogen nearly fills the nucleus of cells that specialize in glycogen storage.
(e) Glycogen storage occurs in the form of dense granules in the cytoplasm of cells.

3. How do the multiple enzymes bound to glycogen granules differ from multienzyme complexes, such as pyruvate dehydrogenase?

Reactions of Glycogen Degradation and Synthesis

4. Explain why the phosphorolytic cleavage of glycogen is more energetically advantageous than its hydrolytic cleavage.

5. Which of the following statements about the role of pyridoxal phosphate in the mechanism of action of phosphorylase are correct?

(a) It interacts with orthophosphate.
(b) It acts as a general acid-base catalyst.
(c) It orients the glycogen substrate in the active site.
(d) It donates a proton to the O-4 of the departing glycogen chain.
(e) It binds water at the active site.

6. Match the enzymes that degrade glycogen in the left column with the appropriate properties from the right column.

(a) Phosphorylase 1, 6, 4

(b) α-1,6-Glucosidase ____________

(c) Transferase ____________

(1) Is part of a single polypeptide chain with two activities
(2) Cleaves α-1,4 glucosidic bonds
(3) Releases glucose
(4) Releases glucose 1-phosphate
(5) Moves three sugar residues from one chain to another
(6) Requires ATP.

7. The phosphoglucomutase reaction is similar to the phosphoglyceromutase reaction of the glycolytic pathway. Which of the following properties are common to both enzymes?

(a) Both have a phosphoenzyme intermediate.
(b) Both use a glucose 1,6-bisphosphate intermediate.
(c) Both contain pyridoxal phosphate, which donates its phosphate group to the substrate. ✓
(d) Both transfer the phosphate group from one position to another on the same molecule. ✓

8. The activity of which of the following enzymes is *not required* for the release of large amounts of glucose from liver glycogen?

(a) Glucose 6-phosphatase
(b) Fructose 1,6-bisphosphatase
(c) α-1,6-Glucosidase
(d) Phosphoglucomutase
(e) Glycogen phosphorylase

9. Which of the following are common features of both glycogen synthesis and glycogen breakdown?

(a) Both require UDP–glucose.
(b) Both involve glucose 1-phosphate.
(c) Both are driven in part by the hydrolysis of pyrophosphate.
(d) Both occur on cytoplasmic glycogen granules.
(e) Both use the same enzyme for branching and debranching.

10. If glycogen synthase can add a glucose residue to a growing glycogen molecule only if the glucose chain is at least four units long, how does a new glycogen molecule start?

____

____

11. Why is the existence of distinct biosynthetic and catabolic pathways for glycogen important for the metabolism of liver and muscle cells?

____

____

12. Is it true or false that branching in the structure of glycogen increases the rates of its synthesis and degradation? Explain.

___

___

13. Answer the following questions about the enzymatic degradation of amylose, a linear α-1,4 polymer of glucose that is a storage form of glucose in plants:

(a) Would phosphorylase act on amylose? Explain.

___

___

___

(b) Would the rates of glucose 1-phosphate release from an amylose molecule by phosphorylase relative to that from a glycogen molecule having an equivalent number of glucose monomers be equal? Explain.

___

___

___

(c) If the amylose were first treated with an endosaccharidase that cleaved some of its internal glycosidic bonds, how might the rate of production of glucose 1-phosphate be affected?

___

___

___

14. Starting from a glucose residue in glycogen, how many net ATP molecules will be formed in the glycolysis of the residue to pyruvate?

(a) 1 (b) 2 (c) 3 (d) 4 (e) 5

Control of Glycogen Metabolism, and Structural Features of the Enzymes of Glycogen Metabolism

15. Consider the diagram of the different conformational states of muscle glycogen phosphorylase in Figure 23-2. Answer the following questions:

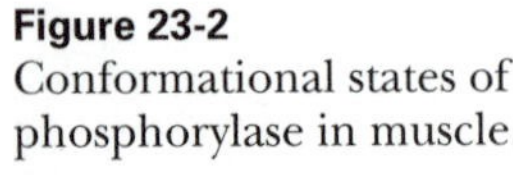

Figure 23-2
Conformational states of phosphorylase in muscle.

(a) Which are the active forms of phosphorylase?

A D

___

(b) Which form requires high levels of AMP to become activated?

B

___

(c) Which conversion is antagonized by ATP and glucose 6-phosphate?

____

(d) What enzyme catalyzes the conversion of C to B?

____

16. How does the regulation of phosphorylase in the liver differ from the scheme for phosphorylase regulation in muscle shown in Figure 23-2?

____

____

17. Indicate which of the following substances have binding sites on phosphorylase. For those that do, give their major roles or effects.

(a) Calmodulin _______________

(b) Glycogen _______________

(c) Pyridoxal phosphate _______________

(d) Ca^{2+} _______________

(e) AMP _______________

(f) P_i _______________

(g) ATP _______________

(h) Glucose _______________

18. Explain the roles of protein kinase A and calmodulin in the control of phosphorylase kinase in the liver.

____

____

19. Which of the following statements about glycogen synthase are correct?

(a) It is activated when it is dephosphorylated.
(b) It is activated when it is phosphorylated.
(c) It is activated when it is phosphorylated and in the presence of high levels of glucose 6-phosphate.
(d) It is activated when it is phosphorylated and in the presence of high levels of AMP.

Reaction Cascades in the Phosphorylation and Dephosphorylation of Glycogen Synthase and Phosphorylase

20. What would increased epinephrine do to protein phosphatase 1 (PP1) in muscle, and how would muscle glycogen metabolism be affected?

____

____

21. Place the following steps of the reaction cascade of glycogen metabolism in the proper sequence:

(a) Phosphorylation of protein kinase _______________

(b) Formation of cyclic AMP by adenylate cyclase _______________

(c) Phosphorylation of phosphorylase *b* ______

(d) Hormone binding to target cell receptors ______

(e) Phosphorylation of glycogen synthase *a* and phosphorylase kinase ______

22. Explain the effect of insulin on the activity of protein phosphatase 1 and the subsequent effects on glycogen metabolism.

______

______

23. Why are enzymatic cascades, such as those that control glycogen metabolism and the clotting of blood, of particular importance in metabolism?

______

______

24. Which of the following statements about the hormonal regulation of glycogen synthesis and degradation are correct?

(a) Insulin increases the capacity of the liver to synthesize glycogen.
(b) Insulin is secreted in response to low levels of blood glucose.
(c) Glucagon and epinephrine have opposing effects on glycogen metabolism.
(d) Glucagon stimulates the breakdown of glycogen, particularly in the liver.
(e) The effects of all three of the regulating hormones are mediated by cyclic AMP.

Regulation of Blood Glucose Levels, and Glycogen-Storage Diseases

25. Which of the following are effects of glucose on the metabolism of glycogen in the liver?

(a) The binding of glucose to phosphorylase *a* converts this enzyme to the inactive T form.
(b) The T form of phosphorylase *a* becomes susceptible to the action of phosphatase.
(c) The R form of phosphorylase *b* becomes susceptible to the action of phosphorylase kinase.
(d) When phosphorylase *a* is converted to phosphorylase *b*, the bound phosphatase is released.
(e) The free phosphatase dephosphorylates and activates glycogen synthase.

26. Explain how a defect in phosphofructokinase in muscle can lead to increased amounts of glycogen having a normal structure. Patients with this defect are normal except for having a limited ability to perform strenuous exercise.

______

______

27. For the defect in question 26, explain why there is not a massive accumulation of glycogen.

____

____

ANSWERS TO SELF-TEST

Introduction

1. (a) A and G

   (b) G and E

   (c) All the bonds are α-1,4 glycosidic linkages except for the one between residues G and E.

   (d) No. The two branches are too short for phosphorylase cleavage. Phosphorylase stops cleaving four residues away from a branch point.

   (e) Yes. Residue G can be hydrolyzed by α-1,6-glucosidase (the debranching enzyme).

   (f) No. The branching enzyme transfers a block of around 7 residues from a nonreducing end of a chain at least 11 residues long. Furthermore, the new α-1,6 glycosidic linkage must be at least four residues away from a preexisting branch point at a more internal site. The fragment of glycogen in Figure 23-1 does not fulfill these requirements.

2. b, c, d

3. Unlike the enzymes in such multienzyme complexes as pyruvate dehydrogenase, the enzymes in glycogen granules are not present in defined stoichiometric ratios. Rather, they bind to a subset of the many available sites on glycogen granules, and their relative amounts vary.

Reactions of Glycogen Degradation and Synthesis

4. The phosphorolytic cleavage of glycogen produces glucose 1-phosphate, which can enter into the glycolytic pathway after conversion to glucose 6-phosphate. These reactions do not require ATP. On the other hand, the hydrolysis of glycogen would produce glucose, which would have to be converted to glucose 6-phosphate by hexokinase, requiring the expenditure of an ATP. Therefore, harvesting the free energy stored in glycogen by phosphorolytic cleavage rather than a hydrolytic one is more efficient because it decreases the ATP investment.

5. a, b

6. (a) 2, 4 (b) 1, 3 (c) 1, 2, 5. None of these enzymes requires ATP.

7. a

8. b

9. b, d

10. The primer required to start a new glycogen chain is formed by the enzyme glycogenin, which has a glucose residue covalently attached to one of its tyrosine residues. Glycogenin uses UDP–glucose to add approximately 8 glucose residues to itself to generate a primer that glycogen synthase can extend.

11. The separate pathways for the synthesis and degradation of glycogen allow the synthesis of glycogen to proceed despite a high ratio of orthophosphate to glucose 1-phosphate, which energetically favors the degradation of glycogen. In addition, the separate pathways allow the coordinated reciprocal control of glycogen synthesis and degradation by hormonal and metabolic signals.

12. True. Since degradation and synthesis occur at the nonreducing ends of glycogen, the branched structure allows simultaneous reactions to occur at many nonreducing ends, thereby increasing the overall rates of degradation or biosynthesis.

13. (a) Yes, phosphorylase would act on amylose by removing one glucose residue at a time from the nonreducing end.

(b) No, the rate of degradation of amylose would be much slower than that of glycogen because amylose would have only a single nonreducing end available for reaction, whereas glycogen has many ends. See self-test question 12.

(c) The increased the number of ends available to phosphorylase as a result of cleaving the chain into more pieces with the endosaccharidase would allow a more rapid production of glucose 1-phosphate by phosphorylase.

14. c. A glucose molecule that is degraded in the glycolytic pathway to two pyruvate molecules yields two ATP; however, the formation of glucose-1-P from glycogen does not consume the ATP that would be required for the formation of glucose-6-P from glucose. Thus, the net yield of ATP for a glucose residue derived from glycogen is three ATP.

Control of Glycogen Metabolism, and Structural Features of the Enzymes of Glycogen Metabolism

15. (a) A and D (c) B to A

(b) B (d) Protein phosphatase 1

The phosphorylated form of glycogen phosphorylase is phosphorylase *a*, which is mostly present in the active conformation designated as D in Figure 23-2. In the presence of high levels of glucose, phosphorylase *a* adopts a strained, inactive conformation, designated C in the figure. The dephosphorylated form of the enzyme is called phosphorylase *b*. Phosphorylase *b* is mostly present in an inactive conformation, labeled B in the figure. When AMP binds to the inactive phosphorylase *b*, the enzyme changes to an active conformation, designated A in the figure. The effects of AMP can be reversed by ATP or glucose 6-phosphate.

16. AMP doesn't activate liver phosphorylase (the B to A conversion shown in Figure 23-2), and glucose shifts the equilibrium between the activated phosphorylase *a* toward the inactivated form (the D to C conversion).

17. (b) Glycogen, as the substrate, binds to the active site; there is also a glycogen particle binding site that keeps the enzyme attached to the glycogen granule.

(c) Pyridoxal phosphate is the prosthetic group that positions orthophosphate for phosphorolysis and acts as a general acid-base catalyst.

(e) AMP binds to an allosteric site and activates phosphorylase *b* in muscle.

(f) P_i binds to the pyridoxal phosphate at the active site and attacks the α-1,4 glycosidic bond. Another P_i is covalently bound to serine 14 by phosphorylase kinase. This phosphorylation converts phosphorylase *b* into active phosphorylase *a*.

(g) ATP binds to the same site as AMP and blocks its effects in muscle.

(h) Glucose inhibits phosphorylase *a* in the liver by changing the conformation of the enzyme to the inactive T form.

Answers (a) and (d) are incorrect because calmodulin and Ca^{2+} bind to phosphorylase kinase rather than to phosphorylase.

18. Protein kinase A, which is itself activated by cAMP, phosphorylates phosphorylase kinase to activate it. Phosphorylase kinase can also be activated by the binding of Ca^{2+} to its calmodulin subunit. Upon binding Ca^{2+}, calmodulin undergoes conformational changes that activate the phosphorylase kinase. The activated kinase in turn activates glycogen phosphorylase. These effects lead to glycogen degradation in active muscle.

19. a, c

Reaction Cascades in the Phosphorylation and Dephosphorylation of Glycogen Synthase and Phosphorylase

20. Increased epinephrine activates PKA, which phosphorylates the G subunit of PP1 and thus reduces the ability of PP1 to act on its protein targets. Furthermore, inhibitor 1 is also phosphorylated by PKA so that it too decreases PP1 activity, albeit by a different mechanism. Inactivated PP1 leads to increased levels of activated (phosphorylated) phosphorylase and inactivated (phosphorylated) glycogen synthase. Glycogen breakdown would be stimulated under these conditions.

21. d, b, a, e, c

22. Insulin results in the activation of PP1. The hormone activates an insulin-sensitive protein kinase that phosphorylates the G subunit of PP1, rendering the phosphatase more active. The activated phosphatase dephosphorylates phosphorylase, protein kinase, and glycogen synthase. These

changes result in a decrease in glycogen degradation and the stimulation of glycogen synthesis.

23. Enzymatic cascades lead from a small signal, due to a few molecules, to a large subsequent enzymatic response. Thus, small biological signals can be amplified in a short time, and their effects can be regulated at various levels of the cascade.

24. a, d

Regulation of Blood Glucose Levels, and Glycogen-Storage Diseases

25. a, b, d, e

26. Since a defect in phosphofructokinase does not impair the ability of muscle to synthesize and degrade glycogen normally, the structure of glycogen will be normal. However, the utilization of glucose 6-phosphate in the glycolytic pathway is impaired, and it equilibrates with glucose 1-phosphate; therefore, some net accumulation of glycogen will occur. The inability to perform strenuous exercise is likely due to the impaired glycolytic pathway in muscle and the diminished production of ATP.

27. Although the impaired use of glucose 6-phosphate in glycolysis will lead to the storage of extra glycogen, it will not become excessive, because the increased concentration of glucose 6-phosphate will inhibit hexokinase and hence the sequestering of glucose in muscle.

PROBLEMS

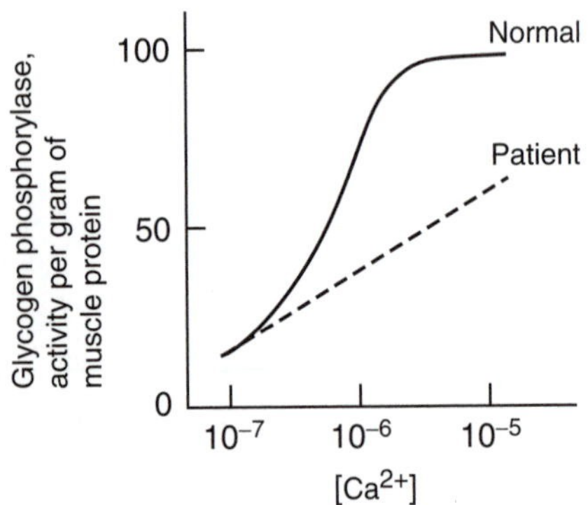

Figure 23-3
Response of glycogen phosphorylase to calcium ion in a patient and in a normal person.

1. A patient can perform nonstrenuous tasks but becomes fatigued upon physical exertion. Assays from a muscle biopsy reveal that glycogen levels are slightly elevated relative to normal. Crude extracts from muscle are used to determine the activity of glycogen phosphorylase at various levels of calcium ion for the patient and for a normal person. The results of those assays are shown in Figure 23-3. Briefly explain the clinical and biochemical findings for the patient.

2. A certain strain of mutant mice is characterized by limited ability to engage in prolonged exercise. After a high carbohydrate meal, one of these mice can exercise on a treadmill for only about 30 percent of the time a normal mouse can. At exhaustion, blood glucose levels in the mutant mouse are quite low, and they increase only marginally after rest. When liver glycogen in fed mutant mice is examined before exercise, the polymers have chains that are highly branched, with average branch lengths of about 10 glucose residues in either α-1,4 or α-1,6 linkage. Glycogen from exhausted normal mice has the same type of structure. Glycogen from exhausted mutant mice is still highly branched, but the polymer has an unusually large number of single glucose residues with α-1,6 linkages. Practically all the chains with α-1,4 linkages are still about 10 residues in length, although some have 13 to 15 residues. Explain the metabolic and molecular observations for the mutant mice.

3. Your colleague discovers a fungal enzyme that can liberate glucose residues from cellulose. The enzyme is similar to glycogen phosphorylase in that it utilizes inorganic phosphate for the phosphorolytic cleavage of

glucose residues from the nonreducing ends of cellulose. Why would you suspect that other types of cellulases may be important in the rapid degradation of cellulose?

4. Consider a patient with the following clinical findings: fasting blood glucose level is 25 mg per 100 ml (normal values are from 80 to 100 mg per 100 ml); feeding the patient glucose results in a rapid elevation of blood glucose level, followed by a normal return to fasting levels; feeding the patient galactose or fructose results in the elevation of blood glucose to normal levels; the administration of glucagon fails to generate hyperglycemia; biochemical examination of liver glycogen reveals a normal glycogen structure.

(a) Which of the enzyme deficiencies described on page 598 of Stryer's text could account for these clinical findings?
(b) What additional experiments would you conduct to provide a specific diagnosis for the patient?

5. Vigorously contracting muscle often becomes anaerobic when the demand for oxygen exceeds the amount supplied through the circulation. Under such conditions, lactate may accumulate in muscle. There is some evidence that under anaerobic conditions a certain percentage of lactate can be converted to glycogen in muscle. One line of evidence involves the demonstration of activity for malic enzyme, which can use CO_2 to convert pyruvate to malate, using NADPH as an electron donor.

(a) Why is lactate produced in muscle when the supply of oxygen is insufficient?
(b) In muscle, pyruvate carboxylase activity is very low. How could malic enzyme activity facilitate the synthesis of glycogen from lactate?
(c) Why would you expect the conversion of lactate to glycogen to occur only after vigorous muscle contraction ceases?
(d) Is there any energetic advantage to converting lactate to glycogen in muscle rather than sending the lactate to the liver, where it can be resynthesized to glucose and then returned to muscle for glycogen synthesis?

6. Cyclic nucleotide phosphatases are inhibited by caffeine. What effect would this have on glycogen metabolism when epinephrine levels are dropping in the blood?

7. During the degradation of branched chains of glycogen, a transferase shifts a chain of three glycosyl residues from one branch to another, exposing a single remaining glycosyl residue to α-1,6-glucosidase activity. Free glucose is released, and the now unbranched chain can be further degraded by glycogen phosphorylase.

(a) Estimate the free-energy change of the transfer of glycosyl residues from one branch to another.
(b) About 10 percent of the glycosyl residues of normal glycogen are released as glucose, whereas the remainder are released as glucose 1-phosphate. Give two reasons why it is desirable for cells to convert most of the glycosyl residues in glycogen to glucose 1-phosphate.

(c) Patients who lack liver glycosyl transferase have been studied. Why would you expect liver extracts from such people also to lack α-1,6-glucosidase activity?

8. You are studying a patient with McArdle's disease, which is described on page 599 of Stryer's text. Explain what you would expect to find when you carry out each of the following analyses:

(a) Fasting level of blood glucose
(b) Structure and amount of liver glycogen
(c) Structure and amount of muscle glycogen
(d) Change in blood glucose levels upon feeding the patient galactose
(e) Change in blood lactate levels after vigorous exercise
(f) Change in blood glucose levels after administration of glucagon
(g) Change in blood glucose levels after administration of epinephrine

9. An investigator has a sample of purified muscle phosphorylase *b* that he knows is relatively inactive.

(a) Suggest two in vitro methods that could be employed to generate active phosphorylase from the inactive phosphorylase *b*.
(b) After the phosphorylase is activated, the investigator incubates the enzyme with a sample of unbranched glycogen in a buffered solution. He finds that no glycosyl residues are cleaved. What else is needed for the cleavage of glycosyl residues by active phosphorylase?

10. Arsenate can substitute for many reactions in which phosphate is the normal substrate. However, arsenate esters are far less stable than phosphate esters, and they decompose to arsenate and an alcohol:

$$R-OAsO_3^{2-} + H_2O \rightarrow R-OH + AsO_4^{2-}$$

(a) In which of the steps of glycogen metabolism could arsenate be used as a substrate?
(b) What are the energetic consequences of utilizing arsenate as a substrate in glycogen degradation?

11. As described on page 597 of Stryer's text, the ratio of glycogen phosphorylase to protein phosphatase 1 is approximately 10 to 1. Suppose that in some liver cells the overproduction of the phosphatase results in a ratio of one to one. How will such a ratio affect the cell's response to an infusion of glucose?

12. A young woman cannot exercise vigorously on the treadmill without leg pains and stiffness. During exercise, lactate levels do not increase in her serum, in contrast to results of exercise in normal subjects. As is the case with normal subjects, no significant hypoglycemia is observed when the patient exercises or fasts. Analyses of muscle biopsy samples show that glycogen content is about 10 times greater than normal in the young woman, but the level of muscle phosphorylase activity is normal. Other experiments with biopsy samples show that rapid incorporation of ^{14}C from radioactive glucose into fructose 6-phosphate and glycogen is observed, but very little incorporation of radioisotope into lactate is seen. When ^{14}C-pyruvate is incubated with another sample of the homogenate, the radioisotope is readily incorporated into glycogen.

What specific deficiency in a metabolic pathway could contribute to the observations described above? Propose two additional studies that could confirm your conclusion.

13. In 1952, Dr. D.H. Andersen described a seriously ill infant with an enlarged liver as well as cirrhosis. When epinephrine was administered, a relatively low elevation in the patient's blood glucose levels was noted. Several days later when the infant was fed galactose, normal elevation of glucose was observed in the circulation. The infant died at the age of 17 months, and at autopsy Dr. Andersen found that glycogen from the liver, while present in unusually high concentration, was relatively insoluble, making it difficult to extract. She sent a sample of the liver glycogen to Dr. Gerty Cori. In an experiment designed to characterize the glycogen, Dr. Cori incubated a sample with orthophosphate (P_i) and two normal liver enzymes, active glycogen phosphorylase and debranching enzyme. She found that the ratio of glucose 1-phosphate to glucose released from the glycogen sample was 100:1, while the ratio from normal glycogen is 10:1.

(a) What enzyme of glycogen metabolism is most likely to be deficient in the liver tissue of the infant? Write a concise explanation for your answer, and relate it to the relative insolubility of the glycogen in the autopsy sample.

(b) Dr. Andersen, aware that a number of enzyme deficiencies might cause a glycogen-storage disease, sought to rule out a deficiency of a particular enzyme in the infant by studying the elevation of glucose levels after feeding galactose. What is that enzyme, and how does normal elevation of blood glucose after galactose feeding rule out a deficiency of that enzyme in the infant?

14. While muscle cells in tissue culture can be stimulated to break down glycogen only minimally when incubated in a solution containing cyclic AMP, they are more readily stimulated by compounds like dibutyryl cyclic AMP, whose structure is shown below. Explain the difference in response of cells to these two substances.

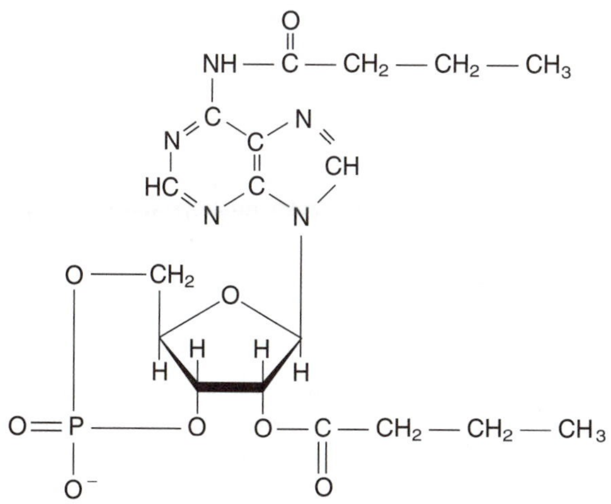

NH — C(=O) — CH_2 — CH_2 — CH_3
O — CH_2
O=P(O⁻) — O   O — C(=O) — CH_2 — CH_2 — CH_3

Dibutyryl-cyclic AMP

15. One method for the analysis of glycogen involves incubating a sample with methyl iodide, which methylates all free hydroxyl groups. Acid hydrolysis of exhaustively methylated glycogen yields a mixture of methyl

glucosides, which can be separated and analyzed. Considering the various types of glycogen-storage diseases listed in Table 23-1 of Stryer's text, which of them could be diagnosed using exhaustive methylation and acid hydrolysis of glycogen?

16. Propose a scheme to identify the specific lysine residue in glycogen phosphorylase that is in Schiff base linkage with pyridoxal phosphate.

17. Patients with Cori's disease lack debranching enzyme, and therefore the structure of liver and muscle glycogen is unusual, with short outer branches. Design an assay that would enable you to demonstrate the presence of short branches in glycogen from one of these patients. Also explain how you would demonstrate that the debranching enzyme is deficient in these patients.

18. Table 23-1 in Stryer's text lists eight diseases of glycogen metabolism, all of which affect the level of glycogen in muscle and liver or the structure of the polysaccharide in one or both of those tissues. Another rare disease of glycogen metabolism is caused by a deficiency in liver glycogen synthase. After fasting, affected subjects have low blood glucose. Hyperglycemia and high blood lactate are observed after a meal.

(a) Briefly explain how these symptoms could be caused by glycogen synthase deficiency.
(b) Under normal nutritional conditions, glycogen comprises about 4% of the wet weight of liver tissue in normal subjects. What proportion of glycogen in liver would you expect in a patient who lacks liver glycogen synthase?

ANSWERS TO PROBLEMS

1. Calcium ion normally activates muscle phosphorylase kinase, which in turn phosphorylates muscle phosphorylase. In the patient, glycogen phosphorylase activity is less responsive to Ca^{2+} than it is in the normal subject. It is likely that Ca^{2+} cannot activate phosphorylase kinase in the patient, perhaps because the δ subunit of the enzyme is altered in some way. As a result, there are too few molecules of enzymatically active glycogen phosphorylase to provide the rate of glycogen breakdown that is needed to sustain vigorous muscle contraction. Elevated levels of muscle glycogen should be expected when glycogen phosphorylase activity is lower than normal.

2. The longer chains of glucose residues in α-1,4 linkage and the unusually high number of single glucose residues in α-1,6 linkage suggest that while transferase activity is present, α-1,6-glucosidase activity is deficient in the mutant strain. For such chains, far fewer ends with glucose residues are available as substrates for glycogen phosphorylase. (Recall that glycogen phosphorylase cannot cleave α-1,6 linkages.) This limited ability to mobilize glucose residues means that less energy is available for prolonged exercise.

3. Cellulose is an unbranched polymer of glucose residues with β-1,4 linkages. Therefore, each chain has only one nonreducing end that is

available for phosphorolysis by the fungal enzyme. Compared with the rate of breakdown of molecules of glycogen, whose branched chains provide more sites for the action of glycogen phosphorylase, the generation of glucose phosphate molecules from cellulose via the fungal enzyme alone could be quite slow. Therefore, you should expect to find other cellulases that generate additional nonreducing ends in cellulose chains.

In order for fungal cells to degrade cellulose, they must secrete extracellular enzymes that can degrade cellulose, a very large and insoluble macromolecule. It seems unlikely that a phosphorylase would be the enzyme of choice for cellulose degradation, because glucose phosphate molecules, which are negatively charged at neutral pH, would be unable to cross cellular membranes and enter the cytosol. In addition, concentrations of phosphate outside the cell could be too low to drive phosphorolysis of cellulose.

4. (a) A low fasting blood glucose level indicates a failure either to mobilize glycogen or to release glucose from the liver. However, the elevations in blood glucose levels after feeding the patient glucose, galactose, or fructose indicate that the liver can release glucose derived from the diet or formed from other monosaccharides. The lack of response to glucagon indicates that the enzymatic cascade for glycogen breakdown is defective. Therefore, you would suspect a deficiency of liver glycogen phosphorylase or phosphorylase kinase. Of the diseases described in Table 23-1 of Stryer's text, both type VI and type VIII could account for the findings, which include normal glycogen structure.

   (b) The direct assay of the activities of glycogen phosphorylase and phosphorylase kinase would enable you to make a specific diagnosis. For these purposes, a liver biopsy would be necessary.

5. (a) Muscle cells produce lactate from pyruvate under anaerobic conditions in order to generate NAD^+, which is required to sustain the activity of glyceraldehyde 3-phosphate dehydrogenase in the glycolytic pathway.

   (b) The low activity of muscle pyruvate carboxylase means that other pathways for the synthesis of oxaloacetate must be available. The formation of malate, which is then converted to oxaloacetate (see p. 573 of Stryer's text), enables the muscle cell to carry out the synthesis of glucose 6-phosphate via gluconeogenesis. Glycogen can than be synthesized through the conversion of glucose 6-phosphate to glucose 1-phosphate, the formation of UDP–glucose, and the transfer of the glucose residue to a glycogen primer chain.

   (c) Energy for vigorous muscle contraction under anaerobic conditions is derived primarily from the conversion of glycogen and glucose to lactate. The simultaneous conversion of lactate to glycogen would simply result in the unnecessary hydrolysis of ATP.

   (d) In liver, the conversion of two lactate molecules to a glucose residue in glycogen through gluconeogenesis requires seven high-energy phosphate bonds; six are required for the formation of glucose 6-phosphate from two molecules of lactate, and one is needed for the synthesis of UDP–glucose from glucose 1-phosphate. The conversion of lactate to glycogen in muscle requires two fewer high-energy bonds because the formation of oxaloacetate through the action of malic enzyme does not require ATP. Recall that pyruvate carboxylase requires ATP for the synthesis of oxaloacetate from pyruvate.

6. When epinephrine levels in the blood decrease, the synthesis of cyclic AMP decreases. Existing cyclic AMP is degraded by cyclic nucleotide phosphatases. The inhibition of these enzymes by caffeine prolongs the degradation of glycogen because the remaining cyclic AMP continues to activate protein kinase, which in turn activates phosphorylase kinase. Glycogen phosphorylase is, in turn, activated by phosphorylase kinase. Sustained activation of phosphorylase results in continued mobilization of glucose residues from glycogen stores in liver.

7. (a) Because the bonds broken and formed during the transferase action are both α-1,4 glycosidic bonds, the free-energy change is likely to be close to zero.

(b) The generation of glucose 1-phosphate rather than glucose means that one less ATP equivalent is required for the conversion of a glucose residue to two molecules of lactate. In addition, the phosphorylation of glucose ensures that the molecule cannot diffuse across the cell membrane before it is utilized in the glycolytic pathway.

(c) The glucosidase and the transferase activities are both found on the same 160-kd polypeptide chain. A significant alteration in the structure of the domain for glucosidase could impair the functioning of the transferase domain.

8. (a) In McArdle's disease, muscle phosphorylase is deficient, but liver phosphorylase is normal. Therefore, you would expect glucose and glycogen metabolism in the liver to be normal and the control of blood glucose by the liver also to be normal.

(b) Normal glyogen metabolism in the liver means that both the amount of liver glycogen and its structure would be the same as in unaffected people.

(c) Defective muscle glycogen phosphorylase means that glycogen breakdown is impaired. Moderately increased concentrations of muscle glycogen could be expected, although the structure of the glycogen should be similar to that in unaffected people.

(d) Galactose can be converted to glucose 6-phosphate in the liver, which can then export glucose to the blood. Because defective muscle phosphorylase has no effect on galactose metabolism, you would expect similar elevations in blood glucose after the ingestion of galactose in normal and affected people.

(e) During vigorous exercise, blood lactate levels normally rise as muscle tissue exports the lactate generated through glycogen breakdown. The defect in muscle phosphorylase limits the extent to which glycogen is degraded in the muscle. This in turn reduces the amount of lactate exported during exercise, so the rise in blood lactate levels would not be as great in the affected person.

(f) Glucagon exerts its effects primarily on liver, not muscle. In patients with McArdle's disease, blood glucose levels increase normally in response to glucagon.

(g) There may be a slight increase in blood glucose concentration after epinephrine administration, because liver is somewhat responsive to this hormone. Epinephrine does have a greater glycogenolytic effect on muscle, but you would not expect to see any change in blood glucose concentration when it is administered. The reason is that even if glycogen breakdown is accelerated (which is unlikely to occur in patients with McArdle's disease), the glucose 6-phosphate produced cannot be converted to glucose for export into the circulation, because muscle lacks the enzyme glucose 6-phosphatase.

9. (a) The investigator can activate the phosphorylase by adding AMP to the sample or by using active phosphorylase kinase and ATP to phosphorylate the enzyme.

(b) Inorganic phosphate is also required for the conversion of glycosyl residues in glycogen to glucose 1-phosphate molecules.

10. (a) Arsenate can substitute for inorganic phosphate in the glycogen phosphorylase reaction, generating glucose arsenate esters.

(b) When P_i is used as a substrate for glycogen phosphorylase, glucose 1-phosphate is generated. The glucose 1-arsenate esters that are generated when arsenate is used as a substrate spontaneously degrade to yield glucose and arsenate. The conversion of glucose to pyruvate requires one more ATP equivalent than does the conversion of glucose 1-phosphate to pyruvate.

11. The normal 10-to-1 ratio means that glycogen synthase molecules are activated only after most of the phosphorylase *a* molecules are converted to the inactive *b* form, which ensures that the simultaneous degradation and synthesis of glycogen does not occur. A phosphorylase to phosphatase ratio of one to one means that, as soon as a few phosphorylase molecules are inactivated, phosphatase molecules that are no longer bound to phosphorylase begin to convert glycogen synthase molecules to the active form. Glycogen degradation and synthesis then occur simultaneously, resulting in the wasteful hydrolysis of ATP.

12. From the clinical observations it appears that the pathway from pyruvate to glucose 6-phosphate and on to glycogen is functional and that gluconeogenesis is working normally in liver (there is no hypoglycemia during fasting or exercise, when demands for glucose increase). Although muscle glycogen content is higher, normal phosphorylase activity indicates that glycogen could be phosphorylized normally. You should then consider whether there is a deficiency in the glycolytic pathway, because lactate does not accumulate during exercise and it is not labeled when ^{14}C-glucose is administered. Labeled fructose 6-phosphate can be made from radioactive glucose in the biopsy sample, but knowledge about subsequent glycolytic reactions is not available. There could be a significant block at the level of phosphofructokinase or beyond. Such a deficiency would mean that while normal demands for glucose can be taken care of, a high rate of glycolytic activity during vigorous exercise cannot be accommodated. You should consider analyzing for additional radioactive glycolytic intermediates when glucose is administered, then testing for deficiency of

one or more glycolytic enzymes using biopsy tissues. The description of the disorder corresponds most closely to a known condition for a deficiency in muscle phosphofructokinase. One might also argue that lactate dehydrogenase could be absent, explaining why no lactate is generated during exercise. However, in cases in which muscle lactate dehydrogenase is defective, affected subjects cannot exercise vigorously, but they have no accumulation of glycogen in their muscle tissue.

13. (a) The branching enzyme was deficient in the infant. This enzyme removes blocks of glucosyl residues from a chain of α-1 $\rightarrow$ 4-linked residues and transfers them internally to form a branch with an α-1 $\rightarrow$ 6 link to a polymer chain. The most important clue to the deficiency is found in the ratio of glucose 1-phosphate to glucose, which is 10 times higher in glycogen from the affected infant than from a normal polymer sample. Recall that glucose 1-phosphate is produced through the action of phosphorylase, which phosphorylizes α-1 $\rightarrow$ 4 linkages, while glucose is produced when the glycogen debranching enzyme hydrolyzes a glucose in α-1 $\rightarrow$ 6 linkage at a branch point. Normal glycogen has a branch at every 10 or so glycosyl residues, so that treatment with a mixture of normal phosphorylase and debranching enzyme will yield a 10:1 ratio of glucose 1-phosphate to glucose. The autopsy sample yielded a ratio of 100:1, suggesting that there are far fewer branches in the sample. This conclusion is consistent with the relative insolubility of the infant's glycogen, which, with fewer branches, is more like amylopectin, a linear glucosyl polymer which has limited solubility in water.

(b) The pathway for galactose metabolism includes its conversion, through steps that include epimerization, to glucose 6-phosphate. Thus feeding galactose should result in an increased concentration of glucose 6-phosphate in the liver cell. If glucose 6-phosphatase were deficient, glucose 6-phosphate would not be converted to glucose, so that the levels of blood glucose would not be elevated after galactose feeding. Dr. Andersen considered a glucose 6-phosphatase deficiency because of the limited increase in blood glucose levels after administration of epinephrine, so she used galactose feeding to increase glucose 6-phosphate levels in liver cells. When glucose levels rose in the blood, she concluded that glucose 6-phosphatase levels were normal. She subsequently considered other deficiencies that would result in storage of abnormal amounts of liver glycogen.

14. Like other nucleotides, cyclic AMP is polar and negatively charged at neutral pH. It therefore crosses plasma membranes at a relatively low rate. The presence of two hydrophobic acyl chains on the molecule make it much more hydrophobic, so that it can more easily dissolve in the bilayer and more readily enter the cytosol.

15. Type IV glycogen-storage disease, in which glycogen with a much lower number of α-1,6 glycosidic linkages is produced, could be analyzed using methylation and hydrolysis. Any glucose residue derived from a branch point will have methyl groups at C-2 and C-3, while all other residues (with one exception) will emerge from hydrolysis as 2,3,6-*O*-

trimethyl glucose molecules. The glucose at the reducing end of the glycogen molecule will be converted to a tetramethlyglucoside. In normal subjects, the ratio of trimethylglucose to dimethyl glucose should be about 10 to 1, while glycogen from a person with a deficiency in the branching enzyme will have a much higher ratio.

16. Treat the protein with sodium borohydride, which reduces the Schiff base linkage between pyridoxal phosphate and the lysine residue. Then convert the protein to peptide fragments using various proteases or chemical methods. This is a time-consuming task because the protein is composed of two identical chains, each containing 842 amino acids. Every fragment that is known to contain lysine can be subjected to acid hydrolysis. A lysine residue from one of the fragments will be covalently attached to pyridoxamine (the phosphate group is usually removed during acid hydrolysis). Analysis of phosphorylase by these methods shows that Lys 680 is in Schiff base linkage with pyridoxal phosphate.

17. A short outer branch in a glycogen molecule has only a small number of α-1,4-glucosyl residues on the nonreducing side of a branch or an α-1,6 link. Incubating such a glycogen molecule with active phosphorylase and P_i will liberate only limited amounts of glucose 1-phosphate, compared with the number liberated from normal glycogen. Recall that phosphorylase cannot free glucose molecules that are within four residues of a branch point in glycogen. To demonstrate that phosphorylase action is limited by short outer branches, you can incubate another sample with purified debranching enzyme and phosphorylase, and you would expect to see an increase in production of glucose 1-phosphate. To demonstrate a debranching enzyme deficiency in a patient, you could treat normal glycogen with active muscle phosphorylase and muscle extracts from a patient with Cori's disease. If debranching enzyme activity is low, only limited amounts of glucose 1-phosphate will be produced. Larger numbers of glucose 1-phosphate molecules will be released from a normal glycogen sample treated with active phosphorylase and muscle cell extracts from a normal person.

18. (a) Lack of glycogen synthase implies that the liver's ability to store glucose as glycogen is impaired. After fasting, when blood glucose concentrations are low, liver glycogen is normally converted to glucose 6-phosphate, which is converted to glucose and exported to the blood. Low glycogen levels in liver tissue would make it impossible for liver to maintain proper glucose levels in the blood. After a meal containing carbohydrates, the liver would be unable to convert glucose to glycogen. Even though glucokinase may convert glucose to glucose 6-phosphate, the high concentration of that substrate may cause accelerated conversion back to glucose through the action of glucose 6-phosphatase. Glucose levels would then increase in the circulation. The elevation of lactate levels in blood suggests that any glucose metabolized in the liver is preferentially converted to lactate rather than to glycogen.

(b) As discussed above, liver cells deficient in glycogen synthase would be unable to synthesize high levels of glycogen. You would therefore expect the percentage of glycogen in affected people to be lower. In those few patients with the disorder, glycogen makes up less than 1% of liver tissue.

EXPANDED SOLUTIONS TO TEXT PROBLEMS

1. Recall that galactose enters metabolism by reacting with ATP in the presence of galactokinase to yield galactose 1-phosphate and subsequently glucose 1-phosphate (Stryer, Chapter 19, p. 492). On the way to glycogen the latter reacts with UTP to give UDP–glucose. Hence:

$$\text{Galactose} + \text{ATP} + \text{UTP} + H_2O + (\text{glycogen})_n \rightarrow (\text{glycogen})_{n+1} + \text{ADP} + \text{UDP} + 2\ P_i + H^+$$

2. In normal glycogen, branches occur about once in 10 units. Therefore, degradation of this glycogen is expected to give a ratio of glucose 1-phosphate to glucose of about 10:1. An increased ratio (100:1) indicates that the glycogen has a much lower degree of branching, suggesting a deficiency of the branching enzyme.

3. The enzymatic defect in von Gierke's disease is the absence of liver glucose 6-phosphatase. The resulting high concentrations of glucose 6-phosphate allosterically (p. 594) activate the inactive glycogen synthase *b*, causing a net increase in liver glycogen.

4. Glucose is an allosteric inhibitor of phosphorylase *a*. Hence, crystals grown in its presence are in the T state. The addition of glucose 1-phosphate, a substrate, shifts the R $\leftrightarrow$ T equilibrium toward the R state. The conformational differences between these states are sufficiently large that the crystal shatters unless it is stabilized by chemical cross-links. The shattering of a crystal caused by an allosteric transition was first observed by Haurowitz in the oxygenation of crystals of deoxyhemoglobin.

5. Since glucose 1,6-bisphosphate is an intermediate in the reaction catalyzed by phosphoglucomutase and phosphorylates this enzyme during catalysis (p. 584), it is reasonable to expect that it can phosphorylate the dephosphoenzyme. Glucose 1,6-bisphosphate is formed from glucose 1-phosphate and ATP by phosphoglucokinase.

6. Water is excluded from the active site to prevent hydrolysis. The entry of water could lead to the formation of glucose rather than glucose 1-phosphate. A site-specific mutagenesis experiment is revealing in this regard. In phosphorylase, Tyr 573 is hydrogen-bonded to the 2′-OH of a glucose residue. The ratio of glucose 1-phosphate to glucose product is 9000:1 for the wild-type enzyme, and 500:1 for the Phe 573 mutant. Model building suggests that a water molecule occupies the site normally filled by the phenolic OH of tyrosine and occasionally attacks the oxocarbonium ion intermediate to form glucose. See D. Palm, H.W. Klein, R. Schinzel, M. Buehner, and E.J.M. Helmreich. Biochemistry 29(1990): 1099.

7. When two soluble enzymes catalyze consecutive reactions, the product formed by the first enzyme must leave and diffuse to the second enzyme. Catalytic efficiency is substantially increased if both active sites are in close proximity in the same enzyme molecule. A similar advantage is obtained when consecutive enzymes are held close to each other in multienzyme complexes.

8. (a) Muscle phosphorylase *b* will be inactive even when the AMP level is high. Hence, glycogen will not be degraded unless phosphorylase is converted into the *a* form by hormone-induced or Ca^{2+}-induced phosphorylation.

(b) Phosphorylase *b* cannot be converted into the much more active *a* form. Hence, the mobilization of liver glycogen will be markedly impaired.

(c) The elevated level of the kinase will lead to the phosphorylation and activation of glycogen phosphorylase. Little glycogen will be present in the liver because it will be persistently degraded.

(d) Protein phosphatase 1 will be continually active. Hence, the level of phosphorylase *b* will be higher than normal, and glycogen will be less readily degraded.

(e) Protein phosphatase 1 will be much less effective in dephosphorylating glycogen synthase and glycogen phosphorylase. Consequently, the synthase will stay in the less active *b* form, and the phosphorylase will stay in the more active *a* form. Both changes will lead to increased degradation of glycogen.

(f) The absence of glycogenin will block the initiation of glycogen synthesis. Very little glycogen will be synthesized in its absence.

9. The slow phosphorylation of the a subunits of phosphorylase kinase serves to prolong the degradation of glycogen. The kinase cannot be deactivated until its α subunits are phosphorylated. The slow phosphorylation of α assures that the kinase and, in turn, phosphorylase stay active for a defined interval. See H.G. Hers. *Ann. Rev. Biochem.* 45(1976):167.

CHAPTER

24

Fatty Acid Metabolism

In his discussion of the generation and storage of metabolic energy, Stryer has thus far focused on the carbohydrates (Chapters 19, 22, and 23). In Chapter 24, he turns to the fatty acids as metabolic fuels. After describing the nomenclature of fatty acids, he explains why fatty acids are the most concentrated energy stores. The pathway of the oxidation of fatty acids, which liberates the energy of fatty acids and makes it available to the cell, is then presented. The oxidation of unsaturated fatty acids is described, and the formation and role of the ketone bodies as acetyl transport molecules in circulation are discussed. Stryer then describes how both saturated and unsaturated fatty acids are synthesized. The energetics of the oxidation and synthesis of fatty acids are given, and an outline of the control of these processes is provided. Stryer concludes this chapter with a discussion of eicosanoid hormones derived from fatty acids. A review of Chapter 11 will remind you of the structural role of lipids in membranes and of the effect of the fatty acids in determining membrane fluidity.

LEARNING OBJECTIVES

When you have mastered this chapter, you should be able to complete the following objectives:

Introduction (Stryer pages 603–604)

1. List the four major physiologic functions of *fatty acids.*

2. Derive the structure of a *saturated* or *unsaturated fatty acid* from its systematic name. Specify the α, β, and ω carbon atoms and designate the position of a double bond in a fatty acid, given either its Δ or ω number.

3. Recognize the structures of *palmitate, stearate, palmitoleate, oleate, linoleate,* and *linolenate.* Note their relative melting points.

Fatty Acid Oxidation (Stryer pages 604–612)

4. Explain why *triacylglycerols* are highly concentrated forms of *stored metabolic energy.*

5. Describe the *lipolysis* of triacylglycerols by *lipases.* Explain the role of *cyclic AMP* in the regulation of lipase in *adipose cells.* Outline the conversion of glycerol to *glycerol 3-phosphate* and *dihydroxyacetone phosphate.*

6. Recount the experiments of Franz Knoop, which showed that fatty acid catabolism occurs by *β-oxidation.*

7. Describe the reaction that links *coenzyme A (CoA)* to a fatty acid.

8. Explain the involvement of *carnitine* in the transport of fatty acids from the cytoplasm into mitochondria.

9. List the four reactions of the β-oxidation pathway and identify their substrates and products. Explain the function of *NAD^+* and *FAD* in these reactions.

10. Calculate the *energy yield,* in terms of ATP molecules, for the β-oxidation of a given fatty acid.

11. Indicate the two reactions, in addition to those of the β-oxidation pathway, used to oxidize naturally occurring unsaturated fatty acids.

12. Outline the oxidation reactions of an *odd-numbered fatty acid.*

Ketone Bodies (Stryer pages 612–613)

13. Explain the consequences of limiting *oxaloacetic acid* concentrations on the oxidation of fatty acids. Name and identify the structures of the *ketone bodies.*

14. Describe the synthesis and normal catabolism of the ketone bodies. Explain why only the liver exports *acetoacetate* and *3-hydroxybutyrate.*

15. Describe the effect of high levels of *acetoacetate* on fat metabolism in adipose tissue.

16. Provide the biochemical basis for the inability of animals to convert fatty acids into glucose.

Fatty Acid Synthesis (Stryer pages 614–621)

17. Contrast fatty acid oxidation and *fatty acid synthesis.*

18. List the substrates and products of the committed step in fatty acid synthesis and describe its catalytic mechanism.

19. Name the common component of *acyl carrier protein* (*ACP*) and CoA, give its functions. Describe the overall functions of ACP and CoA in fatty acid metabolism.

20. Describe the four reactions of the elongation cycle of fatty acid synthesis. Explain how *malonyl CoA* provides the driving force for the condensation of acetyl units with the growing acyl chain.

21. Calculate the energy cost of the synthesis of a given fatty acid.

22. Contrast the enzymatic machinery for fatty acid biosynthesis in bacteria with that in eukaryotes. Outline the movements of the elongating acyl chain on the mammalian *fatty acid synthetase* dimer during fatty acid biosynthesis.

23. Describe the transport of acetyl CoA across the inner mitochondrial membrane in the form of *citrate.* Account for the synthesis of *NADPH* during the conversion of oxaloacetate into pyruvate in the cytosol.

Control of Fatty Acid Metabolism. Derivatives of Fatty Acids (Stryer pages 621–625)

24. Discuss the different modes of regulation of *acetyl CoA carboxylase.* Explain the reciprocal control of fatty acid synthesis and degradation.

25. Describe the elongation and desaturation reactions that can occur on preformed fatty acids. Explain why linoleate and linolenate are essential in the diet.

26. List the different kinds of *eicosanoid hormones.* Outline their metabolic relationships and biological functions.

27. Describe the effects of *acetylsalicylate* (*aspirin*) on the synthesis of eicosanoids.

SELF-TEST

Introduction

1. Which of the following are major physiologic functions of free fatty acids?
   (a) They stabilize the structure of membranes.
   (b) They serve as precursors of phospholipids and glycolipids.
   (c) They serve as fuel molecules.
   (d) They are precursors of triacylglycerols.
   (e) They are precursors of certain hormones and intracellular messengers.

2. For each of the following four naturally occurring fatty acids, give the systematic name, the common name, and the abbreviations. Use the Δ convention for the name and abbreviations of the unsaturated compounds. Also indicate the position of the double bond closest to the methyl end of the chain using the ω convention.

(a) $CH_3(CH_2)_{14}CO_2H$

(b) $CH_3(CH_2)_7\overset{H}{\overset{|}{C}}=\overset{H}{\overset{|}{C}}(CH_2)_7CO_2H$

(c) $CH_3(CH_2)_4\overset{H}{\overset{|}{C}}=\overset{H}{\overset{|}{C}}CH_2\overset{H}{\overset{|}{C}}=\overset{H}{\overset{|}{C}}(CH_2)_7CO_2H$

(d) $CH_3CH_2\overset{H}{\overset{|}{C}}=\overset{H}{\overset{|}{C}}CH_2\overset{H}{\overset{|}{C}}=\overset{H}{\overset{|}{C}}CH_2\overset{H}{\overset{|}{C}}=\overset{H}{\overset{|}{C}}(CH_2)_7CO_2H$

3. Arrange the following fatty acids in the order of increasing melting point:

(a) Oleate
(b) Stearate
(c) Linoleate
(d) Palmitate
(e) Linolenate

Fatty Acid Oxidation

4. What two properties make triacylglycerols more efficient than glycogen for the storage of metabolic energy?

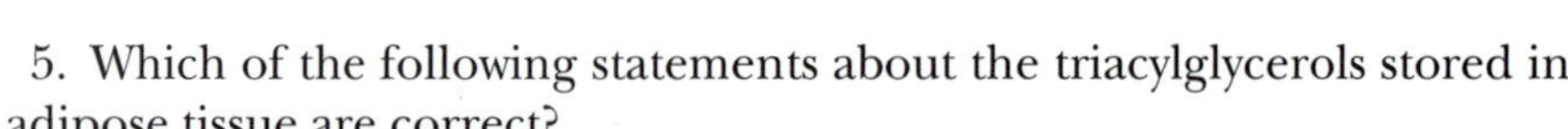

5. Which of the following statements about the triacylglycerols stored in adipose tissue are correct?

(a) They are hydrolyzed to form fatty acids and dihydroxyacetone.
(b) They are hydrolyzed by a lipase that is activated by covalent modification.
(c) They release fatty acids that can be oxidized to CO_2 and H_2O to provide energy to the cell.
(d) They can yield a precursor of glucose.
(e) They are mobilized by epinephrine or glucagon.

6. Predict the residual acids resulting from the oxidation of phenylpentanoic and phenylhexanoic acids in a dog. What is the biochemical meaning of this finding, which was made by Knoop?

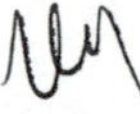

7. (a) Draw a thioester bond and explain its role in the β-oxidation of fatty acids.

___

___

(b) What is the $\Delta G^{\circ\prime}$ value for the hydrolysis of acetyl coenzyme A? What is the significance of this value with respect to fatty acid metabolism?

___

___

(c) Describe the mechanism for the formation of acyl CoA.

___

___

(d) How is pyrophosphatase involved in the activation of fatty acids for β-oxidation?

___

___

8. Place the following incomplete list of reactions or relevant locations during the β-oxidation of fatty acids in the proper order.

(a) Reactions with carnitine 3

(b) Fatty acid in cytosol 1

(c) Activation of fatty acid by joining to CoA 2

(d) Hydration 6

(e) NAD^+-linked oxidation 7

(f) Thiolysis 8

(g) Acyl CoA in mitochondrion 4

(h) FAD-linked oxidation 5

9. Explain the involvement of carnitine in the β-oxidation of fatty acids.

___

___

10. Calculate the yield in ATP molecules of the complete oxidation of hexanoic acid (C6:0).

___

___

11. In the process of hydrogenating certain oils to saturate the double bonds in their fatty acids, some of the cis double bonds are converted to the trans conformation. Predict what would happen if a monoenoic fatty acid with a trans-Δ^{10} bond were produced, ingested, and degraded by the β-oxidation pathway. If another of the ingested fatty acids contained a cis-Δ^{11} double bond, what would be the outcome of these processes? What

effect would the presence of the double bond have on the yield of ATP obtained by the β-oxidation of these fatty acids?

___

___

Ketone Bodies

12. Indicate whether the following statement is true or false and explain your answer:

After a meal rich in carbohydrates, acetyl CoA levels rise and ketone body synthesis increases.

___

___

13. Which of the following statements about acetoacetate and 3-hydroxybutyrate are correct?

(a) They are normal fuels for heart muscle and the renal cortex.
(b) They are synthesized in the liver.
(c) They can give rise to acetone.
(d) They contain four carbon atoms and require three acetyl CoA molecules for their synthesis.
(e) They can be regarded as water-soluble, transportable forms of citrate in the blood.

Fatty Acid Synthesis

14. Match the reactant or characteristic in the right column with the appropriate pathway in the left column.

(a) Fatty acid oxidation ___
(b) Fatty acid synthesis ___

(1) Acyl CoA
(2) Occurs in the cytosol
(3) Uses NAD^+
(4) D-3-Hydroxyacyl derivative involved
(5) Pantetheine involved
(6) Malonyl CoA
(7) Single polypeptide with multiple activities involved
(8) Uses FAD

15. Explain the requirement for bicarbonate in fatty acid biosynthesis.

___

___

16. Which of the following statements about citrate are correct?

(a) It transports reducing power from the mitochondria into the cytosol.
(b) It inhibits gluconeogenesis.
(c) It activates the first enzyme of fatty acid biosynthesis.
(d) It transports acetyl groups from the mitochondria into the cytosol.
(e) It supplies the CO_2 required for formation of malonyl CoA.

17. The fatty acid synthase of mammals is a dimer consisting of identical subunits, each of which contains all the activities necessary to synthesize fatty acids from malonyl CoA and acetyl CoA. Why is a single subunit unable to carry out the reactions?

---

18. Possible advantages of multifunctional polypeptide chains, that is, polypeptide chains having more than one active site, include which of the following?

(a) Enhanced stability beyond that expected for a noncovalent complex of the same activities on separate polypeptide chains.
(b) Fixed stoichiometric relationships among the different enzymatic activities because of their coordinate synthesis.
(c) Enhanced specificity and decreased side reactions because the product of each active site is in the immediate vicinity of the active site carrying out the next reaction in the sequence.
(d) Enhanced versatility because the product of any one active site could be used by any other active site in its immediate vicinity to generate a variety of products.
(e) Accelerated overall reaction rate because of the proximity of the active sites.

19. Which of the following answers completes the sentence correctly? The major product of the fatty acid synthase complex in mammals is

(a) oleate.
(b) stearate.
(c) stearoyl CoA.
(d) linoleate.
(e) palmitate.
(f) palmitoyl CoA.

20. Calculate the ATP and NADPH requirements for the synthesis of lauric acid (C12:0) from acetyl CoA.

Control of Fatty Acid Metabolism. Derivatives of Fatty Acids

21. Which of the following statements about acetyl CoA carboxylase are correct?

(a) It is active in the phosphorylated form.
(b) It is partially active in the phosphorylated form in the presence of citrate.
(c) It is phosphorylated by a cAMP-dependent protein kinase.
(d) It is stimulated by a high energy charge.

22. Which of the following enzymes are allosterically regulated by palmitoyl CoA?

(a) Pyruvate carboxylase
(b) Citrate translocase
(c) Carnitine acyltransferase
(d) Glucose 6-phosphate dehydrogenase
(e) Acetyl CoA carboxylase

23. Match the appropriate compound or process in the right column with its ability to control the process in the left column.

(a) β-Oxidation of fatty acids 3, 4, 5, 6

(b) Fatty acid biosynthesis 1, 2, 4, 7

(1) Citrate
(2) Palmitoyl CoA
(3) Acetyl CoA
(4) NADH
(5) Malonyl CoA
(6) Hormones stimulating the formation of cyclic AMP in adipose tissue
(7) Adaptive control through the synthesis and degradation of enzymes

24. Which of the following statements about desaturases in humans are correct?

(a) They cannot introduce double bonds into a fatty acid that already contains a double bond.
(b) They cannot introduce double bonds between the Δ^9 position and the ω end of the chain.
(c) They convert the essential fatty acid linoleate into arachidonate.
(d) They use an isozyme of the FAD-linked dehydrogenase of the β-oxidation cycle to form double bonds.

25. Which of the following statements about the eicosanoid hormones are incorrect?

(a) The major classes of eicosanoid hormones include prostaglandins, leukotrienes, thromboxanes, and prostacyclin.
(b) The eicosanoid hormones are derived from arachidonic acid.
√ (c) The eicosanoid hormones are very potent and exert global effects because they are widely distributed by the circulatory system.
(d) The prostaglandins have a variety of physiologic effects.
(e) Prostaglandins are derived from phospholipids.

26. Explain why aspirin is a potent anti-inflammatory agent.

____

____

ANSWERS TO SELF-TEST

Introduction

1. b, c, d, e

2. (a) Hexadecanoic acid; palmitic acid; C16:0

(b) *cis*-Δ^9-Octadecenoic acid; oleic acid; C18:1 *cis*-Δ^9; ω-9

(c) *cis,cis*-Δ^9, Δ^{12}-Octadecadienoic acid; linoleic acid; C18:2 *cis,cis*-$\Delta^{9,12}$; ω-6

(d) *cis,cis,cis-* (or all *cis-*) $\Delta^{9,12,15}$-Octadecatrienoic acid; linolenic acid; C18:3 *cis,cis,cis*-$\Delta^{9,12,15}$; ω-3

Note that question 2 shows the undissociated form of the fatty acids, whereas Table 24-1 in Stryer names their ionized or carboxylate forms.

3. e, c, a, d, b. The order of e, c, a, b is determined by the content of double bonds of the C_{18} fatty acids. Double bonds decrease intermolecular interactions and packing in the solid state and thus depress the melting points of fatty acids. The order of (d) relative to (b) is determined by the chain length, C_{16} versus C_{18}, of the saturated fatty acids.

Fatty Acid Oxidation

4. Triacylglycerols contain a high proportion by weight of fatty acids. Fatty acids are highly reduced and have a higher energy content (9 kcal/g) than glycogen (4 kcal/g), which is composed of carbohydrate residues containing numerous oxygen atoms. In addition, fats are anhydrous, whereas glycogen is hydrated, so on the basis of actual storage weight, triacylglycerols contain six times more calories per gram than glycogen does.

5. b, c, d, e. The hydrolysis of triacylglycerols yields glycerol, not dihydroxyacetone, and glycerol can be converted into glyceraldehyde 3-phosphate, which can ultimately give rise to glucose. Several hormones affect the hormone-sensitive lipase of adipose tissue via a cyclic AMP–modulated phosphorylation that activates the enzyme.

6. The compound with the phenyl group bearing the odd-numbered fatty acid chain would give rise to benzoic acid. In contrast, phenylhexanoic acid would be degraded to phenylacetic acid. These results suggest that the fatty acids are being metabolized from the carboxyl terminus and, more importantly, that the carbon atoms are being removed two at a time. Thus, oxidation of the β carbon atom of the fatty acid was a likely step in the degradation.

7. (a) A thioester bond is shown in the margin. The thioester linkage joins the fatty acid to CoA, which acts as a tag or handle by which the enzymes of the β-oxidation path can recognize, bind, and act on the saturated alkane chains of the fatty acids. Furthermore, since the thioester linkage is a "high-energy" bond, it can transfer the acyl group to carnitine—a reaction that is necessary to deliver the acyl group from the cytosol to the mitochondrial matrix for oxidation.

$$R-\overset{O}{\overset{\|}{C}}-SR'$$

Thioester bond

(b) $\Delta G^{\circ\prime} = -7.5$ kcal/mol (see p. 451 in Stryer). The relatively large and negative value for the free energy of hydrolysis for acetyl CoA indicates that energy must be supplied to synthesize it and that, conversely, it is an "activated" donor of acetyl groups.

(c) The carboxyl group of the fatty acid is first activated by reaction with ATP to form an acyl adenylate, which contains a mixed anhydride linkage between the carboxylate and the 5′-phosphate of AMP, with the release of PP_i. In a second step, also catalyzed by acyl CoA synthetase, the acyl group is transferred to the sulfhydryl group of CoA to form the thioester bond and release AMP.

(d) The hydrolysis of PP_i couples the cleavage of a second high-energy bond to the formation of the thioester bond to make its formation exergonic.

8. b, c, a, g, h, d, e, f

9. Acyl CoA is formed in the cytosol, and the enzymes of the β-oxidation pathway are in the matrix of the mitochondrion. The mitochondrial inner membrane is impermeable to CoA and its acyl derivatives. However, a translocase protein can shuttle carnitine and its acyl derivatives across the inner mitochondrial membrane. The acyl group is transferred to carnitine on the cytosol side of the inner membrane and back to CoA on the matrix side. Thus, carnitine acts as a transmembrane carrier of acyl groups.

10. After activation of hexanoic acid to hexanoyl CoA, two rounds of β-oxidation are required to produce 3 acetyl CoA molecules. The two cycles also produce 2 $FADH_2$ and 2 NADH molecules. Each acetyl CoA yields 10 ATP upon complete oxidation, each $FADH_2$ produces 1.5 ATP, and each NADH makes 2.5 ATP—a total of 38 ATP. Activation uses 2 ATP molecules because AMP is one of the products of the reaction, so the net yield is 36 ATP.

11. Four rounds of β-oxidation of a fatty acid with a trans-Δ^{10} double bond would yield a trans-Δ^2-enoyl CoA derivative. This compound is the natural intermediate formed by an acyl CoA dehydrogenase. It would be hydrated by enoyl CoA hydratase to form the L-3-hydroxyacyl CoA derivative. For the fatty acid with a cis-Δ^{11} double bond, four rounds of β-oxidation would produce a cis-Δ^3 double bond, which would not serve as a substrate for enoyl CoA hydratase. An isomerase would convert this bond into the trans-Δ^2 configuration to allow subsequent metabolism. Since the double bond already exists in the fatty acids and does not arise from β-oxidations, one less $FADH_2$ would be formed. Consequently, 1.5 fewer ATP would be produced for each preexisting double bond.

Ketone Bodies

12. False. When carbohydrates are abundant, oxaloacetate levels are high and condensation of acetyl CoA and oxaloacetate produces citrate. Citrate is used for energy production as well as for fatty acid biosynthesis. Ketone bodies are produced when acetyl CoA is abundant but oxaloacetate is depleted.

13. a, b, c, d. Choice (e) is incorrect because ketone bodies are the transportable form of acetyl units in blood.

Fatty Acid Synthesis

14. (a) 1, 3, 5, 8 (b) 1, 2, 4, 5, 6, 7. Acyl CoA is involved in both the synthesis and the oxidation of fatty acids.

15. The irreversible and committed step of fatty acid biosynthesis is the formation of malonyl CoA from acetyl CoA and HCO_3^- by acetyl CoA car-

boxylase. HCO_3^- is fixed to form a dicarboxylic acid at the expense of an ATP cleavage. This facilitates the subsequent condensation reactions with activated acyl groups to form an acetoacyl–ACP by releasing CO_2 to help drive the reaction.

16. c, d

17. The reactions of elongation require the interactions of domains from different subunits of the dimer in order to form active sites at the interfaces of the subunits.

18. a, b, c, e. (For a discussion relevant to the correct answers, see p. 618 in Stryer.)

19. e

20. For a C12:0 fatty acid, 6 acetyl CoA molecules are required. One serves as a primer forming the ω end of the chain and five undergo condensation reactions as their malonyl CoA derivatives. Each malonyl CoA requires 1 ATP, and each cycle of elongation uses 2 NADPH molecules. Thus, 5 ATP and 10 NADPH are required.

Control of Fatty Acid Metabolism. Derivatives of Fatty Acids

21. b, d. Acetyl CoA carboxylase is inactivated by phosphorylation. This effect is partially abolished by citrate. Phosphorylation-dephosphorylation of this enzyme is under the control of hormones, whose action is mediated by a protein kinase that is not dependent on cAMP. High energy charge stimulates acetyl CoA carboxylase, while low energy charge, i.e., high AMP levels, inhibits it.

22. b, d, e. Answer (a) is incorrect because pyruvate carboxylase is controlled by acetyl CoA, and answer (c) is incorrect because carnitine acyltransferase is regulated by malonyl CoA.

23. (a) 3, 4, 5, 6. Answer 6 is correct because, although indirect, it is a mechanism that supplies fatty acids for oxidation.

(b) 1, 2, 6, 7

24. b, c. For (c) additional elongations as well as desaturations are required.

25. c. Answer (c) is incorrect because the eicosanoid hormones have very short half-lives and therefore exert local rather than global effects. Answer (e) is correct because phospholipids supply the arachidonate for prostaglandin synthesis.

26. Aspirin acetylates a specific Ser residue in the cyclooxygenase component of prostaglandin synthase. Thus aspirin inhibits the synthesis of prostaglandins, thromboxanes, and prostacyclin, which mediate the inflammatory response.

PROBLEMS

1. Many plants have enzyme systems that catalyze the formation of a cis double bond in oleic acid at one or more positions between C-9 and the terminal methyl group. The fact that these enzyme systems exist in plants is of great significance to animals. Why?

2. (a) Suppose that the normal mechanism for the oxidation of fatty acids in mammals were through oxidation at the α carbon rather than the β carbon. What results would Knoop have obtained in his experiments?

   (b) Stumpf and his colleagues have described an α-oxidation system in plant leaves and seeds. Molecular oxygen is used in the α-oxidative decarboxylation of a free fatty acid, which yields a fatty aldehyde that is one carbon shorter than the original fatty acid. The fatty aldehyde is in turn oxidized to the corresponding fatty acid, with NAD^+ serving as an electron acceptor. These steps are repeated, resulting in the complete oxidation of the fatty acid. Suppose that the NADH generated through the α-oxidation of palmitate is reoxidized in the mitochondrial electron transport chain. Compare the yield of ATP generated by the α-oxidation of palmitate with that generated by β-oxidation of the same fatty acid. Assume that the products of the final round of oxidation are carbon dioxide and acetic acid.

   (c) If fatty acid oxidation occurs via the α-oxidation route, will odd-numbered fatty acids be glucogenic? Why?

3. Although most components of the diet contain fatty acids with unbranched chains, some plant tissues contain fatty acids with methyl groups at odd-numbered carbons in the acyl chain. These fatty acids cannot be broken down through β-oxidation.

   (a) Which step in β-oxidation is likely to be blocked when branched-chain fatty acids are substrates?
   (b) Some tissues, including brain tissue, can carry out the limited α-oxidation of a fatty acid with one or more methyl groups at odd-numbered carbons. Using the pathway discussed in problem 2, show how one round of α-oxidation enables a cell to bypass the block to β-oxidation. Use as your substrate a molecule of palmitate with a methyl branch at C-3.

4. The microbial oxidation of long-chain alkanes, which are found in crude oil, is the subject of intense study because of concern about oil spills. In many bacteria, alkane oxidation occurs within the outer membrane. A monooxygenase enzyme uses molecular oxygen and an oxidizable substrate, such as NADH, to convert an alkane to a primary alcohol. Studies show that three additional reactions are required for the primary alcohol to undergo β-oxidation. Propose a pathway for the conversion of a long-chain primary alcohol to a substrate that can undergo β-oxidation. Include cofactors and electron acceptors that may be required.

5. Malonyl CoA, labeled with ^{14}C in the methylene carbon, is used in excess as a substrate in an in vivo system for the synthesis of palmitoyl CoA,

which is catalyzed by a yeast fatty acid synthase complex. Acetyl CoA and other substrates are also present in the system, but acetyl CoA carboxylase is not. Which carbons in palmitoyl CoA will be labeled?

6. A deficiency of carnitine acyltransferase I in human muscle causes cellular damage and recurrent muscle weakness, especially during fasting or exercise. A deficiency of the enzyme in the liver causes an enlarged and fatty liver, hypoglycemia, and a reduction in the levels of ketone bodies in blood. Explain the causes of these symptoms.

7. One intermediate in the conversion of propionyl CoA to succinyl CoA is methylmalonyl CoA, the structure of which is shown in the margin. This compound is an analog of malonyl CoA. In people who are unable to convert propionyl CoA to succinyl CoA, high levels of methylmalonyl CoA are observed. What effect could such levels of methylmalonyl CoA have on fatty acid metabolism?

COO^-
H_3C—C—H
C—S—CoA
‖
O

L-Methylmalonyl CoA

8. Animals cannot synthesize glucose from even-numbered fatty acids, which make up the bulk of the fatty acids in their diet.

(a) How can odd-numbered fatty acids be used for the net synthesis of glucose in animals?
(b) Triacylglycerols can be used as precursors of glucose. Give two reasons why this is possible.
(c) Why are most of the fatty acids found in animal tissues composed of an even number of carbon atoms?
(d) Some bacteria synthesize odd-numbered fatty acids. What CoA derivative is required, in addition to acetyl CoA and malonyl CoA, for the synthesis of an odd-numbered fatty acid?

9. (a) Describe how malonyl CoA affects the balance between the rates of synthesis and β-oxidation of fatty acids in a liver cell.

(b) Show that failure to regulate these two processes reciprocally results in the wasteful hydrolysis of ATP.

10. Plant seeds contain triacylglycerols in organelles called *spherosomes.* During germination, lipases located in the spherosome membrane convert triacylglycerol to monoacylglycerols, free fatty acids, and glycerol. Both free fatty acids and monoacylglycerols enter the glyoxysome, whereas most of the glycerol is metabolized in the plant cell cytosol. A membrane-bound lipase in the glyoxysome converts monoacylglycerols to free fatty acids and glycerol.

(a) Describe two possible metabolic fates of glycerol in the cytosol.
(b) What is the fate of fatty acids in the glyoxysome?
(c) When a germinating plant begins to carry out photosynthesis, the number of glyoxysomes in the germinating plant decreases rapidly. Why?
(d) Plant tissues with high numbers of mitochondria also have high concentrations of carnitine, but there is little correlation between numbers of glyoxysomes and carnitine concentrations in germinating tissue. What does this observation suggest about the role of carnitine in fatty acid metabolism in these two organelles?
(e) Another difference between plant glyoxysomes and plant mitochondria is that glyoxysomes cannot oxidize acetyl CoA, whereas

mitochondria can. How is this observation related to the metabolism of fatty acids in these two organelles?

11. People concerned about weight must pay attention not only to triacylglyceride intake but also to the consumption of starch, glucose, and other carbohydrates. Although carbohydrates can be converted to glycogen in liver, muscle, and other tissues, only about 5 percent of the energy stored in the body is present as glycogen. What happens to most carbohydrates that are consumed in caloric excess?

12. Wakil's pioneering studies on fatty acid synthesis included the crucial observation that bicarbonate is required for the synthesis of palmitoyl CoA. He was surprised to find that very low levels of bicarbonate could sustain palmitate synthesis; that is, there was no correlation between the amount of bicarbonate required and the amount of palmitate produced. Later he also found that ^{14}C-labeled bicarbonate is not incorporated into palmitate. Explain these observations.

13. Many of the enzymes of the β-oxidation pathway have relatively broad specificities for fatty acyl chain lengths. Why is this important for the economy of the cell?

14. Liver tissue carries out the synthesis of ketone bodies from fatty acids. Suppose a liver cell converts palmitic acid to acetoacetate and then exports it to the circulation. How many molecules of ATP per molecule of palmitate converted to acetoacetate are available to the liver cell?

15. In tissue culture, cells that are deficient in $NADP^+$-linked malate enzyme can be isolated. They exhibit a slightly lower rate of fatty acid synthesis when compared with normal cells. However, cells lacking citrate lyase are very difficult to isolate. Why?

16. An unusual sphingolipid contains a 22-carbon, polyunsaturated fatty acid called *clupanodonic acid,* or *7,10,13,16,19-docosapentaenoic acid.* In mammals, both the mitochondrial and endoplasmic reticular acyl chain elongation and desaturation systems can synthesize clupanodonate from linolenate.

(a) What steps are required to synthesize clupanodonate from linolenate?
(b) Why are mammals unable to synthesize clupanodonate from linoleate?

17. You are conducting some experiments with the mammalian fatty acid synthase complex, and you wish to use radioactively labeled precursors to produce labeled palmitoyl CoA. You have two samples of ^{14}C-labeled malonyl CoA, one labeled in the carboxyl carbon and one labeled in the methylene carbon. Explain where palmitate will be labeled when each of the radioactive malonyl CoA samples is used with fatty acid synthase.

18. Certain desert mammals can survive long periods of drought by consuming plants and seeds and then generating water by metabolizing those foods.

(a) Briefly describe how water is generated through intermediary metabolism. Include the sources of oxygen and hydrogen and describe reactions that lead to the formation of water.

(b) While parts of mature plants are a reliable source of carbohydrates and proteins, plant seeds contain high quantities of triacylglycerols and free fatty acids. Would plants or seeds be better for generating metabolic water? Why?

(c) Suppose that a desert rat metabolizes 30 g of palmitoyl CoA from seeds. How many milliliters of water can be generated from the process?

19. Explain why the metabolism of a C_{15} fatty acid can lead to the net synthesis of glucose, but the metabolism of a C_{16} fatty acid cannot.

20. Compare the effects of high levels of intracellular citrate on pathways of fatty acid and carbohydrate metabolism. Explain how its transport from the mitochondrion to the cytosol is essential for the action of citrate on both sets of pathways.

21. You are examining mitochondria from muscle cells of an infant who has a deficiency in one of the enzymes in the fatty acid oxidative pathway. Those mitochondria consume oxygen normally when incubated with pyruvate and malate, with succinate, or with palmitoyl CoA (in the presence of carnitine), but the rate of oxygen utilization is decreased when the mitochondria are incubated with linoleoyl CoA in the presence of carnitine. Blood levels of carnitine in the patient are low, while the levels of an unusual acylcarnitine derivative are present in blood and urine. Analysis of this acylcarnitine species using mass spectroscopy reveals that it is *trans*-Δ^2, *cis*-Δ^4 decadienoyl (C10:2)-acylcarnitine. The infant suffers from hypotonia (lack of muscle tone), poor feeding, and little weight gain.

What enzyme is deficient in the cells of the infant? Explain the observed symptoms on the basis of such a deficiency and how you might treat such a disorder.

22. In mammals, acetyl CoA from fatty acid oxidation cannot be used for the net synthesis of pyruvate or oxaloacetate, which in turn means that net glucose synthesis from acetyl CoA is not possible. However, glucose can be radioactively labeled when ^{14}C-labeled acetate is introduced into human tissue culture cells and converted to acetyl CoA by acetyl CoA synthetase. Radioactive fatty acids can also be used to label glucose. Why? If the methyl carbon of acetate is labeled, where will glucose be labeled?

ANSWERS TO PROBLEMS

1. Because they lack an enzyme that can introduce double bonds beyond the C-9 position in a fatty acid, animals cannot synthesize linoleate and linolenate. These unsaturated fatty acids are precursors for a number of other important fatty acids as well as the prostaglandins. Animals therefore rely on their diet as the source of linoleate and linolenate, which are synthesized only in plants.

2. (a) Because the α-oxidation of a fatty acid results in the removal of a single terminal carbon atom during each round of oxidation, Knoop would have found that the degradation of both phenylbutyrate and phenylpropionate yields benzoate.

(b) The net yield from the β-oxidation of palmitate is 106 molecules of ATP, as discussed on page 611 of Stryer's text. If 1 molecule of NADH is generated for 15 of the 16 carbons of palmitate, then the yield of ATP is 2.5×15, or 37.5. For α-oxidation, activation of the acetate molecule requires 2 ATP. Subsequent oxidation of acetyl CoA generates 10 ATP molecules. Thus 45.5 molecules of ATP are generated by the α-oxidation of a molecule of palmitate.

(c) In β-oxidation of odd-numbered fatty acids, the products include propionyl CoA, which can be converted to succinyl CoA, a glucogenic substrate. However, α-oxidation of an odd-numbered fatty acid would yield carbon dioxide as well as a single molecule of acetate or acetyl CoA, neither of which is glucogenic.

3. (a) As shown in Figure 24-1, the oxidation of a fatty acid with a methyl group at C-3 proceeds to the formation of the L-hydroxymethylacyl CoA derivative. Subsequent oxidation of the β carbon to the ketoacyl derivative is blocked by the methyl group. Compare this pathway with the one shown in Figure 24-5 (p. 609) of Stryer's text.

(b) The oxidation of the α carbon in the palmitate derivative followed by decarboxylation of the molecule yields a fatty acid that has a methyl group at an even-numbered carbon. Activation of the fatty acid to form an acyl CoA derivative followed by oxidation at the β carbon results in the generation of propionyl CoA and a shortened acyl derivative, lauroyl CoA (see Figure 24-2).

$$R-CH_2-CH(CH_3)-CH_2-C(=O)-S-CoA$$

Branched-chain acyl CoA

FAD → $FADH_2$ Oxidation

$$R-CH_2-C(CH_3)=CH-C(=O)-S-CoA$$

Enoyl CoA

H_2O Hydration

$$R-CH_2-C(CH_3)(OH)-CH_2-C(=O)-S-CoA$$

L-Hydroxymethylacyl CoA

Figure 24-1
Formation of L-hydroxymethylacyl CoA through β-oxidation of branched-chain acyl CoA.

4. The normal route for β-oxidation in bacteria utilizes acyl CoA derivatives, which are formed from free fatty acids. To convert a primary alcohol to a free fatty acid, two oxidative steps are needed, each requiring an electron acceptor. In *Corynebacterium*, NAD^+-dependent dehydrogenases catalyze the sequential conversion of a primary alcohol to a fatty acid, with the corresponding aldehyde as an intermediate. Conversion of the free fatty acid to an acyl CoA derivative requires two equivalents of ATP (because ATP is converted to AMP and PP_i), as well as coenzyme A. The reaction is catalyzed by acyl CoA synthase.

5. As shown in Figure 24-13 on page 617 of Stryer's text, acetyl–ACP and malonyl–ACP condense to form acetoacetyl–ACP. Carbons 4 and 3 of acetoacetyl–ACP are not labeled, because they are derived from acetyl CoA. These two carbons will become carbons 15 and 16 of palmitate. Only C-2 of acetoacetyl–ACP will be labeled because it is derived from the methylene carbon of malonyl–ACP. When the second round of synthesis begins, butyryl–ACP condenses with a second molecule of methylene-labeled malonyl–ACP, which contributes C-1 and C-2 of the newly formed six-carbon ACP derivative. In this compound, C-2 and C-4 will be labeled. Chain elongation continues until palmitoyl–ACP is formed. Each even-numbered carbon atom, except for carbon 16 (at the ω end), will be labeled.

6. Carnitine acyltransferase I facilitates the transfer of long-chain fatty acids into the mitochondrion by catalyzing the formation of fatty acyl carnitine molecules. The failure to form such molecules means that long-chain fatty acids are not available for cellular oxidation. In muscle, exercise or fasting increases dependence on fatty acids as a source of energy, so the inability to metabolize them interferes with cellular functions, causing cramps, weakness, and muscle damage. Liver cells also require forma-

$$CH_3(CH_2)_{12}-CH(CH_3)-CH_2-COO^- \xrightarrow[\alpha\text{-Oxidation}]{\frac{1}{2}O_2} CH_3(CH_2)_{12}-CH(CH_3)-CH(OH)-COO^- \xrightarrow[\text{Oxidative decarboxylation}]{NAD^+ \to NADH,\ CO_2} CH_3(CH_2)_{12}-CH(CH_3)-CHO$$

$$\xrightarrow[\text{Oxidation}]{NAD^+ \to NADH + H^+} CH_3(CH_2)_{12}-CH(CH_3)-COO^- \xrightarrow[\text{Activation}]{ATP + CoA \to AMP + PP_i} CH_3(CH_2)_{12}-CH(CH_3)-CO-S-CoA$$

$$\xrightarrow{\beta\text{-Oxidation}} CH_3(CH_2)_{10}-CO-S-CoA + CH_3-CH_2-CO-S-CoA$$

Acyl CoA **Propionyl CoA**

Figure 24-2
The α-oxidation and oxidative decarboxylation of a branched-chain fatty acid allow generation of intermediates that can enter normal oxidative pathways.

tion of fatty acyl carnitine molecules to oxidize fats in mitochondria. If fatty acids cannot be utilized, they will remain in the cytosol, where high concentrations of them cause cell enlargement and interfere with other functions. Liver cells then must use glucose as a source of energy instead of exporting it to other cells. Because liver cells use acetyl CoA, which is derived primarily from fatty acid oxidation, as a precursor of ketone bodies, the failure to oxidize fatty acids will result in a reduction in the rate of ketone body synthesis. This in turn will exacerbate the symptoms of hypoglycemia because tissues, such as cardiac muscle, that normally use ketone bodies as a source of energy will rely more heavily on glucose as a source of energy.

7. Because malonyl CoA is a substrate for fatty acid synthase, competition from methylmalonyl CoA could cause a decrease in the rate of palmitoyl CoA synthesis in the cytosol, which could in turn lead to an increase in the amount of acetyl CoA carboxylase activity (recall that palmitoyl CoA inhibits acetyl CoA carboxylase). In addition, high levels of methylmalonyl CoA could interfere with transport of long-chain fatty acyl chains into mitochondria by inhibiting carnitine acyltransferase, as does malonyl CoA. Thus, both the synthesis and the oxidation of fatty acids could be inhibited by methylmalonyl CoA.

8. (a) The oxidation of an odd-numbered fatty acid yields a number of acetyl CoA molecules as well as one molecule of propionyl CoA, which can be converted to succinyl CoA, a component of the citric acid cycle. Although two-carbon compounds like acetyl CoA cannot be used for the net synthesis of glucose, succinyl CoA can contribute net carbons to the citric acid cycle, enabling oxaloacetate and, ultimately, glucose to be formed through gluconeogenesis.

(b) Triacylglycerols are converted to glycerol and three free fatty acids through the action of lipases. Glycerol can be converted to glucose via dihydroxyacetone phosphate. Odd-numbered fatty acids found in triacylglycerols can also be used for net synthesis of glucose, whereas even-numbered fatty acids cannot.

(c) During fatty acid synthesis, most organisms use acetyl CoA as a source of the ω carbon and its adjacent carbon in the acyl chain. Two of the three carbons of malonyl CoA are incorporated during each cycle of acyl chain elongation. Thus the resulting fatty acid will contain an even number of carbon atoms.

(d) To produce an odd-numbered fatty acid, at least one odd-numbered CoA intermediate must be incorporated in its entirety during fatty acid synthesis. Propionyl CoA can be used by certain bacteria for the initial condensation step with malonyl CoA in fatty acid synthesis. The resulting five-carbon acyl intermediate is then extended in two-carbon units to yield an odd-numbered fatty acid.

9. (a) Malonyl CoA is a key substrate for the synthesis of fatty acids; when it is abundant, synthesis is stimulated. In addition, high levels of this intermediate inhibit carnitine acyltransferase I, thereby limiting the entry of fatty acyl chains into the mitochondrion, where they are oxidized. A decrease in the concentration of malonyl CoA leads to a decrease in the rate of fatty acid synthesis and an increase in the rate of fatty acid oxidation in the mitochondrion.

(b) The overall equation for the synthesis of palmitoyl CoA is

$$\text{8 Acetyl CoA} + \text{7 ATP} + \text{14 NADPH} \rightarrow \text{palmitoyl CoA} + \text{14 NADP}^+ + \text{7 CoA} + 7\ H_2O + \text{7 ADP} + 7\ P_i$$

The overall equation for the oxidation of palmitoyl CoA is

$$\text{Palmitoyl CoA} + \text{7 FAD} + \text{7 NAD}^+ + \text{7 CoA} + 7\ H_2O \rightarrow \text{8 acetyl CoA} + 7\ FADH_2 + \text{7 NADH} + 7\ H^+$$

Assuming that NADPH is equivalent in reducing power to NADH, that a molecule of $FADH_2$ yields 1.5 ATP during electron transport and oxidative phosphorylation, and that a molecule of NADH yields 2.5 ATP, then 42 ATP molecules are required to synthesize a molecule of palmitoyl CoA, whereas 28 ATP are generated by the conversion of palmitoyl CoA to 8 molecules of acetyl CoA. There is a net loss of 14 ATP molecules if the two processes occur simultaneously.

10. (a) Glycerol is converted to dihydroxyacetone phosphate, which in turn can serve as a source of glucose or can be converted to acetyl CoA.

(b) Fatty acids serve as a source of acetyl CoA, which is used in the glyoxylate cycle and gluconeogenesis.

(c) The primary function of glyoxysomes is to utilize fatty acids from triacylglycerols for the synthesis of glucose, which is used as a source of other molecules by the developing plant. Once leaf development enables the plant to generate glucose by photosynthesis, glyoxysomes are no longer needed.

(d) Carnitine functions in the transport of long-chain fatty acids from the cytosol to the interior of the mitochondrion. The observation suggests that, although carnitine may be important in mitochondrial transport of fatty acyl chains, the compound is not involved in the movement of fatty acyl chains into the glyoxysome. It is also possible that glyoxysomes metabolize fatty acids with shorter acyl chains, for which transport facilitated by carnitine is not necessary.

(e) The fate of fatty acids is different in glyoxysomes and in mitochondria. Both organelles carry out β-oxidation of fatty acids to acetyl CoA; however, in glyoxysomes, acetyl CoA is a precursor of glucose, whereas mitochondria oxidize acetyl CoA to CO_2 and H_2O to generate ATP.

11. Carbohydrates consumed in caloric excess are converted to acetyl CoA, which in turn serves as a source of fatty acids. The concurrent synthesis of glycerol from carbohydrates such as glucose and fructose provides the second precursor needed for the synthesis of triacylglycerols, which are the primary storage form of energy in humans.

12. Bicarbonate is a source of carbon dioxide for the reaction catalyzed by acetyl CoA carboxylase, in which malonyl CoA is formed. Malonyl CoA is then used as a source of two-carbon units for fatty acyl chain elongation, and the carbon atom derived originally from bicarbonate is released as CO_2. Carbon dioxide is then rapidly converted to bicarbonate, which is used again for the synthesis of another molecule of malonyl CoA. Thus, the carbon atom derived from bicarbonate can be used many times for the production of malonyl CoA, but it is never incorporated into the growing acyl chain, so it does not appear in palmitate.

13. If each enzyme could operate only on fatty acyl CoA derivatives of a particular chain length, then as many as eight sets of enzymes would be required to carry out the β-oxidation of palmitate. The fact that most enzymes of the β-oxidation pathway can use acyl CoA molecules of different chain lengths as substrates means that the cell needs to synthesize fewer types of proteins in order to carry out fatty acid oxidation.

14. In order to synthesize acetoacetate from palmitate, liver cells must carry out β-oxidation of the 16-carbon acyl chain, generating 8 molecules of acetyl CoA, which will in turn generate 4 molecules of acetoacetate. A total of 7 NADH and 7 $FADH_2$ molecules are generated per molecule of palmitate converted to acetyl CoA. The 14 reduced cofactors are equivalent to 28 ATP molecules. Because 2 molecules of ATP are needed to activate palmitate, the net yield of ATP per palmitate is 26.

15. Both malate enzyme and citrate lyase are part of the shuttle system that transports two-carbon units from the mitochondrion to the cytosol. Malate enzyme also generates reducing power in the form of NADPH, which is used for fatty acid synthesis; however, the phosphogluconate pathway also serves as a source of NADPH, so that fatty acid synthesis can continue even if malate enzyme is deficient. Recall from page 573 of Stryer's text that malate can cross the mitochondrial membrane. Citrate lyase is more critical to fatty acid synthesis because it is required to generate acetyl CoA from

citrate in the cytosol. Without cytosolic acetyl CoA, fatty acid synthesis cannot take place, and the cells cannot grow and divide.

16. (a) To synthesize clupanodonate from linolenate, the acyl chain must be elongated from 18 to 22 carbons, and two new double bonds must be introduced into the chain. Although the details of the various mammalian desaturation systems are not completely understood, it appears that a double bond at C-6 can be introduced when a double bond at C-9 is available and a double bond at C-5 can be introduced when one at C-8 is available. Thus, the probable sequence of reactions includes the introduction of a double bond at C-6 of linolenate (yielding a 18:4 *cis*-$\Delta^6,\Delta^9,\Delta^{12},\Delta^{15}$-acyl chain) followed by chain elongation to a 20-carbon derivative. The introduction of a double bond at C-5 then gives an acyl chain denoted as 20:5 *cis*-$\Delta^5,\Delta^8,\Delta^{11},\Delta^{14},\Delta^{17}$. The final reaction required to yield clupanodonate is chain elongation to the 22-carbon fatty acyl chain.

(b) Linoleate has cis double bonds at C-9 and C-12. Elongation to a 22-carbon chain would yield an acyl chain with double bonds at C-13 and C-16. To form clupanodonate, a double bond at C-19 is needed, but mammals lack the enzymes required to introduce double bonds beyond C-9. Thus, linoleate cannot be used for the synthesis of clupanodonate in mammals.

17. When malonyl CoA condenses with the growing acyl chain during palmitate synthesis, its carboxyl carbon is lost as carbon dioxide or bicarbonate ion. Thus the label from carboxyl-labeled malonyl CoA will not be transferred to the growing fatty acyl chain.

When the methylene-labeled species is used, seven carbon atoms (14, 12, 10, 8, 6, 4, and 2) in palmitoyl CoA will be labeled. During palmitoyl CoA synthesis, 7 two-carbon units from malonyl CoA molecules are incorporated into the growing chain, beginning with a unit that condenses with acetyl–ACP on the fatty acyl synthase complex. The first methylene carbon joins the acetyl–ACP molecule to become what will be carbon 14 of palmitate. Succeeding methylene carbons are incorporated into even-numbered positions in the growing molecule. Recall that the fatty acid is numbered from the carboxyl end, which is added last during biosynthesis.

18. (a) The source of oxygen for formation of water during respiration is atmospheric oxygen, while the sources of hydrogen include oxidizable foodstuffs such as carbohydrates and fats. These substances are oxidized to generate energy-rich electrons, which are in turn used to reduce oxygen to generate water. The principal terminal reaction in the process occurs in the mitochondrion, where electrons are transferred from cytochrome *c* to oxygen to generate oxidized cytochrome *c* and water. Also important in water generation is the formation of ATP from ADP and inorganic phosphate, where a molecule of water is generated during the formation of each ATP molecule.

(b) The more reduced the carbons of a substrate, the larger the number of electrons available during metabolism and the more water generated. Most carbon atoms of fatty acids are saturated and therefore highly reduced, so that they are a better source of available electrons. Carbohydrate molecules like glucose, whose carbons are at the alcohol level of oxidation or, in the case of the C-1 atom, at the aldehyde level, provide fewer electrons during terminal oxidation. Seeds, which contain a high percentage of fats, are therefore a better source than mature plants for the generation of water.

(c) First determine the number of moles of palmitate that are converted to CO_2 and water. The molecular weight of the molecule ($C_{16}H_{31}O_2$) is 255.

$$30 \text{ g}/255 \text{ g mol}^{-1} = 0.12 \text{ mole palmitate oxidized}$$

Then calculate the number of moles of water produced by the complete oxidation of palmitate.

For palmitoyl CoA, Stryer shows on page 611 that oxidation of the molecule gives 7 $FADH_2$, 7 NADH, and 8 acetyl CoA molecules, utilizing 7 molecules of water.

On page 513, Stryer shows that oxidation of acetyl CoA in the citric acid cycle yields 3 NADH, 1 $FADH_2$, and 1 GTP, equivalent to ATP, utilizing 2 water molecules. Thus the 8 acetyl CoA molecules produced from palmitoyl CoA give 24 NADH, 8 $FADH_2$, and 8 ATP equivalents, utilizing 16 water molecules.

The total number of reduced electron carriers from the oxidative process is 31 NADH and 15 $FADH_2$. Recall that 2.5 ATP molecules are produced when NADH is oxidized in the electron transport chain, and 1.5 ATP are generated from $FADH_2$ oxidation.

The overall equation for the production of NAD^+, ATP, and water from palmitoyl CoA is

$$31 \text{ NADH} + 15.5 \text{ O}_2 + 77.5 \text{ ADP} + 77.5 \text{ P}_i + 108.5 \text{ H}^+ \rightarrow 31 \text{ NAD}^+ + 77.5 \text{ ATP} + 108.5 \text{ H}_2\text{O}$$

And for $FADH_2$ it is

$$15 \text{ FADH}_2 + 7.5 \text{ O}_2 + 22.5 \text{ ADP} + 22.5 \text{ P}_i + 37.5 \text{ H}^+ \rightarrow 15 \text{ FAD} + 22.5 \text{ ATP} + 37.5 \text{ H}_2\text{O}$$

The total number of water molecules produced is 146, and the net water produced is $(146 - 23) = 123$ molecules of water per palmitate oxidized or 123 moles of water per mole of palmitate. The molecular weight of water is 18.0.

30 g palmitoyl CoA is equivalent to 0.12 mole of palmitate, which generates $0.12 \times 123 = 14.8$ moles of water when oxidized. At 18 g/mol, 14.8 moles of water equals 266 g, or 266 ml, of water.

19. The oxidation of a C_{16} fatty acid (palmitate) leads to the formation of eight molecules of acetyl CoA. This compound, which contains two carbon atoms, is oxidized to two CO_2 in the citric acid cycle, so that the net number of carbons entering and leaving the cycle is zero. Thus no net carbons are available to enter the gluconeogenic pathway. On the other hand, oxidation of a C_{15} fatty acid generates seven acetyl CoA molecules, plus one molecule of propionyl CoA. This compound is converted by carboxylation, epimerization, and conversion to succinyl CoA, a *four*-carbon compound that is an intermediate in the citric acid cycle. Succinyl CoA contributes two extra carbons to the gluconeogenic pathway, leading to the net synthesis of glucose.

20. High levels of citrate signal that glucose utilization is no longer necessary and that adequate carbon atoms are available for synthesis of palmitoyl CoA. Citrate inhibits phosphofructokinase 1 activity, decelerating the rate of glycolysis. On the other hand, citrate stimulates the activity of acetyl CoA carboxylase, so that increased production of malonyl CoA leads to stimulation of fatty acid synthesis. The transport of citrate from the mitochondrial matrix to the cytosol is important because both phosphofructokinase 1 and acetyl CoA carboxylase are located in the cytosol.

21. The most likely deficiency is a lack of 2,4-dienoyl CoA reductase, an enzyme that is essential for the degradation of unsaturated fatty acids with double bonds at even-numbered carbons. Such fatty acids include linoleate (9-*cis*,12-*cis* 18:2). Four rounds of oxidation of linoleoyl CoA generates a 10-carbon acyl CoA that contains a trans-Δ^2 and a cis-Δ^4 double bond. This intermediate is a substrate for the reductase, which converts the 2,4-dienoyl CoA to *cis*-Δ^3-enoyl CoA. A deficiency of 2,4-dienoyl reductase leads to an accumulation of *trans*-Δ^2,*cis*-Δ^4-decadienoyl CoA molecules in the mitochondrion. The observation that carnitine derivatives of the 2,4-dienoyl CoA are found in blood and urine provides evidence that these molecules accumulate in the mitochondrion and are then attached to carnitine. Formation of carnitine decadienoate allows the acyl molecules to be transported across the inner mitochondrial membrane into the cytosol, and then into the circulation.

Mitochondria from the patient function normally, taking up oxygen as they carry out oxidation of various substrates including palmitate, a saturated fatty acid. However, incubation of those mitochondria with linoleate results in reduced oxygen uptake, because the absence of the reductase molecule allows only a limited number of rounds of β-oxidation to occur before the 2,4-dienoyl molecule is formed. Lack of muscle tone could mean that there are difficulties in oxidizing fuel molecules needed to provide energy for muscle contraction. If carnitine levels in cells are lower because many of them are esterified to decadienoate molecules, the result is a virtual deficiency of carnitine. The cell's ability to transport other long-chain fatty acids across the inner mitochondrial membrane is limited under these conditions. Impairment of fatty acid oxidation means that fewer ATP molecules are available for muscular activity.

One immediate strategy for dealing with this disorder is to limit linoleate in the diet. However, linoleate is a starting point for other unsaturated fatty acids including arachidonate, a precursor of eicosanoid hormones. Limiting linoleate in the diet of a person with the reductase deficiency would have to be carried out carefully, in order to avoid a deficiency of an essential fatty acid.

22. Radioactive acetyl CoA can be generated by direct synthesis from ^{14}C-acetate or from β-oxidation of radioactive fatty acids, such as uniformly labeled palmitate. Examination of the reactions of the citric acid cycle reveal that neither of the two carbons that enter citrate from acetate are removed as carbon dioxide during the first pass through the cycle. Labeled carbon from ^{14}C-methyl-labeled acetate appears in C-2 and C-3 of oxaloacetate, because succinate is symmetrical, with either methylene carbon in that molecule labeling C-2 or C-3 of oxaloacetate. The conversion of oxaloacetate to phosphoenolpyruvate yields PEP labeled at C-2 or C-3 as well. Formation of glyceraldehyde 3-phosphate and its isomer dihydroxyacetone phosphate gives molecules, both labeled at carbons 2 and 3. Condensation by aldolase gives fructose 1,6-bisphosphate radioactively

labeled at carbons 1, 2, 5, and 6. The corresponding four carbons will then be labeled in glucose 6-phosphate or glucose.

EXPANDED SOLUTIONS TO TEXT PROBLEMS

1. Glycerol + 2 NAD^+ + P_i + ADP $\rightarrow$

$$\text{pyruvate} + \text{ATP} + H_2O + 2\ \text{NADH} + H^+$$

Glycerol kinase and glycerol phosphate dehydrogenase are required. Remember, glycerol enters glycolysis as dihydroxyacetone phosphate; hence, the need for a kinase and a dehydrogenase. For the conversion of dihydroxyacetone phosphate to pyruvate, see Chapter 19.

2. Stearate + ATP + $13\frac{1}{2}$ H_2O + 8 FAD + 8 NAD^+ $\rightarrow$

$$4\tfrac{1}{2}\text{acetoacetate} + 14\tfrac{1}{2}H^+ + 8\ FADH_2 + 8\ \text{NADH} + \text{AMP} + 2\ P_i$$

Note that this equation is the sum of the following three equations:

(1) Stearate + CoA + ATP + H_2O $\rightarrow$

$$\text{stearoyl CoA} + \text{AMP} + 2\ P_i + 2H^+$$

(2) Stearoyl CoA + 8 FAD + 8 NAD^+ + $3\frac{1}{2}$CoA + 8 H_2O $\rightarrow$

$$4\tfrac{1}{2}\text{acetoacetyl CoA} + 8\ FADH_2 + 8\ NADH^+ + 8\ H^+$$

(3) $4\frac{1}{2}$Acetoacetyl CoA + $4\frac{1}{2}H_2O$ $\rightarrow$

$$4\tfrac{1}{2}\text{acetoacetate} + 4\tfrac{1}{2}H^+ + 4\tfrac{1}{2}\text{CoA}$$

(See Stryer, pp. 607, 611.)

3. (a) Oxidation in mitochondria, synthesis in the cytosol.

(b) CoA in oxidation, acyl carrier protein for synthesis.

(c) FAD and NAD^+ in oxidation, NADPH for synthesis.

(d) L isomer of 3-hydroxyacyl CoA in oxidation, D isomer in synthesis.

(e) Carboxyl to methyl in oxidation, methyl to carboxyl in synthesis.

(f) The enzymes of fatty acid synthesis, but not those of oxidation, are organized in a multienzyme complex.

4. Because mammals lack the enzymes to introduce double bonds at carbon atoms beyond C-9 but can increase the length of the fatty acid chain at the carboxyl end, the easiest way to determine which unsaturated fatty acid is the precursor is to note the number of carbons from the ω end (CH_3 end) to the nearest double bond. Thus, in (a) this number is seven carbons; hence, palmitoleate is the precursor. In (b) it is six carbon atoms; hence, linoleate. In (e) it is nine carbon atoms; hence, oleate; etc. (See Stryer, p. 623.)

5. During fatty acid biosynthesis, the carbon chain grows two carbons at a time by the condensation of an acyl–ACP with malonyl–ACP, with the malonyl–ACP becoming, in every case, the carboxyl end of the new acyl-ACP. Thus, the chain grows from methyl to carboxyl. Since ^{14}C-labeled malonyl CoA was added a short time before synthesis was stopped, the fatty acids whose synthesis was completed during this short period will be heavily labeled toward the carboxyl end (the last portion synthesized) and less heavily labeled, if at all, on the methyl end.

6. The enolate anion of one thioester attacks the carbonyl carbon atom of the other thioester to form a C–C bond.

7. Decarboxylation drives the condensation of malonyl–ACP and acetyl–ACP. In contrast, the condensation of two molecules of acetyl–ACP is energetically unfavorable. In gluconeogenesis, decarboxylation drives the formation of phosphoenolpyruvate from oxaloacetate (p. 573).

8. Adipose-cell lipase is activated by phosphorylation. Hence, overproduction of the cAMP-activated kinase will lead to accelerated breakdown of triacylglycerols and depletion of fat stores.

9. The mutant enzyme would be persistently active because it could not be inhibited by phosphorylation. Fatty acid synthesis would be abnormally active. Such a mutation might lead to obesity.

10. Carnitine translocase deficiency and glucose 6-phosphate transporter deficiency. For an explanation, see Stryer, pp. 608, 598–599.

11. In the fifth found of β-oxidation, *cis*-Δ^2-enoyl CoA is formed. Hydration by the classic hydratase yields D-3-hydroxyacyl CoA, the wrong isomer for the next enzyme in β-oxidation. This dead end is circumvented by a second hydratase that removes water to give *trans*-Δ^2-enoyl CoA. Addition of water by the classic hydratase then yields L-3-hydroxyacyl CoA, the appropriate isomer. Thus, hydratases of opposite stereospecificities serve to *epimerize* (invert the configuration of) the 3-hydroxyl group of the acyl CoA intermediate. See J.K. Hiltunen, P.M. Palosaari, and W-H. Kunau. *J. Biol. Chem.* 264(1989):13536.

12. The probability of synthesizing an error-free polypeptide chain decreases as the length of the chain increases. A single mistake can make the entire polypeptide ineffective. In contrast, a defective subunit can be spurned in forming a noncovalent multienzyme complex; the good subunits are not wasted.

CHAPTER

25

Amino Acid Degradation and the Urea Cycle

The organism derives energy from both stored and exogenous fuels. Stryer described how carbohydrates (Chapter 19) and fats (Chapter 24) are catabolized to form NADH and $FADH_2$ (Chapter 20). Electron transport and oxidative phosphorylation harvest some of the free energy of these compounds to form transmembrane proton-motive gradients and ATP (Chapter 21). In Chapter 25, Stryer explains how proteins fit into this scheme. Although proteins are not stored as fuels per se as carbohydrates and fats are, supplies of proteins in excess of those needed to provide biosynthetic precursors are degraded for energy or are converted into fats or carbohydrates. Each amino acid contains at least one nitrogen atom. Stryer describes how the α-amino groups of most amino acids are transferred to α-ketoglutarate to form glutamate by transamination and how the α-amino group of glutamate is converted to ammonia by an oxidative deamination. In addition, he describes the formation of aspartate by transamination between oxaloacetate and glutamate. He then presents the urea cycle, in which ammonia, the α-amino group of aspartate, and CO_2 are condensed to form urea—a nontoxic excretory product of nitrogen in higher animals. He also explains that some other organisms excrete nitrogen as uric acid or ammonia.

Because there are 20 amino acids, the catabolic pathways of their carbon skeletons are numerous and of varied types. Stryer describes how the carbon atoms of each amino acid are funneled into one or more of seven primary products. Two of these, acetyl CoA and acetoacetyl CoA, can be converted to ketone bodies (Chapter 24), and the remaining five can be converted into glucose (Chapter 22). The two groups of products lead to the glycogenic-ketogenic classification of the amino acids. Throughout the chapter, Stryer explains the roles of several cofactors, including pyridoxal phosphate, tetrahydrobiopterin, and vitamin B_{12}, in the mechanisms of particular enzymes involved in amino acid metabolism. The pathological consequences of defects or deficiencies in some of the enzymes involved in catabolism of amino acids and the synthesis of urea are also described.

LEARNING OBJECTIVES

After mastering this chapter, you should be able to complete the following objectives:

Introduction (Stryer pages 629)

1. Describe the fates of exogenously supplied amino acids that are not used for protein synthesis.

Amino Group Metabolism (Stryer pages 630–635)

2. Name the major organ of *amino acid* degradation in mammals.

3. Describe the reactions catalyzed by the *aminotransferases* (*transaminases*) and state the major function of these reactions. Explain how α-amino groups are removed from serine and threonine.

4. Describe the reaction catalyzed by *glutamate dehydrogenase,* outline its regulation, and state its major function. Note that the participation of NAD^+ in the reaction links nitrogen metabolism and energy generation.

5. Recognize the structures of *pyridoxal phosphate* (*PLP*) and *pyridoxamine phosphate* (*PMP*), indicate the reactive functional groups on each, and name the dietary precursor of PLP.

6. Describe the aminotransferase reaction mechanism and explain the involvement of a *Schiff base* and PLP. List the other kinds of reactions catalyzed by PLP and describe the common features of PLP catalysis. Define *stereoelectronic control.*

Urea Synthesis (Stryer pages 634–637)

7. Define the terms *ureotelic, uricotelic,* and *ammonotelic.*

8. List the immediate precursors of the nitrogen and carbon atoms of *urea* and name the carrier molecule on which urea is assembled during the *urea cycle.*

9. Name the enzymes and the other components of the urea cycle, note their intracellular locations, and indicate the molecular connection between this cycle and the citric acid cycle. Account for the ATP requirement of the cycle.

10. Explain how deficiencies in several different enzymes of the urea cycle give rise to *hyperammonemia.*

Carbon Atom Metabolism (Stryer pages 638–648)

11. State the strategy used by humans for catabolizing the carbon atom skeletons of amino acids and name the seven major metabolic products formed.

12. Describe the basis for the *glycogenic-ketogenic* designation of the amino acids and classify each amino acid accordingly. Appreciate the limitations of this classification.

13. List the amino acids that give rise to pyruvate. Consider the role of NAD^+ in these conversions.

14. List the amino acids that give rise to *oxaloacetate.*

15. List the amino acids that give rise to *fumarate.*

16. List the amino acids that give rise to *α-ketoglutarate.* Describe the role of *tetrahydrofolate* in one of the conversions.

17. List the amino acids that give rise to *succinyl CoA.* Describe the roles of CoA, biotin, and *coenzyme B_{12}* in these conversions.

18. Describe the biological source and the structure of coenzyme B_{12} and outline its formation from B_{12a}. Indicate its general function in the three types of reactions it catalyzes.

19. Explain the cause of *pernicious anemia,* and outline the normal sequence of events in the absorption of *cobalamin* and its conversion to coenzyme B_{12}.

20. List the amino acids that give rise to *acetyl CoA* and to *acetoacetyl CoA.* Describe the roles of biotin, CoA, NAD^+, FAD, thiamine pyrophosphate, lipoate, and *tetrahydrobiopterin* in these conversions.

21. Distinguish between *monoxygenases* and *dioxygenases.* Explain the role of O_2 in the catabolism of aromatic amino acids.

Inborn Errors of Metabolism (Stryer pages 648–650)

22. Describe *alcaptonuria.* Relate Garrod's hypothesis concerning this disorder and appreciate the historical significance of his analysis.

23. Explain the biochemical bases of *phenylketonuria, methylmalonic aciduria,* and *maple syrup urine disease.* Describe some of the consequences of these diseases.

SELF-TEST

Introduction

1. Which of the following answers complete the sentence correctly? Surplus dietary amino acids may be converted into

(a) proteins.
(b) fats.
(c) ketone bodies.
(d) glucose.
(e) a variety of biomolecules for which they are precursors.

all

Amino Group Metabolism

2. Which of the following compounds serves as an acceptor for the amino groups of many amino acids during catabolism?

(a) Glutamine
(b) Asparagine
(c) α-Ketoglutarate
(d) Oxalate

3. Which of the following answers completes the sentence correctly? The removal of α-amino groups from amino acids for conversion to urea in animals may occur by

(a) transamination.
(b) reductive deamination.
(c) oxidative deamination.
(d) transamidation.

4. Which of the following amino acids have their α-amino groups removed by dehydratases?

(a) Histidine
(b) Tryptophan
(c) Serine
(d) Glutamine
(e) Threonine

5. Which of the following answers completes the sentence correctly? The products of an aminotransferase-catalyzed reaction between pyruvate and glutamate would be

(a) aspartate and oxaloacetate.
(b) aspartate and α-ketoglutarate.
(c) alanine and oxaloacetate.
(d) alanine and α-ketoglutarate.

6. Explain the role of pyridoxal phosphate in aminotransferase reactions. Be sure to describe the Schiff base and the ketimine that are involved in the mechanism.

Urea Synthesis

7. Considering all forms of life, which of the following are major excretory forms of the α-amino groups of amino acids?

(a) Urea
(b) Uracil
(c) Ammonia
(d) Uric acid

8. How many moles of ATP are required to condense 2 moles of nitrogen and 1 mole of CO_2 into 1 mole of urea via the urea cycle? How many

high-energy bonds are used in this process? Do both atoms of nitrogen enter the cycle as NH_4^+?

____

____

9. What would be the net effect of linking the following two transamination reactions, and why would such coupling be useful when excess proteins were being catabolized for energy generation?

$$\text{alanine} + \alpha\text{-ketoglutarate} \rightleftharpoons \text{pyruvate} + \text{glutamate}$$

$$\text{oxaloacetate} + \text{glutamate} \rightleftharpoons \text{aspartate} + \alpha\text{-ketoglutarate}$$

____

____

10. Describe the role of ornithine in the urea cycle.

____

____

11. Explain how hyperammonemia, the increased concentration of NH_4^+ in the serum, can arise from defects in more than one enzyme of the pathway that forms urea.

____

____

Carbon Atom Metabolism

12. Match the catabolic products in the right column with the amino acids in the left column from which they can be derived.

(a) Alanine ______________
(b) Aspartate ______________
(c) Glutamine ______________
(d) Phenylalanine ______________
(e) Leucine ______________
(f) Valine ______________

(1) Succinyl CoA
(2) Acetoacetyl CoA
(3) α-Ketoglutarate
(4) Oxaloacetate
(5) Pyruvate
(6) Acetyl CoA
(7) Fumarate

13. Classify the following amino acids as glycogenic (G), ketogenic (K), or both (GK):

(a) Leucine ______________
(b) Alanine ______________
(c) Tyrosine ______________
(d) Serine ______________
(e) Histidine ______________
(f) Isoleucine ______________
(g) Aspartate ______________
(h) Phenylalanine ______________

14. What common feature is shared by the catabolism of fatty acids having an odd number of carbon atoms and the catabolism of the amino acids isoleucine, methionine, and valine?

15. Cobalt is a component of which of the following coenzymes?

(a) FAD
(b) NAD^+
(c) Dihydrobiopterin
(d) CoA
(e) Coenzyme B_{12}
(f) Pyridoxal phosphate

16. Catabolism of which of the following amino acids requires the direct involvement of O_2?

(a) Histidine
(b) Phenylalanine
(c) Tyrosine
(d) Isoleucine
(e) Glutamine

17. Describe the reactions of the oxygenases and distinguish between monoxygenases and dioxygenases.

18. Match the functions from the right column with the appropriate coenzymes, listed in the left column, that are involved in amino acid catabolism.

(a) Pyridoxal phosphate ____________
(b) Coenzyme B_{12} ____________
(c) Tetrahydrobiopterin ____________
(d) NAD^+ ____________
(e) Biotin ____________

(1) Carries electrons
(2) Provides free radicals
(3) Carries amino groups
(4) Carries CO_2

Inborn Errors of Metabolism

19. Match the enzyme or protein in the right column with the disorder in the left column that often produces a deficiency of the enzyme or protein.

(a) Methylmalonic aciduria ____________
(b) Pernicious anemia ____________
(c) Hyperammonemia ____________
(d) Phenylketonuria ____________
(e) Alcaptonuria ____________
(f) Maple syrup urine disease ____________

(1) Carbamoyl phosphate synthetase
(2) Homogentisate oxidase
(3) Intrinsic factor
(4) Phenylalanine hydroxylase
(5) Branched-chain α-keto acid dehydrogenase
(6) Transferase that converts B_{12s} into deoxyadenosylcobalamin

Introduction

1. All are correct. The amino acids that are needed for protein synthesis and as precursors for other biomolecules are used directly for those purposes. The carbon skeletons of any in excess can be converted into acetyl CoA or glucose, depending on the particular amino acid, and thus into products that are derivable from these two basic molecules.

Amino Group Metabolism

2. c. The transamination of several different amino acids with α-ketoglutarate forms glutamate and α-keto acids that can subsequently be catabolized.

3. a, c

4. c, e. Serine and threonine are deaminated by dehydratases that take advantage of the β-hydroxyl of these amino acids to carry out a dehydration followed by a rehydration to release NH_4^+.

5. d. A five-carbon amino acid will yield a five-carbon α-keto acid as a result of a transamination; in this case, glutamate yields α-ketoglutarate. The other partner in the reaction, pyruvate, will yield alanine as a product.

6. Pyridoxal phosphate (PLP) acts as an amino carrier in transamination reactions. It is covalently bound to the ε-amino group of a lysine residue in the enzyme by a Schiff base or aldimine bond, that is, by a carbon-nitrogen double bond between the ε-amino group of the lysine and a carbon of PLP (see p. 631 of Stryer). The enzyme catalyzes a displacement of the ε-amino group of the lysine of the enzyme and forms an analogous aldimine bond with the α-amino group of an amino acid. Through a deprotonation and a reprotonation, the aldimine is isomerized to a ketimine in which the carbon-nitrogen bond is between the α-amino nitrogen and the α-carbon of the amino acid (see Figure 25-2 in Stryer). The addition of H_2O to the ketimine releases the carbon skeleton of the amino acid as an α-keto acid and leaves the coenzyme in the enzyme-bound pyridoxamine form, which now contains the α-amino group of the amino acid. After dissociation of the α-keto acid, a different α-keto acid binds to the enzyme to form a new ketimine, and the overall process reverses, resulting in the transfer of the enzyme-bound amino group to the keto acid to form a new amino acid and regenerate the pyridoxal phosphate.

Urea Synthesis

7. a, c, d. Uracil is a pyrimidine component of RNA and is not a nitrogen excretory product. Uric acid is a purine derivative excreted by uricotelic organisms.

8. Three moles of ATP are directly involved in the synthesis of urea. Two are converted to ADP and P_i by carbamoyl phosphate synthetase and one is converted to AMP and PP_i by argininosuccinate synthetase. Two ATP would be required to convert AMP back into ATP, so a total of four high-

energy bonds are used. Only one molecule of nitrogen enters as NH_4^+; the other enters in the α-amino group of aspartate, which can be formed by a transamination between oxaloacetate and glutamate.

9. The sum of the two reactions would be

$$\text{alanine} + \text{oxaloacetate} \rightleftharpoons \text{aspartate} + \text{pyruvate}$$

and the net effect would be that the α-amino group of alanine would appear as the α-amino group of aspartate, one of the two direct donors of nitrogen into the urea cycle—the other being carbamoyl phosphate. The α-amino groups of many other amino acids can be similarly collected on aspartate and fed into the cycle.

10. Ornithine serves as the carrier on which the urea molecule is constructed. The δ-amino group of ornithine has a carbamoyl group added to it by ornithine transcarbamoylase to form citrulline. Subsequent steps of the cycle add aspartate to bring in the second nitrogen atom and also regenerate ornithine when arginase cleaves arginine to form urea.

11. A defect in carbamoyl phosphate synthesis would cause increased NH_4^+ concentrations as would a defect in any of the four reactions that condense carbamoyl phosphate with ornithine and regenerate the ornithine. Essentially, blocking a biochemical pathway at any of its steps may lead to increased concentrations of any of the precursors or members of the pathway.

Carbon Atom Metabolism

12. (a) 5 (b) 4, 7; aspartate can be converted to fumarate via the urea cycle. (c) 3 (d) 2, 7 (e) 2, 6 (f) 1

13. (a) K (b) G (c) GK (d) G (e) G (f) GK (g) G (h) GK

14. Both give rise to methylmalonyl CoA, which can, in turn, be converted into succinyl CoA and ultimately into glucose (See Stryer, p. 641 and p. 612).

15. e

16. b, c. Monoxygenases and dioxygenases are involved in the conversion of phenylalanine to tyrosine and the subsequent opening of the aromatic ring during tyrosine catabolism.

17. Oxygenases use O_2 as a substrate and incorporate one or both of the atoms into the oxidized product. Monoxygenases, also called *mixed-function oxygenases,* incorporate one atom of the oxygen into the product; the other appears in water. Dioxygenases incorporate both atoms of O_2 into the product. Monoxygenases are often involved in hydroxylation, and dioxygenases often participate in the opening of aromatic rings.

18. (a) 3 (b) 2 (c) 1 (d) 1 (e) 4

Inborn Errors of Metabolism

19. (a) 6 (b) 3 (c) 1 (d) 4 (e) 2 (f) 5

1. Birds require arginine in their diet. Would you expect to find the production of urea in these animals? Explain.

2. Which would you expect to have a greater effect on the rate of urea biosynthesis, a defect in fumarase activity or a defect in alanine aminotransferase?

3. Pyridoxal phosphate or related metabolites are required growth factors for *Lactobacillus* species. For example, when amino acids such as alanine or glutamate are used as the sole source of nutrition, these bacilli do not grow nor do they generate metabolic energy unless pyridoxal phosphate or its metabolites are supplied. Explain these observations.

4. A male infant six weeks of age exhibited symptoms of pronounced hyperammonemia, which included vomiting, fever, irritability, and screaming episodes interspersed with periods of lethargy. The infant had low levels of blood urea, elevated serum transaminase, and generalized hyperaminoacidemia and aminoaciduria. The levels of citrulline, argininosuccinic acid, and arginine were relatively low in both blood and urine. Enzymatic assays of liver tissue established that the level of mitochondrial carbamoyl phosphate synthetase (CPS) was approximately 20% of normal; the enzyme was active only in the presence of relatively high concentrations of *N*-acetylglutamate. All other urea cycle enzyme levels were relatively normal. The infant was treated with a supplement containing arginine, pyridoxine, and α-keto analogs of essential amino acids (those that cannot be synthesized in humans). As part of the therapy, dietary protein was also restricted.

(a) Why were blood levels of ammonia high in this infant?
(b) Explain why infants with CPS deficiency normally exhibit relatively low blood levels of citrulline, argininosuccinic acid, arginine, and urea.
(c) Why is the administration of supplemental arginine recommended for this infant?
(d) Why were α-keto analogs of essential amino acids and pyridoxine administered?
(e) Why was dietary protein restricted?
(f) Why would you expect to see elevated concentrations of glutamate, glutamine, and alanine in the blood and urine of this infant?

5. Why is glutamate dehydrogenase a logical point for the control of ammonia production in cells?

6. During the process of glomerular filtration in the kidney, amino acids, as well as other metabolites, enter the lumen of the kidney tubule. Normally, a large portion of these amino acids are reabsorbed into the blood through the action of membrane-bound carrier systems that are specific for different classes of amino acids. Cystinuria is a disorder whose symptoms include urinary excretion with unusually high concentrations of cystine as well as excess amounts of ornithine, lysine, and arginine. Patients with this disorder often have urinary tract stones, which are caused by the limited solubility of cystine. A related disorder found in other people is characterized

by the appearance of ornithine, lysine, and arginine in the urine, although the levels of urinary cystine are normal.

(a) What is the most likely source of cystine in cells?
(b) What common structural feature of the four amino acids—cystine, ornithine, lysine, and arginine—is recognized by the carrier in the kidney tubule membrane?
(c) How many carrier systems may exist for these molecules?
(d) Other amino acidurias are due to a deficiency in one or more of the enzymes in the catabolic pathway for an amino acid. This leads to higher concentrations of the amino acid in the blood and a corresponding increase in the concentrations in the glomerular filtrate. In this case, the capacity of the reabsorption system is surpassed, causing some amino acid to be lost in the urine. How could you distinguish between a defect in amino acid metabolism and a defect in a renal transport system?

7. Why is the catabolism of isoleucine said to be both glucogenic and ketogenic?

8. Brain cells take up tryptophan, which is then converted to 5-hydroxytryptophan by tryptophan hydroxylase, an enzyme whose activity is similar to that of phenylalanine hydroxylase. Aromatic amino acid decarboxylase then catalyzes the formation of the potent neurotransmitter 5-hydroxytryptamine, also called *serotonin.* In the blood, tryptophan is bound to serum albumin, with an affinity such that about 10% of the tryptophan is freely diffusable. The rate of tryptophan uptake by brain cells depends on the concentration of free tryptophan. In these cells, tryptophan concentration is normally well below that of the K_M for tryptophan hydroxylase. Aspirin and other drugs displace tryptophan from albumin, thereby increasing the concentration of free tryptophan.

(a) What cofactor is required for the activity of tryptophan hydroxylase?
(b) Dietary deficiencies in pyridoxin and related metabolites can induce a number of symptoms, including those that appear to be related to derangements in serotonin metabolism. What enzyme could be affected by a deficiency of vitamin B_6?
(c) What effect does aspirin have on tryptophan metabolism in brain cells?

9. Sodium borohydride reduces imines to stable secondary amines, as shown in the following general reaction:

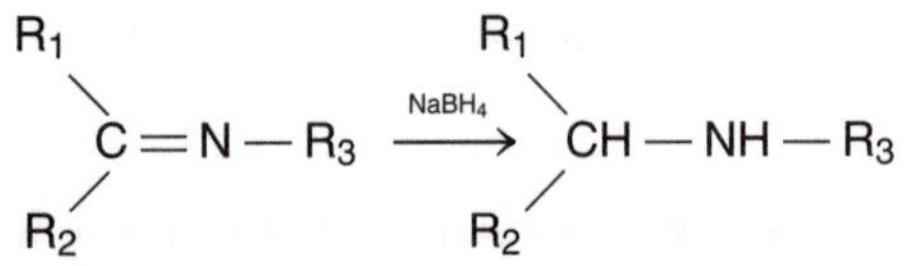

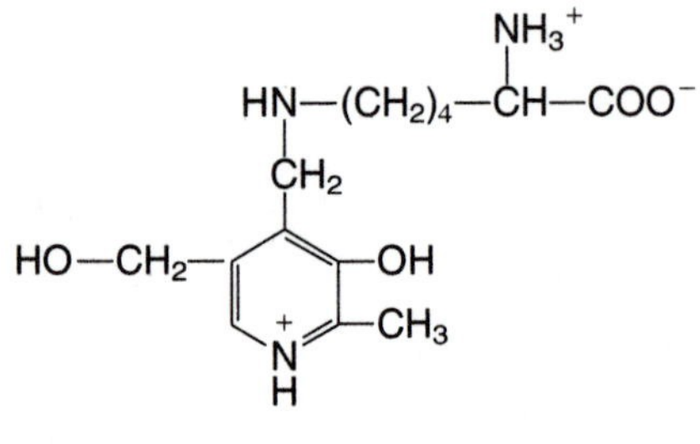

$$R_1R_2C{=}N{-}R_3 \xrightarrow{NaBH_4} R_1R_2CH{-}NH{-}R_3$$

A purified sample of alanine aminotransferase is treated with sodium borohydride and is then subjected to partial acid hydrolysis. Among the amino acids isolated is the fluorescent derivative shown in the margin. Identify the derivative, and describe the role of its parent compound in the catalytic action of the aminotransferase.

NH_3^+ / $HN{-}(CH_2)_4{-}CH{-}COO^-$, CH_2 on N, attached to pyridinium ring with $HO{-}CH_2{-}$, $-OH$, $-CH_3$, $\overset{+}{N}H$

10. In many microorganisms, glutamate dehydrogenase (GDH) participates in the catabolism of glutamate by generating ammonia and α-ketoglutarate, which undergoes oxidation in the citric acid cycle. However, when *E. coli* is grown with glutamate as the sole source of carbon, the synthesis of GDH protein is strongly repressed. Under these conditions, aspartase, an enzyme that catalyzes the removal of ammonia from aspartate to form fumarate, is required in order for the cell to grow in glutamate. Propose a cyclic pathway for the catabolism of glutamate that includes aspartate.

When *E. coli* is grown in glucose and ammonia, GDH synthesis is accelerated and the enzyme is active. Under these conditions, what role does GDH play in bacterial metabolism?

11. The formation of the aldimine linkage between an amino acid substrate and the pyridoxal phosphate in alanine aminotransferase can be portrayed as a condensation. However, when the enzyme reacts with its substrate in ^3H_2O, less than 1% of tritium is incorporated into the product. What does this observation tell you about the substrate-enzyme interaction?

12. Certain forms of maple syrup urine disease respond to the administration of thiamine at 50 to 100 times the normal recommended level. Why?

13. After an overnight fast, muscle tissue proteolysis generates free amino acids, many of which pass into the blood. Among the amino acids that are found in the blood are alanine, glutamate, and glutamine, all of which are rapidly taken up by the liver. What happens to these amino acids when they enter hepatic cells?

14. Propionyl CoA and methylmalonyl CoA both inhibit *N*-acetylglutamate synthase activity in slices of liver tissue. What clinical symptom would you expect to see in patients suffering from methylmalonic aciduria as a result of this inhibition?

15. Dialysis of purified glutamate aminotransferase can be used to remove pyridoxal phosphate from the enzyme, but the dissociation of the cofactor from the enzyme is very slow. Why would the addition of glutamate to the enzyme solution increase the rate of dissociation of the cofactor from the enzyme?

16. Most enzymes that use pyridoxal phosphates as a cofactor lose their catalytic activity when they are reduced by sodium borohydride. For example, aspartate aminotransferase is rapidly inactivated by sodium borohydride reduction. Why? However, glycogen phosphorylase, which contains two pyridoxal phosphate molecules in each of its subunits, is not inactivated by sodium borohydride. Why?

17. Early work by Esmond Snell on the enzymes which employ pyridoxal phosphate included experiments in which free pyridoxal was heated with amino acids like glutamate. Snell found that the α-amino group of glutamate was transferred to pyridoxal, generating pyridoxamine. Why was this an important clue to the function of pyridoxal as a cofactor?

18. The enzyme serine dehydratase employs pyridoxal phosphate (PLP) in the direct deamination of serine (see Stryer, p. 634). In this reaction, the

α carbon of serine undergoes a two-electron oxidation through β-elimination. Show how PLP participates in the process by writing a mechanism for serine dehydration and deamination.

19. A small number of infants who have phenylketonuria have normal levels of phenylalanine hydroxylase activity, but on normal diets they continue to accumulate phenylalanine as well as other metabolites, including phenylpyruvate, phenyllactate, and phenylacetate. They also have high levels of quinonoid dihydrobiopterin.

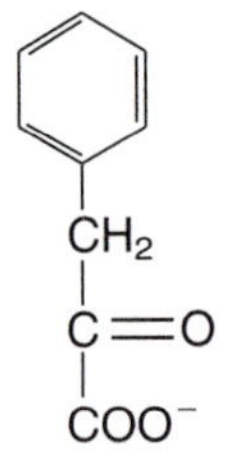

$$\text{C}_6\text{H}_5\text{—CH}_2\text{—C(=O)—COO}^-$$

Phenylpyruvate

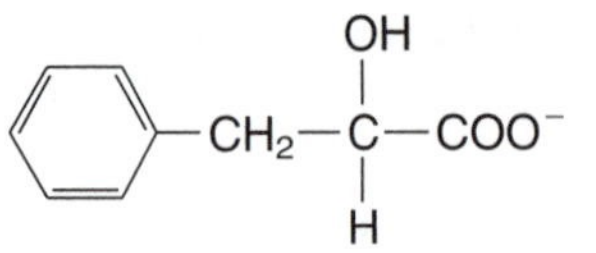

$$\text{C}_6\text{H}_5\text{—CH}_2\text{—CH(OH)—COO}^-$$

Phenyllactate

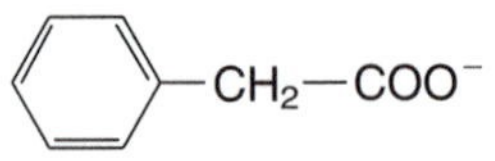

$$\text{C}_6\text{H}_5\text{—CH}_2\text{—COO}^-$$

Phenylacetate

(a) What is the probable enzyme deficiency in these infants? Rationalize the deficiency with the observed clinical symptoms.
(b) Write brief pathways for the formation of the phenylalanine metabolites found in these infants.

ANSWERS TO PROBLEMS

1. In the urea cycle, arginine is cleaved to yield urea and ornithine. The fact that birds require arginine in their diet indicates that they are unable to synthesize it for utilization in protein synthesis. As a result, they are also unable to synthesize urea to dispose of ammonia; instead, they synthesize uric acid. Birds do have carbamoyl phosphate synthetase activity; however, it is located in the cytosol, and it catalyzes the formation of carbamoyl phosphate, which is then utilized for pyrimidine synthesis.

2. Fumarase activity has an effect on the urea cycle because it is needed, along with malate dehydrogenase, for the regeneration of oxaloacetate, which in turn undergoes transamination to form aspartate. The amino group of aspartate contains one of the two nitrogen atoms that are used to synthesize urea. Alanine aminotransferase is one of a number of aminotransferases that can transfer amino groups from amino acids to α-ketoglutarate to generate glutamate. Subsequent deamination of glutamate provides ammonia for the urea cycle. If all the other aminotransferases in the cell are active, then alanine aminotransferase would not be particularly essential. Thus, a defect in fumarase activity would have the greater effect on the rate of urea biosynthesis.

3. In order to utilize amino acids as sources of oxidative energy or to generate glucose through gluconeogenesis, the bacilli must carry out transamination reactions to dispose of ammonia (or to use it for the biosynthesis of other nitrogen-containing compounds) as well as to generate α-keto acids that can be used in the citric acid cycle or other pathways. Pyridoxal phosphate is a required cofactor for the aminotransferase enzymes, and in bacteria in which it cannot be synthesized, it must be derived from the growth medium. Otherwise, amino acids cannot be metabolized. In this case, where an amino acid is the only source of carbon and of nitrogen, the

bacilli will not be able to introduce the amino acid into any catabolic pathway. Pyridoxal phosphate also functions as a cofactor for a large number of other enzymes, including decarboxylases, racemases, aldolases, and deaminases. Therefore, deficiencies in pyridoxal phosphate would also adversely affect a large number of other pathways.

4. (a) Carbamoyl phosphate synthetase utilizes ammonia and bicarbonate to synthesize carbamoyl phosphate, which enters the urea cycle. A deficiency in the activity of this enzyme leads to an accumulation of ammonia in blood and in urine.

(b) Because carbamoyl phosphate condenses with ornithine to form citrulline, it is in effect a precursor of all the components of the urea cycle. When its synthesis is depressed, the rate of synthesis of other urea cycle components will be decreased.

(c) Arginine is needed as a component of most proteins; it must therefore be supplied to avoid arginine deficiency, which would result in a decrease in the rate of protein synthesis. In the urea cycle, arginine serves as a precursor of ornithine, which in turn serves as an acceptor of the relatively small number of carbamoyl groups that enter the urea cycle. Thus, supplemental arginine could accelerate the rate of urea synthesis in an attempt to drive the reaction catalyzed by the partially active CPS enzyme toward the net formation of carbamoyl phosphate. Finally, recent findings suggest that arginine is a feed-forward activator for the synthesis of *N*-acetyl glutamate, which activates CPS. Recall that the deficient enzyme appeared to be heavily dependent on levels of *N*-acetyl glutamate. Supplying arginine indirectly activates CPS, thereby increasing the rate of ammonia utilization.

(d) The administration of α-keto analogs of essential amino acids serves two functions. First, these substrates undergo transamination with glutamate as the amino donor. The corresponding increase in the synthesis of glutamate, through the action of glutamate dehydrogenase, utilizes more ammonia, removing at least some of it from the blood. Second, the transamination of the α-keto acids generates essential amino acids, which are used for protein synthesis. This is especially important in a situation in which dietary protein intake is restricted (see the following).

(e) When dietary protein is hydrolyzed to its component amino acids, subsequent catabolic steps include transamination and deamination, which produce free ammonia. The less protein consumed, the lower the production of ammonia from the breakdown of amino acids. A reduction in the blood ammonia level is the primary goal of the therapy for this infant.

(f) Increased levels of ammonia in the cells will drive the reaction catalyzed by glutamate dehydrogenase toward the net formation of glutamate and will also stimulate the synthesis of glutamine, which is in effect a carrier of two molecules of ammonia. With the high concentrations of glutamate, the equilibria for most transamination reactions would be shifted toward the net formation of other amino acids, such as alanine, a transaminated form of pyruvate.

Concentrations of these and other amino acids would therefore be increased in both the blood and the urine.

5. The deamination of amino acids occurs through the action of transaminases as well as through the action of glutamate dehydrogenase (GDH). GDH is the only enzyme that catalyzes the oxidative deamination of an L amino acid. It deaminates glutamate, whose precursor, α-ketoglutarate, is the ultimate acceptor of amino groups from almost all the amino acids. In addition, while the reactions catalyzed by the transaminases are freely reversible, the GDH reaction is far from equilibrium; it is therefore a logical activity to control because large changes in its velocity can be achieved with small changes in the concentrations of allosteric effectors like ATP or NADH. Thus, GDH is the enzyme of choice for the control of ammonia synthesis.

6. (a) Cystine is a dibasic amino acid, composed of two cysteine molecules joined by a disulfide linkage. Disulfide linkages exist in many proteins, and when they are hydrolyzed by proteases to yield free amino acids, cystine is often one of the products.

(b) Like cystine, the other three amino acids—arginine, ornithine, and lysine—have two basic groups. The carrier systems probably recognize and bind these groups for transport.

(c) From the disorders described, it is likely that there are two transport systems. One carries all four dibasic acids; when it is defective, reabsorption of all four species fails, allowing all four to spill over into the urine. Another system transports ornithine, arginine, and lysine but not cystine; its failure to function accounts for the appearance of these three amino acids but not cystine in the urine.

(d) A defect in the catabolic pathway for a particular amino acid causes elevation in the concentration of that single amino acid in blood and in urine as well, unless other amino acids carried by the same kidney transport system are lost to the urine. A defect in renal reabsorption means that all those amino acids that share the affected carrier system will be lost to the urine. Their concentration in blood will be lower than normal, while their concentration in urine will be higher.

7. The catabolic pathway for isoleucine leads to the formation of acetyl CoA and propionyl CoA. Acetyl CoA can be utilized for the net synthesis of fatty acids or ketone bodies, but it cannot be used for the net synthesis of glucose; thus, it is said to be ketogenic. In contrast to acetyl CoA, propionyl CoA is converted to succinyl CoA, which can be utilized through part of the citric acid cycle and the gluconeogenic pathway to give the net formation of glucose. The distinction between the two types of substrates is somewhat arbitrary, however. For example, succinyl CoA can also be converted via pyruvate and acetyl CoA to citrate, which, when transported to the cytosol, serves as the source of carbons for the synthesis of fatty acids. Thus, glucogenic substrates can, under certain conditions, be ketogenic; however, ketogenic substrates cannot be glucogenic, unless a cell has a functional glyoxylate pathway.

8. (a) Tetrahydrobiopterin is utilized as a reductant by many hydroxylase enzymes, including phenylalanine hydroxylase and tryptophan hydroxylase.

(b) Pyridoxal phosphate, which is derived from dietary pyridoxine, is a cofactor for a number of enzymatic reactions that occur at the α carbon of an amino acid. In this case, the cofactor participates in the decarboxylation of 5-hydroxytryptophan. A deficiency of vitamin B_6 would lead to a reduction in the rate of synthesis of serotonin.

(c) The higher the concentration of free tryptophan, the greater the rate of uptake of the amino acid by the brain cells. Because the normal concentration of tryptophan in these cells is below that of the K_M for tryptophan hydroxylase, an influx of more tryptophan into the cells provides more substrate for the enzyme. You would therefore expect an increase in the level of 5-hydroxytryptophan production.

9. The fluorescent derivative is ε-pyridoxyllysine, in which lysine is covalently linked to a reduced pyridoxyl group. At the active site of the native enzyme, the parent compound is pyridoxyl phosphate in Schiff base linkage with the ε-amino group of a specific lysine residue. When an amino acid enters the active site, the ε-amino group of lysine is displaced and a new Schiff base is formed between PLP and the α-amino group of the amino acid. The formation of the new Schiff base initiates the transamination reaction.

10. A possible catabolic pathway for glutamate that includes aspartate is as follows:

Glutamate + Oxaloacetate ⟶ Aspartate + α-Ketoglutarate

Aspartate ⟶ Fumarate + NH_4^+ ; α-Ketoglutarate ⟶ Oxidation in the citric acid cycle

Fumarate + H_2O ⟶ Malate ; Malate + NAD^+ ⟶ Oxaloacetate + NADH + H^+

Glutamate undergoes transamination with oxaloacetate to generate α-ketoglutarate and aspartate. Oxidation of α-ketoglutarate is carried out in the citric acid cycle, whereas aspartate is cleaved to yield ammonia and fumarate. Fumarate is converted to malate, which is then oxidized to oxaloacetate in the citric acid cycle so that α-ketoglutarate can be regenerated to serve as an acceptor of the amino group from glutamate. It should be noted that glutamate will also be used as a source of amino groups by the aminotransferases and other enzymes involved in biosynthesis.

In cells grown in glucose and ammonia, GDH catalyzes the assimilation of ammonia by incorporating it into oxaloacetate to yield glutamate, which serves as a source of amino groups for other biosynthetic reactions.

11. The fact that very little hydrogen exchange occurs when alanine binds to the enzyme suggests that a direct transaldiminization occurs. Pyridoxal phosphate is linked by a Schiff base to a specific lysine residue at the active site. Alanine displaces the cofactor, with a direct exchange of the Schiff base linkage to the α-amino group of alanine. It appears that there is no intermediate formation of free aldehyde during the binding of the substrate. The small amount of tritium that is incorporated into the product may be due to the ionization of lysine.

12. Maple syrup urine disease is characterized by the appearance of large quantities of leucine, valine, and isoleucine and their corresponding α-keto acids in the blood and urine. This is due to a deficiency of branched-chain α-keto acid dehydrogenase, the enzyme that catalyzes the oxidative decarboxylation of the three aliphatic amino acids. The reaction is analogous to those catalyzed by pyruvate dehydrogenase and α-ketoglutarate dehydrogenase. These enzymes utilize thiamine pyrophosphate as a cofactor. A response to the administration of thiamine indicates that the deficient enzyme does not bind thiamine as efficiently as a normal enzyme does. Higher concentrations of the cofactor are therefore required for optimal enzyme activity.

13. In the liver, alanine, glutamate, and glutamine are utilized as sources of carbon for gluconeogenesis. Glutamine is deaminated to yield glutamate and ammonia. Glutamate undergoes oxidative deamination to form ammonia and α-ketoglutarate, a substrate for gluconeogenesis. Pyruvate is generated from the transamination of alanine, and carboxylation of the α-keto acid yields oxaloacetate, another source of carbon for gluconeogenesis. The ammonia generated by the conversion of the amino acids to their corresponding α-keto acids is used for the synthesis of urea. Glucose synthesized by liver can be returned through the blood to the muscle, where it serves as a source of energy. Thus, muscle uses the amino acids as a means of contributing to the generation of glucose in liver as well as a means of transporting ammonia to the liver, where the synthesis of urea can be carried out.

14. The activity of mitochondrial carbamoyl phosphate synthetase (CPS) depends on the availability of *N*-acetylglutamate, which is generated from glutamate and acetyl CoA. A reduction in the availability of the activating molecule will lead to a decrease in the activity of CPS, which utilizes ammonia and bicarbonate for the synthesis of carbamoyl phosphate. This in turn leads to an increase in the level of ammonia in blood and urine. Over two-thirds of patients with methylmalonic aciduria are hyperammonemic.

15. In the native enzyme, the dissociation of the cofactor from the enzyme during dialysis is very slow because PLP is covalently bound to a lysine residue. The addition of glutamate, a substrate for the transamination reaction, leads to the formation of a Schiff base between glutamate and PLP, which means that the cofactor is no longer covalently attached to the enzyme molecule. Although PLP is bound to the enzyme by noncovalent forces, such as those shown in Figure 25-3 of the text, these forces are not as strong as a covalent bond, so during dialysis, the rate of dissociation of the cofactor from the enzyme is increased.

16. Many of the reactions in which pyridoxal phosphate takes part involve rearrangements, in which the cofactor, initially in Schiff base linkage with an ε-amino group of an enzyme, forms a new Schiff base link with an arriving substrate (see Stryer, p. 631). Thus during transamination of aspartate, pyridoxal phosphate is transiently in Schiff base linkage with aspartate, forming an external aldimine, which undergoes hydrolysis to form oxaloacetate and pyridoxamine phosphate. Then the cofactor forms a second external aldimine with an incoming α-ketoacid (α-ketoglutarate), and the reaction is reversed. Glutamate is generated and pyridoxal phosphate is again linked to the enzyme. Treatment of an enzyme that functions by formation of internal and external aldimines with

sodium borohydride generates a stable C–N bond between pyridoxal phosphate and the lysine residue. Under these conditions, external aldimine linkages cannot be established with an incoming substrate, and the enzyme is inactivated. However, glycogen phosphorylase employs pyridoxal phosphate as a proton donor and proton acceptor, as the 5′-phosphate of PLP acts in tandem with orthophosphate when the carbonium ion intermediate is attacked to form glucose 1-phosphate. (See Stryer, p. 589.) Although PLP is in Schiff base linkage with Lys 680 of the enzyme, the catalytic cycle does not involve formation of external aldimine links with incoming substrates. Thus stabilization of the Schiff base link by sodium borohydride does not inactivate glycogen phosphorylase.

17. Snell's observations of the formation of pyridoxamine by heating with α-amino acids suggested that pyridoxal is involved in transamination reactions. As an enzyme cofactor it could transfer an α-amino group from an amino acid to the α-keto group of an α-keto acid. It is now established that the action of pyridoxal phosphate in aminotransferase enzymes includes formation of pyridoxamine phosphate during the catalytic cycle (see Stryer, pp. 631 and 632).

18. In the accompanying figure, PLP in Schiff base linkage with a lysine residue in the enzyme forms a new Schiff base link with serine. A hydrogen atom is removed from the α carbon, and then the hydroxyl group is removed by elimination from the β-carbanion, generating aminoacrylate attached to pyridoxal phosphate. Hydrolysis of the Schiff base to give aminoacrylate and PLP is followed by tautomerization to the imino form. This compound hydrolyzes spontaneously to form pyruvate and ammonia. PLP is once again linked covalently to the enzyme.

ENZ
$(CH_2)_4$
N
H—C
OH
PLP
+
$H-O$ H
$CH_2-C-COO^-$
NH_3^+
Serine
H^+
H—O
$CH_2-C-COO^-$
N H
H—C O$^-$
Carbanion
OH^-
β-Elimination
$H_2C{=}C-COO^-$
+N H
H—C O$^-$
H_2O
$H_2C{=}C-COO^-$
NH_2
Aminoacrylate
+
PLP
$H_3C-C-COO^-$
$^+NH_2$
Imino Form
H_2O
Pyruvate
$CH_3-C-COO^-$
O
+
Ammonia NH_3

19. (a) The enzyme that is deficient in these infants is dihydropteridine reductase, which converts quinonoid dihydrobiopterin to tetrahydrobiopterin, using NADH as a substrate. In the phenylalanine hydroxylase reaction, tetrahydrobiopterin, the reductant in the conversion of phenylalanine to tyrosine, is oxidized to quinonoid dihydrobiopterin. The reductase enzyme regenerates tetrahydrobiopterin so that it can be used for further use in tyrosine formation. Cells that are deficient in the reductase cannot carry out

efficient conversion of phenylalanine to tyrosine because they cannot regenerate tetrahydrobiopterin.

(b) Phenylpyruvate can be generated from phenylalanine by transamination, and reduction of phenylpyruvate by NADH or NADPH generates phenyllactate, in a reaction similar to that catalyzed by lactate dehydrogenase. Phenylacetate can be generated by oxidative decarboxylation of phenylpyruvate, in a reaction reminiscent of the conversion of pyruvate to acetyl CoA, catalyzed by pyruvate dehydrogenase.

EXPANDED SOLUTIONS TO TEXT PROBLEMS

1. (a) Pyruvate, (b) oxaloacetate, (c) α-ketoglutarate, (d) α-ketoisocaproate, (e) phenylpyruvate, and (f) hydroxyphenylpyruvate

2. (a) Aspartate + α-ketoglutarate + GTP + ATP + 2 H_2O + NADH + H^+ $\rightarrow$ ½ glucose + glutamate + CO_2 + ADP + GDP + NAC^+ + $2P_i$

The glucogenic route for aspartate involves transamination to oxaloacetate, conversion of the latter to phosphoenolpyruvate, which is then converted to glucose (Stryer, Chapter 22). PLP and NADH participate as coenzymes in the conversion of aspartate to glucose.

(b) Aspartate + CO_2 + NH_4^+ + 3 ATP + NAD^+ + 4 H_2O $\rightarrow$

oxaloacetate + urea + 2 ADP + 4 P_i + AMP + NADH + H^+

This equation represents the summation of the stoichiometry of urea synthesis (p. 636), the hydrolysis of PP_i, and the conversion of fumarate to oxaloacetate in the citric acid cycle.

3. (a) Label the methyl carbon atom of L-methylmalonyl CoA with ^{14}C. Determine the location of ^{14}C in succinyl CoA. The group transferred is the one bonded to the labeled carbon atom.

It should be noted that one must distinguish between the free and thioester carboxyls in succinyl CoA. In free succinic acid, the carboxyls are indistinguishable. A possible strategy for distinguishing them is to convert succinyl CoA to the half amide of succinate (thioester $\rightarrow$ amide) and then run a Hoffmann degradation.

(b) The proton that is abstracted from the methyl group of L-methylmalonyl CoA is directly transferred to the adjacent carbon atom.

4. Thiamine pyrophosphate

5. It acts as an electron sink. See C. Walsh, *Enzymatic Reaction Mechanisms* (W.H. Freeman, 1979), p. 178.

6. Deuterium is abstracted by the radical form of the coenzyme. The methyl group rotates before hydrogen is returned to the product radical. (See Stryer, p. 643.)

7. A carbanion or a carbonium ion. (See Stryer, p. 643.)

8. $CO_2 + NH_4^+ + 3\ ATP + NAD^+ + 3H_2O$ + glutamate $\rightarrow$

$$\text{urea} + 2\ ADP + 2\ P_i + AMP + PP_i + NADH + H^+ + \alpha\text{-ketoglutarate}$$

The answer in Stryer is for the conversion of fumarate to oxaloacetate rather than to aspartate. The equation above gives the correct stoichiometry for the synthesis of urea from NH_4^+ and glutamate. It includes a non-energy requiring transamination reaction. Hence, the number of ~P spent remains at 4. Note that aspartate does not appear in this equation, since it is resynthesized. Thus aspartate can be considered a nitrogen-carrying cofactor in the synthesis of urea.

9. The mass spectrometric analysis strongly suggests that three enzymes —pyruvate dehydrogenase, α-ketoglutarate dehydrogenase, and the branched-chain α-keto dehydrogenase—are deficient. Most likely, the common E3 component of these enzymes (p. 517) is missing or defective. This proposal could be tested by purifying these three enzymes and assaying their capacity to catalyze the regeneration of lipoamide.

10. Benzoate, phenylacetate, and arginine would be given in a protein-restricted diet. Nitrogen would emerge in hippurate, phenylacetylglutamine, and citrulline. See S.W. Brusilow and A.L. Horwich in C.R. Scriver, A.L. Beaudet, W.S. Sly, and D. Valle (eds.), *The Metabolic Basis of Inherited Disease,* 6th ed. (McGraw-Hill, 1989), pp. 629–663, for a detailed discussion of the therapeutic rationale.

This therapy is designed to facilitate nitrogen excretion in the absence of urea synthesis. Note that much of the nitrogen would be excreted as glycine (hippurate is benzoylglycine), glutamine, and citrulline.

11. Aspartame, a dipeptide ester (aspartylphenylalanine methyl ester), is hydrolyzed to L-aspartate and L-phenylalanine. High levels of phenylalanine are harmful in phenylketonurics.

12. *N*-acetylglutamate is synthesized from acetyl CoA and glutamate. Once again, acetyl CoA serves as an activated acetyl donor. This reaction is catalyzed by *N*-acetylglutamate synthase.

Photosynthesis

To this point, Stryer has dealt with the mechanisms by which organisms obtain energy from their environment by oxidizing fuels to generate ATP and reducing power. In this chapter, he describes how light energy is transduced into the same forms of chemical energy and how CO_2 is converted into carbohydrate by photosynthetic organisms.

In the first part of the chapter Stryer recounts the experimental discoveries that led to the formulation of the basic equation of photosynthesis. He then describes the overall structures, components, and reactions of photosystems II and I, and the cytochrome *bf* complex, including the absorption of light, charge separation, electron transport events, and the evolution of O_2. Stryer explains how these light reactions lead to the formation of proton gradients, and the synthesis of ATP and NADPH. Next, Stryer turns to the dark reactions, which use the ATP and NADPH produced in the light reactions to fix CO_2 into carbohydrate. He describes the Calvin cycle, the competing carboxylase and oxygenase reactions of rubisco, the C_4 pathway of tropical plants, and the coordinated regulation of the light and dark reactions.

A review of the basic concepts of metabolism in Chapter 17; mitochondrial structure, redox potentials, the proton-motive force, and free-energy changes in Chapter 21; and the pentose phosphate pathway in Chapter 22 will help you to understand this chapter.

LEARNING OBJECTIVES

When you have mastered this chapter, you should be able to complete the following objectives:

The Chloroplast, Chlorophylls, and Basic Reactions of Photosynthesis (Stryer pages 653–657)

1. Write the basic equation of photosynthesis.

2. Provide an overview of the roles of light, *chlorophylls, reaction centers, photosystems I* and *II,* and the *Calvin cycle* in photosynthesis.

3. Describe the structure of the *chloroplast.* Locate the *outer, inner,* and *thylakoid membranes;* the *thylakoid space;* the *granum;* and the *stroma.* Associate these structures with the functions they perform.

4. Describe the properties of the thylakoid membrane.

5. Relate the findings of Priestley, Ingenhousz, Senebier, de Saussure, and Mayer to our present understanding of photosynthesis.

6. List the structural components of chlorophylls *a* and *b,* and explain why chlorophylls are effective *photoreceptors.*

7. Explain the significance of the observation that the rate of photosynthesis reaches a limiting value as the intensity of the incident light increases.

8. Distinguish between the *light* and *dark reactions* of photosynthesis.

9. Define the *photosynthetic unit* and describe the function of the *reaction centers.* Explain how the components of the photosynthetic unit interact to funnel light to the reaction centers.

The Reactions of Photosystems II and I (Stryer pages 657–664)

10. Name the molecular source of O_2 evolved in photosynthesis, and write the general formula for the *splitting of the hydrogen donor* proposed by Van Niel.

11. Write the equation of the *Hill reaction,* and relate its significance to our understanding of photosynthesis.

12. Describe the *red drop* and the *enhancement phenomenon,* and recount the proposal of Emerson to explain them.

13. List the *oxidants* and *reductants* produced by the light reactions of *photosystems I* and *II.*

14. Diagram photosystem II and identify its major components. Describe the roles of P680, *pheophytin,* and *plastoquinone* in the absorption of light, *separation of charge,* and electron transfer in photosystem II.

15. Explain the significance of the two plastoquinone binding sites, Q_A and Q_B in photosystem II.

16. Explain the function of the *manganese center* in the H_2O-splitting reaction.

17. Describe the composition and function of the *cytochrome* bf *complex*, and outline the roles of *plastocyanin*, Cu^{2+}, and *Fe-S centers* in the formation of a *transmembrane proton gradient.*

18. Diagram photosystem I. Indicate the components and reactions of photosystem I, including the roles of P700, A_0, *ferredoxin, FAD, NADPH,* and plastocyanin(Cu^+) in these processes.

19. Write the net reaction carried out by the combined actions of photosystem II, the cytochrome *bf* complex, and photosystem I.

Photophosphorylation and ATP Synthesis (Stryer pages 664–671)

20. Explain how photosystem I can synthesize ATP without forming NADPH or O_2.

21. Contrast the formation of ATP by *photophosphorylation* and by *oxidative phosphorylation.* Describe the structure of the *ATP synthase* of chloroplasts.

22. Rationalize the differences in the *photosynthetic assemblies* in the *stacked* and *unstacked* regions of the thylakoid membranes.

23. Explain the molecular and ecological functions of *phycobilisomes* in *cyanobacteria* and *red algae.*

24. Compare and contrast the photosynthetic reaction center of *Rhodopseudomonas viridis* (a purple sulfur bacterium) with photosystem II of green plants. Note the site of action of urea-derived herbicides.

25. Summarize the common features of diverse photosynthetic reaction centers.

The Formation of Hexoses by CO_2 Fixation (Stryer pages 671–679)

26. Describe *Calvin's experiments* with *Chlorella* using $^{14}CO_2$.

27. Describe the formation of *3-phosphoglycerate* by *ribulose 1,5-bisphosphate carboxylase* (*rubisco*). Note the activator and substrate roles of CO_2.

28. Outline the formation of *phosphoglycolate* by the *oxygenase reaction* of rubisco, and follow its subsequent metabolism. Define *photorespiration.*

29. Outline the conversion of 3-phosphoglycerate into fructose 6-phosphate and the *regeneration of ribulose 1,5-bisphosphate.*

30. Describe the formation of *starch* and *sucrose.*

31. Write a balanced equation for the *Calvin cycle,* and account for the ATP and NADPH expended to form a hexose molecule. Estimate the efficiency of photosynthesis.

32. Describe the role of *thioredoxin* in coordinating the light and dark reactions of photosynthesis.

33. Describe the C_4 *pathway* and its adaptive value to tropical plants. Explain how CO_2 transport suppresses the oxygenase reaction of rubisco.

SELF-TEST

The Chloroplast, Chlorophylls, and Basic Reactions of Photosynthesis

1. Write the basic reaction for photosynthesis in green plants.

____

____

2. Assign each function or product from the right column to the appropriate structure or pathway in the left column.

(a) Chlorophyll ____________
(b) Light-harvesting complex ____________
(c) Photosystem I ____________
(d) Photosystem II ____________
(e) Calvin cycle and gluconeogenic pathway ____________

(1) O_2 generation
(2) ATP synthesis
(3) Light collection
(4) NADPH synthesis
(5) Separation of charge
(6) 3-Phosphoglycerate formation
(7) Light absorption
(8) Hexose formation
(9) Transmembrane proton gradient

3. Thylakoid membranes contain which of the following?

(a) Light-harvesting complexes
(b) Reaction centers
(c) ATP synthase
(d) Electron transport chains
(e) Galactolipids
(f) Sulfolipids
(g) Phospholipids

4. For each observation about photosynthesis in the left column, indicate the investigator from the right column who made it.

(a) CO_2 is consumed. ____________
(b) Plants revitalize air in which candles have been burned. ____________
(c) H_2O is a reactant. ____________
(d) Sunlight is needed to revitalize air in which candles have been burned. ____________
(e) Light energy is converted into chemical energy. ____________

(1) Priestley
(2) Ingenhousz
(3) Senebier
(4) De Saussure
(5) Mayer

5. Which of the following are constituents of chlorophylls?

(a) Substituted tetrapyrrole
(b) Plastoquinone
(c) Mg^{2+}
(d) Fe^{2+}
(e) Phytol
(f) Iron porphyrin

6. Why do chlorophylls absorb visible light efficiently?

7. Which of the following statements about the photosynthetic unit are true?

(a) It is a chlorophyll molecule.
(b) It collects light energy through the absorption of light by chlorophyll molecules.
(c) It contains a reaction center with a specialized chlorophyll pair that contributes to the transduction of light energy into chemical energy.
(d) It contains chlorophyll molecules that transfer energy from one to another by direct electromagnetic interactions.
(e) It is the product of Planck's constant *h* and the frequency of the incident light ν.

The Reactions of Photosystems II and I

8. Write the equation with which Van Niel indicated that H_2O is split by light energy in green plants to provide hydrogen atoms for CO_2 reduction. What did studies with $H_2{}^{18}O$ reveal about this phenomenon?

9. Which of the following statements about the Hill reaction are true?

(a) It confirmed that H_2O is the source of the O_2 evolved during photosynthesis.
(b) It demonstrated the central role of the chloroplast in photosynthesis.
(c) It demonstrated that electron acceptors other than CO_2 can be used in the generation of O_2.
(d) It revealed the role of Fe^{2+} in photosynthesis.
(e) It connected light absorption to an endergonic oxidation-reduction reaction.

10. Describe the observations that suggested that two interacting light reactions occur during photosynthesis and that they use light of different wavelengths. What are the molecular bases of the two light reactions?

11. Which of the following statements about photosystem II are correct?

(a) It is a multimolecular transmembrane assembly containing several polypeptides, more than 200 chlorophyll *a* and *b* molecules, antenna chlorophylls, a special chlorophyll (P680), pheophytin, and plastoquinones.
(b) It transfers electrons to photosystem I via the cytochrome *bf* complex.

(c) It uses light energy to create a separation of charge whose potential energy can be used to oxidize H_2O and to produce a reductant, plastoquinol.
(d) It uses an Fe^{2+}-Cu^{+} center as a charge accumulator to form O_2 without generating potentially harmful hydroxyl radicals, superoxide anions, or H_2O_2.

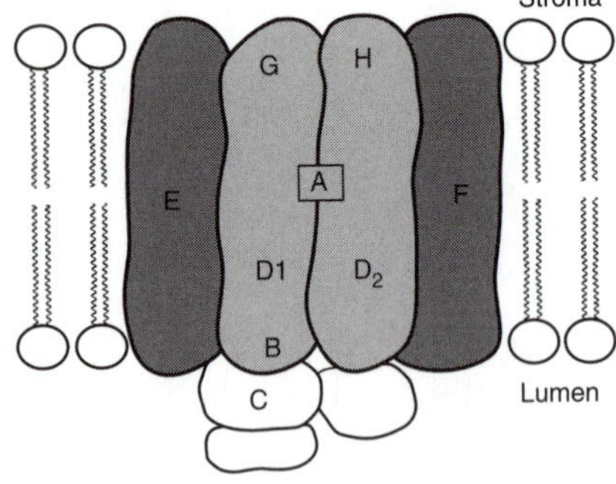

12. Using the diagram of photosystem II in the margin, identify the appropriate components, sites, and functions listed below:

(a) Q_A plastoquinone site ________
(b) Q_B plastoquinone site ________
(c) Chlorophyll P680 ________
(d) Mn^{2+} site ________
(e) Antennae chlorophylls ________
(f) Photon absorption ________
(g) H_2O splitting ________
(h) Reaction center ________
(i) Electron transfer chain ________
(j) Tyrosyl radical ________
(k) Reaction site of urea-derived herbicides ________

13. Which statement about the Mn center of photosystem II is incorrect?

(a) The Mn center has four possible oxidation states.
(b) Electrons are transferred from the Mn center to $P680^{+}$.
(c) A tyrosine residue on the D1 protein is an intermediate in electron transfer.
(d) The O_2 released by the Mn center comes from the oxidation of water.
(e) Each photon absorbed by the reaction center leads to the removal of an electron from the Mn cluster.

14. The interaction of photosystems I and II in photosynthesis by green plants can be represented as follows:

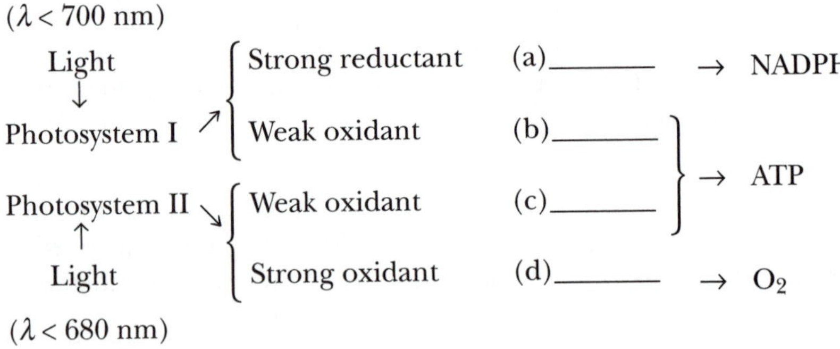

From among the following, identify the oxidants and reductants in the scheme:

(1) $\text{Ferredoxin}_{\text{reduced}}$
(2) Plastoquinol
(3) Chlorophyll $P680^{+}$
(4) $\text{Ferredoxin}_{\text{oxidized}}$
(5) Plastoquinone
(6) Plastocyanin(Cu^{2+})

15. Explain how plastocyanin and plastoquinol are involved in ATP synthesis.

16. Write the net equation of the reaction catalyzed by photosystem I, and describe how NADPH is formed. What is the role of FAD in this process?

Photophosphorylation and ATP Synthesis

17. Which of the following statements about cyclic photophosphorylation are correct?

(a) It doesn't involve NADPH formation.
(b) It uses electrons supplied by photosystem II.
(c) It is activated when $NADP^+$ is limiting.
(d) It does not generate O_2.
(e) It leads to ATP production via the cytochrome *bf* complex.
(f) It involves a substrate-level phosphorylation.

18. Describe the experiment by which Jagendorf showed that chloroplasts could synthesize ATP in the dark when an artificial pH gradient was created across the thylakoid membrane. What have subsequent studies shown concerning the stoichiometry of protons transported per ATP synthesized?

19. Which of the following statements about the thylakoid membrane are correct?

(a) It contains photosystem I and ATP synthase in the unstacked regions.
(b) It contains the cytochrome *bf* complex in the unstacked regions only.
(c) It contains photosystem II mostly in the stacked regions.
(d) It facilitates communication between photosystems I and II by the circulation of plastoquinones and plastocyanins in the thylakoid space.
(e) It allows direct interaction between P680* and P700* reaction centers through its differentiation into stacked and unstacked regions.
(f) It can respond to environmental signals by varying its content, the degree of stacking, and the proportions of different photosynthetic assemblies.

20. Cyanobacteria and red algae are photosynthetic algae that live more than a meter underwater, where little blue or red light reaches them. Explain how they can carry out photosynthesis.

21. Match the photosystems of the purple sulfur bacterium or of green plants with the appropriate properties listed in the right column.

(a) Reaction center of *Rhodopseudomonas viridis* ____________

(b) Photosystem II of green plants ____________

(1) Contains a Mn center
(2) Contains two binding sites for plastoquinones
(3) Absorbs light of > 900 nm
(4) Energy conserving event is separation of charge from a chlorophyll$^+$ to pheophytin
(5) Transfers electrons from QH_2 to a cytochrome
(6) Special-pair chlorophyll$^+$ reduced by electrons from H_2O
(7) Special-pair chlorophyll$^+$ reduced through a cytochrome and four covalently attached hemes
(8) Can be inhibited by several urea derivatives

22. Explain why the multiple redox centers are closely spaced in reaction centers.

____

____

The Formation of Hexoses by CO_2 Fixation

23. The observation that the incubation of photosynthetic algae with $^{14}CO_2$ in the light for a very brief time (5 s) led to the formation of ^{14}C-labeled 3-phosphoglycerate suggested that the $^{14}CO_2$ was condensing with some two-carbon acceptor. That acceptor was in fact which of the following?

(a) Acetate
(b) Acetyl CoA
(c) Acetyl phosphate
(d) Acetaldehyde
(e) Glycol phosphate
(f) None of the above

24. Which of the following statements about ribulose 1,5-bisphosphate carboxylase (rubisco) are correct?

(a) It is present at low concentrations in the chloroplast.
(b) It is activated by the addition of CO_2 to the ε-amino group of a specific lysine to form a carbamate that then binds a divalent metal cation.
(c) It catalyzes, as one part of its reaction sequence, an extremely exergonic reaction, the cleavage of a six-carbon diol derivative of arabinitol to form 2 three-carbon compounds.
(d) It catalyzes a reaction between ribulose 1,5-bisphosphate and O_2 that decreases the efficiency of photosynthesis.
(e) It catalyzes the carboxylase reaction more efficiently and the oxygenase reaction less efficiently as the temperature increases.

25. The rubisco-catalyzed reaction of O_2 with ribulose 1,5-bisphosphate forms which of the following?

(a) 3-Phosphoglycerate
(b) 2-Phosphoacetate
(c) Phosphoglycolate
(d) Glycolate
(e) Glyoxylate

26. Describe photorespiration, and explain why it decreases the efficiency of photosynthesis.

____

____

27. Outline the synthesis of fructose 6-phosphate from 3-phosphoglycerate.

____

____

28. Match the two major storage forms of carbohydrates, starch and sucrose, with the appropriate items listed in the right column.

(a) Starch ____________

(b) Sucrose ____________

(1) Contains glucose
(2) Contains fructose
(3) Is a polymer
(4) Is synthesized in the cytosol
(5) Is synthesized in chloroplasts
(6) Is synthesized from UDP–glucose
(7) Is synthesized from CDP–, GDP–, or ADP–glucose

29. Which of the following statements about the Calvin cycle are true?

(a) It regenerates the ribulose 1,5-bisphosphate consumed by the rubisco reaction.
(b) It forms glyceraldehyde 3-phosphate, which can be converted to fructose 6-phosphate.
(c) It requires ATP and NADPH.
(d) It is exergonic because light energy absorbed by the chlorophylls is transferred to rubisco.
(e) It consists of enzymes, several of which can be activated through reduction of disulfide bridges by reduced thioredoxin.
(f) It is controlled, in part, by the rate of the rubisco reaction.
(g) Its rate decreases as the level of illumination increases because both the pH and the level of Mg^{2+} of the stroma decrease.

30. How many moles of ATP and NADPH are required to convert 6 moles of CO_2 to fructose 6-phosphate?

____

____

31. Which statement is correct about thioredoxin?

(a) It contains a heme that cycles between two oxidation states.
(b) Its oxidized form predominates while light absorption is taking place.

(c) It activates some biosynthetic enzymes by reducing disulfide bridges.
(d) It activates some degradative enzymes by reducing disulfide bridges.
(e) Oxidized thioredoxin is reduced by plastoquinol.

32. It is said that the C_4 pathway increases the efficiency of photosynthesis. What is the justification for this statement when more than 1.6 times as much ATP is required to convert 6 moles of CO_2 to a hexose when this pathway is used in contrast with the pathway used by plants lacking the C_4 apparatus? Account for the extra ATP molecules used in the C_4 pathway.

____

____

ANSWERS TO SELF-TEST

The Chloroplast, Chlorophylls, and Basic Reactions of Photosynthesis

1. The basic reaction for photosynthesis in green plants is

$$H_2O + CO_2 \xrightarrow{\text{Light}} (CH_2O) + O_2$$

where (CH_2O) represents carbohydrate.

2. (a) 3, 5, 7 (b) 3, 7 (c) 2, 3, 4, 5, 7, 9 (d) 1, 2, 3, 5, 7, 9 (e) 6, 8. Chlorophylls are involved in light absorption, light collection in the antennae, and reaction center chemistry. Photosystems I and II cooperate to generate a transmembrane proton-motive force that can synthesize ATP. The enzymes of the Calvin cycle and the gluconeogenic pathway combine to form hexoses.

3. a, b, c, d, e, f, g

4. (a) 3 (b) 1 (c) 4 (d) 2 (e) 5

5. a, c, e

6. The polyene structure (alternating single and double bonds) of chlorophylls causes them to have strong absorption bands in the visible region of the spectrum. Their peak molar absorption coefficients are higher than $10^5\ cm^{-1}\ M^{-1}$.

7. b, c, d

The Reactions of Photosystems II and I

8. Van Niel's equation for photosynthesis in green plants is

$$CO_2 + 2\ H_2O\ \text{Light} \xrightarrow{\text{Light}} (CH_2O) + O_2 + H_2O$$

Photosynthesis carried out in ^{18}O-labeled H_2O generated $^{18}O_2$, showing that water was oxidized to oxygen by the process.

9. a, b, c, e

10. When the rate of photosynthesis as a function of the wavelength of the incident light was examined, the rate declined precipitously for wavelengths longer than 680 nm even though chlorophyll continues to absorb light in the range from 680 to 700 nm. If there were only a single species of photoreceptor, the quantum yield of the process should remain constant over its entire absorption band. The fact that it did not suggested that more than one species of photoreceptor was involved in accumulating light energy. A second observation was that light at a shorter wavelength, such as 600 nm, could augment light in the 680- to 700-nm range such that the rate of photosynthesis in the presence of both was greater than the sum of the rates obtained with light of each wavelength separately. This red drop and enhancement phenomenon indicated that there were two separate but interacting light reactions in photosynthesis.

Photosystems I and II have different molecular compositions (e.g., different chlorophylls) and thus absorb light maximally at different wavelengths. They catalyze the two different light reactions, but interact with each other through the cytochrome *bf* complex.

11. a, b, c. Answer (d) is incorrect because a cluster of four manganese ions serves as a charge accumulator by interactions with the strong oxidant $P680^+$ and H_2O to form O_2.

12. (a) H (b) G (c) A (d) B (e) E, F (f) A, E, F (g) B (h) D1, D2 (i) D1, D2 (j) D1 (k) G

13. a. Answer (a) is incorrect because the Mn center contains four Mn atoms. The Mn center can adopt five oxidation states.

14. (a) 1 (b) 6 (c) 2 (d) 3

15. Two electrons from the weak reductant plastoquinol (QH_2) are transferred to two molecules of the weak oxidant plastocyanin (PC) in a reaction catalyzed by the transmembrane cytochrome *bf* complex; in the process, two protons are pumped across the thylakoid membrane to acidify the thylakoid space with respect to the stroma. The transmembrane proton gradient is used to synthesize ATP.

16. The net reaction catalyzed by photosystem I is

$$\text{PC(Cu}^+) + \text{ferredoxin}_{\text{oxidized}} \longrightarrow \text{PC(Cu}^{2+}) + \text{ferredoxin}_{\text{reduced}}$$

where PC is plastocyanin. Reduced ferredoxin is a powerful reductant. Two reduced ferredoxins reduce $NADP^+$ to form NADPH and two oxidized ferredoxins in a reaction catalyzed by ferredoxin-$NADP^+$ reductase. FAD is a prosthetic group on the enzyme that serves as an adapter to collect two electrons from two reduced ferredoxin molecules for their subsequent transfer to a single $NADP^+$ molecule.

Photophosphorylation and ATP Synthesis

17. a, c, d, e. Answer (b) is incorrect because photosystem I provides the electrons for photophosphorylation.

18. In Jagendorf's experiment, chloroplasts were equilibrated with a buffer at pH 4 to acidify their thylakoid spaces. The suspension was then rapidly

brought to pH 8, and ADP and P_i were added. The pH of the stroma suddenly increased to 8, whereas that of the thylakoid space remained at 4, resulting in a pH gradient across the thylakoid membrane. Jagendorf observed that ATP was synthesized as the pH gradient dissipated and that the synthesis occurred in the dark. Later studies of the stoichiometry of ATP synthesis revealed that approximately three protons traverse the membrane for each ATP synthesized.

19. a, c, d, f. Answer (b) is incorrect because the cytochrome *bf* complex is uniformly distributed throughout the thylakoid membrane. Answer (e) is incorrect because the differentiation into stacked and unstacked regions probably prevents direct interaction between the excited reaction center chlorophylls P680* and P700*.

20. Cyanobacteria and red algae contain protein assemblies called *phycobilisomes* that absorb some of the green and yellow light that reaches them to perform photosynthesis.

21. (a) 2, 3, 4, 5, 7, 8 (b) 1, 2, 4, 5, 6, 8

22. The rate of electron transfer decreases very rapidly with distance (see Figure 26-29 in Stryer). Therefore electron donors and acceptors must be less than 15 Å apart for efficient electron transfer.

The Formation of Hexoses by CO_2 Fixation

23. f. The $^{14}CO_2$ actually condenses with the five-carbon compound ribulose 1,5-bisphosphate to form a six-carbon compound that is rapidly hydrolyzed to form two molecules of 3-phosphoglycerate, one of which is ^{14}C-labeled.

24. b, c, d. Answer (e) is incorrect because the rate of the oxygenase reaction increases relative to that of the carboxylase reaction as the temperature increases; the altered ratio of the two reaction rates decreases the efficiency of photosynthesis as the temperature increases.

25. a, c

26. The oxygenase reaction of rubisco and the salvage reactions that convert two resulting phosphoglycolate molecules into serine are called *photorespiration* because CO_2 is released and O_2 is consumed in the process. Unlike genuine respiration, no ATP or NADPH is produced by photorespiration. Ordinarily, no CO_2 is released during photosynthesis, and all the fixed CO_2 can be used to form hexoses. During photorespiration, no CO_2 is fixed, and the products into which ribulose 1,5-bisphosphate is converted by the oxygenase reaction of rubisco cannot be completely recycled into carbohydrate because of the loss of CO_2 in the phosphoglycolate salvage reactions.

27. Phosphoglycerate kinase converts 3-phosphoglycerate, the initial product of photosynthesis, to the glycolytic intermediate 1,3-bisphosphoglycerate, which is then converted to glyceraldehyde 3-phosphate (G-3-P) by an NADPH-dependent G-3-P dehydrogenase in the chloroplast. Triosephosphate isomerase converts G-3-P to dihydroxyacetone phosphate, which aldolase can condense with another G-3-P to form fructose 1,6-bisphosphate. The phosphate ester at C-1 is hydrolyzed to give fructose

6-phosphate. The result of this pathway, which is functionally equivalent to the gluconeogenic pathway, is the conversion of the CO_2 fixed by photosynthesis into a hexose. (See Figure 26-36, in Stryer.)

28. (a) 1, 3, 5, 7 (b) 1, 2, 4, 6

29. a, b, c, e, f

30. Eighteen moles of ATP and twelve moles of NADPH are required to fix six moles of CO_2. Two moles of ATP are used by phosphoglycerate kinase to form two moles of 1,3-pisphosphoglycerate, and one mole of ATP is used by ribulose 5-phosphate kinase to form one mole of ribulose 1,5-bisphosphate per mole of CO_2 fixed. Two moles of NADPH are used by G-3-P dehydrogenase to form two moles of G-3-P per mole of CO_2 incorporated. Therefore, three moles of ATP and two moles of NADPH are used for each mole of CO_2 fixed. (See Figure 26-37, in Stryer.)

31. c. Thioredoxin contains cysteine residues that cycle between two oxidation states. It is reduced by ferredoxin while the light reactions are proceeding. It activates biosynthetic enzymes and inhibits degradative enzymes by reducing their disulfide bridges.

32. Plants lacking the C_4 pathway cannot compensate for the relative increase in the rate of the oxygenase reaction of rubisco with respect to the rate of the carboxylase reaction that occurs as the temperature rises. Plants with the C_4 pathway increase the concentration of CO_2 in the bundle-sheath cell, where the Calvin cycle occurs, thereby increasing the ability of CO_2 to compete with O_2 as a substrate for rubisco. As a result, more CO_2 is fixed and less ribulose 1,5-bisphosphate is degraded into phosphoglycolate, which cannot be efficiently converted into carbohydrate. Thus, the Calvin cycle functions more efficiently in these specialized plants under conditions of high illumination and at higher temperatures than it would otherwise.

The concentration of CO_2 is increased by an expenditure of ATP. The collection of one CO_2 molecule and its transport on C_4 compounds from the mesophyll cell into the bundle-sheath cell is brought about by the conversion of one ATP to AMP and PP_i in a reaction in which pyruvate is phosphorylated to PEP. The PP_i is hydrolyzed, and two ATP are required to resynthesize ATP from AMP. Thus, an *extra* ATP/CO_2 $\times$ 6 CO_2/hexose = 12 ATP/hexose are used by the C_4 pathway.

PROBLEMS

1. Suppose you were designing spectrophotometric assays for chlorophyll *a* and chlorophyll *b*. What wavelengths would you use for the detection of each? (See Figure 26-5, p. 656, of Stryer.) Explain your answer very briefly.

2. Green light has a wavelength of approximately 520 nm. Explain why solutions of chlorophyll appear to be green. (See Figure 26-5, p. 656, of Stryer.)

3. If you were going to extract chlorophylls *a* and *b* from crushed spinach leaves, would you prefer to use acetone or water as a solvent? Explain your answer briefly.

4. The spectrophotometric absorbance A of a species is given by the Beer-Lambert law:

$$A = E \times l \times [c]$$

where [c] is the molar concentration of the absorbing species, l is the length of the light path in centimeters, and E is the absorbance of a 1 M solution of the species in a 1-cm cell. Suppose that a mixture of chlorophylls *a* and *b* in a 1-cm cell gives an absorbance of 0.2345 at 430 nm and an absorbance of 0.161 at 455 nm. Calculate the molar concentration of each chlorophyll in the mixture. Use the values in the margin for E. (Hint: The absorbance of a mixture is equal to the sum of the absorbances of its constituents.)

	E_{430}	E_{455}
Chlorophyll *a*	1.21×10^5	0.037×10^5
Chlorophyll *b*	0.53×10^5	1.55×10^5

5. Predict what product or products will be rapidly labeled with ^{18}O when the following additions are made to actively photosynthesizing systems.

(a) $H_2{}^{18}O$ is added to green plants.
(b) $C^{18}O_2$ is added to green plants.
(c) $C^{18}O_2$ is added to green sulfur bacteria.

6. Would you expect oxygen to be evolved when $NADP^+$ is added to an illuminated suspension of isolated chloroplasts? Explain briefly.

7. Would your answer to problem 6 change if the chloroplasts were illuminated with extremely monochromatic light of 700 nm? Explain the basis for your answer.

8. $NADP^+$ and A_0 are two components in the electron transport chain associated with photosystem I (see Stryer, Figure 26-15, p. 662). A_0^- carries a single electron, whereas NADPH carries two electrons. Write the overall reaction that occurs, and calculate $\Delta E'_0$ and $\Delta G^{\circ\prime}$ for the reduction of $NADP^+$ by A_0^- using the thermodynamic data presented by Stryer on pages 532, 533, and 663.

9. Calculate the maximum free-energy change $\Delta G^{\circ\prime}$ that occurs as a pair of electrons is transferred from photosystem II to photosystem I. Estimate the E'_0 values from Figure 26-15 on page 662 of Stryer. Then compare your answer with the free-energy change that occurs in mitochondria as a pair of electrons is transferred from NADH + H^+ to oxygen. (See Stryer, p. 533.)

10. Explain the defect or defects in the hypothetical scheme for the light reactions of photosynthesis depicted in Figure 26-1.

11. The following overall equation may be written for photosynthesis in green plants:

$$6\ CO_2 + 6\ H_2O + n(Nh\nu) = \text{glucose} + 6\ O_2 \qquad \Delta G^{\circ\prime} = +686\ \text{kcal/mol}$$

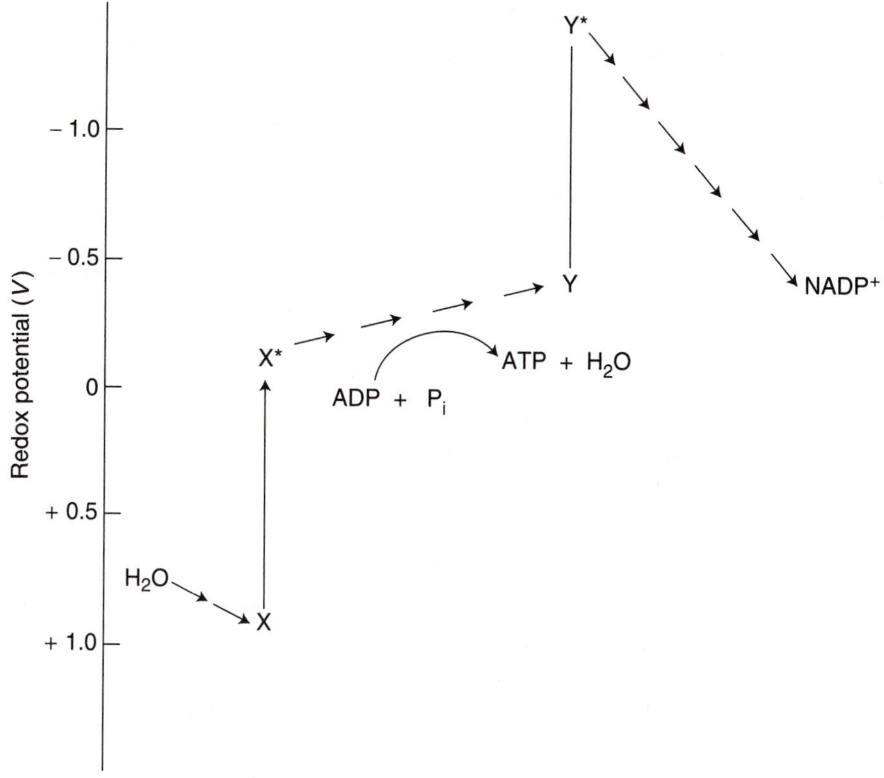

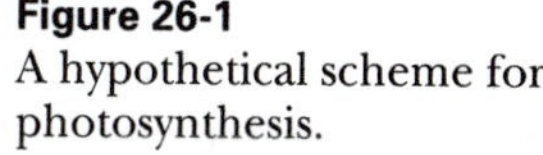
Figure 26-1
A hypothetical scheme for photosynthesis.

where n is the number of moles of photons required, N is Avogadro's number, and Nhv is the energy of 1 mole of photons.

(a) Give the minimum value for n, and explain your answer briefly.

(b) The energy of a photon is given by the relationship

$$E = hv = h(c/\lambda)$$

where h is Planck's constant, v is the frequency, λ is the wavelength, and c is the velocity of light. Calculate the thermodynamic efficiency of photosynthesis when light of 550 nm is used to drive the synthesis of 1 mole of glucose. Then repeat the calculation when light of 720 nm is used. Which has the greater thermodynamic efficiency? Use the physical constants given in Appendix A on page 1012 of Stryer.

(c) In practice, using extremely monochromatic light of 720 nm to drive photosynthesis would not be a good idea. Why not?

(d) In view of the fact that neither chlorophyll *a* nor chlorophyll *b* absorbs light at 550 nm, explain how any photosynthesis may occur when light of 550 nm is used. (See Stryer, Figure 26-5, p. 656.)

12. Explain why ATP synthesis requires a larger pH gradient across the thylakoid membrane of a chloroplast than across the inner membrane of a mitochondrion.

13. When a short pulse of $^{14}CO_2$ is added to systems that are actively carrying out photosynthesis, in which position or positions of glucose will the label first appear? Explain.

14. If $^{18}O_2$ were added to C_3 plants on a bright, sunny day, would you expect glycine subsequently isolated from the leaves to be labeled? Explain.

15. (a) Write a balanced equation for the conversion of oxaloacetate to malate in the mesophyll cell of a C_4 plant.

(b) What serves as the likely source of reducing power?

16. Give the source of the 12 NADPH used in the biosynthesis of glucose in a bundle-sheath cell of a C_4 plant.

17. Dinitrophenol (DNP) is a weak lipophilic acid that passes readily through membranes (see Figure 21-34 in Stryer's text). In a pH gradient, DNP binds a proton on the acidic side of the membrane, diffuses through, and releases the proton on the alkaline side. Suppose chloroplasts are incubated in the presence of light and CO_2 and are carrying out carbon fixation at a high rate. DNP is added to the incubation solution. Within 30 s after DNP is added, several processes related to photosynthesis are affected. Explain each of the changes listed below.

(a) The rate of photophosphorylation decreases.
(b) The rate of electron transport increases.
(c) The rate of NADPH generation increases.
(d) The overall rate of carbon fixation decreases.

18. Plants synthesize starch via pathways that are reminiscent of the pathway for glycogen synthesis in mammals, but there are certain important differences. Refer to page 676 of Stryer and then write a pathway for the formation of a glucosyl residue in starch from two molecules of ribulose 5-phosphate. Include the name of the enzyme catalyzing each reaction and specify other molecules needed, including CO_2, NADPH, and ATP.

19. The formation of a carbamate residue in rubisco is essential for its conversion into a catalytically active form. The carbamate adduct binds Mg^{2+} or Mn^{2+} to form a positively charged active site.

(a) Formation of the carbamate adduct is faster at alkaline pH. How do light-induced pH changes facilitate rubisco activation?
(b) In considering the reaction pathway catalyzed by rubisco, describe steps in which a divalent cation like Mg^{2+} or Mn^{2+} might be involved.

20. Certain photosynthetic bacteria produce acetone during the production of carbohydrate units from carbon dioxide. What substrate are they oxidizing during photosynthesis? Write a balanced equation for the process, in the form shown in Stryer, page 657.

1. You would use 430 nm for chlorophyll *a* and 455 nm for chlorophyll *b*. These are the wavelengths of maximum absorbance, so they would provide the most sensitive spectrophotometric assays.

2. Chlorophyll appears to be green because it has no significant absorption in the green region of the spectrum and therefore transmits green light.

3. Acetone is the preferred solvent. Because of the hydrophobic porphyrin ring and the very hydrophobic phytol tail of the chlorophylls, they are soluble in organic solvents like acetone but are insoluble in water.

4. The total absorbance at each wavelength is the sum of the absorbances due to chlorophylls *a* and *b* separately. At 430 nm,

$$0.2345 = (1.21 \times 10^5)[\text{chl } a] + (0.53 \times 10^5)[\text{chl } b]$$

At 455 nm,

$$0.161 = (0.037 \times 10^5)[\text{chl } a] + (1.55 \times 10^5)[\text{chl } b]$$

Solving these two equations for the two unknowns yields

$$[\text{chl } a] = 1.5 \times 10^{-6}\ \text{M and } [\text{chl } b] = 1.0 \times 10^{-6}\ \text{M}$$

5. (a) $^{18}O_2$ will be evolved. The oxygen produced by green plants comes from water split by photosystem II.

(b) Carbohydrate and water will both be labeled with ^{18}O. (See the Van Niel equation on p. 657 of Stryer.) The oxygen liberated derives from water. Therefore, of the two oxygen atoms of CO_2, one is incorporated into carbohydrate and the other into water.

(c) Again, both carbohydrate and water will be labeled with ^{18}O.

6. Oxygen would be evolved. In this reaction, $NADP^+$ would be the counterpart of the artificial electron acceptors used by Hill (See Stryer, p. 658).

7. Yes. Little oxygen would be evolved when 700-nm light is used. This is the red drop (see Stryer, p. 658, including Figure 26-9). Oxygen is evolved by photosystem II, which contains P680 and is therefore not maximally excited by 700-nm light.

8. A_0^- is the stronger reductant in the system because it is associated with the half-cell having the more negative standard reducing potential. A_0^- will therefore reduce $NADP^+$ to NADPH under standard conditions. Following the convention for redox problems presented by Stryer on page 533, we first write the partial reaction for the reduction involving the

weaker reductant (the half-cell with the more positive standard reducing potential):

$$\text{NADP}^+ + \text{H}^+ + 2\,e^- \longrightarrow \text{NADPH} \qquad E'_0 = -0.32\text{ V} \quad (1)$$

Next, we write, *again as a reduction,* the partial reaction involving the stronger reductant (the half-cell with the more negative standard reducing potential):

$$\text{A}_0 + e^- \longrightarrow \text{A}_0^- \qquad E'_0 = -1.1\text{ V} \quad (2)$$

To get the overall reaction that occurs, we must equalize the number of electrons by multiplying equation 2 by 2. (We do not, however, multiply the half-cell potential by 2.)

$$2\,\text{A}_0 + 2\,e^- \longrightarrow 2\,\text{A}_0^- \qquad E'_0 = -1.1\text{ V} \quad (3)$$

Then we *subtract* equation 3 from equation 1. This yields

$$\text{NADP}^+ + \text{H}^+ + 2\,\text{A}_0^- \longrightarrow \text{NADPH} + 2\,\text{A}_0 \qquad \Delta E'_0 = +0.78\text{ V}$$

In calculating $\Delta E'_0$ values, do not make the mistake of multiplying half-cell reduction potentials by factors used to equalize the number of electrons. Remember that $\Delta E'_0$ is a *potential difference* and hence, at least for our purposes, is independent of the *amount* of electron flow. For example, in a house with an adequate electrical power supply, the potential difference measured at the fuse box is approximately 117 V regardless of whether the house is in total darkness or all the lights are turned on.

To get the free-energy change for the overall reaction, we start with the relationship given on page 533 of Stryer:

$$\Delta G^{\circ\prime} = -nF\,\Delta E'_0$$

Substitution yields

$$\Delta G^{\circ\prime} = -2 \times 23.06 \times 0.78 = -36.0\text{ kcal/mol}$$

9. In this process, electrons are transferred down an electron transport chain from P680* to P700. The E'_0 value for P680* is approximately −0.8 V, and that for P700 is approximately 0.4 V; $\Delta E'_0$ is therefore +1.2 V. The free-energy change is calculated from the relationship given on page 533 of Stryer:

$$\begin{aligned}\Delta G^{\circ\prime} &= -nF\,\Delta E'_0\\ &= -2 \times 23.06 \times 1.2\\ &= -55.3\text{ kcal/mol}\end{aligned}$$

In the mitochondrial electron transport chain, the free-energy change as a pair of electrons is transferred from NADH + H^+ to oxygen is −52.6 kcal/mol (see p. 533 of Stryer).

10. In the scheme in Figure 26-1, electrons are shown flowing "uphill" from X* to Y as ATP is being formed. This is a thermodynamic impossibility. For electrons to flow spontaneously from X* to Y, the redox potential of X* must be more negative than that of Y. For ATP to be formed as elec-

tron transfer occurs, the free-energy change must be of sufficient magnitude to allow for ATP biosynthesis. In order to make electrons flow from X* to Y as depicted in the hypothetical scheme, ATP would be consumed, not generated.

11. (a) The minimum value of $n = 48$. In each turn of the Calvin cycle two NADPH are required and one CO_2 is fixed (see Figure 26-37, p. 675 of Stryer). Therefore, for 6 CO_2 to be reduced to the oxidation level of carbohydrate, 12 NADPH are required. Each NADPH generated has received two electrons from photosystem I, requiring the absorption of two photons. The electrons donated by photosystem I are replaced by photosystem II, requiring the absorption of another two photons for a total of four photons for each NADPH.

(b) The energy of a mole of 550-nm photons is given by

$$E = Nh(c/\lambda)$$

where N is Avogadro's number. Substitution yields

$$E = (6.022 \times 10^{23}\ \text{mol}^{-1})(1.584 \times 10^{-34}\ \text{cal s})\ \frac{2.998 \times 10^{17}\ \text{nm s}^{-1}}{550\ \text{nm}}$$

$$= 52\ \text{kcal}$$

Since 48 moles of photons are required for the synthesis of 1 mole of glucose, the total energy input is $48 \times 52 = 2496$ kcal, and the efficiency is $(686/2496) \times 100\% = 27\%$.

When light of 720 nm is used, the total energy input is $39.5 \times 48 = 1896$ kcal, and the efficiency is $(686/1896) \times 100\% = 36\%$. Therefore, light of longer wavelength gives a higher thermodynamic efficiency.

(c) Even though photosynthesis is more efficient thermodynamically with 720-nm light, the rate of photosynthesis under these conditions would be negligible owing to the red drop (see Stryer, p. 658).

(d) Accessory pigments absorb light of 550 nm and transfer the light energy to the reaction centers. Note that the quantum yield of photosynthesis is significant at 550 nm. (See Stryer, p. 658, Figure 26-9. Also see the discussion of light antennae in cyanobacteria and red algae on pp. 667-668.)

12. The synthesis of ATP in both the chloroplast and the mitochondrion is driven by the proton-motive force across the membrane. The proton-motive force Δp is given by the equation

$$\Delta p = E_m - \Delta\text{pH}$$

where E_m is the membrane potential, and ΔpH is the pH gradient across the membrane. In mitochondria, a membrane potential of 0.14 V is established during electron transport. In chloroplasts, the light-induced potential is close to 0. Therefore, there must be a greater pH gradient in the chloroplast to give the same free-energy yield. (See Stryer, pp. 545 and 665–666.)

13. The label will first appear in positions 3 and 4 of glucose. Phosphoglycerate labeled in the carboxyl carbon will be formed, and this carbon becomes carbon atoms 3 and 4 of glucose.

14. Yes. C_3 plants photorespire, and oxygen is incorporated into the carboxyl group of phosphoglycolate as shown in Figure 26-33 on page 673 of Stryer. Since the carboxyl group of phosphoglycolate will become the carboxyl group of glycine, the glycine will be labeled. (See Figure 26-34 on p. 674 of Stryer and the accompanying text.)

15. (a) The equation for the conversion of oxaloacetate to malate is

$$\text{Oxaloacetate} + \text{NADPH} + \text{H}^+ \quad \text{malate} + \text{NADP}^+$$

(b) The NADPH is furnished by the light reactions of photosynthesis.

16. Six of the twelve NADPH are provided by the oxidative decarboxylation of malate:

$$6\ \text{Malate} + 6\ \text{NADP} + \longrightarrow 6\ \text{pyruvate} + 6\ \text{NADPH} + 6\ \text{H}^+ + 6\ \text{CO}_2$$

The other six NADPH come from the light reactions of photosynthesis. Note, however, that the malate is originally produced in the mesophyll cell by the reduction of oxaloacetate by NADPH + H^+ coming from photosystem I. (See the answer to problem 15.) Thus, all the NADPH required for the reduction of CO_2 ultimately arises from the light reactions.

17. (a) The proton gradient is dissipated by DNP. ATP synthase activity depends on movement of protons through the enzyme stalk in order to drive the phosphorylation of ADP. When the proton gradient collapses, the rate of ATP synthesis decreases.

(b) When the proton gradient is dissipated by DNP, there is then no concentration gradient against which protons must be pumped during electron transport. The rate of electron transport will increase as the proton gradient collapses.

(c) Generation of NADPH depends on transport of electrons through photosystems II and I to reduce ferredoxin, which then reduces $NADP^+$. An increase in the rate of electron transport as the pH gradient collapses will allow more electrons to be used to reduce $NADP^+$, so that NADPH production will increase.

(d) ATP is required for the conversion of carbon dioxide to glucose (it is specifically required for the reaction catalyzed by 3-phosphoglycerate kinase, but this precise answer is not necessary). Since the rate of photophosphorylation and ATP production decreases when DNP is introduced, the rate of carbon fixation decreases.

18. See the accompanying figure. The most direct path from ribulose 1,5-phosphate to a glucosyl unit in starch includes formation of two molecules of 3-phosphoglycerate (3-P-G) from ribulose 1,5-phosphate and a molecule of carbon dioxide. Phosphorylation of 3-P-G to form 1,3-bisphospho-

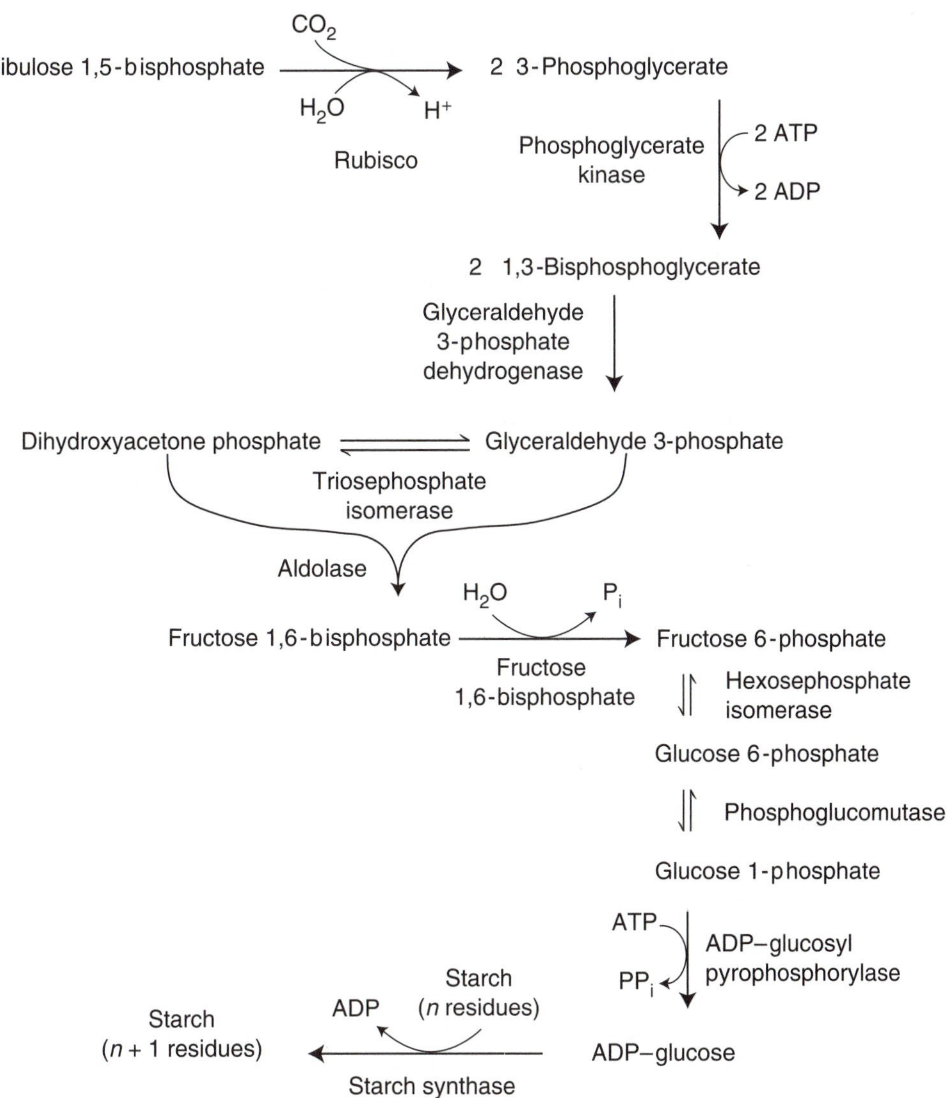

glycerate requires ATP. The reduction of 1,3-bisphosphoglycerate, catalyzed by glyceraldehyde 3-phosphate dehydrogenase, requires NADPH, generated in the light reactions of photosynthesis. Isomerization of glyceraldehyde 3-phosphate (3-P-Gal) yields dihydroxyacetone phosphate, which is condensed with 3-P-Gal to form fructose 1,6-bisphosphate, which is then converted to fructose 6-phosphate by fructose 1,6-bisphosphatase. The sequential actions of hexose 6-phosphate isomerase and phosphoglucomutase yield glucose 1-phosphate, a substrate for ADP–glucosyl pyrophosphorylase, which uses ATP as a nucleotidyl donor. Note that starch synthesis differs from glycogen synthesis, in which UDP–glucose is an intermediate. In plants, activated precursors for starch synthesis include ADP–glucose, CDP–glucose, or GDP–glucose but not UDP–glucose. The addition of a glucose residue to the nonreducing end of a starch molecule is catalyzed by starch synthase.

19. (a) Rubisco is found in the stroma, where hydrogen ion concentrations are lowered when photosystem I, stimulated by light absorption, pumps protons into the thylakoid membrane. As the pH in the stroma becomes more alkaline, the formation of carbamate adducts increases. Divalent cations also enter the stroma as proton

concentration increases in the thylakoid lumen. An increase in the rate of carbamate formation as well as in divalent cation concentration accelerates the conversion of rubisco to the active form.

(b) Divalent cations at the active site of rubisco may help in stabilizing a deprotonated enediol intermediate as it adds carbon dioxide, and also act as an electron sink in the cleavage of the carbon-carbon bond that generates two molecules of glyceraldehyde 3-phosphate.

20. They are oxidizing isopropanol, which is converted to acetone during photosynthesis. The balanced equation is

$$CO_2 + 2\ CH_3-\underset{\underset{H}{|}}{\overset{\overset{OH}{|}}{C}}-CH_3 \longrightarrow (CH_2O) + H_2O + 2\ CH_3-\overset{\overset{O}{\|}}{C}-CH_3$$

Isopropanol **Acetone**

EXPANDED SOLUTIONS TO TEXT PROBLEMS

1. $\Delta E'_0 = -0.32 - (-0.43) = +0.11$ V. $\Delta G°'$ (to reduce 1 mol of $NADP^+$) $= -2 \times 23.06 \times 0.11 = -5.08$ kcal/mol. (See Stryer, p. 533.)

2. DCMU inhibits electron transfer between Q and plastoquinone in the link between photosystems II and I. O_2 evolution can occur in the presence of DCMU if an artificial electron acceptor such as ferricyanide can accept electrons from Q.

3. (a) The energy of photons is inversely proportional to the wavelength. Since 600-nm photons have an energy content of 47.6 kcal/einstein, 1000-nm light will have an energy content of $600/1000 \times 47.6 = 28.7$ kcal/einstein. Since 1 cal = 4.184 joule, 28.7 kcal × 4.184 = 120 kJ/einstein.

(b) −28.7 kcal/mol ($\Delta G°'$) $= -1 \times 23.06 \times V$. Therefore, $V = -28.7/-23.06 = 1.24$ volts.

(c) If 1000-nm photons have a free-energy content of 28.7 kcal/einstein and ATP has a free-energy content of 12 kcal/mol, then 1000-nm photon has the free-energy content of 28.7/12, or 2.39 ATP. Therefore, a minimum of 0.42 (1/2.39) photon is needed to drive the synthesis of an ATP.

4. The formation of O_2 is a four-electron process. Each brief light pulse provides one electron for reduction, which is accumulated by the manganese center.

5. Aldolase participates in the Calvin cycle, whereas transaldolase participates in the pentose phosphate pathway.

6. The conversion of ribulose 1,5-bisphosphate to 3-phosphoglycerate does not require ATP, so it will continue until the ribulose 1,5-bisphosphate is largely depleted.

7. When the concentration of CO_2 is drastically decreased, the rate of conversion of ribulose 1,5-bisphosphate to 3-phosphoglycerate will greatly decrease, whereas the rate of utilization of 3-phosphoglycerate will not be diminished.

8. Phycoerythrin, the most peripheral protein in the phycobilisome.

9. (a)

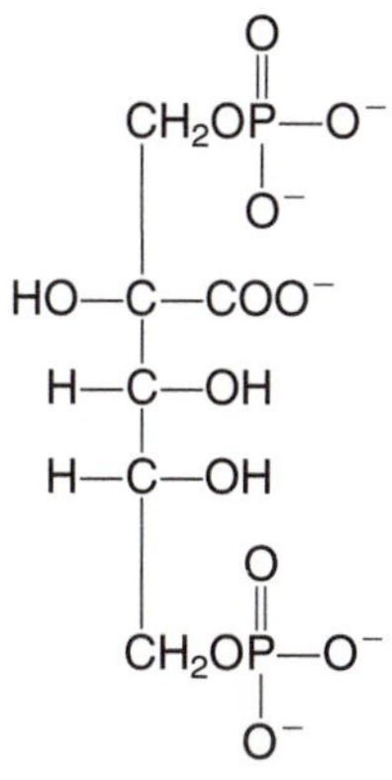

```
         O
         ‖
   CH2OP—O⁻
   |     |
   |     O⁻
HO—C—COO⁻
   |
 H—C—OH
   |
 H—C—OH
   |     O
   |     ‖
   CH2OP—O⁻
         |
         O⁻
```

2-Carboxyarabinitol 1,5-bisphosphate (CABP)

(b) CABP resembles the addition compound formed in the reaction of CO_2 and ribulose 1,5-bisphosphate (p. 672).

(c) As predicted, CABP is a potent inhibitor of rubisco.

10. Aspartate + glyoxylate $\longrightarrow$ oxaloacetate + glycine

CHAPTER

27

Biosynthesis of Membrane Lipids and Steroids

This chapter describes the biosynthesis of membrane lipids, steroids, and other important lipid molecules, such as bile salts, vitamin D, and polymers of isoprene units. As background material for this chapter, you should review the earlier chapters on cell membranes (Chapter 11) and fatty acid metabolism (Chapter 24), paying particular attention to the structure and properties of lipids and the central role of acetyl CoA in the metabolism of lipids. Stryer begins Chapter 27 with a discussion of the formation of triacylglycerols, phosphoglycerides, and sphingolipids from the simple precursors glycerol 3-phosphate, fatty acyl CoAs, certain polar alcohols (e.g., choline), serine, and sugars. He then describes the synthesis of cholesterol from acetyl CoA via the important intermediate isopentenyl pyrophosphate. The regulation of a key enzyme in the biosynthetic pathway is also outlined. Cholesterol is the precursor of bile salts as well as steroid hormones. The cholesterol and triacylglycerols synthesized in the liver and intestines are transported by lipoproteins to peripheral tissues. Cholesterol from dietary sources is moved from the intestine to the liver by lipoproteins. Therefore, the classification, the properties, and the mechanisms by which the lipoproteins deliver lipids to cells are discussed next. Finally, Stryer describes the synthesis of steroid hormones and vitamin D from cholesterol and introduces a variety of isoprenoid lipids that are derived from isopentenyl pyrophosphate.

LEARNING OBJECTIVES

When you have mastered this chapter, you should be able to complete the following objectives:

Synthesis of Triacylglycerols and Phosphoglycerides (Stryer pages 685–689)

1. Describe the roles of *glycerol 3-phosphate, lysophosphatidate,* and *phosphatidate* in the synthesis of *triacylglycerols* and *phosphoglycerides.*

2. Describe the formation of triacylglycerols from phosphatidate.

3. Contrast the biosynthesis of phosphoglycerides in bacteria and mammals. Note the significance of *CDP–diacylglycerol* and *CDP–choline* or *CDP–ethanolamine,* the activated precursors in these biosyntheses.

4. Compare the biosynthesis of *phosphatidyl serine, phosphatidyl ethanolamine, phosphatidyl choline,* and *phosphatidyl inositol* in bacteria and mammals.

5. State the physiologic role of phosphatidyl inositol and its degradation products, *inositol 1,4,5-trisphosphate* and *diacylglycerol.*

6. Compare the structures and biosynthetic pathways of the *glyceryl ether phospholipids,* including *platelet-activating factor* and *plasmalogens,* with those of the *glyceryl ester phospholipids.*

7. List the *phospholipases* together with their specificities for the ester bonds of phosphoglycerides. Describe the biological roles of phospholipases.

Synthesis of Sphingolipids (Stryer pages 689–691)

8. Summarize the steps in the biosynthesis of *sphingosine* from *palmitoyl CoA* and *serine.*

9. Outline the synthesis of *sphingomyelin, cerebrosides,* and *gangliosides* from sphingosine. Note the use of activated sugars and acidic sugars.

10. Discuss the general degradation pathway of gangliosides and the biochemical basis of *Tay-Sachs disease.*

Synthesis of Cholesterol (Stryer pages 691–697)

11. Describe the physiologic roles of *cholesterol.*

12. List the major stages in cholesterol biosynthesis and give the key intermediates.

13. Compare the synthetic paths leading from acetyl CoA to *mevalonate* and to the *ketone bodies.* Note the role of *3-hydroxy-3-methylglutaryl CoA reductase* (*HMG CoA reductase*) as the major regulatory enzyme in cholesterol biosynthesis.

14. Describe the conversion of mevalonate into *isopentenyl pyrophosphate.*

15. Outline the condensation reactions leading from isopentenyl pyrophosphate to *squalene.* Describe the mechanisms of these condensation reactions.

16. Discuss the *cyclization* of squalene and the formation of cholesterol from *lanosterol.* Note the role of O_2 in the formation of cholesterol.

17. Describe the physiologic roles and the general structures of the *bile salts.*

18. List the sources of cholesterol and outline the mechanisms of regulation of cholesterol biosynthesis.

Lipoproteins (Stryer pages 697–702)

19. List the various classes of *lipoproteins* together with their lipid and protein components. Describe their lipid transport functions.

20. Summarize the steps in the delivery of cholesterol to cells via the *low-density-lipoprotein (LDL) receptor.* Discuss the regulation of cellular functions by the *LDL pathway.*

21. Describe the proposed domain structure of the LDL receptor derived from the primary sequence of this protein. Define the term *mosaic protein.*

22. Discuss the biochemical defects of the LDL receptor that result in *familial hypercholesterolemia.*

23. Summarize approaches used to reduce serum cholesterol.

Synthesis of Steroid Hormones (Stryer pages 702–707)

24. Give the numbering scheme for the carbon atoms of cholesterol and its derivatives, and distinguish between the α and β*-oriented* groups and *cis* or *trans ring fusions.*

25. List the five major classes of *steroid hormones,* their physiologic functions, and their sites of synthesis. Outline their biosynthetic relationships.

26. Describe the *hydroxylation reactions* involving *cytochrome P450.* Indicate the role of these *monooxygenase reactions* in steroid biosynthesis, the *detoxification* of *xenobiotic compounds,* and the generation of *carcinogens.*

27. Describe the synthesis of *pregnenolone* from cholesterol, the conversion of pregnenolone into *progesterone,* and the subsequent reactions leading to *cortisol* and *aldosterone.*

28. Outline the synthesis of *androgens* and *estrogens* from progesterone.

29. Explain the biochemical basis for both the virilization and the Na^+ loss observed in *21-hydroxylase deficiency.*

Vitamin D and Derivatives of Isopentenyl Pyrophosphate (Stryer pages 707–709)

30. Discuss the synthesis and the physiologic role of *vitamin D.*

31. Give examples of biomolecules that contain *isoprene units.*

SELF-TEST

Synthesis of Triacylglycerols and Phosphoglycerides

1. Which of the following reactions are sources of glycerol 3-phosphate that is used in lipid synthesis?

(a) Reduction of dihydroxyacetone phosphate
(b) Oxidation of glyceraldehyde 3-phosphate
(c) Phosphorylation of glycerol
(d) Dephosphorylation of 1,3-bisphosphoglycerate
(e) Reductive phosphorylation of pyruvate

2. Match the lipids in the left column with the major synthetic precursors or intermediates listed in the right column.

(a) Triacylglycerol ________
(b) Phosphatidyl ethanolamine (bacteria) ________
(c) Phosphatidyl ethanolamine (mammals) ________

(1) Phosphatidate
(2) Diacylglycerol
(3) Acyl CoA
(4) Glycerol 3-phosphate
(5) CDP–diacylglycerol
(6) CDP–ethanolamine

3. Explain the role of the CDP derivatives in the synthesis of phosphoglycerides.

____

____

4. Calculate the number of "high-energy" phosphate bonds that are expended in the formation of phosphatidyl choline from diacylglycerol and choline in mammals.

____

____

5. Which of the following is a common reaction used for the formation of phosphatidyl ethanolamine in bacteria?

(a) Decarboxylation of phosphatidyl serine
(b) Reaction of CDP–ethanolamine with a diacylglycerol
(c) Demethylation of phosphatidyl choline
(d) Reaction of ethanolamine with CDP–diacylglycerol
(e) Reaction of CDP–ethanolamine with CDP–diacylglycerol

6. Which of the following is a lipid with a signal-transducing activity?

(a) Phosphatidyl choline
(b) Phosphatidyl serine
(c) Plasminogen activator
(d) Phosphatidyl inositol 4,5-bisphosphate
(e) Phospholipase A_2

7. For the lipid classes listed in the left column, select the characteristic structural components or properties from the right column.

(a) Glyceryl ester phospholipids__________

(b) Plasmalogens __________

(c) Platelet-activating factor __________

(1) Two long hydrocarbon chains
(2) Acetyl group
(3) Phosphate group
(4) Ether linkage
(5) α, β-double bond
(6) Glycerol group
(7) Long fatty acyl chain
(8) Relatively high solubility in water

8. Which of the following phospholipases would *not* cleave the phospholipid shown below?

(a) Phospholipase A_1
(b) Phospholipase A_2
(c) Phospholipase C
(d) Phospholipase D

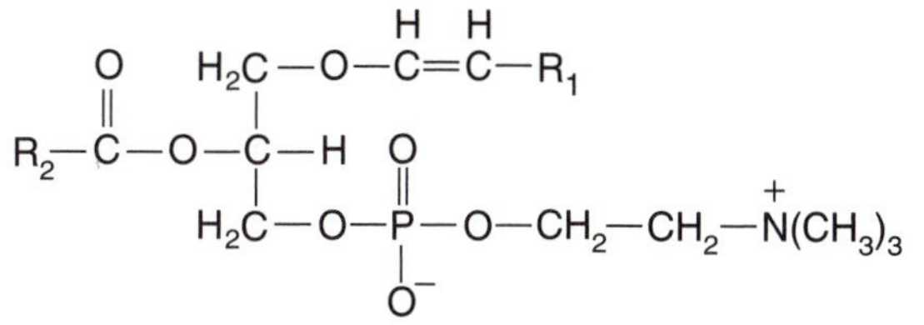

```
                 H  H
  O     H2C—O—C=C—R1
  ‖        |
R2—C—O—C—H   O
           |     ‖              +
        H2C—O—P—O—CH2—CH2—N(CH3)3
                 |
                 O⁻
```

Figure 27-1
A phospholipid.

Synthesis of Sphingolipids

9. Which of the following is *not* a precursor or intermediate in the synthesis of sphingomyelin?

(a) Palmitoyl CoA
(b) Lysophosphatidate
(c) CDP–choline
(d) Acyl CoA
(e) Serine

10. Match the lipids in the left column with the appropriate activated precursors from the right column.

(a) Sphingomyelin__________

(b) Ganglioside__________

(c) Phosphatidyl serine__________

(1) Acyl CoA
(2) CDP–choline
(3) CDP–diacylglycerol
(4) CMP–*N*-acetylneuraminate
(5) UDP–sugar

11. In which compartment of the cell does ganglioside G_{M2} accumulate in Tay-Sachs patients? What is the biochemical defect?

____

____

Synthesis of Cholesterol

12. From the following compounds, identify the intermediates in the synthesis of cholesterol and list them in their proper sequence:

(a) Geranyl pyrophosphate ____________
(b) Squalene ____________
(c) Isopentenyl pyrophosphate ____________
(d) Mevalonate ____________
(e) Cholyl CoA ____________
(f) Farnesyl pyrophosphate ____________
(g) Lanosterol ____________

13. Which of the following are common features of the syntheses of mevalonate and ketone bodies?

(a) Both involve 3-hydroxy-3-methylglutaryl CoA (HMG CoA).
(b) Both require NADPH.
(c) Both require the HMG CoA cleavage enzyme.
(d) Both occur in the mitochondria.
(e) Both occur in liver cells.

14. Select the appropriate characteristics from the right column for the three stages in the synthesis of cholesterol in the left column.

(a) Mevalonate to isopentenyl pyrophosphate ____________
(b) Isopentenyl pyrophosphate to squalene ____________
(c) Squalene to cholesterol ____________

(1) Releases PP_i
(2) Requires NADPH
(3) Requires O_2
(4) Releases CO_2
(5) Requires ATP

15. Yeast cells growing aerobically synthesize sterols and incorporate them into membranes. However, under anaerobic conditions yeast cells do not survive unless they are provided with an exogenous source of sterols. Explain the metabolic basis of this nutritional requirement.

____

____

16. The physiologic roles of bile salts include which of the following?

(a) They aid in the digestion of lipids.
(b) They aid in the digestion of proteins.
(c) They facilitate the absorption of sugars.
(d) They facilitate the absorption of lipids.
(e) They provide a means for excreting cholesterol.

17. Which of the following are common features in the structures of cholesterol and glycocholate?

(a) Both have three hydroxyl groups.
(b) Both contain four fused rings.
(c) Both have a hydrocarbon side chain.
(d) Both contain a carboxylate group.
(e) Both contain double bonds.
(f) Both contain a sulfur atom.

18. Explain the structural characteristics of bile salts that make them effective biological detergents.

____

____

19. The key step in cholesterol biosynthesis is the conversion of 3-hydroxy-3-methylglutaryl CoA to mevalonate. Which of the following are ways in which this reaction can be modulated?

(a) Covalent modification HMG CoA reductase through phosphorylation
(b) Controlling the rate of translation of the mRNA encoding HMG CoA reductase
(c) Controlling the rate of transcription of the gene encoding HMG CoA reductase
(d) Proteolytic degradation of HMG CoA reductase
(e) Deletion and duplication of the gene encoding HMG CoA reductase

Lipoproteins

20. Match the appropriate components or properties in the right column with the lipoproteins in the left column.

(a) Chylomicron __________
(b) VLDL__________
(c) LDL__________
(d) HDL__________

(1) Contains apoprotein B-100
(2) Contains apoprotein B-48
(3) Contains apoprotein A-1
(4) Transports endogenous cholesterol esters
(5) Transports dietary triacylglycerols
(6) Transports endogenous triacylglycerols
(7) Is degraded by lipoprotein lipase
(8) Is taken up by cells via receptor-mediated mechanisms
(9) Is a precursor of LDL
(10) May remove cholesterol from cells

21. Which of the following events occur in the LDL pathway in fibroblasts? Place them in their proper sequential order.

(a) Breakdown of LDL in lysosomes __________
(b) Endocytosis of LDL along with LDL receptors __________
(c) Degradation of LDL receptors in lysosomes __________
(d) Binding of LDL to LDL receptors __________
(e) Return of LDL receptors to the plasma membrane __________

22. Exons in the gene for the LDL receptor give rise to structurally diverse domains. What is the likely function of the cysteine-rich amino-terminal domain, which contains a cluster of negatively charged side chains?

(a) Carbohydrate binding
(b) Membrane attachment
(c) Lipoprotein binding
(d) Growth-factor binding
(e) Clathrin binding

23. Explain how LDL regulates the cholesterol content in fibroblasts.

___

___

___

24. Assume that LDL is produced normally in a patient but that the apoprotein B-100 domain that recognizes the receptor is functionally defective, which prevents the binding of LDL to its receptor. What outcome would this defect have on cholesterol metabolism in peripheral cells?

___

___

Synthesis of Steroid Hormones

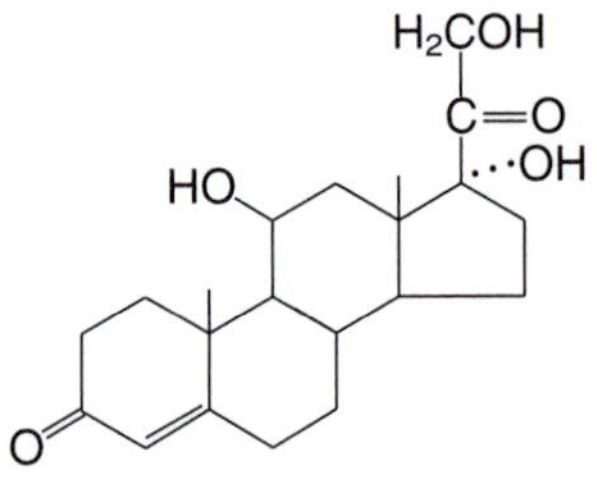

H$_2$COH
C=O
OH
HO
O

Figure 27-2
A sterol.

25. For the sterol structure in the margin, complete the following:

(a) Name this sterol: ___

(b) It is synthesized from ___ via three hydroxylation reactions.

(c) It has how many fewer carbon atoms than cholesterol?

___

(d) Its concentration will be diminished if there is a deficiency of 21-hydroxylase—true or false?

___

Explain why.

___

___

26. Hydroxylation reactions involving cytochrome P450 have which of the following characteristics?

(a) They require a proton gradient.
(b) They involve electron transport from NADPH to O_2.
(c) They activate O_2 by binding it to adrenodoxin.
(d) They transfer one oxygen atom from O_2 to the substrate and form water from the other oxygen atom.
(e) They occur in adrenal mitochondria and liver microsomes.

27. Explain how foreign aromatic compounds are detoxified and excreted by mammals. Describe a possible deleterious effect of this process.

___

___

28. Match the steroid hormones in the left column with the characteristics in the right column that distinguish them from one another.

(a) Aldosterone____________
(b) Estrogen____________
(c) Testosterone____________

(1) Has 18 carbon atoms
(2) Has 19 carbon atoms
(3) Has 21 carbon atoms
(4) Contains an aromatic ring
(5) Contains an aldehyde group at C-18

29. Name the principle form of excreted cholesterol.

____

____

Vitamin D and Derivatives of Isopentenyl Pyrophosphate

30. Which of the following statements about active vitamin D are incorrect?

(a) It has the same fused ring system as cholesterol.
(b) It requires hydroxylation reactions for its synthesis from cholecalciferol.
(c) It is important in the control of calcium and phosphorus metabolism.
(d) It can be synthesized from cholesterol in the presence of uv light.
(e) It can be derived from the diet.

31. Which of the following lipids does not contain isoprene units?

(a) Coenzyme Q
(b) Carotene
(c) Vitamin K
(d) Arachidonate
(e) Phytol side chain of chlorophyll

ANSWERS TO SELF-TEST

Synthesis of Triacylglycerols and Phosphoglycerides

1. a, c. Reduction of dihydroxyacetone phosphate is the primary route.

2. (a) 1, 2, 3, 4 (b) 1, 3, 4, 5 (c) 1, 2, 3, 4, 6

3. CDP–diacylglycerol and the CDP–alcohols are activated intermediates that allow the formation of phosphate ester bonds in phosphoglycerides, a process that is otherwise highly exergonic. ATP supplies the energy to form these compounds. Recall that UDP–sugars are used in a similar manner in the synthesis of carbohydrates (see Stryer, p. 586).

4. Summing the individual reactions:

$$\text{Choline} + \text{ATP} \longrightarrow \text{phosphorylcholine} + \text{ADP}$$
$$\text{Phosphorylcholine} + \text{CTP} \longrightarrow \text{CDP–choline} + \text{PP}_i$$
$$\text{CDP–choline} + \text{diacylglycerol} \longrightarrow \text{CMP} + \text{phosphatidyl choline}$$
$$\text{Choline} + \text{ATP} + \text{CTP} + \text{diacylglycerol} \longrightarrow \text{ADP} + \text{PP}_i + \text{CMP} + \text{phosphatidyl choline}$$

Two high-energy bonds (from ATP and CTP) are directly consumed in these reactions. In addition, pyrophosphate is hydrolyzed by pyrophosphatase, driving the net reaction farther to the right. A total of three high-energy bonds would be consumed to regenerate ATP and CTP from ADP and CMP.

5. a

6. d

7. (a) 1, 3, 6, 7 (b) 1, 3, 4, 5, 6, 7 (c) 2, 3, 4, 6, 8

8. a. Phospholipases are specific for *ester* bonds; therefore, phospholipase A_1 will not cleave the *ether* bond on the C-1 carbon of the given plasmalogen.

Synthesis of Sphingolipids

9. b

10. (a) 1, 2 (b) 1, 4, 5 (c) 1, 3

11. The degradative enzymes for gangliosides are located in lysosomes; therefore, ganglioside G_{M2} will accumulate in the lysosomes of Tay-Sachs patients. The enzyme that removes the terminal sugar, GalNAc, from the ganglioside is deficient in these people.

Synthesis of Cholesterol

12. All the compounds given are intermediates in the biosynthesis of cholesterol except for (e) cholyl CoA, which is a catabolic derivative of cholesterol and a precursor of bile salts. The proper sequence is d, c, a, f, b, g.

13. a, e

14. (a) 4, 5 (b) 1, 2 (c) 2, 3

15. A key intermediate in the biosynthesis of cholesterol and related sterols is squalene, an open-chain isoprenoid hydrocarbon. It is converted to squalene 2,3-epoxide, which in turn is converted to lanosterol. The conversion of squalene to the 2,3-epoxide is catalyzed by a monooxygenase, and molecular oxygen is a required component for this reaction. Under anaerobic conditions, yeast cells cannot synthesize sterols because they lack oxygen, a substrate for the monooxygenase reaction.

16. a, d, e

17. b, e

18. Bile salts are effective detergents because they contain both polar and nonpolar regions. They have several hydroxyl groups, all on one side of the ring system, and a polar side chain that allow interactions with water. The ring system itself is nonpolar and can interact with lipids or other nonpolar substances. Bile salts are planar, amphipathic molecules, in contrast with such detergents as sodium dodecyl sulfate (see Stryer, p. 275), which are linear.

19. a, b, c, d

Lipoproteins

20. (a) 2, 3, 5, 7 (b) 1, 6, 7, 9 (c) 1, 4, 8 (d) 3, 4, 10

21. All the events except (c) occur in the LDL pathway. The proper sequence is d, b, a, e.

22. c. The negatively charged cluster of amino acids are thought to interact with a positively charged site on the apoB-100 molecule.

23. The main source of cholesterol for cells outside the liver and intestine is from circulating LDL. Cholesterol released during the degradation of LDL suppresses the formation of HMG CoA reductase, the enzyme that controls the endogenous synthesis of cholesterol. In addition, the synthesis of new LDL receptors is suppressed, decreasing the uptake of exogenous cholesterol by the cell.

24. A defect in apoprotein B-100 that prevents the binding of LDL to the cell-surface receptor would result in the stimulation of the synthesis of endogenous cholesterol and LDL receptors and a decrease in the synthesis of cholesterol esters via the ACAT reaction. Indeed, the cellular and physiologic consequences of such a mutation may be similar to those seen in familial hypercholesterolemia.

Synthesis of Steroid Hormones

25. (a) Cortisol

(b) Progesterone

(c) Six

(d) True. A deficiency of 21-hydroxylase will impair hydroxylation at C-21 of progesterone, which will prevent the normal synthesis of cortisol and mineralocorticoids from progesterone.

26. b, d, e

27. In mammals, foreign aromatic molecules are hydroxylated by the cytochrome P450–dependent monooxygenases that are present in the endoplasmic reticulum of the liver cells. The hydroxylated derivatives are more water-soluble and have functional groups for the attachment of very polar substances, such as glucuronate, that allow them to be excreted in

urine. The action of the cytochrome P450 system sometimes converts potential carcinogenic compounds into highly carcinogenic derivatives.

28. (a) 3, 5 (b) 1, 4 (c) 2

29. The water-soluble bile salt glycocholate is a major cholesterol breakdown product.

Vitamin D and Derivatives of Isopentenyl Pyrophosphate

30. a

31. d

PROBLEMS

1. Why is de novo cholesterol synthesis dependent on the activity of citrate lyase?

2. An infant has an enlarged liver and spleen, cataracts, and anemia and exhibits general retardation of development. Mevalonate is found in the urine. Investigation reveals a deficiency of mevalonate kinase, which catalyzes the formation of 5-phosphomevalonate from mevalonate.

   (a) Why is urinary excretion of mevalonate consistent with a deficiency of mevalonate kinase?
   (b) How would a deficiency of mevalonate kinase affect cholesterol synthesis in this infant?
   (c) What level of activity, relative to normal, would you expect to find for HMG CoA reductase in cells isolated from the infant? Briefly explain your answer.

3. Normally, most of the bile acids that are secreted into the intestine undergo reabsorption and are returned to the liver. Cholestyramine is a positively charged resin that binds bile acids in the intestinal lumen and prevents their reabsorption.

   (a) To examine the effects of cholestyramine on LDL metabolism, two fractions of LDL were prepared: one was covalently labeled on tyrosine residues with ^{125}I; the other was labeled with ^{131}I and treated with cyclohexanedione, which interferes with binding to the LDL receptor. When rabbits were given cholestyramine, hepatic uptake of ^{125}I-labeled LDL was enhanced relative to normal, whereas the uptake of ^{131}I-labeled LDL was unchanged relative to that in rabbits that had not been given cholestyramine. Briefly explain the relationship between the action of cholestyramine and LDL uptake in the liver.
   (b) The administration of cholestyramine usually results in a 15 to 20% reduction in levels of circulating LDL, whereas the administration of a combination of cholestyramine and mevinolin can often yield a 30 to 40 percent reduction. Why?

4. The presence of apoprotein E in lipoproteins enables them to be taken up by hepatic cells. Provide a brief explanation for each of the following observations made of a person with a deficiency in apoprotein E synthesis.

(a) Elevated levels of plasma triacylglycerols and cholesterol, coupled with the presence of chylomicron remnants and IDL. These latter particles persist in the bloodstream much longer than in normal people.
(b) Abnormally low levels of LDL in the blood.
(c) Abnormally high levels of LDL receptors in liver cells.
(d) A marked reduction in levels of circulating chylomicron remnants and IDL when the diet is low in cholesterol and fat.

5. Pregnant women often have increased rates of triacylglycerol breakdown and, as a result, have elevated levels of ketone bodies in their blood. Why do they also often exhibit an increase in plasma lipoprotein levels?

6. Hopanoids, previously described in this student companion in problem 6 of Chapter 11, are pentacyclic molecules that are found in bacteria and in some plants. These organisms utilize a pathway similar to that of cholesterol synthesis to make hopanoids. The hopane biosynthetic pathway includes the formation of squalene, followed by the synthesis of hopane, a C_{30} compound. Hopane is similar to lanosterol, but lacks the hydroxyl group.

(a) How many molecules of mevalonic acid are required for the synthesis of hopane?
(b) Squalene can undergo concerted cyclization to form hopane in a reaction that is catalyzed by a unique type of squalene cyclase. The reaction is initiated by a proton and does not require oxygen. Compare this step with the formation of lanosterol from squalene. Why could it be argued that the synthesis of hopanoids preceded the synthesis of sterols in evolution?

7. Your colleague has discovered a compound that is a very powerful inhibitor of HMG CoA reductase, and she has evidence that the drug will completely block the synthesis of mevalonate in liver. Why is this compound unlikely to be useful as a drug?

8. The liver is the site of the synthesis of plasma phospholipids and lipoproteins. Rats maintained on a diet deficient in choline often develop fat deposits in liver tissue. How could choline deficiency be related to this aberration in lipid metabolism?

9. Among the sugar residues found in a blood group ganglioside is fucose. Experiments utilizing isolated Golgi membranes and ribonucleoside triphosphates show that fucose can be incorporated into the ganglioside only when GTP is available. What is the role of GTP in fucose incorporation?

10. Suppose a cell is deficient in phosphatidate phosphatase, which catalyzes the formation of diacylglycerol from phosphatidate. What effects on lipid metabolism would you expect?

11. (a) Would you expect virilization among patients who have desmolase deficiency?

(b) Why is enlargement of the adrenal glands common among such patients?

12. In the adult form of Gaucher's disease, glucosylcerebrosides accumulate in liver, spleen, and bone-marrow cells. Although the common galactosylceramides and their derivatives are found in the tissues of affected patients, accumulations of galactosylcerebrosides or their metabolites are not found, nor do ceramides accumulate. What enzyme activity is probably deficient in patients with Gaucher's disease?

13. Cells of the adrenal cortex have very high concentrations of LDL receptors. Why?

14. At low concentrations of phospholipid substrates in water, the reaction catalyzed by a phospholipase occurs at a rather low rate. The reaction rate accelerates when the concentrations of the phospholipid substrates increase to the point that micelles are formed. How is this property of phospholipases related to their activity in the cell?

15. Glucagon has been shown to reduce the activity of HMG CoA reductase. Why is this observation consistent with the overall effect of glucagon on cellular metabolism?

16. People who have elevated levels of low-density lipoproteins (LDL) in their serum can be treated in a number of ways. These include restriction of dietary intake of cholesterol, ingestion of positively charged resin polymers that inhibit intestinal reabsorption of bile salts, and administration of lovastatin, a competitive inhibitor of 3-hydroxy-3-methylglutaryl CoA reductase.

(a) Briefly explain how each of these treatments reduces serum LDL levels.
(b) None of the above treatments should be used for patients who are homozygous for a defect in LDL receptors? Why?

17. Currently there are two established methods for dealing with hypercholesterolemia in patients homozygous for LDL receptor deficiency.

(a) The first is plasma apheresis, whereby the plasma and blood cells of a patient are separated in a continuous flow device, and the plasma is passed over a column that removes lipoproteins containing apoprotein B-100. Which lipoproteins are removed by this procedure, and how could their removal lower plasma cholesterol levels?
(b) A more extreme method of dealing with extreme hypercholesterolemia in FH homozygotes is liver transplantation. The rationale for this procedure is based on the observation that over 70% of total body LDL receptors are in the liver. In the small group of patients who have undergone liver transplants, LDL cholesterol levels are substantially reduced. In one particular case, the rate of LDL turnover increased about threefold, and one patient became responsive to lovastatin after the transplant. Why did increased LDL turnover and lovastatin response indicate that the transplantation procedure was successful?

18. Provide a physiologic rationale for each of the following responses to an increase in the rate of transport of unesterified cholesterol into a mammalian cell:

(a) Stimulation of the synthesis of cholesteryl oleate esters.
(b) Suppression of the activity of HMG CoA reductase.
(c) Suppression of the synthesis of LDL receptors.

19. Glycerol phosphate acyltransferase can convert 3,4-dihydroxybutyl-1-phosphonate, an analog of glycerol 3-phosphate, to diacylbutyl-1-phosphonate, which is an analog of phosphatidate. Would this analog be more likely to inhibit triglyceride synthesis or CDP–diacylglycerol synthesis? Briefly explain your answer. See the accompanying marginal figure.

H_2C-OH, $HO-C-H$, $CH_2-CH_2-P(=O)(O^-)-$

3, 4–Dihydroxybutyl–1–phosphonate

20. During the uptake of low-density lipoprotein (LDL) by a liver cell, LDL–receptor protein complexes are internalized by endocytosis. The endosomes then fuse with lysosomes, where protein components of LDL are hydrolyzed to free amino acids, while cholesterol esters are hydrolyzed by a lysosomal acid lipase. The LDL receptor itself is not affected by lysosomal enzymes.

(a) Briefly describe what would happen to cholesterol metabolism in a cell deficient in lysosomal acid lipase.
(b) Why is it important that LDL receptors are not degraded by lysosomal enzymes?

21. Glycerol kinase catalyzes the conversion of free glycerol into glycerol 3-phosphate, using ATP as a phosphoryl donor. While liver tissue has high levels of the enzyme, the activity of glycerol kinase in adipose tissue is low. How do these differences contribute to the balance between carbohydrate and triglyceride metabolism in mammals?

22. Niemann-Pick disease is an inherited disorder of sphingomyelin breakdown due to a deficiency of sphingomyelinase. This enzyme, found in all tissues and easily assayed in white blood cells collected from a patient, converts sphingomyelin to ceramide and phosphocholine. Phosphocholine is highly soluble in water, while sphingomyelin is more soluble in chloroform. Assuming that you can obtain sphingomyelin labeled with ^{14}C in any desired carbon atoms, design an assay that would allow you to confirm a diagnosis of Niemann-Pick disease.

ANSWERS TO PROBLEMS

1. Citrate lyase catalyzes the formation of acetyl CoA in the cytosol (see p. 620 of Stryer's text). Acetyl CoA is utilized for the synthesis of HMG CoA, which gives rise to mevalonate for the synthesis of cholesterol in the cytosol.

2. (a) A deficiency of mevalonate kinase activity means that mevalonate cannot be utilized as a precursor of 5-phosphomevalonate. If there are no other pathways where mevalonate can be used, its concentration in the liver will increase until it spills over into the blood

and then into the urine. Furthermore, because the activity of HMG CoA reductase is increased in this infant (see answer c), the rate of mevalonate synthesis will be stimulated.

(b) You would expect the rate of cholesterol synthesis to be depressed because the pathway is blocked at the step in which 5-phosphomevalonate is formed.

(c) You would expect to find a higher-than-normal level of HMG CoA reductase activity. A depressed rate of cholesterol synthesis lowers the amount of cholesterol in the cell. HMG CoA reductase activity increases because inhibition of its synthesis and its activity by cholesterol is reduced.

3. (a) The experiments show that rabbits given cholestyramine have higher rates of removal of LDL from the blood and that hepatic uptake of LDL depends on the ability of the lipoprotein to bind to the LDL receptor. One explanation for this is that cholestyramine interferes with the return of bile acids to the liver, stimulating the synthesis of more bile acids from cholesterol. An increased demand for cholesterol stimulates the synthesis of LDL receptors, which take up more cholesterol-containing LDL particles from the blood.

(b) Although cholestyramine stimulates the hepatic uptake of cholesterol-containing LDL, it has no direct effect on de novo cholesterol synthesis in the liver. Mevinolin inhibits HMG CoA reductase, thereby depressing the rate of cholesterol biosynthesis. The subsequent requirement for cholesterol leads to a further increase in the number of LDL receptors, which in turn can take up more LDL from the circulation.

4. (a) Both chylomicron remnants and IDL normally contain apoprotein E. A deficiency of the apoprotein means that hepatic uptake of chylomicron remnants and IDL is impaired, so these particles persist in the circulation. Because both of these types of particles contain triacylglycerols and cholesterol, circulating levels of these compounds are also elevated.

(b) Both chylomicron remnants and IDL particles serve as precursors of VLDL in the liver. When the uptake of the VLDL precursors by hepatic tissue is impaired by an apoprotein E deficiency, the rate of synthesis and export of VLDL particles is reduced. Since VLDL are LDL precursors in circulation, LDL are reduced.

(c) As discussed in answer (b), VLDL synthesis in the liver is impaired. Additional VLDL receptors are synthesized because their synthesis is no longer repressed by VLDL-derived cholesterol.

(d) A diet low in cholesterol and fat will reduce the rate of formation of chylomicrons, which are precursors of chylomicron remnants.

5. Increased levels of ketone bodies, such as acetoacetate, imply that the levels of acetyl CoA and HMG CoA, both precursors of cholesterol, are also elevated. Cholesterol synthesis is stimulated by an increase in the availability of these substrates. The subsequent decreased demand for dietary cholesterol results in an elevation in cholesterol-containing lipoproteins.

6. (a) Mevalonic acid, a six-carbon compound, is a precursor of isopentenyl pyrophosphate (IPP), which contains five carbon atoms. IPP serves as the basic unit for the formation of squalene, a 30-carbon compound. Six molecules of IPP are needed for the synthesis of a molecule of squalene, which is in turn the precursor of hopane. Thus, six molecules of mevalonic acid are required.

(b) Aerobic processes, such as the synthesis of sterols, probably evolved later than anaerobic processes and only after free oxygen became available. Thus, the synthesis of hopane from squalene, an anaerobic process, probably preceded the synthesis of sterols, such as lanosterol, from squalene.

7. The synthesis of mevalonate is required not only for the synthesis of cholesterol but also for the synthesis of a number of other important compounds derived from isopentenyl pyrophosphate, including ubiquinone, an important component of the electron transport chain. Therefore, the complete blockage of mevalonate synthesis, even if adequate cholesterol is available in the diet, is not advisable.

8. Choline, which is synthesized only to a limited extent in mammals, is a constituent of phosphatidyl choline, an important component of membranes and lipoproteins. A deficiency in choline could interfere with the synthesis and export of lipoproteins like VLDL, which is a carrier of triacylglycerols to peripheral tissues. Failure to export fats such as triacylglycerols leads to their accumulation in the liver.

9. Nucleotide sugars, such as UDP–glucose, serve as donors during the incorporation of sugar residues into gangliosides. In this case, it appears that the donor of fucose residues is GDP–fucose, which is synthesized from fucose and GTP.

10. You would expect to see reduced rates of synthesis of triacylglycerols, which use diacylglycerols as acceptors of activated acyl groups. In addition, phosphatidyl choline synthesis is dependent on the availability of diacylglycerols as acceptors of choline phosphate from CDP–choline.

11. (a) You would not expect desmolase deficiency to lead to virilization. An increase in androgen production, which causes virilization in both males and females, is due to elevated levels of 17α-hydroxyprogesterone. The pathway from cholesterol to 17α-hydroxyprogesterone includes a step catalyzed by desmolase, which cleaves the bond between C-20 and C-22 in 20α, 22β dihydroxycholesterol to form pregnenolone (see Stryer, p. 705). A deficiency of desmolase would therefore decrease the rate of androgen synthesis.

(b) Pregnenolone is a precursor of the glucocorticoids, which exert feedback control on the activity of the adrenal cortex. Desmolase deficiency leads to diminished production of glucocorticoids. Failure of the normal feedback mechanism leads to increased ACTH production and to enlargement of the adrenal glands.

12. Because glucosylcerebrosides accumulate but galactosylcerebrosides do not, you would suspect that the defect involves ganglioside breakdown rather than ganglioside synthesis. The defect involves the step that removes glucose from the cerebroside to yield free ceramide, or *N*-acyl

sphingosine. The enzyme that carries out this step is a glycosyl hydrolase; it is also called *β-glucosidase.*

13. Cells of the adrenal cortex utilize cholesterol for the synthesis of a number of steroid hormones, including cortisol. Although these cells can themselves synthesize cholesterol, it is often also necessary for additional cholesterol to be obtained from plasma lipoproteins. A high concentration of LDL receptors enables cortical cells to take up LDL, which contains cholesterol, rapidly.

14. In the cell, a phospholipase would most often encounter substrates that are part of an aggregate, such as those phospholipids found in membranes. Thus, the enzyme should be expected to function at a higher rate with aggregates or assemblies of lipid molecules.

15. The presence of glucagon is a signal that carbohydrate and triacylglycerol catabolism is needed to generate energy in the organism. Under such conditions, one would expect biosynthetic reactions to be suppressed. Glucagon stimulates cAMP production, which in turn activates a protein kinase. Phosphorylation of HMG CoA reductase by the kinase suppresses cholesterol synthesis.

16. (a) Restricting the level of dietary cholesterol lowers the input of exogenous cholesterol into lipoproteins, so that fewer LDL molecules are present in serum. Compounds that inhibit bile salt reabsorption stimulate additional synthesis of bile acids from cholesterol in the liver, decreasing concentrations of the sterol in liver cells and, by causing an increase in the number of receptors on the cell surface, stimulating LDL uptake from the circulation. Finally, mevinolin reduces the rate of de novo cholesterol biosynthesis, which can also stimulate uptake of LDL from the bloodstream. Any of these treatments could help in reducing circulating levels of cholesterol as LDL.

(b) Homozygotes have virtually no functional LDL receptors. Such people are unable to internalize significant amounts of LDL, which means that circulating levels of that lipoprotein are elevated in the blood. In addition, an absence of LDL receptors means that there is no introduction of endogenous cholesterol to the liver cell to suppress de novo synthesis. Dietary restriction could reduce exogenous LDL levels somewhat but would not prevent formation of LDL and other lipoproteins arising from cell turnover of cholesterol. Bile sequestrants could cause some depletion of liver cell cholesterol but again would not stimulate LDL uptake from the circulation. Mevinolin will suppress de novo cholesterol synthesis, but this would again not be compensated for by uptake from the circulation. Homozygotes with two nonfunctional receptor genes are therefore resistant to compounds that inhibit LDL synthesis and stimulate uptake. Thus none of these measures have a great effect on reducing circulating LDL levels in these patients.

17. (a) Protein B-100 is found in VLDL, IDL, and LDL, and all three lipoproteins are removed from the plasma. Each contains cholesterol, and, in addition, VLDL and IDL are regarded as precursors of LDL, so that total cholesterol concentration in the blood would be lowered by apheresis. This method can reduce LDL choles-

terol levels by 70%. Treatment must be repeated about every two weeks, and the effects of long-term apheresis are unknown.

(b) Increased LDL turnover indicates that the transplanted liver is producing normal LDL receptors which can take up LDL from the circulation and can facilitate its conversion to other lipoproteins. A response to lovastatin also indicates that functional LDL receptors are available to accelerate uptake in response to diminished de novo cholesterol synthesis, which is inhibited by lovastatin. It should be noted that liver transplant operations are hazardous, especially in FH homozygotes who usually have advanced atherosclerosis. The fact that normal liver cells can contribute functional LDL receptors has stimulated interest in gene therapy designed to target a normal LDL gene to liver cells of FH homozygotes.

18. (a) Formation of cholesteryl esters provides the cell with a means of storing cholesterol until it is needed for membrane biosynthesis or other purposes.

(b) Suppression of the activity of HMG CoA reductase leads to a decrease in de novo cholesterol synthesis; this occurs when the cell has sufficient endogenous cholesterol so that it does not need to synthesize the steroid on its own.

(c) Suppression of LDL receptor synthesis leads to a gradual reduction in the number of LDL receptors in the cell, because the receptors undergo a relatively constant rate of degradation. Reduction in LDL receptor levels means that fewer LDL particles will be able to enter the cell, resulting in a decrease in the entry of endogenous cholesterol.

19. The conversion of phosphatidate to a triacylglycerol is initiated by hydrolysis of the phosphate group and the formation of diacylglycerol. The C-P bond in the phosphonate is less likely to be cleaved by the phosphatase (also known as *phosphatidate phosphohydrolase*), so it is likely that triglyceride synthesis could be impaired by the analog. On the other hand, formation of CDP–diacylglycerol involves the formation of an anhydride bond between the phosphates of phosphatidic acid and cytidylic acid (see Stryer, p. 686). Since the phosphonate is not cleaved in this reaction, formation of the phosphonyl analog of CDP–diacylglycerol seems possible, or at least the synthesis of normal substrates would not be impaired. Phosphonolipids are known in trace amounts in mammals, but are found more extensively in some invertebrates, including the protozoan *Tetrahymena,* where they may represent up to 25% of total phospholipid [see D.E. Vance and J. Vance (eds.), *Biochemistry of Lipids, Lipoproteins and Membranes.* (Elsevier, 1991), p. 205].

20. (a) Both the LDL receptor gene and the gene for HMG CoA reductase contain sterol regulatory elements that are responsive to free cholesterol. A reduction in free cholesterol release from the lysosome leads to an increase in LDL receptor production and to increased HMG CoA reductase activity. Both these consequences lead to an increase in cholesterol concentrations in the cell, through an increased rate of LDL entry and accelerated cholesterol synthesis. Lysosomal accumulation of cholesteryl esters and triglycerides can eventually destroy the cell. One form of this

disorder, termed *Wolman disease,* is characterized by liver enlargement, digestive difficulties, and enlargement and deterioration of the adrenal glands. The disease is usually fatal within a year after birth.

(b) LDL receptors, after their release from lysosomes, return to the cell surface, where they take up other LDL particles and bring them back to lysosomes. A round trip takes about 10 minutes. Destruction of LDL receptors by lysosomal enzymes would make it necessary to synthesize the 115-kd glycoprotein at a much faster and energetically wasteful rate.

21. Free glycerol in mammals is produced in adipocytes when triglycerides are converted to free fatty acids and glycerol by hormone-sensitive lipases. Hormone-responsive lipolysis occurs when glucose levels are low and fatty acids are needed as fuels. Under these conditions, the low activity of glycerol kinase in adipocytes prevents unnecessary resynthesis of triacylglycerols from free fatty acids and glycerol 3-phosphate, via phosphatidic acid. Triacylglycerol synthesis is more likely to occur when glucose levels are high. Adipocytes can then synthesize glycerol 3-phosphate from dihydroxyacetone phosphate produced from glucose during glycolysis. Thus adipocytes are unable to synthesize triglycerides unless glucose or another suitable carbon source is available.

22. Extracts of white cells from a person suspected of having a sphingomyelinase deficiency are incubated in a buffered solution with radioactive sphingomyelin labeled with ^{14}C in the methyl groups of the phosphocholine moiety. Then the incubation mixture is extracted with chloroform. Any radioactive phosphocholine liberated by the enzyme will remain in the upper aqueous layer, while intact radioactive sphingomyelin will be extracted into the lower chloroform layer. Using white cells from patients with Neimann-Pick disease, incubation with cell extracts followed by chloroform extraction results in little or no radioactivity in the aqueous phase, confirming the deficiency of sphingomyelinase activity.

EXPANDED SOLUTIONS TO TEXT PROBLEMS

1. $$\text{Glycerol} + 4\ \text{ATP} + 3\ \text{fatty acids} + 4\ H_2O \longrightarrow \text{triacylglycerol} + \text{ADP} + 3\ \text{AMP} + 7\ P_i + 4\ H^+$$

One ATP is used in the formation of glycerol 3-phosphate and three ATPs are used to convert three fatty acids to acyl CoAs (see p. 607 in Stryer). The three PPi formed during fatty acid activation are converted to P_i, hence, the total of seven P_i in the equation above.

2. $$\text{Glycerol} + 3\ \text{ATP} + 2\ \text{fatty acids} + 2\ H_2O + \text{CTP} + \text{serine} \longrightarrow \text{phosphatidyl serine} + \text{CMP} + \text{ADP} + 2\ \text{AMP} + 6\ P_i + 3\ H^+$$

The equation above is a summation of equations shown in Stryer, pp. 686–687.

3. (a) CDP–diacylglycerol, (b) CDP–ethanolamine, (c) acyl CoA, (d) CDP–choline, (e) UDP–glucose or UDP–galactose, (f) UDP–galactose, and (g) geranyl pyrophosphate.

4. (a and b) None, because the label is lost as CO_2.

5. (a) No receptor is synthesized.

(b) Receptors are synthesized but do not reach the plasma membrane, because they lack signals for intracellular transport or do not fold properly.

(c) Receptors reach the cell surface, but they fail to bind LDL normally because of a defect in the LDL-binding domain.

(d) Receptors reach the cell surface and bind LDL, but they fail to cluster in coated pits because of a defect in their carboxyl-terminal region.

6. Deamination of cytidine to uridine changes CAA (Gln) into UAA (stop).

7. Benign prostatic hypertrophy can be treated by inhibiting the 5α-reductase. *Finasteride,* the 4-aza steroid analog of dihydrotestosterone, competitively inhibits the reductase but does not act on androgen receptors. Patients taking finasteride have a markedly lower plasma level of dihydrotestosterone and a nearly normal level of testosterone. The prostate becomes smaller, whereas testosterone-dependent processes such as fertility, libido, and muscle strength appear to be unaffected (see E. Stoner, *Steroid Biochem. Molec. Biol.* 37(1990):375–378). Genetic deficiencies of 5α-reductase are discussed by J.E. Griffin and J.D. Wilson in C.R. Scriver, A.L. Beaudet, W.S. Sly, and D. Valle (eds.), *The Metabolic Basis of Inherited Disease,* 6th ed. (McGraw-Hill, 1989), pp. 1919–1944.

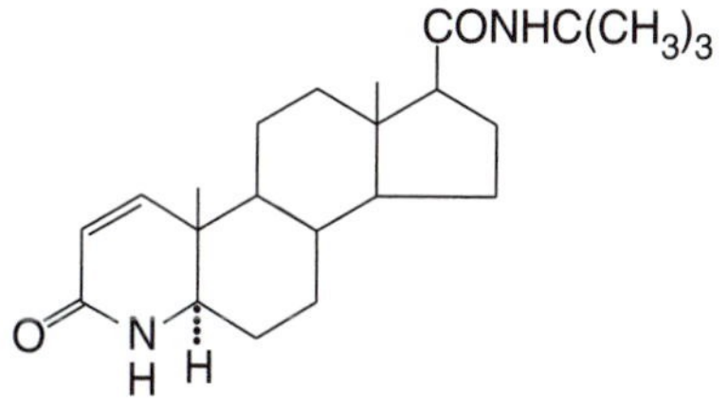

$CONHC(CH_3)_3$

O N H H

Finasteride

8. Patients who are most sensitive to debrisoquine have a deficiency of a liver P450 enzyme encoded by a member of the *CYP2* subfamily. This characteristic is inherited as an autosomal recessive trait. The capacity to degrade other drugs may be impaired in people who hydroxylate debrisoquine at a slow rate because a single P450 enzyme usually handles a broad range of substrates. See W.B. Pratt and P. Taylor, *Principles of Drug Action: The Basis of Pharmacology,* 3rd ed. (Churchill Livingstone, 1990), pp. 496–500; and F.J. Gonzalez and D.W. Nebert. *Trends Genet.* 6(1990):182.

9. Deficiencies of 11β-hydroxylase, 17α-hydroxylase, 3β-dehydrogenase, and desmolase also lead to adrenal hyperplasia. See M.I. New, P.C. White, S. Pang, B. Dupont, and P.W. Speiser in C.R. Scriver, A.L. Beaudet, W.S. Sly, and D. Valle (eds.), *The Metabolic Basis of Inherited Disease,* 6th ed. (McGraw-Hill, 1989), pp. 1181–1917.

10. Many hydrophobic odorants are deactivated by hydroxylation. O_2 is activated by a cytochrome P450 monooxygenase. NADPH serves as the reductant. One oxygen atom of O_2 goes into the odorant substrate, whereas the other is reduced to water.

CHAPTER

28

Biosynthesis of Amino Acids and Heme

In this chapter, Stryer explains the biosynthetic origins of the amino acids. He deals first with how atmospheric nitrogen in the form of N_2 is converted to NH_3 by the process of nitrogen fixation involving a molybdenum-iron cofactor. He then explains how NH_4^+ is incorporated into the covalent structures of the amino acids glutamate and glutamine. Next, a major portion of the chapter discusses how the individual carbon atom skeletons of most of the amino acids are constructed from a few common intermediates and how nitrogen from glutamate and glutamine is incorporated into their structures, primarily in α-amino groups. He explains that the lack of some biosynthetic pathways in humans has led to the dietary requirement for nine amino acids. The roles of tetrahydrofolate and *S*-adenosylmethionine as carriers of single carbon atoms during metabolism are also explained. Stryer includes a general discussion of the control of metabolic pathways and uses examples from amino acid metabolism to illustrate the relevant principles.

Stryer emphasizes the important role of the amino acids as precursors of many biomolecules by listing several such derivatives. He explains the synthesis of glutathione, a sulfhydryl buffer and detoxifying agent, and nitric oxide, a short-lived signal molecule. He concludes the chapter with a discussion of the formation and degradation of another important nitrogen containing molecule—heme. A review of the amino acid structures in Chapter 2, heme structure in Chapter 7, enzyme control in Chapter 10, and amino acid catabolism in Chapter 25 will facilitate your understanding of this chapter.

LEARNING OBJECTIVES

When you have mastered this chapter, you should be able to complete the following objectives:

The Conversion of N_2 into Organic Nitrogen (Stryer pages 714–717)

1. Define *nitrogen fixation* and name the groups of organisms that can carry out this conversion.

2. Describe the *nitrogenase complex* and explain the roles of its *reductase* and *nitrogenase* components. Note the function of the *FeMo-cofactor.*

3. Explain the energy requirement for nitrogen fixation and write the equation giving the *stoichiometry* of the overall reaction.

4. Outline the key roles of *glutamate* and *glutamine* in the assimilation of NH_4^+ into amino acids and describe the reactions of *glutamate dehydrogenase, glutamine synthetase,* and *glutamate synthase.* Recognize the functions of *ATP* and *NADPH* in these processes.

Amino Acid Biosyntheses and One-Carbon-Atom Metabolism (Stryer pages 717–726)

5. Identify the *essential* amino acids for humans and explain why they are essential.

6. Classify the amino acids into six *biosynthetic families* and identify their *seven precursors.* Name the metabolic pathways from which these precursors originate.

7. Describe the single-step biosyntheses of alanine, aspartate, asparagine, and tyrosine.

8. Outline the syntheses of glutamate, glutamine, proline, and arginine from *α-ketoglutarate.*

9. Outline the syntheses of serine, glycine, and cysteine from *3-phosphoglycerate.*

10. Explain the roles of *pyridoxal phosphate, tetrahydrofolate,* and S-*adenosylmethionine* in amino acid biosyntheses.

11. Identify the structure of tetrahydrofolate and indicate the reactive part of the molecule. Describe the sources of this cofactor in humans.

12. Draw the structures of the single-carbon groups that can be carried on tetrahydrofolate and provide examples of reactions that generate and use them.

13. Draw the structure of *S*-adenosylmethionine and describe its synthesis. Indicate the reactive part of the molecule and describe the basis of its *high methyl group–transfer potential.*

14. Outline the *activated methyl cycle* and describe the roles of *methylcobalamin* and *ATP* in the cycle. Give examples of important derivatives from S-adenosylmethionine.

15. Describe the synthesis of cysteine from *homocysteine* and serine.

16. Outline the biosyntheses of phenylalanine, tyrosine, and tryptophan in *E. coli.* Describe the roles of *phosphoenolpyruvate, erythrose 4-phosphate,* and *phosphoribosylpyrophosphate* in these reactions.

17. Describe the structure of *tryptophan synthetase* and the role of *substrate channeling* in its catalytic reaction.

Control of Amino Acid Biosynthesis (Stryer pages 727–730)

18. Define the *committed step* of a metabolic pathway and recognize that it is often the target of *feedback regulation.* Note the main features of control by *sequential feedback, enzyme multiplicity, concerted feedback,* and *cumulative feedback.*

19. Describe the cumulative feedback control of *glutamine synthetase* from *E. coli.* Explain the mechanisms and functions of the *reversible covalent modifications* and describe the advantage of employing an *enzymatic cascade* in regulating this reaction.

Amino Acids as Versatile Precursors of Biomolecules (Stryer pages 730–736)

20. Provide examples of important biomolecules that are derived from amino acids.

21. Draw the structure of *glutathione* and describe its synthesis. Indicate the functions of glutathione and describe the involvement of *selenium* in the glutathione peroxidase reaction.

22. Outline the synthesis of *nitric oxide* (*NO*) and explain its function.

23. Name the two molecular precursors of the *porphyrins* in mammals and outline the synthesis of *heme.* Identify the key regulatory points and mechanisms in heme biosynthesis.

24. Explain the molecular defect in *congenital erythropoietic porphyria.*

25. Outline the catabolism of heme to *bilirubin* and describe the involvement of *UDP–glucuronate* in converting bilirubin to an excretory form in humans. Explain why *biliverdin* is reduced to bilirubin in mammals.

SELF-TEST

The Conversion of N_2 into Organic Nitrogen

1. Define nitrogen fixation and explain why it is crucial to the maintenance of life on earth.

____

____

2. Place the following components, reactants, and products of the nitrogenase complex reaction in their correct sequence during the electron transfers of nitrogen fixation:

(a) Oxidized ferredoxin ________

(b) Reductase component ________

(c) Nitrogenase component ________

(d) NH_3 ________

(e) N_2 ________

(f) Reduced ferredoxin ________

(g) Electron source ________

3. Write the net equation for nitrogen fixation and describe the sources of the electrons and ATP.

____

____

4. Match the structural components or features in the right column with the appropriate component of the nitrogenase reaction.

(a) Reductase component ________

(b) Nitrogenase component ________

(1) MoFe-cofactor
(2) [4Fe-4S] cluster
(3) ATP-ADP binding site
(4) N_2-binding site
(5) $\alpha_2\beta_2$ tetramer
(6) Dimer of identical subunits

5. Match the enzyme with the reaction it catalyzes.

(a) Glutamine synthetase ________

(b) Glutamate dehydrogenase ________

(c) Glutamate synthase ________

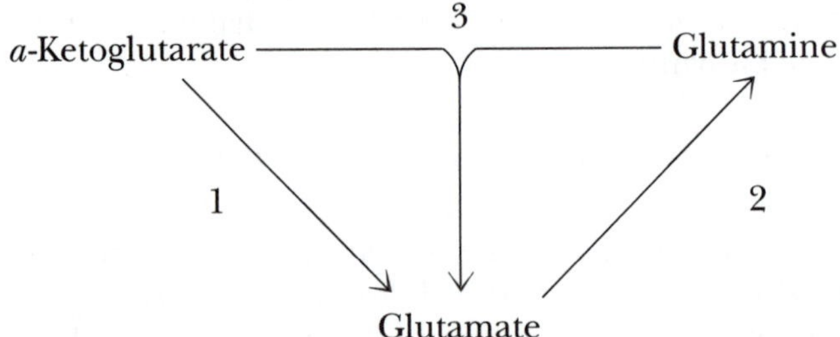

6. Which of the reactions shown in question 5 require the following?

(a) NH_4^+ ________
(b) ATP ________
(c) NADH ________
(d) NADPH ________

7. All organisms can incorporate NH_4^+ into glutamate and glutamine using glutamate dehydrogenase and glutamine synthetase. Why do pro-

karyotes have an additional enzyme, glutamate synthase, to perform this function?

Amino Acid Biosynthesis and One-Carbon-Atom Metabolism

8. Propose a simple pathway to generate ^{14}C-labeled aspartate from pyruvate that has all its carbon atoms radiolabeled.

---

9. Which of the following amino acids are essential dietary components for an adult human?

(a) Alanine
(b) Aspartate
(c) Histidine
(d) Tryptophan
(e) Leucine
(f) Phenylalanine
(g) Glutamine
(h) Asparagine
(i) Glutamate
(j) Threonine
(k) Methionine

10. Which of the following amino acids are derived from pyruvate?

(a) Phenylalanine
(b) Alanine
(c) Tyrosine
(d) Histidine
(e) Valine
(f) Leucine
(g) Cysteine
(h) Glycine

11. Which of the following amino acids are derived from α-ketoglutarate?

(a) Glutamate
(b) Proline
(c) Cysteine
(d) Aspartate
(e) Glutamine
(f) Arginine
(g) Ornithine
(h) Serine

12. Which of the following compounds provide the carbon skeletons of the six biosynthetic families of amino acids?

(a) Pyruvate
(b) Oxaloacetate
(c) α-Ketoglutarate
(d) Succinate
(e) 2-Deoxyribose
(f) 3-Phosphoglycerate
(g) Ribose 5-phosphate
(h) Glucose 6-phosphate
(i) Phosphoenolpyruvate
(j) Erythrose 4-phosphate
(k) α-Ketobutyrate

13. Refer to question 12. Name the metabolic pathways from which each of the precursor compounds originates.

---

14. Three coenzymes are involved in carrying activated one-carbon units. Match the activated group in the right column with the appropriate coenzyme in the left column.

(a) Tetrahydrofolate _______________
(b) *S*-Adenosylmethionine __________
(c) Biotin _______________________

(1) $-CH_3$
(2) $-CH_2-$
(3) $-CHO$
(4) $-CHNH$
(5) $-CH=$
(6) $-CO_2^-$

15. Which of the following answers completes the sentence correctly? The major source of one-carbon units for the formation of the tetrahydrofolate derivative N^5,N^{10}-methylenetetrahydrofolate is the conversion of

(a) methionine to homocysteine.
(b) deoxyuridine 5′-phosphate to deoxythymidine 5′-phosphate.
(c) 3-phosphoglycerate to serine.
(d) serine to glycine.

16. Why does *S*-adenosylmethionine have a higher methyl group–transfer potential than N^5-methyltetrahydrofolate?

____

____

17. *S*-adenosylmethionine is involved directly in which of the following reactions?

(a) Methyl transfer to phosphatidyl ethanolamine
(b) Synthesis of glycine from serine
(c) Synthesis of polyamines
(d) Conversion of homocysteine into methionine
(e) Synthesis of ethylene in plants
(f) Methyl transfer to glutamate residues of bacterial chemotaxis receptors

18. How many high-energy bonds are expended during the synthesis of *S*-adenosylmethionine from ATP and methionine?

____

____

19. The conversion of homocysteine into methionine involves which of the following cofactors?

(a) N^5-Methyltetrahydrofolate
(b) N^5,N^{10}-Methylenetetrahydrofolate
(c) Methylcobalamin
(d) Pyridoxal phosphate

20. Fill the empty boxes in the pathways outlined in Figure 28-1 using the following compounds:

(1) *S*-Adenosylhomocysteine
(2) Serine
(3) N^5,N^{10}-Methylenetetrahydrofolate
(4) N^5-Methyltetrahydrofolate
(5) Cysteine
(6) *S*-Adenosylmethionine
(7) ATP

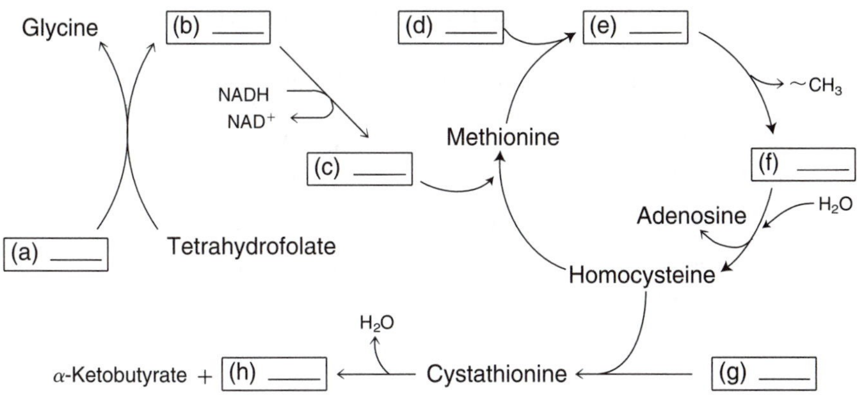

Figure 28-1
Some reactions of one-carbon metabolism.

21. Which of the following are aromatic amino acids?

(a) Leucine
(b) Phenylalanine
(c) Tryptophan
(d) Tyrosine
(e) Threonine

22. What is the ultimate source of most of the essential amino acids required by humans?

____

____

23. Which of the following are intermediates in the pathway for the biosynthesis of both phenylalanine and tryptophan?

(a) Anthranilate
(b) Chorismate
(c) Shikimate
(d) Prephenate

24. The two binding sites for indole on tryptophan synthetase subunits α and β are about 25 Å apart. Explain how indole is transferred between these sites.

____

____

Control of Amino Acid Biosynthesis

25. In the following biosynthetic pathway

$$A \longrightarrow B \longrightarrow C \longrightarrow D \longrightarrow E \longrightarrow F \longrightarrow G$$

which is likely to be the committed step?

____

26. In the biosynthetic pathway in question 25, which compound is likely to inhibit the committed step?

____

27. Since glutamine is an important source of nitrogen in biosynthetic reactions, the enzyme that synthesizes it is carefully regulated. Which of the following compounds act as inhibitors of glutamine synthetase in *E. coli*?

(a) Tryptophan
(b) Histidine
(c) Carbamoyl phosphate
(d) Glucosamine 6-phosphate
(e) AMP
(f) CTP
(g) Alanine
(h) Glycine

28. How does covalent modification contribute to the regulation of glutamine synthetase in *E. coli*?

____

____

Amino Acids as Versatile Precursors of Biomolecules

29. Glutathione is composed of which of the following amino acids?

(a) Glutamine
(b) Glutamate
(c) Methionine
(d) Cysteine
(e) Glycine

30. Which of the following answers complete the sentence correctly? Glutathione

(a) cycles between oxidized and reduced forms in the cell.
(b) is involved in the detoxification of H_2O_2 and organic peroxides.
(c) donates amide groups from its γ-glutamyl residue during biosynthetic reactions.
(d) contains a Se atom.

31. Which of the following statements about nitric oxide (NO) is correct?

(a) It is used as an inhalation anesthetic.
(b) It is a long-lived signal molecule.
(c) It is produced from asparagine.
(d) Its synthesis requires NADPH and O_2.
(e) Its synthesis requires ATP and NH_4^+.

32. Which of the following are intermediates or precursors in the synthesis of heme?

(a) δ-Aminolevulinic acid
(b) Bilirubin
(c) Porphobilinogen
(d) Biliverdin
(e) Glycine
(f) Succinyl CoA
(g) Fe^{2+}

33. Which reaction of heme metabolism releases carbon monoxide?

____

____

34. Which of the following compounds are effective antioxidants in blood?

(a) Biliverdin
(b) Bilirubin
(c) Protoporphyrin IX
(d) Ascorbate
(e) Urate

ANSWERS TO SELF-TEST

The Conversion of N_2 into Organic Nitrogen

1. Nitrogen fixation is the process by which nitrogen present in the atmosphere as N_2 is enzymatically converted to NH_3 by some bacteria and blue-green algae. This process is crucial to all other organisms because they can use only NH_4^+, and not N_2, as the source of nitrogen for biosynthesis.

2. g, a, f, b, c, e, d

3. The net equation for nitrogen fixation is

$$N_2 + 8\,e^- + 16\,\text{ATP} + 16\,H_2O \longrightarrow 2\,NH_3 + 16\,\text{ADP} + 16\,P_i + 8\,H^+ + H_2$$

The eight electrons needed to reduce N_2 are supplied by oxidative processes in nonphotosynthetic nitrogen-fixing organisms and by light energy from the sun in photosynthetic nitrogen-fixing organisms. The ATP requirement is met by the usual oxidative or photosynthetic mechanisms of the cells.

4. (a) 2, 3, 6 (b) 1, 2, 4, 5

5. (a) 2 (b) 1 (c) 3

6. (a) 1, 2 (b) 2 (c) None (d) 1, 3. Glutamate dehydrogenase uses NADPH when catalyzing reductive aminations and NAD^+ when carrying out oxidative deaminations.

7. Glutamate synthase catalyzes the reductive amination of α-ketoglutarate in a reaction with glutamine to form two glutamates. Glutamate can also be made from NH_4^+ and α-ketoglutarate using glutamate dehydrogenase. However, this route requires high concentrations of NH_4^+ because of the high K_M of the enzyme for NH_4^+. Prokaryotes can use glutamine synthetase, which has a low K_M for NH_4^+, to form glutamine when NH_4^+ concentrations are low. Thus, by using an additional enzyme, they can form glutamate from glutamine and α-ketoglutarate when NH_4^+ is scarce.

8. Pyruvate is converted to acetyl CoA by pyruvate dehydrogenase. Acetyl CoA condenses with oxaloacetate and passes through the TCA cycle to form labeled oxaloacetate, which undergoes a transamination reaction with glutamate to form labeled aspartate. Recall that the label in the acetyl portion of acetyl CoA is not eliminated as $^{14}CO_2$ in a single pass through the cycle. Alternatively, pyruvate carboxylase (Stryer, p. 571 forms oxaloacetate directly from pyruvate. Again, transamination with glutamate leads to labeled aspartate, but in this case, with a different pattern of labeled atoms.

9. c, d, e, f, j, k

10. b, e, f

11. a, b, e, f, g. Recall that ornithine is a precursor of arginine in the urea cycle and is derived from α-ketoglutarate.

12. a, b, c, f, g, i, j

13. Glycolysis: pyruvate (a)
3-phosphoglycerate (f)
phosphoenolpyruvate (i)

Citric acid cycle: oxaloacetate (b)
α-ketoglutarate (c)

Pentose phosphate pathway: ribose 5-phosphate (g)
erythrose 4-phosphate (j)

14. (a) 1, 2, 3, 4, 5 (b) 1 (c) 6

15. d. Furthermore, since 3-phosphoglycerate can give rise to serine, you can see how carbohydrates can provide activated one-carbon units via a glucose ⟶ 3-phosphoglycerate ⟶ serine ⟶ glycine pathway.

16. The positive charge on the sulfur atom of *S*-adenosylmethionine activates the methyl sulfonium bond and makes methyl group transfer from *S*-adenosylmethionine energetically more favorable than from N^5-methyltetrahydrofolate.

17. a, c, e, f. The cofactor for reactions (b) and (d) is tetrahydrofolate.

18. Three high-energy bonds are expended. The adenosyl group of ATP is condensed with methionine to form a carbon-to-sulfur bond with the release of P_i and PP_i, which is hydrolyzed to 2 P_i.

19. a, c. Homocysteine transmethylase uses a vitamin B_{12}–derived cofactor.

20. (a) 2 (b) 3 (c) 4 (d) 7 (e) 6 (f) 1 (g) 2 (h) 5

21. b, c, d

22. The essential amino acids in the diet of humans are ultimately derived from plants.

23. b, c

24. Tryptophan synthetase contains a 25-Å-long channel between the active sites of adjacent α and β subunits. This channel allows the diffusion of the intermediate, indole, through the protein from one binding site to the other.

Control of Amino Acid Biosynthesis

25. A → B. Control of the first step conserves the first compound, A, in the sequence and also saves metabolic energy by preventing subsequent reactions in the pathway.

26. G. The end product of a biosynthetic pathway often controls the committed step.

27. All the choices are correct. When all eight compounds are bound to the enzyme, it is almost completely inactive. The control of this enzyme is an excellent example of cumulative feedback inhibition.

28. Glutamine synthetase can be covalently modified by the attachment of an AMP to each of its 12 subunits. The more adenylylated the enzyme becomes, the more susceptible it is to feedback inhibition by the compounds listed in question 27. Thus, covalent modification modulates the sensitivity of the enzyme to its effectors. An added level of control exists in this system; adenylyltransferase, the enzyme that adenylylates glutamine synthetase, is itself covalently modified. See pages 729–730 of Stryer for a description of how a cascade of modifications such as this can amplify the signals embodied in the changing levels of the effectors.

Amino Acids as Versatile Precursors of Biomolecules

29. b, d, e

30. a, b. Answer (d) is incorrect because the enzyme glutathione peroxidase, rather than glutathione, contains a Se analog of cysteine.

31. d. Note that (c) is incorrect because arginine rather than asparagine is the precursor of NO.

32. a, c, e, f, g

33. Heme oxygenase uses O_2 and NADPH to open the tetrapyrrole heme structure and, in the process, releases one of the methene-bridge carbon atoms as CO. This is the first step in a pathway that leads to the formation of a diglucuronide derivative of bilirubin that is excreted in the bile.

34. b, d, e

PROBLEMS

1. The *glyA*⁻ mutation in Chinese hamster ovary cells in tissue culture makes these cells partially dependent on glycine. The mutation affects the mitochondrial form of serine transhydroxymethylase, which catalyzes the conversion of serine to glycine, with tetrahydrofolate serving as an acceptor of the hydroxymethyl group.

(a) Would you expect heme synthesis to be adversely affected in *glyA*⁻ mutants? Why?
(b) These cells appear to grow and divide much faster when choline is added to the medium. Why?

2. Hereditary coproporphyria (HCP) is a form of hereditary hepatic porphyria that is caused by a partial deficiency of coproporphyrinogen oxidase, which converts coproporphyrinogen III to protoporphyrin IX. The symptoms include abdominal pain, vomiting, and neuropsychiatric disorders.

(a) Why are patients with a total deficiency of coproporphyrinogen oxidase unlikely to be found?
(b) Which of the intermediates in heme biosynthesis would you expect to accumulate in patients with this disorder? What products would have lower than normal concentrations? Explain your answer not only in terms of the enzyme deficiency but also in terms of the regulation of heme synthesis.
(c) Why would you expect patients with HCP to be extremely sensitive to light?
(d) Why should patients with HCP not be given barbiturates or steroid drugs?

3. In early nutritional studies, cysteine was thought to be an essential amino acid. In 1937, Abraham White and E.F. Beach showed that cysteine could be removed from protein hydrolysates with cuprous mercaptide. Rats fed on such treated hydrolysates could grow, provided that sufficient methionine was supplied in the diet.

(a) What did this result reveal about cysteine metabolism?
(b) What would you expect the result to be if the rats were fed homocysteine along with the treated hydrolysates?

4. The Crigler-Najjar syndrome is characterized by the lack of glucuronyl transferase, which catalyzes the attachment of glucuronate units to the two propionate side chains of bilirubin. Infants with this disorder have high concentrations of bilirubin and its breakdown products in their skin, and their serum concentrations of bilirubin are 20 to 50 times above normal, but no bilirubin is found in the bile of these infants. How would you explain these observations?

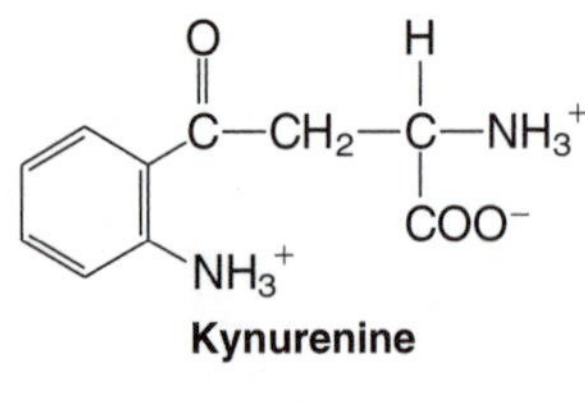

Kynurenine

3-Hydroxykynurenine

5. Deficiencies in pyridoxine or in pyridoxal phosphate (PLP) are relatively rare because these compounds are widespread in the normal diet. Patients that have one of these deficiencies frequently exhibit symptoms related to pellagra, a disease often ascribed to an insufficiency of niacin or nicotinamide. In addition, there is often an accumulation of the compounds kynurenine and hydroxykynurenine in people with PLP deficiency. (The structure of these two compounds is shown in the margin.) Describe the relationships among pyridoxal phosphate, kynurenine, and pellagra.

6. Blood from rabbits that have been made anemic in order to produce a high percentage of reticulocytes can be used to study the biosynthesis of heme in vitro. Isotopically labeled 3-phosphoglycerate with ^{14}C at C-1, C-2, or C-3 can be prepared; the three possible structures are shown in the margin. Assuming that reticulocytes can synthesize the appropriate precursors for heme synthesis from glycolytic intermediates, which labeled form of 3-phosphoglycerate will yield a ^{14}C-labeled, nitrogen-containing compound that will be incorporated into heme with the least dilution of the isotope?

$^{14}COO^-$ — H—C—OH — CH_2 — OPO_3^{2-}

1-^{14}C-3-Phosphoglycerate

COO^- — H—^{14}C—OH — CH_2 — OPO_3^{2-}

2-^{14}C-3-Phosphoglycerate

COO^- — H—C—OH — $^{14}CH_2$ — OPO_3^{2-}

3-^{14}C-3-Phosphoglycerate

7. The essential amino acids are those that cannot be synthesized de novo in humans. Given an abundance of other amino acids in the diet, the α-keto acid analogs that correspond to the essential amino acids can substitute for these compounds in the diet.

(a) What do these observations tell you about the steps in the synthesis of essential amino acids that may be missing in humans?
(b) If ^{15}N-labeled alanine is supplied in the diet, many other amino acids in the body will contain at least a small amount of the label within 48 hours. What enzymes are primarily responsible for this observation?

8. In muscle, glutamine synthetase is very active, catalyzing the formation of glutamine from glutamate and ammonia at the expense of a molecule of ATP. In the liver, the rate of formation of glutamine is very low, but a high level of glutaminase activity, which generates ammonia and glutamate, is observed. How would you explain the difference in the levels of enzyme activity in these two organs?

9. Consider three forms of bacterial glutamine synthetase: GS, the deadenylylated form; GS–(AMP)$_1$, a form with one AMP unit per 12 subunits; and GS–(AMP)$_{12}$, the fully adenylated form.

(a) Which of these forms is most sensitive to feedback inhibition by several of the final products of glutamine metabolism, such as tryptophan or histidine? Why is it important that the activity of the most sensitive form not be *completely* inhibited by tryptophan?
(b) Which form has the lowest K_M for ammonia?
(c) Why is it important that adenylyl transferase not carry out adenylylation and deadenylylation of glutamine synthetase at the same time?
(d) Glutamine synthetase in mammals is not subject to the same type of complex regulation that is seen in bacteria. Why?

10. For every acetyl CoA introduced into the citric acid cycle, two molecules of CO_2 are released during a complete turn of the cycle. Yet, ^{14}C-labeled acetyl CoA can be used in an in vitro system for the synthesis of radioactive heme. Why?

11. Which of the carbon atoms of glucose must be labeled with ^{14}C in order to synthesize ^{14}C-labeled N^5,N^{10}-methylenetetrahydrofolate that can in turn be used to generate ^{14}C-methyl-S-adenosylmethionine?

12. In addition to catalyzing the reduction of diatomic nitrogen to ammonia, nitrogenase carries out a number of other reactions, including the conversion of acetylene to ethylene, the conversion of cyanide to methane and ammonia, and the conversion of azide to diatomic nitrogen and ammonia.

(a) Write balanced equations for these reactions, showing how many electrons are involved in each.

(b) One of the requirements for nitrogenase activity is ATP. In bacterial organisms like *Rhizobium,* ATP is generated by electron transport and oxidative phosphorylation, both of which are processes that require oxygen. However, nitrogenase is extremely sensitive to oxygen. How could the presence of leghemoglobin in the bacterial membrane of *Rhizobium* resolve this paradox?

13. In many bacteria, all the enzymes of the histidine pathway are coordinately synthesized when the culture medium is deficient in histidine. Conversely, the synthesis of these enzymes is coordinately repressed when histidine is added to the medium. The genes for all 10 enzymes are closely linked to one another on the bacterial chromosome. In many cases, a stop codon mutation in the gene of one of these enzymes not only results in a defective enzyme but also reduces the rate of synthesis of those enzymes coded for by genes farther along the chromosome.

(a) What are the advantages of the coordinate synthesis and repression of the enzymes of the histidine pathway in bacteria?

(b) Briefly explain the observations concerning stop codon mutations.

14. Most of the proteins synthesized in mammals contain all 20 common amino acids. As stated on page 717 of the text, more protein is degraded than is synthesized when even one essential amino acid is missing from the diet.

(a) Under such conditions, how could an increase in the rate of protein degradation provide the missing amino acid?

(b) How does an increase in the rate of protein degradation contribute to increased levels of nitrogen excretion?

15. The diagram below outlines the biosynthesis of a compound that is required for the oxidation of fatty acids in the mitochondrion.

A: $^{+}NH_3-CH_2-CH_2-CH_2-CH_2-CH(^{+}NH_3)-COO^-$

$\xrightarrow{3\ B}$

C: $(H_3C)_3N^+-CH_2-CH_2-CH_2-CH_2-CH(^{+}NH_3)-COO^-$

$\longrightarrow \longrightarrow \longrightarrow$

$(H_3C)_3N^+-CH_2-CH_2-CH_2-COO^-$

$\xrightarrow[O_2]{\alpha\text{-Ketoglutanate} \quad \text{Succinate} + CO_2}$

D: $(H_3C)_3N^+-CH_2-CH(OH)-CH_2-COO^-$

(a) Name compound D and briefly explain its role in fatty acid metabolism.

(b) Name compound A. Why is it considered essential in human diets?

(c) Three molecules of compound B are required for the formation of compound C. Its synthesis depends on the availability of an essential amino acid. Name that amino acid and then name compound B and compound C.

16. A pathway for the synthesis of ornithine from glutamate is shown in Figure 28-2.

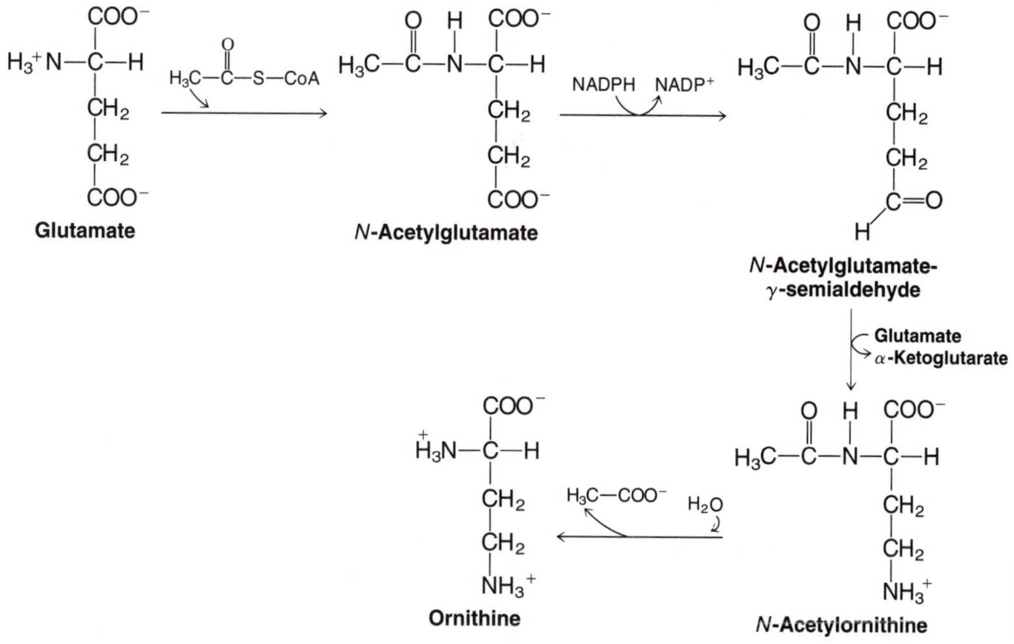

Glutamate $\xrightarrow{H_3C-C(=O)-S-CoA}$ *N*-Acetylglutamate $\xrightarrow{NADPH \rightarrow NADP^+}$ *N*-Acetylglutamate-γ-semialdehyde $\xrightarrow{\text{Glutamate} \rightarrow \alpha\text{-Ketoglutarate}}$ *N*-Acetylornithine $\xrightarrow{H_2O \rightarrow H_3C-COO^-}$ Ornithine

Figure 28-2
Biosynthesis of ornithine from glutamate.

(a) Why can this pathway also be considered to be part of the de novo pathway for the synthesis of arginine?
(b) Inspect the pathway for proline biosynthesis given on page 719 of the text, and then explain why the *N*-acetylation of glutamate is required for the synthesis of ornithine.

17. Elevated levels of ammonia in blood can result from deficiencies in one or another of the enzymes of the urea cycle. Measures taken to relieve hyperammonemia have included limiting intake of dietary proteins, administering α-keto analogs of several of the naturally occurring L-amino acids, or administering other compounds designed to exploit pathways of nitrogen metabolism and excretion.

(a) In trying to determine why a patient has hyperammonemia, which organ should you check first for normal function? Why?
(b) Why would limiting protein intake assist in relieving chronic hyperammonemia? Why would eliminating dietary proteins *altogether* (without any other supplement) probably increase the level of hyperammonemia?
(c) Write a brief rationale for using α-keto acid analogs in treating hyperammonemia, mentioning a particular group of enzymes essential to your explanation. Would it be better to use α-keto analogs of essential or nonessential amino acids? Why?

COO^-
|
CH_2
|
NH
|
CH_2
|
$^-O{-}P{=}O$
|
O^-

Glyphosate

18. Plants synthesize all 20 common amino acids de novo. Glyphosate, a weed killer sold under the trade name "Roundup," is an analog of phosphoenolpyruvate that specifically inhibits 3-enolpyruvylshikimate 5-phosphate synthase, a key enzyme of the pathway for chorismate biosynthesis. This compound is a very effective plant herbicide, but has virtually no effect on mammals. Why?

19. Mutant *B. subtilis* that lack tryptophan synthetase can grow on minimal medium only when supplemented with exogenous tryptophan. Under these conditions, none of the intermediates in the tryptophan biosynthetic pathway from anthranilate to indole 3-glycerol phosphate are produced. However, when the bacteria have depleted the medium of tryptophan, the levels of those intermediates increase, even though there is no net production of tryptophan. Why?

20. Both genetic and biochemical methods have been used to establish the biosynthetic pathways for essential amino acids in bacteria and other microorganisms. One classic approach is isotope competition, which begins with the use of radioactive glucose as the sole source of carbon for growing bacteria. Under these conditions, all the intermediates in a particular pathway will be uniformly labeled, but if a nonradioactive intermediate in the pathway is added to the growing cells, it will reduce or dilute the radioactivity of that intermediate and others following it in a pathway.

Britten and his co-workers used isotope competition to examine the biosynthetic pathway for threonine, methionine, and other amino acids. They grew *E. coli* in a minimal medium containing labeled glucose and nonlabeled homoserine, which was known to relieve auxotrophy for several amino acids. After growth, the bacterial cells were harvested, washed, and disrupted. Cell proteins were hydrolyzed with concentrated HCl. The neutralized amino acids from the hydrolysate were then separated by paper chromatography, located by ninhydrin reaction, and assayed for radioactivity. Isoleucine, threonine, and methionine had little or no radioactivity under these conditions. The respective radioactivities of aspartate and lysine from protein hydrolysates were not reduced by the use of nonradioactive homoserine.

In a similar experiment, nonradioactive aspartate was used in the growth medium; aspartate from protein hydrolysates was virtually nonradioactive, as were lysine, methionine, threonine, and isoleucine.

When nonlabeled threonine was used along with radioactive glucose, only threonine and isoleucine from protein hydrolysates had reduced radioactivity. And when either isoleucine or methionine was used, they affected only their own levels of radioactivity in protein hydrolysates.

On the basis of these observations, write an outline of the biosynthetic pathway for the amino acids isoleucine, threonine, homoserine, methionine, and lysine.

ANSWERS TO PROBLEMS

1. (a) Glycine is an obligatory precursor of heme; in the reaction catalyzed by δ-aminolevulinate synthase, glycine condenses with succinyl CoA to form δ-aminolevulinate. The reduction in the concentration of glycine in the cell caused by the *glyA*$^-$ mutation will cause a decrease in the rate of heme synthesis.

(b) Choline cannot be synthesized as a free molecule in animal cells; however, it can be synthesized as part of a phosphatide by the successive methylation of ethanolamine, which is in phosphate linkage with diacylglycerol (see p. 687 of Stryer's text). The methyl donor for the formation of phosphatidyl choline is *S*-adenosylmethionine, which is converted to *S*-adenosylhomocysteine. Regeneration of *S*-adenosylmethionine depends on the availability of N^5-methyltetrahydrofolate, which in turn depends on the generation of one or another of the methylated forms of tetrahydrofolate. Serine transhydroxymethylase activity, which generates N^5,N^{10}-methylenetetrahydrofolate as well as glycine, is therefore required for the continued entry of methyl groups into tetrahydrofolate. A decrease in the rate of synthesis of methylated folate will in turn depress the rate of synthesis of compounds that require one or more methyl transfers. These compounds include choline, which must be added as a supplement to ensure that sufficient levels of the molecule are available for optimal membrane synthesis.

2. (a) The heme molecule is required for a large number of important cellular functions, including the transport of oxygen in hemoglobin and the transport of electrons in the cytochromes. The total absence of any one of the enzymes in the heme biosynthetic pathway, including coproporphyrinogen oxidase, would result in a failure to synthesize heme, a condition that is fatal.

(b) Because heme synthesis is blocked at a later step in the pathway, intermediates synthesized in the steps preceding the block, such as uroporphyrinogen III and coproporphyrinogen III, will accumulate, while those synthesized in the steps past the block will be present in low concentrations. Among these scarce substances will be heme itself, since the small quantity produced is rapidly incorporated into hemoglobin and other substances that require heme. Because the cellular concentration of heme is relatively low, both the synthesis and activity of δ-aminolevulinate synthetase are no longer under control. The synthesis of δ-aminolevulinate and other early intermediates in the pathway will be stimulated.

(c) As noted in the answer to 2(b), at least two types of porphyrin molecules accumulate as a result of the block in the pathway. These intermediates, which accumulate in the skin and teeth, strongly absorb light, resulting in extreme photosensitivity.

(d) Barbiturates and steroid drugs undergo hydroxylation in liver cells in a process mediated by cytochrome P450, a heme-containing protein (see p. 704 of Stryer's text). Induction of cytochrome P450 synthesis in response to an increase in drug concentrations will deplete the already low number of heme molecules available. A decrease in heme concentration will stimulate the synthesis and activity of δ-aminolevulinate synthetase, resulting in the synthesis of additional precursors of heme. This will in turn exacerbate the symptoms of HCP.

3. (a) The studies led White and Beach to conclude that methionine, which is another sulfur-containing amino acid, is a biosynthetic precursor of cysteine. Later work elucidated the roles of methion-

ine in the active methyl cycle and in the synthesis of cysteine and confirmed their conclusion.

(b) In the active methyl cycle, *S*-adenosylhomocysteine is cleaved to yield adenosine and homocysteine. Although homocysteine can be converted to methionine, it can also condense with serine to form cystathionine which is then cleaved to yield cysteine and α-ketobutyrate. Feeding rats homocysteine along with a treated hydrolysate will make supplementation with methionine unnecessary.

4. Bilirubin is an intermediate in the breakdown of heme. When it enters the circulation, it is immediately complexed with serum albumin, which serves as a carrier for relatively insoluble compounds like bilirubin and free fatty acids. In the liver, glucuronyl transferase makes bilirubin more soluble through the attachment of two glucuronate groups; bilirubin diglucuronide is then secreted into bile and eliminated. Unconjugated bilirubin cannot be secreted into bile. As the concentration of bilirubin in the circulation increases, the capacity of serum albumin to absorb the molecule is surpassed, and the excess diffuses into the tissues. Such a condition is known as *jaundice*. In extreme cases, unconjugated bilirubin can enter nerve tissues, where it disrupts normal neurological function.

5. Examination of the structures of kynurenine and 3-hydroxykynurenine shows they are both related to tryptophan. As mentioned on page 730 of the text, the nicotinamide ring of NAD^+ is synthesized from tryptophan. The fact that a deficiency in PLP results in an insufficiency of this compound suggests that one or more of the steps in the pathway from tryptophan to nicotinamide involve PLP or one of its metabolites. The enzyme kynureninase is a PLP-dependent enzyme that catalyzes the cleavage of the side chain of 3-hydroxykynurenine to generate alanine and 3-hydroxyanthranilate:

3-Hydroxykynurenine $\xrightarrow{H_2O}$ 3-Hydroxyanthranilate + Alanine

The latter compound is a necessary intermediate in NAD^+ synthesis. Failure to synthesize 3-hydroxyanthranilate leads to an insufficiency of nicotinamide, which in turn can lead to pellagra.

6. The most appropriate nitrogen-containing compound for the synthesis of heme is glycine, which condenses with succinyl CoA to form δ-aminolevulinate. Glycine is synthesized from 3-phosphoglycerate through a pathway that involves serine, as described on page 719 of the text. When serine is converted to glycine, the β carbon of serine, which corresponds to C-3 in 3-phosphoglycerate, is transferred to tetrahydrofolate. The carboxyl group of glycine, which is derived from C-1 of 3-phosphoglycerate, is released as CO_2 when the amino acid condenses with succinyl CoA. The fact that only the α carbon of glycine is incorporated in δ-aminolevulinate means that you should choose 2-^{14}C-3-phosphoglycerate as a means of labeling heme.

7. (a) The fact that α-keto acid analogs can substitute for essential amino acids means that the carbon skeletons of the essential amino acids are not synthesized in humans. Many studies have shown that one or more of the enzymes needed for the synthesis of these structures are missing.

(b) The enzymes that are primarily responsible for the distribution of the ^{15}N label among the other amino acids are the aminotransaminases, which catalyze the interconversions of amino acids and their corresponding α-keto acids. The redistribution of the label begins with the transamination of alanine, with α-ketoglutarate serving as the amino acceptor to yield pyruvate and glutamate. Glutamate then serves as an amino donor for other α-keto acids. In order for an essential amino acid to be labeled, you must postulate the transamination of that amino acid to yield the corresponding α-keto acid analog, followed by the donation of a labeled amino group from glutamate or another donor of amino groups.

8. Ammonia, which is generated as part of the process of amino acid catabolism in muscle, is toxic and must be removed from the cells. This could be done through the synthesis of urea, but that process occurs only in the liver. In muscle cells, therefore, glutamine synthetase catalyzes the formation of glutamine, which is an efficient and nontoxic carrier of ammonia. This accounts for the high activity of that enzyme in muscle. The glutamine is transported by the blood to the liver, where glutaminase and aspartate aminotransferase work together to generate aspartate and two molecules of ammonia from glutamine, hence, the high activity of glutaminase in the liver. Aspartate and ammonia are both used by the liver for the synthesis of urea, a nontoxic and disposable form of ammonia.

9. (a) The fully adenylylated form of glutamine synthetase is the most sensitive to molecules like tryptophan and histidine. Because glutamine is utilized for the synthesis of a variety of compounds, complete inhibition of the enzyme by only one of those products, such as tryptophan, would inappropriately inhibit the synthesis of all the others. Thus, the enzyme is cumulatively inhibited by at least eight different nitrogen-containing compounds.

(b) The deadenylylated form, which is not subject to cumulative feedback inhibition, is generated in response to increases in the cellular concentrations of α-ketoglutarate (the precursor of glutamate) and ATP. These molecules signal the need for glutamine synthesis, even when other nitrogen-containing compounds are present. Under these conditions, the deadenylylated form of the enzyme binds ammonia even when ammonia concentrations are relatively low; that is, it has a relatively low K_M for ammonia.

(c) The simultaneous adenylylation and deadenylylation of glutamine synthetase would result in a loss of feedback control of the enzyme because the adenylylated form is subject to cumulative inhibition whereas the deadenylylated form is not. In addition, it would lead to the wasteful hydrolysis of ATP, since every round of adenylylation and deadenylylation generates AMP and inorganic pyrophosphate from ATP.

(d) Mammals acquire many nitrogen-containing compounds, such as tryptophan and histidine, in their diet rather than through de novo biosynthesis, so glutamine synthetase does not play so prominent a role in the nitrogen metabolism of mammals as it does in that of bacteria. Complex regulation of the enzyme is therefore not needed in mammals.

10. In the citric acid cycle, the two radioactive carbons of acetyl CoA combine with oxaloacetate to form citrate. As the cycle continues, two carbons are lost as carbon dioxide, but these atoms are not the two radioactive atoms that are derived directly from acetyl CoA (see page 513 of the text); the two radioactive carbons are lost in later turns of the cycle, when they appear in a different position in citrate. Thus, if radioactive succinyl CoA is generated directly from radioactive citrate, it could be incorporated immediately into δ-aminolevulinic acid, the precursor of heme. Inspection of the pathway for heme synthesis, which is outlined on pages 732–734 of the text, shows that many of the carbons in heme can indeed be labeled using radioactive acetyl CoA.

11. The methyl group of *S*-adenosylmethionine is derived from the methylene group of N^5,N^{10}-methylenetetrahydrofolate, which is in turn derived from the hydroxymethyl side chain of serine. Serine is synthesized from glucose through the glycolytic pathway. The series of steps include the generation of 3-phosphoglyceric acid (3-PGA) from glucose, followed by oxidation, transamination, and phosphorolysis to yield serine (see p. 719 of the text). In order to generate serine with radioactive carbon in the hydroxymethyl side chain, 3-^{14}C-3-PGA is needed. Such a molecule can be generated by labeling the C-1 or C-6 of glucose, as shown in Figure 28-3.

12. (a) The reduction of acetylene requires two electrons and two hydrogen ions, whereas the conversion of cyanide to methane and ammonium ion requires six electrons along with eight hydrogen ions. The generation of diatomic nitrogen and ammonium ion requires two electrons and four hydrogen ions.

$$\underset{\textbf{Acetylene}}{HC{\equiv}CH} + 2\,H^+ + 2\,e^- \longrightarrow \underset{\textbf{Ethylene}}{H_2C{=}CH_2}$$

$$\underset{\textbf{Cyanide}}{CN^-} + 8\,H^+ + 6\,e^- \longrightarrow \underset{\textbf{Methane}}{CH_4} + \underset{\textbf{Ammonia}}{NH_4^+}$$

$$\underset{\textbf{Azide}}{N_3^-} + 4\,H^+ + 2\,e^- \longrightarrow \underset{\textbf{Nitrogen}}{N_2} + \underset{\textbf{Ammonia}}{NH_4^+}$$

The role of these reactions in nitrogen-fixing organisms is not understood. It is worth noting that an agricultural field test for nitrogen fixation has been developed that takes advantage of the ability of nitrogenase to reduce acetylene.

(b) The presence of leghemoglobin in the bacterial membrane of organisms like *Rhizobium* functions in two ways. First, it binds oxygen with high affinity, so it thereby simply prevents oxygen from entering the cell and denaturing the nitrogenase molecule. Second, leghemoglobin supplies oxygen as an electron acceptor for reduced electron carriers in the bacterial membrane, allowing the ATP required for nitrogen fixation to be generated by oxidative phosphorylation.

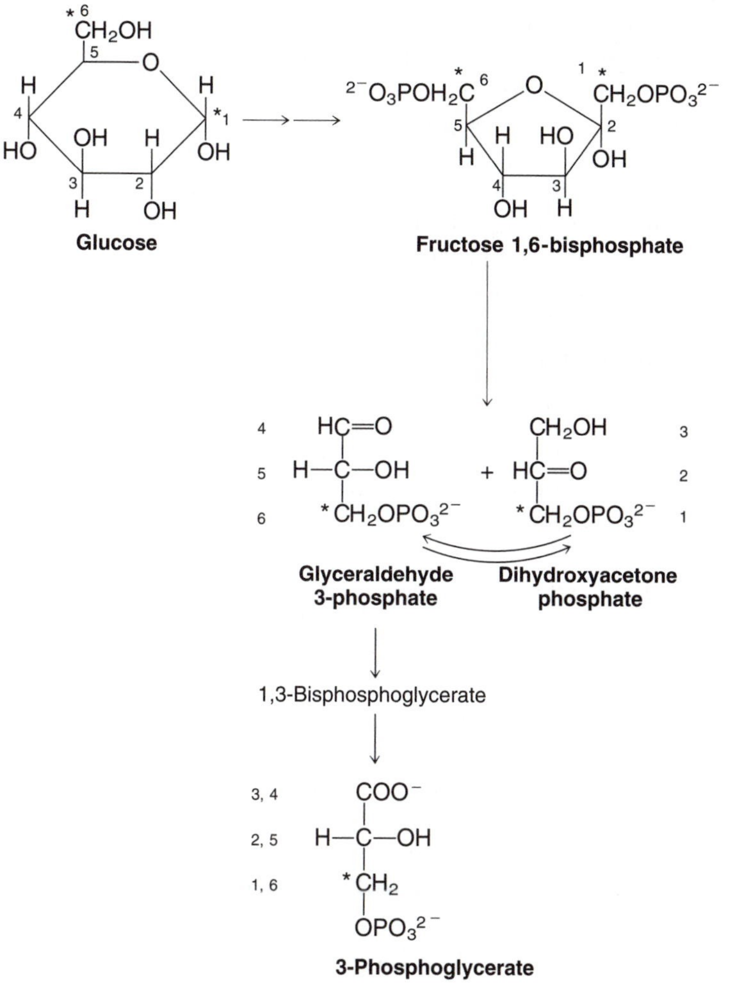

Glucose

Fructose 1,6-bisphosphate

Glyceraldehyde 3-phosphate

Dihydroxyacetone phosphate

1,3-Bisphosphoglycerate

3-Phosphoglycerate

Figure 28-3
Synthesis of 3-^{14}C-3-phosphoglycerate from glucose labeled at C-1 or C-6. Asterisks denote labeled carbon atoms, and numbers beside the three-carbon compounds correspond to the numbered carbons in glucose and in fructose 1,6-bisphosphate.

13. (a) The 10 enzymes of the histidine pathway catalyze the formation of intermediates that are needed only for the synthesis of histidine. The coordinate synthesis of all the histidine pathway enzymes ensures that roughly equivalent numbers of enzyme molecules will be produced so that all the enzymes will be available as a group when the biosynthesis of histidine is required. On the other hand, if histidine is available, none of the enzymes are needed, so coordinate repression shuts off the synthesis of all the enzymes simultaneously.

    (b) The effects of stop codon mutations, coupled with the fact that all 10 genes are closely linked, suggest that the genes for the enzymes of the histidine pathway are transcribed as a single message, which may be then used for protein synthesis without further processing. For such a message, the ribosomes begin translation at one end of the molecule and carry out the synthesis of a number of proteins as they proceed toward the other end. When the ribosomes encounter a stop codon mutation, they terminate translation prematurely, and they cannot reinitiate translation farther downstream. This could explain why enzymes having genes downstream from a stop codon mutation are not synthesized.

14. (a) Many experiments have shown that under normal conditions cells continuously synthesize and degrade proteins. Although both

essential and nonessential amino acids are continuously recycled during these processes, reutilization is not completely efficient; thus, additional amino acids are needed. In mammals, there are no reservoirs of free amino acids; the only sources of essential amino acids are dietary proteins or the proteins of the body tissues. If an essential amino acid is not available from the diet, cells appear to accelerate the hydrolysis of their own proteins, in order to generate the missing essential amino acid. How the rate of cellular proteolysis is accelerated in response to a deficiency of an essential amino acid is not understood.

(b) An increased rate of protein degradation generates a higher concentration of free amino acids. During the oxidation of those amino acids not used for synthesis of other proteins, ammonia will be produced. An elevation in ammonia concentration in the body stimulates the formation of urea, causing the level of nitrogen excretion to increase.

15. (a) Compound D is carnitine, which, when esterified to the acyl group of a long-chain fatty acid, shuttles it from the cytosol to the matrix of the mitochondrion, where fatty acid oxidation takes place.

(b) Compound A is the essential amino acid lysine; it is termed *essential* because it cannot be synthesized de novo in humans. Lysine and other essential amino acids must be obtained from the diet.

(c) Compound C is trimethyllysine, and the methyl groups that are attached to lysine are likely to be derived from compound B, *S*-adenosylmethionine, the major donor of methyl groups in biosynthetic reactions. The methyl group of *S*-adenosylmethionine is derived from methionine, an essential amino acid.

16. (a) Ornithine is a precursor of arginine, as part of the pathway for the synthesis of urea. Thus the pathway for the synthesis of ornithine from glutamine, along with part of the urea cycle pathway, can together be considered as a de novo pathway for the synthesis of arginine. Arginine can in turn be used for the synthesis of urea, or it can serve instead as one of the amino acids used for polypeptide synthesis.

(b) In the pathway for proline biosynthesis, glutamic-γ-semialdehyde cyclizes with the loss of water to form Δ^1-pyrroline-5-carboxylate. However, in the pathway for the formation of ornithine shown in Figure 28-2, an *N*-acetylated derivative of the semialdehyde molecule is formed. The *N*-acetyl group blocks the condensation of the amino group with the aldehyde group, thereby preventing the formation of the pyrroline ring. This allows the pathway to proceed toward the synthesis of ornithine.

17. (a) You should assess liver function, because enzymes of the urea cycle are found primarily in this organ. In addition, liver takes up amino acids such as alanine, glutamate, and glutamine, which are in effect nontoxic forms of ammonia generated by muscle and other tissues. Amino groups of these compounds, along with carbon dioxide from carbamoyl phosphate, are precursors of urea in the liver.

(b) During digestion, dietary proteins are hydrolyzed to their component amino acids. Those amino acids that are not needed immediately for protein synthesis or for the biosynthesis of other nitrogen-containing compounds are degraded. One of the products of amino acid degradation is ammonia. Usually ammonia is metabolized through conversion to nitrogen carriers such as alanine, glutamate, and glutamine, and it is ultimately utilized for the synthesis of urea when the urea cycle is operating. Thus limiting protein intake in a patient with a deficiency in urea synthesis would be expected to reduce ammonia production in liver and other tissues.

Protein turnover and amino acid degradation constantly take place in the tissues, and essential amino acids (those that cannot be synthesized de novo in human tissues) must be generated either from dietary sources or from additional breakdown of body proteins. A complete restriction of dietary protein would accelerate body protein breakdown and would exacerbate the condition of hyperammonemia.

(c) The α-keto acid analogs can serve as acceptors for amino groups from glutamate and other amino acids, in reactions catalyzed by transaminases or aminotransferases. Nonessential amino acids formed as the result of this process may be themselves eliminated, degraded (often generating more ammonia), or else used for biosynthesis of other nitrogen-containing compounds. Employing analogs of essential amino acids might be preferable, because tissues are more likely to require them for protein synthesis or other biosyntheses.

18. Chorismate is an intermediate in the biosynthesis of the aromatic amino acids tryptophan, phenylalanine, and tyrosine. Mammals do not synthesize these amino acids from chorismate. Instead, they obtain the essential aromatic amino acids tryptophan and phenylalanine from the diet, and they can synthesize tyrosine from phenylalanine. Glyphosate is an effective herbicide because it prevents synthesis of aromatic amino acids in plants. But the compound has no effect on mammals because they have no active pathway for de novo aromatic amino acid synthesis.

19. In *B. subtilis,* the pathway from chorismate to tryptophan is feedback-inhibited by tryptophan, which suppresses anthranilate synthase activity. Exogenous tryptophan from the medium, therefore, inhibits the production of intermediates in the biosynthetic pathway. When exogenous tryptophan is depleted, intracellular levels of tryptophan decrease and anthranilate synthase activity is no longer inhibited. An increase in anthranilate production leads to an increase in the production of other intermediates in the pathway until equilibria are established among those compounds. The block at the step catalyzed by tryptophan synthetase prevents large accumulations of the intermediates, but under these conditions their concentrations are higher than in the presence of exogenous tryptophan.

20. These experiments are described in R.B. Roberts, P.H. Abelson, D.B. Cowie, E.T. Bolton, and R.J. Britten, *Studies in Biosyntheses in* Escherichia coli. Carnegie Institution of Washington, Pub. No. 607, 1955. The experiments show that aspartate is a key precursor for the five amino acids, because when it is included in the growth medium with labeled glucose, radioactivity of each of the amino acids is reduced. Homoserine affects

the labeling of isoleucine, threonine, and methionine. Thus, homoserine is an intermediate in the pathway for those three amino acids but not lysine or aspartate. Dilution of threonine and isoleucine from cells grown in nonradioactive threonine suggests that those two amino acids are on the same pathway, separated from those of the other three. Methionine and isoleucine are affected only when they are used in growth experiments, indicating that they are not on other pathways (and, for isoleucine, that threonine must precede it in a biosynthetic pathway).

The overall pathway below shows that lysine and homoserine derived from aspartate. Threonine and isoleucine are both derived from aspartate, and homoserine must be an intermediate in their synthesis, as well as in the synthesis of methionine.

$$\begin{array}{ccccccc} \text{Aspartate} & \rightarrow & \text{Homoserine} & \rightarrow & \text{Threonine} & \rightarrow & \text{Isoleucine} \\ \downarrow & & \downarrow & & & & \\ \text{Lysine} & & \text{Methionine} & & & & \end{array}$$

EXPANDED SOLUTIONS TO TEXT PROBLEMS

1. Glucose + 2 ADP + 2 P_i + 2 NAD^+ + 2 glutamate → 2 alanine + 2 α-ketoglutarate + 2 ATP + 2 NADH + H^+

Glucose is converted to pyruvate via glycolysis, and pyruvate is converted to alanine by transamination.

2. $N_2 \rightarrow NH_4^+ \rightarrow$ glutamate → serine → glycine → δ-aminolevulinate → porphobilinogen → heme

3. (a) N^5,N^{10}-Methylenetetrahydrofolate, (b) N^5-methyltetrahydrofolate

4. γ-Glutamyl phosphate is a likely reaction intermediate.

5. The administration of glycine leads to the formation of isovalerylglycine. This water-soluble conjugate, in contrast with isovaleric acid, is excreted very rapidly by the kidneys. See R.M. Cohn, M. Yudkoff, R. Rothman, and S. Segal. *New Engl. J. Med.* 299(1978):996.

6. They carry out nitrogen fixation. The absence of photosystem II provides an environment in which O_2 is not produced. Recall that the nitrogenase is very rapidly inactivated by O_2. See R.Y. Stanier, J.L. Ingraham, M.L. Wheelis, and P.R. Painter, *The Microbial World,* 5th ed. (Prentice-Hall, 1986), pp. 356–359, for a discussion of heterocysts.

7. The cytosol is a reducing environment, whereas the extracellular milieu is an oxidizing environment.

8. The synthesis of δ-aminolevulinate requires succinyl CoA, an intermediate in the citric acid cycle, which occurs in the mitochondrial matrix. Thus it makes sense to have the first step in porphyrin biosynthesis occur in the matrix.

9. The reduction of N^5,N^{10}-methylenetetrahydrofolate to N^5-methyltetrahydrofolate is essentially irreversible. Tetrahydrofolate is normally recovered in the conversion of homocysteine to methionine, a methylcobalamin-dependent reaction. In pernicious anemia, vitamin B_{12} deficiency blocks this methyl transfer reaction. Hence, the folate pool is trapped in the form of N^5-methyltetrahydrofolate. Most one-carbon transfers are thereby blocked.

10. CO, which is formed in the breakdown of heme, and HCN, which is formed in the synthesis of ethylene, are two plausible candidates. Both bind to heme, CO to the ferrous form and CN^- to the ferric form. A heme protein could serve as a receptor for either. It will be interesting to learn whether CO and HCN have messenger roles.

CHAPTER

29

Biosynthesis of Nucleotides

In this chapter, Stryer completes his treatment of the biosyntheses of the major classes of macromolecular precursors by describing the synthesis of the purine and pyrimidine nucleotides. Besides being the precursors of RNA and DNA, these compounds serve a number of other important roles that are reviewed in the opening paragraph of the chapter. You should review the nomenclature of the nucleotides, which is given on page 76 of Stryer.

Stryer begins with the synthesis of the purine nucleotide inosine 5′-monophosphate (IMP). He lists its precursors and describes the synthesis of the activated form of ribose 5-phosphate, 5-phosphoribosyl-1-pyrophosphate (PRPP). He then describes the steps in the construction of the purine ring on ribose 5-phosphate and the conversion of IMP, the initial product, to AMP or GMP. The reactions that allow cells to salvage free purines and the control of purine biosynthesis are then presented. Next, Stryer describes the synthesis of the free pyrimidine orotate and its attachment to ribose 5-phosphate through reaction with PRPP to form the pyrimidine nucleotide. The reactions leading to the uracil and cytosine nucleotide derivatives are then given, followed by the reactions that interconvert the mono-, di-, and triphosphate derivatives of the nucleotides. The regulation of pyrimidine synthesis is also covered at this point.

Two reactions that are required to form the precursors of DNA are described in detail: ribonucleotide reductase converts ribonucleotides to deoxyribonucleotides, and thymidylate synthase methylates dUMP to form dTMP. Stryer presents the mechanisms and cofactors of these enzymes and explains how two important anticancer drugs and an antibiotic inhibit the synthesis of dTMP and thus the growth of cells. Nucleotides also serve important roles as constituents of NAD^+, $NADP^+$, FAD, and coenzyme A (CoA), so the syntheses of these cofactors are described. Stryer concludes the chapter with an explanation of how the purines are catabolized and a discussion of two pathological conditions that arise from defects in the catabolic pathway of the purines.

LEARNING OBJECTIVES

When you have mastered this chapter, you should be able to complete the following objectives:

Introduction (Stryer pages 739–740)

1. List the major biochemical roles of the *nucleotides.*

2. Distinguish among the *purine* and *pyrimidine nucleosides* and *nucleotides.*

Biosynthesis of the Purine Nucleotides (Stryer pages 740–745)

3. List the precursors that provide the carbon and nitrogen atoms of the purine ring. Note the numbering of the ring atoms.

4. Describe the substrates and products of the *ribose phosphate pyrophosphokinase* (*PRPP synthetase*) reaction and explain the importance of *5-phosphoribosyl-1-pyrophosphate* (*PRPP*) in purine as well as amino acid metabolism.

5. Outline the synthesis of the purine base *hypoxanthine* as it occurs on ribose 5-phosphate to form *inosine 5′-monophosphate* (*IMP*), noting the sources of the atoms and the cofactors involved. List the steps that require ATP.

6. Describe the synthesis of *adenylate* (*AMP*) and *guanylate* (*GMP*) from IMP. List the cofactors and intermediates of the reactions.

7. Explain the mechanism for replacing a carbonyl oxygen with an amino group, and list the potential sources of the nitrogen atom in these reactions.

8. Describe the synthesis of purine nucleotides by the *salvage reactions* and explain why these reactions are energetically advantageous.

9. Outline the regulation of the biosynthesis of the purine nucleotides and name the committed step in the pathway.

Biosynthesis of the Pyrimidine Nucleotides (Stryer pages 745–748)

10. List the precursors that provide the carbon and nitrogen atoms of the pyrimidine ring. Note the numbering of the ring atoms.

11. Explain the role of *carbamoyl phosphate* in pyrimidine biosynthesis, and compare the two *carbamoyl phosphate synthetase* reactions that occur in eukaryotic cells.

12. Describe the *aspartate transcarbamoylase* reaction and outline the remaining reactions that form *orotate.* Outline the conversion of orotate to *uridylate* (*UMP*).

13. Contrast the enzymes of pyrimidine biosynthesis in *E. coli* with those of higher organisms and list the potential advantages of *multifunctional enzymes.*

14. Explain how nucleoside *mono-* and *diphosphate kinases* interconvert the nucleoside mono-, di-, and triphosphates.

15. Describe the reaction that converts UTP to *CTP* and note the different nitrogen atom sources used in bacteria and mammals.

16. Outline the regulation of pyrimidine biosynthesis and name the committed step. Note the steps that require ATP.

Formation of the Deoxyribonucleotides (Stryer pages 748–754)

17. Describe the *ribonucleotide reductase* reaction and outline its mechanism and regulation. Explain the roles of *NADPH* and *thioredoxin* (or *glutathione* and *glutaredoxin*) in the reaction. Compare the reaction in *E. coli* and mammals with that in *Lactobacillus leichmanni* and note the *radical* mechanism used in all cases.

18. Describe the *thymidylate synthase* reaction. Account for the source of the methyl group and describe the change in the oxidation state of the transferred carbon atom that occurs during the reaction.

19. Relate the reactions of ribonucleotide reductase and thymidylate synthase to the hypothesis that RNA was the primordial informational molecule.

20. Explain the role of *dihydrofolate reductase* in the synthesis of deoxythymidylate.

21. Account for the ability of *fluorouracil* to act as a *suicide inhibitor.* Describe the inhibitory mechanism of *methotrexate* and *aminopterin.* Explain how these three compounds interfere with the growth of cancer cells. List the mechanisms by which a cell could become resistant to methotrexate. Explain the antibiotic activity of *trimethoprim.*

Biosynthesis of NAD$^+$, FAD, and CoA (Stryer pages 754–755)

22. Describe the synthesis of *NAD$^+$* and account for the sources of the *nicotinamide* portion of the molecule.

23. Outline the syntheses of *FAD* and *CoA* and explain the role of the PP_i that is released during their syntheses and that of NAD$^+$.

Degradation of the Purines (Stryer pages 755–758)

24. Describe the reactions of the *nucleotidases* and *nucleoside phosphorylases.*

25. Outline the conversions of AMP and *guanine* to *uric acid.* Describe the role of *xanthine oxidase* in these processes.

26. Describe the major clinical findings in patients with *gout* and explain the rationale for the use of *allopurinol* to alleviate the symptoms of the disease. Relate the functioning of *hypoxanthine-guanine phosphoribosyl transferase* (*HGPRT*) to the rate of de novo purine biosynthesis in these patients.

27. Describe the *antioxidant* role of *urate.*

28. Give the biochemical lesion that leads to the *Lesch-Nyhan syndrome* and describe the symptoms of the disease.

SELF-TEST

Introduction

1. Describe the physiologic roles of the nucleotides.

____

____

2. Which of the following answers completes the sentence correctly? Cytosine is a

(a) purine base.
(b) pyrimidine base.
(c) purine nucleoside.
(d) pyrimidine nucleoside.

3. Which of the following answers completes the sentence correctly? Guanosine is a

(a) purine base.
(b) pyrimidine base.
(c) purine nucleoside.
(d) pyrimidine nucleoside.

4. Which of the following are nucleotides?

(a) Deoxyadenosine
(b) Cytidine
(c) Deoxyguanylate
(d) Uridylate

Biosynthesis of the Purine Nucleotides

5. Which of the following compounds directly provide atoms to form the purine ring?

(a) Aspartate
(b) Carbamoyl phosphate
(c) Glutamine
(d) Glycine
(e) CO_2
(f) N^5,N^{10}-Methylenetetrahydrofolate
(g) N^{10}-Formyltetrahydrofolate
(h) NH_4^+

6. Which of the following statements about 5-phosphoribosyl-1-pyrophosphate (PRPP) are true?

(a) It is an activated form of ribose 5-phosphate.
(b) It is formed from ribose 1-phosphate and ATP.
(c) It has a pyrophosphate group attached to the C-1 atom of ribose in the α configuration.
(d) It is formed in a reaction in which PP_i is released.

7. Which of the following answers completes the sentence correctly? The first product of purine nucleotide biosynthesis that contains a complete purine ring (hypoxanthine) is

(a) AMP.
(b) GMP.
(c) IMP.
(d) xanthylate (XMP).

8. The conversion of IMP to AMP requires which of the following?

(a) ATP
(b) GTP
(c) Aspartate
(d) Glutamine
(e) NAD^+

9. The conversion of IMP to GMP requires which of the following?

(a) ATP
(b) GTP
(c) Aspartate
(d) Glutamine
(e) NAD^+

10. Describe the general mechanism used by cells to replace a carbonyl group with an amino group in nucleotide biosynthesis.

____

____

11. Which of the following reactants and products are involved in the salvage reactions of purine biosynthesis?

(a) IMP → AMP
(b) IMP → GMP
(c) Adenine → AMP
(d) Guanine → GMP

12. During a purine salvage reaction, what is the source of the energy required to form the C–N glycosidic bond between the base and ribose?

____

____

13. Show which of the nucleotides in the right column regulate each of the conversions in the left column.

(a) Ribose 5-phosphate → PRPP __________	(1) AMP
(b) PRPP → phosphoribosylamine __________	(2) GMP
(c) Phosphoribosylamine → IMP __________	(3) IMP
(d) IMP → adenylosuccinate __________	
(e) IMP → xanthylate (XMP) __________	

14. How does the purine repressor, PurR, function to regulate the amount of GMP and AMP in *E. coli*?

____

____

Biosynthesis of the Pyrimidine Nucleotides

15. Which of the following statements about the carbamoyl phosphate synthetase of mammals, which is used for pyrimidine biosynthesis, are true?

(a) It is located in the mitochondria.
(b) It is located in the cytosol.
(c) It uses NH_4^+ as a nitrogen source.
(d) It uses glutamine as a nitrogen source.
(e) It requires *N*-acetylglutamate as a positive effector.

16. What is the committed step in pyrimidine biosynthesis?

17. Would orotidylate 5′-monophosphate (OMP) or UMP move more rapidly toward the anode during electrophoresis at pH 8? Explain.

18. How is orotate, a free pyrimidine, converted into a nucleotide? Is this reaction considered to be a salvage reaction or a biosynthetic one?

19. Why might covalently linked (multifunctional) enzymes, such as those of the pyrimidine biosynthetic pathway of mammals, be advantageous to an organism?

20. Which of the following enzymes are involved in converting the nucleoside 5′-monophosphate (NMP) products of the purine or pyrimidine biosynthetic pathways into their 5′-triphosphate (NTP) derivatives?

(a) Purine nucleotidase
(b) Nucleoside diphosphate kinase
(c) Nucleoside monophosphate kinases
(d) Nucleoside phosphorylase

21. How is the exocyclic amino group on the N-4 position of cytosine formed?

22. Which of the following compounds are involved in the control of the pyrimidine biosynthetic pathway?

(a) UTP
(b) UMP
(c) CTP
(d) CMP

Formation of the Deoxyribonucleotides

23. Which of the following statements about ribonucleotide reductase are true?

(a) It converts ribonucleoside diphosphates into 2′-deoxyribonucleoside diphosphates in humans.
(b) It catalyzes the homolytic cleavage of a bond.

(c) It accepts electrons directly from $FADH_2$.
(d) It receives electrons directly from either thioredoxin or glutaredoxin.
(e) It contains two kinds of allosteric regulatory sites; one for control of overall activity and another for control of substrate specificity.

24. Select from the following those compounds that are precursors of 2′-deoxythymidine-5-triphosphate (dTTP) in mammals and place them in their correct biosynthetic order:

(a) OMP ________ (f) dUDP ________

(b) UMP ________ (g) dUTP ________

(c) UDP ________ (h) dTMP ________

(d) UTP ________ (i) dTDP ________

(e) dUMP ________ (j) dTTP ________

25. Define *suicide inhibitor* and give an example from pyrimidine biosynthesis.

____

____

26. Methotrexate, a folate antagonist, interferes with nucleic acid biosynthesis. Would you expect it to inhibit purine or pyrimidine biosynthesis or both processes? Explain.

____

____

27. Methotrexate and trimethoprim are both inhibitors of dihyrofolate reductase. Why is trimethoprim the drug of choice in treating a human microbial infection?

____

____

Biosynthesis of NAD$^+$, FAD, and CoA

28. Which of the following can serve as precursors of NAD$^+$ in humans?

(a) Riboflavin
(b) Pantothenate
(c) Tyrosine
(d) Tryptophan
(e) Niacin

Degradation of the Purines

29. Which of the following compounds would give rise to urate if they were catabolized completely in humans?

(a) ADP–glucose
(b) GDP–mannose
(c) CDP–choline
(d) UDP–galactose
(e) CoA
(f) FAD
(g) UMP

30. Which of the following statements about patients with gout are true?
   (a) They have elevated levels of serum urate.
   (b) They have elevated levels of serum PRPP.
   (c) They sometimes have a partial deficiency of hypoxanthine-guanine phosphoribosyl transferase.
   (d) They sometimes have abnormally high levels of PRPP synthetase.
   (e) They can be treated with allopurinol to decrease de novo purine biosynthesis.

31. Explain why allopurinol is not effective in decreasing de novo purine biosynthesis in patients with Lesch-Nyhan syndrome when it does so in many patients with gout?

____

____

ANSWERS TO SELF-TEST

Introduction

1. The nucleotides (1) are the activated precursors of DNA and RNA; (2) are the source of derivatives that are activated intermediates in many biosyntheses; (3) include ATP, the universal currency of energy in biological systems, and GTP, which powers many movements of macromolecules; (4) include the adenine nucleotides, which are components of the major coenzymes NAD^+, FAD, and CoA; and (5) serve as metabolic regulators.

2. b

3. c. Nucleosides consist of a purine or pyrimidine base attached to a ribose or 2′-deoxyribose.

4. c, d. Nucleotides are nucleosides that contain one or more phosphate substituents on their ribose or deoxyribose moieties.

Biosynthesis of the Purine Nucleotides

5. a, c, d, e, g

6. a, c

7. c

8. b, c

9. a, d, e

10. A carbonyl oxygen is converted into a phosphoryl ester or a substituted phosphoryl ester through a reaction with a high-energy phosphate (ATP or GTP) to form a mono- or diphosphate ester. The phosphate or pyrophosphate group can then be readily displaced by the nucleophilic attack of the

nitrogen atom from NH_3, the side-chain amide group of glutamine, or the α-amino group of aspartate. The resulting adduct is an amino group or can be converted into one. See Stryer, Figure 29-8.

11. c, d

12. The activated form of ribose 5-phosphate (R-5-P), PRPP, reacts with the purine base to form the nucleotide and release PP_i. The displacement and subsequent hydrolysis of PP_i drives the formation of the *N*-glycosyl bond. ATP ultimately provides the energy through its reaction with R-5-P to form PRPP.

13. (a) 1, 2, 3 (b) 1, 2, 3 (c) none (d) 1 (e) 2

14. When the purines hypoxanthine and guanine are abundant, they are bound by PurR, causing it to act as a specific transcriptional repressor. Genes encoding enzymes in the de novo biosynthetic pathway of purines are thereby not expressed, and the amount of these enzymes decreases. Consequently, the amounts of AMP and GMP that can be synthesized are decreased.

Biosynthesis of the Pyrimidine Nucleotides

15. b, d. The mitochondrial carbamoyl phosphate synthetase used for urea synthesis is activated by *N*-acetylglutamate and uses NH_4^+ as the nitrogen source.

16. The formation of *N*-carbamoylaspartate by the aspartate transcarbamoylase (ATCase) reaction is the committed step in pyrimidine biosynthesis. See pages 238–241 of Stryer for a detailed description of the regulation.

17. Since OMP contains a carboxyl substituent on the pyrimidine ring that is lacking in UMP, it has an extra negative charge at pH 8, so it will move toward the anode more rapidly.

18. Orotate condenses with PRPP in a reaction catalyzed by orotate phosphoribosyl transferase to form the nucleotide orotidylate (OMP). Orotidylate decarboxylase converts OMP to the more abundant nucleotide UMP. The reaction occurs during de novo pyrimidine biosynthesis and is therefore not a salvage reaction.

19. The clustering of two or more enzymes (active sites) in a single polypeptide chain ensures that their synthesis is coordinated and helps assure that they will assemble into a coherent complex. Also, the proximity of the active sites means that side reactions are minimized as substrates are channeled from one active site to another. Finally, a multifunctional complex with covalently linked active sites is likely to be more stable than a complex formed by noncovalent interactions.

20. b, c. Several different specific nucleoside monophosphate kinases phosphorylate dNMPs and NMPs, using ATP as the phosphoryl donor. A single enzyme, nucleoside diphosphate kinase, uses the phosphorylation potential of ATP to convert the dNDPs and NDPs to dNTPs and NTPs. The ubiquitous adenylate nucleotides are interconverted by adenylate kinase (myokinase).

21. CTP is formed by amination of UTP. The carbonyl oxygen at C-4 of UTP is replaced with an amino group via the formation of an enol phosphate ester intermediate (see the answer to question 10). In *E. coli,* NH_4^+ serves as the source of the nitrogen atom that displaces the phosphate group, whereas the amide group of glutamine serves this purpose in mammals.

22. b, c. ATCase is regulated through feedback inhibition by CTP, the end product of the pyrimidine pathway. See the answer to question 16. Carbamoyl phosphate synthetase, which forms a substrate for ATCase, is inhibited by UMP.

Formation of the Deoxyribonucleotides

23. a, b, d, e. NADPH provides electrons via thioredoxin or glutaredoxin. See pages 750 and 751 of Stryer.

24. a, b, c, f, g, e, h, i, j. In mammals, NDPs are converted to dNDPs by ribonucleotide reductase. Thus, UDP is converted to dUDP, which is converted to dUTP by nucleoside diphosphate kinase. A specific pyrophosphatase hydrolyzes dUTP to dUMP, which is then converted to dTMP by thymidylate synthase. The dTMP is converted to dTTP. A priori, you might have expected the dUDP product of the ribonucleotide reductase reaction to be converted directly to dUMP. However, in fact, cells contain a dUTP pyrophosphatase to prevent dUTP from serving as a DNA precursor, and it is this enzyme that functions in the dTTP biosynthetic pathway.

25. A suicide inhibitor is a substrate that is converted by an enzyme into a substance that is capable of reacting with and inactivating the enzyme. In pyrimidine biosynthesis, thymidylate synthase converts fluorouracil into a derivative that becomes covalently attached to the enzyme and thereby inactivates it.

26. Methotrexate and aminopterin, a similar compound, are analogs of dihydrofolate (DHF) and inhibitors of dihydrofolate reductase, an enzyme that converts DHF to tetrahydrofolate (THF). The thymidylate synthase reaction converts N^5,N^{10}-methylenetetrahydrofolate to DHF in the process of methylating dUMP to form dTMP. In the presence of one of the inhibitors, this reaction functions as a sink that reduces the THF level of the cell by converting THF to DHF. Since THF derivatives are substrates in two reactions of purine metabolism and one of pyrimidine metabolism, both pathways are affected by the inhibitor.

27. Since trimethoprim binds to the mammalian dihydrofolate reductase much less tightly than to the enzyme of susceptible microorganisms, it causes fewer deleterious effects to humans than does methotrexate.

Biosynthesis of NAD$^+$, FAD, and CoA

28. d, e. Niacin becomes a dietary requirement if the supply of the essential amino acid tryptophan is inadequate.

Degradation of the Purines

29. a, b, e, f. Each of these compounds contains a heterocyclic purine base.

30. a, c, d, e

31. Allopurinol is converted to a ribonucleotide by hypoxanthine-guanine phosphoribosyl transferase (HGPRT). Allopurinol ribonucleotide inhibits the conversion of PRPP to phosphoribosylamine and thus inhibits de novo purine biosynthesis. Since patients with Lesch-Nyhan syndrome are deficient in HGPRT, they cannot form the inhibitor ribonucleotide, so de novo purine biosynthesis is not diminished.

PROBLEMS

1. In the early 1950s, Rose and Schweigert carried out experiments that involved the injection of radioactive nucleosides into animals. They found that when cytidine with tritium-labeled cytosine and ^{14}C-labeled ribose were injected, the deoxyribonucleotide later appeared with the same ratio of specific activities in DNA isolated from the tissues of the injected animals. Why was this an important clue to biochemists studying the biosynthesis of deoxyribonucleotides?

2. Mammalian lymphocytes that lack adenosine deaminase (see problem 11, p. 761, of Stryer's text) neither grow nor divide. The level of dATP in these cells is 100 times higher than that in normal lymphocytes, and the synthesis of DNA in the cells is impaired.

(a) How is adenosine converted to dATP? Assume that the first step is catalyzed by a specific nucleoside kinase.
(b) How does the elevation in dATP concentration in the abnormal lymphocytes affect the synthesis of DNA?

3. Clinicians who use F-dUMP and methotrexate *together* in cancer treatment find that the combined effects on cancer cells are not synergistic. Suggest how the administration of methotrexate could interfere with the action of F-dUMP.

4. Elevated levels of ammonia in the blood can be caused by a deficiency of mitochondrial carbamoyl phosphate synthetase or a deficiency of any of the urea cycle enzymes. These two types of disorders can be distinguished by the presence of orotic acid or related metabolites in the urine.

(a) Why is it possible to determine the basis of hyperammonemia in this way?
(b) Why would a deficiency of cytoplasmic carbamoyl phosphate synthetase not cause hyperammonemia? What problems would such an enzyme deficiency cause? How would you treat a patient who has a deficiency in cytoplasmic carbamoyl phosphate synthetase?

5. The degradation of thymine yields β-aminoisobutyrate, as shown in the figure in the margin, and the compound can be converted to succinyl CoA and then degraded in the citric acid cycle. What cofactors are needed to convert β-aminoisobutyrate to succinyl CoA?

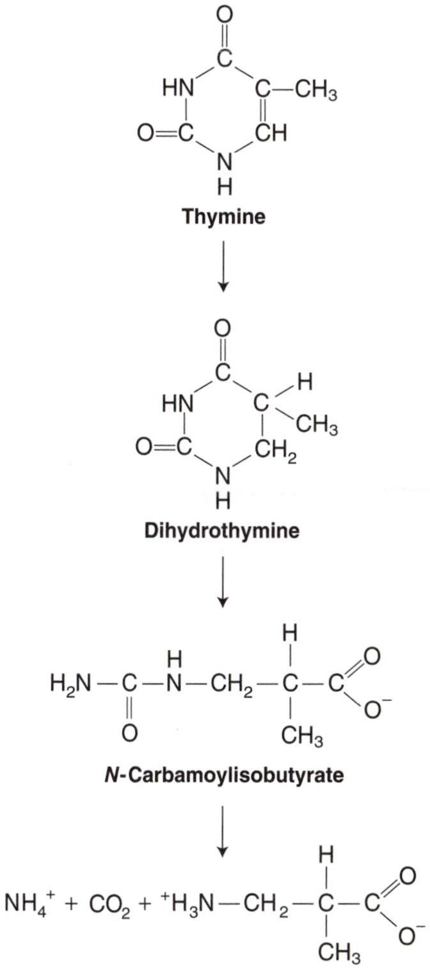

Thymine

↓

Dihydrothymine

↓

$H_2N-C(=O)-NH-CH_2-CH(CH_3)-COO^-$

N-Carbamoylisobutyrate

↓

$NH_4^+ + CO_2 + {}^+H_3N-CH_2-CH(CH_3)-COO^-$

β-Aminoisobutyrate

6. In order to study the enzymes of the salvage reactions for nucleotide synthesis in *E. coli,* you plan to grow a large batch of cells in a liquid

medium. You have a choice between a simple glucose-salts medium and a medium that includes a hydrolyzed extract prepared from yeast cells. Which of these media should you choose? Why?

7. You wish to prepare ^{14}C-labeled purines by growing bacteria in a medium containing a suitably labeled precursor. The only precursors available are amino acids that are all uniformly labeled to the same specific activity per carbon atom. Which of the amino acids would you use to obtain purine rings that are labeled to the highest specific activity?

8. A deficiency of glucose 6-phosphatase in the liver is responsible for the clinical disorder known as von Gierke's syndrome, as is discussed on page 598 of Stryer's text. Among the clinical findings in victims of this disease is hyperuricemia, an elevation of urate levels in the blood. Several explanations for this particular finding have been offered.

(a) Lactate interferes with the secretion of urate by the kidney tubules. Why are lactate concentrations often elevated in patients with von Gierke's syndrome, and how could interference with the tubular secretion of urate cause hyperuricemia?
(b) The incorporation of 1-^{14}C-labeled glycine into urate is higher in patients with von Gierke's disease than in normal people. This indicates that the rate of synthesis of purines is elevated in these patients. Can you suggest why?

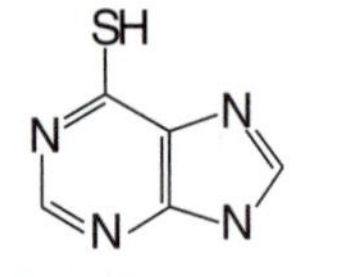

6-Mercaptopurine

9. 6-Mercaptopurine, after its conversion to the corresponding nucleotide (6-thioinosine-5′-monophosphate) through salvage reactions, is a potent inhibitor of purine biosynthesis. It is therefore a clinically useful anticancer agent. The chemotherapeutic effectiveness of 6-mercaptopurine is enhanced when it is administered with allopurinol.

(a) Briefly describe the salvage reactions required to convert 6-mercaptopurine to the corresponding nucleotide.
(b) What step in the biosynthesis of purines is likely to be inhibited by 6-thioinosine-5′-monophosphate?
(c) Explain why allopurinol enhances the effectiveness of 6-mercaptopurine as a chemotherapeutic agent.

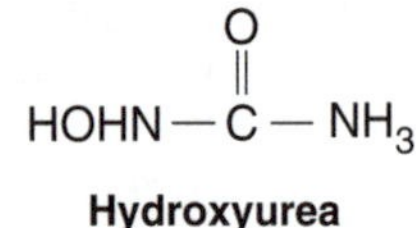

Hydroxyurea

10. Hydroxyurea, a potent chelator of ferric ions, has been shown to interfere with DNA synthesis, and it is used as an antitumor agent. What is the target enzyme for hydroxyurea?

11. Infants with hereditary orotic aciduria type I are deficient in the enzymes orotate phosphoribosyl transferase and orotidylate decarboxylase. Among the clinical findings in these patients are developmental retardation, anemia, and orotic acid in the urine. Treatment includes the oral administration of a source of pyrimidine nucleotides.

(a) Would you choose cytosine, uridine, or UMP as a source of pyrimidine nucleotides for affected infants? Why?
(b) Red blood cells from patients with orotic aciduria type I exhibit increased activities of aspartate transcarbamoylase and dihydroorotase. These activities return to a more normal level when a source of pyrimidine nucleotides is provided. Can you explain these findings?

12. Nucleoside phosphorylases catalyze the interconversion of bases and nucleosides through the following reactions:

$$\text{Ribose 1-phosphate} + \text{base} \rightleftharpoons \text{ribonucleoside} + P_i$$

or

$$\text{Deoxyribose 1-phosphate} + \text{base} \rightleftharpoons \text{deoxyribonucleoside} + P_i$$

The equilibrium constant for each of these reactions is close to 1.

(a) The pathway for the incorporation of radioactive thymine into bacterial DNA includes a step catalyzed by nucleoside phosphorylase. It has often been observed that the incorporation of thymine into DNA is enhanced when deoxyadenosine or deoxyguanosine is added to the medium. Can you explain this observation? Why might deoxyguanosine be preferable to deoxyadenosine?

(b) In cells that cannot carry out de novo synthesis of IMP, inosine can be utilized to produce IMP but only through an indirect salvage route because of the absence of inosine kinase. Suggest an alternative pathway for the formation of IMP from inosine. Among the enzymes you will need are nucleoside phosphorylase and phosphoribomutase, which isomerizes ribose 1-phosphate to ribose 5-phosphate.

13. In contrast to the normal bacterial and mammalian enzymes, the ribonucleotide reductase coded for by bacteriophage T4 is *activated* by dATP. Why?

14. (a) Write out the steps for the formation of glycinamide ribonucleotide from glycine and 5-phosphoribosyl-1-amine, and show how ATP participates.

(b) Suppose that you incubate glycinamide ribonucleotide synthase with glycinamide ribonucleotide, ADP, and inorganic phosphate labeled with ^{18}O. Will you recover any ^{18}O-labeled glycine? Why?

15. Many multivitamin preparations contain nicotinamide. Most mammalian cells contain cytosolic enzymes that convert nicotinamide directly to NAD^+. What other substrates are required for the formation of NAD^+ from nicotinamide? How could PRPP and ATP be used as sources of those substrates?

16. Which carbons of glucose must be radioactively labeled to provide 5-methyl ^{14}C-thymidylate?

17. Hypoxanthine-guanine phosphoribosyl transferase (HGPRT), a salvage enzyme of nucleotide metabolism, uses 5′-phosphoribosylpyrophosphate (PRPP) to convert hypoxanthine to IMP and guanine to GMP. A deficiency of this enzyme can lead to an increased level of purine synthesis, excess formation of uric acid, and hyperuricemia, or gout.

(a) How might a deficiency in HGPRT stimulate purine synthesis?

(b) Under what conditions might one expect a deficiency of hypoxanthine-guanine phosphoribosyl transferase to affect the rate of *pyrimidine* nucleotide synthesis? How could you estimate the rate of pyrimidine nucleotide synthesis in humans?

18. While free purine and pyrimidine bases generated by nucleotide degradation are recycled through salvage pathways, ribose 5-phosphate molecules can be used as sources of metabolic energy. Outline a metabolic route from ribose 5-phosphate to pyruvate in a liver cell.

19. In mammals, the committed step for pyrimidine synthesis is catalyzed by carbamoyl phosphate synthetase, while in bacteria, the committed step is the formation of *N*-carbamoylaspartate, catalyzed by aspartate transcarbamoylase.

(a) Account for these differences in mammals and bacteria.
(b) Bacterial carbamoyl phosphate synthetase is only partially inhibited by UMP. Why?

20. The synthesis of deoxythymidylate can proceed not only from dUMP but also from dCMP. The route from dCMP begins with the formation of dCDP from CDP, catalyzed by ribonucleotide reductase, followed by dephosphorylation to dCMP, and the deamination of dCMP to form dUMP, catalyzed by dCMP deaminase.

$$\text{dCMP} + H_2O + H^+ \longrightarrow \text{dUMP} + NH_3^+$$

dCMP deaminase is an allosteric enzyme that is stimulated by dCTP and inhibited by dTTP. Account for these effects and relate them to the regulation of ribonucleotide reductase by deoxynucleoside triphosphates.

21. In vigorously contracting muscle, AMP levels increase as a result of ATP hydrolysis and the activity of adenylate kinase. Some of the AMP produced is introduced into a pathway called the *purine nucleotide cycle,* which includes AMP deaminase as well as two enzymes of the purine biosynthetic pathway. In this cycle, AMP is deaminated to IMP by AMP deaminase, and then IMP is converted to AMP by the sequential action of adenylosuccinate synthetase and adenylosuccinate lyase. The importance of the cycle is demonstrated by the observation that people deficient in AMP deaminase suffer from fatigue and cramping during moderate exercise.

(a) Outline the purine nucleotide cycle, including all substrates. What products are generated in a turn of the cycle? How could any of these products compensate for the fact that muscle tissue has only a low level of pyruvate carboxylase?
(b) Explain why people deficient in AMP deaminase are unable to carry out sustained muscular contraction.

ANSWERS TO PROBLEMS

1. The experiments showed that ribonucleosides are converted *intact* to deoxyribonucleotides. Had cleavage occurred to produce the free sugar and base, the ribose and cytosine would have been incorporated into pools that are likely to be of different sizes. The subsequently synthesized deoxyribonucleotides would then contain deoxyribose and cytosine with a different ratio of specific activities. These results also suggested that free ribose is

not converted to deoxyribose before deoxyribonucleotides are synthesized. As a result of these experiments, biochemists began to search for enzymatic activities that convert intact ribonucleotides to deoxyribonucleotides. Their work led to the discovery of ribonucleotide reductase, which reduces ribonucleoside diphosphates to deoxyribonucleoside diphosphates.

2. (a) Adenosine is phosphorylated to AMP, with ATP serving as the phosphoryl donor, in a reaction carried out by a specific nucleoside kinase. The conversion of AMP to ADP through the action of a specific nucleoside monophosphate kinase is accomplished, with ATP again utilized as a phosphate donor. Ribonucleotide reductase catalyzes the reduction of ADP to dADP, which is then converted to dATP by nucleotide diphosphokinase.

   (b) High concentrations of dATP displace ATP from the overall activity site on ribonucleotide reductase, which lowers the rate of synthesis of all four deoxyribonucleoside diphosphates. This in turn leads to a depletion of deoxyribonucleoside triphosphates, which are the substrates for DNA synthesis.

3. Methotrexate blocks the regeneration of tetrahydrofolate from dihydrofolate, which is produced during the synthesis of thymidylate. The failure to regenerate tetrahydrofolate means that those biochemical reactions in the cell that depend on one-carbon metabolism cannot be carried out. One of the products of tetrahydrofolate is methylenetetrahydrofolate, which is used as a substrate by thymidylate synthetase and is required for inhibition of the enzyme by F-dUMP. A deficiency of methylenetetrahydrofolate means that F-dUMP cannot irreversibly inactivate thymidylate synthetase. Conversely, F-dUMP prevents the formation of dihydrofolate, thereby abolishing the adverse effects caused by the depletion of tetrahydrofolate in the cell.

4. (a) The presence of orotic acid, a precursor of pyrimidines, in the urine suggests that carbamoyl phosphate synthesized in mitochondria is not utilized there. Instead, carbamoyl phosphate enters the cytosol, where it stimulates an increase in the rate of synthesis of precursors of pyrimidines, including orotic acid. An excess of carbamoyl phosphate arises in mitochondria whenever any of the urea cycle enzymes are deficient. Such a condition will lead to hyperammonemia, as well as to the accumulation of carbamoyl phosphate. Although a deficiency in mitochondrial carbamoyl phosphate synthetase leads to hyperammonemia, it cannot lead to an accumulation of mitochondrial carbamoyl phosphate and therefore does not stimulate pyrimidine synthesis in the cytosol.

   (b) Cytoplasmic carbamoyl phosphate synthetase is involved primarily in the pathway for pyrimidine synthesis, not for the assimilation of ammonia. Recall that, in the cytosol, the substrate for the formation of carbamoyl phosphate is glutamine, not ammonia. A deficiency of carbamoyl phosphate synthesis in the cytosol would cause a depletion of pyrimidines. Such a deficiency is treated by administration of uracil or uridine, which are precursors of UMP and CMP.

5. The transamination of β-aminoisobutyrate to form methylmalonate semialdehyde requires pyridoxal phosphate as a cofactor. This reaction is

similar to the conversion of ornithine to glutamate γ-semialdehyde. Then NAD^+ serves as an electron acceptor for the oxidation of methylmalonate semialdehyde to methylmalonate. The conversion of methylmalonate to methylmalonyl CoA requires coenzyme A. The final reaction, in which methylmalonyl CoA is converted to succinyl CoA, is catalyzed by methylmalonyl CoA mutase, an enzyme that contains a derivative of vitamin B_{12} as its coenzyme.

6. You should choose the medium that includes the extract prepared from yeast cells because it contains nucleotides, nucleosides, and purine and pyrimidine bases. All these compounds can be utilized by the enzymes of the bacterial salvage pathways to synthesize the nucleotides needed for growth and division, thereby making the de novo synthesis of purines and pyrimidines less necessary. Bacteria growing in a simple glucose-salts medium are required to carry out de novo synthesis of the nucleotides; in these bacteria the activities of the enzymes of the salvage reactions would be lower because, in rapidly growing cells, the concentrations of the breakdown products of nucleotide metabolism would be lower.

7. Examination of the pathway for purine synthesis shows that only glycine is incorporated intact into the purine ring at the C-4 and C-5 positions. Therefore, glycine is a good choice as the radiolabeled precursor. Serine can also be considered because it is a precursor of glycine and the ultimate donor of C-1 groups to tetrahydrofolate, and activated tetrahydrofolate derivatives participate in two reactions in the formation of purines. Whether serine is a better choice than glycine depends on the relative amounts of the two unlabeled amino acids in the cell.

8. (a) A deficiency of glucose 6-phosphatase in the liver leads to increased concentrations of glucose 6-phosphate in that organ. This causes an increase in the rate of glycolysis, one result of which is the elevation of both lactate and pyruvate concentrations. Because lactate interferes with the tubular secretion of urate into the urine, increased levels of urate are found in the blood.

(b) Among the alternative pathways that the excess glucose 6-phosphate in the liver can enter is the pentose phosphate pathway, and among the products of that pathway is ribose 5-phosphate. One explanation for the increased rate of purine synthesis is that an elevation in the level of ribose 5-phosphate results in an increase in the synthesis of PRPP. Since PRPP is a precursor of the purines, their synthesis is also stimulated. This in turn leads to higher concentrations of urate, which is a breakdown product of the purine nucleotides.

9. (a) 6-Mercaptopurine is converted to the mononucleotide through the action of hypoxanthine-guanine phosphoribosyl transferase, which uses PRPP to add 5′-phosphoribose to the purine ring. The resulting compound is 6-thioinosine-5′-monophosphate, an analog of IMP.

(b) In the purine biosynthetic pathway, IMP is used for the synthesis of AMP via adenylosuccinate and for the synthesis of XMP, a precursor of GMP. Either step in IMP utilization could be impaired by 6-thioinosine-5′-monophosphate, an analog of IMP. In either case, the tissue would be deprived of an essential purine nucleotide.

(c) Allopurinol is also converted to a nucleotide through the action of hypoxanthine-guanine phosphoribosyl transferase, and the consumption of the corresponding substrate, PRPP, could make less substrate available for the formation of IMP and the purines. This could make 6-thioinosine-5′-monophosphate more effective at lower concentrations. Allopurinol also inhibits amidophosphoribosyl transferase, one of the early enzymes in purine synthesis; this might also allow the thio analog to be more effective at lower concentrations in inhibiting purine biosynthesis. Another possibility is that allopurinol inhibits purine degradation by inhibiting xanthine and hypoxanthine oxidation. A decrease in the rate of degradation of the thiopurine would also potentiate its action.

10. Hydroxyurea inhibits ribonucleotide reductase. By sequestering ferric ions, hydroxyurea destabilizes the organic free radical in the B2 subunit of the enzyme. The inhibition of enzyme activity leads to a depletion of deoxyribonucleoside diphosphates, which are normally converted to deoxyribonucleoside triphosphates, the substrates for DNA synthesis.

11. (a) You should choose uridine, an uncharged molecule, that can pass across plasma membranes into the cytoplasm, where it can be converted to UMP through the action of a nucleoside kinase. The synthesis of CMP is dependent on the availability of UMP. While cytosine can be used to generate CMP in the cell, it cannot be used to generate UMP. The administration of UMP is not recommended because its phosphate group is charged at neutral pH, making its transport across the plasma membrane difficult.

(b) The observations suggest that pyrimidine nucleotides control either enzyme synthesis or catalytic activities of preexisting enzymes. Recall that carbamoyl phosphate synthetase, aspartate transcarbamoylase, and dihydroorotase are covalently joined on a single polypeptide chain, which in turn suggests that the activities are coded for by a single gene. Regulation of their synthesis by UTP offers a convenient way to regulate pyrimidine synthesis, as does allosteric control by UTP or another product of the pyrimidine pathway. Although it is known that UTP inhibits carbamoyl phosphate synthetase activity, more work needs to be done to establish the molecular basis for regulation of the pyrimidine biosynthetic pathway.

12. (a) Deoxyribonucleosides such as deoxyadenosine can be converted to the free base and deoxyribose 1-phosphate by nucleoside phosphorylase. Increased levels of deoxyribose 1-phosphate are then available for the formation of deoxythymidine from thymine in the reverse reaction catalyzed by nucleoside phosphorylase. Deoxyguanosine might be preferable to deoxyadenosine because the conversion of elevated levels of deoxyadenosine to dAMP and then to dATP could lead to the inactivation of ribonucleotide reductase, which is sensitive to the concentration of dATP.

(b) Inosine is cleaved to produce hypoxanthine and ribose 1-phosphate through the action of nucleoside phosphorylase; note that inorganic phosphate is required for this reaction. Hypoxanthine-guanine phosphoribosyl transferase converts free hypoxanthine to IMP by condensation with PRPP. PRPP can be derived from ribose 1-phosphate in two steps: (1) the conversion of ribose 1-phosphate to ribose 5-phosphate, which is catalyzed by phosphoribomutase, and (2) the formation of PRPP from ribose 5-phosphate and ATP, which is catalyzed by PRPP synthetase.

13. In most organisms, the inhibition of ribonucleotide reductase is desirable when an adequate number of deoxyribonucleotides have been synthesized. In bacteria and in mammals the enzyme is inhibited by high concentrations of dATP. However, the goals of a viral infection include the rapid synthesis of large quantities of deoxyribonucleotides so that many copies of the viral genome can be synthesized. Evidently the need for the unrestricted synthesis of nucleotides can account for the fact that dATP stimulates, rather than inhibits, the production of deoxyribonucleoside diphosphates.

14. (a) The pathway for the synthesis of glycinamide ribonucleotide is shown in Figure 29-1. It includes the formation of glycinyl phosphate, an acyl phosphate. The amino group of phosphoribosylamine then displaces the phosphate group of glycinyl phosphate as an amide linkage is formed.

(b) The reverse of the reaction that produces glycinamide ribonucleotide would yield ^{18}O-labeled glycinyl phosphate, which could then react with ADP to form glycine and ATP. One of the oxygens of the carboxylate group in glycine is thus derived from ^{18}O-labeled inorganic phosphate. Experiments of this type were performed to show that glycinyl phosphate is an intermediate in the pathway to glycinamide ribonucleotide.

Ribose-adenine

ATP

Glycine

Glycinyl phosphate

Ribose-adenine

ADP

Glycinyl phosphate

Phosphoribosylamine

Glycinamide ribonucleotide

P_i

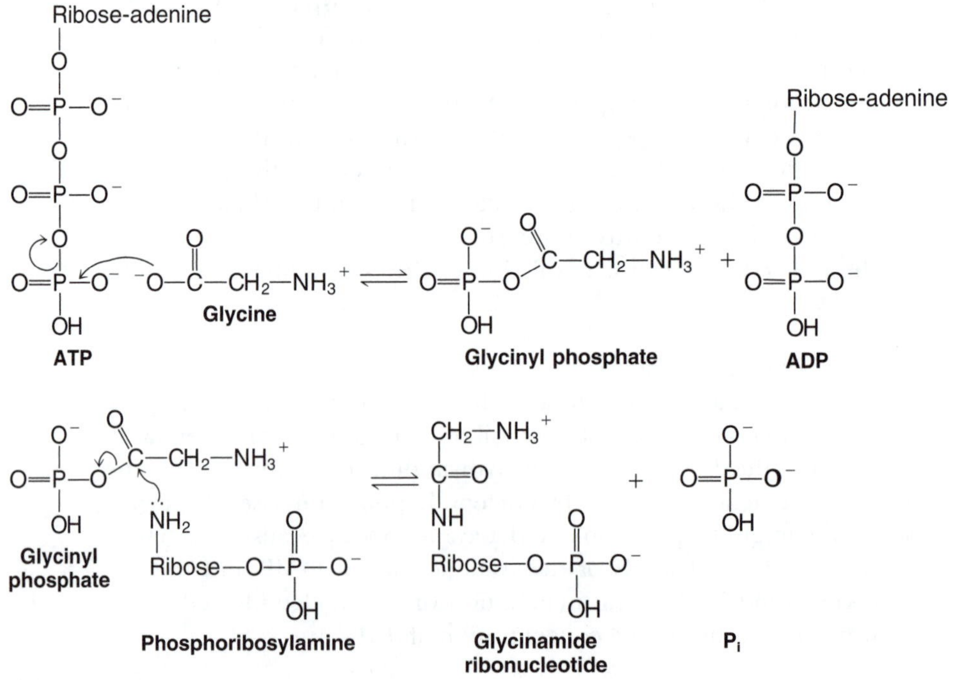

Figure 29-1
Formation of glycinamide ribonucleotide from glycine and 5-phosphoribosyl-1-amine. Glycinyl phosphate is an intermediate in the reaction.

15. Both ribose phosphate and an AMP moiety must be added to nicotinamide in order to convert it to NAD^+. The ribose phosphate is derived from PRPP when a phosphoribosyl transferase catalyzes the formation of nicotinamide ribonucleotide or nicotinamide mononucleotide. The final step utilizes ATP, which serves as an adenyl donor for the formation of the dinucleotide NAD^+.

16. When glucose is converted to 2 molecules of 3-phosphoglycerate by the glycolytic pathway, C-1 and C-6 of glucose become C-3 of 3-phosphoglycerate, which is subsequently converted to C-3 of serine. This carbon is then transferred to tetrahydrofolate in a serine transhydroxymethylase–catalyzed reaction that produces 5,10-methylenetetrahydrofolate, which in turn provides the 5-methyl group of thymidylate.

17. (a) When active, HGPRT consumes PRPP as it catalyzes the synthesis of GMP and IMP. Decreased flux through this reaction raises the steady-state level of PRPP, thereby increasing the activity of PRPP amidotransferase, which catalyzes the initial step in purine synthesis. Increased activity may make the amidotransferase resistant to feedback inhibition by AMP and GMP, the end products of the purine biosynthetic pathway.

(b) As noted above, decreased activity of HGPRT increases the concentration of PRPP. This increases the rate of pyrimidine synthesis at the orotate phosphoribosyl transferase reaction, if PRPP levels are normally subsaturating for that enzyme. Orotate incorporation into nucleotide pools or into nucleic acids would give a reasonable estimate of de novo pyrimidine nucleotide synthesis.

18. Ribose 5-phosphate molecules are converted to pyruvate through the nonoxidative branch of the pentose phosphate pathway and the glycolytic pathway from glyceraldehyde 3-phosphate and fructose 6-phosphate. The initial steps involve epimerization of some molecules of ribose 5-phosphate to xylulose 5-phosphate, required for transketolase reactions that convert the two pentose phosphates to sedoheptulose 7-phosphate and glyceraldehyde 3-phosphate. Other reactions involving transketolase as well as transaldolase generate additional molecules of glyceraldehyde 3-phosphate and fructose 6-phosphate. Fructose 6-phosphate can be converted to glyceraldehyde 3-phosphate through the action of phosphofructokinase, aldolase, and triosephosphate isomerase. The steps that convert glyceraldehyde 3-phosphate to pyruvate are those of the glycolytic pathway.

19. (a) Bacteria use a single form of carbamoyl phosphate synthetase not only for the synthesis of pyrimidines but also for the synthesis of arginine. Arginine biosynthesis begins with glutamate and includes formation of citrulline from ornithine and carbamoyl phosphate, a pathway that resembles urea formation in mammals. Two forms of carbamoyl phosphate synthetase, a cytoplasmic form for pyrimidine synthesis and a mitochondrial form for arginine and urea synthesis, are employed in mammals. Because the two pathways are compartmentalized, the formation of carbamoyl phosphate in the cytosol is regarded as the committed step for pyrimidine synthesis.

(b) UMP does not completely inhibit bacterial carbamoyl phosphate synthetase because that inhibition would interfere with arginine production.

20. An increase in dCTP levels signals that the cell has ample deoxynucleotides for DNA synthesis and that there is a need for thymidylate synthesis. An increase in dTTP levels signals that the activity of thymidylate synthase can be decreased, and the inhibition of dCMP deaminase by dTTP reduces the input of dUMP into the pathway. While ribonucleotide reductase is subject to regulation by other deoxynucleotides, it is not subject to allosteric regulation by dCTP. Instead it appears that regulation of dCMP deaminase provides a second control point for the generation of deoxynucleotides in the cell.

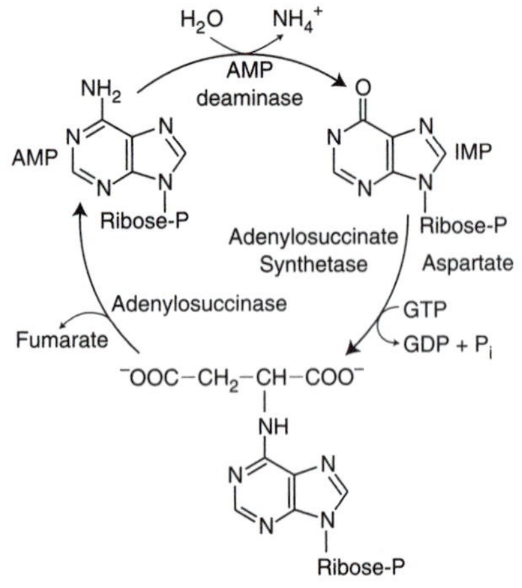

21. (a) An outline of the purine nucleotide cycle is shown in the margin. The net products of the cycle are ammonia and fumarate, an intermediate in the citric acid cycle. A low level of activity for pyruvate carboxylase in muscle means that little oxaloacetate can be generated from pyruvate when there is increased demand for an increase in the rate of the citric acid cycle. Such an increase would be required during muscle contraction to generate additional ATP molecules. Fumarate from the purine nucleotide cycle provides an alternative source of carbons that can serve to increase oxaloacetate levels as well as those of other citric acid cycle intermediates and to increase the rate of flux through the cycle.

(b) A deficiency in AMP deaminase causes a reduction in the generation of ammonia and IMP during muscle contraction. In turn, fumarate synthesis is inhibited because fewer IMP molecules are available as acceptors for aspartyl groups in the formation of adenylosuccinate. The flux of intermediates through the citric acid cycle cannot be increased if no additional fumarate molecules are available, so that an increase in ATP synthesis in response to muscle contraction cannot be realized.

EXPANDED SOLUTIONS TO TEXT PROBLEMS

1. Glucose + 2 ATP + 2 $NADP^+$ + H_2O $\longrightarrow$
PRPP + CO_2 + ADP + AMP + 2 NADPH + H^+

This equation is the summation of the following conversions: Glucose $\longrightarrow$ glucose 6-phosphate $\longrightarrow$ ribose 5-phosphate $\longrightarrow$ PRPP (see Chapters 19 and 22 and p. 741 in Stryer).

2. Glutamine + aspartate + CO_2 + 2 ATP + NAD^+ $\longrightarrow$
orotate + 2 ADP + 2 P_i + glutamate + NADH + H^+

This equation is the summation of reactions shown on page 746 in Stryer.

3. (a, c, d, e) PRPP, (b) carbamoyl phosphate

4. PRPP and formylglycinamide ribonucleotide accumulate because they are reactions 1 and 4 in the first stage of purine biosynthesis (Stryer, p. 742). If these glutamine-requiring amidotransferase reactions are inhib-

ited, the precursor molecules will accumulate. However, since the synthesis of formylglycinamide requires that PRPP be converted to phosphoribosylamine, probably only PRPP will accumulate significantly.

5. dUMP + serine + NADPH + $H^+ \longrightarrow$ dTMP + $NADP^+$ + glycine

This is the summation of reactions shown in Stryer, Figure 29-18. Note that the coenzyme, tetrahydrofolate, is regenerated; it acts catalytically and does not appear in the equation.

6. There is a deficiency of N^{10}-formyltetrahydrofolate (see Figure 29-6 in Stryer). Sulfanilamide inhibits the synthesis of folate by acting as an analog of *p*-aminobenzoate, one of the precursors of folate.

7. PRPP is the activated intermediate in the synthesis of (a) phosphoribosylamine in the de novo pathway of purine formation, (b) purine nucleotides from free bases by the salvage pathway, (c) orotidylate in the formation of pyrimidines, (d) nicotinate ribonucleotide, (e) phosphoribosyl-ATP in the pathway leading to histidine, and (f) phosphoribosylanthranilate in the pathway leading to tryptophan.

8. The small subunit releases ammonia from glutamine. The nascent ammonia then reacts with an activated form of CO_2 that is formed by the large subunit. The bicarbonate-dependent ATPase activity implies that this activated species is carbonyl phosphate. Reaction of this carbonic-phosphoric mixed anhydride with NH_3 yields carbamate, which would then react with ATP to give carbamoyl phosphate. See C. Walsh, *Enzymatic Reaction Mechanisms* (W.H. Freeman, 1979), pp. 150–154. The holoenzyme probably contains a channel for the diffusion of nascent NH_3 from the small subunit to the large one. Recall that tryptophan synthetase (Stryer, p. 726) contains a channel for the diffusion of indole from the α to the β subunit. An intriguing possibility is that the catalytic activities of the subunits of carbamoyl phosphate synthetase are coordinated: NH_3 production by the small subunit is stimulated when carbonyl phosphate is formed by the large subunit.

9. Analogous reactions occur in the urea cycle—from citrulline to argininosuccinate, and then to arginine (Stryer, p. 635).

10. In vitamin B_{12} deficiency, methyltetrahydrofolate cannot donate its methyl group to homocysteine to regenerate methionine. Because the synthesis of methyltetrahydrofolate is irreversible (Stryer, p. 720), the cell's tetrahydrofolate will ultimately be converted into this form. No formyl or methylene tetrahydrofolate will be left for nucleotide synthesis. Pernicious anemia illustrates the intimate connection between amino acid metabolism and nucleotide metabolism.

11. (a) *S*-Adenosylhomocysteine hydrolase activity is markedly diminished in ADA-deficient patients because of reversible inhibition by adenosine and suicide inactivation by 2′-deoxyadenosine.

(b) The activated methyl cycle (Stryer, p. 722) is blocked in these patients. *S*-Adenosylhomocysteine is a potent inhibitor of methyl transfer reactions involving *S*-adenosylmethionine. For a discussion of adenosine deaminase deficiency, see the article by N.M. Kredich and M.S. Hershfield in C.R. Scriver, A.L. Beaudet, W.S. Sly, and D. Vale (eds.), *The Metabolic Basis of Inherited Disease,* 6th ed. (McGraw-Hill, 1989), pp. 1045–1075.

12. (a) Cell A cannot grow in a HAT medium, because it cannot synthesize dTMP either from thymidine or from dUMP. Cell B cannot grow in this medium, because it cannot synthesize purines by either the de novo pathway or the salvage pathway. Cell C can grow in a HAT medium because it contains active thymidine kinase from cell B (enabling it to phosphorylate thymidine to dTMP) and hypoxanthine-guanine phosphoribosyl transferase from cell A (enabling it to synthesize purines from hypoxanthine by the salvage pathway).

(b) Transform cell A with a plasmid containing foreign genes of interest and a functional thymidine kinase gene. The only cells that will grow in a *HAT* medium are those that have acquired a thymidylate kinase gene; nearly all these transformed cells will also contain the other genes on the plasmid.

13. These patients have a high level of urate because of the breakdown of nucleic acids. Allopurinol prevents the formation of kidney stones and blocks other deleterious consequences of hyperuricemia by inhibiting the formation of urate (Stryer, p. 757).

14. Though often termed *binding constants,* the values given are really *dissociation constants.* Therefore, to calculate the free energy of binding, one uses the reciprocals of the values given. Thus, for the wild type,

$$\Delta G^{\circ\prime} = -1.36 \log \frac{1}{7 \times 10^{-11}} = -1.36 \times 10.15$$

$$= -13.8 \text{ kcal/mol}$$

Similar calculations for Asn 27 and Ser 27 will give the results in Stryer's answer.

15. The cytosolic level of ATP in liver falls and that of AMP rises above normal in all three conditions. The excess AMP is degraded to urate. See C.R. Scriver, A.L. Beaudet, W.S. Sly, and D. Vale (eds), *The Metabolic Basis of Inherited Disease,* 6th ed. (McGraw-Hill, 1989), pp. 984–988, for an illuminating discussion.

CHAPTER

30

Integration of Metabolism

This chapter, which concludes the two major sections of the text devoted to metabolism, provides an integrated view of mammalian metabolism and a review of the principal themes of metabolism. Stryer starts with a recapitulation of the roles of ATP, NADPH, and the building-block molecules derived from fuels in biosynthesis and cellular processes; he also stresses the regulatory mechanisms that control metabolism. (Chapters 10 and 17 contain important background material for these sections.) Stryer then reviews the major metabolic pathways, from glycolysis through fatty acid metabolism (Chapters 19 through 24), and stresses the roles of glucose 6-phosphate, pyruvate, and acetyl CoA as key intermediates at junctions between the various metabolic pathways. The metabolic characteristics of the major organs are presented next, followed by a description of the major hormonal effects on their metabolism. The special role of the liver in buffering blood glucose levels is also considered. Finally, the adaptations of metabolism to intense exercise and to prolonged starvation and diabetes are described.

LEARNING OBJECTIVES

When you have mastered this chapter, you should be able to complete the following objectives:

Strategy and Regulation of Metabolism (Stryer pages 764–766)

1. Review the sources of *ATP, NADPH,* and the *building-block molecules* and their roles in biosyntheses.

2. List the general mechanisms for the *regulation* of *metabolism.* Give examples of metabolic regulation in the major metabolic pathways.

3. Locate the enzymes of the major metabolic pathways in the *compartments* of a eukaryotic cell.

Major Metabolic Pathways and Key Intermediates (Stryer pages 766–770)

4. Describe the roles of *glycolysis,* name its products, and describe the regeneration of *NAD^+* under aerobic and anaerobic conditions.

5. Outline the regulation of *phosphofructokinase* in the liver and in muscle.

6. Discuss the roles of the *citric acid cycle,* the *electron transport chain,* and *oxidative phosphorylation* in the oxidative degradation of fuels. Explain the control of these pathways by the availability of *ADP.*

7. Describe the roles and the regulation of the *pentose phosphate pathway.*

8. Discuss the physiologic role of *gluconeogenesis* and the reciprocal regulation of gluconeogenesis and glycolysis.

9. Outline the *synthesis* and *degradation* of *glycogen* and their coordinated control by *phosphorylation* and *dephosphorylation* and allosteric effectors.

10. Summarize the metabolism of *fatty acids.* Describe the regulation of *fatty acid β-oxidation* and *fatty acid synthesis.* Explain how compartmentation is involved in these processes.

11. Describe the fates of *glucose 6-phosphate* in cells. Outline the pathways that give rise to glucose 6-phosphate, and discuss the release of glucose into the blood.

12. Describe the sources and the key metabolic products of *pyruvate.*

13. Discuss the sources and fates of *acetyl CoA.*

Metabolic Profiles of the Major Organs, Hormonal Regulators, and the Control of Blood Glucose Levels (Stryer pages 770–775)

14. Summarize the fuel requirements of the *brain* and note the relationship of blood to brain glucose levels.

15. Explain the use of fuels by resting and active *muscle.*

16. Describe the synthesis and turnover of *triacylglycerols* by *adipose tissue.* Indicate the role of glucose in this tissue.

17. Discuss the role of the *liver* in providing glucose to other tissues and in regulating lipid metabolism. List the fuels used by the liver for its own needs.

18. Explain and contrast the metabolic effects of *insulin, glucagon,* and *epinephrine* or *norepinephrine.*

19. Describe how the *blood glucose level* is controlled by the liver in response to glucagon and insulin. Discuss the contributions of muscle and adipose tissue to the regulation of the blood glucose level.

Metabolic Adaptations in Starvation, Intense Exercise, and Diabetes
(Stryer pages 775–781)

20. List the approximate *fuel reserves* of a 70-kg man in *kilocalories.*

21. Describe the metabolic changes that occur after one and three days of *starvation.*

22. Discuss the metabolic adaptations that occur after prolonged starvation; note especially the shift in brain fuels and the decreased rate of protein degradation.

23. Explain the role of large *fat deposits* in migratory birds.

24. Discuss the different patterns of fuel use in short- and long-distance running.

25. Describe the metabolic derangements in *diabetes mellitus* resulting from relative insulin insufficiency and glucagon excess.

26. Describe the formation of hemoglobin A_{Ic} and its use in monitoring blood glucose in diabetic patients.

SELF-TEST

Strategy and Regulation of Metabolism

1. Biosynthetic pathways that require NADPH include which of the following?

(a) Gluconeogenesis
(b) Fatty acid biosynthesis
(c) Ketone body formation
(d) Cholesterol biosynthesis
(e) Tyrosine biosynthesis

2. Match the biomolecules in the left column with their precursors from the right column.

(a) Phospho-glycerides __________
(b) Porphyrins __________
(c) Cholesterol __________
(d) Triacylglycerols __________
(e) Nucleic acids __________

(1) *S*-Adenosylmethionine
(2) Acetyl CoA
(3) Ribose 5-phosphate
(4) Dihydroxyacetone phosphate
(5) Tetrahydrofolate
(6) Succinyl CoA

3. Consider the following examples of metabolic regulation:

(1) Fatty acid oxidation in mitochondria is diminished when fatty acid biosynthesis in the cytosol is active due to the inhibition of carnitine acyltransferase I by malonyl CoA.
(2) The synthesis of HMG CoA reductase in various cells is inhibited by low-density lipoproteins.
(3) Glucose 6-phosphatase is present in the liver and kidneys but not in muscle.
(4) Amidophosphoribosyl transferase, the enzyme that catalyzes the committed step in the biosynthesis of purine nucleotides, is inhibited by all purine nucleotides.
(5) The enzyme that catalyzes the synthesis and degradation of fructose 2,6-bisphosphate is phosphorylated and dephosphorylated in response to hormonal signals.

Indicate which of these examples apply to each of the following modes of metabolic regulation:

(a) Allosteric interactions __________
(b) Covalent modifications __________
(c) Enzyme levels __________
(d) Compartmentation __________
(e) Metabolic specialization of organs __________

4. In which compartment of a liver cell are ketone bodies synthesized?

(a) Plasma membrane
(b) Cytosol
(c) Lysosomes
(d) Mitochondria
(e) Endoplasmic reticulum

Major Metabolic Pathways and Key Intermediates

5. Match each metabolic pathway in the left column with its *major* role in metabolism from the right column.

(a) Glycolysis __________
(b) Gluconeo-genesis __________
(c) Pentose phosphate pathway __________
(d) Glycogen synthesis __________
(e) Fatty acid degradation __________

(1) Control of glucose levels in blood
(2) Formation of NADH and $FADH_2$
(3) Storage of fuel
(4) Synthesis of NADPH and ribose 5-phosphate
(5) Production of ATP and building blocks of biomolecules

6. The control of phosphofructokinase in the liver and in muscle is different. Both epinephrine and glucagon initiate responses to low glucose levels, yet epinephrine stimulates glycolysis in muscle, whereas glucagon inhibits glycolysis in the liver. Explain this fact.

____

____

7. Explain how it is possible to have high $NADPH/NADP^+$ and low $NADH/NAD^+$ ratios in the cytosol of the same cells.

____

____

8. Which of the following answers completes the sentence correctly? Regulation of fatty acid biosynthesis occurs at the enzymatic step catalyzed by

(a) carnitine acyltransferase I.
(b) acetyl CoA carboxylase.
(c) pyruvate carboxylase.
(d) citrate synthase.
(e) citrate-malate translocase.

9. Match the three key metabolic intermediates in the left column with their major products from the right column. Indicate the most direct relationships, that is, those not separated by other key intermediates.

(a) Glucose 6-phosphate ____________
(b) Pyruvate ____________
(c) Acetyl CoA____________

(1) Ketone bodies
(2) Oxaloacetate
(3) Pyruvate
(4) Glycogen
(5) CO_2
(6) Lactate
(7) Ribose 5-phosphate
(8) Fatty acids
(9) Alanine
(10) Cholesterol
(11) Acetyl CoA

Metabolic Profiles of the Major Organs, Hormonal Regulators, and the Control of Blood Glucose Levels

10. In adipose tissue, glucose 6-phosphate is not converted into which of the following?

(a) Pyruvate
(b) Glycogen
(c) Glucose
(d) Ribose 5-phosphate

11. In the liver, the major fates of pyruvate include the formation of which of the following?

(a) Acetyl CoA
(b) Lactate
(c) Oxaloacetate
(d) Alanine

12. Which of the following statements about the metabolism of the brain are incorrect?

(a) It uses fatty acids as fuel in the fasting state.
(b) It uses about 60% of the glucose consumed by the whole body in the resting state.

(c) It lacks fuel reserves.
(d) It can use acetoacetate and 3-hydroxybutyrate under starvation conditions.
(e) It releases lactate during periods of intense activity.

13. Which of the following tissues converts pyruvate to lactate most effectively?

(a) Liver
(b) Muscle
(c) Adipose tissue
(d) Brain
(e) Kidney

14. Adipose cells constantly break down and resynthesize triacylglycerols, but synthesis cannot proceed without an external supply of glucose. Explain why.

____

____

15. Which of the following statements about the metabolism of adipose tissue are correct?

(a) It has an active pentose phosphate pathway.
(b) It contains a hormone-sensitive lipase that hydrolyzes triacylglycerols.
(c) It uses ketone bodies as its preferred fuel.
(d) It releases fatty acids to the blood as triacylglycerols that are packaged in VLDL.
(e) It is the most abundant source of stored fuel.

16. Select the statements from the right column that best describe the metabolism of each organ, tissue, or cell in the left column.

(a) Brain ____________
(b) Muscle ____________
(c) Adipose tissue ____________
(d) Liver ____________
(e) Red blood cell ____________

(1) Releases glycerol and fatty acids into the blood during fasting periods
(2) In a normal nutritional state, utilizes glucose as the exclusive fuel
(3) Synthesizes ketone bodies when the supply of acetyl CoA is high
(4) Can release lactate into the blood
(5) Utilizes α-keto acids from amino acid degradation as an important fuel
(6) Can store glycogen but cannot release glucose into the blood
(7) Can synthesize fatty acids, triacylglycerols, and VLDL when fuels are abundant

17. Which of the following statements about ketone bodies is incorrect?
    (a) Thiolase is required for the utilization of ketone bodies as fuels.
    (b) Ketone bodies are produced in the liver.
    (c) The adipose tissue normally uses ketone bodies as fuels.
    (d) Ketone bodies are the blood transport forms of acetyl units.
    (e) During starvation, the brain derives approximately two-thirds of its fuel needs from ketone bodies.

18. When fuels are abundant, the liver does not degrade fatty acids; rather, it converts them into triacylglycerols for export as VLDL. Explain how β-oxidation of fatty acids and the formation of ketone bodies from fatty acids are prevented under these conditions.

____

____

19. Match each hormone in the left column with its properties, *major* target organ, and metabolic effects from the right column.

(a) Insulin________________
(b) Glucagon______________
(c) Epinephrine____________

(1) A catecholamine hormone
(2) A polypeptide hormone
(3) Secreted by the α cells of the pancreas
(4) Effects are mediated by cAMP
(5) Target organ is the liver
(6) Target organ is muscle
(7) Signals the fed state
(8) Secreted in response to low blood glucose levels
(9) Promotes the storage of fuels
(10) Promotes the breakdown of stored fuels

20. Use an "S" to indicate the following metabolic processes that are stimulated and an "I" to indicate those that are inhibited by the action of insulin.
    (a) Gluconeogenesis in liver ________
    (b) Entry of glucose into muscle and adipose cells ________
    (c) Glycolysis in the liver ________
    (d) Intracellular protein degradation ________
    (e) Glycogen synthesis in liver and muscle ________
    (f) Uptake of branched-chain amino acids by muscle ________
    (g) Synthesis of triacylglycerols in adipose tissue ________

21. Explain the allosteric effects of glucose on glycogen metabolism.

____

____

22. The blood glucose level of a normal person, measured after an overnight fast, is approximately 80 mg/100 ml. After a meal rich in carbohydrate, it rises to about 120 mg/100 ml and then declines to the fasting level.

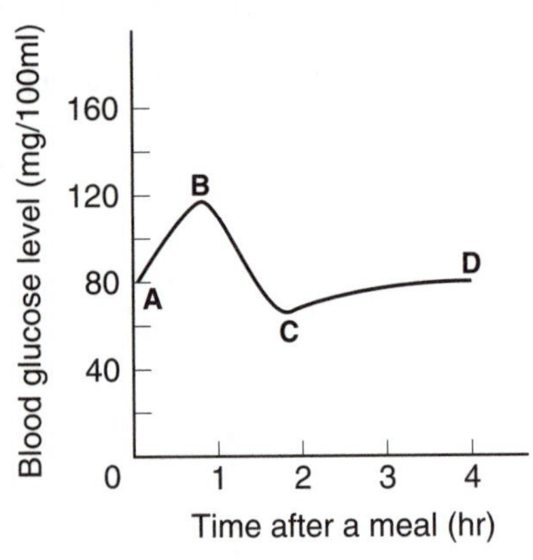

Figure 30-1
Blood glucose levels after a meal rich in carbohydrate.

The approximate time course of these changes and the inflection points are shown in Figure 30-1. After examining the figure, complete the following sentences:

(a) The increase in the glucose level from A to B is due to

____

(b) The decrease in the glucose level from B to C is due to

____

(c) The leveling off of the glucose level from C to D is due to

____

(d) The slight overshoot that is sometimes observed at C can be explained by

____

Metabolic Adaptations in Starvation, Intense Exercise, and Diabetes

23. Estimate the ratio of glucose produced by gluconeogenesis starting with 1 g of protein to that produced starting with 1 g of triacylglycerol. Assume that the protein is made up entirely of Ala residues, which have a molecular mass of 72 daltons, that the molecular mass of the triacylglycerol is 890 daltons, and that the mass contributed by the glycerol moiety is 89 daltons. Also assume complete conversion.

24. Match the fuel storage forms in the left column with the most appropriate characteristics from the right column.

(a) Glycogen ______________
(b) Triacylglycerols______________
(c) Protein ______________

(1) Largest storage form of calories
(2) Most readily available fuel during muscular activity
(3) Major source of precursors for glucose synthesis during starvation
(4) Depleted most rapidly during starvation
(5) Not normally used as a storage form of fuel

25. Relative to the well-fed state, fuel utilization after three days of starvation shifts in which of the following ways?

(a) More glucose is consumed by the brain.
(b) Adipose tissue triacylglycerols are degraded to provide fatty acids to most tissues.
(c) The brain begins to use ketone bodies as fuels.
(d) Proteins are degraded in order to provide three-carbon precursors of glucose.
(e) Glycogen is stored as a reserve fuel.

26. Metabolic adaptations to prolonged starvation include which of the following changes relative to the metabolic picture after three days of starvation?

(a) The rate of lipolysis in the adipose tissue increases.
(b) The glucose output by the liver decreases.
(c) The ketone body output by the liver decreases.
(d) The utilization of glucose by the brain decreases as the utilization of ketone bodies increases.
(e) The rate of degradation of muscle protein decreases.

27. List the following metabolic pathways or sources in the order of decreasing ATP production rate during strenuous exercise.

(a) Muscle glycogen to CO_2
(b) Liver glycogen to CO_2
(c) Muscle glycogen to lactate
(d) Adipose tissue fatty acids to CO_2
(e) Muscle creatine phosphate

28. Given the energy sources in question 27, list them in decreasing order of total, potentially available ATP production.

____

____

29. From the energy sources given in question 27, select the ones that provide most of the energy in

(1) A 100-meter race ________________________
(2) A 1000-meter race ________________________
(3) A marathon race ________________________

30. Show the changes in blood glucose levels you would expect to see for an insulin-dependent diabetic patient after a meal rich in carbohydrate by plotting their time course on Figure 30-1 (see question 22). Explain your answer.

____

____

31. Which of the following occur in people with untreated diabetes?

(a) Fatty acids become the main fuel for most tissues.
(b) Glycolysis is stimulated and gluconeogenesis is inhibited in the liver.
(c) Ketone body formation is stimulated.
(d) Excess glucose is stored as glycogen.
(e) Triacylglycerol breakdown is stimulated.

32. Is it true or false that in diabetes the brain shifts to ketone bodies as its major fuel? Explain.

____

____

33. Which of the following statements about hemoglobin A_{Ic} are true?
    (a) It is a mutant hemoglobin that occurs in diabetic patients.
    (b) It is absent in normal individuals.
    (c) It involves the covalent modification of hemoglobin A by glucosyl transferase.
    (d) It is useful in monitoring glucose levels in blood.
    (e) It is an example of how reducing sugars at high concentrations may damage proteins.

ANSWERS TO SELF-TEST

Strategy and Regulation of Metabolism

1. b, d, e. Answer (e) is correct because phenylalanine hydroxylase requires O_2 and NADPH in the conversion of phenylalanine to tyrosine.

2. (a) 1, 2, 4 (b) 6 (c) 2 (d) 2, 4 (e) 3, 5. Regarding the answer to (a), (1) is correct because the synthesis of phosphatidylcholine in bacteria requires *S*-adenosylmethionine.

3. (a) 1, 4 (b) 5 (c) 2 (d) 1 (e) 3

4. d

Major Metabolic Pathways and Key Intermediates

5. (a) 5, (b) 1, (c) 4, (d) 3, (e) 2, 5

6. The different effects of glucagon and epinephrine in the liver and in muscle are due to the different properties of the kinase and phosphatase enzymes that catalyze the synthesis and degradation of fructose 2,6-bisphosphate in these organs. In the liver, the cAMP cascade leads to inhibition of the kinase and activation of the phosphatase. This decreases fructose 2,6-bisphosphate levels and inhibits glycolysis. In muscle, the phosphorylation of a homologous enzyme activates the kinase, thus stimulating the formation of fructose 2,6-bisphosphate, the activity of phosphofructokinase, and glycolysis.

7. Because NADPH and NADH are distinct compounds that bind specifically to different enzymes and are not readily interconverted, it is possible to have high NADPH/$NADP^+$ ratios and low NADH/NAD^+ ratios in the same compartment of a cell.

8. b

9. (a) 3, 4, 7 (b) 2, 6, 9, 11 (c) 1, 5, 8, 10

Metabolic Profiles of the Major Organs, Hormonal Regulators, and the Control of Blood Glucose Levels

10. b, c

11. a, c. The conversions of pyruvate into lactate or alanine mainly occur in the muscles and red cells. In the liver, pyruvate is mostly used in gluconeogenesis or for lipid synthesis.

12. a, e

13. b

14. The synthesis of triacylglycerols requires glycerol 3-phosphate, which is derived from glucose. The glycerol that is released during triacylglycerol hydrolysis cannot be reutilized in adipose cells because they lack glycerol kinase. Thus, externally supplied glucose is required.

15. a, b, e

16. (a) 2 (b) 4, 6 (c) 1 (d) 3, 5, 7 (e) 2, 4

17. c

18. The selection of the pathway depends on whether or not the fatty acids enter the mitochondrial matrix, the compartment of β-oxidation and ketone body formation. When citrate and ATP concentrations are high, as in the fed state, the activity of acetyl CoA carboxylase is stimulated. The resulting malonyl CoA, which is a precursor for fatty acid synthesis, inhibits carnitine acyltransferase I, which translocates fatty acids from the cytosol into the mitochondria for oxidation.

19. (a) 2, 6, 7, 9 (b) 2, 3, 4, 5, 8, 10 (c) 1, 4, 6, 10

20. (a) I (b) S (c) S (d) I (e) S (f) S (g) S

21. Phosphorylase *a* binds glucose, which makes this enzyme susceptible to the action of phosphatase. The resulting phosphorylase *b* is inactive; therefore, glycogen degradation is decreased. Since phosphorylase *b* does not bind phosphatase, phosphatase is released and activates glycogen synthase by dephosphorylating it, leading to the production of glycogen.

22. (a) The increase in the glucose level from A to B is due to the absorption of dietary glucose.

(b) The decrease in the glucose level from B to C is due to the effects of insulin, which is secreted in response to increased blood glucose. Glucose is removed from the blood by the liver, which synthesizes glycogen, and by muscle and adipose tissue, which store glycogen and triacylglycerols.

(c) The leveling off of the glucose level from C to D is due to the increased secretion of glucagon and the diminished concentration of insulin. Glucagon maintains blood glucose levels by promoting gluconeogenesis and glycogen degradation in the liver and by promoting the release of fatty acids, which partially replace glucose as the fuel for many organs.

(d) The slight overshoot that is sometimes observed at C can be explained by the continued effects of insulin, which are not yet balanced by the metabolic effects of glucagon.

Metabolic Adaptations in Starvation, Intense Exercise, and Diabetes

23. The entire protein is converted by proteolysis to Ala molecules, which by transamination give rise to pyruvate. Gluconeogenesis from two pyruvate molecules yields one molecule of glucose. The triacylglycerol contains 10% of glycerol by weight, which is the only part of the lipid that can give rise to glucose. Glycerol enters glycolysis as dihydroxyacetone phosphate, and two molecules of this triose phosphate yield one molecule of glucose. Therefore, the glucose produced from 1 g of protein is

$$\frac{1\text{ g}}{72\text{ g/mol}} \times \frac{1}{2}$$

and the glucose produced from 1 g of triacylglycerol is

$$\frac{0.1\text{ g}}{89\text{ g/mol}} \times \frac{1}{2}$$

The ratio of the glucose from protein to the glucose from triacylglycerol is thus

$$\frac{(1\text{ g})/(72\text{ g/mol})}{(0.1\text{ g})/(89\text{ g/mol})} = \frac{1}{0.1} \times \frac{89}{72} = 12.4$$

24. (a) 2, 4 (b) 1 (c) 3, 5

25. b, c, d

26. b, d, e. Answer (a) is incorrect because lipolysis remains essentially constant.

27. e, c, a, d, b. Liver glycogen and adipose tissue fatty acids as fuels for active muscle are the slowest and are about equivalent in terms of the maximal rate of ATP production (6.2–6.7 mmol/s). This rate is probably limited by the slow transport of the fuels from the storage sites to the muscle.

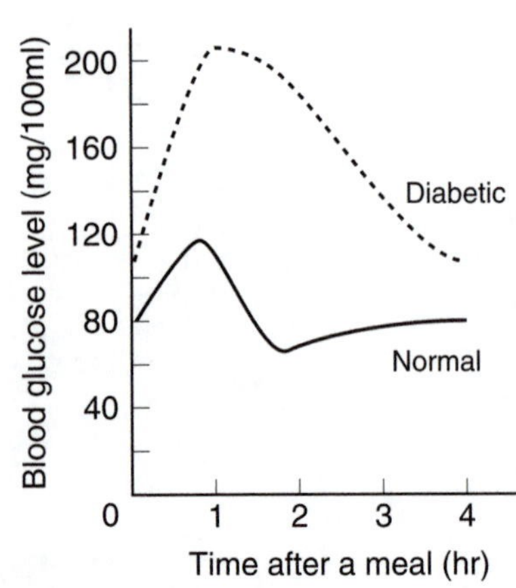

Figure 30-2
Blood glucose levels for a diabetic patient and a normal person after a meal rich in carbohydrate.

28. d, a, b, c, e. The anaerobic conversion of muscle glycogen to lactate yields much less ATP than its complete oxidation to CO_2.

29. (1) c, e (2) a, c, e (3) a, b, d

30. See Figure 30-2. In diabetic patients, the level of insulin is too low and that of glucagon is too high, so after a meal the glucose levels will reach higher values than in a normal person. Also the removal of glucose from blood will be slower, and the fasting glucose levels in the blood may remain higher.

31. a, c, e

32. False. Although ketone body concentrations in blood may become high in diabetes, glucose is even more plentiful. Therefore, the brain continues to use glucose as its major fuel.

33. d, e

PROBLEMS

1. Cardiac muscle exhibits a high demand for oxygen, and its functioning is severely impaired when coronary circulation is blocked.

   (a) Considering the energy-generating substrates available to and used by the heart under normal circumstances, why is oxygen required by heart muscle?
   (b) Suppose that in an intact animal you can measure the concentrations of biochemical metabolites in the arteries leading to the cardiac muscle and in the veins carrying blood away from heart tissue. If the supply of oxygen to heart tissue is reduced, what differences in arterial and venous glucose concentrations will you observe? What metabolite will be elevated in heart muscle that has an insufficient supply of oxygen?
   (c) Patients suffering from cardiac damage due to oxygen insufficiency exhibit an increase in the levels of certain enzymes in the plasma; among these are the H_4 form of lactate dehydrogenase and several types of aminotransferase enzymes. During recovery from a heart attack, the plasma levels of these enzymes decrease. Explain these findings.

2. An infant suffering from a particular type of organic acidemia has frequent attacks of vomiting and lethargy, which are exacerbated by infections, fasting, and the consumption of protein or fat. During these episodes, the patient suffers from hypoglycemia, which can be alleviated by injections of D-3-hydroxybutyrate. In addition, concentrations of ketone bodies in the blood are extremely low. The patient also has elevated concentrations of a number of organic acids in both blood and urine. Among these acids are 3-hydroxy-3-methylglutarate, β-methylglutaconate, and isovalerate.

   (a) What enzyme deficiency could cause an elevation in the concentration of these organic acids?
   (b) How could the same enzyme deficiency lead to a reduction in the concentration of ketone bodies?
   (c) How does fasting exacerbate the symptoms of the disorder?
   (d) The consumption of fat causes a noticeable increase in the concentration of 3-hydroxy-3-methylglutarate. Why?
   (e) How can the consumption of protein cause an increase in the levels of organic acids?
   (f) How can the administration of D-3-hydroxybutyrate relieve hypoglycemia?

3. Patients who remain unconscious after a serious surgical operation are given 100 to 150 g of glucose daily through the intravenous administration of a 5% solution. This amount of glucose falls far short of the daily caloric needs of the patient. What is the benefit of the administration of glucose?

4. A biochemist in the Antarctic is cut off from his normal food supplies and is forced to subsist on a diet that consists almost entirely of animal fats. He decides to measure his own levels of urinary ketone bodies, beginning on the day he starts the high-fat diet. What changes in urinary ketone body levels will he find?

5. In liver tissue, insulin stimulates the synthesis of glucokinase. What implications does this have for a person who has an insulin deficiency?

6. Within a few days after a fast begins, nitrogen excretion accelerates to a relatively high level. After several weeks, the rate of nitrogen excretion falls to a lower level. The excretion of nitrogen then continues at a relatively constant rate until the body is depleted of triacylglycerol stores; then the rate of urea and ammonia excretion again rises to a very high level.

(a) What events trigger the initial surge of nitrogen excretion?
(b) Why does the nitrogen excretion rate decrease after several weeks of starvation?
(c) Explain the increase in nitrogen excretion that occurs when lipid stores are exhausted.

7. Some bacterial species can grow and divide with either alanine or leucine as a sole carbon source. Others exhibit a strong preference for alanine and grow poorly on leucine. Can you explain the difference between the two types of cells?

8. Suppose that genetic engineering techniques enable you to transfer the genes for the enzymes of the glyoxalate pathway to human tissues and to express those genes in mitochondria. Why might those wishing to lose weight rapidly be interested in such a system?

9. Among the difficulties caused by prolonged fasting are metabolic disorders caused by vitamin deficiencies. What vitamins are needed during starvation to ensure that cells can continue to carry out the metabolic adaptations discussed on pages 776 and 777 of Stryer's text?

10. Describe the general fate of each of the following compounds in the mitochondria and in the cytosol of a liver cell:

(a) Palmitoyl CoA
(b) Acetyl CoA
(c) Carbamoyl phosphate
(d) NADH
(e) Glutamate
(f) Malate

11. Young men who are championship marathon runners have levels of body fat as low as 4%, whereas most casual runners have levels ranging from 12 to 15%. Why would marathoners be at greater risk during prolonged fasting?

12. Assume that a typical 70-kg man expends about 2000 kcal of energy per day. If the energy for his activities were all derived from ATP, how many grams of ATP would have to be generated on a daily basis? How many grams of glucose would be required to drive the formation of the needed amount of ATP? The molecular weight of ATP is 500, and that of glucose is 180. One mole of glucose generates 686 kcal of energy when

completely oxidized, and 7.3 kcal are required to drive the synthesis of one mole of ATP. Assume that 40% of the energy from the oxidation of glucose can be used for ATP synthesis.

13. Suppose that the concentration of mitochondrial oxaloacetate increases dramatically in a liver cell. Briefly describe how the increase would affect each of the pathways below, and justify your answer:

(a) Gluconeogenesis
(b) Biosynthesis of palmitoyl CoA
(c) Biosynthesis of cholesterol
(d) Degradation of acetyl CoA

14. The primary site of ethanol metabolism in humans is the liver. Ethanol is first oxidized to acetaldehyde, then further to acetate. Note that NAD^+ serves as an electron acceptor, so that the intracellular ratio of NADH to NAD^+ is elevated as ethanol is oxidized. Formation of acetyl CoA from acetate permits carbon atoms from ethanol to be further oxidized to carbon dioxide in the citric acid cycle.

In chronically malnourished people or in healthy people who have missed one or two meals, hypoglycemia develops rapidly with the ingestion of moderate amounts of ethanol. A reduced rate of hepatic glucose synthesis is observed, along with increases in intracellular ratios of lactate to pyruvate, of glycerol 3-phosphate to dihydroxyacetone phosphate, of glutamate to α-ketoglutarate, and of D-3-hydroxybutyrate to acetoacetate. In a well-fed person whose liver contains normal amounts of glycogen, ethanol infusion is less likely to induce hypoglycemia. The rate of hepatic glucose production is relatively normal. Increases in intracellular ratios of the pairs of compounds named above do occur, however.

(a) How do elevated intracellular ratios of NADH to NAD^+ in response to ethanol infusion lead to an increase in ratios of the pairs of compounds named above?
(b) Briefly describe how an increase in the ratios of any of the pairs of compounds could impair glucose synthesis in the liver of a malnourished person.
(c) Why is ethanol infusion less likely to impair hepatic glucose production when liver glycogen content is normal?

15. Humans cannot synthesize biotin and are dependent on exogenous sources. Some free biotin arises in the intestine from microorganisms or from the diet. However, most dietary biotin is covalently linked to enzyme protein. It must be cleaved from peptides after digestion in order to be incorporated into newly synthesized enzymes. Human serum contains an enzyme, biotinidase, which cleaves free biotin from peptides and which also may act as a carrier of free biotin. Deficiencies of biotinidase are inherited as an autosomally recessive disorder.

Affected people are normal at birth, but within six months they may have seizures, skin rash, hair loss, and hearing deficiencies. In these infants, elevated levels of lactic and pyruvic acids, several amino acids, organic acids (including propionic, methylcrotonic, D-3-hydroxybutyric and acetoacetic) are found in the serum. Daily administration of 150 mg of free biotin restores normal levels of each group of compounds listed above and releases the clinical symptoms.

Explain how a biotinidase deficiency leads to an increase in concentration of each of the groups of biochemical compounds mentioned above as well as how administration of biotin acts to restore those concentrations to normal levels.

16. During starvation, the rate of liver cell lipolysis accelerates, and the concentration of acetyl CoA increases. These changes are accompanied by an increase in the activity of pyruvate carboxylase, which converts pyruvate to oxaloacetate. Give two reasons why it is desirable to increase oxaloacetate concentrations in starved cells. Also name a source of pyruvate during starvation, when glucose availability in liver is low.

17. "The duration of starvation compatible with life is mainly determined by the size of the triacylglycerol depot" (Stryer, p. 777). Briefly explain this statement.

18. For several hours after birth, premature infants are particularly susceptible to hypoglycemia and are also unable to rapidly generate ketone bodies. Describe how each of the characteristics below would contribute to hypoglycemia, low circulating levels of ketone bodies, or both:

(a) A large brain to body weight ratio
(b) A small store of liver glycogen
(c) Low specific activity of cytosolic carnitine long-chain acyl CoA transferase in liver
(d) Very low levels of liver phosphoenolpyruvate carboxykinase

19. The Cori cycle is especially important during early phases of starvation, in which lactate molecules generated in the peripheral tissues are sent to the liver for use in gluconeogenesis.

(a) Why is it important for fatty acid oxidation to occur in the liver while the Cori cycle is operating?
(b) Suppose that lactate from muscle were completely oxidized to carbon dioxide and water in the liver. How would this make it more difficult for that organ to maintain glucose homeostasis during the early phases of starvation?
(c) Give two advantages of the alanine cycle in muscle and liver, in comparison with the Cori cycle.

ANSWERS TO PROBLEMS

1. (a) Under normal conditions, heart muscle consumes acetoacetate and D-3-hydroxybutyrate, both of which are converted to two molecules of acetyl CoA. Oxygen is required for terminal oxidation of acetyl CoA in the citric acid cycle.

(b) Cardiac cells deprived of oxygen are unable to generate metabolic energy by oxidizing acetyl CoA. An alternative source of energy is glucose, which can be converted to lactate under anaerobic conditions, with the subsequent generation of ATP. Thus, under anaerobic conditions, glucose uptake by heart muscle increases; the concentration of glucose in cardiac veins will decrease relative to glucose levels in coronary arteries. The concentration of lactate in the coronary veins is also elevated, compared with the levels in the coronary arteries.

(c) Under anaerobic conditions, the limited amount of ATP produced by the glycolytic pathway is insufficient to maintain normal cell functions, and many of the cells in anaerobic or ischemic tissues will die. Cell lysis permits release of cytoplasmic enzymes into the circulation. The H_4 form of lactate dehydrogenase, the form found in cardiac cells, and the aminotransferases are relatively easy to assay and are therefore used as indicators of cellular damage. During recovery, as heart tissue undergoes some degree of repair, the release of cytoplasmic enzymes slows, so there is a concomitant decrease in the plasma concentrations of these enzymes.

2. (a) The organic acids that are elevated in the blood and urine are those that are normal intermediates in the degradation of leucine. Examination of the pathway for leucine degradation (see Stryer's text, pp. 645–646) shows that the deficient enzyme is probably the enzyme that converts 3-hydroxy-3-methylglutaryl CoA (HMG CoA) to acetoacetate and acetyl CoA. This enzyme is known as HMG CoA cleavage enzyme, and its reduced activity in the infant leads to a reduction in the rate of leucine catabolism.

(b) The production of acetoacetate and D-3-hydroxybutyrate is also dependent on the activity of HMG CoA cleavage enzyme, which is deficient in the infant. Thus, the production of ketone bodies from HMG CoA in response to increases in the concentration of acetyl CoA is impaired because of the enzyme deficiency.

(c) Fasting causes an increase in acetyl CoA production through the increased rates of lipolysis that occur in the attempt to generate sources of metabolic energy. Any increase in acetyl CoA concentration stimulates HMG CoA synthesis, and the inability to convert HMG CoA to acetoacetate leads to an increase in concentrations of 3-hydroxy-3-methylglutarate, which is excreted by the cells into the plasma.

(d) Consumption of fats such as triacylglycerols leads to the increased production of acetyl CoA, from the β-oxidation of fatty acids. As noted in the answer to 2(c), elevation in the level of acetyl CoA stimulates the production of HMG CoA, leading to an increase in the concentration of 3-hydroxy-3-methylglutarate.

(e) Most proteins contain leucine, and the fraction of leucine that is not required for the synthesis of other proteins will undergo oxidation. As noted in the answer to 2(a), the catabolism of leucine generates a number of intermediates that accumulate as a result of the deficiency of HMG CoA lyase.

(f) Because the liver is unable to generate normal levels of ketone bodies, those tissues that normally utilize them are required to use other substrates as metabolic fuels. Since glucose is the substrate of choice in such situations, increased demand for blood glucose results in hypoglycemia. The administration of D-3-hydroxybutyrate provides an alternative source of metabolic energy for organs such as the heart, and glucose is thereby conserved.

3. The limited amount of glucose is given to prevent the hydrolysis of muscle protein during fasting. It serves as a source of energy for the brain and blood cells; otherwise, during fasting, body proteins are hydrolyzed to provide carbon atoms for the generation of glucose by gluconeogenesis in the liver and kidney.

4. A high-fat diet will stimulate ketone body formation because the oxidation of fatty acids causes an increase in acetyl CoA concentration, which in turn stimulates the production of ketone bodies. The overall profile of ketone body production may resemble that found during starvation because a source of carbon for glucose production will be lacking. However, dietary fats will serve as a source of energy instead of the triacylglycerols stored in body tissue.

5. Under normal conditions, glucokinase acts to phosphorylate glucose when concentrations of the hexose are high. The failure to synthesize sufficient quantities of glucokinase means that the liver cannot control the levels of glucose in blood, compounding the other difficulties caused by insulin deficiency described on pages 779–780 of Stryer's text.

6. (a) During the first few days of starvation, the brain continues to utilize glucose. Glycogen stores are exhausted, so the primary source of carbon atoms for gluconeogenesis is amino acids. Because the concentration of free amino acids in the tissues is limited, body proteins are broken down to provide the amino acids to support gluconeogenesis. Nitrogen excretion increases because the amino groups of those amino acids are eliminated as urea.

(b) After several weeks of fasting, the brain adapts to the utilization of ketone bodies as a source of energy, so less glucose is required. The resulting reduction in gluconeogenesis means a reduction in the rate of oxidation of amino acids and in the production of ammonia and urea.

(c) When triacylglycerol stores are depleted, the body relies on body proteins not only as a source of glucose for the brain but also as a source of energy for all other tissues. These requirements cause a great increase in the rate of body protein catabolism, with corresponding increases in amino acid oxidation and nitrogen excretion. Often more than a kilogram per day in weight is lost, indeed causing a threat to life.

7. Bacteria that lack a glyoxylate pathway are unable to carry out gluconeogenesis if leucine, a ketogenic amino acid, is the sole source of carbon. They can utilize alanine for the production of pyruvate, which in turn can be used as a source of carbon for the synthesis of glucose. Bacteria having the enzymes of the glyoxylate pathway can use either leucine or alanine for the de novo synthesis of glucose.

8. One of the difficulties caused by fasting is that body proteins are hydrolyzed to provide a source of carbon atoms for gluconeogenesis. Normally, fatty acids cannot serve as a source of glucose because acetyl CoA is not gluconeogenic. The enzymes of the glyoxylate pathway provide a means for acetyl CoA to be used for the synthesis of glucose. Such a system in human cells could conceivably be used to prevent the hydrolysis of body proteins during fasting.

9. Most of the vitamins and cofactors discussed in previous chapters of Stryer's text would be needed during starvation because many of the essential metabolic pathways must continue to operate. Among the most obvious vitamins needed for those pathways are pyridoxal phosphate (for the transamination of amino acids), niacin and riboflavin (for electron transport), thiamin (for the oxidative decarboxylation of pyruvate, α-ketoglutarate, and the branched-chain amino acids), biotin (for the carboxylation of pyruvate), and cobalamin (for the conversion of methylmalonyl CoA to succinyl CoA).

10. (a) In the cytoplasm, the acyl chain of palmitoyl CoA can be esterfied to glycerol as part of the process of triacylglycerol formation in the liver cytosol, or the palmitoyl chain can be transferred to carnitine for subsequent transport to the mitochondria. In mitochondria, palmitoyl CoA is oxidized to CO_2 and H_2O.

(b) In mitochondria, acetyl CoA is oxidized to carbon dioxide and water, or it can undergo carboxylation to form oxaloacetate. It can also be used for the synthesis of ketone bodies. In the cytosol acetyl CoA serves as a precursor of malonyl CoA, as well as a precursor of HMG CoA and cholesterol.

(c) In the mitochondria, carbamoyl phosphate combines with ornithine to form citrulline in the urea cycle, whereas in the cytosol, it serves as a precursor of pyrimidines when it condenses with aspartate to yield carbamoyl aspartate.

(d) In mitochondria, NADH is reoxidized to NAD^+ in the electron transport chain. In the cytosol, it can be reoxidized to NAD^+ through the action of lactate dehydrogenase or NADPH dehydrogenase or by means of the glycerol-phosphate or malate-aspartate shuttles.

(e) In mitochondria, glutamate undergoes oxidative deamination, yielding ammonia and α-ketoglutarate. It can also serve as a source of amino groups for a number of aminotransferase enzymes. In the cytosol, it can also serve as an amino donor for aminotransferases. Glutamate can also be used in the cytosol as a precursor of glutamine, and it can of course be incorporated into newly synthesized protein.

(f) Malate is converted to oxaloacetate in the citric acid cycle, which takes place in the mitochondria. In the cytoplasm, as a component of the malate-aspartate shuttle, it serves as an electron carrier to transfer electrons from NADH to the inner mitochondrial membrane. Malate can also be used as a source of electrons for the generation of NADPH in the reaction catalyzed by malic enzyme.

11. The greater the percentage of body fat, the larger the reserves of triacylglycerols, which are the primary source of metabolic energy during fasting. Once these reserves are depleted, the body accelerates the breakdown of muscle protein as an energy source. Extensive hydrolysis of muscle tissue threatens many vital body functions and can lead to death.

12. Since 7.3 kcal are required to drive the synthesis of 1 mole of ATP, the caloric value of the energy in each mole of ATP is 7.3 kcal. Therefore, the amount of ATP needed per day is

$$\frac{2000 \text{ kcal/day}}{7.3 \text{ kcal/mol ATP}} = 274 \text{ mol ATP per day}$$

$$274 \text{ mol ATP/day} \times 500 \text{ g/mol} = 137 \text{ kg ATP per day}$$

If the oxidation of 1 mole of glucose yields 686 kcal of energy, and if 40% can be used to drive the synthesis of ATP, glucose yields (0.4)(686 kcal/mol) = 274 kcal/mol of usable energy. Therefore, the amount of glucose needed per day is

$$\frac{2000 \text{ kcal/day}}{274 \text{ kcal/mol glucose}} = 7.3 \text{ mol glucose per day}$$

$$7.3 \text{ mol glucose/day} \times 180 \text{ g/mol} = 1.31 \text{ kg glucose per day}$$

Normal fuel stores available to the typical 70-kg man include about 250 g of glycogen and approximately 25 g of glucose. These figures make it evident that humans can rely on stores of carbohydrate for only a short time.

13. (a) Oxaloacetate is a source of carbon atoms for the synthesis of phosphoenolpyruvate, an intermediate in gluconeogenesis. An increase in oxaloacetate concentration will therefore stimulate gluconeogenesis.

(b) In mitochondria, oxaloacetate combines with acetyl CoA to form citrate. When citrate levels increase, it is shuttled across the mitochondrial membrane where it is a source of cytosolic acetyl CoA. Formation of malonyl CoA from acetyl CoA allows synthesis of palmitoyl CoA through the action of the fatty acyl CoA synthase complex. An increase in oxaloacetate concentration will therefore stimulate synthesis of palmitoyl CoA.

(c) Cholesterol synthesis occurs in the cytosol, beginning with the utilization of acetyl CoA. As discussed in (b) above, an increase in oxaloacetate concentration results in an increase in cytosolic concentrations of acetyl CoA, which could drive the formation of additional cholesterol.

(d) The synthesis of citrate in the mitochondrion is stimulated by higher concentrations of oxaloacetate, which serves as an acceptor of acetyl units from acetyl CoA. An increase in mitochondrial citrate concentration accelerates the rate of the reactions of the citric acid cycle, the function of which includes the oxidation of acetyl CoA. Therefore an increase in oxaloacetate concentration will stimulate oxidation or degradation of acetyl CoA.

14. (a) Each pair of compounds referred to in the problem is interconvertible through oxidation-reduction reactions linked to the NADH/NAD^+ redox pair. An increased NADH/NAD^+ ratio will therefore suppress net oxidation reactions because the NAD^+ needed to serve as an electron acceptor (for conversion of lactate to pyruvate, for example) is limited in concentration. Thus an ethanol-induced increase in NADH will suppress conversion of lactate to pyruvate, of glycerol 3-phosphate to dihydroxyacetone phosphate, of glutamate to α-ketoglutarate, and of D-3-hydroxybutyrate to acetoacetate.

(b) During starvation the liver carries out gluconeogenesis using lactate, amino acids, and α-glycerol phosphate (from triacylglycerols) as initial substrates. These compounds are converted via NAD^+-linked oxidation to compounds like pyruvate, α-ketoglutarate, and dihydroxyacetone phosphate, all directly in the pathway to glucose synthesis. The inability of the cell to produce these and other compounds (like oxaloacetate) because of the lack of NAD^+ would result in a low level of glucose production. The acetoacetate/D-3-hydroxybutyrate pair is not involved in gluconeogenesis; these are ketone bodies. However, the inability to convert D-3-hydroxybutyrate to acetoacetate would interfere with the terminal oxidation of these compounds as well.

(c) In well-nourished people, ethanol infusion will elevate intracellular NADH levels, and gluconeogenesis through the liver's utilization of amino acids and other compounds will be somewhat impaired. However, stored hepatic glycogen serves as a source of glucose in response to any drop in blood sugar levels. Because liver glycogen levels are virtually depleted after 24 to 36 hours of starvation, ethanol-induced hypoglycemia can develop rapidly in malnourished people.

15. The deficiency in biotin leads to a reduction in activity of carboxylase enzymes, and pathways that depend on carboxylases would be adversely affected. Lactic and pyruvic acids are elevated because pyruvate carboxylase (converting pyruvate to oxaloacetate) is impaired, especially during gluconeogenesis. Several amino acids depend on propionyl CoA carboxylase for their oxidation—these include valine, isoleucine, and methionine. A reduction in the ability of the enzyme to convert propionyl CoA to D-methylmalonyl CoA leads to accumulation of these amino acids and their metabolites (like methylcrotonate) as well as of propionate. Elevation of levels of the ketone bodies D-3-hydroxybutyrate and acetoacetate results from a deficiency of acetyl CoA carboxylase, which normally converts acetyl CoA to malonyl CoA. In this case, elevation of acetyl CoA levels drives ketone body synthesis (not fatty acid synthesis, which requires malonyl CoA).

16. Increased concentrations of oxaloacetate are needed to provide more acceptors of acetyl groups from acetyl CoA to form citrate, ensuring that the citric acid cycle can operate at higher capacity. This makes it able to oxidize the increasing amounts of acetyl CoA present in the cell. Elevated levels of oxaloacetate are also required to provide more molecules that can serve as precursors for gluconeogenesis. During starvation, liver cells increase their rate of glucose formation through glycogenolysis and gluconeogenesis, in order to provide more glucose to peripheral tissues. Gluconeogenesis depends primarily on the availability of oxaloacetate molecules, which are converted first to phosphoenolpyruvate and then ultimately to glucose. One source of pyruvate during starvation is alanine, produced by the degradation of muscle tissue proteins. Alanine, along with glutamine, serves as a carrier of carbon atoms and nitrogen from muscle to liver. Alanine is converted to pyruvate by aminotransferase enzymes, which use another α-keto acid as an acceptor of the amino group from alanine.

17. Once triacylglycerols are depleted during starvation, the only source of fuels for the ~1800 kcal/day needed by the body is protein, and $2\frac{1}{2}$ g of

protein are needed to produce the same number of calories produced by 1 g of fat. Thus body wasting occurs at a higher rate than when triacylglycerols are available.

18. (a) Because brain tissue preferentially uses glucose as a fuel, the demand for glucose in neonates is disproportionately high compared with that of an older person with a lower brain to body weight ratio.

    (b) Low reserves of liver glycogen mean that the ability of liver to export glucose synthesized from glycogen is limited.

    (c) Neonates are unable to generate sufficient levels of ketone bodies as an alternative fuel during hypoglycemia because the low specific activity of carnitine acyl transferase limits transport of long-chain fatty acids across the mitochondrial membrane. The depletion of fatty acids in the mitochondrion means that only limited amounts of acetyl CoA from fatty chain oxidation are available for synthesis of acetoacetyl CoA and β-hydroxybutyrate.

    (d) Liver cannot effectively carry out gluconeogenesis, because the activity of phosphoenolpyruvate carboxykinase (which carries out synthesis of PEP from oxaloacetate), a key enzyme in glucose synthesis, is present only at a very low level.

19. (a) Because most of the ATP used for gluconeogenesis in the liver is generated by β-oxidation of fatty acids.

    (b) Compounds of the Cori cycle, such as lactate and pyruvate, provide a readily available source of carbon for gluconeogenesis in liver. If those molecules were unavailable, additional carbon atoms from proteolysis would be required to sustain the level of glucose required through production by gluconeogenesis.

    (c) Formation of alanine in muscle allows transport of an atom of nitrogen in nontoxic form to the liver, where it can be disposed of as urea. Conversion of pyruvate to alanine means that the electrons in NADH normally consumed in the conversion of pyruvate to lactate can now be sent to the mitochondrion to be used to drive ATP synthesis.

EXPANDED SOLUTIONS TO TEXT PROBLEMS

1. Liver contains glucose 6-phosphatase, whereas muscle and brain do not. Hence, muscle and brain, in contrast with liver, do not release glucose. Another key enzymatic difference is that liver has little of the transferase needed to activate acetoacetate to acetoacetyl CoA. Consequently, acetoacetate and 3-hydroxybutyrate are exported by the liver for use by heart muscle, skeletal muscle, and brain.

2. (a) Adipose cells normally convert glucose to glycerol 3-phosphate for the formation of triacylglycerols. A deficiency of hexokinase would interfere with the synthesis of triacylglycerols.

    (b) A deficiency of glucose 6-phosphatase would block the export of glucose from liver following glycogenolysis. This disorder (called von Gierke's disease) is characterized by an abnormally high content of glycogen in the liver and a low blood glucose level.

(c) A deficiency of carnitine acyltransferase I impairs the oxidation of long-chain fatty acids. Fasting and exercise precipitate muscle cramps in these people.

(d) Glucokinase enables the liver to phosphorylate glucose even in the presence of a high level of glucose 6-phosphate. A deficiency of glucokinase would interfere with the synthesis of glycogen.

(e) Thiolase catalyzes the formation of two molecules of acetyl CoA from acetoacetyl CoA and CoA. A deficiency of thiolase would interfere with the utilization of acetoacetate as a fuel when the blood sugar level is low.

(f) Phosphofructokinase will be less active than normal because of the lowered level of F-2,6-BP. Hence, glycolysis will be much slower than normal.

3. (a) A high proportion of fatty acids in the blood are bound to albumin. Cerebrospinal fluid has a low content of fatty acids because it has little albumin.

(b) Glucose is highly hydrophilic and soluble in aqueous media, in contrast with fatty acids, which must be carried by transport proteins such as albumin. Micelles of fatty acids would disrupt membrane structure.

(c) Fatty acids, not glucose, are the major fuel of resting muscle.

4. (a) A watt is equal to 1 joule (J) per second (0.239 calorie per second). Hence, 70 W is equivalent to 0.07 kJ/s or 0.017 kcal/s.

(b) A watt is a current of 1 ampere (A) across a potential of 1 volt (V). For simplicity, let us assume that all the electron flow is from NADH to O_2 (a potential drop of 1.14 V). Hence, the current is 61.4 A, which corresponds to 3.86×10^{20} electrons per second (1 A = 1 coulomb/s = 6.28×10^{18} charges/s).

(c) About 2.5 ATP are formed per NADH oxidized (two electrons). Hence, one ATP is formed per 0.80 electrons transferred. A flow of 3.86×10^{20} electrons per second therefore leads to the generation of 4.83×10^{20} ATP per second or 0.8 mmol per second.

(d) The molecular weight of ATP is 507. The total body content of ATP of 50 g is equal to 0.099 mol. Hence, ATP turns over about once per 125 seconds when the body is at rest.

5. (a) The stoichiometry of complete oxidation of glucose is

$$C_6H_{12}O_6 + 6\ O_2 \longrightarrow 6\ CO_2 + 6\ H_2O$$

and that of tripalmitoylglycerol is

$$C_{51}H_{98}O_6 + 72.5\ O_2 \longrightarrow 51\ CO_2 + 49\ H_2O$$

Hence, the RQ values are 1.0 (6/6) and 0.703 (51/72.5), respectively.

(b) An RQ value reveals the relative usage of carbohydrate and fats as fuels. The RQ of a marathon runner typically decreases from 0.97 to 0.77 during the race. The lowering of the RQ reflects the shift in fuel from carbohydrate to fat.

6. One gram of glucose (molecular weight 180.2) is equal to 5.55 mmol, and one gram of tripalmitoylglycerol (molecular weight 807.3) is equal to 1.24 mmol. The reaction stoichiometries (see problem 5) indicate that 6 mol of H_2O are produced per mole of glucose oxidized, and 49 mol of H_2O per mole of tripalmitoylglycerol oxidized. Hence, the H_2O yields per gram of fuel are 33.3 mmol (0.6 g) for glucose, and 60.8 mmol (1.09 g) for tripalmitoylglycerol. Thus, complete oxidation of this fat gives 1.82 times as much water as does glucose. Another advantage of triacylglycerols is that they can be stored in essentially anhydrous form, whereas glucose is stored as glycogen, a highly hydrated polymer (p. 605 in Stryer). A hump consisting mainly of glycogen would be an intolerable burden—far more than the straw that broke the camel's back!

7. A typical macadamia nut has a mass of about 2 g. Because it consists mainly of fats (~9 kcal/g), a nut has a value of about 18 kcal. The ingestion of 10 nuts results in an intake of about 180 kcal. As was discussed in problem 4, a power consumption of 1 W corresponds to 0.239 calorie per second, and so 400-W running requires 95.6 cal/s, or .0956 kcal/s. Hence, one would have to run 1882 s, or about 31 min, to spend the calories provided by 10 nuts.

8. A high blood glucose level would trigger the secretion of insulin, which would stimulate the synthesis of glycogen and triacylglycerols. A high insulin level would impede the mobilization of fuel reserves during the marathon.

DNA Structure, Replication, and Repair

In Part V of the text, Stryer returns to the topic of the flow of genetic information and considers the detailed biochemical mechanisms underlying this complex process. Chapters 4 and 5 introduced you to DNA and RNA and outlined the storage, duplication, and expression of genetic information. In Chapter 6, the techniques used to analyze, construct, and clone DNA were presented. You should review these chapters to prepare for studying Chapter 31. Pay particular attention to DNA structure, the supercoiling of DNA, and DNA polymerase I in Chapter 4 and to the restriction endonucleases and DNA ligase in Chapter 6.

Chapter 31 covers the biochemistry of the flow of genetic information from parent to progeny by describing DNA and the enzyme systems that replicate and maintain it. Stryer expands his earlier coverage of DNA with a more thorough description of the A, B, and Z structures it can assume and the underlying chemical determinants of these structural variations. The mechanisms by which proteins interact with specific DNA sequences are exemplified by EcoRV restriction endonuclease. The mechanisms of DNA ligases are also given. Stryer then provides a detailed description of the topology of covalently closed circular DNAs and the topoisomerases that modulate their linking numbers. He expands on the previous description of the biochemistry of DNA polymerase I from *E. coli* and introduces DNA polymerases II and III. He then describes the replication fork; replication initiation; and RNA-primed, semidiscontinuous DNA elongation. The structure and important role of DNA polymerase III in replication is described in detail. The special problems of replication due to the structure of eukaryotic chromatin will be covered in Chapter 37. Stryer concludes the chapter with a description of mutations—their nature, causes, consequences, and repair. He describes examples of pathological repair deficiencies in humans, states the relationship of repair impairments and mutation to carcinogenesis, and describes a test system for detecting potential carcinogens through their mutagenic action on bacteria.

LEARNING OBJECTIVES

When you have mastered this chapter, you should be able to complete the following objectives:

DNA Structure and Protein-DNA Interactions (Stryer pages 787–794)

1. Contrast the structural information provided by the *x-ray diffraction analysis* of *DNA fibers* and *DNA crystals.*

2. Explain why *local DNA structure* depends on *base sequence.* Contrast deformations in the DNA helix due to *smooth bending* and *kinking.*

3. Describe the *major* and *minor grooves* of the *B-DNA helix.* Distinguish between the four possible *base pairs* ($A \cdot T$, $T \cdot A$, $C \cdot G$, and $G \cdot C$) in terms of the unique arrays of *hydrogen bond acceptors* and *donors* and *methyl groups* they present in the grooves of the DNA.

4. Indicate the role of the *puckering* of deoxyribose in determining the structural differences between *A–DNA* and B–DNA helices. Compare the structure of *double-stranded RNA* and *RNA–DNA hybrids* with that of the A–DNA helix.

5. Describe *Z-DNA* in terms of the *handedness of the helix* and the shape of the path traced by its phosphodiester backbone.

6. Describe the *specific interactions* between *EcoRV endonuclease* and its *palindromic recognition sequence.* Outline the *scanning* process by which it locates its target sequence.

7. List the substrates and outline the reaction mechanisms of the *DNA ligases.*

DNA Topology and Topoisomerases (Stryer pages 794–799)

8. Define the *linking number* (Lk) of a circular DNA molecule and relate *supercoiling* to the *electrophoretic* and *centrifugal mobility* of the molecule.

9. Write the equation relating the linking number to the *twisting number* (Tw) and *writhing number* (Wr) of a DNA *topoisomer.* Give the equation for the *supercoiling density* (σ) of a topoisomer and explain how a negative *superhelix density* can facilitate the unwinding of the helix. Describe the partitioning of the free energy of a negatively supercoiled molecule into Tw and Wr.

10. Describe the three steps catalyzed by *topoisomerases;* distinguish between *type I* and *type II* topoisomerases; and describe the substrates, products, and mechanisms of *topoisomerase I* and *DNA gyrase.*

DNA Polymerases (Stryer pages 799–803)

11. Outline the key features of the reactions catalyzed by *DNA polymerase I* of *E. coli.* Relate its $3' \rightarrow 5'$ and $5' \rightarrow 3'$ *exonuclease activities* to the *fidelity* of the polymerization reaction and the physiologic functions of the enzyme. Distinguish between the small and large fragments produced by limited proteolysis of DNA polymerase I.

12. Recount the significance of the discovery of a *mutant* of *E. coli* that was deficient in DNA polymerase I, and describe the reactions catalyzed by *DNA polymerase III.*

DNA Replication (Stryer pages 803–809)

13. Draw a *replication fork,* and describe the reactions and the movements of the DNA strands that occur during replication.

14. Define *continuous replication* and *discontinuous replication* and relate these terms to the *leading* and *lagging strands.* Describe an *Okazaki fragment.*

15. Describe the function and features of the nucleotide sequence of *oriC* and note that it is the unique site of *bidirectional replication initiation* in *E. coli.* List the proteins that interact with the DNA in this region of the chromosome, and give the reactions they catalyze.

16. Explain the roles of *RNA* in DNA replication. Describe the *primosome,* and name the enzyme activity that removes *RNA primers* from the genome.

17. List the distinctive features of *DNA polymerase III holoenzyme,* and describe how an asymmetric dimer of the enzyme, along with other proteins, synthesizes the leading and lagging strands of the daughter duplexes.

18. Summarize the reactions and identify the proteins at the replication fork that carry out DNA replication.

Mutations—Their Nature, Consequences, and Repair (Stryer pages 809–814)

19. Distinguish between *substitution, insertion,* and *deletion mutations,* and relate them to changes in DNA. Distinguish between *transversion* and *transition* substitution mutations and explain the origin of mutations that alter the reading frame in translation.

20. Give examples of some *chemical mutagens* and their mechanisms. Describe the structure of the *pyrimidine dimer* formed by *ultraviolet light.* Outline the mechanisms for the enzymatic repair of damaged DNA.

21. Explain why *thymine* rather than *uracil* is used in DNA.

22. Describe *xeroderma pigmentosum,* its cause, pathological consequences, and relationship to *DNA repair.*

23. Describe the relationship between *hereditary nonpolyposis colorectal cancer* and DNA repair.

24. Relate *mutagens* and *carcinogens.* Outline the *Salmonella mutagen assay.*

SELF-TEST

DNA Structure and Protein-DNA Interactions

1. Which of the following statements about the double-helical structure of DNA are correct?

   (a) It has adenine paired with thymine and cytosine paired with guanine.
   (b) It can assume many different forms including A-DNA, B-DNA, and Z-DNA.
   (c) In B-DNA, all the hydrogen-bonded base pairs lie in a plane perpendicular to the helix axis.
   (d) It can be deformed by both smooth bends and kinks.
   (e) It is rigid and static.

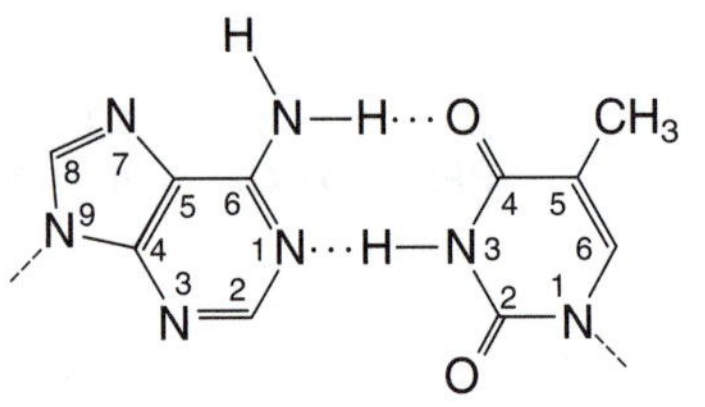

A·T

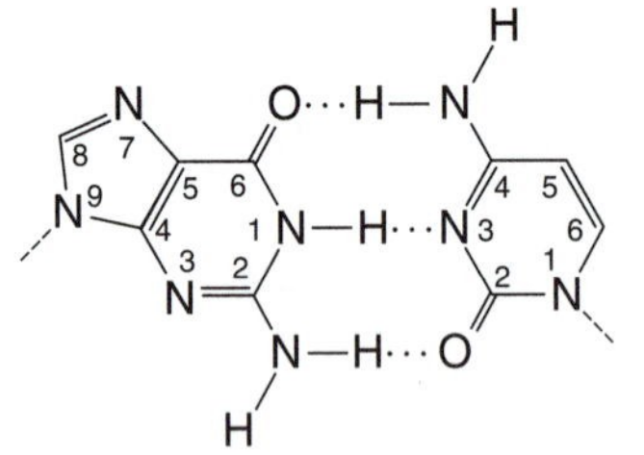

G·C

Figure 31-1
Watson-Crick base pairs.

2. Figure 31-1 shows the structures of the adenine and thymine (A·T) and guanine and cytosine (G·C) base pairs in B-DNA. Use the figure to answer the following questions:

(a) Which hydrogen bond donors or acceptors of the A·T base pair are in the major groove?

(b) Which hydrogen bond donors or acceptors of the G·C base pair are in the minor groove?

(c) Using *n* to represent heterocyclic ring nitrogen atoms, *o* for oxy groups, and *h* for protons, give the patterns of hydrogen bond acceptors and donors in the major groove for a G·C and a C·G base pair.

(d) Using the same symbols as in (c), give the patterns of hydrogen bond acceptors and donors in the minor groove for an A·T and a T·A base pair.

(e) Explain the possible biological significance of the unique arrays of hydrogen bond donors and acceptors in the grooves of B-DNA.

3. X-ray diffraction studies of DNA held at low relative humidity revealed the existence of A-DNA. Since such arid conditions presumably never occur in the cell, what is the significance of the structure of this DNA?

4. Match the features or characteristics in the right column with the type of DNA helix in the left column.

(a) A-DNA ____________

(b) B-DNA ____________

(c) Z-DNA ____________

(1) Phosphates in the backbone are zigzagged
(2) Formation is favored by negative supercoiling
(3) Has a relatively wide and deep major groove
(4) Has a right-handed helix
(5) Has 10.4 base pairs per turn
(6) Has a structure similar to that of double-stranded RNA
(7) Strands in the helix have opposite polarities

5. How does EcoRV endonuclease recognize the sequence it cleaves? What are the symmetry relationships in the interaction between the enzyme and the substrate?

___

___

6. Which of the following statements about DNA ligase are correct?

(a) It forms a phosphodiester bond between a 5′-hydroxyl and a 3′-phosphate in duplex DNA.
(b) It requires a cofactor, either NAD^+ or ATP, depending on the source of the enzyme, to provide the energy to form the phosphodiester bond.
(c) It catalyzes its reaction by a mechanism that involves the formation of a covalently linked enzyme adenylate.
(d) It catalyzes its reaction by a mechanism that involves the activation of a DNA phosphate through the formation of a phosphoanhydride bond with AMP.
(e) It is involved in DNA replication, repair, and recombination.

DNA Topology and Topoisomerases

7. The topological features of circular DNA may affect which of the following?

(a) The electrophoretic mobility of the DNA
(b) The sedimentation properties of the DNA
(c) Its affinities toward proteins that bind to the DNA
(d) The susceptibility of the strands of the DNA to unwinding
(e) The susceptibility of the DNA to the action of DNA ligase

8. Which of the following statements about DNA molecules that are topoisomers are correct?

(a) They are bound to topoisomerases.
(b) They differ from one another topologically only in that they have different linking numbers.
(c) The values of their supercoiling densities are unequal.
(d) They have identical molecular weights.
(e) They are topological or spatial isomers.

9. Which of the following statements about topoisomerases are correct?

(a) They alter the linking numbers of topoisomers.
(b) They break and reseal phosphodiester bonds.
(c) They require NAD^+ as a cofactor to supply the energy to drive the conversion of a supercoiled molecule to its relaxed form.
(d) They form covalent intermediates with their DNA substrates.
(e) They can, in the case of a particular topoisomerase, use ATP to form negatively supercoiled DNA from relaxed DNA in *E. coli.*

DNA Polymerases

10. Which of the following statements about DNA polymerase I are correct?

(a) It adds deoxyribonucleotide units to the 3′-hydroxyl of a primer.
(b) It uses the template strand to select which deoxyribonucleotide unit to add to the growing DNA chain.

(c) It contains a $3' \rightarrow 5'$ nuclease that cleaves phosphodiester bonds to yield 3′-dNMPs and 3′-phosphate-terminated DNA.
(d) It contains two nuclease activities in the same polypeptide chain that contains the polymerase active site.
(e) It can be cleaved with a protease into two fragments, each of which has a nuclease activity.

11. Match the properties or functions in the right column with the DNA polymerase in the left column.

(a) DNA polymerase I__________
(b) DNA polymerase II__________
(c) DNA polymerase III__________

(1) Involved in replication
(2) Requires a primer and a template
(3) Involved in DNA repair
(4) Makes most of the DNA during replication
(5) Removes the primer and fills in gaps during replication

DNA Replication

12. Which of the following statements about DNA replication in *E. coli* are correct?

(a) It occurs at a replication fork.
(b) It starts at a unique locus on the chromosome.
(c) It proceeds with one replication fork per replicating molecule.
(d) It is bidirectional.
(e) It involves discontinuous synthesis on the leading strand.
(f) It uses RNA transiently as a template.

13. Which of the following statements about DNA polymerase III holoenzyme from *E. coli* are correct?

(a) It elongates a growing DNA chain approximately 100 times faster than does DNA polymerase I.
(b) It associates with the parental template, adds a few nucleotides to the growing chain, and then dissociates before initiating another synthesis cycle.
(c) It maintains a high fidelity of replication, in part by acting in conjunction with a subunit containing a $3' \rightarrow 5'$ exonuclease activity.
(d) When replicating DNA, it is a molecular assembly composed of at least 10 different kinds of subunits.

14. Explain how the β_2 subunit of DNA polymerase III holoenzyme contributes to the processivity of the DNA synthesis machinery.

____

____

15. Why is RNA synthesis essential to DNA synthesis in *E. coli*?

____

____

16. Match the functions or features related to DNA replication in *E. coli* listed in the right column with the molecules or structures in the left column.

(a) Replication fork ____

(b) *oriC* ____

(c) Lagging strand ____

(d) Leading strand ____

(e) Okazaki fragments ____

(f) DnaB helicase ____

(g) Single-strand binding protein ____

(h) DNA gyrase ____

(i) Primase ____

(j) DNA polymerase III holoenzyme ____

(k) ε subunit of DNA polymerase III ____

(l) DNA polymerase I ____

(m) DNA ligase ____

(1) Synthesis direction is opposite that of replication fork movement
(2) Unwinds strands at the origin of replication in association with dnaA and dnaC proteins
(3) Is synthesized continuously
(4) Synthesizes most of DNA
(5) Is synthesized discontinuously
(6) Relieves positive supercoiling
(7) Is the locus of DNA unwinding
(8) Hydrolyzes ATP to reduce the linking number of DNA
(9) Binds dnaA, dnaB, and dnaC proteins
(10) Fills in gaps where RNA existed
(11) Is the point of initiation of synthesis
(12) Joins lagging strand pieces to each other
(13) Contains a $5' \rightarrow 3'$ exonuclease that removes RNA primers
(14) Is an RNA polymerase
(15) Performs "proofreading" on most of the DNA synthesized
(16) Stabilizes unwound DNA
(17) Uses NAD^+ to form phosphodiester bonds

17. Why are the antibiotics novobiocin and nalidixic acid, which inhibit DNA gyrase, useful in treating bacterial infections in humans?

____

____

Mutations—Their Nature, Consequences, and Repair

18. Which of the following nucleotide substitutions are transition mutations?

(a) G for A
(b) A for C
(c) C for T
(d) T for G

19. Which of the substitutions in question 19 are transversion mutations?

____

____

20. How could the tautomerization of a keto group on a guanine residue in DNA to the enol form lead to a mutation?

____

____

____

21. Explain why most nucleotides that have been misincorporated during DNA synthesis in *E. coli* do not lead to mutant progeny.

____

____

22. Match the type of mutation or physiologic consequence in the right column with the appropriate mutagen in the left column.

(a) 5-Bromouracil ________

(b) 2-Aminopurine ________

(c) Hydroxylamine ________

(d) Acridines ________

(e) Nitrous acid________

(f) Ultraviolet light ________

(1) Transversion
(2) Transition
(3) Insertion or deletion
(4) Translational frameshift
(5) Block in replication

23. What property of DNA allows the repair of some residues damaged through the action of mutagens?

____

____

24. How does the repair machinery of *E. coli* identify a DNA strand that has recently misincorporated a noncomplementary nucleotide during replication in order to repair it?

____

____

25. Which of the following enzymes or processes can be involved in repairing DNA in *E. coli* damaged by uv light-induced formation of a thymine dimer?

(a) DNA ligase seals the newly synthesized strand to undamaged DNA to form the intact molecule.
(b) The UvrABC enzyme (excinuclease) hydrolyzes phosphodiester bonds on both sides of the thymine dimer.
(c) DNA polymerase I fills in the gap created by the removal of the oligonucleotide bearing the thymine dimer.

(d) The UvrABC enzyme recognizes a distortion in the DNA helix caused by the thymine dimer.
(e) A photoreactivating enzyme absorbs light and cleaves the thymine dimer to re-form two adjacent thymine residues.

26. Given that T requires more energy to synthesize than U, and A pairs equally well with U or T, why does DNA contain A·T base pairs instead of A·U base pairs?

---

27. Explain how mutations in genes encoding proteins likely to be involved in DNA repair, such as those defective in xeroderma pigmentosum and hereditary nonpolyposis colorectal cancer, may contribute to the onset of cancer.

---

28. Explain how some strains of *Salmonella* are used to detect carcinogens. How is an extract from human liver involved in this test?

ANSWERS TO SELF-TEST

DNA Structure and Protein-DNA Interactions

1. a, b, d. Answer (c) is incorrect because B-DNA has local variations from the average structure. The hydrogen-bonded base pairs are often twisted and tilted out of the plane that is perpendicular to the helix axis.

2. (a) In the major groove, N-7 of A is an acceptor, H-6 on the 6-exocyclic (i.e., not in the ring) amino group of A is a donor, and O-4 on T is an acceptor.

(b) In the minor groove, N-3 of G is an acceptor, H-2 on the 2-exocyclic amino group of G is a donor, and O-2 on C is an acceptor.

(c) The patterns in the major groove are *noh* for G·C and *hon* for C·G.

(d) The patterns in the minor groove are *no* for A·T and *on* for T·A with no donors being present.

(e) The patterns of hydrogen bond donors and acceptors in the major and minor grooves of B-DNA, recognized by complementary hydrogen bond donors and acceptors on the amino acid side chains of proteins, facilitate the sequence-specific interaction of proteins and DNA. Using such interactions proteins can recognize specific sequences in DNA without disrupting the DNA helix. Note that A·T base pairs have two symmetrically related acceptor atoms in the minor groove and thus have less "information" in this groove than do G·L base paris. The methyl groups on T may also participate in this process through hydrophobic interactions with proteins.

3. Double-stranded RNA, RNA–DNA hybrids, and some short sequences of double-stranded DNA embedded in B-DNA have structures like that of A-DNA. Thus, knowledge about the helical structure of A-DNA contributes to an understanding of other similar helices that are of physiologic importance.

4. (a) 4 ,6, 7 (b) 3, 4, 5, 7 (c) 1, 2, 7. Answer 2 is correct for (c) because the conversion of B-DNA to Z-DNA would require the complete unwinding of the right-handed helix to form the left-handed one. Negative supercoiling promotes unwinding of B-DNA and thus will facilitate the conversion of a segment of B-DNA into Z-DNA.

5. The enzyme dimer interacts specifically with the palindromic DNA sequence GATATC by forming 12 specific hydrogen bonds between side chains on the enzyme and the groups exposed in the major groove of the DNA (see question 2). Only the sequence GATATC offers the correct array of hydrogen bond acceptors and donors for the 12 hydrogen bonds to form. When the target sequence is encountered by the enzyme, the K_M for Mg^{2+} decreases and phosphodiester bond hydrolysis occurs. In addition, there are electrostatic interactions between the enzyme and the DNA. The twofold symmetry of the recognition sequence implies a similar symmetry in the enzyme, which is, in fact, a homodimer.

6. b, c, d, e. Answer (a) is incorrect because the enzyme joins a 3′-hydroxyl to a 5′-phosphate.

DNA Topology and Topoisomerases

7. a, b, c, d. Regarding answer (e), supercoiling is ordinarily a property of covalently closed circular DNA—that is, of DNA in which there are no discontinuities in either strand of the helix. Hence, these molecules are not substrates for DNA ligase, because they lack ends.

8. b, c, d, e. The supercoiling density, $\sigma = (Lk - Lk_0)/Lk_0$, where Lk is the linking number of the topoisomer and Lk_0 is the linking number of the relaxed molecule. Since, by definition, two topoisomers will differ in Lk values but will have identical Lk_0 values under the same conditions, σ will necessarily be different for the two topoisomers.

9. a, b, d, e. Although all topoisomerases break and reseal phosphodiester bonds, an external energy source is not always required. Relieving the torsional stress in a negatively supercoiled DNA molecule by relaxing it with topoisomerase I is exergonic and requires no energy input, whereas introducing negative supercoils with DNA gyrase is endergonic and must be coupled to ATP hydrolysis. The particular catalytic mechanisms of given topoisomerases determine whether they are coupled to ATP hydrolysis.

DNA Polymerases

10. a, b, d, e. The 3′ → 5′ exonuclease cleaves a 5′-dNMP from the 3′-hydroxyl of the primer, leaving a 3′-hydroxyl group on the residual DNA.

The two exonucleases of DNA polymerase I can be separated from each other by cleaving the enzyme with a protease.

11. (a) 1, 2, 3, 5 (b) 2, 3 (c) 1, 2, 4

DNA Replication

12. a, b, d. Although not explicitly stated in Stryer, answer (c) is incorrect because the replicating *E. coli* chromosome has two replication forks that synthesize the DNA bidirectionally from the unique *oriC* origin. Answer (e) is incorrect because only the lagging strand is synthesized discontinuously. Answer (f) is incorrect because RNA serves as a primer and not as a template.

13. a, c, d. Answer (b) is incorrect because DNA polymerase III holoenzyme is a highly processive enzyme that synthesizes extensively before dissociating from its template.

14. The β_2 subunit forms a torus, with the duplex DNA in its aperture. The β_2 ring acts as a sliding clamp that holds the replication machinery on the DNA.

15. Because DNA polymerases are unable to initiate DNA chains de novo and because they require a primer with a 3′-hydroxyl group, short RNA chains are used as primers to start DNA replication at the origin of replication and to initiate the Okazaki fragments of the lagging strand. RNA polymerases can start RNA chains by adding a nucleotide to an initiating NTP, but they do so with relatively low accuracy. RNA-initiated DNA chains facilitate high-fidelity replication at the beginning sequences of new chains because they allow DNA polymerase I to replace the ribonucleotides with DNA, using its polymerase activity and both of its exonucleases in a nick translation reaction (see Stryer p. 809).

16. (a) 7 (b) 7, 9, 11 (c) 1, 5 (d) 3 (e) 1 (f) 2 (g) 16 (h) 6, 8 (i) 14 (j) 4 (k) 15 (l) 1, 10, 13 (m) 12, 17. Answer (5) is not a correct match with (e), because each Okazaki fragment is synthesized continuously.

17. These compounds interfere with the essential function of DNA gyrase in bacterial DNA replication. By inhibiting its action, they prevent the gyrase from relieving the positive supercoils that build up ahead of the moving replication fork. Human cells lack an enzyme similar to gyrase, and thus they are relatively unharmed by these antibiotics.

Mutations—Their Nature, Consequences, and Repair

18. a, c. Transition mutations are substitutions of a purine for a purine or a pyrimidine for a pyrimidine.

19. b, d. Transversion mutations substitute a purine for a pyrimidine or vice versa.

20. The rare enol tautomer of G could base-pair with a T in the template to allow its incorporation into a growing DNA strand during replication. If the proofreading process missed this erroneous incorporation, the resulting daughter DNA duplex would contain a G · T base pair. During the

next round of replication, the T would direct the incorporation of an A into its complementary daughter strand. The final result would be the substitution of an A·T base pair for the original G·C base pair.

21. The proofreading $3' \rightarrow 5'$ nuclease of the ε subunit of DNA polymerase III holoenzyme removes most of the misincorporated nucleotides that do not form a base pair with the template. The polymerase activity of the enzyme then has a second chance to incorporate the correct nucleotide.

22. (a) 2 (b) 2 (c) 2 (d) 3, 4 (e) 2; (f) 5. None of these mutagens commonly cause transversions.

23. Since DNA is double-stranded, damage to one strand of the DNA can often be repaired using the undamaged complementary strand as a template to direct the incorportion of the correct nucleotide.

24. The authentic parental template has its GATC sequences methylated by a sequence-specific DNA methyltransferase (*dam* methylase). The newly synthesized strand has not yet had time for its GATC sequences to become methylated. The methyl groups on the parental strand serve as "tags" to direct the repair enzymes (MutS, MutL, and MutH) to cut the unmethylated strand containing the wrong nucleotide, and remove the mispaired nucleotide and a variable number of neighboring nucleotides. DNA repair synthesis then replaces these nucleotides to form a corrected strand, which DNA ligase joins to form intact DNA.

25. a, b, c, d, e. Although not mentioned, the uvrD protein is a helicase that removes the 12-nucleotide-long oligonucleotide bearing the thymine dimer.

26. C spontaneously deaminates to form U in DNA. This change would lead to a mutation during replication, since U would pair with A changing what was a C·G base pair into an A·U base pair during the next replication of the DNA. The repair machinery of a cell that used U in its DNA would be unable to distinguish the U in an A·U base pair arising from a C deamination from one formed during replication. The methyl group on T distinguishes it from undesirable uracils formed by deamination.

27. The inability to effectively repair mutagenic lesions in DNA may lead to their accumulation. As a consequence, genes regulating cellular proliferation may malfunction, causing cancer.

28. Special strains of *Salmonella* have been developed to detect substitution, insertion, and deletion mutations in their DNA as a result of exposure to exogenously supplied chemicals. Mutagens can alter the DNA in these strains and thus convert them from auxotrophs, which are unable to grow in the absence of histidine, to prototrophs. The revertants can grow on media lacking histidine and are detected with high sensitivity. Since there is a correlation between mutagenicity and carcinogenicity, these strains are used as an inexpensive initial test of the carcinogenic potential of a compound. Because animals sometimes metabolize innocuous compounds and convert them to carcinogens, incubation of a suspect chemical with a human liver extract before using the bacterial test can sometimes mimic what would happen to the chemical in vivo. This adjunct to the test expands its capacity to detect potential human carcinogens.

1. The following are regional nucleotide sequences on one of the two strands of B-DNA. Which sequence might induce kinking of the helix? Briefly explain your answer.

(a) ACTGGUTT
(b) GGTAAAGT
(c) CTTTTTAG
(d) GAACTCGC
(e) GATCGATC

2. As discussed by Stryer on page 789, the major and minor grooves of B-DNA contain groups that are potential hydrogen bond donors and acceptors and hence might be important in specific interactions between proteins and DNA. Suppose that a given region of B-DNA is G·C-rich. Will the number of potential hydrogen bond-forming groups in the major groove and the minor groove differ from the number that would be present if the region were A·T-rich? Explain.

3. Which of the following nucleotide sequences might allow for the formation of Z-DNA? Explain your answer.

(a) CCTAGTTA
(b) TATATATA
(c) CGTGTACA
(d) TGAATTCA

4. Many proteins that bind to DNA consist of two α-helical recognition units that are related by a twofold axis of symmetry and are separated by a distance of 34 Å.

(a) Are these proteins likely to bind to double-stranded or to single-stranded DNA? Explain.
(b) Would the proteins bind preferentially to A-DNA, B-DNA, or Z-DNA? Explain.
(c) Describe how these proteins likely interact with DNA.

5. Suppose that two polynucleotide chains are joined by DNA ligase in a reaction mixture to which ATP labeled with ^{32}P in the α-phosphoryl group has been added as an energy source. What products of the reaction would be expected to carry the radioactive label? Explain.

6. Suppose that negatively supercoiled DNA with $Lk = 23$, $Tw = 25$, and $Wr = -2$ is acted on by topoisomerase I. After one catalytic cycle, what would be the values of *Lk, Tw,* and *Wr*?

7. Suppose that negatively supercoiled DNA with $Lk = 23$, $Tw = 25$, and $Wr = -2$ is acted on by DNA gyrase and ATP. After one catalytic cycle, what would be the values of *Lk, Tw,* and *Wr*?

8. What property of DNA polymerase I leads to the observation that *polA1* mutants are more sensitive to ultraviolet light than are their wild-type counterparts? (See p. 803 of Stryer.)

9. Suppose that a single-stranded circular DNA with the base composition 30% A, 20% T, 15% C, and 35% G serves as the template for the synthesis of a complementary strand by DNA polymerase.

(a) Give the base composition of the complementary strand.
(b) Give the overall base composition of the resulting double-helical DNA.

10. Suppose that a plasmid containing only genes A, B, C, and D begins to replicate rapidly at time $t = 0$. At $t = 1$, there are twice as many copies of genes B and C as there are copies of genes D and A. Is it possible to establish the order of the four genes on the plasmid? Explain.

11. Suppose that a bacterial mutant is found to replicate its DNA at a very low rate. Upon analysis, it is found to have entirely normal activity of DNA polymerases I and III, DNA gyrase, and DNA ligase. It also makes normal amounts and kinds of dnaA, dnaB, dnaC, and SSB proteins. The *oriC* region of its chromosome is found to be entirely normal with respect to nucleotide sequence. What defect might account for the abnormally low rate of DNA replication in this mutant? Explain briefly.

12. Which of the following mutations in a polypeptide chain could have been induced by a single hit of the mutagen 5-bromouracil on DNA? (See Stryer, p. 810. Also see the genetic code on p. 109 of Stryer.)

(a) Phe → Glu
(b) Asp → Ala
(c) Phe → Leu
(d) Met → Lys

13. Which of the following mutations could result from the action of a single hit of the mutagen hydroxylamine on DNA?

(a) Gln → Asn
(b) Glu → Lys
(c) His → Tyr
(d) Gly → Asp

14. The drug fluorouracil is widely used as an anticancer agent (see Stryer, p. 753). It irreversibly inactivates the enzyme thymidylate synthase. Explain how this treatment retards the growth of tumor tissue. Will the growth of normal cells be affected as well?

15. Mammalian cells of two differing genotypes can be fused together, usually in the presence of Sendai virus, to form multinucleate cells (heterokaryons) containing nuclei of both genotypes (see Stryer, p. 276). When fibroblasts from two patients suffering from xeroderma pigmentosum were fused, the resulting heterokaryons showed no deficiency in DNA repair. What conclusions can be drawn from this observation? Explain.

16. Physical studies on the interaction of the β_2 subunit of DNA polymerase III holoenzyme show that the β_2 subunit binds much more tightly to circular than to linear DNA molecules.

(a) Propose an explanation for this observation.
(b) What do you think would happen if the circular DNA were treated with an endonuclease?

17. As you will learn in Chapter 37 of Stryer, eukaryotic DNA is highly methylated at the C-5 position of cytosine. The degree of methylation is inversely correlated with gene expression. Although the exact role of C-5 methylation in gene expression is not known, it is known that these C-5-methylated cytosines can cause mutations. How?

18. Acyclovir is an antiviral agent used to reduce the pain and promote the healing of skin lesions resulting from adult chicken pox. Its structure is shown in Figure 31-2.

(a) What nucleoside does this drug resemble?
(b) Why does the drug need to be administered in dephosphorylated form?
(c) Acyclovir has very few side effects because it inhibits DNA replication only in herpes-infected cells. This is because all herpes viruses encode a thymidine kinase gene that is able to activate the drug. What is this activating reaction?
(d) Once acyclovir is activated, how does it inhibit DNA replication?
(e) Why are cells uninfected by virus relatively unaffected by acyclovir?
(f) The herpes virus can become insensitive to acyclovir therapy by mutations in either of two genes. What might they be?

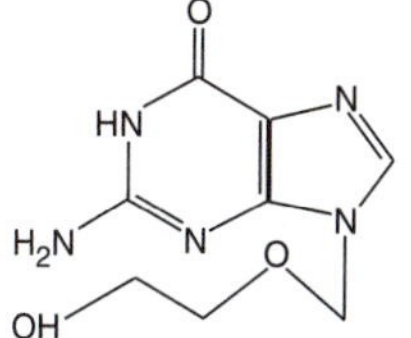

Figure 31-2
Structure of acyclovir.

ANSWERS TO PROBLEMS

1. A run of at least four adenine residues can induce kinking (see Stryer, p. 789). Therefore, sequence (c) might induce kinking, since a run of five adenines would be present on the complementary strand.

2. Both G·C and A·T base pairs of B-DNA have one hydrogen bond donor and two hydrogen bond acceptors in the major groove. However, a G·C pair has one donor and two acceptors in the minor groove, whereas an A·T pair has only two acceptors. Thus, a G·C-rich region on DNA differs from an A·T-rich region with respect to the number of hydrogen bond–forming groups in the minor groove.

3. Both (b) and (c) are alternating sequences of purines and pyrimidines and are therefore sites of possible Z-DNA formation (see Stryer, p. 791).

4. (a) They bind to a region of double-stranded DNA that also has a twofold axis of symmetry, like a palindromic region.

(b) B-DNA. The distance between adjacent major grooves is 34 Å. Also these grooves are wide enough to accommodate the recognition helix of the binding protein.

(c) The helical recognition regions bind to two adjacent major grooves of B-DNA.

5. In the overall reaction, ATP is hydrolyzed to AMP and pyrophosphate. Only AMP would be labeled. The phosphates involved in the formation of phosphodiester bonds are furnished by the incoming polynucleotide chains and do not arise from ATP.

6. *Lk* = 24, *Tw* = 25, and *Wr* = −1. Topoisomerase I increases the linking number of DNA by 1 each catalytic cycle (see Stryer, p. 797). This increase comes about at the expense of unwinding the negative supercoil.

7. *Lk* = 21, *Tw* = 25, and *Wr* = −4. DNA gyrase catalyzes a reaction in which both DNA strands are broken, the linking number is decreased by 2, and the number of negative supercoils is correspondingly increased by 2.

8. The *polA1* mutants are deficient in DNA polymerase I, having only 1% of the activity of their wild-type counterparts. They can, however, replicate their DNA at normal rates because DNA polymerase III is the enzyme that is primarily responsible for DNA replication. They are extraordinarily sensitive to ultraviolet irradiation because they lack the $5' \rightarrow 3'$ exonuclease activity of DNA polymerase I and are therefore deficient at DNA repair.

9. (a) 30% T, 20% A, 15% G, 35% C

(b) 25% A, 25% T, 25% C, 25% G. The base composition of the double strand is the average of that of the two single strands.

10. The order cannot be unambiguously established. Two possibilities are shown in Figure 31-3. Other solutions exist if one assumes unidirectional, rather than bidirectional, replication.

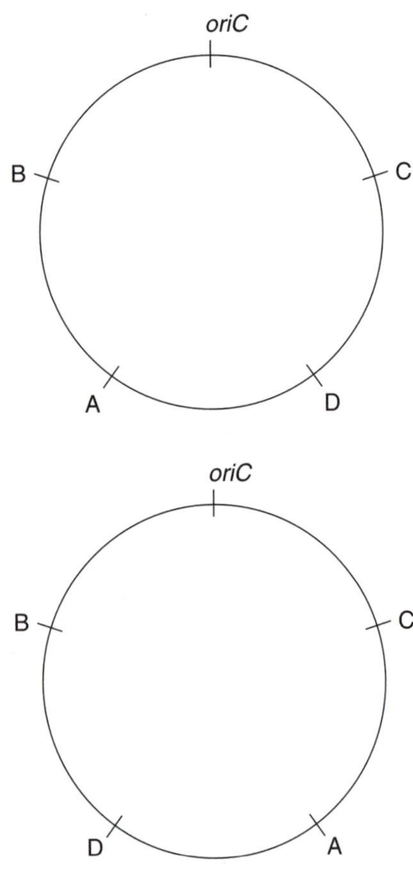

Figure 31-3
Two possible gene arrangements for problem 10.

11. A decrease in the activity of primase would account for the low rate of DNA replication. DNA replication requires the prior synthesis of RNA primers. Also, decreased rates of dNTP synthesis would slow replication.

12. The mutagen 5-bromouracil changes A·T pairs to G·C pairs or G·C pairs to A·T pairs. The mutation in (c) could be induced by 5-bromouracil. For example, the DNA sequence AAA, which codes for phenylalanine, could be changed to the sequence AAG, which codes for leucine. The other mutations could not arise from treatment with 5-bromouracil. Remember that the genetic code presented in Stryer is expressed in terms of RNA. The sequence UUU on RNA corresponds to the sequence AAA on the informational strand of DNA. Leucine is encoded by the sequence CUU on RNA, which corresponds to the sequence AAG on the informational strand of DNA. Remember also that, unless otherwise specified, nucleotide sequences are written in the $5' \rightarrow 3'$ direction.

13. Hydroxylamine is a mutagen that causes the unidirectional change of C·G pairs to T·A pairs. The mutation in (a) cannot result from the action of hydroxylamine. Those in (b), (c), and (d) might. In (b), TTC (Glu) could change to TTT (Lys). In (c), ATG (His) could be converted to ATA (Tyr). In (d), ACC (Gly) could change to ATC (Asp), or GCC (Gly) could change to GTC (Asp).

14. Because thymidylate synthase is inactivated, the supply of 5′-dTTP is insufficient to support the synthesis of DNA at normal rates. If DNA synthesis is suppressed, so will be the rate of division of the tumor cells. This type of treatment takes advantage of the fact that tumor cells divide more rapidly than do normal cells. The dosage of the drug is adjusted so that it will primarily affect more rapidly dividing cells. However, the division of some normal cells may be retarded as well.

15. The fibroblasts from the two patients show complementation, so it is likely that the two patients suffer from different genetic variants of xeroderma pigmentosum. The action of several genes is likely responsible for the excision and subsequent repair of damaged DNA. One patient, for example, could have produced a normal nuclease that excises damaged DNA but could have been deficient in a ligase. The other patient could have produced normal ligase but could have been deficient in nuclease activity. There are at least seven different complementation groups among xeroderma patients.

16. (a) A possible explanation is that the β_2 subunit falls off the end of a linear molecule.

(b) Treatment of circular DNA with an endonuclease would convert it to a linear DNA, and the affinity for β_2 protein would correspondingly decrease.

17. C-5 cytosine can deaminate just as easily as cytosine. When C-5 cytosine deaminates, it forms thymidine, not uracil. Therefore the uracil *N*-glycosylase will not recognize this product of deamination as an inappropriate base and will not remove it from the DNA, causing a transition mutation.

18. (a) guanosine

(b) The phosphorylated form could not cross cell membranes.

(c) The activation involves building phosphates onto the distal –OH group at the expense of ATP to produce a compound resembling 5′-GTP. (Although the enzyme responsible for the activation is called *thymidine kinase,* it can activate other nucleosides as well.)

(d) The triphosphate of acyclovir will serve as a substrate for DNA polymerase, and acyclovir nucleotide residues will be incorporated into growing polynucleotide chains in the place of guanosine residues. Acyclovir will, however, cause premature termination of nascent polynucleotide chains because it lacks a free –OH group onto which further nucleotides can be linked by phosphodiester bonds, that is, it lacks the analog of a 3′-OH group.

(e) Thymidine kinase is encoded by viral, not host, genes. Therefore uninfected cells will lack the enzyme.

(f) Mutations leading to inactive thymidine kinase or viral DNA polymerase that does not incorporate acyclovir would render infected cells insensitive to the agent.

EXPANDED SOLUTIONS TO TEXT PROBLEMS

1. DNA polymerase I uses deoxyribonucleoside triphosphates; pyrophosphate is the leaving group. DNA ligase uses DNA-adenylate (AMP joined to the 5′-phosphate) as a reaction partner; AMP is the leaving group. Topoisomerase I uses a DNA-tyrosyl intermediate (5′-phosphate linked to the phenolic OH); the tyrosine residue of the enzyme is the leaving group.

2. FAD, CoA, and $NADP^+$ are plausible alternatives. Note that, like NAD^+ and ATP, all three of these molecules contain a pyrophosphate (PP) linkage.

3. DNA ligase relaxes supercoiled DNA by catalyzing the cleavage of a phosphodiester bond in a DNA strand. The attacking group is AMP, which becomes attached to the 5′-phosphoryl group at the site of scission. AMP is required because this reaction is the reverse of the final step in the joining of pieces of DNA (see Figure 31-14 on p. 793 in Stryer).

4. Positive supercoiling resists the unwinding of DNA. The melting temperature of DNA increases in going from negatively supercoiled to relaxed to positively supercoiled DNA. Positive supercoiling is probably an adaptation to high temperature, where the unwinding (melting, denaturing) of DNA is markedly increased.

5. (a) Long stretches of each occur because the transition is highly cooperative.

(b) B-Z junctions are energetically highly unfavorable.

(c) A-B transitions are less cooperative than B-Z transitions because the helix stays right-handed at an A-B junction but not at a B-Z junction.

6. ATP hydrolysis is required to release DNA gyrase after it has acted on its DNA substrate. Negative supercoiling requires only the binding of ATP, not its hydrolysis.

7. (a) Pro (CCC), Ser (UCC), Leu (CUC), and Phe (UUC). Alternatively, the last base of each of these codons could be U.

(b) These C $\rightarrow$ U mutations were produced by nitrous acid.

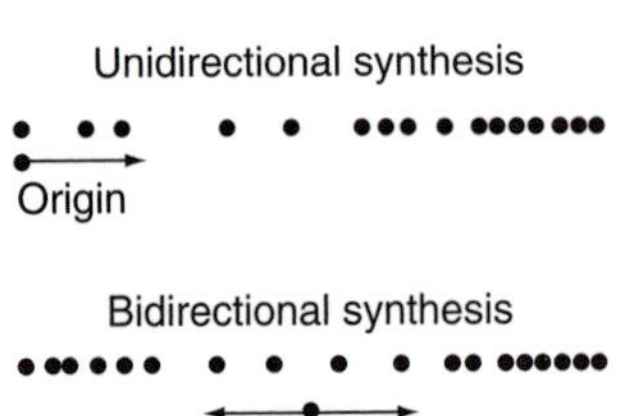

8. If replication were unidirectional, tracks with a low grain density at one end and a high grain density at the other end would be seen. On the other hand, if replication were bidirectional, the middle of a track would have a low density, as shown in the margin. For *E. coli,* the grain tracks are denser on both ends than in the middle, indicating that replication is bidirectional.

9. Potentially deleterious side reactions are avoided. The enzyme itself might be damaged by light if it could be activated by light in the absence of bound DNA harboring a pyrimidine dimer. The DNA-induced absorption band is reminiscent of the glucose-induced activation of the phosphotransferase activity of hexokinase (p. 499 in Stryer).

10. (a) 1000 nucleotides/s divided by 10.4 nucleotides/turn for B-DNA gives 96.2 revolutions per second (see Stryer p. 807).

(b) 0.34 μm/s. (1000 nucleotides/s corresponds to 3400 Å/s because the axial distance between nucleotides in B-DNA is 3.4 Å.)

(c) The rate of rotation of DNA at a replication fork (96.2 rps) is nearly the same as that of a bacterial flagellum (100 rps) (see Chapter 15). Skeletal muscle sarcomere shortening is about 15-fold as rapid as polymerase movement at a replication fork (5 μm/s, compared with 0.34 μm/s) (see Chapter 15). Smooth muscle sarcomere shortening (0.4 μm/s) occurs at about the same speed as replication fork movement.

CHAPTER

32

Gene Rearrangements

New sequences of nucleotides can be generated in DNA not only by mutation (Chapter 31) but also by recombination between two different DNA molecules or by the transposition of a segment of DNA within a chromosome. The breakage and joining of fragments of DNA also rearranges gene order within the chromosome and is a mechanism for repairing damaged DNA and regulating gene expression. Review the material on plasmids and bacteriophages in Chapter 6 in preparation for studying this chapter.

Stryer first describes homologous (general) recombination, which requires extensive sequence similarity between the interacting DNA duplexes. He describes the DNA strand movements that occur and explains how the recA and recBCD enzymes and other proteins carry out the recombination. Stryer then introduces recombination systems that do not require extensive regions of sequence similarity. Plasmids are discussed in terms of their roles as mobile genetic elements. F factor plasmids, which allow the exchange of DNA between bacteria, are described, as is their use to map the gene order of the bacterial genome when they are integrated into the bacterial chromosome to form an Hfr cell. Specialized R factor plasmids, which contain genes that confer antibiotic resistance on bacteria that harbor them, are presented next. Stryer continues with a general discussion of the nature of transposons and of the kinds of changes they induce in DNA sequences as a result of their movements. Site-specific recombination occurs at unique sites not necessarily determined by sequence similarities between the reacting duplexes. Site-specific recombination of λ phage into bacterial DNA, recombination of *V, D,* and *J* genes in the immune system, and retrovirus replication and integration into the host genome are described at the end of the chapter.

LEARNING OBJECTIVES

When you have mastered this chapter, you should be able to complete the following objectives:

Introduction (Stryer pages 819–820)

1. List the different kinds of *gene rearrangements.*

2. Distinguish between *homologous (general) recombination* and *transposition* in terms of the requirement for *sequence similarity* between the reacting DNA molecules.

3. List the *biological functions* of gene rearrangement and recombination.

Homologous Recombination (Stryer pages 820–826)

4. Describe the *Holliday model* for general recombination. Explain how the *Holliday recombination intermediate* can form two distinct recombinant products.

5. Draw the *chi form* and relate it to the Holliday model for recombination.

6. Outline the interactions of the *recA protein* with DNA that give rise to a *D loop.* List the substrates, intermediates, and products of the reaction and explain the roles of ATP.

7. Describe the consequences of the vectorial assembly of the *recA–single-stranded DNA filament* on recombination.

8. Relate the functioning of the *recBCD complex* to the generation of single-stranded DNA. Explain the roles of ATP and the *chi sequence* in this process.

9. Explain the functions of *single-strand binding protein (SSB), topoisomerase I, ruv A, B, C proteins,* and *DNA ligase* in homologous recombination.

10. Outline the *SOS response* to *DNA damage.* Describe the role of *rec*A and the repressor and protease roles of the *lex*A protein in this process.

Mobile Genetic Elements (Stryer pages 826–831)

11. Define *allele* and distinguish between the rearrangement of alleles and the rearrangement of genes. Assign these events either to homologous recombination or to the action of *mobile genetic elements (transposons).*

12. List the properties of *plasmids.* Describe *conjugation,* and explain the roles of *sex pili, pili receptors,* and *F factors* in this process.

13. Describe how an F factor can create an *Hfr cell.* Describe gene mapping using Hfr cells. Distinguish between an F factor and an *F′ factor.*

14. Explain how the F factor and *lysogenic bacteriophages* facilitate *bacterial evolution.*

15. Describe the acquisition of *simultaneous multiple-drug resistance* by a bacterial population. Explain the role of *R factor plasmids* in this process, and

indicate the functions of the *resistance transfer factor* (*RTF*) and r-*genes.* Distinguish between *simple* and *complex* R factor plasmids.

16. Describe the structure of *insertion sequences* (*IS*) and note the functions of their *inverted terminal repeats* and *gene products.* Outline the events that occur during the insertion of a *transposon* into a chromosome.

17. Explain how transposons can lead to *gene activation* or *inactivation* and, depending on *IS orientation,* to the *inversion, deletion,* or *duplication* of genes.

18. Distinguish between *simple* and *complex transposons.*

Site-Specific Recombination and Retroviruses (Stryer pages 831–837)

19. Distinguish site-specific recombination from homologous recombination.

20. Outline the lytic and lysogenic pathways in the life cycle of the λ bacteriophage and describe its recombination with the bacterial chromosome.

21. Explain how site-specific recombination generates the vast diversity of genes for antibody molecules and T-cell receptors. Describe the DNA sequences required for the recombination of *V, D,* and *J genes* (see Chapter 14 in Stryer).

22. Outline the life cycle of a *retrovirus.*

23. Describe the roles of *reverse transcriptase* and explain how it replicates the entire (+) *RNA strand* of retroviruses.

24. Outline the process of recombination of retroviral DNA with host DNA.

25. Provide examples of *retrotransposons* (*retroposons*) and describe their role in eukaryotic gene rearrangements.

SELF-TEST

Introduction

1. Which of the following statements about gene rearrangements are correct?

(a) They generate new combinations of genes.
(b) They can move a segment of DNA from one chromosome to another.
(c) They are mediated by the breakage of DNA and the rejoining of the resulting fragments.
(d) They generate genome sequence variability, upon which natural selection can act.
(e) They can regulate gene expression.
(f) They always require extensive regions of sequence similarity between the interacting DNA molecules in order to juxtapose the recombining sites.

2. DNA was isolated from *E. coli* that had been infected, in the presence of a DNA-synthesis inhibitor, with a mixture of bromouracil-labeled and ^{32}P-labeled T4 bacteriophages. If this DNA were centrifuged to equilibrium in a CsCl gradient, how would you expect it to distribute in the density gradient?

____

____

3. Place the following events in the order in which they would occur during homologous recombination between two DNA molecules:

(a) Strand exchange occurs between duplexes via branch migration at the crossover point. ________
(b) A pair of strands with similar sequences is cleaved in each duplex. ________
(c) Two duplexes align at a region of sequence similarity. ________
(d) Each invading strand becomes covalently joined to its corresponding strand in the other duplex. ________
(e) Strands of the recombinant intermediate are cleaved at or near the crossover point and are joined to their corresponding strands in each duplex. ________
(f) The end of each single strand invades the other duplex and forms base pairs with its complementary strand. ________

4. Which of the following statements about structures like those at the nodes or crossover points of chi-form DNA molecules are correct?

(a) They are formed in high yields when bacteria bearing certain plasmids are grown in the presence of chloramphenicol, a protein-synthesis inhibitor.
(b) They are intermediates that are formed during plasmid replication.
(c) They are formed from plasmids in *E. coli* strains that have mutations in the *rec* genes.
(d) When they are derived from plasmids, they are composed of two duplexes that are connected via four single strands of DNA.
(e) When they are derived from plasmids, they can be resolved in two different ways; one resolution yields two unit-length circular molecules, and the other produces a single, two-unit-long circular molecule.

5. When two circular plasmids containing a single restriction endonuclease site recombine and the recombination intermediates are isolated and cut with an endonuclease specific for that site, chi-shaped products with pairs of arms of equal length are formed. What do the structures of the products reveal about the sequences at the site of the crossover that forms the node of the chi form?

____

____

6. Which of the following statements about the recA protein of *E. coli* are correct?

(a) It is an ATP-dependent nuclease that generates single-stranded DNA.
(b) It catalyzes an ATP-dependent strand-assimilation reaction in which a single-stranded DNA molecule associates with duplex DNA.
(c) It hydrolyzes ATP to promote branch migration.
(d) It binds to single-stranded DNA to form a filament.
(e) It facilitates the search of duplex DNA for regions with sequence similarity to the invading single-stranded DNA.

7. Considering that duplex DNA is a stable molecule—that is, it ordinarily exists as a duplex in the cell—explain how the recA protein is capable of catalyzing the net insertion of a single-stranded DNA molecule into the duplex.

____

____

8. Explain how the single-stranded DNA required for general recombination is generated, using recBCD and SSB.

____

____

9. Homologous recombination is likely to require which of the following?

(a) DnaB protein
(b) Topoisomerase I
(c) RecA protein
(d) RecBCD complex
(e) ATP
(f) NAD^+
(g) Single-strand binding protein
(h) DNA-dependent RNA polymerase
(i) DNA polymerase I
(j) DNA ligase
(k) dATP

10. Which of the following statements about the lexA protein of *E. coli* are correct?

(a) It is cleaved by the recBCD complex.
(b) It represses the synthesis of recA protein.
(c) It is a protease as well as a repressor.
(d) It represses the synthesis of more than 15 different proteins.
(e) It is involved in the SOS response to DNA damage.

Mobile Genetic Elements

11. Which of the following are transposable elements in *E. coli*?

(a) Insertion sequences
(b) Bacteriophage mu (μ)
(c) R factor plasmids
(d) RecBCD complex genes
(e) Bacteriophage lambda (λ)
(f) Chi sequences
(g) F factor plasmids
(h) F′ factor plasmids

12. Which of the following pairs of polypeptides has genes that are alleles?
    (a) Hemoglobin α chain and hemoglobin β chain
    (b) Hemoglobin β chain and hemoglobin β^S chain (found in Hb S, sickle-cell hemoglobin)
    (c) Hemoglobin α chain and hemoglobin γ chain (found in Hb F, fetal hemoglobin)
    (d) Hemoglobin α chain and myoglobin
    (e) Hemoglobin β chain and myoglobin
    (f) Lactate dehydrogenase (LDH) isozyme M (found in muscle) and LDH isozyme H (found in the heart)

13. Which of the following statements about plasmids are correct?
    (a) They can serve as accessory chromosomes in a bacterium.
    (b) They can be essential for bacterial chromosomal replication when the *oriC* sequences of both the plasmid and the bacterium are nearly identical.
    (c) They can sometimes recombine with the chromosome of the bacterium harboring them.
    (d) They can replicate independently of the chromosome of their bacterial host.
    (e) They can confer drug resistance on the bacterium harboring them.
    (f) They can promote the transfer of genetic information between bacteria.

14. Which of the following are necessary for two bacteria to conjugate successfully?
    (a) Both must have pili.
    (b) Both must have pili receptors.
    (c) One must contain an F factor plasmid.
    (d) DNA synthesis must occur in the donor.
    (e) DNA synthesis must occur in the acceptor.

15. What is an Hfr cell?

    ____

    ____

16. Which of the following statements about the F′ factor are correct?
    (a) Its formation requires the integration of an F factor into the chromosome of a bacterium.
    (b) It can give rise to a bacterium that is diploid for some genes.
    (c) It is an F factor without a gene specifying a pilus protein.
    (d) It is functionally equivalent to the transducing phages in that chromosomal DNA can associate with its DNA and be transferred between cells.
    (e) It can produce a bacterium that is auxotrophic for an amino acid.

17. Which of the following answers complete the sentence correctly? A transmissible or infectious drug-resistance plasmid can be created when
    (a) an *r*-gene integrates into an R factor plasmid bearing a resistance transfer factor (RTF).
    (b) an R factor plasmid loses several *r*-genes.

(c) an RTF plasmid integrates into the chromosome of its host bacterium.
(d) a simple R factor plasmid excises from the host bacterial chromosome.

18. Which of the following statements about transposons are correct?

(a) They contain insertion sequences.
(b) They contain inverted terminal repeat sequences.
(c) They contain one or more genes specifying one or more enzymes that catalyze the transposition event.
(d) They require the *rec* gene products to complete their movements between or within genomes.
(e) They can lead to the duplication of short sequences of DNA in the recipient genome.

19. How could the insertion of a transposon into a bacterial chromosome enhance the expression of a bacterial gene?

____

____

Site-Specific Recombination and Retroviruses

20. Arrange the following list of events in the λ bacteriophage's lysogenic cycle in the correct chronological sequence.

(a) Integrase cuts both strands of the O sequences. ________
(b) Integrase and integration host factor bind to specific host and phage DNA sequences. ________
(c) Host DNA ligase forms a circular λ DNA molecule. ________
(d) Stable prophage exists in a lysogenic *E. coli* cell. ________
(e) Excisionase and integrase catalyze the breaking of B-P and P-B sequences. ________

21. Match the three proteins found in lysogenic *E. coli* with the appropriate items in the right column.

(a) Integrase ____________
(b) Excisionase ____________
(c) Integration host factor ____________

(1) Is encoded by *E. coli* DNA
(2) Is encoded by λ phage
(3) Recognizes P-O-P and B-O-B attachment sites
(4) Bends the P-O-P sequence
(5) Is required for the formation of prophage
(6) Is required for the release of λ DNA from lysogenic *E. coli*.

22. Which of the following answers completes the sentence correctly? Site-specific recombination of *V*- to *V*-segment genes of immunoglobulins does not occur because

(a) there is not an extensive sequence homology between them.
(b) they are flanked by recognition sequences lacking a spacer sequence.

(c) they lack an AT-rich nonamer sequence.
(d) their recognition sequences, which include a 23-bp spacer sequence, cannot recombine with each other.
(e) The recA enzyme does not bind to their palindromic heptamer sequences.

23. The life cycle of a retrovirus includes the following steps. List them in their correct sequence.

(a) Reverse transcriptase synthesizes the (−) and (+) strands of the viral DNA and digests the viral (+) RNA. ________
(b) The virion binds to a specific receptor on the cell surface. ________
(c) Integration of retroviral DNA into host DNA occurs. ________
(d) Viral (+) RNA is uncoated. ________
(e) Transcription of retroviral DNA occurs. ________

24. Explain the role of the tRNA bound to the (+) RNA viral genome.

____

____

25. Which statements about the recombination of retroviral DNA with host DNA are correct?

(a) It requires recognition of long terminal repeats by integration protein (IN).
(b) It occurs in a few specific sites of the host genome.
(c) It requires ATP hydrolysis.
(d) It involves the duplication of 4 to 6 nucleotides at the host integration site.
(e) It requires that the 3′-OH ends of the viral DNA become phosphorylated before attack of the host DNA.

ANSWERS TO SELF-TEST

Introduction

1. a, b, c, d, e. Answer (f) is incorrect because transposition, unlike homologous recombination, doesn't require extensive sequence similarity between the regions of DNA that react with each other.

Homologous Recombination

2. In addition to heavy, bromouracil-labeled DNA and light, ^{32}P-labeled DNA derived from unaltered parental genomes, you would expect to find recombinant molecules of intermediate density containing both labels.

3. c, b, f, d, a, e. (See Figure 32-2, p. 820, in Stryer.)

4. a, d, e. Answer (b) is incorrect because chi-form nodes are intermediates in recombination, not replication. Answer (c) is incorrect because their formation depends on the *rec* gene products. For answer (e), see Figure 32-4, p. 821, of Stryer.

5. Pairs of arms of equal length show that the crossover occurs at regions of sequence homology between the two joined molecules. Crossover at regions of sequence similarity will give rise to intermediates in which the single restriction site is uniquely located with respect to the joint. If the crossover occurred randomly, chi forms would be likely to have arms of unequal length because the unique restriction sites would be randomly distributed with respect to the node. Because the pairs of arms on chi structures can have any length, recombination can occur at almost any location on the DNA. (See Figure 32-5, p. 822, in Stryer.)

6. b, c, d, e. Answer (a) is incorrect because the recA protein lacks nuclease activity.

7. The recA protein binds both single-stranded and duplex DNA and facilitates interactions between them at regions that share sequence similarity. ATP plays several roles in the overall recombination reaction. It is required for the formation of the extended filament of recA on single-stranded DNA. Also ATP hydrolysis releases recA from DNA in a cycle reminiscent of the activities of actin and myosin in muscle contraction. Since recA filament assembly has polarity, ATP hydrolysis and recA release are coupled to the extension of the recA–DNA filament in the $5' \rightarrow 3'$ direction only, leading to unidirectional branch migration. In addition, ATP hydrolysis allows the invading single-stranded DNA to proceed past regions of duplex DNA that do not share sequence similarity.

8. The products of the *recB, recC,* and *recD* genes form the recBCD complex, an ATP-driven, site-specific nuclease and helicase that unwinds duplex DNA and hydrolyzes it to create free, single-stranded DNA. The recBCD complex binds to the end of a linear duplex and acts as a helicase to unwind the DNA as it invades the duplex. The unwound DNA rewinds behind the moving complex more slowly than it is unwound, leading to the formation of two single-stranded loops. When the complex encounters a chi site, which is a specific sequence, it cuts the strand. The complex moves on, thus forming single-stranded DNA with a free end (see Figure 32-11, p. 825, in Stryer). The single-stranded DNA is trapped by SSB so that it can then interact with the recA protein to initiate a strand-assimilation reaction with a homologous duplex.

9. b, c, d, e, f, g, i, j, k. ATP is required for the actions of recA and the recBCD complex; dATP and the other three dNTPs, for DNA synthesis by DNA polymerase I; and NAD^+, for the DNA ligase reaction.

10. b, c, d, e. The lexA protein is a repressor that regulates several genes whose products are involved in DNA repair via the SOS pathway. Single-stranded DNA, which arises from the damaged DNA, binds to recA protein, and the resulting ssDNA–recA complex can activate associated lexA. The activated lexA then acts as a protease to cleave neighboring lexA proteins, allowing the proteins involved in the repair of DNA to be synthesized.

Mobile Genetic Elements

11. a, b, c, e, g, h. Answer (f) is incorrect because chi sequences promote general homologous recombination, which does not ordinarily result in the rearrangement of the order of genes on the chromosome. See Table 32-1 of Stryer.

12. b. Only in choice (b) are the two proteins derived from two different or alternative forms of the same gene. For all the other pairs, the two proteins are derived from different genes. Homologous recombination generates new combinations of alleles, whereas transposition can reorder the positions of genes.

13. a, c, d, e, f

14. c, d, e. Answers (a) and (b) are incorrect because the male bacterium must have pili and the female must have pili receptors.

15. Bacteria that harbor an F factor plasmid in their chromosomes are called *Hfr cells* because they exhibit a high frequency of recombination. An Hfr cell is capable of transferring its entire chromosome into a female cell by conjugation. The F factor contains the genes for the formation of the sex pili and the other components required for conjugation.

16. a, b, d, e. The integration of an F factor is a necessary prerequisite to the formation of an F′ factor. When an F′ factor carries a chromosomal gene into a bacterium that duplicates one already present, the recipient cell becomes diploid for that gene. If an F′ factor integrates into the bacterial chromosome such that a chromosomal gene encoding an essential enzyme needed to synthesize an amino acid is disrupted, the bacterium would then have a nutritional requirement, that is, it would be auxotrophic, for that amino acid.

17. a. An infectious plasmid bearing genes that confer drug resistance on a bacterium can be formed by recombination between two plasmids: an R factor plasmid containing an RTF and a plasmid bearing one or more drug-resistance genes. The excision of a simple R factor plasmid would not produce a transmissible plasmid, because it would lack the RTF genes.

18. a, b, c, e. A transposon can be composed of as little as an insertion sequence (IS) and of as much as the IS plus several genes, including those necessary for transposition. That is, a transposon may contain only a transposase, or it may contain a transposase plus a resolvase plus additional genes that specify drug resistance as well as other functions. Transposition does not require the *rec* gene products, but it does use DNA polymerase I and DNA ligase.

19. A promoter on the transposon might start a transcript that extends into a neighboring bacterial gene. This transcript would lead to enhanced expression of the neighboring gene relative to its expression before transposition.

Site-Specific Recombination and Retroviruses

20. c, b, a, d, e. See Figures 32-23 and 32-24 in Stryer.

21. (a) 2, 3, 5, 6 (b) 2, 6 (c) 1, 4, 5, 6

22. d. Answer (a) is incorrect because site-specific recombination does not require extensive sequence homology between the DNA strands. See Figure 32-25 in Stryer for the sequences of the two types of recognition sites. Recombination occurs only between different classes of sites as shown in Figure 32-26 in Stryer.

23. b, d, a, c, e

24. The 3′-OH end of the tRNA acts as the primer for DNA synthesis by reverse transcriptase on the (+) RNA template. See Figure 32-29 in Stryer.

25. a, d. Answer (b) is incorrect because integration can occur nearly anywhere in the host DNA. Answers (c) and (e) are incorrect because the integration reaction does not require ATP and the 3′-OH group of viral DNA directly attacks a phosphodiester bond in the target DNA.

PROBLEMS

1. Suppose that one parental DNA has the genes *A, B,* and *C,* in that order, and that the other parental DNA has the alleles *a, b,* and *c.*

   (a) Give all the recombinant genotypes that may be formed.
   (b) Which recombinant genotype or genotypes will be produced least frequently? Why?

2. Suppose that an Hfr strain of *E. coli* that is sensitive to streptomycin (str^S) and is able to synthesize amino acids A and B is mixed together in a culture medium with an F^- strain that is resistant to streptomycin and is unable to synthesize amino acids A and B. After some time, mating is interrupted by placing the mixture in a Waring blender, and then samples of the bacteria are transferred to other culture media to test whether or not conjugation has occurred.

   (a) When the bacteria are transferred to a medium containing streptomycin but lacking amino acids A and B, some bacterial growth occurs. Explain this result.
   (b) Suppose that you carry out a similar experiment but place the bacterial mixture into the blender after a shorter time. Explain why some bacteria might grow in a medium containing only streptomycin and A.
   (c) Studies of conjugation in bacteria always involve following more than one marker. In this experiment, for example, three markers, *A, B,* and str^s are followed. Why do you think this is necessary? What possibility could not be excluded if only gene *A* were followed? (Hint: Recombination in bacteria is a rare event.)

3. An early and very attractive mechanism proposed for genetic recombination was the copy-choice model. It suggested that recombination between two parental DNA duplexes occurs during DNA replication when DNA polymerase switches from one parental duplex to the other so as to

produce a recombinant daughter DNA duplex. The copy-choice model is now known to be incorrect. One experimental finding inconsistent with the copy-choice model is the observation that when *E. coli* bacteria are infected by T4 bacteriophage of two different genotypes, whose DNAs are distinctly marked, one by ^{32}P and the other by bromouracil, recombinant DNAs containing both markers are found under conditions in which DNA synthesis is blocked. How are these findings inconsistent with the copy-choice model, but consistent with the breakage-reunion model?

4. When one thinks about genetic recombination from a classical point of view, one may envision that the farther apart two genes are from each other on a chromosome, the greater the likelihood of recombination between them will be. However, the recombination frequency between two given nucleotides is not always a function of the internucleotide distance separating them. Briefly explain how recombination may not occur, for example, between two sites in the *E. coli* chromosome that are 3000 nucleotide residues apart.

5. When transposons having long, inverted repeat sequences are denatured and are allowed to self-anneal, they form a characteristic structure that may be seen through electron microscopy. Sketch the structure that would be expected for a transposon having the sequence *ABCDEFGHIJKL-GFEDCBA*.

6. Sketch the product or products that would be expected when recombination between *A* and *B* occurs in the insertion sequence *ABCWXYZcba*. What happens to the sequence *WXYZ*?

7. Sketch the product or products that would be expected when recombination between *B* and *C* occurs in the insertion sequence *ABCDEFGH-abcd*. What happens to the sequence *EFGH*?

8. The kind of recombination displayed by transposons, sometimes called recA-independent recombination, has been regarded as having far greater evolutionary significance than does general, or recA-dependent, recombination. Why do you think this may be the case?

9. Suppose that a DNA-containing bacterial virus has five genes, *A, B, C, D,* and *E,* in that order on its genome. The viral DNA is known to exist during infection as a concatamer in which the sequence *ABCDE* is repeated many times. After replication, viral DNA is stuffed into preformed heads and is snipped by an endonuclease once the head is filled. The capacity of the head is a length of DNA that contains six genes. Although the DNA of this virus is a linear duplex, when genetic mapping studies are conducted, the DNA maps as a circle. Explain how this might come about.

10. The viral-mediated transfer of host DNA from one bacterial cell to another is known as *transduction.* Bacteriophage P1 may carry a wide variety of *E. coli* genes from one bacterial cell to another. In contrast, bacteriophage λ carries only *gal* or *bio* genes from one cell to another. Explain the difference in the transducing behavior of the two phages.

11. What addition must be made to a lexA preparation to make it proteolytically active?

12. Suppose the gene for resistance to antibiotic X is carried on a small R factor plasmid that cannot be transmitted by conjugation. What genetic

engineering manipulations would you carry out so that the gene for resistance to X may be transmitted by conjugation?

13. Suppose that in the coding region for the heavy chains of an antibody molecule there were 230 *V* genes, 15 *D* genes, and 5 *J* genes. Calculate the number of different kinds of H genes that could be formed, assuming that *V* and *J* can be formed in three frames. (You might wish to refer to p. 374 of Stryer.)

14. How do the observations that methotrexate and fluorodeoxyuridylate block infection by RNA tumor viruses implicate DNA synthesis as a requirement for multiplication of those viruses within a host?

15. Infectious disease specialists use plasmid profiles to determine whether antibiotic-resistant bacteria isolated from different patients represent different resistant strains of the same species, or the same strain colonizing all the patients. In addition, they determine the antibiotic resistance profile of each isolate. Using these techniques, epidemiologists have documented the spread of triple antibiotic resistance genes from one species of bacteria to another, which presents great difficulties in patient care. As an infectious disease specialist, how would you show that the latest strain of penicillin-resistant streptococcus had acquired its penicillin resistance from another species of bacteria?

16. Explain four different ways that an antibiotic resistance gene might move from the genomic DNA of one bacterium to the genomic DNA of another.

17. The copy-choice model of recombination does not occur in *E. coli,* but does occur in retroviral replication. Explain this observation, using as a basis your understanding of the mechanisms by which *E. coli* and retroviruses replicate.

ANSWERS TO PROBLEMS

1. (a) The possible recombinant genotypes are *Abc* and *aBC,* resulting from recombination between *A* and *B* only; *ABc* and *abC,* resulting from recombination between *B* and *C* only; and *AbC* and *aBc,* resulting from simultaneous recombination between *A* and *B* and between *B* and *C*.

   (b) The genotypes AbC and aBc will be produced least frequently because they involve the simultaneous occurrence of rare events.

2. (a) Neither parent strain could grow in such a medium. The Hfr parent could not survive, because the medium contains streptomycin, and the F^- parent would be killed because the medium lacks amino acids A and B. The result is explained by conjugation followed by genetic recombination. The Hfr strain could have donated DNA containing information for the production of enzymes that synthesize amino acids A and B to the F^- strain during mating. Provided that there was recombination between the newly

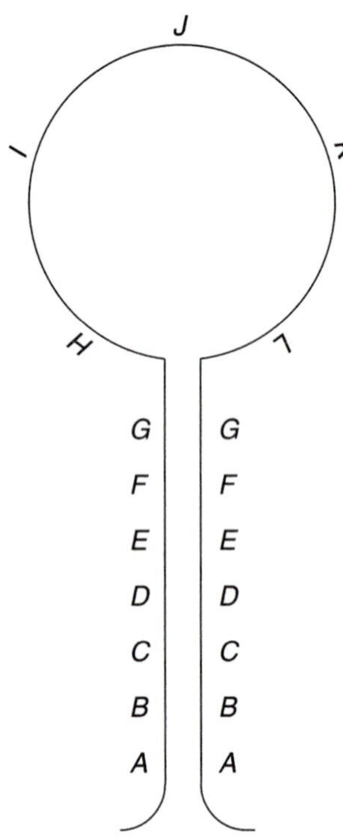

Figure 32-1
Structure of a transposon with long, inverted repeats following denaturation and self-annealing.

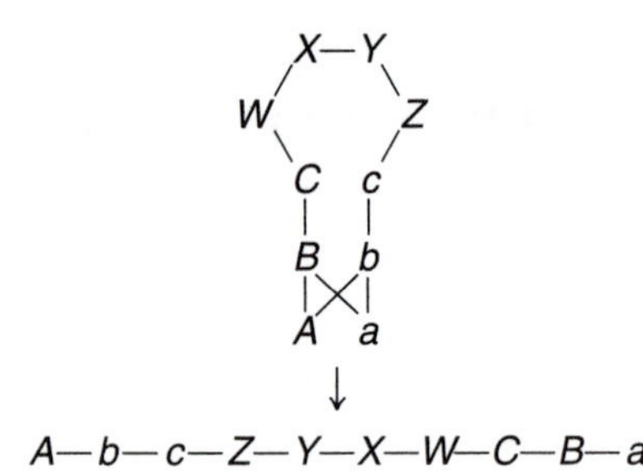

Figure 32-2
Solution to problem 6.

introduced DNA and the DNA of the F$^-$ chromosome, the resulting progeny would be able to synthesize amino acids A and B and would be resistant to streptomycin.

(b) This would occur if *B* were the first gene to be transferred during mating and *A* were the second. If mating were interrupted after the transfer of *B* but before the transfer of *A*, the result would be F$^-$ bacteria that are streptomycin-resistant but require the addition of *A* for growth.

(c) In an F$^-$ strain that is unable to synthesize *A* and *B*, it is always possible that regaining the ability to make either *A* or *B* could result from mutation and not from conjugation. (The frequencies of occurrence of mutation and recombination are similar.) However, it is highly improbable that mutation would simultaneously reactivate the genes for producing both *A* and *B*.

3. First, if copy-choice were a correct model, no recombinant phage should be produced in the absence of new DNA synthesis. Second, according to that model, no recombinant DNA duplexes should contain both bromouracil and ^{32}P. However, the Holliday model for homologous recombination (Stryer, Figure 32-2, p. 820) does account for the observations.

4. The mechanism of recombination mediated by the recA protein requires that breaks be introduced into one strand of a DNA duplex by the recBCD complex (see Figure 32-11 on p. 825 of Stryer). The break occurs near a specific sequence called *chi*, which occurs on the average of once for each 4000 nucleotide residues in *E. coli* (see p. 825 of Stryer).

5. The structure is a "lollipop," as shown in Figure 32-1.

6. The new sequence is *AbcZYXWCBa*, as shown in Figure 32-2. The original sequence *WXYZ* is inverted.

7. The two products are shown in Figure 32-3. The sequence *EFGH* is deleted from one of the products.

8. RecA-dependent recombination can occur only between homologous segments of chromosomes. RecA-independent recombination allows for rather drastic rearrangements of chromosomes, including the insertion, deletion, and duplication of genes, as shown in Figure 32-20 on page 830 of Stryer. Without recA-independent recombination, evolutionary possibilities would be far more limited.

9. The concatameric DNA of this virus has the following gene sequence:

ABCDEABCDEABCDEABCDEABCDEABCDEABCDE . . . .

Because the viral DNA is stuffed into preformed heads, and because the maximum capacity of each head is six genes, the first virion will have the sequence *ABCDEA;* the second, *BCDEAB;* the third, *CDEABC;* and so forth. When genetic mapping experiments are conducted with these viruses, the genetic map will be circular because, for example, gene A will be equally close to genes B and E. Thus, concatameric linear DNA behaves genetically as if it were circular.

10. Bacteriophage λ is integrated as a prophage at a specific site between the *gal* and *bio* genes on the *E. coli* chromosome. If it carries any host-cell DNA with it as it breaks out of the host chromosome, it will be the DNA immediately adjacent to its site of integration. In contrast, bacteriophage P1 is a lytic, not a temperate, phage whose preformed capsids are filled with DNA. When a host-cell DNA fragment is packaged into a preformed capsid, the host-cell DNA will be carried to a cell subsequently infected by the virus particle. (A virus particle carrying host-cell DNA to the exclusion of its own DNA could not be infective.) The host DNA carried in the capsid may be a fragment of any part of the host chromosome.

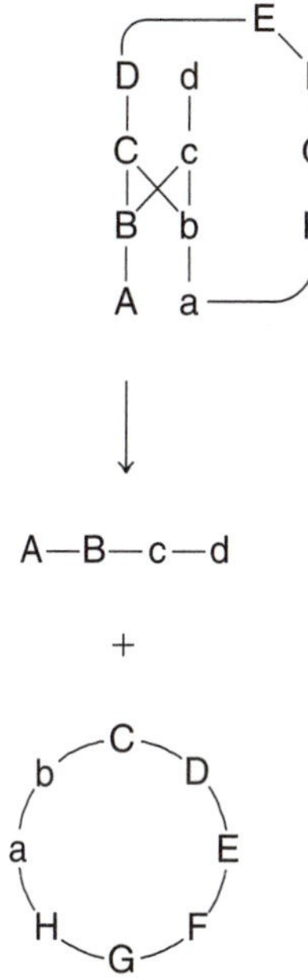

Figure 32-3
Solution to problem 7.

11. ssDNA–recA must be added.

12. Cut the small plasmid with a restriction endonuclease, and cut a larger plasmid containing an RTF region by the same endonuclease. Mix the two linear DNAs with their complementary sticky ends together. Some larger plasmids will re-form containing both the gene for resistance to X and the RTF region. Those recombinant plasmids may be transmitted by conjugation.

13. The number is $230 \times 15 \times 5 \times 3 = 51{,}750$.

14. Fluorodeoxyuridylate and methotrexate both result in the decreased synthesis of dTMP from dUMP, although they act at different points in the pathway of dTMP synthesis (see Stryer, Figure 29-18, p. 753). If dTMP cannot be synthesized, neither can DNA, because of the lack of one of the four nucleotide precursors, dTTP, which is produced from dTMP. Thus, the ability of these agents to inhibit proliferation of RNA tumor viruses suggests that DNA synthesis is required.

15. You would need to show the presence of a new plasmid that appears whenever the gene conferring penicillin resistance is present. Furthermore, this plasmid could not be present when a species was found to be penicillin-sensitive. Although this would probably serve as ample proof, to be absolutely sure one could do the following: Take one of the penicillin-resistant strains and remove or otherwise inactivate the plasmid in question and show conversion to penicillin sensitivity. Then reintroducing that plasmid into a sensitive strain should make it resistant. Note: This approach would not work if the resistance were spread through phage transduction.

16. (1) Movement might occur through generalized phage transduction, as occurs with bacteriophage P1. (2) The gene could be carried on a prophage. (3) The gene could be introduced by an Hfr cross. (4) The gene could be part of a mobile genetic element such as a retrotransposon. The mobile fragment of DNA could hitch a ride to the next cell on a prophage or an Hfr plasmid and then hop into the new bacterial genome.

17. During *E. coli* replication the DNA polymerase remains attached to the same DNA template throughout the entire process. Therefore the only recombination that can take place is through breaking and rejoining preformed pieces of DNA. However, retroviral replication has an obligatory template-switching step as part of its replication cycle (see Stryer, p. 836). Furthermore, each viral particle contains two RNA molecules, and one could argue that copy-choice recombination occurs during every cycle of retroviral replication.

EXPANDED SOLUTIONS TO TEXT PROBLEMS

1. Helicase participates in DNA replication. The recB component of the recBCD complex generates single-stranded DNA in general recombination. The recA protein binds to single-stranded DNA and catalyzes its invasion of duplex DNA and a switch in base pairing.

2. (a) The recA–DNA complex contains 18.6 base pairs per turn, compared with 10.4 for B-DNA. Hence, the linking number (*Lk*) is 56, compared with 100 for the relaxed circle formed in the absence of recA protein (1040/18.6 = 56, and 1040/10.4 = 100; see p. 795 in Stryer).

   (b) Recall that about 70% of the change in *Lk* in an isolated circular DNA molecule is expressed in *Wr* (writhing) and 30% in *Tw* (twisting) (p. 796 in Stryer). Hence, after removal of recA, *Wr* is 30.8 less than that of the relaxed circle [$(100 - 56) \times 0.70 = 30.8$], which means that the duplex is negatively supercoiled (i.e., right-handed). The degree of supercoiling σ is −0.44, compared with −0.06 for most naturally occurring DNA. Thus, closure in the presence of recA generates a highly supercoiled molecule, which is much more compact than its relaxed counterpart and will sediment much more rapidly. To calculate sigma we use the equation on page 796 in Stryer as follows: $\sigma = (56 - 100)/100 = -0.44$.

3. Proteolysis of the lexA protein at the onset of the SOS response relieves the repression of synthesis of lexA mRNA. DNA repair then leads to a decrease in the amount of the ssDNA–recA complex, the coprotease. Hence, the level of intact lexA increases, which in turn terminates the SOS response.

4. (a) In the presence of ATPγS, recA protein binds to DNA, but its dissociation is blocked because this analog is not readily hydrolyzed. Strand exchange does not take place.

   (b) RecB protein binds to DNA in the presence of this analog, but its release is blocked. Consequently, single-stranded DNA is not formed.

5. Negative supercoiling promotes recombination by inducing strand separation (negative *Wr* can be changed into negative *Tw*). Furthermore, negatively supercoiled DNA occupies a much smaller volume than does relaxed DNA. Hence, the effective concentration of sites undergoing recombination is increased by negative supercoiling. For an illuminating discussion, see R. Kanaar and N.R. Cozzarelli. *Curr. Opin. Struct. Biol.* 2(1992):369.

6. Homologous recombination between a genomic DNA sequence and a similar cloned DNA sequence introduced into a cell makes it possible to transfer any modification of the cloned gene into the genome of a cell. For example, a stop codon could be placed early in a sequence encoding a protein of interest. See M.R. Capecchi. *Science* 244(1989):1288.

7. Viruses make the most of a small genome. The encoding of a different second subunit would require about 2 additional kb of RNA. Replication would be slower and packaging of the nucleic acid more difficult.

CHAPTER

33

RNA Synthesis and Splicing

The conversion of DNA nucleotide sequences into RNA sequences is an early step in the expression of genetic information. DNA-dependent RNA polymerases catalyze this reaction, and the product RNA transcript must sometimes be cleaved and modified in various ways before it becomes functional. You were introduced to these topics in Chapter 5; in this chapter Stryer describes them in detail in both prokaryotes and eukaryotes. Reread Chapter 5, paying particular attention to the nomenclature, structures, and kinds of RNA found in cells; RNA polymerases, promoters, and terminators; and introns and RNA splicing. Also refresh your understanding of the relationship between the mRNA nucleotide sequence and the protein amino acid sequence, which is outlined in Chapter 5. Review the material on DNA polymerase I on pages 88–89 and 799–800 of Stryer in order to be better able to contrast its properties with those of the RNA polymerases. In addition, reread pages 230–232 in Stryer on catalytic RNA before studying the section on self-splicing RNA in this chapter.

Stryer begins this chapter by describing RNA polymerase from *E. coli.* He describes the subunit structure of the enzyme and divides the RNA synthesis reaction into initiation, elongation, and termination stages. The nature of prokaryotic promoters is given, and the function of the σ subunit of RNA polymerase in specific transcript initiation is explained. Stryer next describes the structure of the ternary elongating complex consisting of the template DNA, RNA polymerase, and product RNA. Finally, two mechanisms of transcription termination, one of which requires the ρ protein, are detailed. Stryer outlines the cleavage and modification reactions involved in the maturation of ribosomal and transfer RNAs. In addition, he describes two antibiotics, rifampicin and actinomycin D, that inhibit transcription by different mechanisms.

Stryer next turns to the more complex process of transcription in eukaryotes. He points out that the coupling of transcription and translation that occurs in prokaryotes is impossible in eukaryotes because they occur in different cellular compartments, namely, the nucleus and cytoplasm, respectively. He describes the three RNA polymerases that carry out this process and relates them to the kinds of RNA they synthesize. Eukaryotic promoters and enhancers and some of the proteins that bind them are then presented. He explains how multiple combinations of transcription factors binding to specific sites on the DNA can give rise to temporally regulated gene expression. The reactions that modify the 5′ and 3′ ends of typical eukaryotic mRNA transcripts to cap and add a poly(A) tail to them are also described. He also shows how the information content, that is, the nucleotide sequence of certain mRNAs can be modified by base alterations and insertions. The molecular machinery and reactions that remove RNA segments from the interior of eukaryotic RNA precursors, that is, the splicing reactions that remove introns, are outlined. The discovery of catalytic RNA and its catalytic versatility is presented next. Descriptions of the mechanistic and possible evolutionary relationships between three different splicing mechanisms, two of which do not require the participation of proteins, close the chapter.

LEARNING OBJECTIVES

When you have mastered this chapter, you should be able to complete the following objectives:

RNA Polymerase from *E. coli* (Stryer pages 841–845)

1. Describe the *subunit structure* of *RNA polymerase* from *E. coli* and assign functions to the individual subunits.

2. Name the three *stages* of *RNA synthesis,* and list the functions of RNA polymerase in these processes.

3. Define *promoter,* and recount how RNA polymerase protects promoters from hydrolysis by DNase. Explain how *footprinting* can be used to locate promoters.

4. Recognize the convention for numbering the nucleotides in the DNA template with regard to the *transcription start site.* Distinguish between the *template* (or *antisense*) *strand* and the *coding* (or *sense*) *strand* of the duplex DNA template.

5. Note the *consensus sequences* around the *−35* and *−10* positions of *E. coli* promoters. Contrast the rates of transcript initiation on *strong* and *weak promoters* in *E. coli.*

6. Explain how the *σ factor* enables RNA polymerase to *recognize promoters.* Describe the effect of the *unidimensional random walk* of RNA polymerase along the DNA template on the rate constant for the binding of RNA polymerase to the promoter.

7. Distinguish between the σ^{70} and σ^{32} subunits, contrast the sequences of *standard promoters* and *heat-shock promoters,* and provide examples of how σ factors can determine which genes are expressed.

8. Describe how topoisomerase I was used to determine the number of promoter *DNA base pairs that are unwound* upon the binding of RNA polymerase. Relate *negative supercoiling* to promoter efficiency. Distinguish between *closed promoter* and *open promoter complexes.*

Prokaryotic Transcription (Stryer pages 845–852)

9. Draw the structure of the *5′ end* of an RNA transcript in *E. coli.* Explain how [γ–^{32}P]ATP or [γ–^{32}P]GTP can be used to show that RNA polymerase elongates RNA chains in the 5′ → 3′ direction.

10. Describe the model for the *transcription bubble.* State the *number of base pairs in the RNA–DNA hybrid.* Appreciate the *rate of RNA chain elongation* in terms of both the nucleotides added and the distance on the template traversed by RNA polymerase.

11. Contrast the *fidelity of polymerization* of RNA polymerase with that of DNA polymerase, and explain the reason for this difference.

12. Sketch a *ρ-independent transcription termination structure* and state the roles of the *GC-rich hairpin* and *oligo(U) stretch* of the RNA structure in the termination of transcription.

13. Contrast *ρ-dependent* and ρ-independent transcription termination. Outline the mechanisms of the ρ protein and explain the role of *ATP hydrolysis* in its function.

14. Compare the degree of *modification* of mRNA with that of *ribosomal* and *transfer RNA* (*rRNA* and *tRNA*) in prokaryotes. Outline the mechanisms of *intron removal* from some *precursors of eukaryotic tRNA.*

15. Describe the mechanisms of *inhibition of transcription* by *rifampicin* and *actinomycin D.*

Eukaryotic Transcription (Stryer pages 852–860)

16. Note the spatial and temporal differences in transcription and translation between prokaryotes and eukaryotes. Consider the regulatory implications of these differences.

17. Contrast *mRNA processing* in prokaryotes and eukaryotes. Outline the major reactions that process the *primary RNA transcripts* of eukaryotes.

18. Describe the *RNA polymerases of eukaryotes* and locate them within the cell. List the kinds of RNA they synthesize. Compare the structural and catalytic features they share with the RNA polymerase of *E. coli.*

19. Account for the toxic effects of *α-amanitin* and describe how it can be used to differentiate the three eukaryotic RNA polymerases.

20. List the salient *sequence elements* of *eukaryotic promoters.* Contrast the nucleotide sequences and locations of the *TATA box* of eukaryotes and the −10 sequence of prokaryotes.

21. Explain the key role of *TATA box-binding protein* in assembling active *transcription complexes.*

22. Describe *enhancers,* list their functions, and present a molecular model that accounts for their activity. Relate *DNA looping* to enhancer function.

23. Outline the combinatorial activity of *transcription factors* and other DNA-binding regulatory proteins in directing RNA polymerase to specific genes.

24. Draw the structure of the *5′ end* of a typical *eukaryotic mRNA,* and distinguish *caps 0, 1,* and *2.* Outline the reactions required to cap the primary transcript.

25. Contrast eukaryotic and prokaryotic transcription termination. Describe the events leading to the production of mRNA with a *poly(A) tail.*

26. Describe *RNA editing* and provide examples of its effects on gene expression.

RNA Splicing (Stryer pages 860–870)

27. Describe the *splicing of eukaryotic mRNA,* give the *consensus sequences* at the *splice site junctions,* and designate the other nucleotide *sequence elements* involved in the process. List the functions of this *post-transcriptional modification.*

28. Provide an example of *aberrant splicing* that leads to a form of *thalassemia* in humans.

29. Draw the structure of the nucleotide at the *branch site* of an mRNA *lariat intermediate.*

30. Describe the role of *transesterification reactions* in splicing and compare the number of phosphodiester bonds broken and formed during mRNA splicing.

31. Describe the *spliceosome* and detail the structural and catalytic involvement of *small nuclear ribonucleoprotein particles (snRNPs)* in mRNA splicing.

32. Describe the discovery of *self-splicing RNA* and outline the reactions that occur during the conversion of a ribosomal RNA (rRNA) precursor from *Tetrahymena* into the spliced, mature rRNA and other linear and circular products. Explain the role of *guanosine* or a *guanyl nucleotide* in the self-splicing reaction.

33. Describe the activities of *L19 RNA* and explain the significance of its dual activities.

34. Describe the *hammerhead self-catalytic RNA* and explain the significance of its structure and mechanism.

35. Contrast the *group I* and *group II self-splicing* introns. Compare *spliceosome-catalyzed* splicing with self-splicing.

SELF-TEST

RNA Polymerase from *E. coli*

1. Give the subunit composition of the RNA polymerase of *E. coli* for both the holoenzyme and the core enzyme.

____

____

2. Match the subunit of the RNA polymerase of the *E. coli* in the left column with its putative function during catalysis from the right column.

(a) α ____2____
(b) β ____3____
(c) β' ____1____
(d) σ^{70} ____4____

(1) Binds the DNA template
(2) Binds regulatory proteins and sequences
(3) Binds NTPs and catalyzes bond formation
(4) Recognizes the promoter and initiates synthesis

3. Which of the following statements about *E. coli* promoters are correct?

(a) They may exhibit different transcription efficiencies.
(b) For most genes they include variants of consensus sequences.
(c) They specify the start sites for transcription on the DNA template.
(d) They have identical and defining sequences.
(e) They are activated when C or G residues are substituted into their −10 regions by mutation.
(f) Those that have sequences that correspond closely to the consensus sequences and are separated by 17 base pairs are very efficient.

4. The sequence of a duplex DNA segment in a DNA molecule is

5′-ATCGCTTGTTCGGA-3′
3′-TAGCGAACAAGCCT-5′

When this segment serves as a template for *E. coli* RNA polymerase, it gives rise to a segment of RNA with the sequence

5′-UCCGAACAAGCGAU-3′

Which of the following statements about the DNA segment are correct?

(a) The top strand is the coding strand.
(b) The bottom strand is the sense strand.

(c) The top strand is the template strand.
(d) The bottom strand is the antisense strand.

5. Which of the following statements about the σ subunit of RNA polymerase are correct?

(a) It enables the enzyme to transcribe asymmetrically.
(b) It confers on the core enzyme the ability to initiate transcription at promoters.
(c) It decreases the affinity of RNA polymerase for regions of DNA that lack promoter sequences.
(d) It facilitates the termination of transcription by recognizing hairpins in the transcript.

6. The rate constant for the binding of RNA polymerase holoenzyme to a promoter on a long DNA molecule is greater than that for the collision of two small molecules in solution. Since small molecules diffuse through solutions more rapidly than large ones, how can this be true?

7. When growing *E. coli* are subjected to a rapid increase in temperature, a new and characteristic set of genes is expressed. Explain how this alteration in gene expression occurs.

Prokaryotic Transcription

8. A restriction fragment of bacterial DNA is transcribed in vitro by *E. coli* RNA polymerase in the presence of different ^{32}P-labeled ribonucleoside triphosphates (NTPs). In NTPs the innermost phosphate is designated α, the middle phosphate is β, and the outermost phosphate is γ. For which of the following pairs of radiolabeled NTPs would a mixture containing *either* of the radiolabeled NTPs, along with the other three NTPs (unlabeled) necessary to make a complete set of all four NTPs, give rise to a transcript that contains ^{32}P?

(a) [α-^{32}P]ATP or [α-^{32}P]GTP
(b) [β-^{32}P]ATP or [β-^{32}P]GTP
(c) [γ-^{32}P]ATP or [γ-^{32}P]GTP
(d) [α-^{32}P]CTP or [α-^{32}P]UTP
(e) [β-^{32}P]CTP or [β-^{32}P]UTP
(f) [γ-^{32}P]CTP or [γ-^{32}P]UTP

9. Which of the following are components or structures that lie within the transcription bubble formed by *E. coli* RNA polymerase?

(a) The polymerization site of the enzyme
(b) The σ subunit
(c) Approximately 17 nucleotides of the DNA coding strand in the single-stranded form
(d) Approximately 12 nucleotides of the 5′ end of the elongating RNA strand
(e) Approximately 12 bp of RNA–DNA duplex

10. Match the regions of a ρ-independent transcription termination signal in a DNA template in the left column with the structures or the functions performed by the encoded transcript segments in the right column.

(a) GC-rich palindromic region ______

(b) AT-rich region ______

(1) Oligo(U) stretch in RNA
(2) Hairpin in RNA
(3) Promotes the dissociation of RNA–DNA hybrid helix
(4) Causes the enzyme to pause

11. Which of the following statements about the ρ protein of *E. coli* are correct?

(a) It is an ATPase that is activated by binding to single-stranded DNA.
(b) It recognizes specific sequences in single-stranded RNA.
(c) It recognizes sequences in the DNA template strand.
(d) It causes RNA polymerase to terminate transcription at template sites that are different from those that lead to ρ-independent termination.
(e) It acts as a RNA–DNA helicase.

12. To what extent are mRNA, rRNA, and tRNA modified in prokaryotes? List some of the modifications that occur.

______

______

13. Match the functions in the right column with the antibiotic inhibitors of *E. coli* transcription in the left column.

(a) Rifampicin ______

(b) Actinomycin D ______

(1) Interacts with the template
(2) Interacts with nascent mRNA polymerase
(3) Prevents initiation
(4) Prevents elongation
(5) Intercalates into mRNA hairpins

14. Explain how a mutation might give rise to an *E. coli* that is resistant to the antibiotic rifampicin.

______

______

Eukaryotic Transcription

15. Which of the following statements about eukaryotic mRNAs are correct?

(a) They are derived from larger RNA precursors.
(b) They result from extensive processing of their primary transcripts before serving as translation components.

(c) They usually have poly(A) tails at their 5′ ends.
(d) They have a cap at their 3′ ends.
(e) They are often encoded by noncontiguous segments of template DNA.

16. Match the descriptions in the right column with the appropriate eukaryotic DNA-dependent RNA polymerases in the left column.

(a) RNA polymerase I ________
(b) RNA polymerase II ________
(c) RNA polymerase III ________

(1) Is located in the nucleolus
(2) Is located in the nucleoplasm
(3) Makes mRNA precursors
(4) Makes tRNA precursors
(5) Makes 5S rRNA
(6) Makes 18S, 5.8S, and 28S rRNA precursors
(7) Is strongly inhibited by α-amanitin
(8) Synthesizes RNA in the $5' \rightarrow 3'$ direction
(9) Is composed of several subunits
(10) Has a subunit with repeated amino acid sequences subject to phosphorylation

17. List the major sequence features of promoters for eukaryotic mRNA genes.

____

____

18. Describe the role of TATA box–binding protein (TBP, TFIID) in forming the basal transcription apparatus. What properties does TBP confer on this apparatus?

____

____

19. Describe how upstream promoter elements such as those referred to in question 17 might act to facilitate transcription. How might DNA looping be involved in this mechanism?

____

____

20. Which of the following statements about enhancers are correct?

(a) They function as promoters.
(b) They function when in either orientation in the DNA.
(c) They function when on either side of the activated promoter.

(d) They function even when located many base pairs away from the promoter.
(e) They function only in specific types of cells.

21. Which of the following are substrates or products in the capping reactions that form the structure at the 5′ end of mature mRNAs in eukaryotes?

(a) PP_i
(b) P_i
(c) GTP
(d) GDP
(e) GMP
(f) 5′-Triphosphate-terminated RNA
(g) 5′-Diphosphate-terminated RNA
(h) 5′-Monophosphate-terminated RNA
(i) 5′-Hydroxyl-terminated RNA
(j) *S*-Adenosylmethionine (SAM)
(k) *S*-Adenosylhomocysteine (SAH)
(l) N^5,N^{10}-Methylenetetrahydrofolate

22. Which of the following statements about the poly(A) tails that are found on most eukaryotic mRNAs are correct?

(a) They are added as preformed polyriboadenylate segments to the 3′ ends of mRNA precursors by an RNA ligase activity.
(b) They are encoded by stretches of polydeoxythymidylate in the template strand of the gene.
(c) They are added by RNA polymerase II in a template-independent reaction using ATP as the sole nucleotide substrate.
(d) They are added by poly(A) polymerase using dATP as the sole nucleotide substrate.
(e) They are cleaved from eukaryotic mRNAs by a sequence-specific endoribonuclease that recognizes the RNA sequence AAUAAA.

23. What functions are the caps and tails of mRNAs thought to perform?

____

____

24. Which of the following statements about apolipoprotein B (apo B) are correct?

(a) The apo B-48 form is formed by the proteolytic cleavage of the primary (apo B-100) translation product.
(b) Apo B-48 and apo B-100 are formed in different tissues.
(c) Apo B-48 arises from the expression of a form of the gene for apo B-100 that has been shortened by nonhomologous recombination.
(d) The transcript of the apo B-100 gene is spliced to remove a segment and form apo B-48.
(e) A specific nucleotide in the apo B-100 transcript is altered, thereby creating a stop codon in the mRNA.

RNA Splicing

25. Which of the following are important sequence elements in the splicing reactions that produce eukaryotic mRNAs?

(a) Exon sequences located between 20 and 50 nucleotides from the 5′ splice site
(b) Exon sequences located between 20 and 50 nucleotides from the 3′ splice site
(c) Intron sequences located between 20 and 50 nucleotides from the 5′ splice site
(d) Intron sequences located between 20 and 50 nucleotides from the 3′ splice site
(e) Intron sequences at the 5′ splice site
(f) Intron sequences at the 3′ splice site

26. Eukaryotic mRNA splicing involves which of the following?

(a) The formation of 2′ → 5′ phosphodiester bonds
(b) A sequence-specific endoribonuclease that hydrolyzes the phosphodiester bond at the junctions of the intron with the exon
(c) The spliceosome
(d) The coupling of phosphodiester bond formation to ATP hydrolysis
(e) The formation of lariat intermediates

27. What are snRNPs, and how are they involved in the eukaryotic mRNA splicing reaction?

____

____

28. Place the following events in the order in which they occur during the formation of mature rRNA in *Tetrahymena*:

(a) The 3′-hydroxyl of a guanine nucleoside attacks the phosphodiester bond at the 5′ splice site leaving the 5′ (upstream) exon with a free 3′-hydroxyl and attaching the guanine nucleoside to the 5′ end of the intron. ________
(b) The transcript from the rRNA gene folds into a specific structure. ________
(c) The rRNA transcript specifically binds a guanine nucleoside or nucleotide. ________
(d) Self-splicing occurs within the intron to form L19 RNA. ________
(e) The 3′-hydroxyl of the 5′ exon attacks the bonds at the 3′ splice junction to form the spliced rRNA and eliminate the intron. ________

29. What properties does L19 RNA have that indicate it is an enzyme?

____

____

30. Match the descriptions in the right column with the type of splicing in the left column.

(a) Group I splicing ______________
(b) Group II splicing ______________
(c) Nuclear mRNA splicing ______________
(d) Eukaryotic tRNA splicing ______________

(1) A $2' \rightarrow 5'$ phosphodiester bond is involved.
(2) Nuclease and ligase activities are required.
(3) Transesterification reactions break and form bonds.
(4) A guanine nucleoside or nucleotide is required.
(5) Spliceosomes are required.
(6) Lariat intermediates are involved.
(7) snRNPs are involved.

ANSWERS TO SELF-TEST

RNA Polymerase from *E. coli*

1. The holoenzyme has the subunit composition $\alpha_2\beta\beta'\sigma$. The core enzyme lacks the σ subunit.

2. (a) 2 (b) 3 (c) 1 (d) 4

3. a, b, c, f. The promoters of most *E. coli* genes include variants of defining, consensus sequences that are centered at about the −35 and −10 positions. The nearer the sequences of a promoter are to the consensus sequence and the nearer the separation between them is to the optimal 17-bp spacing, the more efficient the promoter. The −10 consensus sequence is TATAAT. The substitution of a C or G into the sequence would likely lower the efficiency of a promoter.

4. b, c. The sense strand of the template DNA has the same sequence as the mRNA.

5. a, b, c. The σ subunit recognizes promoter sites, decreases the affinity of the enzyme for regions of DNA lacking promoter sequences, and facilitates the specific, oriented initiation of transcription. Orienting the binding of the enzyme to the DNA results in only one of the two DNA strands functioning as a template for RNA transcription; that is, it gives rise to asymmetric transcription.

6. RNA polymerase holoenzyme has lower affinity for nonspecific DNA sequences than for promoter sequences. The nonspecific affinity, however, allows the enzyme to bind to "random-sequence" DNA and then "slide" along the molecule in a unidimensional random walk until it encounters a promoter sequence, for which its binding affinity is higher. Diffusion in one dimension is much faster than diffusion in three dimensions, thereby explaining the observed rapid rate constant for the binding of RNA polymerase holoenzyme to promoter sequences. If one measured the encounter of the polymerase with the nonspecific regions of the DNA rather than with promoter sequences, the value of the rate constant would be much lower and would fit our expectations for a three-dimensional, diffusion-limited reaction between macromolecules.

7. The temperature increase induces the synthesis of a new σ factor, σ^{32}, which directs RNA polymerase to promoters that have −10 and −35 sequences which are different from those recognized by σ^{70}. Transcription from these promoters gives rise to characteristic heat-shock proteins.

Prokaryotic Transcription

8. a, b, c, d. The β and γ phosphates of NTPs will be incorporated only as a part of the nucleotide at the 5′ end of the transcript, which bears a 5′-triphosphate terminus. Since almost all transcripts start with purine nucleotides, only choices (b) and (c) are correct for the labeled β and γ phosphates. The α phosphates of all the NTPs will be incorporated as phosphodiester bonds within the transcript, with the β and γ phosphates of these NTPs being eliminated as PP_i.

9. a, c, e. Answer (b) is incorrect because the σ subunit dissociates from the transcription bubble shortly after the initiation of phosphodiester bond formation. The enzyme unwinds approximately 17 bp of DNA, and approximately 12 bp of the template strand are in a duplex with the 3′ end of the growing RNA chain, leaving approximately 17 bp of the coding strand unpaired. The 5′ end of the RNA is free of the transcription bubble. (See Figure 33-10 on p. 846 in Stryer.)

10. (a) 2, 4 (b) 1, 3

11. d, e. The ρ protein recognizes and binds stretches of RNA that are devoid of hairpins and are at least 72 nucleotides long. It acts to hydrolyze ATP and to unwind the RNA–DNA hybrid in the transcription bubble.

12. In prokaryotes, tRNA and rRNA are modified extensively, whereas mRNA undergoes little, if any, alteration. Primary transcripts are cleaved by enzymes such as RNase III and have their ends trimmed by other enzymes, as exemplified by RNase P—an enzyme with a catalytically active RNA subunit. Some tRNA transcripts are also modified by the addition of nucleotides to their 3′ ends; all of them undergo conversion of many of their bases to modified forms, for example, U to T or U to ψ, and the methylation of some of their ribose residues. Base and ribose modifications to the rRNA of prokaryotes also occur.

13. (a) 2, 3 (b) 1, 4. Actinomycin D intercalates only into duplex DNA.

14. Rifampicin must bind to the β subunit of RNA polymerase to inhibit the enzyme. A mutation in the gene encoding this subunit that would interfere with the binding of the antibiotic but not with polymerization would produce a rifampicin-resistant cell.

Eukaryotic Transcription

15. a, b, e. Answers (c) and (d) are incorrect because the poly(A) tail is found at the 3′ end and a cap is found at the 5′ end of typical eukaryotic mRNA. Segments of the primary transcript are discarded by RNA splicing to bring RNA sequences encoded by noncontiguous regions of the template together; that is, introns are removed to form a product smaller than the primary transcript.

16. (a) 1, 6, 8, 9 (b) 2, 3, 7, 8, 9, 10 (c) 2, 4, 5, 8, 9. Both RNA polymerases I and III are involved in rRNA synthesis, with polymerase I synthesizing the 18S, 5.8S, and 28S precursor transcripts and polymerase III synthesizing the 5S rRNA.

17. A TATA box centered at about −25 from the start site of transcription and consisting of a variant of the consensus heptanucleotide sequence *TATAAAA* is essential for promoter activity. In addition, many genes have a CAAT box, a GC box, or both elements located between −40 and −110. These sequences can function in either orientation, that is, be in either DNA strand. Genes that are expressed constitutively, as distinct from those whose expression is regulated, tend to have GC boxes. Activating sequences farther upstream of these promoter elements are necessary for the functioning of most promoters.

18. TBP serves a critical role as an enucleating center for the assembly of the minimal molecular apparatus (the basal transcription assembly) required for transcription by RNA polymerase II. After TBP binds the TATA box, other TFII proteins and RNA polymerase II join the supramolecular complex to render it transcriptionally competent. In addition to enabling the assembly of the apparatus through its ability to recognize and bind the TATA box, TBP binds this asymmetric sequence in one orientation that "points" the RNA polymerase in the right direction, thereby defining the strand of DNA that will serve as the template.

19. Diverse upstream activating sequences (UAS) are individually recognized and bound by specific proteins called *transcription factors*. The transcription factor–DNA complex stimulates transcription through its interaction with the DNA-bound TATA box–binding protein (TBP), which is, itself, a transcription factor. Since the UAS may be some distance from the bound TBP, the DNA may loop to bring the proteins into apposition. You can imagine how other proteins that bend or modulate the bending of DNA could affect transcription in this mechanism.

20. b, c, d, e. Answer (a) is incorrect because enhancer sequences do not serve as promoters per se. They will not, by themselves, enable RNA polymerase II to initiate a transcript, and will function only when the basal transcription apparatus exists.

21. a, b, c, f, g, j, k. PP_i comes from GTP when its α phosphorus atom is attacked by the 5′-diphosphate end of the mRNA, and P_i comes from the triphosphate terminus of the mRNA precursor to form its 5′-diphosphate end (see p. 858 in Stryer). SAM provides the methyl groups that are transferred to the base and the sugars, and SAH is the remnant of SAM (see pp. 721–722 in Stryer).

22. None of the statements are correct. Poly(A) polymerase uses ATP, not dATP, to add a stretch of A residues to the 3′-hydroxyl formed by the cleavage of an mRNA precursor by a specific ribonuclease that recognizes the upstream sequence, AAUAAA, within particular mRNA sequence contexts.

23. The 5′ caps are thought to contribute to the stability and efficiency of translation of the mRNA. The poly(A) tails probably perform the same functions, albeit by different, unknown mechanisms.

24. b, e. The apo B system illustrates one kind of RNA editing. The pre-mRNA transcript has its sequence altered—in this case, a specific cytidine is deaminated to uracil to create the stop codon. Other editing mechanisms involve the addition and removal of U residues, an A-to-G change, and an adenosine to inosine change.

RNA Splicing

25. d, e, f

26. a, c, e. Answers (b) and (d) are incorrect because hydrolysis of phosphodiester bonds is not involved during mRNA splicing up to the point of intron removal and lariat formation; only transesterification reactions occur; consequently, ATP is not required for the synthesis of phosphodiester bonds.

27. The snRNPs are small ribonucleoprotein particles that occur in the nucleus. Each is composed of a small RNA molecule and several characteristic proteins, some of which are common to different snRNPs. Distinct snRNPs recognize and bind to splice junctions and the branch site and are involved in assembling the spliceosome in an ATP-dependent manner. They are requisite components of the splicing apparatus, and the RNA components of some of them are probably catalytically active. The RNAs of some snRNPs form hydrogen bonds with sequences within introns and exons to help to juxtapose properly the reacting splice junctions.

28. b, c, a, e, d

29. It catalyzes substrate-specific reactions at rates far faster than occur spontaneously, it displays Michaelis-Menton kinetics, and it can be competitively inhibited (see Stryer, pp. 231 and 867).

30. (a) 3, 4 (b) 1, 3, 6 (c) 1, 3, 5, 6, 7 (d) 2. The splicing of eukaryotic tRNAs requires protein nucleases to cut the phosphodiester bonds at the intron-exon junctions and protein RNA ligases to form the bonds joining the exons.

PROBLEMS

1. The technique of footprinting described by Stryer on page 843 involves, in part, the digestion of unprotected DNA by DNase I.

   (a) Would you expect a restriction endonuclease such as EcoRI to be useful in footprinting experiments? Explain.
   (b) Why is it necessary for digestion with the cleavage enzyme to be brief and not extensive?

2. One would expect an analog of 5′-ATP that lacks an oxygen at the 3′ position of its ribose (3′-deoxy-5′-ATP; see Figure 33-1) to interrupt RNA formation, since it cannot form phosphodiester bonds at its 3′ position. Could such a compound be used to ascertain the direction of chain growth in RNA synthesis? Explain. (Refer to Figure 33-9 on p. 846 of Stryer.)

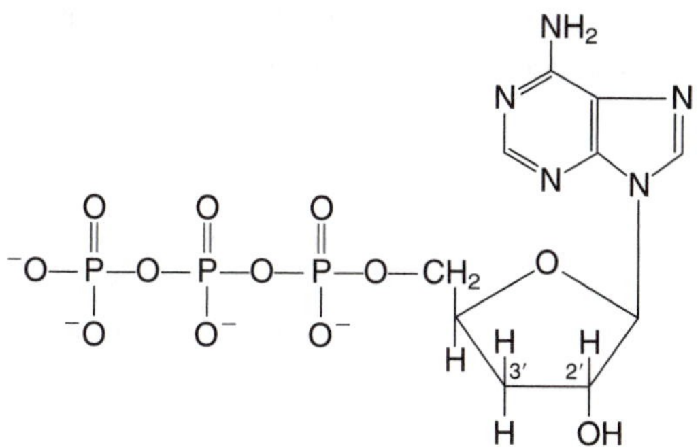

Figure 33-1
Structure of 3′-deoxy-5′-ATP.

3. The following is a partial nucleotide sequence of a recombinant plasmid containing a T7 promoter region and sites for cleavage by two restriction enzymes, EcoRI and BamHI.

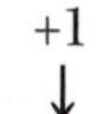

5′-ATAGGGAGGAGGATTACCCCTCGAATTCGAGCTCGCCCGGGGATCCTC-3′

3′-TATCCCTCCTCCTAATGGGGAGCTTAAGCTCGAGCGGGCCCCTAGGAG-5′

T7 promoter

The top chain is the sense strand, and the bottom strand is the template strand, and the first nucleotide to be transcribed is designated +1. It is possible to cleave the DNA with one of the restriction enzymes and then use the resulting cleavage fragment as a template for the preparation of "run-off" transcripts by T7 RNA polymerase.

(a) Give the RNA sequence that would be expected when the DNA is first cleaved with EcoRI and is then transcribed with T7 RNA polymerase. The cleavage pattern for EcoRI is 5′-GAATTC-3′.
(b) Give the sequence expected when BamHI is used to prepare the template. The cleavage pattern for BamHI is 5′-GGATCC-3′.
(c) Such run-off experiments can be used to identify the start site of transcription. How? Would you need to use both restriction enzymes to locate the start site?

4. The technique of nearest-neighbor analysis was a valuable tool in the characterization of RNA transcripts in the days before sequencing techniques for nucleic acids became available. The technique involves the addition of a nucleotide triphosphate labeled with ^{32}P in the α position to a reaction mixture in which RNA polymerase is active. Under such conditions, the label will be incorporated into the growing polynucleotide chain because phosphodiester bond formation involves the α-phosphoryl group of the incoming nucleotides. The newly formed RNA can then be cleaved with a mixture of RNA-hydrolyzing enzymes that cleave RNA completely to 3′-mononucleotides. Separation of these nucleotides and the determination of their ^{32}P content can be used to establish the identity of nucleotide residues that were on the 5′ side of the labeled nucleotides in the RNA.

(a) Suppose that α-labeled UTP is used to prepare the transcripts in problem 3(a) and that the resulting RNA is hydrolyzed to 3′-mononucleotides. Identify the 3′-mononucleotides that would be labeled and give their relative frequencies.
(b) Give the kinds and proportions of labeled 3′-mononucleotides that would be expected if α-labeled ATP had been used instead.

5. State whether each of the following cleavage fragments would be a major product of the digestion of the ribonucleotide pApG by alkali:

(a) pAp
(b) pG

(Hint: Alkaline digestion of RNA produces a mixture of 2′- and 3′-nucleoside monophosphates. Cyclic 2′-3′-phosphodiesters are intermediates in the digestion, and they are randomly cleaved by a base at both the 2′ and 3′ positions to yield 3′- and 2′-monophosphates, respectively.)

6. The ciliated protozoan *Tetrahymena* contains an enzyme that can synthesize 5′-pseudouridine monophosphate from a mixture of PRPP and uracil. For a time it was thought that this enzyme was instrumental in the synthesis of transfer RNAs in *Tetrahymena.* Explain why this is not the case.

7. When mammalian genes are cloned, a strategy that is frequently followed involves the isolation of mRNA rather than DNA from a cell and the preparation of a complementary DNA (cDNA) by the enzyme reverse transcriptase (see p. 136 of Stryer). Suppose that mRNA isolated from a cell specialized for the production of protein X is used as a template for the production of cDNA. What major difference or differences would you expect to find between the structure of that cDNA and genomic DNA for protein X?

8. Rifampicin specifically inhibits the initiation of transcription in prokaryotes and may therefore be used in humans as a therapeutic antibacterial agent. Would you expect actinomycin D to be useful in antibacterial therapy? Why or why not?

9. The mRNAs produced by mammalian viruses undergo modification at the 5′ and 3′ ends in a fashion similar to that of eukaryotic mRNA. Why do you think this is the case?

10. Sketch the most stable secondary structure that could be assumed by the oligonucleotide AAGGCCCUACGGGGCCG.

11. Suppose that human DNA is cleaved into fragments approximately the size of a given mature human messenger RNA and that RNA–DNA hybrids are then prepared. The corresponding procedure is then carried out for *E. coli.* When the RNA–DNA hybrids from each species are examined using an electron microscope, which will show the greater degree of hybridization? Explain.

12. How is it possible for a DNA binding protein to stimulate or repress transcription in both prokaryotes and eukaryotes without interacting with other transcription factors or RNA polymerase?

13. A form of β-thalassemia results from base-change mutations in or near the CAAT box for the β-globin gene.

(a) Explain how such a change could result in thalassemia.
(b) Characterize the mRNA and β-globin protein in an affected person in terms of amount and nucleotide or amino acid sequence.

14. Figure 33-35 on page 861 of Stryer gives an example of how a base-change mutation within an intron can produce a 3′ splice site upstream

from the normal 3′ splice site. The result is that a small extra amino acid sequence is inserted into the β chain before synthesis is terminated by a stop codon. (See problem 8 on p. 874 of Stryer, together with its answer on p. 1031.) Corresponding mutants in exons are also known in which base change mutations introduce 5′ splice sites upstream from the normal 5′ sites. What changes in amino acid sequence would be expected to result?

15. In an attempt to determine whether a given RNA was catalytically active in the cleavage of a synthetic oligonucleotide, the following experimental results were obtained. When the RNA and the oligonucleotide were incubated together, cleavage of the oligonucleotide occurred. When either the RNA or the oligonucleotide was incubated alone, there was no cleavage. When the RNA was incubated with higher concentrations of the oligonucleotide, saturation kinetics of the Michaelis-Menten type were observed. Do these results demonstrate that the RNA has catalytic activity? Explain. (You may wish to refer to the discussion of catalytic RNA on pp. 230–232 of Stryer.)

ANSWERS TO PROBLEMS

1. (a) EcoRI would not be very useful in footprinting experiments, because it is highly specific for the rather rare nucleotide sequence GAATTC. DNase I, however, is rather nonspecific and will cleave duplex DNA into short, random fragments. Thus potential cleavage sites would exist on either side of the protecting protein.

   (b) In principle, one wants to time the digestion with DNase so that there is, on the average, one cut per DNA molecule. Since there are many potential cleavage sites on each DNA molecule, the result will be a population of molecules of varying lengths (see Figure 33-3 on p. 843 of Stryer). Extensive digestion with DNase would produce a population of very short fragments.

2. A 3′-deoxy analog of ATP could be used to establish the direction of chain growth. In 5′ → 3′ growth the analog would donate a nucleotide containing 3′-deoxyadenosine, and the polynucleotide chain would be terminated as a result. No additional nucleotides could be added, because of the lack of a 3′-OH group on the terminal 3′-deoxyadenosine. In 3′ → 5′ growth the nucleotide could not be added to the growing polynucleotide chain because of the lack of the 3′-OH group.

3. (a) The following sequence would result:

   5′-GGGAGGAGGAUUACCCCUCGAAUU-3′

   Remember that EcoRI puts a staggered cut into DNA so that the template strand is longer than the coding strand in this instance.

   (b) The following sequence would result:

   5′-GGGAGGAGGAUUACCCCUCGAAUUCGAGCUCGCCCGGGGAUC-3′

(c) The start site of transcription is determined by measuring the exact length of the RNA product. This is done by carrying out electrophoresis on a DNA sequencing gel along with standard molecular weight markers that could be the products of a DNA sequencing reaction. The start site of transcription is deduced from the location of the EcoRI site and the length of the RNA product. In this example, the RNA product is 27 nucleotides long and the start site of transcription is 27 nucleotides away from the EcoRI site. However, you cannot tell if the start site is on the 5′ or the 3′ side of the EcoRI site. To answer this question, the start site needs to be mapped relative to another restriction site. In this case, the BamHI site would give an RNA product 45 nucleotides long.

4. (a) Each nucleotide to the left of U in the sequence will be labeled by this technique. The kinds and proportions of labeled mononucleotides are 2A:2U:1C. The U-containing segment of the transcript, with the phosphates indicated, is as follows:

$$\text{pA}\overset{*}{\text{p}}\text{U}\overset{*}{\text{p}}\text{UpApCpCpCpC}\overset{*}{\text{p}}\text{UpCpGpApA}\overset{*}{\text{p}}\text{U}\overset{*}{\text{p}}\text{U—OH}$$

The phosphates marked with asterisks were donated by UTP when the RNA was synthesized, but they will depart with the neighbor on the left (the 5′ side) when the RNA is cleaved to a mixture of 3′-mononucleotides.

(b) The kinds and proportions of labeled mononucleotides are 4G:1A:1U.

5. (a) The fragment pAp would be produced by the cleavage of pApG between p and G. Guanosine is also formed.

(b) The fragment pG (5′-GMP) would not be produced, because the cyclic 2′,3′-phosphodiester could not be an intermediate.

6. Nascent polynucleotides formed by RNA polymerases contain only the four usual bases. Subsequently, some of the bases are chemically modified. Were unusual nucleotides to be incorporated into a growing RNA chain, this would in turn require the presence of unusual bases on DNA. The pseudouridine found in transfer RNAs is formed by breaking the nitrogen-carbon bond linking uracil to ribose and forming a carbon-carbon bond instead. Only certain uracils are modified in this manner, owing to their position in the three-dimensional structure of the RNA and to the specificity of the enzymes that carry out the modification.

7. The cDNA prepared from mRNA would have a long poly(T) tail, unlike genomic DNA. Remember that the poly(A) tail is added to the 3′ end of mammalian mRNA and that there is no counterpart on DNA. A second striking difference would be that the cDNA would contain no intervening sequences (introns) and would therefore be much shorter than the corresponding sections of genomic DNA. (Remember that most mammalian genes are mosaics of introns and exons.) A third difference would be found if any RNA editing were involved. An edited mRNA could generate a cDNA with nucleotides that did not correspond to those in genomic DNA.

8. In order to be useful as a therapeutic antibacterial agent, a compound must selectively inhibit processes in prokaryotes but leave the correspond-

ing processes in eukaryotes (including those in mitochondria) largely unaffected. Because rifampicin selectively inhibits the initiation of transcription in prokaryotes but not in eukaryotes, it is useful as an antibacterial agent. Actinomycin D is an intercalating agent that binds to DNA duplexes and inhibits both DNA replication and transcription, although it has a greater inhibiting effect on transcription than on replication. It cannot discriminate between the duplex DNA of bacteria and that of humans, however, and will therefore bind to both. Because it disrupts eukaryotic as well as prokaryotic processes, it is not very useful as an antibacterial agent. It is sometimes used as an anticancer agent, however, because of its ability to slow the replication rate of human DNA.

9. Viruses use the host's enzyme system to replicate their DNA and to synthesize their proteins. Since eukaryotic translation systems must synthesize viral protein, the structure of viral mRNAs must mimic that of the host mRNA.

10. The structure is a stem-and-loop ("lollipop") hairpin structure, as shown in Figure 33-2.

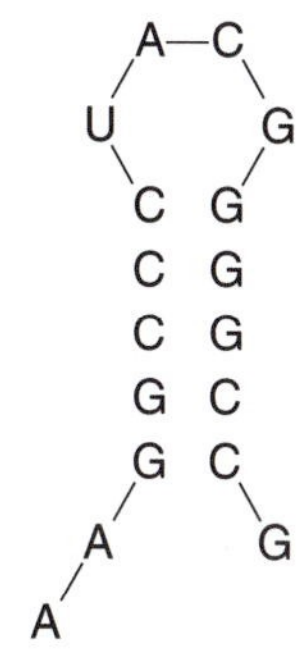

Figure 33-2
Stable secondary structure for the oligonucleotide in problem 10.

11. The RNA–DNA hybrid of *E. coli* will show greater hybridization because it is produced continuously from a DNA template without processing. Because of the presence of intervening sequences in human DNA, there will be regions in the human RNA–DNA hybrids where no base pairing occurs. Also, the bacterial DNA is less complex than the human DNA and will hybridize faster (see Stryer, p. 990).

12. Gene expression in both prokaryotes and eukaryotes can involve DNA looping. The purpose of the looping is to bring together transcription factors that would otherwise be distant from each other as they bind DNA. This looping could be aided or thwarted by a DNA-binding protein that bent the DNA, depending on the direction of the bend.

13. (a) The CAAT box helps to bind RNA polymerase to the promoter and increases transcriptional efficiency. A mutation in that region would lead to decreased production of mRNA for β-globin genes. Decreased mRNA levels within the cell could lead to decreased amounts of protein.

(b) The globin mRNA would be entirely normal in sequence, but decreased in amount, as would be the protein itself.

14. Some amino acids would be omitted from a region of the β chain, the region encoded by nucleotide residues between the newly introduced and normal 5′ splicing sites. Also, the introduction of a new splicing site could lead to changes in the reading frame, and therefore to gross changes in amino acid sequence, polypeptide chain length, or both.

15. These results alone do not establish that the RNA has catalytic activity. A catalyst must be regenerated. It is entirely possible that the results observed could be accounted for by a stoichiometric, as opposed to a catalytic, interaction between RNA and the oligonucleotide, in which the RNA may "commit suicide" as the oligonucleotide is cleaved. In such an interaction, a portion of the RNA would participate chemically in the cleavage of the oligonucleotide, but it would also be cleaved itself as a part of the reaction. Four reaction products would accumulate, two resulting

from the cleavage of RNA and two from the cleavage of the oligonucleotide. To show that this particular RNA was catalytic, it would be necessary to demonstrate that it turns over and is regenerated in the course of the reaction.

EXPANDED SOLUTIONS TO TEXT PROBLEMS

1. The sequence of the coding (+, sense) strand is 5′-ATGGGGAACAG CAAGAGTGGGGCCCTGTCCAAGGAG-3′, and the sequence of the template (−, antisense) strand is 3′-TACCCCTTGTCGTTCTCACCCCGGGAC AGGTTCCTC-5′. Note that the *coding* strand has the same sequence as mRNA (except U → T), whereas the *template* strand is complementary to the coding strand (see Stryer, p. 843).

2. Heparin, a glycosaminoglycan (Stryer, p. 474), is highly anionic. Its negative charges, like the phosphodiester bridges of DNA templates, bind to lysine and arginine residues of β'.

3. This mutant sigma would competitively inhibit the binding of holoenzyme and prevent the specific initiation of RNA chains at promoter sites.

4. The core enzyme without sigma binds more tightly to the DNA template than does the holoenzyme. The retention of sigma after chain initiation would make the mutant RNA polymerase less processive. Hence, RNA synthesis would be much slower than normal.

5. A 100-kd protein contains about 910 residues, which are encoded by 2730 nucleotides. At a maximal transcription rate of 50 nucleotides per second, the protein would be synthesized in 54.6 seconds (Stryer, p. 846).

6. Initiation at strong promoters occurs every two seconds. In this interval, 100 nucleotides are transcribed. Hence, centers of transcription bubbles are 34 nm (340 Å) apart (100×3.4 Å = 340 Å). (See Stryer, p. 846.)

7. (a) The lowest band on the gel will be that of (i), whereas the highest will be that of (v). Band (ii) will be at the same position as (i) because the RNA is not complementary to the nontemplate strand, whereas band (iii) will be higher because a complex is formed between RNA and the template strand. Band (iv) will be higher than the others because strand 1 is complexed to 2, and strand 2 to 3. Band (v) is the highest because core polymerase associates with the three strands.

(b) None, because rifampicin acts before the formation of the open complex (Stryer, p. 851).

(c) RNA polymerase is processive. Once the template is bound, heparin cannot enter the DNA-binding site.

(d) The longest RNA product formed is a 36-mer rather than the full 72-mer. Because GTP is absent, synthesis stops when the first C downstream of the bubble is encountered in the template strand.

8. Note that the DNA strands shown in Figure 33-35 are, by convention, complementary to the DNA strand coding for mRNA. If every T is changed to U, the polarity and base sequence of the DNA strands shown are identical with the mRNA synthesized from the complementary strand. Thus, TCT becomes UCU (Ser), and so forth.

9. A mutation that disrupted the normal AAUAA recognition sequence for the endonuclease could account for this finding. In fact, a change from U to C in this sequence caused this defect in a thalassemic patient. Cleavage occurred at the AAUAAA 900 nucleotides downstream from this mutant AACAAA site. See S.H. Orkin, T-C. Cheng, S.E. Antonarakis, and H.H. Kazazian, Jr. *EMBO J.* 4(1985):453.

10. Mg^{2+} coordinates strongly to oxygen but not sulfur, whereas Mn^{2+} coordinates strongly to sulfur but not oxygen. See E.K. Jaffe and M. Cohn. *J. Biol. Chem.* 253(1978):4823.

11. One possibility is that the 3′ of the poly(U) donor strand cleaves the phosphodiester bond on the 5′ side of the insertion site. The newly formed 3′ terminus of the acceptor strand then cleaves the poly(U) strand on the 5′ side of the nucleotide that initiated the attack. In other words, a U could be added by two transesterification reactions. This postulated mechanism is akin to the one in RNA splicing (Fig. 33-36 in Stryer). See T.R. Cech. *Cell* 64(1991):667.

CHAPTER

34

Protein Synthesis

In Chapter 34, Stryer details the biochemistry of the translation from the language of nucleic acids to that of proteins. You should review the introduction to this process that was given in Chapter 5 of *Biochemistry*.

Before describing the initiation, elongation, and termination stages of polypeptide synthesis, Stryer presents the detailed structures and conformations of tRNAs, the adaptor molecules that recognize both the codons on the mRNAs and the enzymes that attach the corresponding amino acids. Next Stryer explains how amino acids are activated for the subsequent formation of peptide bonds through their attachment to tRNAs by the two classes of aminoacyl-tRNA synthetases. The catalytic mechanism of the first step of the tyrosyl-tRNA synthetase reaction is considered in detail, and the proofreading strategies used by various aminoacyl-tRNA synthetases in recognizing the correct amino acid and tRNA molecules are described. Stryer presents the wobble hypothesis to explain the lack of strict one-to-one Watson-Crick base-pairing interactions between the three nucleotides of the tRNA anticodons and the mRNA codons. A detailed description of prokaryotic ribosomes completes the prelude to the discussion of peptide-bond formation and protein synthesis.

The experiments that showed the polarities of polypeptide formation and the translation of mRNA are presented next. Then initiation is described, and the roles of a specialized initiator tRNA, the mRNA start codon and 16S rRNA sequences, and the protein initiation factors are outlined. The spatial and functional relationships of the sites on the ribosome that bind aminoacyl-tRNAs and peptidyl-tRNAs, the elongation factors, the peptide-bond-forming reaction, the role of GTP, and the mechanism of the translocation of the peptidyl-tRNA from site to site on the ribosome are presented in the description of the elongation stage of protein synthesis. The termination of translation is outlined, and the role of release factors that recognize translation stop codons is described. Stryer also explains the mechanisms of several potent inhibitors of translation.

The chapter closes with a brief overview of translation in eukaryotes that emphasizes the major contrasting features with respect to translation in prokaryotes. Differences in the initiator tRNA, the selection mechanism of the initiator codon, the ribosomes, and the overall complexity of the process are highlighted. Lastly, the mechanism of the bacterial toxin that causes diphtheria is presented.

LEARNING OBJECTIVES

When you have mastered this chapter, you should be able to complete the following objectives:

Introduction (Stryer pages 875–876)

1. Provide an overview of protein synthesis that includes the roles of the *amino acids,* the *tRNAs,* the *amino acid activating enzymes, mRNA,* the *ribosomes,* and *ATP* and *GTP.*

Transfer RNAs (Stryer pages 876–879)

2. Draw the *cloverleaf structure* of a *tRNA* and identify the regions containing the *anticodon* and the *amino acid attachment site.*

3. List the common features of all tRNAs.

4. Relate the two-dimensional cloverleaf representation of the tRNA structure to its three-dimensional configuration.

5. Describe the role of *ribonuclease P* in the formation of tRNA molecules in *E. coli.*

Amino Acid Activating Enzymes (Stryer pages 879–886)

6. Write the two-step reaction sequence of the *aminoacyl-tRNA synthetases.* Enumerate the high-energy phosphate bonds that are consumed in the overall reaction.

7. Draw the *mixed anhydride* and *ester linkages* that occur in the *aminoacyl-adenylate* intermediate and the *aminoacyl-tRNA* product of the activation reaction.

8. Outline the distinguishing properties of *class I* and *class II* aminoacyl-tRNA synthetases.

9. Explain the *delocalized catalytic mechanism* of the *tyrosyl-tRNA synthetase* reaction.

10. Describe the mechanisms of amino acid selection and proofreading that contribute to the accuracy of the attachment of the appropriate amino acid to the correct tRNA.

11. Describe the different modes of recognition of the correct tRNA molecule by aminoacyl-tRNA synthetases.

Codon-Anticodon Interactions (Stryer pages 886–888)

12. Recount the experiment that showed that the tRNA rather than the amino acid in an aminoacyl-tRNA recognizes an mRNA codon.

13. Explain how some *codons are recognized by more than one anticodon,* that is, how they interact with more than one species of aminoacyl-tRNA. List the base-pairing interactions allowed according to the *wobble hypothesis.*

Ribosomes (Stryer pages 888–892)

14. Provide an overview of the roles of the *ribosomes* in protein synthesis. List the kinds and numbers of macromolecular components of the prokaryotic ribosome. Give the *mass, sedimentation coefficient,* and *dimensions* of the *ribosome* of *E. coli.*

15. List the evidence that suggests that the RNA components of ribosomes have active roles in protein synthesis.

16. Give examples of the techniques that have been used to describe the structure and interactions of ribosomes.

17. Outline the three-dimensional structure of a ribosome.

Polypeptide Formation (Stryer pages 892–903)

18. Recount the experiment that established the *polarity of polypeptide synthesis.*

19. Recount the experiment that related the *polarity* of polypeptide synthesis to that of mRNA translation.

20. Define the *polysome* and outline its assembly. Correlate the polarity of ribosome movement with the polarity of the growing polypeptide chain.

21. Distinguish between the *initiator tRNA,* $tRNA_f$, and $tRNA_m$ and outline the conversion of methionine into *formylmethionyl-*$tRNA_f$.

22. Name the major *initiator codon* and the amino acid it encodes. Explain the roles of the nucleotide sequences in *16S rRNA,* mRNA, and tRNA in *selecting the initiation codon* rather than the identical codon that encodes an internal amino acid.

23. List the components of the *70S initiation complex* and indicate the roles of the *initiation factors* (IF) and GTP in its formation.

24. Outline the elongation stage of protein synthesis and describe the roles of the *elongation factors* (EFs) and GTP in the process. Locate the *aminoacyl-tRNAs* and *peptidyl-tRNAs* in the *A* or *P sites of the ribosome* during one cycle of elongation.

25. Describe how the GTP-GDP cycle of *EF-Tu* controls its affinity for its reaction partners.

26. Explain the role of EF-Tu in determining the accuracy and timing of protein synthesis

27. Outline the translocation steps that occur after the formation of a peptide bond and describe the roles of *EF-G* and GTP.

28. Name the *translation stop codons,* describe the termination of translation, and explain the roles of the *release factors* (RFs) in the process.

29. Describe how *puromycin* was used to define the A and P sites of the ribosome.

30. Provide examples of other *antibiotics that inhibit translation,* and describe their mechanisms of action.

Eukaryotic Translation (Stryer pages 903–906)

31. Contrast eukaryotic and prokaryotic ribosomes with respect to composition and size.

32. Contrast the mechanisms of translation initiation in prokaryotes and eukaryotes. Note the different initiator tRNAs, AUG codon selection mechanisms, and numbers of IFs and RFs.

33. Explain the role of the *protein kinases* in the regulation of the initiation of eukaryotic translation.

34. Describe the mechanism by which the *diphtheria toxin* inhibits protein synthesis in eukaryotes.

SELF-TEST

Introduction

1. Which of the following statements about translation are correct?

   (a) Amino acids are added to the amino terminus of the growing polypeptide chain.
   √ (b) Amino acids are activated by attachment to tRNA molecules.
   √ (c) A specific initiator tRNA along with specific sequences of the mRNA ensure that translation begins at the correct codon.
   √ (d) Peptide bonds form between an aminoacyl-tRNA and a peptidyl-tRNA positioned in the A and P sites, respectively, of the ribosome.
   (e) Termination involves the binding of a terminator tRNA to a stop codon on the mRNA.

2. Which of the following statements about prokaryotic translation are correct?

   (a) The formation of a peptide bond is an exergonic reaction.
   √ (b) The protein grows in the amino-to-carboxyl direction.
   (c) The movement of peptidyl-tRNA on the ribosome is directly powered by the hydrolysis of ATP.
   (d) Translation involves the conversion of genetic information in single-stranded DNA into sequences of amino acids.

Transfer RNAs

3. Which of the following statements about functional tRNAs are correct?

   √ (a) They contain many modified nucleosides.
   √ (b) About half their nucleosides are in base-paired helical regions.
   √ (c) They contain fewer than 100 ribonucleosides.
   (d) Their anticodons and amino acid accepting regions are within 5 Å of each other.
   (e) They consist of two helical stems that are joined by loops to form a U-shaped structure.
   (f) They have a terminal AAC sequence at their amino acid accepting end.

4. Explain why tRNA molecules must have both unique and common structural features.

____

____

5. Which statements are correct about ribonuclease P? Ribonuclease P

(a) contains a catalytically active RNA component.
(b) produces several messenger RNA molecules from a long primary transcript.
(c) produces several transfer RNA molecules from a long primary transcript.
(d) cleaves the primary transcript on the 5′ side of the first nucleotide of the RNA products.
(e) trims the exposed 3′ end of each RNA precursor.

Amino Acid Activating Enzymes

6. Which of the following statements about the aminoacyl-tRNA synthetase reaction are correct?

(a) ATP is a cofactor.
(b) GTP is a cofactor.
(c) The amino acid is attached to the 2′- or 3′-hydroxyl of the nucleotide cofactor (ATP).
(d) The amino group of the amino acid is activated.
(e) A mixed anhydride bond is formed.
(f) An acyl ester bond is formed.
(g) An acyl thioester bond is formed.
(h) A phosphoamide (P–N) bond is formed.

7. The $\Delta G^{\circ\prime}$ of the reaction catalyzed by the *aminoacyl-tRNA synthetases* is

(a) ≅0 kcal/mol.
(b) < 0 kcal/mol.
(c) > 0 kcal/mol.

8. Considering the correct answer to question 7, explain how aminoacyl-tRNAs can be produced in the cell.

____

____

9. Match the *class I* and *class II* aminoacyl-tRNA synthetases with the appropriate items in the right column.

(a) Class I ________________
(b) Class II ________________

(1) Generally activate smaller amino acids
(2) Generally activate larger and more hydrophobic amino acids
(3) Acylate the 2′-hydroxyl of the tRNA
(4) Generally acylate the 3′ hydroxyl of the tRNA
(5) Contain β domains at the active site

10. Given the similarity of valine and isoleucine (there is one extra methylene group in the side chain of Ile) and the fivefold higher concentration of Val than Ile in the cell, calculations indicate that isoleucyl-tRNA synthetase should mistakenly incorporate valine into proteins in place of isoleucine once in every 40 reactions. However, this happens only once in every 3000 reactions. Explain.

11. How can some aminoacyl-tRNA synthetases achieve accurate reactions even though they lack a hydrolytic proofreading mechanism?

12. Indicate the possible ways in which different aminoacyl-tRNA synthetases may recognize their corresponding tRNAs.

(a) Recognize the anticodon
(b) Recognize specific base pairs in the acceptor stem
(c) Recognize the 3′ CCA sequence of the tRNA
(d) Recognize both the anticodon and acceptor stem region
(e) Recognize extended regions of the L-shaped molecules

Codon-Anticodon Interactions

13. In an experiment, it was found that Cys-$tRNA^{Cys}$ can be converted to Ala-$tRNA^{Cys}$ and used in an in vitro system that is capable of synthesizing proteins.

(a) If the Ala-$tRNA^{Cys}$ were labeled with ^{14}C in the amino acid, would the labeled Ala be incorporated in the protein in the places where Ala residues are expected to occur? Explain.

(b) What does the experiment indicate about the importance of the accuracy of the aminoacyl-tRNA synthetase reaction to the overall accuracy of protein synthesis?

14. Which of the following answers completes the sentence correctly? The wobble hypothesis

(a) accounts for the conformational looseness of the amino acid acceptor stem of tRNAs that allows sufficient flexibility for the peptidyl-tRNA and aminoacyl-tRNA to be brought together for peptide-bond formation.
(b) accounts for the ability of some codons to be recognized by more than one anticodon.
(c) explains the occasional errors made by the aminoacyl-tRNA synthetases.
(d) explains the oscillation of the peptidyl-tRNAs between the A and P sites on the ribosome.
(e) assumes steric freedom in the pairing of the first (5′) nucleotide of the codon and the third (3′) nucleotide of the anticodon.

15. Assuming that each nucleoside in the left column is in the first position of an anticodon, with which nucleoside or nucleosides in the right column could it pair during a codon-anticodon interaction if each of the nucleosides on the right is in the third position (3′ position) of a codon?

(a) Adenosine ____4____
(b) Cytidine ____3____
(c) Guanosine ____2____
(d) Inosine ____1, 2, 4____
(e) Uridine ____1, 3____

(1) Adenosine
(2) Cytidine
(3) Guanosine
(4) Uridine

16. According to the wobble principle, what is the *minimum* number of tRNAs required to decode the six leucine codons—UUA, UUG, CUU, CUC, CUA, and CUG? Explain.

Ribosomes

17. Which of the following statements about an *E. coli* ribosome are correct?

(a) It is composed of two spherically symmetrical subunits.
(b) It has a large subunit comprising 34 kinds of proteins and 2 different rRNA molecules.
(c) It has a sedimentation coefficient of 70S.
(d) It has two small subunits, one housing the A site and the other the P site.
(e) It has an average diameter of approximately 200 Å.
(f) It has a mass of approximately 270 kd, one-third of which is RNA.

18. It has become evident that the ribosomal RNA components have an active role in protein synthesis. Which of the following are evidence of the activity of ribosomal RNAs?

(a) Ribosomes almost completely depleted of protein are capable of catalyzing the synthesis of peptide bonds.
(b) Most antibiotics that inhibit protein synthesis do so by interacting with aminoacyl-tRNA synthetases.
(c) A sequence in 16S rRNA recognizes the start site in mRNA.
(d) The wobble base of tRNA in the P site of a ribosome is paired to a base in a highly conserved sequence of 16S rRNA.
(e) The 5′ end of each tRNA interacts with a conserved region of ribonuclease P.

19. What is the significance of the reconstitution of a functional ribosome from its separated components?

20. Explain how neutron diffraction experiments revealed the distances between the proteins of the 30S subunit.

21. Describe how chemical modification was used to determine the regions of 16S rRNA that bind each of its associated proteins.

Polypeptide Formation

22. If labeled amino acids are added to an in vitro system synthesizing protein under the direction of a single mRNA species and samples are withdrawn at different times, the following labeling patterns in the *completed* polypeptide chains would be observed. The dash (–) represents unlabeled amino acids, the X represents labeled amino acids, and A and B represent the ends of the intact protein.

Time 1 (early)	A--------------XXB
Time 2	A-----------XXXXB
Time 3	A------- XXXXXXB
Time 4 (late)	A---- XXXXXXXXB

Which of the following statements about these proteins are correct?

(a) The labeled amino acids are added in the B-to-A direction.
(b) The labeled amino acids are added in the A-to-B direction.
(c) A is the amino terminus of the protein.
(d) B is the amino terminus of the protein.

23. Given an in vitro system that allows protein synthesis to start and stop at the ends of any RNA sequence, answer the following questions:

(a) What peptide would be produced by the polyribonucleotide 5′-UUUGUUUUUGUU-3′? (See the table with the genetic code on the inside back cover of Stryer.)

(b) For this peptide, which is the N-terminal amino acid and which is the C-terminal amino acid?

24. What is the role of the vitamin folic acid in prokaryotic translation?

carrier of formyl groups ; substrate

for transformylase rxn
met → (fmet)

25. The methionine codon AUG functions both to initiate a polypeptide chain and to direct methionine incorporation into internal positions in a protein. By what mechanisms are the AUG start codons selected?

____

____

26. Match the functions or characteristics of prokaryotic translation in the right column with the appropriate translation components in the left column.

(a) IF1 ________
(b) IF2 ________
(c) IF3 ________
(d) EF-Tu ________
(e) EF-Ts ________
(f) EF-G ________
(g) Peptidyl transferase ________
(h) RF1 ________
(i) RF2 ________

(1) Moves the peptidyl-tRNA from the A to the P site
(2) Delivers aminoacyl-tRNA to the A site
(3) Binds to the 30S ribosomal subunit
(4) Recognizes stop codons
(5) Forms the peptide bond
(6) Delivers fMet-tRNA$_f^{Met}$ to the P site
(7) Cycles on and off the ribosome
(8) Binds GTP
(9) Prevents the combination of the 50S and 30S subunits
(10) Is involved in the hydrolysis of GTP to GDP
(11) Associates with EF-Tu to release a bound nucleoside diphosphate
(12) Hydrolyzes peptidyl-tRNA
(13) Modifies the peptidyl transferase reaction

27. Which of the following statements about occurrences during translation are correct?

(a) The carboxyl group of the growing polypeptide chain is transferred to the amino group of an aminoacyl-tRNA.
(b) The carboxyl group of the amino acid on the aminoacyl-tRNA is transferred to the amino group of a peptidyl-tRNA.
(c) Peptidyl-tRNA may reside in either the A or the P site.
(d) Aminoacyl-tRNAs are shuttled from the A to the P site by EF-G.

28. A GTP-GDP cycle regulates the association of which of the following proteins with their partners in diverse biological processes.

(a) Stimulatory G protein in hormone action
(b) Succinyl CoA synthetase in the citric acid cycle
(c) Ras protein in growth control
(d) EF-Tu cycle in protein synthesis
(e) Transducin in vision

29. About 5% of the total bacterial protein is EF-Tu. Explain why this protein is so abundant.

30. For each of the following steps of translation, give the nucleotide cofactor involved and the number of high-energy phosphate bonds consumed:

(a) Amino acid activation

(b) Formation of the 70S initiation complex

(c) Delivery of aminoacyl-tRNA to the ribosome

(d) Formation of a peptide bond

(e) Translocation

31. After the formation of the 70S initiation complex, how many high-energy phosphate bonds are expended for each amino acid that is added to the growing polypeptide chain?

32. Explain why the translocation of peptidyl-tRNA from site A to site P occurs in two steps.

33. Which of the following statements about release factors are correct?

(a) They recognize terminator tRNAs.
(b) They recognize translation stop codons.
(c) They cause peptidyl transferase to use H_2O as a substrate.
(d) They are two proteins in *E. coli*, each of which recognizes two mRNA triplet sequences.

34. Many antibiotics act by inhibiting protein synthesis. How can some of these be used in humans to counteract microbial infections without causing toxic side effects due to the inhibition of eukaryotic protein synthesis?

35. Increasing the concentration of which of the following would most effectively antagonize the inhibition of protein synthesis by puromycin?

(a) ATP
(b) GTP
(c) Aminoacyl-tRNAs
(d) Peptidyl-tRNAs
(e) eIF3

36. Which of the following statements about eukaryotic translation are correct?

(a) A formylmethionyl-tRNA initiates each protein chain.
(b) It occurs on ribosomes containing one copy each of the 5S, 5.8S, 18S, and 28S rRNA molecules.
(c) The correct AUG codon for initiation is selected by the base pairing of a region on the rRNA of the small ribosomal subunit with an mRNA sequence upstream from the translation start site.
(d) It is terminated by a release factor that recognizes stop codons and hydrolyzes GTP.
(e) It involves proteins that bind to the 5′ ends of mRNAs.
(f) It can be regulated by protein kinases.

37. Which of the following statements about the diphtheria toxin are correct?

(a) It is cleaved on the surface of susceptible eukaryotic cells into two fragments, one of which enters the cytosol.
(b) It binds to peptidyl transferase and inhibits protein synthesis.
(c) It reacts with ATP to phosphorylate eIF2 and prevent the insertion of the Met-$tRNA_i$ into the P site.
(d) It reacts with NAD^+ to add ADP–ribose to eEF2 and prevent movement of the peptidyl-tRNA from the A to the P site.
(e) One toxin molecule is required for each translation factor inactivated; that is, it acts stoichiometrically.

ANSWERS TO SELF-TEST

Introduction

1. b, c, d. Answer (a) is incorrect because the incoming activated aminoacyl-tRNA, in the A site of the ribosome, adds its free amino group to the activated carboxyl of the growing polypeptide on a peptidyl-tRNA in the P site. Answer (e) is incorrect because termination does not involve tRNAs that recognize translation stop codons but rather protein release factors that recognize these signals and cause peptidyl transferase to donate the growing polypeptide chain to H_2O rather than to another aminoacyl-tRNA.

2. b. Answer (a) is incorrect because energy is required to form a peptide bond. Answer (c) is incorrect because the shuttling of peptidyl-tRNA on the ribosome is powered by the hydrolysis of GTP. Answer (d) is incorrect because translation converts genetic information from nucleotide sequences in RNA into sequences of amino acids.

Transfer RNAs

3. a, b, c. The molecules consist of two helical stems, each of which is made of two stacked helical segments. However, the molecules are L-shaped, and the anticodon and amino acid accepting regions are some 80 Å from each other. Functional tRNAs have a CCA sequence, not an AAC sequence at their 3′ termini.

4. Transfer RNAs need common features for their interactions with ribosomes and elongation factors but unique features for their interactions with the activating enzymes.

5. a, c, d. Ribonuclease P is a ribonucleoprotein enzyme with a catalytically active RNA; it cuts a long precursor of tRNA not of mRNA. Answer (e) is incorrect because ribonuclease D rather than ribonuclease P carries out the trimming function.

Amino Acid Activating Enzymes

6. a, e, f. The carboxyl group of the amino acid is activated in a two-step reaction via the formation of an intermediate containing a mixed anhydride linkage to AMP. The amino acid is ultimately linked by an ester bond to the 2′- or 3′-hydroxyl of the tRNA.

7. a. Since the standard free energy of the hydrolysis of an aminoacyl-tRNA is nearly equal to that of the hydrolysis of ATP, the reaction has a $\Delta G^{\circ\prime} \cong 0$; that is, it has an equilibrium constant near 1.

8. In the cell, the hydrolysis of PP_i by pyrophosphatase shifts the equilibrium toward the formation of aminoacyl-tRNA.

9. (a) 2, 3, 5 (b) 1, 4, 5. Both classes of enzymes contain β sheet domains at their active sites, but they are parallel in class I enzymes and antiparallel in class II enzymes.

10. Isoleucyl-tRNA synthetase has a proofreading mechanism. The hydrolysis of any Val-AMP that is mistakenly formed in the first step of the reaction gives the enzyme a second chance to bind and react with the correct amino acid to form Ile-AMP. The "decision" to hydrolyze the aminoacyl-adenylate intermediate probably depends on the size of the amino acid substituent. If it is as small as valine, the intermediate fits into the hydrolytic site and is destroyed. If it is as large as isoleucine, it does not fit, is not destroyed, and is added to the tRNA.

11. With these enzymes, the appropriate amino acid may not have to compete with an amino acid that closely resembles it. Consequently, the activating enzyme can easily distinguish the correct amino acid from all other amino acids through simple binding interactions. For example, the hydroxyl on tyrosine allows it to be unambiguously distinguished from phenylalanine.

12. a, b, d, e. There are many different ways in which aminoacyl-tRNA synthetases recognize their specific tRNAs. Answer (c) is incorrect because the 3′ CCA sequence is common to all tRNAs and cannot be used to distinguish between them.

Codon-Anticodon Interactions

13. (a) No. The tRNA, acting as an adaptor between the amino acid and mRNA, would associate with Cys codons in the mRNA through base pairing between codon and anticodon. The labeled alanine would incorporate only at sites encoded by the Cys codons and not at those encoded by Ala codons.

(b) The experiment demonstrates that the tRNA and not the amino acid reads the mRNA. Thus, if the activating enzyme mistakenly attaches an incorrect amino acid to a tRNA, that amino acid will be incorporated erroneously into the protein.

14. b. Answer (e) is incorrect because the ambiguities in base pairing occur between the third nucleotide of the codon and the first nucleotide of the anticodon.

15. (a) 4 (b) 3 (c) 2, 4 (d) 1, 2, 4 (e) 1, 3. (See Table 34-2 on p. 887 of Stryer.)

16. A minimum of three tRNAs would be required. One tRNA having the anticodon UAA could decode both UUA and UUG. For the other four codons, which have C in the first position and U in the second, there are two different combinations of two tRNAs each that could decode them. The first combination would be two tRNAs that have anticodons with A in the second position and G in the third, one with I in the first position to decode CUU, CUC, and CUA, and the other with U or C in the first position to decode CUG. The second combination would be two tRNAs that have anticodons with A in the second position and G in the third, one with G in the first position to decode CUU and CUC, and the other with U in the first position to decode CUA and CUG.

Ribosomes

17. b, c, e. Answer (f) is incorrect because two-thirds of the 2700-kd mass of a ribosome is rRNA.

18. a, c, d. Answer (b) is incorrect because most of the antibiotics that inhibit protein synthesis interact with rRNA.

19. It shows that the components themselves contain all the information necessary to form the structure and that neither a template nor any other factors are involved. Thus, the ribosome serves as model from which we might learn the general principles involved in self-assembly. Reassembly allows systematic study of the roles of the individual components through the determination of the effects of substitutions of mutant or altered individual proteins or rRNAs.

20. The principle of neutron diffraction is similar to that of X-ray diffraction, except that neutron diffraction distinguishes between H and D atoms. Thus pairs of ribosomal proteins from bacteria grown in D_2O can be reconstituted into ribosomes and their relative positions determined.

21. Chemical modification of rRNA with small reagents such as dimethylsulfate reveals a footprint of regions protected by specific ribosomal proteins. In this way the interactions between rRNA and proteins can be determined.

Polypeptide Formation

22. b, c. Although longer incubation times result in completed proteins that have labeled polypeptides closer to their amino terminus, the chains actually grow in the amino-to-carboxyl direction. When the labeled amino

acid is introduced into the system, it begins adding to the carboxyl ends of the growing chains that are already present in all stages of completion. The completed chains in samples withdrawn after a short time will have labeled polypeptides only near their carboxyl end. As time passes, more and more polypeptides that began adding labeled polypeptides near their amino terminals will become complete chains.

23. (a) Phenylalanylvalylphenylalanylvaline

(b) Phe is the N-terminal amino acid and Val is the C-terminal amino acid.

24. After folic acid is converted to N^{10}-formyltetrahydrofolate, it acts as a carrier of formyl groups and is a substrate for a transformylase reaction that converts Met-tRNA$_f$ to fMet-tRNA$_f$—the initiator tRNA. (See pp. 894–895 in Stryer.)

25. A purine-rich mRNA sequence, three to nine nucleotides long (called the Shine-Dalgarno sequence), which is centered about 10 nucleotides upstream of (to the 5′ side of) the start codon, base-pairs with a sequence of complementary nucleotides near the 3′ end of the 16S rRNA of the 30S ribosomal subunit. This interaction plus the association of fMet-tRNA$_f$ with the AUG in the P site of the ribosome sets the mRNA reading frame.

26. (a) 3, 7 (b) 3, 6, 7, 8, 10 (c) 3, 7, 9 (d) 2, 7, 8, 10 (e) 11 (f) 1, 7, 8, 10 (g) 5, 12 (h) 4, 7, 13 (i) 4, 7, 13

27. a, c. The aminoacyl-tRNA in the A site becomes a peptidyl-tRNA when it receives the carboxyl group of the growing polypeptide chain from the peptidyl-tRNA in the P site. After the free tRNA leaves, the extended polypeptide on its new tRNA is then moved to the P site by EF-G. Answer (d) is incorrect because transfer RNAs bearing amino acyl derivatives with free amino groups are never found in the P site.

28. a, c, d, e. Answer (b) is incorrect because succinyl CoA synthetase catalyzes substrate level phosphorylation of GDP to GTP in the citric acid cycle, rather than a hydrolysis that affects macromolecular interactions.

29. The large amounts of EF-Tu in the cell bind essentially all of the aminoacyl-tRNAs and protect these activated complexes from hydrolysis.

30. (a) ATP, 2 (b) GTP, 1 (c) GTP, 1 (d) None (e) GTP, 1. With regard to the answer for (d), the formation of a peptide bond per se doesn't require a cofactor. The energy for the exergonic reaction is supplied by the activated aminoacyl-tRNA.

31. Four high-energy phosphate bonds are expended: two to form the aminoacyl-tRNA, one to insert the aminoacyl-tRNA into the A site, and one to move the peptidyl-tRNA from the A to the P site.

32. A single translocation step could lead to the dissociation of the peptidyl-tRNA from the ribosome and its hydrolysis before the polypeptide chain is completed. Translocation in two steps prevents this from happening.

33. b, c, d. Each of the two release factors of *E. coli* recognizes two of the three translation stop codons and interacts with the synthesis machinery

such that peptidyl transferase donates the polypeptide chain to H_2O and thus terminates synthesis by hydrolyzing the ester linkage of the protein to the tRNA.

34. The inhibition of the prokaryotic translation and not that of the eukaryote can result from differences between their respective ribosomes. Some antibiotics interact with the RNA components that are unique to bacterial ribosomes and, consequently, can inhibit bacterial growth without affecting the human cells.

35. c. Puromycin is an analog of aminoacyl-tRNA. It inhibits protein synthesis by binding to the A site of the ribosome and accepting the growing polypeptide chain from the peptidyl-tRNA in the P site and thus terminating polymer growth. Because aminoacyl-tRNAs compete with the puromycin for the A site, increasing their concentration would lessen the extent of inhibition.

Eukaryotic Translation

36. b, d, e, f. Eukaryotic ribosomes usually scan the mRNA from the 5′ end for the first AUG codon, which then serves to initiate the synthesis. Answer (e) is correct because proteins that bind to the cap of the mRNA are involved in the association of the ribosome with the mRNA.

37. a, d. Answer (e) is incorrect because the toxin acts catalytically and is thus extremely deadly; one toxin molecule can inactivate many translocase molecules by modifying them covalently.

PROBLEMS

1. The template strand of DNA known to encode the N-terminal region of an *E. coli* protein has the following nucleotide sequence:

GTAGCGTTCCATCAGATTT

Give the sequence for the first four amino acids of the protein. (Refer to the genetic code on p. 109 of Stryer.)

2. Suppose that the sense strand of the DNA known to encode the amino acid sequence of the N-terminal region of a mammalian protein has the following nucleotide sequence:

CCTGTGGATGCTCATGTTT

Give the amino acid sequence that would result.

3. The nucleotide sequence on the sense strand of the DNA known to encode the carboxy terminus of a long protein of *E. coli* has the following nucleotide sequence:

CCATGCAAAGTAATAGGT

Give the resulting amino acid sequence.

4. List the major components that must be present in an in vitro system for the synthesis of an *E. coli* protein from a mixture of amino acids.

5. The synthesis of a protein molecule requires a considerable expenditure of energy.

(a) Calculate the minimum number of high-energy phosphate bonds that must be expended to synthesize a single prokaryotic protein consisting of 100 amino acid residues from a mixture of the components listed in your answer to problem 4.
(b) Explain why the number of necessary high-energy phosphate bonds you have calculated in part (a) is a minimum.
(c) On the average how many high-energy phosphates are required to add one amino acid residue to a polypeptide chain starting with a free amino acid?
(d) Compare your answer in part (c) with the number required to extend a glycogen molecule by one glucosyl residue starting with free glucose. (See Stryer, p. 588.)

6. Suppose that a particular aminoacyl-tRNA synthetase has a 10% error rate in the formation of aminoacyl-adenylates and a 99% success rate in the hydrolysis of incorrect aminoacyl-adenylates. What percentage of the tRNAs produced by this aminoacyl-tRNA synthetase will be faulty?

7. Students of biochemistry are frequently distressed by "Svedberg arithmetic," that is, for instance, by the fact that the 30S and 50S ribosomal subunits form a 70S particle rather than an 80S particle. Why don't the numbers add up to 80? (See pp. 51–52 of Stryer.)

8. In the chain of events leading from a nucleotide sequence on DNA to the production of protein by ribosomes, where precisely does the process of translation occur? Explain.

9. Studies on random copolymers of nucleotides have been important in research on the genetic code and on the role of tRNAs in protein synthesis. (See Stryer p. 106.) Give the kinds and proportions of amino acids that you would expect to be incorporated into a polypeptide chain when a random copolymer of A and C in the ratio of 3:1 is added as a template to an in vitro system for protein synthesis. (Protein synthesis in vitro may be initiated at random sites along a template lacking AUG, provided that high concentrations of Mg^{2+} are present.)

10. The possible codons for valine are GUU, GUC, GUA, and GUG.

(a) For each of these codons write down all the possible anticodons with which it might pair. (See Stryer, p. 887.)
(b) How many codons could pair with anticodons having I as the first base? How many could pair with anticodons having U or G as the first base? How many could pair with anticodons beginning with A or C.

11. What amino acid will be specified by a tRNA whose anticodon sequence is IGG?

12. Mutations from codons specifying amino acid incorporation to one of the chain-terminating codons, UAA, UAG, or UGA, so-called *nonsense* mu-

tations, result in the synthesis of shorter, usually nonfunctional, polypeptide chains. It was discovered that some strains of bacteria can protect themselves against such mutations by having mutant tRNAs that can recognize a chain-terminating codon and insert an amino acid instead. The result would be a protein of normal length that may be functional, even though it may contain an altered amino acid residue. How can bacteria with such mutant tRNA molecules ever manage to terminate their polypeptide chains successfully?

13. Suppose that the probability of inserting an incorrect amino acid at each position in a protein is 1%. Calculate the probability that a protein consisting of 100 amino acid residues is entirely correct.

14. Change of one base pair to another in a sense codon frequently results in an amino acid substitution. Change of a C·G to a G·U base pair at the 3:70 position of $tRNA^{Cys}$ causes that tRNA to be recognized by alanyl-tRNA synthetase.

(a) What amino acid substitution or substitutions would result?
(b) How does the pattern differ from base substitutions within a codon?

15. Recent studies have implicated rRNA as the peptidyl transferase activity in ribosomes. If this is true, then RNA-catalyzed reactions would be truly ubiquitous. List 11 enzymatic reactions, including peptidyl transferase, in which RNA is the direct catalyst or an active participant in the reaction. (Note: Two of these reactions you have not seen yet. They are discussed in Chapters 35 and 37 of Stryer.)

16. Laboratory studies of protein synthesis usually involve the addition of a radioactively labeled amino acid and either natural or synthetic mRNAs to systems containing the other components that you specified in your answer to problem 4. To observe the formation of protein, advantage is taken of the fact that proteins, but not amino acids, can be precipitated by solutions of trichloroacetic acid. Thus, one can observe the extent to which radioactivity has been incorporated into "acid-precipitable material" as a function of time to estimate the rate of formation of protein. In one such experiment, poly(U) is used as a synthetic mRNA in an in vitro system derived from wheat germ.

(a) What labeled amino acid would you add to the reaction mixture?
(b) What product will be formed?

For each of the following procedures, explain the results observed. Assume that in a complete system 3000 cpm (counts per minute) are found in acid-precipitable material at the end of 30 minutes.

(c) RNase A is added to the complete system; 85 cpm are found at the end of 30 minutes.
(d) Chloramphenicol is added to the complete system; 2900 cpm are found at the end of 30 minutes.
(e) Cyclohexamide is added to the complete system; 300 cpm are found at the end of 30 minutes.
(f) Puromycin is added to the complete system; 640 cpm are found at the end of 30 minutes.

(g) Puromycin and extra wheat germ tRNA are added to the complete system; 1518 cpm are found at the end of 30 minutes.

(h) Poly(A) is used instead of poly(U); 120 cpm are found at the end of 30 minutes.

ANSWERS TO PROBLEMS

1. The sequence of the first four amino acids of the protein is (formyl)Met-Glu-Arg-Tyr. As the name implies, the template (antisense) strand of DNA serves as the template for the synthesis of a complementary mRNA molecule. (Remember that by convention nucleotide sequences are always written in the 5′ to 3′ direction unless otherwise specified). The template strand of DNA and the mRNA synthesized are as follows:

DNA template strand: 5′-GTAGCGTTCCATCAGATTT-3′

mRNA: 3′-CAUCGCAAGGUAGUCUAAA-5′

Remember that the codons of an mRNA molecule are read in the 5′-to-3′ direction. Because this particular nucleotide sequence specifies the N-terminal region of an *E. coli* protein, the first amino acid must be (formyl)-methionine, which may be encoded by either AUG or GUG. Because there is no GUG and only a single AUG in the mRNA sequence, the location of the initiation codon can be established unambiguously. The portion of the mRNA sequence encoding protein and the first four amino acids it encodes are

mRNA: 5′-AUG-GAA-CGC-UAC-3′

Amino acid sequence: (Formyl)Met-Glu-Arg-Tyr

2. The expected amino acid sequence is Met-Leu-Met-Phe. The nucleotide sequences on DNA and mRNA are

Sense strand of DNA: 5′-CCTGTGGATGCTCATGTTT-3′

mRNA: 5′-CCUGUGGAUGCUCAUGUUU-3′

In eukaryotes the first triplet specifying an amino acid is almost always the AUG that is closest to the 5′ end of the mRNA molecule. In this example there are two AUGs, so there will be two Met residues in the polypeptide that is produced. The reading frame and the resulting amino acids are as follows:

mRNA: 5′-CCUGUGG-AUG-CUC-AUG-UUU-3′

Amino acid sequence: Met-Leu-Met-Phe

3. The sequence is His-Ala-Lys. The DNA and mRNA sequences are

Sense strand of DNA: 5′-CCATGCAAAGTAATAGGT-3′

mRNA: 5′-CCAUGCAAAGUAAUAGGU-3′

Since this sequence specifies the carboxyl end of the peptide chain, it must contain one or more of the chain-termination codons: UAA, UAG, or UGA.

UAA and UAG occur in tandem in the sequence, so we can infer the reading frame. The mapping of the amino acid residues to the mRNA is as follows:

mRNA:	5′-C-CAU-GCA-AAG-UAA-UAG-GU-3′
Amino acid sequence:	His-Ala-Lys

4. The major components are as follows: amino acids, tRNAs, activating enzymes, ATP, GTP, ribosomes, mRNA, N^{10}-formyltetrahydrofolate, transformylase, IF1, IF2, IF3, EF-G, EF-Ts, EF-Tu, RF1, RF2, and Mg^{2+}.

5. (a) The minimum number of high-energy phosphate bonds required would be 399. Each of the 100 amino acids must be attached to a tRNA molecule. Since ATP is cleaved to one AMP and two P_i in this process, a total of 200 high-energy phosphates are required for the synthesis of the aminoacyl-tRNAs. In the synthesis of a protein of 100 residues, 99 elongation and 99 translocation steps are involved, each requiring one GTP, for a total of 198 GTPs. Finally, the formation of the single initiation complex requires a GTP. Thus, there are 200 high-energy phosphate bonds from ATP and 199 GTP molecules required, for a total of 399 high-energy phosphates.

(b) The number of high-energy phosphate bonds calculated in (a) is a minimum because it reflects only the direct expenditure of energy. It does not take into account the energy required for the synthesis of ribosomes, activating enzymes, initiation factors, elongation factors, release factors, and so forth.

(c) 4

(d) The number required is 2. The number starting with glucose 6-phosphate is 1 (Stryer, p. 588). An additional ATP is required to convert glucose into glucose 6-phosphate.

6. The percentage of tRNAs that will be faulty is 0.11%. For every 1000 aminoacyl-adenylates that are produced, 100 are faulty and 900 are correct. The 900 correct intermediates will be converted to correct aminoacyl tRNAs because the intermediates are tightly bound to the active site of the aminoacyl-tRNA synthetase. Of the 100 incorrect aminoacyl-adenylates, 99 will be hydrolyzed and will therefore not form aminoacyl tRNAs. Only 1 will survive to become an incorrect aminoacyl tRNA. The fraction of incorrect aminoacyl tRNAs is therefore 1/901, or 0.11%.

7. The Svedberg unit (S) is a sedimentation coefficient, which is a measure of the velocity with which a particle moves in a centrifugal field. It represents a hydrodynamic property of a particle, a property that depends on, among other factors, the size and shape of the particle. When two particles come together, the sedimentation coefficient of the resulting particle should be less than the sum of the individual coefficients because there is no frictional resistance between the contact surfaces of the particles and the centrifugal medium.

8. Translation involves conversion of the language of nucleotides to that of proteins. The agent of translation is the appropriate aminoacyl-tRNA synthetase, which must recognize a particular amino acid and link it to a tRNA containing an anticodon for that amino acid.

9. The eight possible codons on the template, together with their frequencies and the amino acids that they specify, are as follows:

Codon	Frequency	Amino Acid
AAA	$3/4 \times 3/4 \times 3/4 = 27/64$	Lys
AAC	$3/4 \times 3/4 \times 1/4 = 9/64$	Asn
ACA	$3/4 \times 1/4 \times 3/4 = 9/64$	Thr
CAA	$1/4 \times 3/4 \times 3/4 = 9/64$	Gln
ACC	$3/4 \times 1/4 \times 1/4 = 3/64$	Thr
CAC	$1/4 \times 3/4 \times 1/4 = 3/64$	His
CCA	$1/4 \times 1/4 \times 3/4 = 3/64$	Pro
CCC	$1/4 \times 1/4 \times 1/4 = 1/64$	Pro

The total fraction of each amino acid incorporated is as follows:

Amino Acid	Proportion
Lys	27/64
Thr	$9/64 + 3/64 = 12/64$
Asn	9/64
Gln	9/64
His	3/64
Pro	$3/64 + 1/64 = 4/64$
Total	64/64

10. (a) The possible anticodons with which the codons might pair are as follows:

Codon	Possible Anticodon
GUU	AAC, GAC, IAC
GUC	GAC, IAC
GUA	UAC, IAC
GUG	CAC, UAC

(b) Three codons could pair with the anticodon beginning with I; two codons could pair with an anticodon beginning with U or G; only one codon could pair with an anticodon beginning with A or C.

11. Proline. The three codons that will pair with IGG—CCU, CCC, and CCA—all specify proline.

12. If two different legitimate stop codons are present in tandem, it would be extremely improbable that mutant tRNAs would exist for both and would simultaneously bind to each of them and thereby prevent proper chain termination.

13. The probability of any one residue being the correct amino acid is 0.99. The probability that the entire protein consisting of 100 amino acid residues will be correct is therefore $(0.99)^{100}$ or 0.366.

14. (a) The $tRNA^{Cys}$ will become loaded with Ala rather than Cys, and as a result will insert Ala rather than Cys into polypeptide chains.

(b) In the case of a base change within a codon, only a single amino acid of a single polypeptide is changed. In the case of a tRNA recognition mutation, amino acid substitutions at many positions of many polypeptides would occur.

15. (1) Prokaryotic transcription termination, (2) RNAse P cleavage of tRNAs, (3) group I intron removal, (4) group II intron removal, (5) group III intron splicing, (6) hammerhead RNA cleavage, (7) RNA editing of trypanasome mRNA, (8) apolipoprotein B mRNA processing, (9) telomerase, (10) peptidyl transferase, (11) SRP 7SL RNA.

16. (a) Poly(U) codes for the incorporation of phenylalanine. Therefore, labeled phenylalanine must be added to the reaction mixture.

(b) Polyphenylalanine will be formed.

(c) RNase A will digest poly(U) almost completely to 3′-UMP, thus destroying the template for polyphenylalanine synthesis. Also the tRNA will be digested and the ribosomes damaged.

(d) Chloramphenicol inhibits the peptidyl transferase activity of the 50S ribosomal subunit in prokaryotes but has no effect on eukaryotes. (Wheat is a eukaryote.)

(e) Cyclohexamide inhibits the peptidyl transferase activity of the 60S ribosomal subunit in eukaryotes.

(f) Puromycin mimics an aminoacyl-tRNA and causes premature polypeptide chain termination.

(g) The addition of the extra wheat germ tRNA reduces the inhibiting effect of puromycin, since they both compete for the A site on ribosomes.

(h) Poly(A) directs the synthesis of polylysine; since there is no lysine (either labeled or unlabeled) in the system, no product can be detected. The 120 cpm is not significantly above background.

EXPANDED SOLUTIONS TO TEXT PROBLEMS

1. The enzyme-bound Ile-AMP intermediate is necessary for the $^{32}PP_i$ exchange into ATP. Since isoleucine is a requirement, labeled ATP will be formed only in (c).

2. Four bands: light, heavy, a hybrid of light 30S and heavy 50S, and a hybrid of heavy 30S and light 50S. Recall that "ribosomes dissociate into 30S and 50S subunits after the polypeptide product is released." As protein synthesis continues, 70S ribosomes are re-formed from the various heavy and light subunits.

3. About 799 high-energy phosphate bonds are consumed—400 to activate the 200 amino acids, 1 for initiation, and 398 to form 199 pepitide bonds.

4. Type I: b, c, and f; type 2: a, d, and e.

5. A mutation caused by the insertion of an extra base can be suppressed by a tRNA that contains a fourth base in its anticodon. For example, UUUC rather than UUU is read as the codon for phenylalanine by a tRNA that contains 3′-AAAG-5′ as its anticodon.

6. One approach is to synthesize a tRNA that is acylated with a reactive amino acid analog. For example, bromoacetylphenylalanyl-tRNA is an affinity-labeling reagent for the P site of *E. coli* ribosomes. See H. Oen, M. Pellegrini, D. Eilat, and C.R. Cantor. *Proc. Nat. Acad. Sci.* 70(1973):2799.

7. The sequence GAGGU is complementary to a sequence of five bases at the 3′ end of 16S rRNA and is located several bases on the 5′ side of an AUG codon. Hence this region is a start signal for protein synthesis. The replacement of G by A would be expected to weaken the interaction of this mRNA with the 16S rRNA and thereby diminish its effectiveness as an initiation signal. In fact, this mutation results in a tenfold decrease in the rate of synthesis of the protein specified by this mRNA. See J.J. Dunn, E. Buzash-Pollert, and F.W. Studier, *Proc. Nat. Acad. Sci.* 75(1978):2741, for a discussion of this informative mutant.

8. The nitrogen atom of the deprotonated α-amino group of aminoacyl-tRNA is the nucleophile in peptide-bond formation. See Figure 34-33 and the accompanying discussion in Stryer.

9. Proteins are synthesized from the amino to the carboxyl end on ribosomes, and in the reverse direction in the solid-phase method. The activated intermediate in ribosomal synthesis is an aminoacyl-tRNA; in the solid-phase method, it is the adduct of the amino acid and dicyclohexylcarbodiimide.

10. The error rates of DNA, RNA, and protein synthesis are of the order of 10^{-10}, 10^{-5}, and 10^{-4} per nucleotide (or amino acid) incorporated. The fidelity of all three processes depends on the precision of base pairing to the DNA or mRNA template. No error correction occurs in RNA synthesis. In contrast, the fidelity of DNA synthesis is markedly increased by the $3' \rightarrow 5'$ proofreading nuclease activity and by postreplicative repair. In protein synthesis, the mischarging of some tRNAs is corrected by the hydrolytic action of the aminoacyl-tRNA synthetase. Proofreading also takes place when aminoacyl-tRNA occupies the A site on the ribosome; the GTPase activity of EF-Tu sets the pace of this final stage of editing.

11. GTP is not hydrolyzed until aminoacyl-tRNA is delivered to the A site of the ribosome. An earlier hydrolysis of GTP would be wasteful because EF-Tu–GDP has little affinity for aminoacyl-tRNA.

12. EF-Ts catalyzes the exchange of GTP for GDP bound to EF-Tu. In G protein cascades, an activated seven-helix receptor catalyzes GTP-GDP exchange in a G protein. For example, photoexcited rhodopsin triggers GTP-GDP exchange in transducin (in Stryer, p. 336).

13. The translation of an mRNA molecule can be blocked by antisense RNA, an RNA molecule with the complementary sequence. The antisense-sense RNA duplex is degraded by nucleases. Antisense RNA added to the external medium is spontaneously taken up by many cells. A precise quantity can be delivered by microinjection. Alternatively, a plasmid encoding the antisense RNA can be introduced into target cells. For an interesting discussion of antisense RNA and DNA as research tools and drug candidates, see H.M. Weintraub. *Sci. Amer.* 262(January 1990):40.

CHAPTER

35

Protein Targeting

Proteins are not uniformly distributed throughout a cell. In Chapter 35, Stryer describes the biochemistry responsible for the delivery of proteins from their sites of synthesis on the ribosomes to their final locations. In addition, he describes the receptor-mediated import of specific proteins into certain eukaryotic cells and a programmed pathway for protein degradation.

The translocation path for proteins that are synthesized on the rough endoplasmic reticulum (ER) and targeted for the lysosomes or the plasma membrane or for export is described first. Stryer explains the role of signal sequences within the nascent protein in initiating a series of events, involving a small cytoplasmic ribonucleoprotein particle, a GTP-GDP cycle, and integral membrane proteins, resulting in the unidirectional translocation of the polypeptide across the ER membrane into the lumen of the ER. The role of ATP-driven chaperone proteins in binding to the nascent proteins entering the lumen of the ER and in aiding in their correct folding is also described. The ensuing events, in which the protein may be covalently modified by cleavage and glycosylation and is transported through the Golgi complex into transport vesicles that will deliver it to its final destination, are also outlined. Stryer explains how soluble proteins in the Golgi inevitably are transported to constitutive secretory vesicles or into the plasma membrane unless directed to alternative targets. Stryer explains how the molecular marker that targets proteins to the lysosomes was revealed by a human disorder in which an entire class of enzymes is directed to the wrong intracellular location because its members lack a particular carbohydrate "address " label. Stryer also describes the similarities between protein transport pathways via the ER in eukaryotes and protein targeting in prokaryotes.

Proteins synthesized on free cytosolic ribosomes may also be targeted to specific intracellular sites. Mitochondrial, chloroplast, peroxisomal, and nuclear proteins are synthesized in the cytosol and are translocated through or into the membranes of these organelles. Stryer describes how chimeric proteins, produced by recombinant DNA techniques, have been useful in detecting the targeting signals responsible for directing proteins to specific sites. Several mechanisms for the post-translational attachment of lipophilic groups to proteins for the purpose of associating them with the surfaces of membranes are also described.

The mechanism by which cells import specific proteins and their associated ligands is described next. Receptor-mediated endocytosis provides a way for cells to recognize particular extracellular proteins through interactions with receptors located in cell-surface structures called *coated pits.* Some viruses and toxins have appropriated this mechanism to gain entry into the cell. Stryer closes the chapter with a description of a mechanism cells use to designate proteins for degradation.

In preparation for studying this chapter, you should review the structure of the mitochondrion on pages 530–531 and the structure of the chloroplast on page 654 in *Biochemistry.* In addition, reread pages 475–479 to refresh your memory of the relevant carbohydrate structures and the roles of oligosaccharide substituents on proteins. Review the role of nucleotide sugars in the synthesis of glycogen on pages 586–587 to contrast this mechanism of condensing carbohydrate residues with the one involving dolichol phosphate sugar derivatives described in this chapter. The material on page 280 and Figure 11-43, respectively, will remind you of the linkages by which carbohydrates are attached to proteins and of the hydrophobic nature of the regions of integral proteins that span membranes. Membrane proteins should be reviewed by reading pages 275–276. Pages 429–431 should also be read to review the roles of chaperone proteins and isomerases in promoting proper protein folding. Finally, review the introduction to receptor-mediated endocytosis that was given on pages 699–700 during the presentation of the metabolism of cholesterol.

LEARNING OBJECTIVES

When you have mastered this chapter, you should be able to complete the following objectives:

Protein Translocation across the Endoplasmic Reticulum Membrane (Stryer pages 911–919)

1. List the major cell locations to which proteins can be directed in eukaryotes and prokaryotes. Outline the *sorting pathways* and list the major cell structures involved.

2. Distinguish between *ribosomes* that are *free in the cytosol* and those that are *bound to the endoplasmic reticulum* (*ER*) *membrane system* in terms of the destinations of the proteins they synthesize.

3. List the three major classes of proteins that are synthesized on the *rough ER.*

4. State the *signal hypothesis* and recount the evidence that verified it. List the common features of *signal sequences.* Explain how *chimeric proteins* have been used to show the crucial role of signal sequences in protein targeting.

5. Describe the composition of the *signal recognition particle* (*SRP*) and outline its role in translocating proteins across the ER membrane. Sketch the *SRP cycle* and relate it to the *ribosome cycle.* Describe the roles of the *SRP receptor,* the *ribosome receptor/translocon/protein-conducting channel,* the *signal* sequence, and the *signal peptidase* in the translocation mechanism.

6. Explain the roles of the ribosome and SRP in maintaining a nascent polypeptide as an optimal substrate for translocation. Note that some proteins may be translocated across membranes after they have folded, and list the required components for this process.

7. Describe the function of *stop-transfer sequences.*

8. Explain the role of *chaperone proteins, protein disulfide isomerases,* and *peptidyl prolyl isomerase* in promoting protein folding of nascent proteins in the lumen of the ER. Relate the chaperones to the *heat-shock protein family* and its functions.

Protein Modifications during Translocation and the Golgi Complex (Stryer pages 920–929)

9. List the types of *covalent modifications* that proteins undergo in the *lumen* of the *ER.*

10. Describe the structure of *dolichol phosphate* and outline its role in the synthesis of the *pentasaccharide core* of N-*linked oligosaccharides.* Relate the effects of *bacitracin* and *tunicamycin* to dolichol phosphate metabolism.

11. Explain the functions of the repeated addition and removal of glucose from the oligosaccharide of luminal ER proteins and of *calexin* in selecting properly folded proteins for export.

12. Describe the *Golgi complex* and list its major functions. Distinguish among the *cis, medial,* and *trans* compartments of the Golgi.

13. Describe the origin and destination of *transport vesicles.*

14. Explain how *membrane budding* to form vesicles and *vesicle fusion* with the plasma membrane determine the *asymmetry of the membrane* and the *orientation of its integral membrane proteins.* Contrast the physiologic roles of *constitutive* and *regulated secretory vesicles.*

15. Distinguish between *core* and *terminal glycosylation* of glycoproteins and provide an overview of the reactions that occur in the three compartments of the Golgi.

16. List the key proteins in *vesicular transport* and describe their roles.

17. Indicate the destinations of soluble proteins in the lumen or integral membrane proteins in the membrane of the ER if they lack specific target-

ing signals. Explain how some proteins in the lumen of ER escape export by containing a C-terminal KDEL sequence.

18. State the molecular basis of *I-cell disease.* Explain how this disorder revealed the molecular signal that directs hydrolytic enzymes to the lysosome.

19. List the potential destinations of *bacterial* proteins and compare the transmembrane transport of bacterial proteins with the analogous process across the ER membrane in eukaryotes.

Targeting of Mitochondria, Chloroplast, Peroxisome, Nuclear, and Membrane-Surface Proteins (Stryer pages 930–935)

20. Indicate the *origin* of most *mitochondrial proteins,* and list their potential *destinations* in the *genetic* and *cytological* mitochondrion.

21. Describe the salient features of the *presequences* of proteins destined for the mitochondrial matrix. Contrast the *energy sources* for the translocation of proteins into the mitochondria with that used for the translocation of proteins in *E. coli.* Outline the role of *Hsp70 proteins* in mitochondrial protein import to the matrix.

22. Compare protein transport into *chloroplasts* and with that into mitochondria.

23. Describe *peroxisomes* and *glyoxysomes.* Compare protein import into peroxisomes with protein transport into mitochondria and chloroplasts.

24. Describe the structural features of the *nuclear envelope* that are relevant to protein transport, and discuss the effects of *protein size* and *nuclear localization sequences* on the import of proteins into the nucleus.

25. Describe the ways in which proteins can be targeted to the *cytosolic face* or *outer leaflet* of the plasma membrane.

Receptor-Mediated Endocytosis (Stryer pages 935–942)

26. List the roles of *receptor-mediated endocytosis.*

27. Describe the general features of *transmembrane glycoprotein cell-surface receptors* and name the cell-surface regions in which they are found.

28. Describe a *coated pit* and outline the process of receptor-mediated endocytosis from *receptor binding* to the formation of the *endosome.* Describe the *triskelion* structure of *clathrin.*

29. List the possible fates of the internalized proteins of the endosome.

30. Explain the importance of pH changes during receptor-mediated endocytosis and during *lysosome* functioning.

31. Outline the roles of receptor-mediated endocytosis in the life cycle of some *membrane-enveloped viruses* and in the mechanisms of action of some *toxins.*

Programmed Protein Destruction (Stryer pages 942–943)

32. Note the different *half-lives of proteins* in the cell, and explain the role of *ubiquitin* in determining the stability of some proteins.

33. Relate the identity of the *N-terminal amino acid* of a protein to its *intracellular stability* and describe the fate of ubiquinylated proteins.

SELF-TEST

Protein Translocation across the Endoplasmic Reticulum Membrane

1. What feature distinguishes the rough and smooth endoplasmic reticulum?

____

____

2. Membrane-bound ribosomes synthesize which of the following?

(a) Integral membrane proteins of the mitochondria
(b) Proteins spanning the ER membrane
(c) Lysosomal proteins
(d) Secretory proteins
(e) Nuclear proteins

3. Ribosomes can be classified as "free in the cytosol" or "membrane bound." Describe the experiments that showed all ribosomes are intrinsically identical and differ only in their intracellular locations.

____

____

4. When mRNA specifying a secretory protein was translated in vitro in the absence of ER membranes, the product was longer than the form secreted by living cells. Explain why.

____

____

5. Which of the following changes in the gene encoding a lysosomal protein would be expected to impair the translocation of the protein across the ER membrane?

(a) The deletion of the nucleotides specifying the 25 amino acids following the translation initiation codon
(b) The conversion of the sole lysine codon after the translation initiation codon to an arginine codon
(c) The conversion of the sole lysine codon after the translation initiation codon to an aspartic acid codon
(d) The conversion of two leucine codons and one phenylalanine codon near the translation initiation codon to an isoleucine codon, an aspartic acid codon and a glutamic acid codon, respectively

6. Suppose you wish to engineer an enzyme that ordinarily resides in the cytosol so that it will be secreted by a human cell. How would you proceed?

____

____

7. Match the functions or characteristics in the right column with the components in the left column that are involved in the mechanism that delivers ribosomes to the ER and leads to the translocation of proteins across the ER membrane.

(a) 7SL RNA __________

(b) Signal recognition particle (SRP) __________

(c) SRP receptor __________

(d) Ribosome receptor/ translocation machinery __________

(e) Signal peptidases __________

(1) An α, β-heterodimer with an intrinsic GTPase on the cytosolic face of the ER that is involved in the release of SRP from the ribosomes

(2) Arrests elongation of polypeptide chain

(3) Accepts the nascent signal peptide from the SRP on the cytosolic face of the ER membrane

(4) Binds to a ribosome bearing a polypeptide with a signal sequence

(5) A component of SRP

(6) Located on the lumenal face of the ER membrane

(7) Releases SRP and, consequently, relieves translation arrest

8. If you could digest a nascent polypeptide associated with a ribosome during protein synthesis with a voracious and nonspecific protease, how big a fragment would survive? Explain.

____

____

9. Which of the following statements about translocation across the ER membrane are correct?

(a) It is driven by peptidyl transferase.
(b) It involves protein-conducting channels that can be opened by a signal peptide.
(c) It depends on a proton-motive force across the membrane.
(d) It occurs only for unfolded proteins.
(e) It can be stopped by specific sequences of amino acids in the interior of the translocated protein.

10. What important roles do chaperone proteins play in the transport of proteins into the ER?

____

____

11. Where in the cell will a protein containing both an N-terminal signal sequence and a stop-transfer sequence be found at the conclusion of protein sorting?

____

____

Protein Modifications during Translocation and the Golgi Complex

12. Translocated proteins may undergo which of the following modifications in the lumen of the ER?

(a) Signal sequence cleavage
(b) The attachment of dolichol phosphate to form a lipid anchor
(c) Folding, disulfide-bond formation and isomerization, and cis-trans sisomerization of X-Pro peptide bonds
(d) The addition of oligosaccharides to their asparagine residues to form *N*-linked derivatives
(e) The addition of oligosaccharides to their tyrosine residues to form *O*-linked derivatives

13. Which of the following statements about dolichol phosphate are correct?

(a) It serves as an acceptor of monosaccharides.
(b) It serves as a donor of both monosaccharides and oligosaccharides.
(c) It acts as a lipid carrier to facilitate the transfer of sugar residues from the cytosol to the lumen of the ER.
(d) It is converted to dolichol pyrophosphate by a kinase that uses ATP as a phosphate source.
(e) It is produced from dolichol pyrophosphate by a phosphatase.

14. Which of the following statements about the Golgi complex are correct?

(a) It is a stack of flattened proteoglycan sacs.
(b) It carries out core glycosylation of the proteins being transported.
(c) It is the major protein-sorting center of the cell.
(d) It receives proteins from the ER by fusion with transport vesicles.
(e) It forms secretory granules in its *trans* compartment.
(f) The cisternae of the *cis, medial,* and *trans* compartments are connected by pores.

15. If a protein that is targeted to become an integral plasma membrane protein (see question 11) begins its journey by being inserted into the ER membrane such that its amino terminus is in the lumen, will the amino terminus reside inside or outside the cell when it reaches its destination? Explain.

____

____

16. Why don't all the proteins in the lumen of the ER, for example chaperone proteins, follow the default pathway to secretion from the cell?

____

____

17. Which of the following statements about I-cell disease are correct?

    (a) It results from the inability of lysosomes to hydrolyze glycosaminoglycans and glycolipids.
    (b) It results from a chromosomal deletion of the genes specifying at least eight acid hydrolases ordinarily found in the lysosomes.
    (c) It arises from a deficiency in an enzyme that transfers mannose 6-phosphate onto a core oligosaccharide that is normally found on lysosomal enzymes.
    (d) It arises from the absence of a mannose 6-phosphate receptor in the *trans* Golgi complex.

18. Match each protein in the right column with the kind of secretory vesicle in the left column that is likely to translocate it.

    (a) Constitutive vesicle ______________
    (b) Regulated vesicle ______________

    (1) Serum albumin
    (2) Trypsinogen
    (3) ACTH
    (4) Collagen
    (5) Chymotrypsinogen
    (6) Prolactin
    (7) Integral plasma membrane protein

19. Histones, the small proteins associated with DNA, enter the nucleus through the nuclear pores. Does RNA polymerase II enter the nucleus by the same mechanism? Explain.

    ____

    ____

20. What is the fate of soluble proteins in the lumen of the ER if they do not contain signals directing them to specific locations?

    ____

    ____

21. Describe an experiment that indicated that some of the mechanisms involved in the translocation of proteins across membranes share common features in both eukaryotes and prokaryotes.

    ____

    ____

Targeting of Mitochondria, Chloroplast, Peroxisome, Nuclear, and Membrane-Surface Proteins

22. Which of the following statements about mitochondrial matrix proteins are correct?

    (a) They are mostly encoded by mitochondrial DNA.
    (b) They are mostly synthesized on the rough ER.
    (c) They are synthesized with N-terminal targeting sequences.
    (d) They require an electrical potential across the inner mitochondrial membrane and ATP hydrolysis for their import into the mitochondrial matrix.
    (e) They require chaperone proteins in both the cytosol and the matrix for their translocation into the mitochondrion.

23. Match the cellular location in the left-hand column with the entity in the right-hand column that directs the protein bearing it to the target.

(a) Lumen of the ER ________

(b) Outer leaflet of plasma membrane ________

(c) Plasma membrane ________

(d) Peroxisome ________

(e) Cytosolic face of plasma membrane ________

(f) Lysosome ________

(1) C-terminal glycosyl phosphatidyl inositol (GPI) substituent
(2) *S*-Palmitoyl substituent
(3) C-terminal prenyl group
(4) N-terminal myristoyl group
(5) C-terminal SKF sequence
(6) Stop-transfer sequence
(7) N-terminal signal sequence
(8) 20 to 30 amino acid hydrophobic sequence at C-terminal end
(9) KDEL sequence
(10) Mannose 6-phosphate on core of oligosaccharide

Receptor-Mediated Endocytosis

24. Which of the following statements about receptor-mediated endocytosis are correct?

(a) It delivers some metabolites to cells.
(b) It is involved in the actions of some hormones and growth factors.
(c) It participates in the uptake and delivery of proteins to the lysosomes for degradation.
(d) It is used by some bacteriophages to gain entry into bacteria.
(e) When defective, it can cause human disorders.

25. Which of the following statements about clathrin are true?

(a) It is a component of the protein lattice of coated pits and coated vesicles.
(b) It has an extracellular domain containing carbohydrate substituents.
(c) It exists in different forms that are specific for each protein recognized and bound.
(d) It is a trimer comprising six polypeptide chains.
(e) It has a three-legged structure called a triskelion.
(f) It is recycled after being released from coated vesicles by an ATP-dependent uncoating protein.

26. Place the following probable initial steps in the receptor-mediated endocytosis of a protein in their proper order:

(a) Invagination of the coated pit ________
(b) Loss of clathrin ________
(c) Binding of the protein to its receptor in the coated pit ________
(d) Formation of the endosome (or receptosome) ________
(e) Decrease in pH and dissociation of the protein-receptor complex ________
(f) Formation of the coated vesicle ________
(g) Formation of a clathrin lattice around coated pit ________

27. What are the possible fates of the receptor and its protein ligand after they are internalized by receptor-mediated endocytosis?

28. Outline the steps in the infection of a mammalian cell by Semliki Forest virus, a membrane-enveloped virus.

Programmed Protein Destruction

29. Which of the following statements about ubiquitin are correct?
    (a) It is a small and widely distributed protein.
    (b) It has been highly conserved during evolution.
    (c) Its C-terminal carboxyl group is activated as a thioester in an ATP-dependent reaction.
    (d) It is joined to other proteins, depending on the identity of their C-terminal amino acids.
    (e) It marks the proteins to which it attaches for destruction.
    (f) It joins to other proteins by isopeptide linkages.
    (g) It directs proteins to which it is attached to a 26S protease complex.

ANSWERS TO SELF-TEST

Protein Translocation across the Endoplasmic Reticulum Membrane

1. The presence or absence of associated ribosomes distinguishes the two forms of ER. The rough ER has a studded appearance because of the ribosomes that are associated with it, whereas the smooth ER is devoid of ribosomes.

2. b, c, d. Answers (a) and (e) are incorrect because the proteins of the mitochondrion and the nucleus are synthesized on cytosolic ribosomes.

3. Ribosomes of each class were isolated and mixed with appropriate mRNAs in an in vitro protein synthesizing system. Cytosol-derived ribosomes were capable of synthesizing secretory proteins, and, conversely, ribosomes derived from the ER membrane were capable of synthesizing proteins that ordinarily reside in the cytosol.

4. Since the mRNA encoded a protein destined for secretion, it specified a signal sequence at the N-terminus of the protein. This sequence is usually cleaved from the protein by a signal peptidase during translocation across the ER membrane and is therefore not found on normally secreted mature proteins. Because the signal peptidase was absent in the in vitro system, the signal sequence was not cleaved.

5. a, c, d. The signal sequence, which immediately follows the translation initiation codon, is usually from 13 to 36 amino acids long and includes a positively charged amino acid near the N terminus and a stretch of nearby

hydrophobic residues. Either the removal of the positively charged amino acid near the N terminus or the introduction of a charged residue into the hydrophobic region usually impairs translocation.

6. Insert DNA encoding the signal sequence from a secreted protein into the gene of the target enzyme after the site encoding the translation initiation codon by recombinant DNA techniques so that the protein will have an N-terminal signal sequence. If the inserted DNA doesn't disrupt expression of the gene, the encoded protein will be exported upon translation.

7. (a) 5 (b) 2, 4 (c) 1, 7 (d) 3 (e) 6

8. A fragment of about 35 amino acids in length should survive. The growing N terminus of the polypeptide is extruded through a narrow tunnel in the large subunit of the ribosome, and the amino acids within the tunnel should be inaccessible to the protease.

9. b, e. Elongation and translocation are mechanistically distinct. Although unfolded proteins are probably the optimal substrates for the translocation apparatus, fully folded proteins can be transported across the membrane. Stop-transfer or membrane-anchor sequences can arrest translocation and stabilize a transmembrane peptide segment in the membrane.

10. Although folded proteins can be translocated into the ER, unfolded ones are the best passengers for the transport machinery. Once the protein is inside, it must fold or refold. The ATP-dependent chaperone proteins (in their ADP-bound forms) transiently bind a segment of the nascent, unfolded protein. ATP displaces the bound ADP, and the peptide fragment is released. If it folds, the chaperone can no longer bind it. If not, it may be rebound, and the cycle repeated. The net effect is to promote correct folding within the lumen of the ER by inhibiting rapid, incorrect intermolecular interactions. The ATP-bound form of the chaperone is converted to the ADP form by an intrinsic ATPase activity.

11. The protein will be found as a transmembrane protein in the plasma membrane. The signal sequence will initiate its penetration of the ER membrane, but the stop-transfer sequence will anchor it in the membrane. The process of movement through the Golgi, conversion to a secretory vesicle, and fusion with the plasma membrane will ultimately deposit the protein in the plasma membrane.

Protein Modifications during Translocation and the Golgi Complex

12. a, c, d. Answer (e) is incorrect because threonine and serine provide hydroxyls for the formation of *O*-linked oligosaccharides. Answer (b) is incorrect because dolichol phosphate is attached to an oligosaccharide, not a protein.

13. a, b, c, e. Sugar-substituted dolichol phosphates serve both as acceptors of monosaccharides from nucleotide sugars and other dolichol phosphate sugars and as donors of monosaccharides and oligosaccharides to other dolichol phosphate sugar derivatives and proteins. As a result of glycosyl transfer by the dolichol oligosaccharide, dolichol pyrophosphate is formed. This compound must be hydrolyzed to dolichol phosphate by a phosphatase to regenerate the sugar carrier for continued use. (See p. 921 in Stryer.)

14. c, d, e. The Golgi complex carries out terminal glycosylation by modifying and adding to the core oligosaccharides that were constructed in the ER. Answer (f) is incorrect because the compartments of the Golgi are distinct, and components are transferred between them by vesicles.

15. Its amino terminus will reside outside the cell. (See Figure 35-17 on p. 924 in Stryer.) All budding and fusion events with the surfaces of other closed membranous compartments within the cell will leave the N terminus in the lumen of the resulting compartment. The final fusion is with the plasma membrane that encloses the entire cell. This fusion leads to an eversion of the fusing vesicle that places its contents and, in this case, the N terminus of the integral membrane protein on the outside surface of the cell.

16. These proteins have a C-terminal amino acid sequence signal (KDEL) in vertebrates that acts as a retrieval tag. This sequence is bound by receptors in vesicles that return to fuse with the ER. The increase in pH upon entry into the ER lumen releases the KDEL "tail" from the membrane-bound receptor.

17. a. The disease results from a deficiency in a sugar phosphotransferase that initiates a two-step sequence leading to the formation of a mannose 6-phosphate terminus on an oligosaccharide substituent of the eight or more affected lysosomal hydrolases. The phosphotransferase attaches a GlcNAc phosphate to a mannose residue of the oligosaccharide. Removal of the GlcNAc leaves the phosphate on the mannose. The enzymes lacking this mannose 6-phosphate "address" label are erroneously exported from the cell rather than being directed to the lysosomes. (See Figure 35-21 on p. 927 in Stryer.)

18. (a) 1, 4, 7 (b) 2, 3, 5, 6. Serum albumin and collagen are likely to be secreted constitutively and not in response to specific hormonal signals. Their production is regulated, but not at the level of secretion. Integral membrane proteins are not secreted.

19. No. Many of the subunits of RNA polymerase II are too large (see p. 853 in Stryer) to pass through the nuclear pores. Passage of proteins larger than 90 kd requires specific nuclear localization signal sequences that enable coupled ATP hydrolysis to translocate them through the pores into the nucleus.

20. They are exported from the cell by constitutive secretory vesicles.

21. Messenger RNA encoding the secreted bacterial enzyme β-lactamase was translated in an in vitro protein synthesis system derived from a plant, and the protein was recognized by a mammalian SRP. In addition, determinations of the signal sequences and the specificities of mammalian and bacterial signal peptidases showed that they are similar.

Targeting of Mitochondria, Chloroplast, Peroxisome, Nuclear, and Membrane-Surface Proteins

22. c, d, e. Answers (a) and (b) are incorrect because most mitochondrial proteins are encoded by nuclear DNA and are synthesized on cytosolic ribosomes. ATP is involved in the functioning of the chaperone proteins to maintain the optimal unfolded state of the transported proteins (see Figure 35-25 in Stryer).

23. (a) 7, 9 (b) 1, 7, 8 (c) 6, 7 (d) 5 (e) 2, 3, 4 (f) 7, 10

Receptor-Mediated Endocytosis

24. a, b, c, e. Answer (d) is incorrect because receptor-mediated endocytosis is a phenomenon of eukaryotes, not of prokaryotes.

25. a, d, e, f. With regard to answer (a), clathrin is located on the cytosolic face of coated pits and coated vesicles. Choices (b) and (c) are true for the extracellular receptors in coated pits but not for clathrin.

26. c, a, g, f, b, d, e

27. Depending on the particular mode of receptor-mediated endocytosis (see Table 35-2 on p. 938 in Stryer), either component can be recycled, or transported to the lysosomes for degradation, or transported to another location in the cell.

28. Proteins on the surface of the virus are recognized by receptors in coated pits. These receptors probably recognize some other protein during the normal functioning of the cell. The virus is internalized by receptor-mediated endocytosis. In the endosome, the decreased pH leads to the release of the nucleocapsid into the cytosol through fusion of the viral membrane envelope with the endosome membrane or with the membrane of the fused endosome-lysosome (see Figure 35-42 on p. 940 of Stryer). The viral RNA is exposed, translated, and replicated. The newly made viral proteins form a nucleocapsid that exits the cell by budding from the plasma membrane at a site where glycoproteins encoded by the virus have clustered.

Programmed Protein Destruction

29. a, b, c, e, f, g. Answer (d) is incorrect because it is the identity of the N-terminal amino acid of a protein that determines its tagging by ubiquitin and thus its destruction.

PROBLEMS

1. In the classic study of the events involved in the processing of digestive enzymes for export by the endocrine cells of the pancreas, Jamieson and Palade incubated slices of guinea pig pancreas in a medium containing ^3H-leucine for three minutes (the "pulse"). After this period, the slices were transferred to an incubation medium containing unlabeled leucine (the "chase"). At varying intervals after the transfer, the slices were prepared for autoradiography (see Stryer, p. 47 and p. 122) and electron microscopy.

   (a) Explain briefly what the experiment is designed to establish.
   (b) Predict the order in which "tracks" corresponding to silver halide exposed by β particles will appear in the following structures: (1) the cis side of the Golgi, (2) the RER, (3) the trans side of the Golgi, and (4) zymogen granules.

(c) Explain why the chase with unlabeled leucine is indispensible to this experiment.
(d) Identify the contents of the zymogen granules. What is their fate?

2. Trypsinlike enzymes are ubiquitous in cells and are frequently involved in important proteolytic processing events. Is it possible that signal peptidase is a trypsinlike enzyme? Explain by referring to Figure 35-3 on page 913 of Stryer.

3. In one process for the production of human insulin by recombinant DNA techniques, cDNA for insulin is ligated to *E. coli* DNA that carries information for the synthesis of β-lactamase. The recombinant DNA is then added to *E. coli* cells, where it directs the synthesis of a fusion protein containing amino acid sequences for both lactamase and insulin. Despite the fact that the hybrid protein must be cleaved to release the insulin amino acid sequences, this technique offers considerable advantages in the quantitative isolation and purification of the hybrid protein product. Explain why this might be the case.

4. Explain the roles of (a) the phosphate group and (b) the long lipid chain of dolichol phosphate in the transport of polysaccharides across membranes.

5. What feature of the prokaryotic plasma membrane allows proton-motive force, rather than the hydrolysis of high-energy phosphates, to power the translocation of proteins? Why can proton-motive force not serve as an energy source for translocation across the membranes of the ER in eukaryotes?

6. Suppose that glucose 1-phosphate labeled with ^{32}P is added to a cellular system designed to study the synthesis and processing of N-glycosylated proteins. When bacitracin is added to the system, a lipid-soluble intermediate labeled with ^{32}P accumulates. In the absence of bacitracin, the label appears in inorganic phosphate. Explain these results, and identify the lipid-soluble intermediate that accumulates. (Refer to Stryer, p. 920.)

7. What feature of the signal sequence likely prevents the proper identification of proteins targeted for the ER once significant protein folding has begun? Explain.

8. Suppose you are trying to prepare a chimeric protein for the study of the events involved in the transport of proteins into mitochondria. You have two types of cDNAs that you can use to specify the N terminus of the fusion protein, and you also have cDNAs for four proteins.

cDNAs for N terminus	*cDNAs for proteins*
(a) MLRTSSLFTRRVQPSL-FRNILRLQST	(1) Cytochrome c_1
	(2) Succinate dehydrogenase
(b) MRTNSNNVQPNQNRT-NQQLNVANTNT	(3) Glucose 6-phosphate dehydrogenase
	(4) Malate dehydrogenase

Of the eight possible combinations of *cDNAs,* which one is most likely to give satisfactory results? Explain. (Refer to p. 930 of Stryer.)

9. Proteins are targeted for destruction by the formation of peptide bonds between the carboxyl-terminal glycine of ubiquitin and the ε-amino group of lysine residues in the proteins destined for degradation. How many high-energy phosphates are required for the formation of this peptide bond. Compare this number with the number required for the formation of peptide bonds by ribosomes. (Refer to Figure 35-47 on p. 943 of Stryer.)

10. Electron micrographs of mature plasma cells active in the secretion of antibody molecules are strikingly different in appearance from those of immature undifferentiated lymphoid cells. What major structural feature is present in plasma cells, but absent from immature lymphoid cells? How does it relate to the function of plasma cells?

11. Many cell-surface and cytosolic membrane proteins are attached to the cellular membrane through covalent bonds to membrane lipids instead of membrane-spanning α helices. These extra- and intracellular protein classes are not interchangeable. Why not? Compare these two classes of proteins in terms of their pathway to the cell membrane and the amino acid sequence needed to target a protein to either location.

12. Of what possible significance is the fact that the cleavage site for signal peptidase is next to a helix-breaking residue?

13. One generally regards the inner membrane of the mitochondrion as the one controlling the entry rate of substances, and the outer membrane as allowing free permeability? For which class of compounds is this not the case? Explain.

14. Insulin is a polypeptide hormone that combines with a specific glycoprotein receptor on the surface of target cells. The insulin-receptor complexes then migrate to coated pits and are internalized by receptor-mediated endocytosis. The resulting vesicles may then fuse with lysosomes.

(a) Because insulin is a small protein (MW 5700), investigators wondered for years whether it might diffuse into cells and directly modify cellular processes or whether it must combine with a cell-surface receptor, causing the release of an intracellular second messenger that mediates its intracellular effects. Autoradiographic studies involving the use of radioactively labeled insulin appeared to indicate that it had entered cells. Does this weaken the notion that insulin requires a second messenger? Explain.

(b) Divalent antibodies toward insulin receptors can be prepared. These are antibodies that combine with two molecules of insulin receptors. When a fluorescent group is covalently attached to the antibody and the fluorescently labeled antibodies are employed in an attempt to map the location of insulin receptors on the cell surface, it is found that insulin must be added to the cells as well. Why do you think this is the case?

(c) For years it has been known that liver tissue and kidney tissue have the capacity to remove circulating insulin from the blood and degrade it. Outline the steps that you think are probably involved in this process.

15. The targeting sequences for some proteins are quite short, for example, KDEL for endoplasmic reticular proteins and SKF for peroxisomal proteins. If a typical cell contains 10,000 different proteins, how many of them would be guided to the endoplasmic reticulum or peroxisome if these short sequences were generated randomly? How many DNA bases in a restriction site would be needed to provide the same specificity?

16. Figure 14-31 on page 379 of Stryer depicts the assembly of a class I MHC protein with an intracellular peptide before presenting the peptide to the cell surface. After reading this chapter you should be able to fill in some of the details at each step. For example: which complex of proteins degrades the cellular protein to form the peptides that will be associated with the MHC I molecules? Where do these peptides associate with the MHC I protein, and what problem exists in delivering the peptides to the MHC I protein? What molecule delays the folding of the MHC I molecule until it has become associated with the cellular peptide?

ANSWERS TO PROBLEMS

1. (a) The experiment is designed to allow a temporal sequence of processing events to be inferred.

   (b) The tracks indicate the locations of the labeled protein at the time the tissue was fixed and prepared for microscopy. It was found that label appeared in the RER after 3 minutes, in the cis side of the Golgi after 7 minutes, in the trans side of the Golgi after 37 minutes, and in zymogen granules after 117 minutes.

   (c) If unlabeled leucine were not added, one could not establish a temporal sequence with certainty. A temporal sequence can be inferred only when the label disappears from the precursor as it appears in the product. If labeled leucine were left in the incubation medium, the label would appear first in the RER and later in the Golgi. However, one could not be sure whether it appeared later in the Golgi because it was processed first by the RER and then entered the Golgi or because it entered the Golgi directly by a slower process.

   (d) The contents of the zymogen granules are the major zymogens that are manufactured by the pancreas, including trypsinogen, chymotrypsinogen, procarboxypeptidase, and prolipase, to name just a few. They are released by the pancreatic exocrine cells into the pancreatic duct and travel to the small intestine, where they are activated by proteolytic cleavage.

2. Trypsin cleaves peptides on the carboxyl side of Arg and Lys residues. None of the cleavage sites depicted in Figure 35-3 satisfies these criteria. It is therefore reasonable to conclude that signal peptidase is not a trypsinlike enzyme.

3. β-Lactamase is a protein that is ordinarily exported by *E. coli* into the periplasmic space, the space between the inner and outer membranes. The fusion protein contains the amino acid leader sequence that directs β-lactamase to the periplasmic space, where the fusion protein can be conveniently isolated by techniques that rupture the outer but not the inner membrane of the cell.

4. (a) The phosphate group serves as the site for the covalent attachment of sugar residues to the carrier.

(b) The long lipid chain renders the carrier highly hydrophobic and thus membrane-permeable.

5. The assemblies that carry out electron transport are part of the plasma membrane of prokaryotes. Accordingly, a proton gradient is established across them. In eukaryotes the corresponding assemblies are found on the inner membrane of the mitochondrion and not on the membranes of the ER.

6. The lipid-soluble intermediate that accumulates is dolichol pyrophosphate, whose terminal phosphate comes from glucose 1-phosphate. (See Figure 35-12 on p. 920 of Stryer.) Bacitracin is an antibiotic that forms a 1:1 complex with dolichol pyrophosphate, preventing its hydrolysis to dolichol phosphate and inorganic phosphate. Thus, in the presence of bacitracin, the label will remain in dolichol pyrophosphate. In the absence of bacitracin, the terminal phosphate will be released as inorganic phosphate.

7. The signal sequence is rich in hydrophobic amino acids. Once folding has begun, these residues become buried on the inside of the protein structure, where they are stabilized by interactions among themselves and are no longer free to interact with SRP and thus be directed to the ER.

8. The cDNA that specifies amino acid sequence (a) for the N terminus must be used. The transport of proteins into mitochondria requires this particular mitochondrial entry sequence, which is rich in threonine and serine. If our chimeric protein is to be used to demonstrate that the entry sequence is specific for the translocation of proteins into mitochondria, the balance of the protein must be derived from a protein that is not ordinarily present in mitochondria and is readily identifiable. Of the proteins on the list, only glucose 6-phosphate dehydrogenase is not ordinarily found in mitochondria. It has the additional advantage of being readily assayed.

9. Two high-energy phosphates are expended in the attachment of ubiquitin to lysine. An ATP is cleaved to AMP and pyrophosphate, and the pyrophosphate is subsequently cleaved to two molecules of P_i. A total of four high-energy phosphates are required in protein synthesis. (See the answer to problem 5, p. 577.)

10. There is great proliferation of the rough endoplasmic reticulum in plasma cells, as would be expected in a cell specialized for the export of protein. A similar picture is seen in other protein-exporting cells, like the exocrine cells of the pancreas, which are active in the synthesis of digestive enzymes.

11. Interchanging these two classes of proteins would require flipping a lipid molecule along with its associated protein from one side of the cellular membrane to the other. This is energetically highly unfavorable. Experiments have shown that a lipid molecule alone can flip once every few hours. Adding a protein molecule to the process would slow it down considerably. In fact, the flipping of a protein from from one side of the membrane to the other has not been observed. The cytosolic proteins are synthesized on free ribosomes, while the extracellular proteins are synthesized on the rough endoplasmic reticulum. Both classes of proteins use carboxyl-terminal sequences to guide the covalent attachment of a lipid. The extracellular protein must have, in addition, an amino-terminal sequence to direct the protein to the lumen of the endoplasmic reticulum.

12. Enzymes that hydrolyze peptide bonds usually require that the peptide group be in the fully extended, planar conformation to fit into the active site of the protease.

13. Mitochondrial proteins synthesized in the cytosol on free ribosomes must interact with a receptor site on the outer membrane in order to be transported into the mitochondrion.

14. (a) No, it does not weaken the notion that a second messenger is required. Labeled insulin would be internalized by receptor-mediated endocytosis, but that insulin would likely be destroyed by the hydrolytic enzymes of lysosomes. It is important to realize that the interior of internalized vesicles is topologically equivalent to the exterior of the cell, so a case can be made for asserting that the labeled insulin really did not enter the cell at all.

(b) When insulin combines with its receptor, the insulin-receptor complexes aggregate in coated pits. Such aggregates will combine with divalent antibody. Without the presence of insulin, the receptors remain dispersed on the surface of the cell.

(c) The insulin combines with receptors on the surface of the liver or kidney cells. The insulin-receptor complexes are then internalized by receptor-mediated endocytosis, and the resulting vesicles fuse with lysosomes, leading to the hydrolysis of the vesicle contents.

15. If the targeting sequences were restricted to the last three or four carboxyl-terminal amino acids, as they appear to be, then only 1/[(20)(20)(20)(20)], or 1/160,000, would contain KDEL and 1/8,000 would contain SKF. Therefore, on average, only one protein would be guided to the endoplasmic reticulum or the peroxisome. These sequences, though short, are long enough to provide the needed specificity. The number of DNA base pairs needed to provide the equivalent specificity would be between eight and nine for KDEL and 6.5 for SKF.

16. The 26S proteosome degrades cellular proteins where they are transported into the lumen of the endoplasmic reticulum by a specific pump. It is inside the endoplasmic reticulum that these peptides are joined to the MHC I molecule, which is waiting for them and is prevented from being folded by the chaperone on calnexin. You could have guessed the BiP protein would assist with this process, but it is reserved for assisting the folding of immunoglobulin molecules.

1. (a) Cleavable amino-terminal signal sequences are usually 13 to 36 residues long and contain a highly hydrophobic central region 10 to 15 residues long. The amino-terminal part has at least one basic residue. The cleavage site is preceded by small neutral residues.

   (b) Prokaryotic signal sequences are similar to eukaryotic ones. In addition, a stop-transfer sequence is needed to keep the protein in the plasma membrane.

   (c) Mannose 6-phosphate residues direct proteins to lysosomes.

   (d) Integral membrane proteins initially in the ER go to the plasma membrane unless they carry instructions to the contrary. No specific signal is needed.

   (e) An amino-terminal sequence containing positively charged residues, serine, and threonine, in addition to hydrophobic residues.

   (f) An amino-terminal arginine, histidine, isoleucine, leucine, lysine, phenylalanine, or tryptophan.

2. The cytosolic portion of the receptor, which enables it to interact with coated pits, is likely to be altered.

3. (a) Endocytosis mediated by a receptor specific for mannose 6-phosphate residues. Endocytic vesicles containing the added lysosomal enzymes then fuse with lysosomes.

   (b) The addition of mannose 6-phosphate to the extracellular medium should prevent lysosomal enzymes from reaching their destination if the normal pathway involves secretion outside the cell and import back into the cell. However, mannose 6-phosphate does not inhibit normal lysosomal targeting, showing that lysosomal enzymes reach their destination without leaving the cell.

4. (a) The chimeric protein will probably be found in the cytosol. The transmembrane sequence of a membrane-bound immunoglobulin functions as a stop-transfer sequence and not as a signal sequence.

   (b) The chimeric protein will probably be found in the plasma membrane because chymotrypsinogen is synthesized with a signal sequence.

5. Secretory proteins that are erroneously targeted to the cytosol will be rapidly degraded because their amino termini mark them for rapid destruction (Table 35-3, on p. 943 in Stryer). Rapid elimination of mistargeted secretory proteins is important because some of them (e.g., trypsinogen) could wreak havoc in the cytosol.

6. This genetic disorder, known as *Zellweger syndrome,* is caused by defective import of peroxisomal matrix proteins bearing SKL sequences. The mutation is probably in the gene for the SKL receptor or the associated

translocase. See P.B. Lazarow and H.W. Moser in C.R. Scriver, A.L. Beaudet, W.S. Sly, and D. Vale (eds.), *The Metabolic Basis of Inherited Disease,* 6th ed. (McGraw-Hill, 1989), pp. 1479–1509.

7. Methotrexate binds very tightly to DHFR (see Stryer, p. 753) and prevents it from unfolding. Recall that proteins must be partly or completely unfolded during their translocation across the inner mitochondrial membrane as well as other membranes.

8. Mutant ras proteins are oncogenic only if they are membrane-anchored. Farnesylation of ras is essential for membrane anchoring. Hence, specific inhibitors of farnesyl transferase are potentially valuable anticancer agents. See N.E. Kohl, S.D. Mosser, S.J. deSolms, E.A. Giuliani, D.L. Pompliano, S.L. Graham, R.L. Smith, E.M. Scolnick, A. Oliff, and J.B. Gibbs. *Science* 260(1993):1934.

CHAPTER

36

Control of Gene Expression in Prokaryotes

In this chapter, Stryer describes the biochemistry of several mechanisms that control prokaryotic gene expression. The lactose (*lac*) operon is described in detail as an example of a mechanism in which the initiation of transcription is regulated. Negative control is exerted through the binding of a repressor protein to DNA carrying the *lac* operator. Positive control of the *lac* operon is accomplished by a complex of CAP protein and cyclic AMP (cAMP). The complex binds near the same promoter in the absence of the repressor and stimulates the activity of RNA polymerase at the *lac* promoter. The arabinose operon, another operon that encodes the enzymes of a catabolic pathway, illustrates how a single protein can serve as both a negative and a positive regulator, depending on the levels of arabinose and cAMP in the cell. In addition, this operon exemplifies autoregulation.

The regulation of bacteriophage λ in its alternate life cycles is discussed because control of both the initiation and the termination of transcription plays an important role in determining which of the phage genes are expressed and, hence, the fate of the phage. The λ repressor provides an example of the helix-turn-helix structural motif, which several other DNA-binding proteins also use to attain specific association with their target sequences.

Another example of the regulation of gene expression at the level of transcription termination is provided by the attenuation phenomenon in the *trp* operon. This mechanism, which is used by several biosynthetic amino acid operons, depends on the coupling of transcription and translation in prokaryotes. Translational repression, in which proteins bind to the mRNAs that encode them and thus prevent protein synthesis, is exemplified by some ribosomal proteins. Stryer also describes the stringent response, in which the synthesis of ribosomal and transfer RNA is coordinated with protein synthesis. The chapter closes with a description of a system in which site-specific recombination acts as a switch to juxtapose a promoter and a sequence that encodes a repressor protein. This mechanism provides an absolute and reversible switch between two alternative DNA sequences that give rise to two distinct structural proteins in *Salmonella.*

A review of mRNA initiation and termination and protein synthesis initiation on pages 842–844, 847–849, and 892–895 in Stryer will help you to understand the control mechanisms described in this chapter. Reread pages 826–828 on F and F′ factors as background for the genetic experiments on the *lac* operon. Also, read about the life cycles of bacteriophage λ on pages 128–129 and the formation, hydrolysis, and actions of cAMP on pages 245–246 and pages 340–341. Material on page 789 of Stryer describes some of the functional groups of DNA that can serve as determinants for specific interactions with proteins and will aid you in understanding how repressors interact with DNA. Pages 792–793 describe the binding of EcoRV endonuclease to DNA and will enable you to contrast its mode of binding with that used by the repressors described in this chapter. Finally, reread pages 829–831 on transposons as an aid to understanding the section on DNA inversion in *Salmonella.*

LEARNING OBJECTIVES

When you have mastered this chapter, you should be able to complete the following objectives:

The Lactose Operon and Catabolite Repression (Stryer pages 949–957)

1. List the mechanisms bacteria use to *regulate* their *patterns of metabolism.*

2. Specify the primary mechanism for the *regulation of gene expression* in bacteria.

3. Outline the metabolism of lactose in *E. coli.* Draw the structure of *lactose,* describe its entry into the cell, and write the equations for the reactions catalyzed by *β-galactosidase,* naming both the substrates used and the products formed.

4. Recount the genetic experiments that led to the concept of the *lac operon* and its regulation.

5. Draw the *genetic map* of the *lac operon* and outline the functions of the *promoter, repressor, operator,* and *inducer* in controlling the production of *polycistronic* lac *mRNA.*

6. Describe the role of the *lac* repressor in determining whether a bacterium is *inducible* or *constitutive for β-galactosidase.*

7. Describe the *subunit structure* of the *lac* repressor and relate it to the *symmetrical sequence* of the *lac* operator. Compare the affinities of the *lac* repressor for operator DNA and nonspecific DNA. Describe the effect of the binding of *allolactose* by the repressor on its affinity for the operator.

8. Explain the functions of *cyclic AMP* (cAMP) and the *catabolite activator protein* (CAP) in modulating the expression of the *lac* operon. Contrast the functions of cAMP in bacterial and mammalian cells.

9. Diagram the relative positions of cAMP–CAP complex, RNA polymerase, and *lac* repressor on the DNA template. Explain their effects on one another and on the DNA structure.

10. Provide an overview of the *arabinose* (*ara*) *operon* and its expression. Explain the alternative functions of the *araC protein.*

11. List four general principles of gene regulation exemplified by the *ara* operon.

Bacteriophage λ—The Lytic-Lysogenic Switch (Stryer pages 957–964)

12. Outline the alternative fates of *bacteriophage λ* after it has infected *E. coli.* List the stages of the *lytic pathway.*

13. Outline the roles of *antitermination factors N* and *Q* as *positive regulatory proteins* during lytic development.

14. List the three stages of the *lysogenic cycle* and name the only gene expressed in the *prophage* and its product.

15. Distinguish between the patterns of gene expression during the lytic and lysogenic states.

16. Describe the structure of the λ *repressor.* Outline the organization and salient sequence features of the operators to which it binds.

17. Describe the interactions of the λ repressor with its three target sites in the *left* and *right operators* (O_L and O_R). Explain the *autoregulatory mechanism* of the repressor.

18. Explain how λ is released from lysogeny, focusing on the role of the *recA protein* in decreasing the concentration of λ repressor. Describe the function of *Cro protein* in depressing λ repressor synthesis.

19. Describe the *helix-turn-helix motif* of some DNA-binding proteins. Note the existence of other DNA-binding motifs.

The Tryptophan Operon and Transcription Termination (Stryer pages 964–968)

20. Provide an overview of the regulation of the *tryptophan* (*trp*) *operon.* Describe the structure of the *trp repressor* and its activation by the *corepressor* tryptophan.

21. Sketch the secondary structures of the *trp attenuator* (3:4 hairpin) and *antiterminator* (2:3 hairpin) and relate them to control of the *trp* operon.

22. Explain the function of *leader peptides* and the consequences of the *coupling of transcription and translation* in the regulation of several biosynthetic operons.

Translational Repression and Regulation by DNA Inversion (Stryer pages 968–970)

23. Describe the mechanism of *translational repression* in the regulation of ribosomal protein synthesis.

24. Explain the role of the *stringent response* in the coordination of the synthesis of ribosomal and transfer RNA with the synthesis of nonribosomal proteins.

25. Outline *phase variation* in *Salmonella.*

26. Describe the role of *site-specific recombination* in determining which flagellin appears on the surface of the bacterium.

SELF-TEST

The Lactose Operon and Catabolite Repression

1. Which of the following are common mechanisms used by bacteria to regulate their metabolic pathways?

(a) Control of the expression of genes
(b) Control of enzyme activities through allosteric activators and inhibitors
(c) Formation of altered enzymes by the alternative splicing of mRNAs
(d) Deletion and elimination of genes that specify enzymes
(e) Control of enzyme activities through covalent modifications

2. Which of the following statements about β-galactosidase in *E. coli* are correct?

(a) It is present in varying concentrations depending on the carbon source used for growth.
(b) It is a product of a unit of gene expression called an *operon.*
(c) It hydrolyzes the β-1,4-linked disaccharide lactose to produce galactose and glucose.
(d) It forms the β-1,6-linked disaccharide allolactose.
(e) It is allosterically activated by the nonmetabolizable compound isopropylthiogalactoside (IPTG).
(f) Its levels rise coordinately with those of galactoside permease and thiogalactoside transacetylase.

3. Match each feature or function in the right column with the appropriate DNA sequence element of the *lac* operon in the left column.

(a) *i* ____________

(b) *p* ____________

(c) *o* ____________

(d) *z* ____________

(e) *y* and *a* ____________

(f) CAP binding site ____________

(1) Contains a specific binding sequence for the *lac* repressor
(2) Encodes a galactoside permease
(3) Contains a binding sequence for the cAMP–CAP complex
(4) Encodes a protein that interferes with the activation of RNA polymerase
(5) Encodes a protein that binds allolactose
(6) Contains a specific binding sequence for RNA polymerase
(7) Encodes β-galactosidase
(8) Is a regulatory gene
(9) Is the *lac* promoter
(10) Encodes thiogalactoside transacetylase
(11) Is the *lac* operator
(12) Encodes the *lac* repressor

4. Explain the stoichiometric relationship between the concentrations of β-galactosidase, galactoside permease, and thiogalactoside transacetylase in *E. coli.*

____

____

5. When *E. coli* is added to a culture containing both lactose and glucose, which of the sugars is metabolized preferentially? What is the mechanism underlying this selectivity?

____

____

6. What happens when the first sugar is depleted during the experiment described in question 5?

____

____

7. Which of the following statements about the cAMP–CAP complex are correct?

(a) It protects the −87 to −49 sequence of the *lac* operon from nuclease digestion.
(b) It protects the −48 to +5 sequence of the *lac* operon from nuclease digestion.
(c) It protects the −3 to +21 sequence of the *lac* operon from nuclease digestion.

(d) It bends DNA in the direction opposite from RNA polymerase.
(e) Upon binding to the *lac* operon, it creates an additional interaction site for RNA polymerase.

8. Which of the following statements about the C protein encoded by the *ara* operon are correct?

(a) It binds specifically to DNA only when it is complexed with arabinose.
(b) It binds to an operator (O_1) and shuts off its own synthesis.
(c) It binds to an operator (O_2) and shuts off the expression of the structural genes of the operon.
(d) When complexed to arabinose and when cAMP is present, it binds to O_1 and to a region of DNA (*ara I*) adjacent to the promoter for the structural genes.
(e) It brings noncontiguous segments of DNA into proximity by the simultaneous binding of DNA sequences at different sites.

9. It is postulated that the cAMP–CAP complex stimulates the *lac* operon because it creates an additional interaction site for RNA polymerase when it occupies the CAP binding site on the DNA. This site is directly adjacent to the *lac* promoter. Speculate about how the cAMP–CAP complex can accomplish the same positive activating role in the *ara* operon when its binding site on the DNA is separated from the *araBAD* RNA polymerase binding site by the *araI* sequence?

____

____

Bacteriophage λ—The Lytic-Lysogenic Switch

10. Which of the following statements about bacteriophage λ are correct?

(a) It synthesizes only the λ repressor in a lysogenic bacterium.
(b) It escapes from the bacterial chromosome when the λ repressor interacts with single-stranded DNA produced by agents that damage the DNA.
(c) It exists as a prophage during the lysogenic phase of its life cycle.
(d) It maintains its lysogenic state, in the absence of an inducer, because the synthesis of the λ repressor is autoregulated.
(e) In switching from the lysogenic to lytic phase of its life cycle, it turns off the synthesis of the λ repressor because Cro protein binds to the λ operator O_R3.

11. Which of the following statements about bacteriophage λ during lytic development are correct?

(a) It synthesizes three different classes of mRNA, which are designated by the times after infection they appear.
(b) It carries out a programmed synthesis of the proteins necessary to replicate its genome and produce its structural components.
(c) It forms the N and Q gene products, which act as positive regulatory proteins, leading to the sequential production of λ-encoded proteins.
(d) It initially produces two proteins; one acts as an inhibitor of λ repressor synthesis, and the other acts as a transcription termination factor.

12. Which of the following statements about the λ repressor are correct?

(a) It binds to DNA as a monomer.
(b) The monomer contains an N-terminal domain, which interacts specifically with DNA, and a C-terminal domain, which contains the allolactose binding site.
(c) It binds specifically to some of the same sites on DNA as does the Cro protein.
(d) It binds specifically to O_R and O_L operators with a stoichiometry of up to six polypeptide chains per operator.
(e) It shows different affinities to subsites within O_R and O_L.
(f) It binds to a region of DNA that is 17 base pairs long and is imperfectly symmetrical.
(g) It is inactivated through the hydrolysis of one of its peptide bonds by recA.

13. Explain how the λ repressor regulates its own synthesis in a lysogenic cell.

____

____

14. Which of the following statements about the helix-turn-helix motif are correct?

(a) It is an arrangement of secondary-structure elements in some proteins that bind specifically to DNA.
(b) It occurs in two symmetrically disposed arrays in dimeric proteins to facilitate interaction with DNA sequences of inverted symmetry.
(c) It positions the positively charged N terminus of its recognition α helix into the major groove of the DNA to promote electrostatic interactions with the negatively charged phosphodiester backbone of the DNA.
(d) It is arranged in dimers so that two nearly parallel α helices can be positioned 34 Å apart.
(e) It is found in the λ and tryptophan repressors, the Cro protein, and the *met* repressor structures.
(f) It contains a recognition helix of invariant amino acid composition.

15. Describe the sequence of events that follows damage to DNA by uv radiation or chemical agents, resulting in the switch from the lysogenic to the lytic state of λ.

____

____

The Tryptophan Operon and Transcription Termination

16. Regulation of the *trp* operon involves which of the following?

(a) Controlling the amount of polycistronic mRNA formed at the level of transcription initiation
(b) Controlling the amount of polycistronic mRNA at the level of transcription termination

(c) The sequential and coordinate production of five enzymes of tryptophan metabolism from a single mRNA
(d) The sequential and coordinate production of five enzymes of tryptophan metabolism from five different mRNAs produced in equal concentrations
(e) The production of transcripts of different sizes, depending on the level of tryptophan in the cell

17. With regard to the *trp* operon, why is tryptophan called a *corepressor*?

____

____

18. Which of the following statements about the structure of the *trp* repressor are correct?

(a) It has a β sheet DNA-binding motif.
(b) Binding of Trp to the repressor decreases the distance between the reading heads to 26 Å.
(c) Binding of Trp to the repressor increases the distance between the reading heads to 34 Å.
(d) A Trp molecule binds to each of the reading heads.
(e) Water molecules participate in the interaction between DNA and the repressor.

19. Which of the following statements concerning the *trp* operon leader RNA, which has 162 nucleotides preceding the initiation codon of the first structural gene of the operon, are correct?

(a) A deletion mutation in the DNA encoding the 3′ region of the leader RNA gives rise to increased levels of the biosynthetic enzymes forming Trp.
(b) A short open reading frame, containing Trp codons among others, exists within the leader RNA.
(c) The leader RNA encodes a "test" peptide whose capacity to be synthesized monitors the level of Trp-tRNA in the cell.
(d) The leader RNA contains one Shine-Dalgarno ribosome-binding site.
(e) The leader RNA may form two alternative and mutually exclusive secondary structures.
(f) The structure of the leader RNA in vivo depends on the position of the ribosomes translating it.

Translational Repression and Regulation by DNA Inversion

20. Explain how the synthesis of ribosomal RNA and the synthesis of nonribosomal proteins are coordinated.

____

____

21. Describe how the synthesis of ribosomal proteins is coordinated.

____

____

22. Match each feature or requirement in the right column with the appropriate regulated element or mechanism of regulation of gene expression listed in the left column.

(a) *lac* repressor ______________

(b) λ repressor ______________

(c) *Trp* repressor ______________

(d) cAMP–CAP complex ______________

(e) *AraC* protein ______________

(f) N and Q proteins of λ ______________

(g) *Trp* operon attenuation ______________

(h) Ribosomal proteins ______________

(i) Ribosomal and transfer RNAs ______________

(j) *hin* recombinase ______________

(1) Requires concomitant protein synthesis to function
(2) Controlled by translational repression
(3) Inverts the orientation of a promoter with respect to a repressor gene
(4) Interferes with the functioning of RNA polymerase by binding to DNA
(5) Stimulates transcription of several catabolic operons
(6) Binds to DNA sequences with regions of twofold symmetry
(7) Is inactivated by recA protein
(8) Regulatory effects are antagonized by Cro protein
(9) Breaks and rejoins DNA phosphodiester bonds
(10) Depends on the tight coupling of translation and transcription
(11) Regulates its own synthesis
(12) Involves ribosomes
(13) Stimulates transcription by suppressing transcription termination
(14) Concentration depends on glucose levels
(15) Stimulates transcription initiation
(16) Interacts with repeated sequence variants within its operators
(17) Acts as both a positive and a negative regulator
(18) Participates in global regulation
(19) Is affected by the level of aminoacyl-tRNAs in the cell
(20) Involves ppGpp

23. Which of the following statements about site-specific recombination are correct?

(a) When it occurs between two repeated sequences that are inverted with respect to each other on the same DNA molecule, it leads to the deletion of the DNA between the sequences.

(b) It can move a promoter for a gene such that it is "aimed" away from the DNA sequences that encode the mRNA it ordinarily produces.
(c) In the *Salmonella* phase-variation phenomenon, it disrupts an operon such that the repressor it made before the recombination is no longer produced.
(d) In the *Salmonella* phase-variation phenomenon, it selects genes for transcription by two mechanisms—operator-repressor interaction and operon disruption.
(e) It provides an absolute switch that turns on the synthesis of one flagellin or another at any given time in *Salmonella.*

ANSWERS TO SELF-TEST

The Lactose Operon and Catabolite Repression

1. a, b, e. Answer (c) is incorrect because the splicing of mRNA is not common in bacteria.

2. a, b, c, d, f. Although IPTG is an inducer for the synthesis of β-galactosidase, it is neither a substrate nor an allosteric activator of the enzyme, so answer (e) is incorrect.

3. (a) 4, 5, 8, 12 (b) 6, 9 (c) 1, 11 (d) 5, 7 (e) 2, 10 (f) 3. For (d), 5 is correct because allolactose is the product of a reaction catalyzed by β-galactosidase and, as a product, it binds to the enzyme.

4. The genes encoding these three enzymes are transcribed as a polycistronic mRNA, so there are equal numbers of copies of the mRNA sequences specifying each of the enzymes.

5. Glucose is metabolized preferentially because it results in a decrease in the synthesis of cAMP by adenylate cyclase. The lack of cAMP prevents the formation of the cAMP–CAP complex, which is necessary for the efficient transcription of the *lac* operon and other catabolite-repressible operons.

6. When glucose is depleted, the concentration of cAMP rises. The cAMP–CAP complex forms and binds to the CAP binding site just upstream of the RNA polymerase binding site in the *lac* promoter. At the same time, some lactose has entered the cell, has been converted to allolactose by β-galactosidase, and is bound by the *lac* repressor so that it no longer binds to the *lac* operator. RNA polymerase now binds to the *lac* promoter even more effectively because of the presence of the cAMP–CAP complex. The enzymes and permease of the *lac* operon are expressed fully; consequently, lactose readily enters the cell and is efficiently metabolized.

7. a, e. Answers (b) and (c) are incorrect because they correspond to the binding sequences for RNA polymerase and *lac* repressor, respectively. Answer (d) is incorrect because the cAMP–CAP complex promotes RNA polymerase interaction by bending DNA in the same direction.

8. b, c, d, e. Answer (a) is incorrect because the *araC* protein (C protein) binds to specific sequences of DNA both when by itself and when complexed to arabinose—but with different outcomes. The binding of ara-

binose converts the *araC* protein from a repressor to an activator of transcription.

9. The *araC* protein complex with arabinose binds to *araI* and mediates the action of the cAMP–CAP complex. The binding of the cAMP–CAP complex upstream of the binding site for the *araC* protein-arabinose complex affects the way the *araC* protein interacts with the adjacent RNA polymerase so that the transcription is more efficient.

Bacteriophage λ—The Lytic-Lysogenic Switch

10. a, c, d, e. Answer (b) is incorrect because the prophage is excised from the lysogen by a series of events that are initiated by the proteolytic cleavage of the λ repressor by the recA protein. The activation of the proteolytic activity of recA results from the production of single-stranded DNA in the cell, often as a result of agents that damage DNA. Both Cro and λ repressors bind to O_R3 but with different affinities; Cro binds more tightly and, consequently, blocks repressor synthesis. (See Figures 36-23 and 36-24 in Stryer.)

11. a, b, c. Answer (d) is incorrect because, in addition to the Cro protein, which suppresses λ repressor synthesis, the N protein is an *anti*terminator, not a terminator.

12. c, d, e, f, g. Answers (a) and (b) are incorrect because the λ repressor binds to DNA as a dimer, with the C-terminal domains of the monomers interacting with each other to maintain the dimer. Each dimer interacts with a partially symmetrical sequence that is repeated three times in both O_R and O_L. Because the sequence of each of the subsites is different, the repressor displays different affinities for each.

13. Alternative interactions between the λ repressor and the subsites within O_R can either stimulate or repress the activity of RNA polymerase at the promoter for λ repressor synthesis during lysogeny and thus autoregulate its synthesis. At lower levels of repressor, the distal O_R1 and O_R2 subsites are filled, and because they are adjacent to the promoter, they interact with RNA polymerase, stimulating it to transcribe the repressor gene. Conversely, at higher repressor concentrations, the proximal O_R3 subsite is also occupied by the repressor, and RNA polymerase is then precluded from binding to the promoter.

14. a, b, d. Answer (c) is incorrect because the recognition helix fits sideways, not end-on, into the major grove of DNA. Answer (e) is incorrect because the *met* repressor interacts with DNA via a β sheet motif. Answer (f) is incorrect because the sequences of the recognition helices of those proteins that utilize the helix-turn-helix motif vary, specifying the different sequences they recognize.

15. The main event in the switch from lysogeny to the lytic state of λ is destruction of the λ repressor. RecA bound to single-stranded DNA activates the latent protease activity in λ repressor. An activated λ repressor then digests adjacent λ repressor molecules and allows λ to enter the lytic state.

The Tryptophan Operon and Transcription Termination

16. a, b, c, e. When the RNA is not terminated at the attenuator, a single polycistronic mRNA, which encodes five enzymes, is produced, so answer (d) is incorrect.

17. The *trp* repressor does not bind DNA specifically unless it is complexed to Trp. The binding of Trp to the repressor is necessary to induce a conformational change that forms a specific DNA-binding surface on the protein.

18. c, e. The *trp* repressor shares the helix-turn-helix motif with many other DNA-binding proteins. Binding of Trp acts as a wedge to move the reading heads apart to a distance of 34 Å. The reading heads can then bind to the DNA major groove. Specifically bound water molecules participate in the interaction of *trp* repressor with DNA.

19. a, b, c, e, f. The discovery of a deletion mutation in front of the first structural gene of the operon, and not in the operator, was the first clue that a control mechanism operating at the level of transcription termination was involved in regulating the expression of the *trp* biosynthetic enzymes. This deletion altered potential mRNA structures so that the rho-independent transcription-termination structure could no longer form. Answer (d) is incorrect because the leader RNA contains two Shine-Dalgarno ribosome-binding sites: one in front of the initiation codon of the leader peptide and one preceding the first structural gene of the operon.

Translational Repression and Regulation by DNA Inversion

20. Ribosomes are synthesized rapidly during periods of growth when nutrients are in good supply. If amino acids become limiting in the cell, tRNAs cannot be aminoacylated, and uncharged tRNAs bind to the A site of the ribosomes. This event, coupled with the presence of stringent factor, gives rise to the synthesis of guanosine tetraphosphate (ppGpp), which inhibits the synthesis of rRNA at the level of transcription initiation.

21. If the synthesis of ribosomal proteins exceeds that of rRNAs, the excess of particular ribosomal proteins binds to the mRNAs that encode them and the other ribosomal proteins in the same operons. The repression continues until enough rRNAs, which bind ribosomal proteins more tightly than the mRNAs, are available to form ribosomes.

22. (a) 4, 6 (b) 4, 6, 7, 8, 11, 15, 16, 17. The λ repressor acts as both a repressor and a stimulator of its own transcription, depending on its concentration in the cell. In acting as both a positive and a negative regulator, it is like the *araC* protein. (c) 4, 6 (d) 5, 6, 14, 15, 18 (e) 4, 11, 15, 17 (f) 13 (g) 1, 10, 12, 19 (h) 2, 12, 19 (i) 12, 19, 20. The ribosomes are necessary for the formation of ppGpp. (j) 3, 6, 9. The *hin* recombinase binds and reacts with inverted repeats (insertion sequences) to flip the orientation of the intervening DNA (see Stryer, pp. 830–831).

23. b, c, d, e. Answer (a) is incorrect because site-specific recombination between inverted repeats on the same molecule leads to the inversion of the DNA between them (see Stryer, p. 830). In the *Salmonella* phase-variation phenomenon, the inversion occurs in a piece of DNA containing the promoter for a polycistronic mRNA that encodes one type of flagellin and a repressor for a gene that encodes a second type of flagellin. Inverting the DNA disrupts the expression of this operon and, because of the absence of the repressor it encodes, permits the expression of a new mRNA that encodes the second type of flagellin molecule.

1. What property of enzymes makes them more suitable than, say, structural proteins for studies of the genetic regulation of protein synthesis? Explain.

2. When lactose is used as an inducer, there is a lag before the enzymes of the lactose operon are synthesized. With IPTG there is no lag. Explain this observation. (See Figure 36-1.)

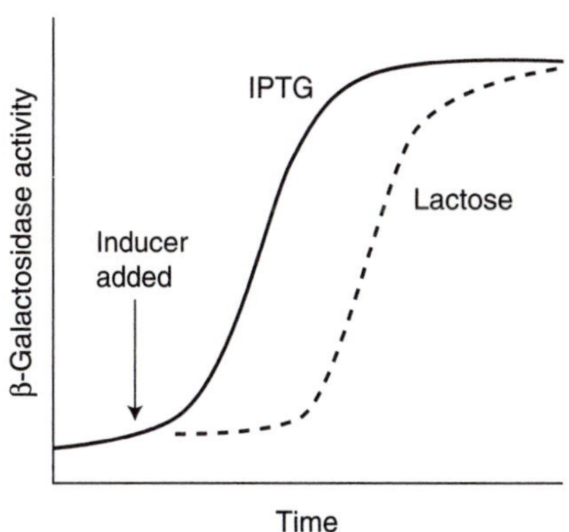

Figure 36-1
Kinetics of β-galactosidase induction by lactose and IPTG. (Assume that each inducer has been removed after an appropriate period.)

3. Some of the known constitutive mutations of the lactose operon occur in the operator sequence rather than the regulator gene.

(a) Would you expect such an o^c mutant to be dominant or recessive to its wild-type o^+ allele? Explain.
(b) Is a constitutive mutation in an operator cis or trans in its effects? Explain.
(c) Design an experiment involving the genes i^+, o^c, o^+, and z^+ that would confirm your answer to part (b). Assume that it is possible to detect whether enzymes are produced in diploid (++) or haploid (+) amounts.

4. Since the permease required for the entry of lactose into *E. coli* cells is itself a product of the lactose operon, how can the first lactose molecules enter uninduced cells? Explain.

5. Design an experiment to show that lactose stimulates the synthesis of new enzyme molecules in *E. coli* rather than fostering the activation of preexisting enzyme molecules.

6. The three known enzymes of the lactose operon in *E. coli* are not produced in equimolar amounts following induction. Rather, more galactosidase than permease is produced, and more permease than transacetylase is produced. Propose a mechanism to account for this that is consistent with known facts about the lactose operon.

7. Assume that the following allelic possibilities exist for the *i* genes and *o* sequence of the lactose operon of *E. coli*:

i^+ = wild-type regulator gene

i^c = regulator constitutive mutation, makes inactive repressor

i^s = repressor substance insensitive to inducer

o^+ = wild-type operator

o^c = operator consitutive mutation

In addition, assume that the mutations z^-, y^-, and a^- lead to nonfunctional enzymes Z, Y, and A, respectively.

For each of the following, predict whether active enzymes Z, Y, and A will or will not be produced. For partially diploid cells, assume semidominance;

that is, the enzyme activity in a diploid cell will be twice that found in a haploid cell. Use the following answer code:

0 = active enzyme absent
+ = active enzyme present in haploid amounts
++ = active enzyme present in diploid amounts

	Without IPTG			With IPTG		
	Z	Y	A	Z	Y	A
(a) $i^+o^+z^+y^+a^+$	____	____	____	____	____	____
(b) $i^co^+z^+y^+a^+$	____	____	____	____	____	____
(c) $i^so^+z^+y^+a^+$	____	____	____	____	____	____
(d) $i^+o^cz^+y^+a^+$	____	____	____	____	____	____
(e) $\frac{i^+o^+z^+y^+a^+}{i^co^+z^+y^+a^+}$	____	____	____	____	____	____
(f) $\frac{i^+o^+z^+y^+a^+}{i^co^+z^-y^+a^+}$	____	____	____	____	____	____
(g) $\frac{i^+o^+z^+y^+a^+}{i^+o^cz^+y^+a^+}$	____	____	____	____	____	____
(h) $\frac{i^so^+z^+y^+a^+}{i^+o^+z^+y^-a^-}$	____	____	____	____	____	____
(i) $\frac{i^so^+z^+y^+a^+}{i^+o^cz^-y^+a^+}$	____	____	____	____	____	____
(j) $\frac{i^+o^+z^+y^+a^+}{i^co^cz^+y^+a^-}$	____	____	____	____	____	____

8. Assume that the structural genes *x* and *y*, which code for the repressible enzymes X and Y, are subject to negative control by a regulator gene *r*, which produces a substance S whose ability to bind to the operator gene *o* is modified by cosubstance T. Assume that the following allelic possibilities exist for genes *r* and *o*:

o^+ = wild-type operator gene

o^0 = operator gene unable to bind S (with or without T)

r^+ = wild-type regulator gene

r^0 = inactive regulator gene product (with or without T)

r^S = regulator gene product always active (with or without T)

In addition, assume that the mutations x^- and y^- in the two structural genes lead, respectively, to nonfunctional enzymes X and Y.

(a) Are enzymes X and Y likely to be biosynthetic or degradative? Justify your answer briefly.
(b) Is substance S active or inactive in the presence of T? Explain.
(c) For each of the following, predict whether active enzymes X and Y will or will not be produced under the specified conditions. For partially diploid cells, assume semidominance; that is, the enzyme

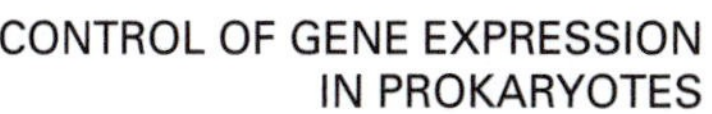

activity in a diploid cell will be twice that found in a haploid cell. Use the following answer code:

0 = active enzyme absent
+ = active enzyme present in haploid amounts
++ = active enzyme present in diploid amounts

		Without T		With T	
		X	Y	X	Y
(1)	$r^+o^+x^+y^+$	___	___	___	___
(2)	$r^+o^+x^-y^+$	___	___	___	___
(3)	$r^So^+x^+y^+$	___	___	___	___
(4)	$r^So^0x^+y^+$	___	___	___	___
(5)	$r^+o^0x^+y^+$	___	___	___	___
(6)	$r^0o^+x^+y^+$	___	___	___	___
(7)	$\frac{r^+o^+x^+y^+}{r^+o^+x^-y^+}$	___	___	___	___
(8)	$\frac{r^+o^+x^+y^+}{r^0o^+x^+y^+}$	___	___	___	___
(9)	$\frac{r^So^+x^-y^-}{r^0o^+x^+y^+}$	___	___	___	___
(10)	$\frac{r^+o^0x^+y^-}{r^0o^+x^+y^+}$	___	___	___	___

9. The kinetics of induction of enzyme X are shown in Figure 36-2. What percentage of total cellular protein is due to enzyme X in induced cells?

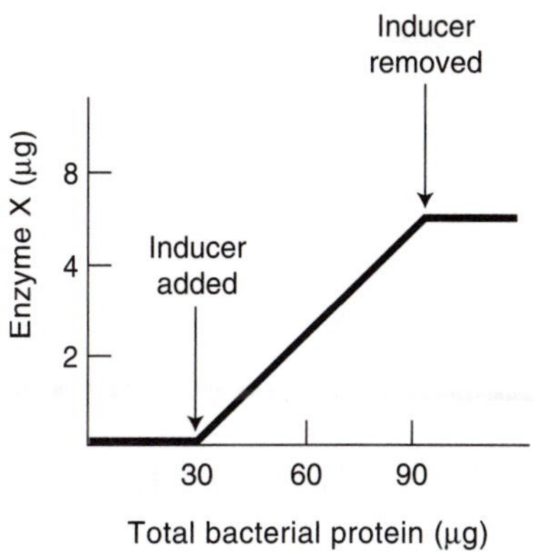

Figure 36-2
Kinetics of induction of enzyme X.

10. Assume that the dissociation constant K for the repressor-operator complex is 10^{-13} M and that the rate constant for association of operator and repressor is 10^{10} M^{-1} s^{-1}. Calculate the rate constant k_{diss} for the dissociation of the repressor-operator complex. What is the $t_{1/2}$ (half-time of dissociation, or half-life) of the repressor-operator complex?

11. In systems of genetic regulation involving positive control, a regulatory gene produces a substance that enhances rather than inhibits transcription. Are there elements of positive control in the lactose operon of *E. coli*? Explain.

12. Suppose that a mutant C protein of the arabinose operon of *E. coli* were found that could not bind arabinose. Predict the results on (a) the level of C protein and (b) the rate of transcription of the mRNA for *BAD* enzymes in the presence of cAMP–CAP complex, explaining your reasoning.

13. What is the minimum number of promoter regions that are necessary for the transcription of the genes for λ repressor, cro protein, and N protein. Where are these regions located? See Figure 36-19 on page 958 of Stryer.

14. Suppose that the relative affinities of the three O_R genes and the three O_L genes for the λ repressor were reversed and that O_R3 and O_L3 have the greatest affinity for a repressor. What would be the likely effect on the mechanism of infection by λ? Explain.

15. An operon for the biosynthesis of amino acid X in a certain bacterium is known to be regulated by a mechanism involving attenuation. What can one confidently predict about the amino acid sequence in the leader peptide for that operon? Explain.

16. In order to prove that regulation by attenuation occurs in vivo, Charles Yanofsky and others studied tryptophan synthesis regulation in a series of *E. coli* mutants. For each mutant described below predict the expression of tryptophan synthesis genes in the presence or absence of tryptophan:

Mutant A: Mutations with decreased, but detectable, tryptophan-tRNA synthetase activity

Mutant B: Mutations in AUG or Shine-Dalgarno sequence of leader peptide sequence

Mutant C: Same as mutant B but with the leader peptide fully expressed on a plasmid

Mutant D: A G·C to A·T transversion mutation that weakens the segment 2/segment 3 interaction without affecting the segment 1/segment 2 region

Mutant E: Mutations replacing the two Trp codons with leu codons

Mutant F: Mutant E with a mutation in the Leu-tRNA synthetase gene

Mutant G: Mutant E that constitutively synthesizes leucine

17. DNA supercoiling can have profound effects on levels of transcription.

(a) Give two ways that negative supercoiling could increase transcription.

(b) The promoter for *E. coli* DNA gyrase is affected by the extent of negative supercoiling in DNA. Would you expect negative supercoiling to increase or decrease the level of DNA gyrase expression?

ANSWERS TO PROBLEMS

1. Enzymes are catalysts, and thus small amounts can be readily detected. An enzyme having a turnover number of 300,000 s^{-1} will provide an assay that is 300,000 times as sensitive as that for a structural protein, which must be assayed stoichiometrically. Many cellular proteins are produced in amounts that are too small to be detected by direct chemical methods.

2. The actual inducer of the lactose operon in vivo is 1,6-allolactose (see Stryer, p. 950). The lag represents the time it takes for lactose to be converted into 1,6-allolactose. IPTG itself directly acts as an inducer. Therefore, no lag is observed.

3. (a) Imagine a partial diploid that has one o^+ and one o^c sequence. The o^+ gene will bind a repressor, so the structural genes on its chromosome will not be expressed. The o^c sequence will not bind a repressor, so the structural genes on its chromosome will always be expressed. Thus, an o^c mutant would be dominant to its wild-type o^+ allele.

(b) Repressors do not bind o^c sequences. Only the structural genes on the same chromosome as the o^c mutant will be affected, a cis effect.

(c) One could prepare a partial diploid with the following genotype

$$\frac{i^+ o^+ z^+}{i^+ o^c z^+}$$

If the effect of the mutation is cis, the haploid amount of enzyme Z will be produced in the absence of inducer. (Its synthesis will be specified by the chromosome containing $i^+ o^c z^+$.) If the effect is trans, the diploid amount of Z will be produced in the absence of inducer.

4. Very low levels of lactose operon enzymes are synthesized even in the absence of an inducer (see Stryer, p. 950).

5. Add IPTG to *E. coli* cells growing in a medium containing a carbon source other than lactose in both the presence and the absence of an inhibitor of prokaryotic protein synthesis, like chloramphenicol. If zymogen activation is involved, chloramphenicol will not inhibit induction. If the synthesis of new protein is involved (as it is), induction will not be observed in the presence of chloramphenicol.

6. Differential expression of the three structural genes in the lactose operon must be at the level of translation and not transcription. Following induction, mRNA transcripts containing genetic information for all three genes are produced. Some ribosomes might drop off the messenger at the end of the structural genes, with a smaller number reading through.

7.						
(a)	0	0	0	+	+	+
(b)	+	+	+	+	+	+
(c)	0	0	0	0	0	0
(d)	+	+	+	+	+	+
(e)	0	0	0	++	++	++
(f)	0	0	0	+	++	++
(g)	+	+	+	++	++	++
(h)	0	0	0	0	0	0
(i)	0	+	+	0	+	+
(j)	+	+	0	++	++	+

8. (a) The enzymes will be biosynthetic. The clue to this is provided by the statement that the enzymes are repressible. (Degradative enzymes are inducible, whereas biosynthetic enzymes are repressible.)

(b) Because this operon is under negative control, substance S must act as a repressor. Substance S is active as a repressor in the presence of T, so T functions as a corepressor.

(c)				
(1)	+	+	0	0
(2)	0	+	0	0
(3)	0	0	0	0
(4)	+	+	+	+
(5)	+	+	+	+
(6)	+	+	+	+
(7)	+	++	0	0
(8)	+	++	0	0
(9)	0	0	0	0
(10)	++	+	+	0

9. From the graph in Figure 36-2, we see that 3 μg of enzyme X is present when the total bacterial protein present is equal to 60 μg. Thus, the percentage of enzyme X is 100% $\times \frac{3}{60} = 5\%$.

10. $$[RO] \rightleftharpoons [R] + [O]$$

$$\frac{[R][O]}{[RO]} = K = 10^{-13}\text{ M} \frac{k_{\text{diss}}}{10^{10}\text{ M}^{-1}\text{ s}^{-1}}$$

$$k_{\text{diss}} = 10^{-3}\text{ s}^{-1}$$

$$t_{1/2} = \frac{0.693}{k_{\text{diss}}} = \frac{0.693}{10^{-3}\text{s}^{-1}} = 693\text{ s} \cong 11.6\text{ min}$$

11. Catabolite activator protein (CAP) is a positive control element. When the level of glucose in cells is low, the level of cAMP is high, leading to the formation of cAMP–CAP complex. The cAMP–CAP complex binds to DNA in the promoter region, creating an entry site for RNA polymerase. The result is the transcription of the lactose operon (providing that no repressor is present). (See Figure 32-10 on p. 954 of Stryer.)

12. (a) The level of C protein would be unaffected because C protein binds to *ara*O_1 in the absence of arabinose, preventing synthesis of C mRNA.

(b) The rate of transcription of mRNA for *BAD* enzymes would be reduced. The maximal transcription rate is achieved only when both cAMP–CAP complex and C protein with bound arabinose are present. (See Figure 36-16 on p. 957 of Stryer.)

13. Three promotor regions must be involved. One within the O_R1 region for the transcription of the *cro* gene, one within O_L1 for the transcription of the *N* gene, and one within O_R3 for the transcription of the *cI* gene.

14. If O_R3 has the greatest affinity for the λ repressor, the result would be abnormally low levels of the λ repressor itself, since it represses its own synthesis by binding to O_R3. Also, transcription of the mRNA for both the N protein and the Q protein would proceed because O_L1 and O_R1 would be unoccupied by the repressor. These effects would lead to lytic, rather than temperate, infection by λ.

15. The leader sequence should be rich in amino acid X. If there is already sufficient X present in the cell, there will be sufficient X-tRNA^X for the synthesis of the leader peptide (as well as for the synthesis of other X-containing proteins in the cell). Therefore, there will be no need to biosynthesize the enzymes needed to produce X.

16. Mutant A: There will be increased transcription of the trp synthesis genes in both the presence and absence of tryptophan. The low levels of Trp-tRNA synthetase will slow down the rate of translation regardless of the levels of tryptophan. It is the stalling or slowing of the ribosome in segment 1 that is important for regulation, not just the inability to synthesize the leader peptide.

Mutant B: Since the ribosome will never start translation, transcription will always terminate regardless of the levels of tryptophan.

Mutant C: The results will be the same as in mutant B. Attentuation relies on the coupling of transcription and translation. Providing the leader peptide in trans would have no effect on the level of attenuation.

Mutant D: Since the mutation weakens the interaction between segments 2 and 3, the interaction between segments 3 and 4 would be increased and the level of Trp synthesis genes would be decreased even in the absence of tryptophan.

Mutant E: Removing the Trp codons would lose all regulation by tryptophan, and Trp synthesis would be regulated by the levels of leucine. In this mutation, the Trp synthesis genes would not be transcribed even in the absence of tryptophan, and this strain, like mutants B–D, would always require tryptophan to grow, unless the strain was grown in the absence of leucine.

Mutant F: There would be constitutive expression of the Trp synthesis genes, even in the presence of tryptophan or leucine.

Mutant G: This mutant would behave like mutants B–E and never express the genes for Trp synthesis.

17. (a) Negative supercoiling would enhance the conversion of RNAP–DNA promoter complexes from closed to open forms, speeding up initiation. Also, supercoiled DNA is highly compact, with distant DNA sequences wrapping around one another. This wrapping could bring transcription factors closer to promoters or enhance any DNA looping needed for transcription initiation.

(b) Extensive negative supercoiling decreases the transcription of DNA gyrase. This control mechanism serves to maintain the level of DNA gyrase and therefore the amount of negative supercoiling in *E. coli* at a fixed level. In general, negative supercoiling increases transcription for the reasons given in part (a), but the promoter for DNA gyrase is an exception.

EXPANDED SOLUTIONS TO TEXT PROBLEMS

1. (a) The *lac* repressor is missing in an i^- mutant. Hence, this mutant is constitutive for the proteins of the *lac* operon.

   (b) This mutant is constitutive for the proteins of the *trp* operon because the *trp* repressor is missing.

   (c) The arabinose operon is not expressed in this mutant because the *araC* protein is needed to activate transcription.

   (d) This mutant is lytic, but not lysogenic, because it cannot synthesize the λ repressor.

   (e) This mutant is lysogenic, but not lytic, because it cannot synthesize the N protein, a positive control factor in transcription.

2. One possibility is that an i^s mutant produces an altered *lac* repressor that has almost no affinity for inducer but normal affinity for the operator. Such a *lac* repressor would bind to the operator and block transcription even in the presence of inducer.

3. This mutant has an altered *lac* operator that fails to bind the repressor. Such a mutator is called O^c (operator constitutive).

4. The cyclic AMP binding protein (CAP) is probably defective or absent in this mutant.

5. An *E. coli* cell bearing a λ prophage contains λ repressor molecules, which also block the transcription of the immediate-early genes of invading λ DNA.

6. (a) Translation of the p_{RE} transcript is from 5 to 10 times as rapid as that of the p_{RM} transcript because it contains the full protein synthesis initiation signal (as discussed on p. 895 in Stryer).

   (b) The more effective translation of the p_{RE} transcript provides a burst of λ repressor molecules needed to establish the lysogenic state. See M. Ptashe, K. Backman, Z. Humagun, A. Jeffrey, R. Maurer, B. Meyer, and R.T. Sauer. *Science* 194(1976):156.

7. The rate constant for association of wild-type *lac* repressor with the operator site is near the diffusion-controlled limit. Hence, a 100-fold increase in binding affinity implies that the mutant dissociates from the

operator site about 100-fold more slowly than does the wild type. The lag between addition of inducer and the initiation of transcription of the *lac* operon will be much longer in the mutant than in the wild type.

8. A repressor finds its target site by first binding to a nonspecific site anywhere in the DNA molecule and then diffusing along the DNA to reach the specific site. The repressor binds less tightly to nonspecific DNA at low ionic strength than at high ionic strength because of increased electrostatic repulsion. In contrast, binding to the operator is nearly independent of ionic strength because the interaction is mediated by hydrogen bond and van der Waals interactions with the bases of the operator site.

9. The mutant protein acts as a repressor even when tryptophan is not bound. Expression of the *trp* operon is permanently repressed in this mutant bacterium. Note that the shape of the mutant aporepressor *without* tryptophan is almost identical with that of the normal aporepressor *containing* bound tryptophan.

10. When RF2 is present, UGA is read as a stop codon, leading to the formation of a truncated 25-residue polypeptide.

- Gly - Tyr - Leu - Stop
GGG UAU CUU UGA ↙ Read by RF2

When the level of RF2 is very low, termination at UGA does not occur. Instead, a shift of the reading frame occurs, resulting in continued protein synthesis with Asp 26 as the next residue. See W. Craigen, R. Cook, W. Tate, and C. Caskey. *Proc. Nat. Acad. Sci.* 82(1985):3619.

- Gly - Tyr - Leu - Asp - Tyr - Asp -
GGG UAU CUU UGAC UAC GAC
↑
Frameshift in absence of RF2. U is skipped. GAC is first codon of new reading frame.

11. Operator sites separated by five or six turns of double helix are oriented so that they can bind to both sites of the dimeric receptor. Only one operator site can bind when the other is 4.5, 5.5, or 6.5 turns away because the second site faces the opposite direction (180 degrees out of phase).

12. Attenuation requires that transcription and translation be tightly coupled. Pausing is a means of synchronizing these two processes. Transcription is halted until a ribosome initiates translation and comes near the paused polymerase. The outcome then depends on whether or not the leader is completely translated (p. 967 in Stryer).

13. The second helix of the HTH motif of both repressors is the major determinant of their specificity. Hence, it is often called the *recognition helix.* See R.P. Wharton and M. Ptashne. *Nature* 316(1985):601. Recall the discussion of the HTH motif (Figure 36-29, p. 963 in Stryer).

CHAPTER

37

Eukaryotic Chromosomes and Gene Expression

In Chapter 37, Stryer elaborates three aspects of the biochemistry of nucleic acids that reflect both the problems and the opportunities afforded by the profusion of DNA in eukaryotic cells. He describes the structure of nucleosomes, which are the basic, repeating units of chromatin. Each nucleosome consists of DNA wrapped about an octameric protein core composed of histones. Individual nucleosomes are linked together by intervening stretches of DNA. The nucleosomes are packed, along with other proteins, to form a highly condensed chromosome. Replication of eukaryotic DNA is also described, with special emphasis on multiple initiations, the DNA polymerases involved, the deposition of newly synthesized histones on only one of the two daughter duplexes, the formation of the telomeres, and the control of the initiation of mitosis by specific protein kinases and protein phosphatases.

The organization of chromosomal DNA is introduced with an explanation of how the kinetics of hybridization of fragmented and denatured eukaryotic DNA revealed the existence of repeated sequences. Stryer provides examples of different arrangements of repeated genes and uses the silk fibroin gene to illustrate the capacity of single-copy genes to produce abundant proteins. Examples of gene amplification and the organization of a developmentally regulated gene are also presented.

Stryer then describes the modifications of chromatin structure that occur during transcription as an introduction to the topic of the regulation of gene expression. He discusses the strategy of using combinations of regulatory proteins that was introduced in Chapter 33 during a discussion of eukaryotic promoters and enhancers (see pages 855–858 in Stryer). He describes the structure of a transcription factor that uses the zinc finger motif to bind DNA specifically and activate the 5S rRNA gene for transcription. Then steroid hormones and morphogens are described. These molecules act by binding receptor molecules that are DNA-binding transcription factors. Therefore, these hormone and morphogen molecules act by directly effecting transcription, rather than working through membrane-associated receptors. The leucine zipper proteins are described as exemplars of a protein structural motif that couples protein-protein interactions with specific protein-DNA interactions to modulate transcription. The chapter concludes with a brief description of the role of the homeo box in the development of the basic architectural plan of higher eukaryotic organisms.

In preparation for studying this chapter, you should review Chapters 31 and 33, focusing on the sections dealing with DNA replication, eukaryotic RNA synthesis and transcription factors, and tRNA and rRNA processing. Also review pages 84–87 on the melting and annealing of DNA as an introduction to the section on the kinetics of DNA hybridization and page 154 on the changing patterns of hemoglobin synthesis during embryonic and fetal development to prepare yourself for the section on the organization of the hemoglobin gene.

LEARNING OBJECTIVES

When you have mastered this chapter, you should be able to complete the following objectives:

The Eukaryotic Chromosome (Stryer pages 976–982)

1. State the relative amounts of *genomic DNA* in the *haploid genomes* of *E. coli, yeast,* the *fruit fly,* and *humans.*

2. Relate the number of chromosomal DNA molecules to the number of chromosomes in a haploid cell, and discuss the empirical bases for arriving at this number.

3. Describe the composition of *chromatin.* List the types of *histones* and describe their general characteristics. Note the evolutionary stability of the sequences of the H3 and H4 histones.

4. Describe the composition and structure of the *nucleosome* and relate the nucleosome to the proposed structure of the *chromatin fiber.*

5. Describe the evidence provided by electron microscopy, x-ray and neutron diffraction analyses, nuclease digestion, and reconstitution studies that led to the current model of the structure of the nucleosome.

6. Distinguish between the DNA associated with the *nucleosome core* and that in the *internucleosome linker.*

7. Assign the types of histones to their locations within the nucleosome. Describe the unique features of histone H1.

8. Explain how the *packing ratio* of 7 arises from the structure of the nucleosome, state the packing ratios that have been estimated for *metaphase* and *interphase chromosomes,* and outline a molecular model that partially accounts for these ratios.

Chromosome Replication and Organelle DNA (Stryer pages 982–989)

9. Compare *DNA replication* in prokaryotes and eukaryotes. Contrast the rates of movement of the *replication fork* in prokaryotes and eukaryotes, and explain how *multiple origins* allow for the rapid duplication of genomes in eukaryotes.

10. List the enzymatic properties and functions of the *DNA polymerases* of eukaryotes. Note the division of labor between *DNA polymerase* α and *DNA polymerase* δ in synthesizing the *lagging* and *leading strands* of the daughter duplexes, respectively, during replication.

11. Explain the difficulty in replicating the ends of a linear DNA duplex, and outline the mechanism for *telomere replication.* Explain the role of the *RNA component* of *telomerase* in determining the sequence of the telomere.

12. Account for the *conservative segregation of parental histones.* Identify the daughter DNA duplex on which newly synthesized histones assemble during replication.

13. Describe the cell cycle and explain how the *cdc2 kinase, cyclin B,* and other specific *protein kinases* and *protein phosphatases* promote mitosis.

14. Describe the phenotype of *petite mutants* in yeast and explain the significance of their independent segregation.

15. Name the *organelles* in plants and animals that contain functional DNA.

16. Describe the human *mitochondrial genome* and contrast its features with those of the nuclear genome.

Repetitive DNA and Gene Organization (Stryer pages 989–994)

17. Write the chemical reaction for the *renaturation of denatured DNA* and the equation that relates the fraction of single-stranded DNA present to the total DNA concentration and the time of hybridization.

18. Draw a C_0t *curve* and describe the significance of the $C_0t_{0.5}$ value with respect to *genome size* and the content of *repeated DNA sequences.*

19. Provide examples of the repeated DNA sequences that are found in mammals. Describe *SINES, LINES,* and *Alu sequences.*

20. Describe *satellite DNA,* locate it in the chromosome, and explain the properties that allow it to be separated from the remaining genomic DNA.

21. Describe the intracellular location and spatial organization of the 18S, 5.8S, and 28S *rRNA genes* and their *spacer elements.*

22. Outline the biochemistry of the expression and maturation of the *rRNA gene cluster.*

23. Explain how the hybridization of histone mRNA to genomic DNA revealed the repetitions of the *histone genes.*

24. List the distinctive features of the histone genes.

Eukaryotic Gene Expression (Stryer pages 994–1006)

25. Explain how a *single-copy gene* can give rise to an *abundant protein,* using the silk fibroin protein from the silkworm as an example.

26. Describe the *amplification* and *selection* of the *dihydrofolate reductase gene* and comment on the implications of this event.

27. Relate the *developmental expression* of α*-globin* and β*-globin* to the chromosomal location and organization of their genes.

28. Relate the globin gene sequences and chromosomal organization to their *evolution.* Distinguish between genes and *pseudogenes.* Note the various classes of eukaryotic DNA.

29. State the fraction of mammalian DNA that is likely to encode proteins. Contrast this value to that for bacteria.

30. Relate *chromosomal puffs* to gene expression.

31. Provide the experimental bases for the correlations between *nuclease susceptibility* and the *degree of methylation* of the chromosome with *transcriptional activity* in mammals. Comment on the meaning of the correlations between enhanced nuclease sensitivity and relative methylation levels with tissue types and development.

32. Describe the usual mode of *positive regulation* of *eukaryotic transcription* and contrast it with the mechanism of regulation usually seen in bacteria.

33. Outline how the *combination of transcription factors* gives rise to specific transcription.

34. Outline the transcriptional activation of the 5S rRNA genes. Describe the *zinc finger structure* of *transcription factor IIIA* and appreciate the ubiquity and functional versatility of zinc finger domains in proteins.

35. Contrast the mechanism of the steroid hormones with hormones initiating their actions through interactions with a transmembrane receptor.

36. List members of the *nuclear receptor superfamily* and describe the functions of their domains. Describe the function of *hormone response elements* in steroid hormone actions. Define *morphogen.*

37. Describe the structures and properties of *leucine zipper proteins* and explain the benefits the cell derives from their ability to form *heterodimers* as well as *homodimers.*

38. Outline the function of *homeotic genes* and the effect of *homeotic mutations.* Describe the relationship between the *homeo box* and the *homeodomain.* Note the structural similarity of the homeodomain to the helix-turn-helix motif in prokaryotic gene regulators.

SELF-TEST

The Eukaryotic Chromosome

1. Which of the following statements about DNA that has been isolated from a eukaryotic chromosome are correct?

   (a) It is resistant to breakage by shearing forces because of its large size.
   (b) It is linear and unbranched.
   (c) It can be sized by a viscoelastic technique, which measures the time the stretched and elongated molecule takes to relax to its normal conformation.
   (d) It is a single molecule.
   (e) It can be more than 100 Mb long.

2. Match the organism listed in the left column with the amount of DNA in its haploid genome from the right column.

   (a) *E. coli* (bacterium) ________
   (b) *S. cerevisiae* (yeast) ________
   (c) *D. melanogaster* (fruit fly) ________
   (d) *H. sapiens* (human) ________

   (1) 4000 Kb
   (2) 170 Mb
   (3) 3900 Mb
   (4) 14 Mb

3. Which of the following statements about histones are correct?

   (a) They are highly basic because they contain many positively charged amino acid side chains.
   (b) They are extensively modified after their translation.
   (c) They can be fractionated by ion-exchange chromatography.
   (d) They can be removed from DNA with dilute acid.
   (e) They account for approximately one-fifth of the mass of a chromosome.

4. Which of the following statements about nucleosomes are correct?

   (a) They constitute the repeating units of a chromatin fiber.
   (b) Each contains a core of eight histones.

(c) They contain DNA that is surrounded by a coating of histones.
(d) They occur in chromatin in association with approximately 200 base pairs of DNA, on average.

5. What did each of the following techniques contribute to our understanding of chromatin structure?

(a) Electron microscopy

___

(b) X-ray and neutron diffraction analyses

___

(c) Nuclease digestions

___

(d) Purification and reconstitution of components

___

6. Describe the model for the structure of the nucleosome.

___

___

7. Does the formation of nucleosomes account for the observed packing ratio of human metaphase chromosomes? Explain.

___

___

Chromosome Replication and Organelle DNA

8. Which of the following statements about eukaryotic DNA are correct?

(a) It is replicated conservatively.
(b) It is replicated bidirectionally.
(c) Its replication originates at numerous initiation points on each chromosome.
(d) Its replication requires primers.
(e) Its replication is template-independent.
(f) It is synthesized by chain growth in the $3' \rightarrow 5'$ direction.

9. Match the DNA polymerase in the left column with the function or property with which it is associated from the right column.

(a) DNA polymerase α ___
(b) DNA polymerase β ___
(c) DNA polymerase γ ___
(d) DNA polymerase δ ___
(e) DNA polymerase ε ___

(1) Involved in chromosomal DNA repair
(2) Involved in chromosomal DNA replication
(3) Involved in mitochondrial DNA replication
(4) Synthesizes leading strand in nucleus
(5) Synthesizes lagging strand in nucleus

10. Describe the problem faced by the biosynthetic apparatus in replication of the DNA at the ends of linear chromosomes.

____

____

11. Explain how the "end" problem referred to in question 10 is solved in eukaryotes.

____

____

12. Which of the following duplex DNA segments is the product of one round in the replication of a eukaryotic chromosome in the absence of any mitotic events, that is, in the absence of recombination. The parental strand is represented by ——, and the newly synthesized daughter strand is represented by ***.

(a) ——— (b) ******* (c) ———
     ———   *******   *******

13. Which of the following statements about histones during DNA replication are correct?

(a) Their number must double.
(b) Those that are present in the G1 phase of the cell cycle initially associate with both daughter DNA duplexes at the replication fork.
(c) They segregate semiconservatively.
(d) Those that are newly synthesized associate with the daughter duplex containing the lagging strand.

14. Which of the following statements about the cell cycle and its regulation are correct?

(a) During the M phase the newly replicated daughter chromosomes are segregated into the daughter nuclei.
(b) The M phase is initiated when the levels of cyclin B are sufficiently high to activate the cdc2 protein kinase.
(c) The M phase begins before the replication of the chromosome is totally completed.
(d) A maturation promoting factor, composed of cdc2 kinase and cyclin B, must form before mitosis can begin.
(e) Specific hydroxyl groups on cdc2 kinase are phosphorylated and dephosphorylated by specific kinases and phosphatases to make the active kinase that actually initiates the M phase.
(f) Entry into the M phase involves the breakdown of the nuclear envelope and the formation of mitotic spindles.

15. Why do the petite mutations of yeast segregate independently of most other mutations?

____

____

16. Which of the following statements about human mitochondrial DNA are correct?

(a) It is a linear, single-stranded molecule.
(b) It encodes both protein and RNA products.
(c) It is responsible for the synthesis of several subunits of proteins that are involved in electron transport and oxidative phosphorylation.
(d) It encodes all the tRNA molecules that are required for mitochondrial protein synthesis.
(e) It contains noncoding spacer DNA between many of its genes.

Repetitive DNA and Gene Organization

17. When the DNA from bacteria is broken into small fragments, thermally denatured, and renatured under defined conditions, which of the following occur?

(a) The formation of duplex DNA follows a time course that is characteristic of a bimolecular reaction.
(b) The fraction of the strands that are renatured is a function of the ratio of the time of renaturation and the concentration of the DNA.
(c) The $C_0t_{0.5}$ value, which corresponds to the reassociation of half the DNA, can be determined.
(d) The rate of reassociation is a function of both the concentration of the DNA fragments and the time of reaction.

18. Under a standard set of conditions and with equimolar concentrations of DNA nucleotides, the *E. coli* genome gives a higher $C_0t_{0.5}$ value than does the genome of bacteriophage T4. Explain.

____

____

19. When DNA from a mouse cell, which contains approximately 1000 times more DNA than an *E. coli* cell, was subjected to a denaturation and renaturation protocol similar to that outlined in question 18, it did *not* yield a $C_0t_{0.5}$ value that was 1000 times that of *E. coli*. Explain.

____

____

20. Which of the following statements about repeated DNA are correct?

(a) It is abundant in all eukaryotic organisms.
(b) It can sometimes be separated from unique DNA sequences.
(c) It is sometimes transcribed.
(d) It can be found at centromeres.
(e) It may have arisen by reverse transcription of RNA sequences followed by the incorporation of the cDNA copies into the chromosome.

21. Which of the following statements about ribosomal RNA genes are correct?

(a) They are localized in the nucleolus.
(b) They are present in the chromosome as single copies driven by the most efficient promoters known so that they can be transcribed rapidly and repeatedly.
(c) They are present in the chromosome as repeated units that are located at highly variable distances from one another.
(d) They exist as units comprising the 18S, 5.8S, and 28S genes plus a spacer region.
(e) They are amplified approximately 4000-fold during oogenesis in the African clawed toad.

22. The rate of hybridization of excess sea urchin histone mRNA to sea urchin DNA revealed what property of histone genes? Explain.

____

____

23. Which of the following statements about histone genes are correct?

(a) The number of copies present per genome likely reflects the need of the organism for rapid histone synthesis.
(b) They exist as clusters of tandemly repeated sets of H1, H2A, H2B, H3, and H4 sequences in some organisms.
(c) They exist as dispersed H1, H2A, H2B, H3, and H4 sequences in some organisms.
(d) They contain introns.
(e) They produce mRNAs without poly(A) tails.

Eukaryotic Gene Expression

24. Which of the following statements about the silk fibroin protein from *Bombyx mori* are correct?

(a) It is an abundant protein during certain stages of development.
(b) It is produced from a gene that is amplified during certain stages of development.
(c) It can be produced in large amounts because its gene exists in multiple copies.
(d) It is an example of an abundant protein that is produced by a single-copy gene.

25. Explain the role of methotrexate, an inhibitor of dihydrofolate reductase, in the induced amplification of the mammalian gene that encodes this enzyme.

____

____

26. Which of the following statements about α-globin and β-globin gene clusters are correct?

(a) They are on different chromosomes.
(b) They are evolutionarily related to each other and to myoglobin.

(c) Each contains globin genes that are arranged in the order the encoded globins appear during the development of the zygote.
(d) Each contains α-globin or β-globin pseudogenes, which are among the first genes to be expressed after the fertilization of the egg.

27. The central dogma of molecular biology summarizes the flow of information in a cell by stating that DNA makes RNA makes protein. What portion of human DNA participates in this pathway? What are the functions of the remaining DNA?

____

____

28. Which of the following statements about transcriptionally active regions of chromosomes are correct?

(a) They are found in the puffed regions of the polytene chromosomes of insects.
(b) They are less sensitive to DNase I digestion than are inactive regions.
(c) They vary with the stage of development of the organism.
(d) They are tissue-specific.
(e) They are undermethylated in comparison with transcriptionally inactive regions of DNA in the same organism.

29. Which of the following statements abut transcription factor IIIA are correct?

(a) It is a sequence-specific DNA-binding protein.
(b) It potentiates the transcription of the genes encoding the 5S rRNA molecule.
(c) It binds to the upstream activating sequence (UAS) of an rRNA gene.
(d) It acts to recruit other transcription factors and RNA polymerase III to the 5S rRNA gene.
(e) It contains several domains comprising 30-residue sequences of amino acids with two Cys and two His side chains disposed such that they chelate a zinc ion.

30. Explain how the use of zinc finger motifs in proteins allows them to recognize different, specific DNA sequences.

____

____

31. Which of the following statements about steroid hormones are correct?

(a) They bind to a seven-helix transmembrane receptor to initiate a series of phosphorylations that culminate in gene transcription.
(b) Upon binding to their specific receptor proteins, they enable the receptors to bind specific DNA sequences.
(c) They activate specific protein kinases and protein phosphatases.
(d) They are recognized by members of the nuclear receptor superfamily of proteins.
(e) They require plasma membrane transporters to go from the blood to the cytosol.

32. Which of the following statements about leucine zipper proteins are correct?

(a) A given monomeric leucine zipper protein can form a heterodimer and thereby bind to a different, specific DNA sequence than it does as a homodimer.
(b) They contain about 35 amino acids in their C-terminal regions that have a leucine residue approximately every 7 residues.
(c) Various combinations of zinc finger elements confer DNA sequence-specific binding properties on the proteins harboring them.
(d) They dimerize to bring their N-terminal regions into proximity so as to serve as specific DNA-binding elements.
(e) They usually act as transcriptional repressors.

33. Which of the following statements about the homeo box are correct?

(a) It is found in prokaryotes and eukaryotes.
(b) It is a group of similar sequences in DNA that is associated with developmental control.
(c) It encodes proteins that bind to DNA with sequence specificity.
(d) It encodes proteins that contain a homeodomain and are localized in the nucleus.

ANSWERS TO SELF-TEST

The Eukaryotic Chromosome

1. b, c, d, e. Answer (a) is incorrect because DNA molecules more than a few kilobases long are sensitive to fragmentation by the shearing forces developed when solutions are stirred.

2. (a) 1 (b) 4 (c) 2 and (d) 3. Note that the length of the bacterial chromosome is expressed in kilobases, not megabases.

3. a, b, c, d. Answer (e) is incorrect because histones make up nearly half the mass of a chromosome.

4. a, b, d. A nucleosome core consists of 140 base pairs of DNA wrapped around a histone octamer. The nucleosome cores are connected by linker DNA, which contains from fewer than 20 to more than 100 base pairs, the exact length depending on the organism and the tissue.

5. (a) Electron microscopy showed that chromatin was a linear array of particles (nucleosomes), with diameters of approximately 100 Å, joined by flexible DNA linkers.

(b) X-ray diffraction analyses also showed 100-Å-diameter particles. Neutron diffraction analyses indicated that the DNA was near the surface of the nucleosome and not buried in its interior.

(c) Limited nuclease digestion with DNase I followed by gel electrophoresis of the products produced a ladder of discrete fragments. The fragments contained multiples of a basic DNA unit of about 200 base pairs. Digestion with micrococcal nuclease revealed that 140 base pairs of DNA were closely associated with the nucleosome core.

(d) When purified histones and purified DNA were mixed, they associated with a stoichiometry of two each of histones H2A, H2B, H3, and H4 and approximately 200 base pairs of DNA to form characteristic beaded chromatin fibers. Histone H1 was not required.

6. The nucleosome core has a disk shape with an average diameter of 100 Å and a thickness of about 55 Å. The octameric core of histones has 140 base pairs of DNA wound about it in approximately 1¾ turns of a left-handed torus. Two copies each of histones H2A, H2B, H3, and H4 are on the inside of the toroidal coil, whereas histone H1 is associated with the DNA that links the cores.

7. No. Each nucleosome is associated with approximately 200 base pairs of DNA. If this DNA were coiled into a sphere with a diameter of approximately 100 Å, it would be condensed from 200 base pairs × 3.4 Å per base pair = 680 Å of linear DNA to 100 Å, which is a packing ratio of about 7. The chromatin fiber must be wound into a helical array, and the resulting 360-Å coils must themselves be looped and folded. Scaffolding proteins, topoisomerases, and small basic molecules, such as the polyamines, also contribute to the ultimate compaction of 10^4 that is observed in metaphase chromosomes.

Chromosome Replication and Organelle DNA

8. b, c, d. Answers (a) and (e) are incorrect because eukaryotic DNA is replicated semiconservatively in a template-dependent reaction that is analogous to the replication process in prokaryotes.

9. (a) 2, 5 (b) 1 (c) 3 (d) 2, 4 (e) 1

10. Since DNA polymerases require primers and can synthesize only in the $5' \rightarrow 3'$ direction, the ends of linear duplex DNA represent a challenge. Semiconservative replication of such a molecule results in a protruding, single-stranded 3′ tail that lacks an associated template strand. (See Figure 37-15 and the text on p. 985 in Stryer.) The genetic information contained in the protruding 3′ tail would be missing in one of the daughter cells if it were not replicated by some other process. Furthermore, increasing amounts of genetic information would be lost with each succeeding cell division.

11. The 3′ tail comprises variable numbers of simple repeated sequences, for example, repeats of GGGGTT, in *Tetrahymena.* The template that forms this sequence is carried as an RNA molecule in a specialized ribonuclear protein DNA polymerase (telomerase) that can add a number of these elements by repeated rounds of synthesis, "slippage," and synthesis. The repeated elements of the 3′ ends resulting from telomerase action probably provide insurance against harm to the information-containing sequences near the ends, if occasional losses of terminal regions should occur.

12. c. Since DNA synthesis is semiconservative, the daughter DNA duplex will contain one strand of parental DNA and one newly synthesized strand.

13. a, d. The old histones remain with the daughter DNA duplex, which is composed of a parental strand and a newly synthesized leading strand. New histones are deposited on the daughter DNA duplex that is formed from the lagging strand and the parental template. The initial association of histones with only one daughter DNA duplex is termed *conservative segregation* and contrasts with the semiconservative segregation of the parental DNA template strands (see question 12).

14. a, b, d, e, f. Answer (c) is incorrect because the replication of the chromosome is completed before mitosis begins.

15. Petite mutations segregate independently with respect to the more common chromosomal mutations because they occur in the mitochondrial DNA, not the nuclear DNA. They segregate with the region of cytosol they occupy when the cells divide.

16. b, c, d. Human mitochondrial DNA is a circular duplex that is densely packed with encoded information for both proteins and RNAs, including the repertoire of tRNAs needed for mitochondrial protein synthesis. Far from having spacer DNA, it often overlaps gene sequences to maximize the density of its information.

Repetitive DNA and Gene Organization

17. a, c, d. Answer (b) is incorrect because the fraction of strands that are renatured is a function of the product, not of the ratio, of the DNA concentration and the time of renaturation.

18. The DNA of both *E. coli* and bacteriophage T4 is composed almost entirely of unique sequences, with the bacterial DNA being approximately 25 times longer than the phage DNA. When equal amounts, in terms of total nucleotides, of bacterial and phage DNA are fragmented, the concentrations of individual fragments will be approximately 25 times lower for the bacterial DNA than for the T4 DNA. Consequently, reassociation for the bacterial DNA will be slower, which causes the $C_0t_{0.5}$ to be higher.

19. Unlike prokaryotic DNA, eukaryotic DNA contains repeated sequences. Consequently, when a given concentration of eukaryotic DNA is fragmented, the concentrations of the repeated sequences are higher than those of the unique sequences, so the repeated sequences will renature more rapidly. Since there are multiple classes of complexity in eukaryotic DNA, the C_0t curve will not be a simple sigmoid as it is with DNA of a single complexity. The curve will be a complex mixture of individual sigmoidal components, each with a characteristic $C_0t_{0.5}$ value. Since some 30 percent of mouse DNA is repeated, the $C_0t_{0.5}$ value is lower than the approximately 15,000 M s that would be expected for DNA containing only unique sequences (see pp. 990–991 in Stryer).

20. a, b, c, d, e. The repeated sequences of rRNA, tRNA, protein-encoding genes (e.g., the histone genes), and some others (e.g., Alu sequences) are transcribed. Other repeated sequences (e.g., satellite sequences) are not transcribed.

21. a, d, e. The ribosomal RNA genes are repeated and exist in tandem clusters. In the toad, the number of copies increases some 4000 times, going from 500 to 2×10^6 copies, during the formation of the egg so that

there are sufficient ribosomes to support protein synthesis during the initial stages of development.

22. The kinetics of hybridization of RNA to DNA reveals information about the number of copies of genes, that is, the copies of particular sequences per DNA molecule, just as the kinetics of hybridization of DNA to DNA does. Because the rate of hybridization of mRNA to DNA was hundreds of times faster than could be accounted for by the concentration of sea urchin DNA assuming a single copy of the histone gene per genome, it was concluded that histone genes were reiterated 300 to 1000 times. The use of excess mRNA ensures that the rate of hybridization is a function of the DNA, not the RNA, concentration.

23. a, b, c, e. Histone genes are atypical in that they lack introns and their mRNAs don't have poly(A) sequences at their 3′ termini.

Eukaryotic Gene Expression

24. a, d. Single-copy genes encode most of the proteins produced by eukaryotes.

25. DHFR is required for the synthesis of DNA because it catalyzes a reaction in the pathway to dTTP. An inhibitor of this enzyme would be expected to retard cell growth. Cells grown in the presence of methotrexate are thus under selective pressure to overproduce DHFR in order to survive. Cells containing spontaneous duplications of the DHFR gene produce more of this enzyme and thus grow better, as do their progeny. Further duplications of the gene lead to more cells with even better survival potentials. The net result of this selection is the amplification of the DHFR gene.

26. a, b, c. Answer (d) is incorrect because pseudogenes are not expressed, since they lack one or more of the features that are required for transcription or translation.

27. Probably only approximately 2% of human DNA encodes proteins. With the exception of the DNA that encodes known RNA products—for example, the ribosomal and snRNP RNAs and telomere-associated RNA sequences—the function of most of the remaining DNA is unknown.

28. a, c, d, e. Answer (b) is incorrect because regions of the chromosome that extend for considerable distances on both sides of transcriptionally active genes are sensitive to digestion by DNase I in vitro. Some regions, usually at the 5′ ends, become hypersensitive to the nuclease; that is, they are more digestible than is naked DNA. With respect to answer (d), tissue specificity of transcription reflects the functional differences between differentiated cells. Common regions of transcriptional activity represent the common functions of cells.

29. a, b, d, e. Transcription factor IIIA binds to a specific sequence within the 5S RNA gene, thus enabling the subsequent binding of other transcription factors and, ultimately, the transcription of the gene by RNA polymerase III.

30. Each zinc finger element "recognizes" a three-base-pair sequence. The combination of various such zinc fingers, each recognizing a particular

three-nucleotide sequence, endows the protein that bears them with the capability of specifically binding longer, specific sequences.

31. b, d. Answers (a) and (c) are wrong because they are properties of a different, more numerous class of hormones.

32. a, b, d. Leucine zipper proteins lack zinc fingers and act as activators rather than repressors.

33. b, c, d. The homeo box sequences are widely distributed among eukaryotes, not prokaryotes, so answer (a) is incorrect.

PROBLEMS

1. Would you expect the interaction between protamines and DNA to be enhanced or diminished in solutions that are highly ionic? Explain the basis for your answer.

2. One measure of the evolutionary divergence between two proteins is the number of amino acid differences between them. It can be argued that a better measure would be the minimum number of mutational events that must have occurred to result in those differences.

(a) The differences in the amino acid sequences of histone H4 between calf thymus and pea seedlings are as follows:

AA position	*Pea seedlings*	*Calf thymus*
60	Ile	Val
77	Arg	Lys

What minimum number of mutational events are necessary to account for these differences. Give the changes that must occur on both mRNA and DNA. (Refer to the genetic code on the inside back cover of Stryer.)

(b) Comment on the nature of the changes, based on your knowledge of amino acid chemistry. What conclusion follows about the function of H4?

3. Norbert Numbskull, a not-too-eminent biochemistry student, and his laboratory partner, Lucy Luckless, decide to fractionate a mixture of chromatin into its component histones. Norbert adds salt to the mixture and pours the resulting solution onto a gel-filtration column. Lucy takes the original mixture and adds it to an ion-exchange chromatography column. Comment on the prospects of success for either of these experiments, giving your reasoning. If necessary, suggest an alternative experiment.

4. Chromatin that is transcriptionally active (*euchromatin*) is disperse in structure, whereas chromatin that is transcriptionally inactive (*heterochromatin*) is compact. When the nuclei of chick globin-producing cells were treated briefly with pancreatic DNase, the adult globin genes were selectively destroyed, but the genes for embryonic globin and ovalbumin remained intact. When the nuclei of oviduct cells were treated with DNase, the ovalbumin genes were destroyed. Explain these results.

5. The chromatin of globin-producing cells can be treated with a micrococcal nuclease that cleaves DNA in the linker regions between nucleosomes. When the resulting nucleosomes are isolated and their DNA is examined, it is found to contain the DNA sequence for the synthesis of globin. Are these results consistent or inconsistent with the explanation for the results in problem 4? Explain.

6. Suppose that the following duplex sequence lies at the end of a eukaryotic chromosome:

5′-ATAGGCC-
3′-CGGTATCCGG-

Suppose further that two blocks of the sequence 5′-TTCCCGG-3′ are added to the overhanging 3′ end. The 3′-OH of the terminal G then loops around and serves as a primer for the synthesis of a complementary 3′ → 5′ sequence. The resulting loop of five nucleotides is then cleaved by a nuclease that excises five nucleotides. Give the sequence of the resulting DNA duplex.

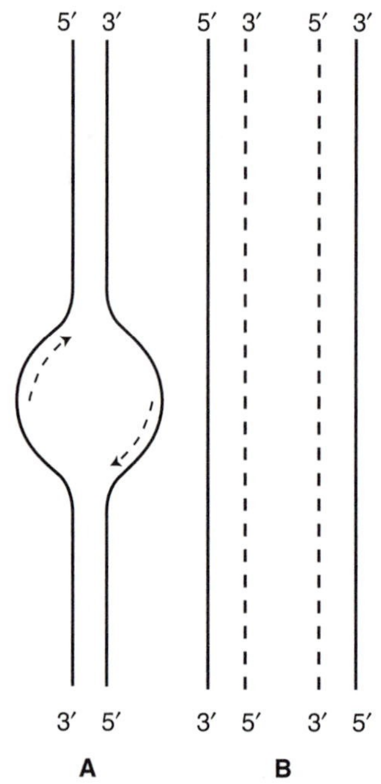

Figure 37-1
A eukaryotic chromosome (A) at the initiation of replication and (B) after replication. The daughter strands are shown as dashed lines.

7. Figure 37-1 depicts the replication of a region of a eukaryotic chromosome. Sketch the expected location of parental histones on the daughter duplexes, using circles to depict nucleosomes.

8. The antibiotic erythromycin binds to the 50S subunit of prokaryotic ribosomes, blocking chain elongation, but it does not bind to the 60S subunit of eukaryotic ribosomes. The tetracyclines bind to the 30S subunit of prokaryotic ribosomes, blocking the A site, but they have no corresponding effect on the 40S subunit of eukaryotic ribosomes. Although erythromycin and the tetracyclines are useful in human antibiosis, one is best advised to be cautious in their administration. Give one good reason why this is the case.

9. Predict whether each of the following manipulations will cause an increase or decrease in the rate at which two separated, complementary DNA sequences reassociate under conditions typical of those used to establish C_0t curves. Briefly state the basis for your prediction.

(a) An increase in salt concentration
(b) An increase in incubation temperature
(c) A decrease in DNA concentration

10. Calculate the relationship that exists between $C_0t_{0.5}$ and k from the following relationship, which is given on page 990 of Stryer:

$$f = \frac{1}{1 + kC_0t}$$

11. As can be seen in Figure 37-21 on page 990 of Stryer, mouse satellite DNA has a $C_0t_{0.5}$ value of approximately 10^{-3} M s. Would you expect a mixture of poly(G) and poly(C) to have a $C_0t_{0.5}$ value that is smaller than, approximately the same as, or greater than that of mouse satellite DNA? Explain your answer.

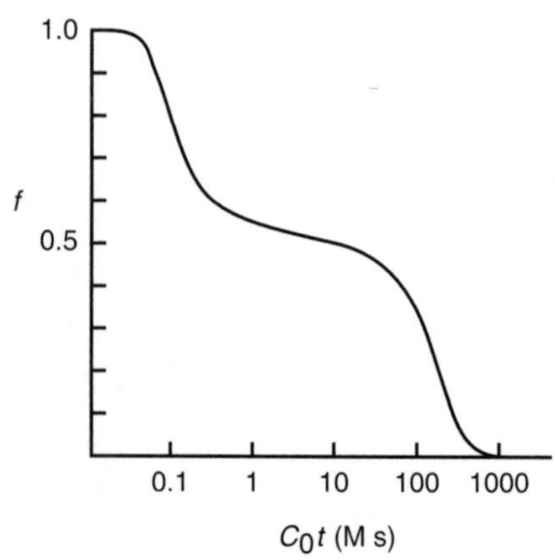

Figure 37-2
C_0t results for problem 12.

12. Suppose that a DNA sample is subjected to C_0t analysis, with the results depicted in Figure 37-2. How would you interpret these results?

13. The nucleotide sequence TCACCCTATAGCCATCG is found on the template (–) strand of mitochondrial DNA known to specify the amino terminal sequence of a mitochondrial protein. Give the amino acid sequence that is specified. (See pp. 109 and 989 of Stryer.)

14. Figure 37-27 on page 994 of Stryer depicts the histone gene cluster of the sea urchin. Do each of the five genes in the cluster produce equal amounts of protein products? Why or why not?

15. Suppose that the DNA specifying the synthesis of an enzyme having a turnover number of 10^4 s^{-1} is transcribed into 10^3 molecules of mRNA and that each of the mRNA molecules is translated into 10^5 molecules of enzyme protein. How many molecules of substrate are converted into product per second for each transcription of the DNA?

16. Would you expect a zinc deficiency in eukaryotes to be associated with any sort of developmental abnormality? Explain.

17. You have isolated the *M* gene, which is involved in muscle cell differentiation, and you wish to study its expression in cultured mammalian cells. In sequencing the region upstream from the open reading frame you notice that there are a few CG-rich regions, from 40 to 150 nucleotides upstream.

(a) When you grow the cells in presence of 5-azacytidine, the *M* gene is expressed, while cells grown in the absence of the analog do not express it. Provide an explanation for these observations, relating them to the CG-rich regions and how they would be affected by 5-azacytidine.

(b) You decide to carry out some experiments to verify your explanation of the experiments with 5-azacytidine. Isolating the *M* gene upstream sequence that contains the CG-rich regions, you attach it on a bacterial plasmid immediately next to the gene for the enzyme chloramphenicol acetyl transferase (CAT), which is not found in mammalian cells. You then amplify this plasmid construct in, and isolate it from, bacteria. When the isolated plasmid molecules are used to transiently infect your cells in culture, CAT is transcribed in the absence of 5-azacytosine. You then isolate the upstream sequence from cell chromosomes as well as from the plasmid, and digest samples from both isolates with the restriction endonucleases HpaII and MspI. The DNA from the digests is separated using gel electrophoresis. What would you expect to see in these experiments?

18. Proteins that interact with a specific sequence of DNA can remain bound under mild conditions (low ionic strength) during electrophoresis in gels containing low percentages of acrylamide. Under these conditions, protein-DNA complexes migrate more slowly than free DNA. Suppose you have a sample of a radioactively labeled DNA fragment that contains an entire promoter sequence. Describe how you could use the sample and gel electrophoresis to isolate transcription factors from cell protein extracts. How could you compensate for the binding of histones to DNA? How could you use unlabeled samples of your promoter fragment to demonstrate the specificity of the interaction between the fragment and a putative transcriptional factor?

19. A group of molecular biologists showed that the three eukaryotic DNA sequences below can activate transcription of a reporter gene, such as the gene encoding chloramphenicol transacetylase.

Sequence X: 5′-TAATTGCGCAATTA-3′
3′-ATTAACGCGTTAAT-5′

Sequence Y: 5′-TAATTGCTACGGTA-3′
3′-ATTAACGATGCCAT-5′

Sequence Z: 5′-TACCGTATACGGTA-3′
3′-ATGGCATATGCCAT-5′

Gel-shift assays, described in problem 18 above, are used to purify *three* dimeric protein activators, each of which binds to one of the three DNA sequences shown above. Further investigation reveals that while three dimeric proteins are purified, only *two* polypeptide monomers have been purified. How is this possible?

20. Unlike bacterial RNA polymerases, eukaryotic polymerases have relatively low affinity for their promoters and therefore often depend on several activator proteins for initiation of transcription. Thus, while many bacterial genes are subject to negative regulation by repressor proteins, eukaryotic genes are more likely to be under positive regulatory control. The reasons for this difference in the mode of regulation may be related to the great difference in genome sizes. For example, the human genome is a thousand times larger than that of *E. coli* and may contain over 100,000 genes. What are the advantages of positive regulation in the control of gene expression in eukaryotes?

21. Simpson and his co-workers digested nucleosomes to generate particles called *chromatosomes,* which contained an octamer of histones, a molecule of H1, and about 160 base pairs of DNA. They then measured the thermal stability of DNA helices in chromatosomes containing H1 as well as in those depleted of that protein. What results do you think they found?

ANSWERS TO PROBLEMS

1. Just as histones can be dissociated from DNA with salt or dilute acid, the interaction between protamines and DNA is diminished in highly ionic solutions. Protamines are arginine-rich proteins whose positively charged guanidinium groups would extend outward in α helices associated with the major groove of a DNA helix (see Stryer, p. 982) in sperm cells. Electrostatic interactions between the guanidinium groups and the negatively charged phosphoryl groups of DNA bind the proteins tightly to the polynucleotide. The positively and negatively charged ions that result from the addition of a salt to an aqueous solution compete with DNA-protamine interactions and, hence, weaken them.

2. (a) There would be a minimum of two mutations involved. The possible codons for Ile and Val are as follows:

Ile	*Val*
AUU	GUU
AUC	GUC
AUA	GUA
	GUG

A single change from A to G in the first position of the codon would give a substitution of Val for Ile. This corresponds to a change from an A·T to a G·C base pair on DNA.

The possible codons for Arg and Lys are as follows:

Arg	*Lys*
CGU	AAA
CGA	AAG
CGG	
CGC	
AGA	
AGG	

Again a single change, from G to A in the second position of the codon, would account for the amino acid difference. This corresponds to a change from a G·C to an A·T base pair on DNA.

(b) These are conservative changes. Both Lys and Arg have positively charged side chains, and both Ile and Val are hydrophobic. Therefore, we would expect virtually no structural or functional difference in the H4 of calf thymus and pea seedlings. However, the two tissues are surely divergent in other respects.

3. The prospects of success in either case are dim. Histones are of sufficiently similar size that a gel-filtration column, which is essentially a molecular sieve, would not be the best choice for separating them. An ion-exchange column would be a better choice for separating the histones, but *only* if the histones were first dissociated from DNA by the addition of salt.

4. In cells that are actively synthesizing adult globins, chromatin is dispersed so that the globin genes may be transcribed into mRNA. Accordingly, this region is sensitive to DNase. In an adult cell specialized for the production of globins, neither embryonic globins nor ovalbumin is produced to any significant extent. Accordingly, the regions of DNA carrying the information for these genes are not dispersed and are therefore not sensitive to DNase.

5. The results are consistent. The genes for globin synthesis are contained within the nucleosomes. When these genes become transcriptionally active, the chromatin becomes dispersed.

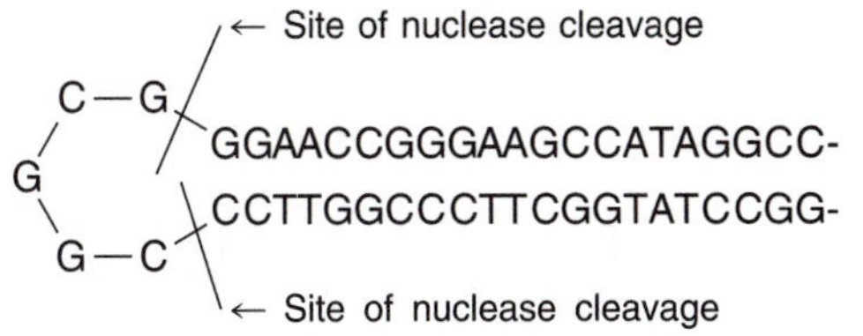

← Site of nuclease cleavage
C—G
G GGAACCGGGAAGCCATAGGCC-
G—C CCTTGGCCCTTCGGTATCCGG-
← Site of nuclease cleavage

Figure 37-3
Cleavage sites on the telomeric nucleotide sequence given in problem 6.

6. The cleavage points are shown in Figure 37-3. The duplex sequence that will result is as follows:

EUKARYOTIC CHROMOSOMES AND GENE EXPRESSION

5′-GGAACCGGGAAGCCATAGGCC-3′

3′-CCTTGGCCCTTCGGTATCCGG-5′

7. The histones will be found on the leading strands of the daughter duplexes, as shown in Figure 37-4. Remember that DNA polymerase reads in the 3′ → 5′ direction and synthesizes polynucleotide chains in the 5′ → 3′ direction.

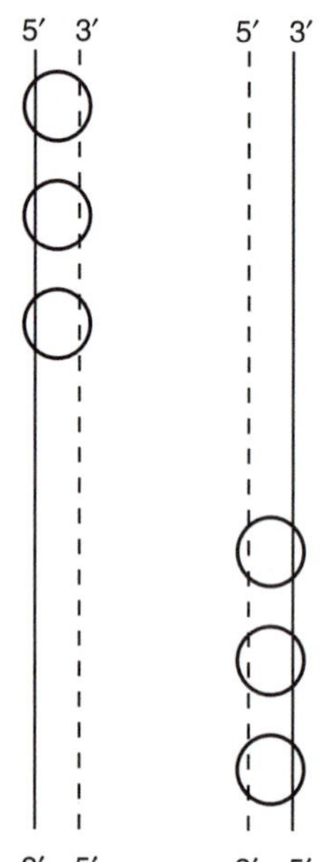

Figure 37-4
Distribution of parental histones on daughter strands for the answer to problem 7.

8. The mitochondria of eukaryotes have ribosomes that resemble those of prokaryotes. Thus, both erythromycin and tetracycline can inhibit protein synthesis in mitochondria as well as in prokaryotes.

9. (a) Increasing the salt concentration will decrease the repulsion between negatively charged phosphoryl groups and, hence, will increase the rate of reassociation.

(b) An increase in the incubation temperature will decrease the stability of the hydrogen bonds that must form between complementary bases and, hence, will decrease the rate of reassociation.

(c) A decrease in DNA concentration will decrease the rate of reassociation. Reassociation involves the coming together of two molecules in a bimolecular process, and the rate is proportional to the product of the concentrations of (+) and (−) strains.

10. When reassociation is half complete, $f = 0.5$ and $C_0t = C_0t_{0.5}$. First, we substitute these values into the original equation:

$$f = \frac{1}{1 + kC_0t}$$

$$0.5 = \frac{1}{1 + kC_0t_{0.5}}$$

We then cross multiply:

$$0.5(1 + kC_0t_{0.5}) = 1$$

$$0.5 + 0.5kC_0t_{0.5} = 1$$

$$0.5kC_0t_{0.5} = 0.5$$

$$C_0t_{0.5} = 1/k$$

Thus $C_0t_{0.5}$ is the reciprocal of k, the second-order rate constant for the association of complementary fragments.

11. One would expect the $C_0t_{0.5}$ value for a mixture of poly(G) and poly(C) to be substantially lower than that for mouse satellite DNA. Although mouse satellite DNA is highly repetitive when compared with DNA from other sources, the sequence of duplex consisting exclusively of G on one strand and C on the other is even more repetitive.

12. At least two frequency classes of DNA must be present in the sample, one of which anneals more quickly than the other.

13. First, the nucleotide sequence of the template strand is used to determine the mRNA sequence:

Template DNA 3′-GCTACCGATATCCCACT-5′

mRNA 5′-CGAUGGCUAUAGGGUGA-3′

The reading frame is established by the occurrence of AUG.

mRNA 5′-CG AUG GCU AUA GGG UGA-3′

Thus, the amino acid sequence of the protein is

Met-Ala-Met-Gly-Trp

Note that AUA codes for Met rather than Ile and UGA codes for Trp rather than chain termination in mitochondria.

14. In nucleosomes there are two copies each of histones H2A, H2B, H3, and H4 but only a single copy of histone H1. Therefore, one would expect only half as much protein product to be produced from histone H1 as from the other four.

15. For each transcription, the number of molecules of substrate that are converted into product per second is given by

$$10^4\ s^{-1} \times 10^3 \times 10^5 = 10^{12}\ s^{-1}$$

16. The transcription of many eukaryotic genes is activated by proteins containing from 1 to 37 zinc fingers, each of which is a 30-residue polypeptide containing two cysteines and two histidines coordinated to a zinc ion. Zinc is also involved in the structure of DNA-binding domains of the glucocorticoid receptor (see Stryer, p. 1002) and may well be a part of the structure of other members of the nuclear receptor superfamily. It is also involved in a large number of other enzyme-catalyzed reactions, including the conversion of acetaldehyde to ethanol, the formation of bicarbonate ion, and the cleavage of peptides by chymotrypsin. One might suspect that zinc deficiencies could lead to significant developmental abnormalities, but there are no documented instances that can be directly related to the normal function of zinc fingers or glucocorticoid receptors.

17. (a) The presence of CG-rich regions in the upstream region of the *M* gene indicates that these are sequences which may be subject to methylation and which can control transcription by influencing promoter activity. In many cells, about three-fourths of CG sequences are methylated, while those sequences in transcriptionally active regions are undermethylated. In your experiments, cells grown in 5-azacytidine incorporate the analog into DNA, and the azacytosine residue cannot be methylated at N-5 by specific methyltransferases (see Stryer, p. 998). The undermethylated region is therefore more susceptible to RNA polymerase binding as well as to the binding of transcription factors. Transcription of the *M* gene leads to expression of the gene in those cells grown in 5-azacytosine. Cells grown in the absence of the analog are more likely to have methylated CG-rich sequences in the promoter region and are less likely to express the gene you are studying.

(b) If your explanation in (a) is correct, you would expect the CG sequences in the upstream sequence to be methylated in samples taken from cells, but not methylated in samples taken from the plasmid grown in bacteria (methylation of C residues occurs frequently in vertebrates). Thus any expression of the gene you observe from the CAT gene must be due to the activity of its reporter. You would expect that both restriction enzymes would cleave your upstream sequence isolated from the bacterial plasmid, while only MspI would cleave the sequence isolated from chromosomes. The fragment would be resistant to cleavage by HpaII, which does not cleave $C^{m}CGG$ sequences.

18. Assays that exploit the differential mobility of protein-DNA complexes versus free DNA are called *gel-shift assays.* In these experiments, specific DNA sequences are allowed to associate with putative DNA-binding proteins from cell extracts or from fractionated samples. They are then subjected to gel electrophoresis along with samples of unincubated labeled samples of the same fragment. Autoradiography of the gel is used to visualize the results. Labeled samples that are retarded on the gel are candidates for sequence-specific protein-DNA complexes. To minimize retardation caused by nonspecific binding of proteins such as histones to the labeled fragment, large amounts of unlabeled, random sequences can be added to the sample to be visualized on the gel. This DNA binds the nonspecific proteins but is not seen on the gel because it is not labeled. Because sequence-specific proteins usually have a much higher affinity for a specific sequence than those binding nonspecifically, the addition of extra DNA does not interfere with the gel-shift assay. Experiments which assess specificity of the DNA-protein interaction can be conducted by including an excess amount of unlabeled DNA which contains the same specific sequence in an analyzed sample. If the interaction is truly sequence-specific, the unlabeled DNA should bind the protein and abolish the shift of the labeled sequence on the gel. A number of research groups have used gel-shift assays to find a large number of DNA-binding proteins. For a good example of use of these assays, see C. Schneidereit, A. Heguy, and R.G. Roeder. *Cell* 51:(1987) 783–793.

19. The most likely possibility is that each of the two monomers can form homodimers, each of which could bind to one of the sequences above. Also the two monomers can associate to form a heterodimer, which could bind to a third sequence. You would expect an activator that is a homodimer to bind to a DNA sequence which has dyad symmetry (or is an inverted repeat) while a heterodimer could bind to an asymmetrical sequence which contains sequences common to each of the symmetric sequences. Inspection of the three sequences above shows that sequence X and Z have dyad symmetry, while sequence Y is asymmetrical, containing sequences that represent half of X and half of Z. Thus one homodimer binds to and activates sequence X, while another homodimer can associate with the symmetric sequence Z. A heterodimer, composed of each of the two monomers, binds to and activates the asymmetric sequence Y, which contains half of each sequence X and Z. This problem illustrates the important principle that heterodimer formation can allow recognition of DNA sequences that do not have dyad symmetry, thereby increasing the potential for regulation of expression.

20. In bacteria, negative regulation requires synthesis of a specific repressor that blocks transcription of a gene. To carry out negative regulation of genes in a human genome, 100,000 repressor proteins would need to be synthesized, which would be a very inefficient means of controlling transcription. Because most eukaryotic genes are normally inactive with regard

to transcription, selective activation through synthesis of a small group of activator proteins can be used to promote transcription of a particular array of genes needed by the cell at a certain time. Another reason for positive control may be related to the fact that a larger genome affords the possibility that a DNA sequence for a regulatory protein could be present in the wrong location, bringing about inappropriate or unneeded gene activation. This can be avoided by requiring that several positive regulatory proteins form a complex that can specifically activate a gene and promote its transcription, thereby reducing the possibility of unneeded gene activation.

21. Simpson found that the DNA in chromatosomes which contain H1 is more resistant to denaturation by an increase in temperature. This suggested that in addition to its interaction with linker DNA, H1 also interacts with DNA wrapped around the histone octamer.

EXPANDED SOLUTIONS TO TEXT PROBLEMS

1. (a) In prokaryotes, translation begins while transcription is still in progress. In eukaryotes, these processes are separate in space and time.

   (b) Eukaryotic transcripts are monogenic, whereas prokaryotic transcripts are typically polygenic.

   (c) In prokaryotes, several proteins can be specified by a single primary transcript, one for each gene encoding the mRNA. In eukaryotes, a primary transcript encoded by a single gene can give rise to multiple proteins through alternative splicing. Cleavage of a polyprotein can also yield multiple proteins in eukaryotes.

   (d) Most of prokaryotic DNA encodes proteins and functional RNAs. In contrast, most of eukaryotic DNA does not encode functional macromolecules.

   (e) Most prokaryotic genes are clustered in operons. Eukaryotes do not have operons.

2. Centromeres, telomeres, and ARS (autonomous replicating sequences serving as origins of replication) are needed to fashion a synthetic yeast chromosome. See A. Murray and J.W. Szostak. *Nature* 305(1983):189.

3. The mitochondrial genome is much smaller than the nuclear genome. Furthermore, a cell contains a large number of mitochondria. A substantial portion of them, say 5%, could be nonfunctional without injuring the cell.

4. Mitochondrial DNA is evolving at a very rapid rate [see R.L. Cann, M. Stoneking, and A.C. Wilson. *Nature* 325(1986):31]. It is possible but not likely that a human mitochondrial gene will be transferred to the nuclear genome in the next 10 million years. The existence of different genetic codes in the mitochondrion and the cytosol is a formidable barrier to gene transfer. In particular, UGA specifies tryptophan in mitochondrial proteins but is a stop signal in cytosolic protein synthesis.

5. One experimental approach is to add histones to linear duplex DNA and then form closed circles using DNA ligase. The histones are then removed from the supercoiled circles. Topoisomerase I is added to an aliquot to form the relaxed counterpart. The melting temperature of the supercoiled DNA is compared with that of the relaxed circle. A left-handed superhelix has a lower melting temperature than the relaxed circle, whereas a right-handed superhelix has a higher melting temperature. Recall that left-handed supercoiled DNA (negatively supercoiled; *Wr* is negative) is poised to be unwound (Stryer, p. 796).

6. 5S RNA has the same base sequence (except for U in place of T) as the coding strand of its gene, the one that strongly binds TFIIIA. Hence, 5S RNA competes with its gene for the binding of TFIIIA. When 5S RNA is abundant, little TFIIIA is bound to the internal control region, and so transcription is slowed. The binding of TFIIIA to 5S RNA, as well as the internal control region of the gene, is the basis of a feedback loop that regulates the amount of 5S RNA.

7. TFIIIA has nine zinc fingers. Some of its zinc fingers could stay bound to the 5S RNA gene, and others could be released to allow RNA polymerase to pass.

8. The leucine zipper symmetrically orients the two basic DNA-binding regions at the half sites of the palindromic DNA target sequence. The insertion of six residues changes the angular relationship of the pair of basic sites. In contrast, the correct angle is preserved when seven residues are inserted because this number corresponds to a heptad repeat (p. 1003 in Stryer). The insertion of four or five residues also destroys activity. See W.P. Pu and K. Struhl. *Proc. Nat. Acad. Sci.* 88(1991):6901.

9. Covalently cross-link a pair of basic DNA-binding regions by forming a flexible disulfide at their C-terminal end. The resulting protein, devoid of a coiled coil, serves as a transcriptional activator. See R.V. Talanian, C.J. McKnight, and P.S. Kim. *Science* 249(1990):769.

10. Retain the DNA-binding domain but delete the activation domain to block the formation of a functional transcription complex.

11. The mutant receptor lacks a nuclear localization signal, which is required for passage through nuclear pores (p. 932 in Stryer).

12. One approach is to synthesize a truncated cDNA that encodes the DNA-binding domain but not the hormone-binding domain of the new nuclear receptor. The DNA-binding domain produced by expressing this cDNA in *E. coli* should then bind with high affinity to target DNA sequences. A complementary approach is to carry out a domain swap experiment. The target cell could be transfected with a gene for a hybrid receptor containing the hormone-binding domain of the glucocorticoid receptor and the DNA-binding domain of the new receptor. The addition of glucocorticoid to the transfected cell would then activate genes normally stimulated by the unknown ligand.

Glossary

α helix a common structural motif in proteins, in which a polypeptide main chain forms the inner portion of a right-handed helix, with the side chains extending outward; the helix is stabilized by intrachain hydrogen bonds between NH and CO groups of the main chain.

acetylcholine receptor a ligand-gated channel which promotes a large inward current of sodium ion and which triggers an action potential; composed of a pentamer of four kinds of polypeptide subunits, the channel opens to allow passage of sodium and potassium ions when two acetylcholine molecules promote transient opening of the channel.

acetylcholinesterase an enzyme in the synaptic cleft that converts acetylcholine to choline and acetate; this enzyme promotes the closing of the acetylcholine receptor membrane by rapidly reducing the concentration of acetylcholine in the region between the presynaptic and postsynaptic membranes.

actin a highly conserved protein found in all eukaryotes; in striated muscle, it forms the thin filaments of the sarcomere and activates the ATPase of myosin.

actinomycin a polypeptide antibiotic from *Streptomyces* that inhibits the RNA synthesis elongation phase by binding to double-helical DNA by intercalating with it, thereby preventing the DNA from serving as a template.

activated carriers small molecules which carry activated functional groups that can be donated to other molecules; for instance, ATP carries activated phosphate and CoA carries activated acyl groups.

activation domain the structural region of a transcription factor that actually facilitates transcription in some manner; see also DNA-binding domain.

active methyl cycle a series of reactions in which *S*-methyl groups from methionine are converted to a biochemically reactive form through insertion into an adenosyl group; an active *S*-methyl group can be transferred from *S*-adenosylmethionine to acceptor molecules such as norepinephrine. The remaining portion of the cycle includes the regeneration of methionine from homocysteine and N^5-methyltetrahydrofolate.

active site a specific region of an enzyme that binds the substrate and carries out catalysis.

active transport transport of an ion or a molecule against a concentration gradient, where ΔG for the transported species is positive, and where the process must be coupled to input of free energy from a source such as ATP, an electrochemical gradient of Na^+ or K^+, or light.

actomyosin a complex formed in vitro between myosin and actin that displays certain properties of muscles, such as contracting in the presence of ATP.

acyl adenylate a mixed anhydride in which the carboxyl group of a molecule is linked to the phosphoryl group of AMP; formation of acyl adenylates is a means of activating carboxyl groups in biochemical reactions, such as the formation of fatty acyl CoA molecules from a free fatty acid and coenzyme A.

acyl carrier protein a bacterial polypeptide which is linked to phosphopantetheine and which acts as a carrier of the growing fatty acyl chain during fatty acid biosynthesis.

adaptation resetting of the sensitivity level of chemoreceptors due to the continued presence of attractants or repellents.

adenylate cyclase cascade a signal transduction pathway that employs cAMP and a series of enzymes to convert an extracellular signal into an intracellular signal.

adipocyte mammalian cells that are the major storage site for triacylglycerols.

A-DNA helix a right-handed double helix made up of antiparallel strands held together by A·T and G·C base pairing; it is wider and shorter than B-DNA and seen in dehydrated DNA as well as in double-stranded regions of RNA and in RNA–DNA helices.

aerobic in the presence of oxygen.

affinity chromatography a protein purification technique based on the high affinity many proteins have for specific chemical groups. Such groups are attached to an inert matrix, and the protein sample is applied; only those with an affinity for the groups will bind.

affinity labeling a means of mapping the active site of an enzyme by using a substrate analog that binds to the active site and forms a covalent bond with a nearby amino acid.

alcoholic fermentation the conversion of glucose into ethanol with a concomitant production of ATP.

aldol condensation the combination of two carbonyl compounds (e.g., an aldehyde and a ketone) to form a β-hydroxy-carbonyl compound, or aldol.

aldose a monosaccharide whose C-1 carbon contains an aldehyde group.

alkaptonuria a relatively harmless hereditary disorder resulting from aberrant breakdown of tyrosine and phenylalanine.

alleles alternative forms of a gene at a particular site on a chromosome.

allosteric control a conformational change and consequent alteration in catalytic activity induced by binding of a small molecule (a ligand) to a site on a protein that may be some distance away from the active site.

Alu sequence a family of 300-bp sequences occurring nearly a million times in the human genome. The Alu sequence is a short interspersed repeat (SINE) thought to be derived from the RNA of the signal recognition particle

amanitin a cyclic octapeptide from the mushroom *Amanita phalloides* (the destroying angel) that is a potent inhibitor of the elongation phase of RNA synthesis catalyzed by RNA polymerase II.

amino acid an organic acid with an α carbon linked to a carboxylic acid, an amino group, a hydrogen and a side chain (the R group). Twenty different amino acids are the building blocks of proteins.

aminoacyl-tRNA an amino acid ester of transfer RNA.

aminoacyl-tRNA synthetase an enzyme that activates an amino acid and then links it to transfer RNA. Also known as *activating enzymes*, aminoacyl-tRNA synthetases are specific for a particular amino acid.

amino sugars sugars that contain an amino group rather than a hydroxyl group at the C-2 position; the most common are glucosamine and galactosamine.

aminotransferase (transaminase) a class of enzymes that transfer an α-amino group from an α-amino acid to an α-keto acid.

ammonotelic characteristic of organisms in which excess ammonia is directly secreted; many aquatic animals are ammonotelic.

ampere a measure of electrical current: the flow of 6.24×10^{18} charges per second.

amphipathic molecule a molecule, such as a membrane lipid, that contains both a hydrophobic and hydrophilic moiety.

amylopectin the branched form of starch, with about one α-1,6 linkage per thirty α-1,4 linkages.

amylose the unbranched form of starch, containing glucose residues in α-1,4 linkage.

amyotrophic lateral sclerosis (also known as *Lou Gehrig's disease*) a debilitating neurological disease, characterized by degeneration of motor neurons, caused by mutations in superoxide dismutase.

amytal a barbiturate that blocks the respiratory chain by inhibiting electron transfer in the NADH-Q reductase complex.

anaerobic in the absence of oxygen.

anaplerotic reaction from the Greek, "fill up," referring to a reaction that replenishes intermediates removed from a metabolic pathway. The most common example is the reaction catalyzed by pyruvate carboxylase, in which the carboxylation of pyruvate produces oxaloacetate, a key component of the citric acid cycle.

Anderson's disease a disease due to an inability to introduce α-1,6 glycosidic bonds during the synthesis of glycogen. Glycogen is present in normal amounts but with long outer branches; clinically, it is characterized by progressive cirrhosis of the liver.

androgens a class of steroid hormones, exemplified by testosterone, that is responsible for the development of male secondary sexual characteristics; synthesized by the testes.

angstrom (Å) a unit of length equal to 10^{-10} meter.

anomers isomers of cyclic hemiacetals or hemiketals, with different configurations only at the carbonyl carbon; that carbon is known as the *anomeric carbon.*

anterograde transport kinesin-driven movement of organelles from the center of a cell to its periphery.

antibody proteins synthesized by an animal in response to the presence of a foreign substance, or antigen; often binds to the antigen, neutralizing it or marking it for destruction.

anticodon three-nucleotide sequence of tRNA that base-pairs with the codon in mRNA.

antigen a foreign substance that elicits the synthesis of an antibody.

antigenic determinant site on an antigen to which an antibody binds; also called an epitope.

antimycin A an antibiotic from *Streptomyces* that inhibits the respiratory chain by blocking electron transfer in the cytochrome reductase complex.

antiporter a transport system in which a molecule is carried across the membrane in the opposite direction of an ion, which in turn is pumped back across the membrane through active transport linked to ATP hydrolysis.

antiserum serum prepared from the blood of an immunized animal containing soluble antibodies specific for a particular antigen.

ascorbate (vitamin C) a water-soluble vitamin that functions as an antioxidant and is required for the hydroxylation of collagen; scurvy results if ascorbate is deficient.

aspartyl protease a class of protein-degrading enzymes whose activity is dependent on an aspartate at the active site. An aspartyl protease is required for HIV replication.

ATP (adenosine 5′-triphosphate) a nucleotide consisting of adenine, ribose, and triphosphate units that serves as the cellular energy currency.

ATP-ADP translocase (adenine nucleotide carrier) a transport protein that carries ADP into the mitochondria and ATP out in a coupled fashion.

ATP synthase molecular assembly of the inner mitochondrial membrane responsible for the respiratory chain–driven synthesis of ATP; also called *mitochondrial ATPase, H^+-ATPase,* or *F_0F_1-ATPase.*

atractyloside a plant glycoside that is an inhibitor of ATP-ADP translocase.

attenuation in bacteria, a mechanism for transcriptional regulation in which a decrease in the rate of translation of an operon mRNA reduces the rate of transcription of that operon.

autoradiography means of detecting radioactive molecules immobilized in a separation medium such as polyacrylamide; the radioactivity of the molecules will blacken x-ray film.

axoneme the fundamental design structure of cilia and flagella; it consists of a bundle of microtubules, enclosed in a membrane, in which nine microtubule doublets surround two microtubule singlets.

β pleated sheet a common structural motif in proteins, in which two or more β strands are associated as stacks of chains, stabilized by interchain hydrogen bonds; a number of β strands running in the same direction form a β pleated sheet, while such strands running in opposite directions form an antiparallel pleated sheet.

β strand an extended polypeptide chain, with an axial distance of 3.5 Å between adjacent amino acids, often found in a β pleated sheet.

β turn a structural element in proteins composed of four amino acids, in which the CO and NH groups of residue 1 are hydrogen-bonded to the NH and CO groups of residue 4, respectively; such a structure forms a hairpin turn, allowing polypeptide chains to reverse their direction.

bacitracin an antibiotic that blocks the transfer of oligosaccarides from dolichol phosphate to proteins.

bacteriorhodopsin a 26-kd integral membrane protein which absorbs light and converts its energy into proton-motive force, which is then used to synthesize ATP.

B-DNA helix a right-handed double helix with the following characteristics: the two strands are antiparallel; the bases are inside the helix and the phosphates and deoxyribose sugars are on the outside; adenine forms hydrogen bonds with thymine and guanine with cytosine; the bases in each pair are coplanar; there are 10.4 residues per turn, with a pitch of 35 Å.

beriberi a neurologic and cardiovascular disorder caused by a dietary deficiency of thiamine (vitamin B_1).

bile salts polar derivatives of cholesterol that are made in the liver, stored in the gall bladder, and released into the small intestine, where they act as detergents to solubilize dietary lipids, facilitating their digestion and absorption.

biotin a vitamin that plays a role in carboxylation and decarboxylation reactions.

Bohr effect the observation made by Christian Bohr that H^+ and CO_2 promote the release of oxygen from oxyhemoglobin.

bongkrekic acid an antibiotic that inhibits the action of the ATP-ADP translocase.

branch migration the ability of a DNA strand partially paired with its complementary strand to move, displacing its homologous resident strand and extending the pairing.

C4 pathway the transport of carbon dioxide in four-carbon compounds from mesophyll cells to bundle-sheath cells, where carbon fixation occurs; the pathway is found in tropical plants and is designed to prevent photorespiration at high environmental temperatures.

CAAT box a component of many eukaryotic promoters with the consensus sequence 5′-GGNCAATCT-3′.

calcium-ATPase an ATP-driven calcium pump that maintains the large electrochemical gradient of calcium ion across the plasma membrane.

calmodulin a ubiquitous protein in vertebrates that, when bound to calcium, stimulates many enzymes and transporters.

calnexin a chaperone protein anchored in the ER membrane that prevents the export of immature or defective glycoproteins by binding glucose residues on the glycoprotein.

calorie the amount of heat required to raise the temperature of 1 gram of water from 14.5°C to 15.5°C.

Calvin cycle in plants, a cyclic metabolic pathway in which carbon dioxide is incorporated into ribulose 1,5-bisphosphate to give compounds that can be used for the synthesis of glucose.

cap a structure at the 5′ end of eukaryotic mRNA that stabilizes the mRNA and enhances its translation. The cap contains a 7-methyl guanylate attached via a triphosphate linkage to the sugar at the 5′ end of the mRNA via a rare 5′-5′ linkage.

capsid the protein coat surrounding virus DNA or RNA.

carbohydrates sugars or saccharides, which are aldehyde or ketone compounds with multiple hydroxyl groups; also defined as organic compounds with the empirical formula $(CH_2O)_n$.

carbonium ion a carbon compound that contains a positively charged carbon atom; a carbonium atom is critical for catalysis by lysozyme

carboxylase an enzyme that catalyzes a carboxyl transfer reaction; biotin is usually required as a coenzyme.

cardiotonic steroids compounds derived from cholesterol that inhibit the Na^+-K^+ pump by blocking the dephosphorylation of the E_2 conformation of the pump protein. Inhibition of the pump in cardiac tissue cells leads to a higher level of sodium ion in the cell, which slows the extrusion of calcium ion and enhances cardiac muscle contractility.

carnitine a zwitterionic compound formed from lysine that acts as a carrier of long-chain fatty acids from the cytosol to the mitochondrial matrix.

cascade a sequence of reactions, in which at each step a product stimulates an ensuing reaction, generating an amplification of a relatively small stimulus or signal.

catalytic group an amino acid or cofactor at an enzyme active site that directly participates in the making or breaking of covalent bonds.

catalytic RNA one of a class of RNA molecules that display enzymatic activity.

cDNA DNA complementary to an mRNA sequence.

CDP–alcohol activated form of an alcohol that can be used in the synthesis of phosphoglycerides by combining with diacylglycerol at the free hydroxyl on glycerol; for instance, CDP–choline can react with diacylglycerol to form the phospholipid phosphatidylcholine.

CDP–diacylglycerol cytidine diphosphodiacylglycerol; formed from phosphatidate and cytidine triphosphate (CTP); activated diacylglycerol used in the synthesis of phosphoglycerides such as phosphatidyl serine.

cellular immune response a system for cellular recognition of foreign substances that employs cell-attached T-cell receptors to eliminate cells infected by a pathogen or to elicit a particular antigenic response by stimulating B-lymphocyte antibody production.

cellulose an unbranched homopolysaccharide in plants, composed of glucose residues in β-1,4 linkage; the major structural polysaccharide in plants.

centromere the attachment site of mitotic spindles in chromosomes.

ceramide (*N*-acyl sphingosine) a sphingosine with a long-chain acyl group attached to the amino group.

cerebroside a sphingolipid in which glucose or galactose is linked to the terminal hydroxyl of a ceramide.

channel a protein passage which is continuous and which allows ions to flow rapidly through a eukaryotic membrane from a compartment of higher to a compartment of lower concentration. Channels (also known as *pores* in bacteria) are generally composed of four to six subunits or domains and are gated by membrane potential, allosteric effectors, or covalent modification.

chaperone proteins slow ATPases that bind newly synthesized protein and assist their proper folding.

chemiosmotic hypothesis the idea that electron transfer in the respiratory chain is used to pump protons across the inner mitochondrial membrane, establishing a proton gradient; this gradient (the proton-motive force) drives the synthesis of ATP by ATP synthase.

chemoreceptors membrane-spanning proteins that detect attractants and repellents in bacteria.

chemotaxis the directed movement of organisms toward certain specific substances and away from others.

chemotrophs organisms that obtain energy by the oxidation of foodstuffs; see also *phototrophs.*

chi intermediates a DNA structure with four arms (resembling the Greek letter chi, χ) generated when a recombination intermediate of two joined circles is cleaved with a restriction endonuclease.

chimeric DNA a recombinant DNA molecule containing unrelated genes.

chloramphenicol an antibiotic that inhibits the peptidyl transferase activity of the prokaryotic 50S ribosomal subunit.

chlorophyll a substituted tetrapyrrole that is the principal photoreceptor in plants.

chloroplast the plant organelle in which photosynthesis takes place.

cholera toxin a toxin that consists of a catalytic and a membrane-penetration domain; the catalytic subunit ADP-ribosylates the G stimulatory protein, persistently activating it, leading to activation of ion pumps and subsequent efflux of large amounts of Na^+ and water into the gut.

cholesterol a sterol that is an important constituent of eukaryotic membranes as well as lipoproteins; also a precursor of steroid hormones.

chromatin the nucleoprotein chromosomal material consisting mainly of DNA and histones.

chromosome walking a technique for analyzing long stretches of DNA by sequential subcloning and rescreening of overlapping segments.

chylomicrons lipoprotein particles that transport dietary triacylglycerols from the intestine to other tissues; apolipoprotein B-48 is a protein component of chylomicrons.

cilia hairlike organelles made of microtubules that protrude from the surface of the cell and move liquid past the cell surface; responsible for movement of many single-celled organisms.

circular dichroism optical activity of a protein resulting from differential absorption coefficients for circularly polarized light due to the dissymetric nature of proteins; can be used to determine a protein's main-chain conformation.

citric acid cycle a cyclic series of metabolic reactions that completely oxidize acetyl units to carbon dioxide; also known as the *tricarboxylic acid cycle* (after citrate) or the *Krebs cycle,* after Hans Krebs, who elucidated the cyclic nature of the pathway.

clathrin a protein which coats the cytosolic side of coated pits and which can form a lattice around the pit, excising it from the membrane to form a coated vesicle.

clonal selection theory the theory of antibody specificity which postulates that each antibody-producing cell makes antibody of a specific kind and that specificity is established before the cell encounters the particular antigen. Proposed by Burnet and others in the 1950s, the theory has essentially been confirmed through work that shows that over 10^8 different antibody types can be randomly generated through a germ-line repertoire of genes for different variable regions of the antibody molecule, combinatorial association of gene regions through somatic recombination, and somatic mutation.

coated pits specialized regions of the plasma membrane containing localized cell-membrane receptors. The cytosolic side of these indentations is coated with the protein clathrin.

coatomer shell a shell of specific proteins, called *coat proteins,* that bind to the surface of a membrane and lead to budding of the membrane and the formation of transport vesicles.

cobalamin (vitamin B_{12}) a complex vitamin involved in a number of reactions including the formation of deoxyribonucleosides from ribonucleosides.

codon nucleotide triplet in mRNA that codes for a particular amino acid.

coenzyme a small organic molecule required for the activity of many enzymes; vitamins are often components of coenzymes.

coenzyme A (CoA) a coenzyme consisting of an adenine nucleotide linked to pantothenate, which is itself linked to mercaptoethylamine; universal carrier of acyl groups, which are bound to the mercaptoethylamine unit.

coenzyme Q a mobile electron carrier that is a component of the respiratory chain; it shuttles between the oxidized ubiquinone form to the reduced ubiquinol form via a semiquinone intermediate; accepts electrons from NADH-Q reductase as well as succinate-Q reductase.

cohesive ends complementary single-stranded ends in double-stranded DNA molecules produced by staggered cuts. Specific cohesive ends in DNA can be produced by certain restriction enzymes like EcoRI and can then be used to join unrelated DNA molecules by annealing and joining with DNA ligase.

colchicine an alkaloid from the autumn crocus that prevents polymerization of tubulin to form microtubules; as a consequence, it inhibits the cell cycle at metaphase.

committed step the first step in a metabolic pathway that is irreversible under physiologic conditions. This step is catalyzed by an allosteric enzyme and commits the product to a particular chemical fate.

compartmentation the location of metabolites, enzymes, or pathways in different membrane-bound organelles in eukaryotic cells.

competitive inhibition the reduction in the rate of enzyme activity observed when the enzyme can bind the substrate or the inhibitor but not both. Many competitive inhibitors resemble the substrate and compete with it for binding to the active site. Relief from inhibition by saturation with substrate is a kinetic hallmark of competitive inhibition.

complement cascade the series of protein activations that leads to the lysis of microorganisms and infected cells; triggered by antigen-antibody complexes and begins with the activation of C1, the first component of the pathway.

cones vertebrate photoreceptor cells that function in bright light and are responsible for color vision.

consensus sequence idealized base sequence which represents common features of a promoter site.

constant region the carboxyl terminal end of an IgG light chain or an IgG heavy chain. For many antibodies these portions of the polypeptide chains have amino acid sequences that are very similar to one another; a light chain has one constant region, whereas a heavy chain has three such regions, each of which specifies a compact domain in the native immunoglobulin molecule.

constitutive genes genes that are not subject to regulation and are constantly transcribed.

core particle particles resulting from micrococcal nuclease digestion of nucleosomes, consisting of 140-bp DNA and the histone octamer of nucleosomes.

Cori cycle a cyclic metabolic pathway in which lactate from active muscle is converted into glucose by the liver, which in turn supplies newly synthesized glucose to muscle and other tissues.

Cori's disease a disease in which the structure of liver and muscle glycogen is abnormal and the amount is increased; it is due to an inability to hydrolyze the α-1,6 glycosidic bonds in glycogen; liver function is compromised.

covalent intermediate an intermediate in a catalytic mechanism in which a component of the substrate is covalently bound to the enzyme.

creatine phosphate (phosphocreatine) a compound found in muscle with a high phosphoryl transfer potential that is used to regenerate ATP from ADP during the initial seconds of muscle contraction.

cryoenzymology a means of studying enzymes by slowing the catalytic process by cooling the enzyme crystal to very low temperatures.

cyclic AMP 3′,5′-adenosine monophosphate; a cyclic nucleotide formed from ATP and an important second messenger in a variety of signaling systems.

cyclic GMP 3′,5′-guanosine monophosphate; a cyclic nucleotide formed from GTP and an important second messenger in vision.

cyclic photophosphorylation in photosynthesis, the generation of ATP without concomitant formation of NADPH. Electron cycling from the reaction center of photosystem I to ferredoxin and then back to the reaction center via cytochrome *bf* and plastoquinone generates a proton gradient that is used to drive ATP formation.

cyclohexamide an antibiotic that inhibits the peptidyl transferase activity of the 60S ribosomal subunit in eukaryotes.

cytochrome an electron-transferring protein that contains a heme prosthetic group whose iron component shuttles between the ferrous (+2) and ferric (+3) state during electron transfer.

cytochrome *c* a water-soluble, highly conserved cytochrome component of the respiratory chain that accepts electrons from cytochrome reductase and is in turn oxidized by cytochrome oxidase.

cytochrome oxidase (Complex IV) the final complex of the respiratory chain, which transfers electrons from cytochrome *c* to molecular oxygen and concomitantly pumps protons across the inner mitochondrial membrane to generate the proton-motive force.

cytochrome P450 system an electron transport chain found in adrenal mitochondria and liver microsomes in which the terminal component is cytochrome P450. This system plays a role in the detoxification of foreign substances by altering them to increase their solubility and facilitate excretion.

cytochrome reductase (Complex III) a component of the respiratory chain that carries electrons from coenzyme Q to cytochrome *c* and, in the process, pumps protons out of the mitochondrial matrix to generate the proton-motive force.

cytoskeleton internal scaffolding of cells, made up of microfilaments, intermediate filaments, and microtubules, which enables cells to transport vesicles, change shape, and migrate.

dalton a unit of mass very nearly equal to that of a hydrogen ion and precisely equal to 1.000 on the atomic mass scale.

dark reactions in chloroplasts, the reactions in which carbon dioxide is fixed into organic compounds, driven by ATP and NADPH.

decarboxylase an enzyme that removes a carbon from a substrate in the form of carbon dioxide.

decorated filaments structure resulting when S1 heads are mixed with actin filaments. Each S1 head binds to the actin filament with its long axis oriented at 45° to the filament. With all S1 units similarly oriented, a structure looking like a set of arrowheads pointing in the same direction results.

deletion a type of mutation in which one or more base pairs are left out during DNA replication; also, the absence of one or more bases from a nucleotide sequence in a gene.

denaturation a disruption in the native conformation of a macromolecule that causes a loss of normal function.

de novo pathway a biosynthetic pathway that builds the final product from simple precursors (see *salvage pathway*).

density-gradient equilibrium sedimentation a centrifugal technique for separating biological molecules according to their density in an aqueous solution of a salt like cesium chloride. When such solutions are centrifuged at high speeds, opposing forces of sedimentation and diffusion create a concentration gradient of the salt; molecules in that gradient are driven by centrifugal force into the region where the solution density is equal to their own.

deoxyhemoglobin conformation of hemoglobin that is unable to bind oxygen; also called the *T,* or *tense, form.*

deoxynucleotide a nucleotide which contains a purine or pyrimidine base covalently linked to 2-deoxyribose which is in turn linked to one or more phosphate groups.

deoxyribose a five-carbon monosaccharide ($C_5H_9O_5$) that constitutes the carbohydrate moiety of deoxynucleotides;

the deoxyribose commonly found in these molecules is 2-deoxyribose.

diabetes a disease characterized by overproduction of glucose by liver and its underutilization by other organs. Type I diabetes is due to insufficiency of insulin production, while type II is due to inability of insulin receptors to respond to the hormone.

diacylglycerol a second messenger in the phosphoinositide cascade that leads to the activation of protein kinase C.

diagonal electrophoresis means of determining location of disulfide bonds in proteins; a mixture of proteins is electrophoresed in a single lane in one direction, disulfide bonds are irreversibly reduced, and the sample is electrophoresed perpendicularly to the original direction. Peptides migrating as a single band in the first direction will migrate as two bands in the second direction if they contained a disulfide bond.

diastereoisomers pairs of molecules with more than one asymmetric center which have opposite configurations at one such center but which are not mirror images of each other; in the aldotetrose series, D-erythrose and D-threose are diastereoisomers.

dicoumarol an antagonist of vitamin K, which is a crucial factor in the carboxylation of γ carbons in glutamate residues in the amino-terminal region of prothrombin; prothrombin synthesized in the presence of dicoumarol cannot bind calcium ion and therefore cannot promote the blood clotting cascade.

difference Fourier method a method to determine the mode of binding of small molecules to a protein crystal using the x-ray reflections of the unbound and bound crystal.

diphtheria toxin a toxin produced by *Corynebacterium diphtheriae* gains entrance to the cell by receptor-mediated endocytosis and kills the cell by ADP-ribosylation of protein synthesis elongation factor 2, thus inhibiting all protein synthesis.

distributive enzyme an enzyme that catalyzes the elongation or degradation of a polymer but which dissociates from the polymer after each catalytic step.

disulfide bond a covalent bond formed by the oxidation of two sulfhydryl groups; oxidation of cysteine residues in a polypeptide yields a disulfide bond linking the two residues.

DNA gyrase a topoisomerase that catalyzes the ATP-driven introduction of negative supercoils into DNA.

DNA ligase an enzyme that catalyzes the formation of a phosphodiester bond between the 3′-OH at the end of one DNA chain and the 5′-phosphate at the end of the other chain; it is involved in the synthesis, repair, and splicing of DNA.

DNA polymerases enzymes that catalyze the template-directed, primer-dependent addition of deoxynucleotide units, using deoxynucleotide triphosphates as substrates, to the 3′ end of the DNA chain; chain growth is in the 5′-to 3′-direction; such enzymes replicate and repair DNA.

DNA probe a radioactively labeled, single-stranded specific base sequence used to locate a complementary sequence among DNA fragments displayed on an electrophoretic gel.

DNA-binding domain the structural region of a transcription factor that recognizes and binds to a particular DNA sequence; see also *activation domain.*

dolichol phosphate a lipid containing approximately 20 isoprene units that serves as a carrier of oligosaccarides destined to be attached to a protein, through a membrane.

domain an independently folded unit in the tertiary structure of a polypeptide chain; may contain a number of supersecondary structures. In multienzyme complexes, each domain may carry out one or more catalytic reactions. Also, in proteins, a compact globular unit of 100 to 400 residues, possibly joined to other domains by a flexible polypeptide segment; often encoded by a specific exon in the gene encoding the protein.

duplication a nucleotide sequence that is repeated in a gene or in a chromosome.

dynein a large protein with ATPase activity that is a component of microtubules; the ATPase activity provides the power for the movement of cilia and flagella; also, in cytoplasm, a motor protein related to the dynein in flagella and cilia that powers retrograde transport

Edman degradation sequential removal of the N-terminal amino acid from a protein as a phenylthiohydantoin derivative; used in sequencing proteins.

EF hand a helix-loop-helix motif that forms a binding site for calcium; found in many calcium-sensitive proteins.

electrophoresis a technique used to separate charged molecules, such as proteins and nucleic acids, that is based on the fact that such molecules will move at differing rates in an electric field, depending on factors such as net charge, size, and shape of the molecules.

electrospray ionization mass spectrometry a means of determining the mass of a protein. A protein sample is sprayed into a mass spectrometer, and the mass of the protein is determined by the mass to charge ratio of proteins differing in the number of bound protons.

ELISA enzyme-linked immunoabsorbant assay; an assay for quantitating the presence of an antigen using an enzyme linked to an antibody to the antigen.

elongation the second of the three stages of all biological polymerization reactions. In this stage, polymerization occurs by repetition of a basic process characteristic of the specific molecule being synthesized.

Embden-Meyerhof pathway glycolysis.

enantiomers pairs of molecules with one or more chiral centers that are mirror images of each other.

endocytosis process of internalization of extracellular material by invagination and budding of the cell membrane.

endoplasmic reticulum (ER) an extensive system of cytoplasmic membranes that comprises about half the total cell membrane. The region of the ER that binds ribosomes is called the *rough ER* while the region that is devoid of ribosomes is called the *smooth ER.*

endosome a component of the receptor-mediated endocytotic pathway in which sorting decisions about the endocytosized material are made; endosomes are derived from coated vesicles that loose the clathrin coat.

energy charge a means of determining the energy status of the cell, equal to the ratio of the concentration of ATP plus ½ the concentration of ADP, all divided by the total adenine nucleotide concentration.

enhancement phenomenon the increase in the rate of photosynthesis in the presence of light at 600 nm and 700 nm, compared with the rates when each wavelength is given separately.

enhancer sequences DNA sequences which have no promoter activity themselves but which can greatly enhance the activity of other promoters. Enhancers can exert their stimulatory effect over a distance of several thousand nucleotides.

entropy a measure of the degree of randomness or disorder in a system; denoted by the symbol S in thermodynamics, the change in entropy (ΔS) increases when a system becomes more disordered and decreases when the system becomes more ordered or less random.

enzymes biological macromolecules that act as catalysts for biochemical reactions; while almost all are composed of protein, catalytically active RNA molecules have also recently been discovered.

enzyme-substrate complex the product of specific binding between the active site of an enzyme and the substrate.

epimer pairs of molecules with more than one asymmetric center that differ in configuration at only one such center; glucose and galactose are epimers, differing only in the configuration at C-4.

epimerases enzymes that catalyze the interconversion of epimers of a compound.

epinephrine a catecholamine released by the adrenal medulla in response to muscle activity or its anticipation that stimulates the breakdown of muscle glycogen..

epitope the specific site on an antigen that is recognized by an antibody; also known as the *antigenic determinant.*

erythromycin an antibiotic that binds to the 50S ribosomal subunit and inhibits peptide translocation in prokaryotes.

essential amino acid an amino acid that cannot be synthesized de novo and therefore must be acquired from the diet; in adult mammals, at least nine amino acids are considered essential.

eubacteria the most common form of extant prokaryotes.

excinuclease (uvrABC enzyme) an enzyme that removes thymidine dimers by hydrolyzing the damaged DNA strand at sites on either side of the dimer site.

exon shuffling a hypothesis that suggests that new proteins arose in evolution by rearranging exons that encoded discrete structural elements.

exons regions of the pre-mRNA that are retained in the mature mRNA.

extrinsic clotting pathway the blood clotting cascade of enzyme activities which is initiated by the activation of factor VII and the release of the lipoprotein tissue factor, both of which are triggered by tissue trauma.

F_{ab} fragment one of two 50-kd polypeptide chains produced when immunoglobulin G is cleaved by the protease papain; F_{ab} fragments bind antigens but cannot cross-link them, because each fragment has only one binding site for an antigen.

FAD/$FADH_2$ (flavin adenine dinucleotide) an important electron carrier in the oxidation of fuel molecules; alternates between the oxidized FAD form and the reduced $FADH_2$. It consists of a flavin moiety and an AMP unit; electrons are carried on the isoalloxazine moiety of the molecule.

familial hypercholesterolemia a disease that results from dysfunctional receptor-mediated endocytosis of cholesterol-bearing lipoprotein particles. With this disease patients have abnormally high blood levels of cholesterol which can lead to arterial narrowing and heart attacks at an early age.

fatty acid synthase the enzyme system that catalyzes the synthesis of saturated long-chain fatty acids from acetyl CoA, malonyl CoA, and NADPH. In bacteria, the constituent enzymes of the synthase complex can be dissociated when cell extracts are prepared; in mammals, all constituent enzyme activities in fatty acid synthase are part of the same polypeptide.

fatty acids carboxylic acids containing long hydrocarbon chains that are an important fuel source as well as a key component of membrane lipids.

F_c fragment one of two 50-kd polypeptide chains produced when immunoglobulin G is cleaved by the protease papain; the F_c portion of an intact IgG molecule cannot participate in antigen binding but can mediate other important biological activities such as complement fixation.

feedback inhibition a mode of enzyme regulation in which the end product of a biosynthetic pathway inhibits the enzyme that catalyzes the first step in that pathway.

fermentation an ATP-generating process in which organic compounds act as both donors and acceptors of electrons. Fermentations can occur in the absence of oxygen.

ferrihemoglobin hemoglobin in which the iron component of the heme prosthetic group is in the ferric (+3) state; ferrihemoglobin cannot bind oxygen.

ferrohemoglobin hemoglobin in which the iron component of the heme prosthetic group is in the ferrous (+2) state; ferrohemoglobin is capable of binding oxygen.

first messenger hormones; molecules that transduce signals from one tissue to another tissue (the target tissue) via the blood.

fertility factor a bacterial plasmid that contains genes required for conjugation.

flagella, eukaryotic hairlike organelles made up of microtubules that protrude from the cell surface and whose movement propels the cell.

flagella, prokaryotic extracellular appendages used to propel bacteria; the required energy is derived directly from a proton-motive force across the cell membrane.

flagellin protein component of bacterial flagella.

flavins electron carriers that use riboflavin in the electron transfer reactions; FAD/FADH2 and FMN/FMNH2 are flavins.

fluorescence photobleaching recovery technique a technique to measure membrane fluidity in which a membrane component is labeled with a fluorescent molecule which is subsequently destroyed (or bleached) by an intense burst from a laser. Fluidity is determined by the rate at which the bleached region fills with fluorescent molecules from adjoining regions of the membrane.

fluorouracil an analog of dUMP that inhibits the methylation of dUMP to form thymidylate; fluorouracil is used in chemotherapy.

footprinting a technique used to determine the binding site of proteins to DNA. DNA is digested in the presence or absence of the protein, and the sequence protected from digestion by the protein is identified. This sequence is the "footprint" of the protein.

free energy form of energy capable of doing work under conditions of constant temperature and pressure. Also, a measure of the usable energy generated in a chemical reaction; denoted by the symbol G in thermodynamics. The change in free energy (ΔG) of a system undergoing transformation at constant pressure is equal to the change in enthalpy (ΔH) minus the product of the absolute temperature (T) and the change in entropy (ΔS).

freeze-etching electron microscopy a technique derived from freeze-fracture microscopy that permits a view of the interior of a native membrane as well as its surfaces.

freeze-fracture electron microscopy a technique for examining the inside of lipid bilayers. A membrane is frozen and fractured by the impact of a microtome knife. Cleavage occurs in the middle of the bilayer, thereby exposing regions within the membrane.

frictional coefficient a characteristic of a molecule that is proportional to the shape of the molecule. This coefficient times the velocity with which a molecule moves through a medium constitutes viscous drag, a measure of resistance to movement of the molecule.

furanose a five-membered heterocyclic ring formed when a monosaccharide cyclizes to form a hemiacetal or a hemiketal; the five-membered oxygen-containing ring is similar to that of furan.

futile cycle a pair of thermodynamically irreversible biochemical reactions that simultaneously produce and consume a pair of metabolic intermediates; also called *substrate cycle;* these paired pathways may amplify metabolic signals and in some cases can also generate heat for maintenance of temperature in tissues.

G protein a type of regulatory protein that is activated by binding GTP and inactivated by the conversion of GTP to GDP.

galactosemia a disease characterized by vomiting, diarrhea, liver dysfunction, and occasionally mental retardation, caused by galactose accumulation due to a deficiency of galactose 1-phosphate uridyl transferase.

ganglioside a ceramide, common in membranes of the nervous system, in which an oligosaccharide is linked to the ceramide by a glucose residue.

GC box a component of many eukaryotic promoters, especially those from constitutively expressed genes. The consensus sequence for the GC box is 5′-GGGCGG-3′.

gel-filtration chromatography a separation technique based on size differences. A sample is applied to a column consisting of porous beads. Large molecules move through the column faster because they cannot enter the beads and, thus, have a shorter path to travel.

general acid catalysis acid catalysis in which the source of the proton is a donor group rather than a free H^+.

general recombination homologous recombination in which DNA strand exchange can occur between any pair of homologous sequences on parental DNA molecules.

genetic code the relationship between nucleic acid sequence information and protein sequence information.

genomic library a collection of DNA fragments, inserted into vector molecules, that represents the entire genome of an organism.

global regulatory network a group of operons controlled by a single regulatory element. In bacteria, a number of inducible catabolic operons are controlled by cAMP concentration.

globin fold a folding structure of a polypeptide chain, exemplified by myoglobin and hemoglobin subunits, that creates an environment for a heme group to reversibly bind oxygen.

glucagon a polypeptide hormone that is secreted by the α cells of the pancreas when blood glucose level is low and leads to glycogen breakdown in the liver and release of glucose to the blood.

glucocorticoids a class of steroid hormones, synthesized by the adrenal cortex and exemplified by cortisol, that promote gluconeogenesis, the formation of glycogen, and the degradation of fats and proteins.

glucogenic describing amino acids whose carbon skeleton, in toto or in part, can be converted into substrates for gluconeogenesis.

glucose 6-phosphatase a hydrolytic enzyme that converts glucose 6-phosphate to free glucose for release into the blood; it is located in the liver and kidney.

glucose transporter an intramembrane, single polypeptide chain that has 12 transmembrane segments and facilitates the movement of glucose across the plasma membrane into the cell.

glycerol phosphate shuttle a pathway that transfers electrons from cytoplasmic NADH into the mitochondria; dihydroxyacetone phosphate (DHAP) is reduced by NADH to glycerol 3-phosphate, which enters the mitochondria and is oxidized to yield $FADH_2$ and DHAP, which leaves the mitochondria.

glycogen a readily mobilized storage form of glucose in which the glucose monomers are linked by α-1,4 glycosidic bonds and with branches (α-1,6 glycosidic bonds) at about every tenth residue

glycogen phosphorylase the enzyme that catalyzes the phosphorolysis of glycogen to yield glucose 1-phosphate; it is an allosteric enzyme whose activity is further regulated by reversible covalent modification.

glycogen storage diseases any of a number of heritable diseases characterized by an inability to store or utilize glycogen appropriately.

glycogen synthase an allosteric enzyme that can also be reversibly covalently regulated and is responsible for the synthesis of glycogen; it transfers glucose from UDP–glucose to the hydroxyl group at a C-4 terminus of glycogen.

glycogenin a protein bearing an oligosaccharide of α-1,4 glucose units which is the primer for glycogen synthase. Glycogenin catalyzes its own autoglycosylation using UDP–glucose.

glycolipids sugar-containing lipids that are derived from sphingosine. The sugar moiety is attached at the alcohol on sphingosine.

glycolysis a sequence of reactions that convert glucose into pyruvate with a concomitant generation of energy.

glycoproteins proteins that have a specific carbohydrate moiety attached.

glycosaminoglycan a heteropolysaccharide made of repeating disaccharide units and containing the amino sugar glucosamine or galactosamine.

glycosidase a class of enzymes that cleave glycosidic bonds; lysozyme is a glycosidase.

glycosidic bond a covalent bond between an aldehyde or ketone function of a monosaccharide and an oxygen, nitrogen, or sulfur atom of another molecule. The most common glycosidic linkages are *O*-links between the anomeric carbon of a sugar and a hydroxyl group of another saccharide.

glyoxylate cycle a metabolic pathway that converts two-carbon units into succinate for energy production and biosyntheses; found primarily in bacteria and plants, the cycle bypasses two decarboxylation steps in the citric acid cycle and allows net formation of glucose and other molecules via oxaloacetate from acetate or acetyl CoA.

glyoxysomes plant organelles in which enzymes of the glyoxalate pathway are present.

Golgi complex a stack of membranous sacks in the cytoplasm which is the major sorting center for proteins which reside in cell membranes and the lumena of organelles.

gout a disease characterized by inflammation of the joints and kidney due to precipitation of abnormally high levels of sodium urate, a breakdown product of purines.

granum a pile or stack of thylakoid membranes in the chloroplast.

Greek key motif the association of four β strands in antiparallel association in a polypeptide, forming an alternating pattern similar to that depicted in ancient Greek ornaments.

hairpin loops loops of nucleic acid formed by duplex formation within a single strand.

hapten a small foreign molecule that can elicit specific antibody formation when attached to a macromolecule. The dinitrophenyl group is effective at eliciting an antigenic response and has been widely used as a haptenic determinant.

Haworth projection a depiction of a cyclic carbohydrate in which the plane of each ring is perpendicular to the plane of the page, and in which ring carbon atoms are not explicitly shown.

HDL (high-density lipoproteins) a lipoprotein that collects cholesterol released into the blood from dying cells or from membranes undergoing turnover.

heat-shock proteins ubiquitous group of proteins which are synthesized in response to stress, such as a heat shock, and which bind unfolded polypeptides and assist in their refolding.

heavy chain a 50-kd polypeptide that is one of two types of paired chains found in the immunoglobulin G molecule; each heavy (H) chain consists of a variable region and three constant regions, and each chain is linked by a disulfide bond to a light chain.

heavy meromyosin along with light meromyosin, one of the tryptic digestion products of myosin; it retains ATPase activity and the ability to bind actin but does not form filaments.

helicase enzymes that catalyze the ATP-driven unwinding of nucleic acids; DNA helicases are important in DNA replication.

helix-turn-helix a recurring motif found in many DNA-binding proteins, in which two helical α-segments are linked by a short hairpin turn; the two segments are 34 Å apart, which allows them to fit into adjoining major grooves in DNA.

heme prosthetic group of myoglobin and hemoglobin as well as other proteins; it consists of an organic constituent, protoporphyrin, and an iron atom.

hemiacetal a compound formed by the reaction of an aldehyde functional group and a hydroxyl group; for example, the C-1 group of the open-chain form of glucose reacts with the C-5 hydroxyl group to form an intermolecular hemiacetal.

hemiketal a compound formed by the reaction of a ketone group and a hydroxyl group; for example, the C-2 keto group of the open-chain form of fructose reacts with the C-5 hydroxyl group to form an intermolecular hemiketal.

hemoglobin an allosteric protein that is the primary oxygen-carrying protein in vertebrates; it also plays a role in transport of CO_2 and H^+.

hemoglobinopathy a genetic disorder resulting from any number of mutations that produce insufficient amounts of normal hemoglobin or normal amounts of defective hemoglobin chains.

hemophilia a general term for a number of disorders of blood clotting, in which one or another protein in the blood clotting pathway is defective or missing.

hereditary nonpolyposis colorectal cancer (HPCC, or Lynch syndrome) a common form of hereditary colon cancer due to defective correction of DNA mismatches.

Hers' disease a disease resulting from a lack of liver glycogen phosphorylase; glycogen is present in increased amounts with mild clinical effects.

heterogeneous nuclear RNA (hnRNA) a class of nuclear RNAs of various sizes, many of which are processed to mRNA.

Hill coefficient a measure of cooperativity derived from a Hill plot; the Hill coefficient increases with the degree of cooperativity, with the maximum value being the number of binding sites capable of cooperation.

Hill reaction the evolution of oxygen without reduction of carbon dioxide in illuminated chloroplasts; a partial reaction of photosynthesis, in which an artificial electron acceptor like ferricyanide can substitute for carbon dioxide.

histones a highly conserved group of small basic proteins found in eukaryotes in association with DNA.

holoenzyme an enzyme that consists of the protein component forming the main body of the enzyme (the apoenzyme) and any necessary, usually small, cofactors.

homeodomain a 60-residue protein domain, corresponding to the homeo box of the DNA sequence, with a high content of basic residues suggestive of DNA binding ability; proteins with such a domain are gene regulators.

homeotic genes genes found in all eukaryotes that control the fundamental architectural plan of the embryo; these genes code for transcription factors.

homologous recombination recombination between homologous segments of two DNA molecules; also called *general recombination.*

hormone response elements (HREs) specific DNA sequences that bind members of the nuclear receptor family of transcription factors. HREs for steroid receptors are palindromic 6-bp sequences separated by a 3-bp spacer.

hormone-binding domain a conserved 240-residue domain in members of the nuclear receptor superfamily of transcription factors that binds a hormone or hormonelike molecule, thus activating the factor.

humoral immune response a system for recognition of foreign substances that employs soluble antibodies to bind to and inactivate such substances.

hybridization a technique used to determine the relatedness of nucleic acids by assaying the ability of single strands of one sample to form a duplex by complementary base pairing to single stands of another sample.

hybridoma cell cell, resulting from the fusion of an antibody-producing cell and a tumor cell, that produces a single antibody and has an unlimited capacity for proliferation.

hydratase an enzyme that adds the elements of water to a carbon-carbon double bond.

hydropathy plot a means of determining transmembrane sequences in proteins by measuring the change in free energy required to move a 20-amino acid segment of the protein from a hydrophobic environment to water. The free-energy change is plotted against the position of the 20-amino acid sequence in the protein.

hyperammonemia a condition characterized by high levels of ammonia in the blood due to deficiencies in the urea cycle, which can result in brain damage and death.

hyperchromism the increase in absorbance of light by unstacking of base pairs when a DNA duplex is melted into single strands.

hypersensitive sites regions of the chromosome that are especially sensitive to digestion by exogenous DNAses; such regions are usually located at the 5′ end of actively transcribed genes.

hyperuricemia excessively high levels of blood urate; hyperuricemia can induce gout.

hypervariable segment polypeptide segments in immunoglobulin L and H chains which display great sequence variability and which are responsible for antigenic specificity.

hypochromicity decrease in the absorbance coefficient as DNA renatures from the single-stranded to the double-stranded form.

I-cell disease a lysosomal storage disease in which certain hydrolyases are missing from the lysosomes owing to a defect in the synthesis of the lysosomal targeting signal, mannose 6-phosphate.

IDL (intermediate-density lipoproteins) lipoprotein particles that remain after much of the triacylglycerides from VLDL are hydrolyzed and absorbed by other tissues; can be absorbed by the liver or converted into LDL.

immotile-cilia syndrome a disease caused by a variety of lesions in microtubule proteins resulting in chronic respiratory disorders due to defective cilia and infertility due to immobile sperm.

immunoglobulin fold a common structural motif for immunoglobulins, in which two broad sheets of antiparallel β strands enclose hydrophobic side chains and complementarity-determining regions of varable domains pair to form an antigen-binding site.

immunoglobulin see *antibody.*

in situ hybridization a technique in which cells are immobilized, their DNA denatured and then hybridized to radioactive RNA probes. These hybrids are then detected by autoradiography.

induced fit the modification of the shape of an active site in an enzyme after the substrate is bound.

inducer a small molecule that binds to a repressor and alters its interaction with an operator.

inhibitor 1 a protein that, when phosphorylated, inhibits the activity of protein phosphatase 1 and thereby sustains glycogen breakdown and inhibits glycogen synthesis.

initiation the first of the three stages common to all biological polymerization reactions. This stage establishes conditions for the beginning of the polymerization process.

initiation complex a complex of an RNA polymerase, promoter elements, and specific transcription factors that permits the initiation of RNA synthesis.

inositol 1,4,5-trisphosphate a second messenger of the phosphoinositide cascade that causes an increase in intracellular calcium levels.

insertion a type of mutation in which one or more base pairs are erroneously inserted into DNA.

insertion sequences a simple DNA transposon, composed of a kilobase sequence which specifies a transposase and which is bounded on either end by inverted terminal repeats. An insertion sequence can insert itself into any site on a bacterial chromosome.

insertional inactivation gene inactivation that occurs when an unrelated DNA fragment is inserted at a restriction site. Inactivation of a gene for antibiotic resistance in a plasmid can be used to detect those plasmids in which a DNA fragment has been successfully inserted.

instructive theory the theory of antibody specificity which postulates that an antigen acts as a specific template to direct the folding of a newly synthesized antibody chain, thereby conferring specificity of the antibody for the antigen present at the time of folding. Proposed by Linus Pauling in the 1940s, the theory was contradicted by experiments which showed that a specific antibody could be unfolded and refolded in the absence of an antigen without loss of specific affinity for that antigen.

insulin a polypeptide hormone secreted by the α-cells of the pancreas that stimulates fuel storage and protein synthesis.

integral membrane proteins proteins found in membranes that interact extensively with the hydrocarbon chains of the membrane lipids and usually span the membrane.

intercalating agents flat, aromatic compounds that can insert between adjacent base pairs in a DNA double helix; these agents, such as ethidium bromide, can cause insertions and deletions.

intermediate filaments filaments approximately 7–10 nm in diameter that can be components of the cell cytoskeleton in epithelial cells; keratins are intermediate filaments.

intrinsic clotting pathway the blood clotting cascade of enzyme activities that is initiated by the activation of Factor XII through contact of activating proteins (kininogen and kallikrein) with abnormal cell surfaces produced by injury; at least six proteins are activated in the pathway to thrombin formation and the continuing sequence that leads to a blood clot.

intrinsic factor a glycoprotein secreted into the gut to bind vitamin B_{12} and assist in its absorption into the intestinal epithelium.

introns regions of the primary transcript that are removed in the mature mRNA; also called *intervening sequences.*

inversion a nucleotide sequence whose normal order is reversed in a gene or in a chromosome.

inverted terminal repeats sequences of 20 or so base pairs at opposite ends of a bacterial insertion sequence.

ion-exchange chromatography a protein purification technique that relies on the charge of proteins. Proteins are applied to an inert matrix to which is attached a charged moiety (e.g., a carboxylate group). Proteins will bind to the matrix with an affinity proportional to their content of the counterion (i.e., positive charges in the case of the carboxylate matrix).

iron-sulfur proteins (also called *nonheme iron proteins*) proteins that contain clusters of iron and sulfur that play a role in electron transfer reactions; iron cycles between the Fe^{2+} and Fe^{3+} state.

irreversible inhibitor an inhibitor that binds very tightly to its target enzyme, either covalently or noncovalently; such an inhibitor dissociates very slowly from the enzyme.

isoelectric focusing a technique for separating proteins. A mixture of proteins is electrophoresed in a pH gradient; each protein will migrate in the electrical field until it reaches its isoelectric point.

isoelectric point (pI) the pH of a protein at which its net charge is equal to zero.

isomerase an enzyme that catalyzes the interconversion of isomeric forms of a compound.

isopentenylpyrophosphate activated isoprene; the basic building block of cholesterol.

isozymes, isoenzymes enzymes in an organism that catalyze the same reactionbut differ in structure; these differences may rang efrom one to several amino acid residues.

joule the amount of energy needed to apply a 1-newton force over a distance of 1 meter. A kilojoule (kJ) is equal to 1000 J, or to 0.239 kcal.

KDEL sequence a protein sequence (Lys-Asp-Glu-Leu) at the carboxyl terminus of proteins that assures that they will be maintained in the ER; the KDEL sequence serves as a retrieval tag to return ER proteins from the Golgi complex to the ER.

ketogenic amino acids whose carbon skeletons, in toto or in part, are degraded into acetyl CoA or acetoacetyl CoA; only leucine and lysine are solely ketogenic.

ketone body refers to acetoacetate, β-hydroxybutyrate, and acetone, produced when acetyl CoA is diverted from the citric acid cycle to formation of acetoacetyl CoA in liver; subsequent reactions generate the three compounds, known as *ketone bodies.*

ketose a monosaccharide that has a ketone group as its most oxidized carbon.

kilobase a unit of length equal to 1000 base pairs of a double-stranded nucleic acid molecule, or 1000 bases of a single-stranded molecule.

kilocalorie a unit of energy equal to 1000 calories, or to 4.184 joules.

kinase an enzyme that catalyzes the attachment of a phosphoryl group to a substrate, using ATP as a phosphoryl donor.

kinesin a protein with ATPase activity that moves cellular organelles along microtubule tracks in anterograde transport.

kinetochores microtubule-binding sites at the centromeres of daughter chromosomes.

Klenow fragment a proteolytic digestion product of DNA polymerase I that retains the polymerase and 3′ to 5′ exonuclease activity.

Krebs cycle see *citric acid cycle.*

L19 RNA a fragment of rRNA intron from *Tetrahymena thermophila* that has ribonuclease and RNA polymerase activity.

lagging strand the newly synthesized strand of DNA at the replication fork that is initially synthesized as Okazaki fragments.

LDL (low density lipoproteins) the major carrier of cholesterol in the blood; consists of a core of esterified cholesterol molecules surrounded by a shell of phospholipids, unesterified cholesterol, and apoprotein B-100; primary source of cholesterol for cells other than the liver or intestine.

leading strand the newly synthesized strand of DNA at the replication fork that is synthesized continuously; see also *lagging strand.*

Leber's hereditary optic neuropathy a maternally inherited form of blindness that is due to mutations in NADH-Q reductase; one of a number of mitochondrial diseases.

lectins plant proteins with a high affinity for specific sugar residues. As such, they are important probes of carbohydrate containing molecules.

leghemoglobin a homolog of hemoglobin found in leguminous plants that also harbor symbiotic nitrogen-fixing bacteria; leghemoglobin binds oxygen, thereby protecting nitrogenase from inactivation.

Lesch-Nyhan syndrome a disease resulting from the loss of a single enzyme in the salvage pathway for purines; marked by mental retardation, extreme hostility and self-mutilation.

ligand-gated channel a transmembrane channel that is opened by the binding of a one or more molecules to a ligand-binding domain of the channel protein.

light chain a 25-kd polypeptide that is one of two types of chains found in immunoglobulin G. Each L chain consists of a varable region and a constant region, and each chain is linked by a disulfide bond to a heavy chain.

light meromyosin along with heavy meromyosin, one of the tryptic digestion products of myosin; forms filaments but has no ATPase or actin-binding activity.

light reactions in chloroplasts, the reactions in which light is used to create reducing potential and to generate oxygen.

LINES long (several kilobase or more) repeats of unknown function interspersed throughout the eukaryotic genome.

linking number a topological property of circular DNA, which is equal to the number of times a strand of DNA winds around the helix axis.

lipid bilayer a bimolecular sheet formed by amphipathic molecules in which the hydrophobic moieties are on the inside of the sheet and the hydrophilic ones are on the aqueous outside.

lipoic acid 6,8-dithiooctanoic acid, an acyl group carrier that functions as a cofactor in dehydrogenase enzymes. Linked covalently to specific lysine residues in enzyme proteins, it can exist as the reduced open-chain form or the closed-ring disulfide form, undergoing interconversion during a catalytic cycle.

lipolysis the enzymatic hydrolysis of triacylglycerols to free fatty acids and glycerol.

lipoprotein particles particles, consisting of a core of hydrophobic lipids surrounded by a shell of polar lipids and specific proteins, that play a role in the transport of cholesterol and triacylglycerols.

liposomes lipid vesicles in which an aqueous region is enclosed by a lipid bilayer

long terminal repeat a sequence that is repeated at either end of a retroviral DNA molecule.

lovastatin (mevinolin) a competitive inhibitor of HMG CoA reductase, the key regulatory enzyme in cholesterol biosynthesis; used therapeutically to lower cholesterol levels.

lysogenic bacteriophage a phage whose DNA is integrated into the host cell by site-specific recombination and whose expression is repressed.

lysogenic bacterium a bacterial cell whose chromosome contains a prophage.

lytic bacteriophage a phage that replicates in and then lyses, or destroys, its host.

magic spot an unusual nucleotide, ppGppp, that mediates the stringent response by inhibiting initiation of transcription of ribosomal RNA genes and by slowing elongation of other transcripts.

main chain the regularly repeating part of the primary structure of a polypeptide; each unit of the chain includes the C_α as well as the CO and NH groups of an amino acid residue.

major groove a 12-Å-wide, 8.5-Å-deep groove in B-DNA resulting from the fact that the glycosidic bonds of a base pair are not diametrically opposite each other.

major histocompatibility complex integral membrane proteins that bind and display on the cell surface peptides derived from the digestion of proteins from the cytosol (class I MHC proteins) or from endosomal compartments (class II MHC proteins). Foreign peptides bound to class I MHC proteins mark them for destruction by killer T cells, while those bound to class II MHC proteins provide a signal for helper T cells, which can in turn stimulate B lymphocyte production.

malate-aspartate shuttle a reversible shuttle, found in the liver and heart, used to transport electrons from cytoplasmic NADH to mitochondrial NAD^+.

maple syrup disease a disease resulting from the inability to oxidatively decarboxylate branch-chain amino acids, characterized by mental and physical retardation and urine that smells like maple syrup.

Maxam-Gilbert method a DNA-sequencing technique that employs preferential chemical cleavage at one of the four nucleotides in a single-stranded molecule; separation of the fragments on electrophoretic gels and comparison of the gel patterns with those from molecules cleaved at each of the other three nucleotides reveal the linear sequence of bases in DNA.

maximal velocity the highest rate of an enzyme-catalyzed reaction, under conditions of constant enzyme concentration and saturating amounts of substrate.

McArdle's disease a disease caused by a lack of muscle glycogen phosphorylase; the glycogen is present in increased amounts but normal structure; clinical characteristics include an inability to perform strenuous exercise.

melting temperature the temperature at which secondary or higher structures of a biological molecule are lost; for a nucleic acid, the melting temperature is defined as the temperature at which half the helical structure is lost.

membrane asymmetry refers to the fact that the two phases of biologically important membranes differ from each other.

membranes sheetlike structures composed of lipids and proteins, usually only a few molecules thick, that form closed boundaries between different compartments. Membranes separate aqueous environments

messenger RNA template for protein synthesis. The base sequence of mRNA is complimentary to a gene in DNA.

metabolism a highly integrated network of chemical pathways that enables a cell to extract energy from the environment and use this energy for biosynthetic purposes.

methotrexate (amethopterin) a competitive inhibitor of dihydro reductase, an enzyme required for the regeneration of the coenzyme tetrahydrofolate, which is necessary for the synthesis of thymidylate. Methotrexate is used as a chemotherapeutic agent.

micelle a globular structure formed by an amphipathic molecules in which the hydrophilic portion is exposed to water and the hydrophobic portion is sequestered inside, away from the water.

Michaelis constant the concentration of substrate at which half the active sites of an enzyme are filled; a ratio of rate constants for the reaction model. $E + S \underset{k_2}{\overset{k_1}{\rightleftharpoons}} ES \overset{k_3}{\rightarrow} E + P$. The constant is defined as $K_M = k_2 + k_3/k_1$. Under conditions where k_2 is much greater than k_3, K_M is equal to the dissociation constant of the enzyme-substrate complex and is a measure of the affinity of the enzyme for the substrate.

Michaelis-Menten equation an equation that expresses the velocity (V) of an enzyme-catalyzed reaction in terms of maximum velocity (Vmax), substrate concentration (S), and the Michaelis-Menten constant (K_M). The equation accounts for the hyperbolic kinetics observed when V is plotted as a function of S; the equation is $V = V_{max}[S]/([S] + K_M)$.

microfilaments filaments of actin approximately 7 nm in diameter; filaments (F-actin) are formed by the polymerization of G-actin monomers; a component of the cytoskeleton.

microsomes closed vesicles formed by self-annealing fragments of the endoplasmic reticulum following cell disruption.

microtubule cytoskeleton element that is a major component of cilia, eukaryotic flagella, and the mitotic spindle; composed primarily of α- and β-tubulin; capable of rapid assembly and disassembly.

microtubule-organizing centers (MTOC) sites of initiation of microtubule growth.

mineralcorticoids a class of steroid hormones, synthesized by the adrenal cortex and exemplified by aldosterone, that act on the kidney to increase absorption of Na^+ and excretion of K^+ and H^+.

minor groove a 6-Å-wide, 7.5-Å-deep groove in B-DNA that arises because the glycosidic bonds of a base pair are not diametrically opposite one another

mitochondria oval-shaped organelles about 2μm in length and 0.5μm in diameter that are the site of oxidative phosphorylation, the enzymes of the citric acid cycle, and the enzymes of fatty acid oxidation.

mitochondrial diseases a set of diseases resulting from mutations in mitochondrial DNA. They are most prevalent in tissues that depend heavily on oxidative phosphorylation, such as the heart and nervous system.

molecular disease a disease whose basis is understood in molecular terms, such as sickle cell anemia.

molten globule an observed intermediate in the transition between the folded and unfolded forms of a protein; the condensed structure contains native secondary but not tertiary structure.

monoclonal antibody an antibody derived from a clone, or a large number of cells which are all descended from the same cell and which have identical properties; normally antigens with a common specificity are heterogeneous because they are produced by a heterogeneous group of cells; fusion of a single antigen-producing cell with an immortal myeloma cell facilitates the production of large amounts of homogeneous antibody protein, these antibodies are valuable analytical and preparative reagents.

monooxygenases enzymes that use O_2 and incorporate one atom of oxygen into a substrate and reduce the other atom to water; these enzymes are important in synthesis of steroid hormones and of tyrosine.

monosaccharides single aldehydes or ketones that have two or more hydroxyl groups; the simplest carbohydrates.

morphogens substances that induce differentiation or control its pattern by associating with specific nuclear receptor transcription factors.

mosaic protein protein encoded by a gene assembled by exon shuffling; the exons code for structural units of the protein.

multienzyme complex a polypeptide chain that contains domains for two or more enzymatic activities.

multifunctional enzymes functionally related enzymes that are covalently linked in some fashion.

mutarotation for carbohydrates, the interconversion of α and β anomers through the open-chain form; usually measured through changes in optical rotation.

mutase an enzyme that catalyzes the intramolecular shift of a chemical group.

myofibrils fibrils inside the cytosol of vertebrate muscle cells that give such cells their striated appearance; the functional unit of a myofibril is a sacromere.

myoglobin a vertebrate oxygen-storage protein found in muscle; structurally, it is similar to an individual hemoglobin subunit.

myosin a protein that forms the thick filaments of striated muscle. It displays ATPase activity at its globular head which, in conjunction with the ability to reversibly bind actin at its fibrous region, provides the power stroke of muscle contraction.

NAD-binding domain a structural motif of NAD^+-linked dehydrogenases that forms a binding site for NAD^+ and consists of four helices and six parallel beta strands.

NADH-Q reductase (NADH dehydrogenase, or Complex I) a large component of the respiratory chain that transfers electrons from NADH to ubiquinone and in the process pumps protons across the inner mitochondrial membrane to generate the proton-motive force.

NAD^+/NADH (nicotinamide adenine dinucleotide) an important electron carrier in the oxidation of fuel molecules; electrons are carried on the nicotinamide moiety of the coenzyme.

$NADP^+$/NADPH (nicotinamide adenine dinucleotide phosphate) the electron donor for reductive biosynthesis; differs from NAD^+/NADH in that a phosphate is attached to the adenine ribose at position 3.

native form the stable, functional conformation of a biological macromolecule.

nerve impulse also known as an *action potential*, the increase in membrane potential and the changes in sodium and potassium conductances that result from alterations in the permeability of the axon membrane to those ions.

neurotransmitter small, diffusible molecules like acetylcholine that mediate the passage of nerve impulses across the synapse.

nicotinate (niacin) vitamin that is a key component of the electron transfer coenzymes NAD^+/NADH and $NADP^+$/NADPH.

nitrogen fixation the conversion of diatomic nitrogen to ammonia; the first step in the flow of nitrogen into amino acids, nucleotides, and other nitrogen-containing compounds in organisms.

nitrogenase an enzyme complex that catalyzes the reduction of diatomic nitrogen to ammonia; found in bacteria and the blue-green algae.

noncompetitive inhibition the reduction in the rate of enzyme activity observed when the enzyme can bind the substrate and the inhibitor simultaneously. Noncompetitive inhibitors decrease the turnover number for an enzyme but do not diminish the proportion of enzyme molecules bound to the substrate; their effects are not overcome by increasing substrate concentration.

Northern blotting a technique analogous to Southern blotting, in which a mixture of RNA fragments is separated by electrophoresis, transferred to a nitrocellulose sheet, hybridized to a radioactively labeled DNA probe complementary to the desired sequence, and visualized by autoradiography. The technique can therefore be used to locate and identify an RNA fragment containing a specific sequence.

nuclear envelope a double membrane, consisting of the inner nuclear membrane and the outer nuclear membrane, which surrounds the nucleus and which is punctuated with openings called *nuclear pores.*

nuclear localization sequence amino acid sequences that direct the protein bearing such sequences into the nucleus.

nuclear magnetic resonance means of determining the structure of a protein in solution based on the ability of certain atoms in a protein to absorb electromagnetic radiation.

nuclear pores a complex protein assembly that provides openings in the nuclear membrane and permits transit of large molecules into and out of the nucleus.

nuclear receptor superfamily a class of transcription factors that bind DNA and activate transcription only in the presence of a specific signal molecule, such as a hormone.

nucleoside a purine or pyrimidine base linked to a sugar.

nucleosomes the repeating unit of chromatin that consists of 200 bp of DNA and two each of the histones H2A, H2B, H3, and H4.

nucleotide a nitrogenous purine or pyrimidine base linked to a sugar that is in turn linked to one or more phosphate groups.

nucleotide kinase enzymes that transfer the phosphoryl group of one nucleotide to another nucleotide, as when ATP reacts with UMP to form UDP and ADP.

Okazaki fragments small fragments of DNA (approximately 1000 nucleotides) that are formed on the lagging strand at the replication fork of DNA synthesis and later joined. They enable $5' \rightarrow 3'$ polymerization at the nucleotide level while overall growth is in the $3' \rightarrow 5'$ direction.

oligomycin an antibiotic that blocks ATP synthesis (and consequently the respiratory chain) by interfering with proton flow through the ATP synthase complex.

oligosaccharides carbohydrates composed of 2 to 10 or 12 monosaccharide units.

one-carbon metabolism biochemical reactions in which tetrahydrofolate derivatives serve as donors of a variety of one-carbon units, ranging from the oxidation level of a methyl group to a formyl group; also includes reactions involving the fully oxidized carbon unit, carbon dioxide, and its carrier, biotin.

operator a DNA segment which is adjacent to a group of structural genes and which is the target sequence for a repressor protein; a unit of gene regulation and expression that includes structural genes and regulatory elements recognized by one or more regulatory gene products.

operon an operator sequence and its associated structural genes.

optical rotation the change in direction of polarization of linearly polarized light on passing through an optically active solution; can be used as an indicator of the main-chain conformation of a protein.

organic fluorophosphates compounds like diisopropylfluorophosphate, which inhibits acetylcholinesterase by forming a stable phosphoryl enzyme complex at a serine residue in the active site.

osteomalacia a clinical condition in adults that is due to vitamin D deficiency and is characterized by softening and weakening of the bones.

β-oxidation oxidation of the C-3 carbon, which is β to a functional group; in the degradation of a fatty acyl CoA molecule, the sequence of oxidation, hydration, and oxidation reactions that converts a methylene group at C-3 to a β-keto group.

oxidative phosphorylation the process in which ATP is formed as the result of the transfer of electrons from NADH or $FADH_2$ to O_2 via a series of electron carriers.

oxygen dissociation curve a plot of the oxygen-binding capacity of a protein versus the partial pressure of oxygen.

packing ratio the degree of condensation of DNA in chromatin. It is the ratio of linear DNA to the length of the packaged DNA. The packing ratio of metaphase human chromosomes is 10^4.

palindrome a word, sentence, or verse that reads the same from right to left as it does from left to right. An example is "*radar*". By extension to biochemistry, a sequence of double-stranded DNA that is the same in each strand when the strands are read in the same direction; that is, it displays a twofold rotational symmetry, for example: 3′-CCTAGG-5′/5′-GGATCC-3′.

pantothenate a vitamin that is a key component of coenzyme A.

passive transport transport of an ion or a molecule down a concentration gradient, where ΔG for the transported species is negative.

Pasteur effect inhibition of glycolysis by respiration, discovered by Louis Pasteur. The rate of glycolysis is lower in the presence of oxygen than under anerobic conditions, a phenomenon largely due to inhibition of phosphofructokinase by ATP and citrate.

patch-clamp technique a method for studying ion channels in which a high-resistance seal is formed between a pipette and a small patch of plasma membrane, allowing the monitoring of the flow of ions through a single channel with high time resolution.

pellagra a disease caused by a dietary deficiency of tryptophan and nicotinate and characterized by dermatitis, diarrhea, and dementia.

pentose phosphate pathway a metabolic pathway that generates NADPH and five-carbon sugars such as ribose 5-phosphate from glucose 6-phosphate; also referred to as the *hexose monophosphate shunt* and the *phosphogluconate pathway.* It includes oxidative reactions which produce NADPH and ribose 5-phosphates as well as nonoxidative reactions which together convert five-carbon sugar phosphates to gluconeogenic precursors of glucose 6-phosphate.

peptide two or more amino acids joined by peptide bonds.

peptidoglycan a macromolecule that consists of linear polysaccharides cross-linked by short peptides. Often found in bacterial cell walls, peptidoglycans confer mechanical support and protect bacterial cells from disruption by osmotic pressure.

peptide bond a covalent linkage formed between the α-carboxyl group of one amino acid and the α-amino group of another; also known as an *amide bond.*

peripheral membrane protein proteins that are associated with the surface of membranes by electrostatic and hydrogen bond interactions.

pernicious anemia a disease in which vitamin B_{12} absorption is impaired owing to a deficiency of intrinsic factor, which results in an inability to synthesize thymine and purines.

peroxidase heme enzymes catalyzing the reduction of an alkyl peroxide to produce an alcohol and water.

peroxisomes small membrane-bound organelles that are present in most eukaryotes and play a role in detoxification, synthesis of plasmologens and bile salts, and β-oxidation of long-chain fatty acids.

phase variation in *Salmonella,* alternative expression of two types of flagellar proteins, which allows the cell to evade the host immune response.

phenylketonuria a disease caused by the inability to convert phenylalanine into tyrosine, which results in excess phenylalanine and its secondary metabolites; the disease is characterized by severe retardation.

phorbol esters polycyclic alcohol derivatives of croton oil that resemble diacylglycerol and inappropriately stimulate protein kinase C; phorbol esters are tumor promoters.

phosphatase an enzyme that catalyzes the removal of a phosphoryl group from a substrate by hydrolysis.

phosphodiesterase an enzyme that converts cyclic nucleotides e.g., cAMP to the noncyclic form e.g., AMP.

phosphoinositide cascade a set of reactions that convert an extracellular signal into an intracellular one. This conversion involves the cleavage of the phospholipid phosphatidyl inositol 4,5-bisphosphate into two second messengers, inositol 1,4,5-trisphosphate and diacylglycerol.

phospholipases a class of enzymes of varying specificity that catalyze the degradation of phospholipids. They can function as digestive enzymes as well as components of signal transduction pathways.

phospholipids important constituents of membranes, composed of three components: a backbone (usually glycerol or sphingosine), two fatty acid chains, and a phosphorylated alcohol.

phosphorolysis the cleavage of a bond by orthophosphate, as in the degradation of glycogen to glucose 1-phosphate.

phosphoryl transfer potential a measure of the tendency of a phosphorylated compound to transfer a phosphate to another compound; presented as the $G^{\circ\prime}$ of hydrolysis of the phosphate compound.

photoreceptors membrane proteins that can convert light to atomic motion and then into a chemical signal.

photorespiration the conversion of organic carbon into carbon dioxide without production of energy-rich metabolites; the result of the oxygenase reaction catalyzed by rubisco and the subsequent regeneration of glucose from two molecules of glycolate, with the release of carbon dioxide and ammonia and the consumption of ATP.

photosynthetic unit a light-harvesting complex that includes about 2500 antenna chlorophyll molecules and a reaction center chlorophyll pair, all in the thylakoid membrane.

photosystem I in chloroplasts, a photosynthetic unit that includes a light-harvesting complex, a reaction center and an electron-transport chain. The system catalyzes the light-driven transfer of electrons from reduced plastoquinone to ferredoxin, which in turn drives the formation of NADPH; it requires light of wavelength shorter than 700 nm.

photosystem II in chloroplasts, a photosynthetic unit that includes a light-harvesting complex, a reaction center, and an electron-transport chain. The system catalyzes the light-driven transfer of electrons from water to plastoquinone, with the concomitant generation of oxygen; it requires light of wavelength shorter than 680 nm.

phototrophs organisms that can meet their energy needs by converting light energy into chemical energy.

plasmalogens phospholipids containing an α,β-unsaturated ether at the C-1 position of the glycerol; abundant in phospholipids of the nervous system

plasmids circular duplex DNA molecules that replicate autonomously and act as accessory chromosomes in bacteria; they carry useful genes but are disposable under certain conditions.

poly(A) tail a long (up to 250 nucleotides) polyadenylate segment that is added post-transcriptionally to the 3′ end of most eukaryotic mRNA.

polycistron an mRNA molecule that codes for more than one protein.

polyclonal antibodies antibodies that are the products of many different populations of antibody-producing cells.

polymerase chain reaction a method for amplifying DNA sequences using DNA polymerase; a series of three-step cycles is employed, in which parental DNA strands are separated by heating, primers to flanking regions of the target sequence are annealed to the separated strands, and the primers are then extended by DNA synthesis.

polymerase an enzyme that catalyzes the step-by-step addition of ribo- or deoxyribonucleotide units to a polynucleotide chain.

polypeptide a series of amino acids, each joined by a peptide bond.

polyprotein a protein that when appropriately cleaved yields several smaller functional proteins.

polysaccharides carbohydrates composed of large numbers of linear or branched monosaccharide units; homopolysaccharides are composed of large numbers of one type of sugar, while heteropolysaccharides contain more than one type.

polysome, polyribosome a group of ribosomes bound to an mRNA molecule and simultaneously carrying out translation.

polytene chromosomes giant chromosomes, for instance, those of *Drosophila* salivary glands, which contain more than a thousand replicated DNA molecules that have not separated.

Pompe's disease a disease affecting all tissues and due to a lack of a lysosomal glycogen-degrading enzyme. The glycogen is structurally normal but present in abnormally large amounts. Death results at an early age from cardiovascular failure.

porin a transmembrane protein that forms channels in the outer mitochondrial membrane, permitting passage of small molecules and ions to the inner membrane space.

porphyrias inherited or acquired metabolic disorders caused by an enzyme deficiency in the biosynthetic pathway for heme; often characterized by the accumulation of one or more pathway intermediates in blood or other tissures, as well as their excretion in urine.

presequences an amino-terminal amino acid sequence that is recognized by receptors on the external face of the outer mitochondrial membrane and leads to the import of the protein bearing it into the matrix.

Pribnow box a promoter component of prokaryotic genes that has the consensus sequence 5′-TATAAA-3′ located at $^{-}$10 bp from the transcription start site.

primary structure usually refers to the linear sequence of amino acids in a protein; more generally, the linear sequence of units that form a polymer.

primase a specialized RNA polymerase that synthesizes the RNA primers for DNA synthesis.

primer in elongation of polymers, the initial segment of the polymer that is to be extended and on which elongation is dependent.

primosome complex of proteins that facilitate the unwinding of DNA and synthesis of RNA primers, thus initiating DNA synthesis.

processive enzyme an enzyme that catalyzes multiple rounds of elongation or digestion of a polymer while the polymer stays bound to the enzyme.

proenzyme a zymogen, or a catalytically inactive precursor of an enzyme; a proenzyme can be converted to the active form by the hydrolysis of one or a few peptide bonds.

progestagen a class of steroid hormone, exemplified by progesterone, that prepares the uterus for implantation of the ovum; synthesized by the corpus leuteum of the ovary.

promoter a specific sequence of DNA, usually just upstream of a gene, that specifies the site and extent of transcription of the associated gene.

propeller twisting the twisting of base pairs in crystals of DNA from the coplanar conformation; this enhances the stacking of bases within a strand.

prophage phage DNA that is integrated into the host chromosome.

prostaglandin a class of short-lived signal molecules that are 20-carbon fatty acids containing a five-membered ring.

protamines a series of arginine-rich proteins that replace histones in packaging DNA in sperm.

protease enzymes that degrade proteins by cleaving peptide bonds.

proteasomes cytosolic assemblies that carry out routine degradation of cellular proteins as well as of those from pathogens. Peptide fragments from proteasome action are eventually transported to the lumen of the endoplasmic reticulum where they combine with nascent class I MHC proteins and are then targeted to the plasma membrane.

protein disulfide isomerase an enzyme that catalyzes the formation of correct disulfide pairings in nascent proteins. Preferentially reacting with peptides that contain cysteine residues but otherwise undiscriminating, the enzyme speeds up the disulfide shuffling required for a protein to find the most thermodynamically stable disulfide pairings among those that can be formed.

protein kinase a class of enzymes that transfer a phosphoryl group from ATP to proteins; protein kinases are frequently found in regulatory pathways.

protein phosphatases enzymes that hydrolyze phosphorylated serine and threonine residues in other proteins. Protein phosphatase 1 reverses the regulatory effects of kinases on glycogen metabolism.

protein a biological macromolecule composed of a linear array of amino acids joined by peptide bonds. Roles of proteins in biological processes include catalysis, transport and storage, motion, mechanical support, immune protection, generation and transmission of nerve impulses, and control of growth and differentiation.

proteoglycan proteins containing one or more covalently linked glycosaminoglycan chains. Cartilage proteoglycan contains keratan sulfate and chondroitin chains linked to a polypeptide backbone.

proton-motive force the energy inherent in the proton gradient established during the functioning of the repiratory chain; consists of a membrane potential as well as a chemical gradient.

protoporphyrin organic constituent of the heme prosthetic group; it consists of four pyrrole rings joined by methylene bridges and contains various side chains.

pseudogenes sequences of DNA that resemble actual genes but do not code for functional products.

pulsed field electrophoresis an electrophoretic technique for separating large DNA molecules. Electric fields, oriented at 120° to each other, are applied in an alternating fashion. Large molecules reorient more slowly than small ones and hence cannot follow the electric field as rapidly.

pulsed hydrogen-deuterium exchange a technique used to assess the pathway for protein folding, in which a protein is unfolded in a D_2O-denaturant solution to convert amide NH groups to ND groups. Dilution in D_2O allows refolding, and many DH groups become inaccessible for exchange, as secondary and tertiary structures are established during refolding; quick dilution in H_2O at pH 9 followed by dropping the pH to 4 some ms later, along with the monitoring of the exchange of specific exchangeable sites using two-dimensional NMR, allows visualization of the refolding pathway, residue by residue.

pump a protein passage in a membrane that can transport a molecule from one compartment to another against a concentration gradient. Also known as *active transporters*, pumps undergo a cycle of conformational changes that alter the affinity of the binding site for the transported molecule, and the eversion of the molecule during the cycle is driven by expenditure of free energy.

purine a nitrogenous base that includes a pyrimidine ring fused with a five-membered imidazole ring; the purine derivatives adenine and guanine are found in nucleotides and nucleic acids.

puromycin an antibiotic that is an analog of the terminal aminoacyl-adenosine portion of aminoacyl-tRNA; during translation, it causes premature chain termination when its amino group joins the carboxyl group of the growing polypeptide chain and the resulting adduct dissociates from the ribosomal complex.

pyranose a six-membered heterocyclic ring formed when a monosaccharide cyclizes to form a hemiacetal or a hemiketal; the six-membered, oxygen-containing ring is similar to that of pyran.

pyridine nucleotides nucleotides, such as NAD^+/NADH and $NADP^+$/NADPH that are employed in redox reactions; the functional group is nicotinamide, a derivative of pyridine.

pyridoxal phosphate prosthetic group derived from vitamin B_6 (pyridoxine) that plays a key role in transamination reactions.

pyrimidine a nitrogenous base that is a six-membered heterocyclic ring containing two nitrogen and four carbon atoms; the pyrimidine derivatives cytosine, uracil and thymine are found in nucleotides and nucleic acids.

pyrimidine dimers a mutation in which exposure to ultraviolet light causes the covalent linkage of two adjacent pyrimidine residues.

pyrrole a five-membered heterocylic diene ring compound in which the hetero atom is nitrogen; it is a building block of the heme group.

Q cycle a set of reactions in which coenzyme Q cycles between the fully reduced state and the fully oxidized state via one-electron transfer reactions in which one of the electrons is temporarily stored in cytochrome *b;* provides a means of passing the two electrons of coenzyme Q to the single electron carrier cytochrome *c*, one electron at a time.

quaternary structure in proteins containing more than one polypeptide chain, the spatial arrangements of those chains (subunits) and the nature of contacts among them.

Ramachandran plot a steric contour diagram that depicts allowed ranges of the angles ϕ and ψ for amino acid residues in polypeptide chains. For each residue, its conformation in the main chain of a polypeptide can be completely defined by ϕ (the degree of rotation at the bond between the nitrogen and the α carbon atoms) and ψ (the degree of rotation at the bond between the α carbon and the carbonyl carbon atoms).

random-coil conformation the disordered, unfolded structure of a denatured macromolecule; many proteins that are not cross-linked lose their native three-dimensional structure as well as their biological function in the presence of unfolding agents such as urea.

reaction center in a photosynthetic unit, a specialized chlorophyll molecule that collects excitation energy from other chlorophylls and mediates the transformation of light into chemical energy.

receptor-mediated endocytosis a means of importing specific proteins into a cell by their binding to plasma membrane receptors and their subsequent endocytosis and inclusion into vesicles.

receptor tyrosine kinase transmembrane receptor proteins that, when bound to the appropriate signal molecules, display intracellular protein kinase activity, phosphorylating proteins on a tyrosine residue.

recombinant DNA technology an array of techniques used to analyze and manipulate DNA; these methods include specific modification of genes as well as construction of new ones, gene cloning and amplification, and expression of new and modified genes to yield protein products.

recombination formation of new arrangements of genes or gene sequences by movement of DNA.

red drop the reduction in the quantum yield of photosynthesis at wavelengths greater than 680 nm.

reductase an enzyme that catalyzes reduction of a functional group, often using NADPH as an electron donor; a type of oxidoreductase.

reductive biosynthesis refers to anabolic pathways that require hydride ion donors to reduce carbon atoms in metabolic intermediates; NADPH serves as an electron donor in many such pathways, including the biosynthesis of palmitoyl CoA.

regulatory proteins proteins that bind to enzymes and regulate their catalytic activity.

relaxed DNA a circular DNA molecule that has no superhelical turns.

release factors proteins that recognize stop codons and terminate translation by releasing the polypeptide chain from the ribosome.

repair the restoration of a normal structure and sequence of a gene after damage from ultraviolet light or chemical agents.

replication fork the site of DNA synthesis where the parental strands are separated and daughter strands complimentary to each parent are synthesized.

replicon a segment of DNA that carries its own origin of replication and can replicate autonomously; bacterial plasmids are replicons.

repressor a protein that binds to an operator sequence and inhibits transcription of the structural genes in the operon.

resistance transfer factor a DNA sequence that allows a plasmid carrying genes for drug resistance to be transferred to other bacteria by conjugation.

respiration an ATP-generating process in which an inorganic compound, such as O_2, serves as the ultimate electron acceptor; the electron donor can be either an organic compound or an inorganic one.

respiratory chain also called the *electron transport chain*; the path that electrons travel in going from NADH or FADH2 to O_2; it consists of three complexes that pump protons as a result of the electron transport and two mobile electron carriers.

respiratory control tight coupling or coordination of the oxidation of reduced cofactors (NADH, $FADH_2$) in the electron transport chain and the phosphorylation of ADP to yield ATP in the mitochondrion; such control ensures that the rate of the citric acid cycle, where reduced cofactors are generated, corresponds to the demand for ATP.

restriction enzyme endonuclease enzymes that recognize specific base sequences in double-stranded DNA and cleave both strands of the duplex at specific places.

restriction-fragment-length polymorphism genetic diversity within a population reflected in mutations within specific sites in DNA; such mutations alter the position of restriction fragments during electrophoretic gel analysis.

retrograde transport cytoplasmic dynein-driven transport of organelles from the periphery to the center of the cell.

retrotransposons (retroposons) transposable elements that use a replicative pathway from DNA to RNA back to DNA to replicate; includes Ty elements in yeast and copia elements in *Drosophila.*

retrovirus a virus which contains an RNA genome but which replicates via a double-stranded DNA intermediate that is integrated into the host-cell genome.

reverse transcriptase an enzyme that synthesizes DNA using an RNA template.

reversible covalent modification the attachment to and removal of chemical groups from an enzyme and the consequent change in the catalytic properties of that enzyme. Catalytic properties of many enzymes are altered by the covalent attachment and removal of phosphoryl groups, while a lesser number of others undergo reversible attachment of AMP units from ATP.

rhodopsin the photoreceptor of rod cells. It is composed of the protein opsin and the prosthetic group 11-*cis*-retinal.

rho factor a protein responsible for the termination of transcription in bacteria.

riboflavin (vitamin B_2) vitamin component of the electron-transfer coenzymes FAD/FADH2 and FMN/FMNH2.

ribonucleoproteins macromolecular complexes whose optimal activity depends on the presence of both specific RNAs and proteins. Ribosomes are ribonucleoproteins.

ribonucleotide a nucleotide which contains a purine or pyrimidine base covalently linked to a ribose which is in turn linked to one or more phosphate groups.
ribose a five-carbon monosaccharide ($C_5H_{10}O_5$) that constitutes the carbohydrate moiety of ATP, other ribonucleosides and ribonucleotides, and cofactors such as NAD and coenzyme A.
ribosomal RNA RNA component of ribosomes, the site of protein synthesis.
ribosome a large ribonucleoprotein assembly that catalyzes the formation of peptide bonds; a molecular machine that coordinates protein synthesis.
ribozymes RNA molecules that display enzymatic activity.
rickets a disease caused by insufficient formation of vitamin D, resulting in inadequate calcification of cartilage and bone.
rifamycin antibiotic isolated from *Streptomyces* that inhibit the initiation of RNA synthesis by blocking the formation of the first phosphodiester bond.
RNA editing the change in the information content of RNA following transcription by processes other than RNA splicing.
RNA enzymes RNA molecules that act as enzymes; also called *ribozymes.*
RNA polymerase a class of enzymes that synthesize RNA molecules complementary to a DNA template.
RNA primers small pieces of RNA that base-pair with the template strand in DNA and serve as primer for primer-dependent DNA synthesis; the RNA is later removed and replaced with DNA.
RNA processing any alteration of a precursor RNA molecule, such as splicing or polyadenylation, that yields a mature RNA molecule.
rods vertebrate photoreceptor cells that function in dim light but are color-insensitive.
rotenone a plant toxin that inhibits electron transfer in the NADH-Q reductase complex; used as a fish and insect poison.
S1 heads proteolytic digestion products of myosin that are the force-generating units of the intact myosin molecule; contain the ATPase and actin-binding sites; S1 heads are smaller than heavy meromyosin.
***Salmonella* test** a simple, rapid means of detecting carcinogens by measuring a chemical's ability to induce mutations in *Salmonella* bacteria; also called the *Ames test.*
salting out a protein purification technique based on the fact that the solubility of most proteins is lowered at higher salt concentrations. Consequently, different proteins will precipitate at varying salt concentrations.
salvage pathway in general, a pathway that synthesizes the final product from preformed components; nucleotides can be synthesized in a salvage pathway by attaching purine bases to PRPP.
Sanger dideoxy method a DNA-sequencing technique that employs controlled interruption of enzymatic replication of the molecule to be analyzed. DNA polymerase I is used with a primer, the four deoxynucleoside triphosphates, and a 2′,3′-dideoxy analog of one of them. Fragments of various lengths are produced in which the dideoxy analog is at the 3′ end. Four sets of chain-terminated fragments (one for each analog) are then displayed by electophoresis and radioautography, so that the base sequence can be read from the four lanes of the gel.
sarcomere the functional unit of a myofibril; its distinct repeating structure is due to overlapping of thick protein filaments, composed of myosin, and thin filaments, composed of actin and other proteins; which contract via a sliding-filament method.
sarcoplasmic reticulum an extensive intracellular compartment in muscle cells that sequesters Ca^{2+} and releases it in response to a nerve impulse, thereby initiating muscle contraction.
satellite DNA highly repetitive DNA, initially resolved by density-gradient centrifugation, that is clustered at centromeres.
Schiff base an intermediate formed when an amine reacts with aldehydes or ketones to from a nitrogen compound analogous to a carbonyl; Schiff bases are intermediates in a number of enzyme-catalyzed reactions.
scRNA small cytoplasmic ribonucleoprotein particles; RNA molecules that are associated with specific proteins to form functional complexes. The signal recognition particle is an scRNP.
scurvy disease caused by a deficiency of ascorbate (vitamin C) which results in formation of unstable collagen.
SDS–polyacrylamide gel electrophoresis Electrophoresis of proteins in a polyacrylamide medium. The detergent SDS renders the movement of the proteins inversely proportional to molecular weight (see *electrophoresis*).
secondary structure in a protein, the spatial arrangement of amino acid residues that are relatively close to one another in the linear sequence; the α helix and the β strand are both elements of primary structure.
second messengers signal molecules that operate within a cell in response to an extracellular (first messenger) signal; cAMP is a second messenger.
sedimentation coefficient the velocity that a macromolecule moves in a centrifugal field divided by the strength of the centrifugal field; usually expressed as Svedberg units.
sedimentation-equilibrium centrifugation an ultracentrifugation technique that can be used to determine the mass of a protein. Centrifugation occurs at a relatively slow speed so that sedimentation is counterbalanced by diffusion.
self-splicing the ability of certain introns to remove themselves from the precursor RNA and assist in the splicing of exons to form the mature RNA.
semiconservative replication the duplication of DNA in which one of the strands of each daughter molecule is newly synthesized while the other is unchanged from the parental DNA double helix.
serine protease a class of protein-degrading enzymes whose activity is dependent on the presence of serine at the active site; chymotrypsin and trypsin are examples.
seven-helix receptor a class of integral membrane proteins typified by transducin in which the intramembrane portion consists of seven helical regions; these receptors are always coupled to G proteins.
severe imunodeficiency disease a genetic disorder in mice in which repair of double-stranded DNA and *VDJ* gene segment recombination are impaired, resulting in the inability to express a diverse group of antibodies.
Shine-Dalgarno sequence in messenger RNA, a purine-rich region about 10 nucleotides on the 5′ side of an initiator codon that pairs with the 3′ end of 16S RNA in the 30S ribosomal subunit; helps to determine where translation is initiated on an mRNA molecule.
sickle-cell anemia a hemoglobinopathy resulting from a point mutation in the β subunit, which results in precipitation of deoxyhemoglobin and deformation of the red blood cells.
side chain the distinctive R group bonded to the α carbon of an amino acid.

siemen the reciprocal of an ohm, which is a measure of resistance to the flow of electrical charge.

signal peptidase an enzyme located in the lumen of the rough endoplasmic reticulum that removes the signal sequence from proteins translocated across the ER.

signal recognition particle (SRP) a ribonucleoprotein particle that acts catalytically to deliver ribosomes attached to nascent proteins with a signal sequence to the rough ER for translocation of the polypeptide across the membrane. Elongation of the protein is halted or slowed when the SRP binds the signal sequence.

signal sequences a sequence of amino acid residues ranging in size from 13 to 36 residues, usually at the amino terminus of the nascent polypeptide chain, that marks the protein for translocation across the rough endoplasmic reticulum.

SINES short (100- to 500-bp) repeats of unknown function interspersed throughout the eukaryotic genome.

sissile bond the bond in a substrate molecule that is cleaved by the enzyme.

site-specific mutagenesis the use of recombinant DNA technology to create specific deletions, insertions, or substitutions in vitro in a particular gene; this technique allows production of proteins having any desired amino acid at any position.

site-specific recombination the exchange of two specific but not always homologous DNA sequences.

small nuclear RNA (snRNA) a class of small RNAs that are confined to the nucleus. Some of these play a role in splicing.

SNAP receptors membrane proteins that bind to SNAPs and subsequently lead to the fusion of a transport vesicle to the target membrane.

SNAPs soluble proteins that attach to a transport vesicle and lead to the fusion of the vesicle with the target membrane.

snRNA see small nuclear RNA.

sodium-potassium pump a protein complex that generates a high intracellular concentration of potassium ion and a low intracellular concentration of sodium, relative to the external medium, through active transport powered by ATP.

solid-phase immunoassay means of detecting a protein of interest in which an antibody to the protein is immobilized on a polymeric support; a sample with the protein of interest is applied to the support, and a different radioactive antibody to the protein is used to detect its presence.

solid-phase method means of synthesizing discrete peptides in which amino acids are added stepwise to a growing peptide chain that is anchored to an insoluble matrix.

Southern blotting a technique used to locate and identify a DNA fragment containing a specific sequence; a mixture of fragments is separated by electrophoresis, transferred to a nitrocellulose sheet, hybridized to a radioactively labeled DNA probe complementary to the desired sequence, and visualized by autoradiography.

special pair in a photosynthetic reaction center, the pair of chlorophyll molecules that collect excitation energy from antenna chlorophylls and then transfer high-energy electrons to other electron acceptors.

sphingomyelin a sphingolipid common in brain tissue in which the terminal hydroxyl group of ceramide has a phosphorylcholine substituent.

sphingosine an amino alcohol containing a long, unsaturated hydrocarbon chain that is a component of the phospholipid sphingomyelin as well as glycolipids; serves a role analogous to glycerol in phosphoglycerides.

spliceosome an assembly of proteins and small nuclear RNAs that splices primary transcripts to mature mRNA.

splicing the removal of introns and the ligation of exons from precursors of RNA to form mature RNA.

spontaneous reaction a process that procedes irreversibly toward equilibrium and is accompanied by an increase in disorder or entropy.

SRP receptor an integral protein of the rough ER that binds the SRP-ribosome complex, frees the signal peptide from the SRP, and directs the nascent protein to the translocation machinery.

stage 1 metabolism stage in which large molecules are broken into smaller units; glycogen, for instance, is converted to glucose.

stage 2 metabolism stage in which small molecules are degraded to a few key intermediates in metabolism, such as acetyl CoA.

stage 3 metabolism stage consisting of the citric acid cycle and oxidative phosphorylation; the final common pathway for the oxidation of fuel molecules.

standard free-energy change the free-energy change of a reaction when it occurs under standard conditions, in which each of the reactants is present at a concentration of 1.0 M.

starch a homopolysaccharide that is a storage form of glucose in plant chloroplasts; amylopectin, the branched form of starch, has approximately one α-1,6 linkage per thirty α-1,4 linkages, while amylose is unbranched, composed of glucose residues in α-1,4 linkage.

start codon the first codon to be translated, usually AUG.

steroid hormones hormones, such as androgen and estrogen, derived from cholesterol.

stop codons a set of codons that are recognized by release factors and mark the end of translation.

stop-transfer sequence (membrane anchor sequence) an amino acid sequence whose presence in a nascent protein in the process of being translocated across the endoplasmic reticulum membrane halts the translocation process. Such sequences are used to imbed integral membrane proteins into the membrane.

streptomycin a highly basic trisaccharide antibiotic that causes both inhibition of translational initiation and misreading of mRNA in prokaryotes.

stringent response in bacteria, the decrease in the rate of ribosomal and transfer RNA synthesis in response to a scarcity of amino acids.

stroma the matrix of the chloroplast, which contains thylakoids as well as soluble enzymes; the stroma is enclosed by the inner membrane of the chloroplast.

submitochondrial particles particles formed by the sonic disruption of the inner mitochondrial membrane, which folds on to itself to form intact vesicles capable of carrying out oxidative phosphorylation.

substrate a reactant in a chemical reaction. An enzyme catalyzes a single chemical reaction or set of closely related reactions, and the components of those reactions are called *substrates.*

substrate-level phosphorylation enzymatic formation of ATP by the transfer of a phosphate from an energy-rich compound to ADP.

subunit any of the polypeptide chains in a protein that contains more than one of such chains.

succinate-Q reductase (Complex II) an integral membrane protein complex of the inner mitochondrial membrane that transfers electrons from the FADH2 formed in the citric acid cycle to coenzyme Q.

suicide inhibition inhibition that results when an enzyme converts a pseudosubstrate into a reactive inhibitor that immediately inactivates its catalytic activity.

supercoiling the ability of closed, circular DNA to coil upon itself.

superhelix the coil of a double-stranded DNA formed when it is twisted so as to cross over its own axis.

supersecondary structure structural motifs in polypeptides that arise by the association of α helices or β strands; these include the antiparallel packing of α helices to form the four-helix bundle found in hemoglobin and myoglobin, and β hairpins or β sheets formed by β polypeptide strands.

Svedberg units unit for measuring the sedimentation of a macromolecule; 1 Svedberg unit (S) is equal to 10^{-13} seconds.

symporter a transport system in which a molecule is carried across the membrane in the same direction as an ion, which in turn is pumped back across the membrane through active transport linked to ATP consumption.

synthase an enzyme catalyzing a synthetic reaction in which two units are joined without the direct participation of a nucleotide triphosphate.

synthetase an enzyme catalyzing a synthetic reaction in which two units are joined with direct participation of ATP or another nucleoside triphosphate.

T-cell receptor the receptor that recognizes peptides displayed by MHC proteins on target cells; the protein consists of two different 43-kd chains joined by a disulfide bond and spanning the plasma membrane; the combination of constant and variable regions among various T-cell receptors allows T cells to recognize a large number of different epitopes.

T-cell-mediated immunity one of the recognition systems of the immune response, in which T cells scan the surfaces of all cells and kill those that exhibit foreign proteins synthesized by intercellular pathogens.

TATA box a promoter element found in nearly all eukaryotic genes giving rise to mRNA. The TATA box is centered between −30 and −90 of the transcription start site and has the consensus sequence 5′ TATAAAA 3′.

taxol a chemical isolated from the Pacific yew tree that stabilizes tubulin in microtubules and promotes polymerization; it is used in chemotherapy to interfere with proliferation of rapidly dividing cells.

Tay-Sachs disease a heritable disorder caused by the accumulation of gangliosides due to the lack of an enzyme (β-*N*-acetylhexosaminidase) responsible for their degradation; clincial characteristics include weakness, mental retardation, dementia, blindness, and death by age three.

telomerase a reverse transcriptase that contains its own template; a highly processive enzyme that elongates the 3′-ending strand of telomeres.

telomeres ends of chromosomes; the DNA at the telomere consists of hundreds of repeats of a hexanucleotide sequence characteristic of the organism.

temperate phage a bacteriophage that can employ a lytic or a lysogenic pathway.

template a sequence of DNA or RNA which directs the production of a complementary sequence.

terminal redundancy the presence of the same nucleotide sequence at both the 3′ and the 5′ end of a viral genome.

termination the final state common to all biological polymerization reactions. The polymerization process is halted in response to a set of termination signals specific for the type of biological molecule being synthesized.

tertiary structure in proteins, the spatial arrangement of amino acid residues that are far from each other in the linear sequence, as well as the pattern of disulfide bonds.

tetracycline an antibiotic which binds to the 30S ribosomal subunit of prokaryotes and which inhibits binding of aminoacyl-tRNA molecules to the ribosomal complex.

tetrodotoxin a highly potent poison from the fugu (puffer) fish that blocks the conduction of nerve impulses along axons and excitable membranes in nerve fibers, leading to respiratory paralysis.

thalassemias genetic disorders characterized by defective synthesis of one or more hemoglobin chains.

thermogenesis the regulated uncoupling of oxidative phosphorylation as a means of generating heat to maintain body temperature.

thermogenin (uncoupling protein) a mitochondrial membrane protein that plays a role in thermogenesis by forming a pathway for the flow of protons into the mitochondria, thereby generating heat without synthesizing ATP.

theta structure structure seen in autoradiographs of replicating circular DNA; it has the form of a closed circle with an internal loop (like the Greek letter θ) and reveals that the DNA maintains its circularity during DNA synthesis.

thiamine (vitamin B_1) vitamin component of the coenzyme thiamine pyrophosphate.

thiamine pyrophosphate the coenzyme form of thiamine (vitamin B_1), composed of a modified thiazole ring linked by a methylene bridge to a substituted pyrimidine; it is a cofactor in enzymatic reactions where bonds to carbonyl carbons are cleaved or synthesized.

thioester an ester in which the noncarbonyl oxygen is replaced with a sulfur. Thioesters are energy-rich intermediates in a number of biochemical reactions.

thiol protease a class of protein-degrading enzymes whose activity is dependent on the presence of a cysteine at the active site. Papain is a thiol protease.

thioredoxin a protein with exposed cysteines that can be reversibly oxidized and reduced; thioredoxin is an important electron carrier in the reduction of ribonucleotides and in photosynthesis.

thylakoid in chloroplasts, a membrane sac or vesicle that contains the energy-transducing machinery of photosynthesis, including light-harvesting proteins, reaction centers, electron transport chains, and ATP synthase.

topoisomerase enzymes that catalyze the interconversion of topoisomers of DNA; topoisomerases can relax supercoiled DNA.

topoisomers molecules of DNA that differ from one another only in their linking number.

transaldolase an enzyme that transfers a three-carbon dihydroxyacetone unit from a ketose to an aldose acceptor; one of the enzymes in the nonoxidative portion of the pentose phosphate pathway.

transamination transfer of an α-amino group from an amino acid to an α-keto acid.

transcription DNA-directed synthesis of RNA catalyzed by RNA polymerase.

transcription bubble the site of RNA synthesis or transcription. It contains the RNA polymerase, a locally melted “bubble” of DNA and helix consisting of the template strand and the newly synthesized RNA.

transcription factor a protein that assists RNA polymerase in the initiation of RNA synthesis. Transcription factors bind to specific promoter elements.

transducin G protein that is the signal coupling protein of visual excitation; transducin is activated by rhodopsin and leads to the activation of cGMP phosphodiesterase, which in turn leads to a nerve impulse.

transferase an enzyme that catalyzes group transfer, often employing a cofactor.

transfer RNA the adapter molecule in protein synthesis. It contains an amino acid recognition site as well as a template-recognition site, or anticodon.

transformation the incorporation of intact foreign DNA into a cell.

transgenic animal an animal that harbors and expresses a foreign gene inserted into the germ line; experiments involving such animals and their offspring show that a foreign gene under the control of a new promoter can be efficiently integrated, expressed, and transmitted.

transition a mutation that results from the substitution of one pyrimidine for another or one purine for another.

transition state a chemical species that has the highest free energy and the lowest concentration of those on the pathway from a substrate to a product.

transition state analogs compounds that have the geometry of the catalyzed transition state before it is bound to the enzyme; often, they are potent inhibitors of the enzyme.

transketolase an enzyme that transfers an activated aldehyde unit from a ketose to an aldose acceptor; one of the enzymes in the nonoxidative portion of the pentose phosphate pathway.

translation cellular protein synthesis, so-named because the four-letter alphabet of nucleic acids is translated into the different amino acid alphabet of proteins.

translational repressor a mechanism for transcriptional regulation in which a protein encoded by an operon binds to mRNA near the initiation site for its own synthesis and blocks synthesis of several proteins encoded by that polygenic message; the synthesis of the 50 or so ribosomal proteins is subject to control by translational repression.

translocase a protein that carries a molecule from one cellular compartment to another; the ATP-ADP translocase facilitates the exchange of ATP and ADP between the mitochondrion and the cytosol.

translocon a multisubunit assembly of integral and peripheral membrane proteins that translocates proteins across the membrane.

transport antibiotics antibiotics that exert their effect by rendering the membranes abnormally permeable.

transport vesicles membrane-bound compartments that mediate the transfer of proteins between the ER and the Golgi, and between the Golgi and subsequent destinations.

transposase a nuclease enzyme encoded by an insertion sequence (IS) or transposon; it makes staggered cuts in donor and recipient sites, facilitating IS insertion into a bacterial gene.

transposition the movement of a gene from one chromosome to another or from one location to another on the same chromosome.

transposons mobile genetic elements in bacteria that enable genes to be transferred among nonhomologous sequences.

transversion a mutation in which a purine is replaced by a pyrimidine or vice versa.

triacylglycerols glycerols that have fatty acyl chains esterified to each of their hydroxyl groups; storage form of fats; also known as *neutral fats* or *triglycerides.*

trimethoprim a competitive inhibitor of dihydrofolate reductase, like methotrexate but specific for bacterial and protozoan enzymes.

tropomyosin protein located in the thin filaments of a sacromere that plays a role in the regulation of muscle contraction by blocking the interaction of myosin with actin at low Ca^{2+} concentrations; regulated by troponin.

troponin complex a set of proteins located in the thin filaments of the sacromere that regulates muscle contraction via allosteric interactions with tropomyosin in response to changes in Ca^{2+} concentrations.

tubulin major protein component of microtubule that exists in two forms, α- and β-tubulin; tubulins display GTPase activity that is vital for the assembly and disassembly of microtubules.

tumor promoters chemicals that promote the proliferation of carcinogenic cells; phorbol esters are tumor promoters.

tunicamycin an antibiotic inhibitor of *N*-glycosylation that is a hydrophobic analog of UDP–*N*-acetylglucosamine; blocks the first step in core oligosaccharide synthesis.

turnover number the number of substrate molecules converted into product by an enzyme molecule in a unit time when the enzyme is fully saturated with substrate; it is equal to the kinetic constant k_3 (see *Michaelis constant*).

ubiquione/ubiquinol see *coenzyme Q.*

ubiquitin a small, highly conserved protein present in all eukaryotes that, when attached to another protein, targets it for proteolytic destruction.

ultracentrifugation high-speed centrifugation that is used to separate biomolecules and determine their mass.

urea cycle cyclic pathway that converts excess ammonia into urea for secretion; it was the first metabolic pathway to be discovered.

ureotelic organisms in which excess ammonia is converted to urea and then excreted; most terrestrial vertebrates are ureotelic.

uricotelic organisms in which excess ammonia is converted into the purine uric acid for secretion; birds and terrestrial reptiles are uricotelic.

variable region the 108-residue amino acid sequence found at the N-terminal end of both the L and H chains of immunoglobulin G; the sequence in this region varies for every antibody type known; a portion of these segments (the hypervariable regions) form the antigen-binding site of the immunoglobulin.

virilization an inherited disorder of steroid hormone synthesis due to abnormally high levels of androgen; clinical characteristics include early sexual development in males, masculinization of external female genitalia, and persistent loss of Na^+ leading to dehydration and hypotension.

virion the complete extracellular form of a virus consisting of DNA or RNA surrounded by a coat.

virus a complex of protein and nucleic acid that can penetrate a cell and replicate itself by coopting the host's metabolism and employing its own as well as the host's gene products; the smallest organism known.

viscoelastic technique a means of measuring the length of DNA molecules in the cell. The cells are lysed, nucleases are inactivated, and protein is removed from the DNA by protease treatment. The DNA molecules are stretched out by solution flow, and then allowed to recoil to normal form. The time for recoiling is proportional to the mass of the molecule.

vitamin A (retinol) fat-soluble vitamin that is the precursor of the light-sensitive pigment retinal, as well as the signal molecule retenoic acid, an activator of certain transcription factors.

vitamin B_{12} (cobalamin) a prosthetic group, consisting of a cobalt atom, a corrin ring, and deoxyadenosine, that plays a role in intramolecular rearrangements, methylations, and reductions of ribonucleotides to deoxyribonucleotides.

vitamin D a fat-soluble vitamin that plays a role in the regulation of calcium and phosphorus metabolism; deficiencies in vitamin D lead to impairment of bone formation.

vitamin E (α-tocopherol) protects unsaturated membrane lipids from oxidation.

vitamin K a fat-soluble vitamin required for blood coagulation.

vitamins organic substances that are required in trace amounts for a number of essential biochemical reactions.

VLDL (very low density lipoproteins) lipoprotein particles, stabilized by apolipoproteins B-100 and E, that transport excess endogenous triacylglycerides and cholesterol from the liver to other tissues.

voltage-gated channel a transmembrane channel that is opened by membrane depolarization; the sodium and potassium channels of axon membranes are good examples.

von Gierke's disease a disease resulting from defective glucose 6-phosphatase that affects liver and kidney; glycogen is normal in structure but present in abnormally large amounts so that there is a massive increase in liver size with resulting damage to liver functions.

Watson-Crick helix see *B-DNA helix.*

Western blotting an immunoassay technique used to detect a specific protein in a cell or in body fluid. A sample is electrophoresed in an SDS-polyacrylamide gel, the resolved proteins are transferred to a polymer sheet, and then an antibody specific for the protein of interest is incubated with the blotted sample; other antibodies or radioactive markers may then be used to help visualize the desired antigen-antibody complex.

wobble steric freedom in the pairing of the third base of an mRNA codon with the anticodon of a transfer RNA, allowing more than one codon to be recognized by a particular tRNA molecule.

x-ray crystallography a technique to determine the three-dimensional structure of protein crystals at atomic resolution by examining the defraction pattern of x-rays striking the crystal.

xeroderma pigmentosa a rare skin disorder, characterized by sensitivity to ultraviolet light and a propensity to skin cancers, which is caused by a defect in excinuclease which plays a role in the removal of pyrimidine dimers.

yeast artificial chromosome a DNA molecule that can be used to clone DNA inserts ranging from 100 to 1000 kb in length; these molecules contain a centromere, an autonomously replicating sequence, a pair of telomeres, selectable marker genes, and an insertion site for the sequence to be cloned.

Z-DNA a left-handed double helix in which the backbone phosphates zigzag; can be formed by oligonucleotides with alternating sequences of purines and pyrimidines.

zinc clusters a DNA-binding motif found in pairs in the nuclear receptor superfamily proteins, in which a zinc ion is coordinated to four cysteine side chains.

zinc finger a eukaryotic DNA-binding motif that consists of 2 cysteines and 2 histidines coordinated to a zinc ion in a 30-residue repeating unit folded into the shape of a finger.

zinc protease a class of protein-degrading enzymes whose catalytic activity is dependent on a zinc ion. Carboxypeptidase A is a zinc protease.

zipper proteins transcription factors that contain a leucine zipper, a structural motif which functions to bring together a pair of DNA-binding molecules by forming a coiled coil stabilized by hydrophobic and van der Waals interactions between regularly repeated leucine residues.

zonal centrifugation an ultracentrifugation technique in which the sample is centrifuged through a gradient of increasing density and the components are separated on the basis of differing sedimentation coefficients.

zwitterion a dipolar ion (from the German *zwitter,* "between"); at neutral pH, amino acids have a protonated amino group and a dissociated carboxyl group and are therefore zwitterions.

zymogen a catalytically inactive precursor of an enzyme.